Troitzsch / Antonatus

Plastics Flammability Handbook

Jürgen Troitzsch
Edith Antonatus

Plastics Flammability Handbook

Principles, Regulations, Testing, and Approval

4th Edition

HANSER

Hanser Publishers, Munich

Hanser Publications, Cincinnati

The Editors:

Dr. Jürgen Troitzsch, 6612 Ascona, Switzerland, jtroitzsch@troitzsch.com

Dipl.-Phys. Edith Antonatus, 78476 Allensbach, Germany, edith.antonatus@gmail.com

Distributed in the Americas by:
Hanser Publications
414 Walnut Street, Cincinnati, OH 45202 USA
Phone: (800) 950-8977
www.hanserpublications.com

Distributed in all other countries by:
Carl Hanser Verlag
Postfach 86 04 20, 81631 Munich, Germany
Fax: +49 (89) 98 48 09
www.hanser-fachbuch.de

Library of Congress Control Number: 2020951740

Editor: Dr. Julia Diaz Luque
Production Management: Der Buchmacher, Arthur Lenner, Windach
Coverconcept: Marc Müller-Bremer, www.rebranding.de, Munich
Coverdesign: Max Kostopoulos
Coverimage: © Southwest Research Institute (SwRI) Fire Technology Department
Typesetting: Eberl & Kœsel Studio GmbH, Krugzell, Germany
Printed and bound by CPI books GmbH, Leck
Printed in Germany

ISBN: 978-1-56990-762-7
E-Book ISBN: 978-1-56990-763-4

Preface to the Fourth Edition

In the past, none of the many publications on the reaction to fire of plastics provided a comprehensive review of the fundamentals, as well as of the relevant regulations and test methods. The *International Plastics Flammability Handbook* was first published in 1983 to fill this gap. However, in the 1980s, major changes occurred in the field of plastics fire behavior ratings on national and (increasingly) international levels. These changes made it necessary to prepare a completely revised 2nd edition of the handbook, which was published in 1990. The 1990s saw a breakthrough in the internationalization of fire testing and classification, particularly for products in buildings, electrotechnical applications, and transportation. At the same time, the perception of how to assess the main parameters governing a fire and the role of combustible materials like plastics was redefined, and led to new approaches particularly in the fields of heat release, smoke development, and toxicity of fire effluents. All these developments required a comprehensive revision and part-rewrite of the handbook, which was published in its 3rd edition in 2004.

Since then, many further changes have taken place in the fire performance requirements for construction products and rail vehicles in the European Union, the USA, and the Asia-Pacific region. In addition, the ongoing developments in the fields of electrical engineering and cables, and furniture and furnishings, as well as the ever-growing interest in smoke and toxicity of fire effluents, made it necessary to prepare a 4th edition of this handbook.

So, more than 15 years since the last edition, it was time to restructure and simplify the contents of the book, and to ask leading experts in the field to revise and to rewrite the various chapters. For this edition, Edith Antonatus agreed to join me as co-editor in order to make this major revision possible, and we want to express our gratitude to the over 30 co-authors whose expertise and commitment have made this project feasible.

The handbook consists of three parts.

Part I, "Fundamentals", begins with an introduction to fire prevention, fires, and fire statistics, and continues with a description of the basic principles of the burning of plastics, the role of flame retardants, their mode of action, and their application to improve the fire safety of plastics. Chapters on the burning behavior of textiles and flame-retardant textiles, smoke development, and smoke suppression follow. In the 3rd edition of this handbook, the chapters on smoke development and toxicity of fire effluents were presented as Part III "Fire effluents" at the end of the book. In this edition, they have been moved to Part I, where, after the chapter on the basics of smoke formation, the chapters on smoke development and toxicity, as well as combustion toxicology of fire effluents, now jointly describe this increasingly important topic.

It is hoped that this will facilitate the reader's introduction to this complex subject, and also provide the background to better understand why the fire test procedures, regulations, and approval criteria covered in the second part of this handbook are necessary to provide adequate fire safety.

Part II, "National and International Fire Protection Regulations and Test Procedures", starts with a general introduction to fire regulations and codes, and the fire test methods needed to satisfy them, followed by a comprehensive overview of international fire safety standardization, with the work of the International Standards Organization (ISO), the European Committee for Standardization (CEN), and the international electrotechnical committees as examples.

The legal provisions regarding the fire performance of plastics and other combustible products form a large part of the book, and are arranged according to a consistent scheme for a selection of countries in the three major regions (North America, Europe, Asia-Pacific). The section on each country commences with an account of the statutory regulations, and continues with a summary of the relevant test methods in the form of diagrams and tables of test specifications. Officially recognized test institutions and procedures for obtaining official product approval are enumerated under the heading "Official Approval". Each section ends with a look at future developments in the relevant country. The reader is thus able to grasp the essentials at a glance. Further details can be taken from the original standards and regulatory documents listed in the bibliography at the end of each section.

The most extensive section of the handbook is devoted to the building sector, for which numerous fire safety regulations and test methods have been developed in all industrialized countries, and where a tremendous harmonization effort has virtually been completed in Europe. But even there, tests and classifications for specific building components such as the fire performance of façades and the external fire exposure of roofs are still not harmonized. To date, the toxicity of construction products is not part of the classification system developed by the European Commission, but subject to intense discussion. In other countries and regions of the world, particularly in Asia, great efforts have been made to improve existing smoke and toxicity regulations and test methods.

In the chapter on transportation (motor vehicles, rail vehicles, ships, and aircraft), the most important change has been the harmonization of railway fire protection standards in Europe, but there have also been many new developments in other parts of the world. Therefore, EU harmonization is covered in some detail, and sections on the USA and Canada, China, Japan, Korea, and Australia have been added.

The chapter on electrical engineering has been completely reworked and simplified in light of the progress made in the international harmonization of regulations and test methods. The growing importance of the fire performance of cables is now presented in a separate comprehensive section.

The chapter on furniture and furnishings concentrates on developments of fire safety regulations and testing in the USA and the UK, and has been revised in light of trends to deregulation caused by concerns regarding the use of flame retardants. A section on measurements and model prediction of the burning rate of furniture has been added.

Part III, "Appendices", contains various lists including a suppliers' index for flame retardants and smoke suppressants, a list of the most common abbreviations for plastics and flame retardants, the chemical names and CAS numbers of the latter, and a choice of journals and books devoted to flame retardancy and fire protection.

It is hoped that this book will be of interest to all those concerned with plastics, flame retardancy, fire testing, and fire protection, and will help the reader to better understand this complex subject.

Jürgen Troitzsch and Edith Antonatus
Ascona and Allensbach, January 2021

List of Contributors

Eleonora Anselmi
Edith Antonatus
Jonas Brandt
Jungmin Choi
Manfred Döring
Marcel Donzé
Jadwiga Fangrat
Mattia Ferraris
Antonio Galán Penalva
Daniela Goedderz
Lara Greiner
Eric Guillaume
Jun Haruhara
Esther Hild
Anja Hofmann
A. Richard Horrocks
Marc Janssens
Xing Jin
Imre Juhász
Baljinder K. Kandola
Torben Kempers
István Móder
Jürgen Pauluhn
Rudolf Pfaendner
Bernhard Schartel
Bart Sette
Björn Sundström
Shuai-Xia Tan
Jürgen Troitzsch
Rudolf van Mierlo
Koichi Wada
Jun-Sheng Wang
Alex Webb
Dieter Werner
Peter Whiting
Hideki Yoshioka

Contents

Preface to the Fourth Edition V

List of Contributors IX

Part I
Fundamentals 1

1 Introduction 3

1.1 Fire Prevention and Fires 3
Jürgen Troitzsch

1.2 Fire Statistics 10
Eric Guillaume

1.2.1 Introduction 10
1.2.2 Lessons Learned from Recent ISO Work on Fire Statistics 10
1.2.3 Current National Fire Statistics: Relevant Data on Specific Issues 12
1.2.3.1 Number of Fires 12
1.2.3.2 Number of Fire Fatalities 14
1.2.3.3 Number of Fire Injuries 17
1.2.3.4 Costs of Fire Losses 19
1.2.4 Future Developments 21

2 The Burning of Plastics 23
Bernhard Schartel

2.1 Production of Volatile Fuel: Heating, Pyrolysis, Charring, Heat of Combustion 27

2.2 Decomposition of Polymers 34

2.3 Ignition and Smoldering 38

2.4 Steady Burning and Flame Spread 42

2.5 Fire Load and Fire Resistance 47
2.6 Conclusions 48

3 Flame Retardants and Flame-Retarded Plastics 53

3.1 Flame Retardants 53
Manfred Döring, Lara Greiner, and Daniela Goedderz
3.1.1 Importance, Development, and Market 53
3.1.2 Mode of Action 54
3.1.2.1 Physical Action 55
3.1.2.2 Chemical Action 55
3.1.3 Important Flame Retardant Classes 57
3.1.3.1 Metal Hydroxides and Mineral Fillers 57
3.1.3.2 Halogenated Flame Retardants 61
3.1.3.3 Phosphorus-Containing Flame Retardants 64
3.1.3.4 Organic Radical-Forming Agents and N-, O- and S-Containing Synergists 72
3.1.3.5 Synergistic Flame Retardant Systems 75
3.1.3.6 Other Flame Retardants 77
3.1.4 Greener Alternatives to Common Flame Retardants 80

3.2 Flame-Retarded Plastics 95
Rudolf Pfaendner
3.2.1 Introduction 95
3.2.2 Thermoplastics 98
3.2.2.1 Polypropylene 98
3.2.2.2 Polyethylene 100
3.2.2.3 Other Polyolefins 103
3.2.2.4 Polystyrenes 104
3.2.2.5 Styrene Copolymers 105
3.2.2.6 Polymethacrylates and Polyacrylates 106
3.2.2.7 Polyoxymethylene and Polyacetals 106
3.2.2.8 Polyamides 107
3.2.2.9 Polyesters 109
3.2.2.10 Polycarbonate and Polycarbonate Blends 111
3.2.2.11 Polyurethanes 112
3.2.2.12 Polyvinylchloride 113
3.2.2.13 Other Halogenated Polymers 114
3.2.2.14 High-Performance Polymers 114
3.2.3 Thermosets 115
3.2.3.1 Epoxy Resins 115
3.2.3.2 Polyurethanes 117

3.2.3.3 Phenolic Formaldehyde Resins ... 118
3.2.3.4 Unsaturated Polyesters ... 118
3.2.4 Elastomers ... 119

4 Textiles ... 129
A. Richard Horrocks and Baljinder K. Kandola

4.1 Introduction ... 129
4.1.1 Burning Behavior of Textile Fibers and Fabrics ... 130
4.1.2 Effect of Fabric and Yarn Structures ... 133

4.2 Flammability Testing of Textiles ... 134
4.2.1 Regulations ... 134
4.2.2 Test Categorization ... 136

4.3 Flame-Retardant Textiles ... 142
4.3.1 Cellulosics ... 143
4.3.1.1 Flame-Retardant Cottons ... 143
4.3.1.2 Flame-Retardant Viscose ... 144
4.3.1.3 Flame-Retardant Cellulosic Blends ... 144
4.3.2 Flame-Retardant Wool and Blends ... 145
4.3.3 Flame-Retardant Synthetic Fibers ... 146

4.4 Inherently Flame-Retardant Synthetic Fibers ... 147

4.5 High Heat- and Flame-Resistant Synthetic Fibers and Textiles .. 148

4.6 Intumescent Applications to Textiles ... 149

5 Smoke Development and Suppression ... 153

5.1 Smoke Development and Measurement ... 153
Eric Guillaume
5.1.1 Introduction ... 153
5.1.2 Smoke Generation ... 154
5.1.2.1 Definitions ... 154
5.1.2.2 Principles ... 155
5.1.2.3 Nucleation and Surface Growth ... 157
5.1.2.4 Coagulation and Agglomeration ... 159
5.1.2.5 Oxidation ... 160
5.1.2.6 Liquid Aerosols ... 160
5.1.3 Smoke Production from the Most Important Polymers .. 161
5.1.4 Smoke Measurement ... 162
5.1.4.1 Smoke Opacity ... 162
5.1.4.2 Other Smoke Parameters ... 166
5.1.5 Visibility Models ... 169
5.1.5.1 Definition of Visibility ... 169

5.1.5.2 Simplified Models (1950–1980) 169
5.1.5.3 Jin and Yamada Models 170
5.1.5.4 ISO 13571 Model 171
5.1.6 Fire Threat Related to Smoke 171

5.2 Smoke Suppressants 173
Rudolf Pfaendner
5.2.1 Introduction 173
5.2.2 Smoke Suppressants for PVC 174
5.2.2.1 Molybdates 175
5.2.2.2 Stannates 175
5.2.2.3 Borates 176
5.2.3 Smoke Suppressants in Combination with Halogenated Flame Retardants 177
5.2.4 Smoke Suppressants in Combination with Halogen-Free Flame Retardants 178
5.2.5 Summary and Outlook 179

6 Smoke and Toxicity from Fire Effluents 185
Eric Guillaume

6.1 Preliminary Remarks 185

6.2 Acute Effects of Fire Effluents 186
6.2.1 Asphyxiants 187
6.2.2 Irritants 188

6.3 Sub-Acute Effects of Fire Effluents 188

6.4 Chronic Effects of Fire Effluents 188

6.5 Environmental Effects of Fire Effluents 189

6.6 Quantitative Risk Assessment of Fire Effluents 190
6.6.1 Introduction 190
6.6.2 Direct and Indirect Methods 190
6.6.2.1 Direct Methods 191
6.6.2.2 Indirect Methods 192

6.7 Generation of Fire Effluents: Physical Fire Models Used for Toxicity Testing 193
6.7.1 Introduction to Physical Fire Models 193
6.7.2 What is a Physical Fire Model? 193
6.7.3 Important Characteristics to be Reproduced by the Fire Model 195
6.7.3.1 Parameters Influencing the Nature and Proportion of Gas Release 195

6.7.3.2 Combustion Conditions 195
6.7.3.3 Thermal Conditions of Degradation and Equivalence Ratio 196
6.7.3.4 Parameters Influencing the Preparation for Analysis 197
6.7.3.5 Parameters Influencing the Scale Effects 198
6.7.3.6 Parameters Influencing the Quality of the Fire Model 202

6.8 Chemical Analysis of Fire Effluents 204
6.8.1 Gases of Interest 204
6.8.2 Sampling 204
6.8.2.1 From Fire Source to Analysis 204
6.8.2.2 Probes 205
6.8.2.3 Filters 206
6.8.2.4 Sampling Line 207
6.8.2.5 Pumping System 208
6.8.2.6 Sampling Conditions 208
6.8.2.7 Gas-Solution Absorbers 208
6.8.2.8 Solid Sorption Tubes 210
6.8.2.9 Gas Bags 211
6.8.3 Selection of Analytical Methods 211
6.8.4 Validation of Analytical Methods 215
6.8.4.1 Evaluation of a Method 215
6.8.4.2 Selectivity 215
6.8.4.3 Sensitivity 215
6.8.4.4 Concentration Ranges 215
6.8.4.5 Detection and Quantification Limits 216
6.8.4.6 Accuracy of the Method 216
6.8.4.7 Repeatability and Reproducibility of the Method 217

7 Combustion Toxicology 219
Jürgen Pauluhn

7.1 Introduction 219

7.2 Toxic Hazards 221
7.2.1 Inhalation Toxicity of HCN and CO 224
7.2.2 Inhalation Toxicity of CO_2 226
7.2.3 Respiratory Tract Irritants 229

7.3 Toxicological Risk Assessment 234
7.3.1 Definition of Uncertainty 234

7.3.2 Analysis of Time-Dose-Response Relationships 235
7.3.2.1 Asphyxiants 235
7.3.2.2 Irritants 238
7.4 Application of the Fractional Effective Dose (FED) Model 240
7.5 Conclusions 242

Part II
National and International Fire Protection Regulations and Test Procedures **247**

8 Regulations and Testing **249**
Edith Antonatus and Jürgen Troitzsch
8.1 Introduction 249
8.2 Sets of Regulations and Codes 251
8.2.1 Types and Basis 251
8.2.2 Application 252
8.2.3 The Role of Standards 253
8.3 Test Methods 253
8.3.1 Laboratory-Scale Tests 254
8.3.1.1 Test Specimens 254
8.3.1.2 Ignition Sources 254
8.3.1.3 Parameters Measured 255
8.3.2 Large-Scale Tests 256

9 International Standardization **257**
9.1 Introduction 257
Edith Antonatus
9.1.1 International Standardization 257
9.1.2 Regional Standardization 259
9.1.3 International and Regional Fire-Related Standards for Transportation 259
9.2 International Organization for Standardization (ISO) 261
Björn Sundström
9.2.1 Introduction 261
9.2.2 Scope of ISO/TC 92 Fire-Related Activities 262
9.2.3 ISO/TC 92/SC1: Fire Initiation and Growth 264
9.2.3.1 General Use of Test Data and Fire Safety Engineering 268
9.2.3.2 Non-Combustibility 268
9.2.3.3 Ignition and Flame Spread 269

9.2.3.4 Heat Release and Smoke Production 273
9.2.3.5 Large-Scale 276
9.2.3.6 Calibration and Supporting Procedures 280
9.2.4 Future Developments 280

9.3 European Committee for Standardization (CEN) 282
Edith Antonatus
9.3.1 Introduction 282
9.3.2 Development of Standards 282
9.3.3 Fire-Related Activities in Building 284

10 Building 287

10.1 Introductory Remarks 287
Edith Antonatus
10.1.1 Classification and Testing of Fire Performance of Building Materials and Components 289
10.1.1.1 Non-Combustible Materials 289
10.1.1.2 Combustible Materials - Laboratory Tests 290
10.1.1.3 Roofs 291
10.1.1.4 Large-Scale Tests 291
10.1.2 Official Approval 292
10.1.3 Present Situation and Future Developments 292

10.2 Americas 293
Marc Janssens
10.2.1 United States of America 293
10.2.1.1 Statutory Regulations 293
10.2.1.2 Consensus Standards 297
10.2.1.3 Test Methods Referenced in U.S. Building and Fire Safety Codes 300
10.2.1.4 Summary of Small-Scale Ignition and Combustibility Tests 303
10.2.1.5 Summary of Small-Scale Reaction-to-Fire Tests 309
10.2.1.6 Summary of Intermediate and Large-Scale Reaction-to-Fire Tests 315
10.2.1.7 Summary of Fire Tests for Furnishings and Contents 325
10.2.1.8 Summary of Fire Resistance Tests 325
10.2.1.9 Summary of Exterior Fire Exposure Tests 329
10.2.1.10 Future Developments 334
10.2.2 Canada 335
10.2.2.1 Statutory Regulations 335

10.2.2.2 Testing of the Fire Performance of Building Materials and Components 336
10.2.2.3 Summary of Selected Material Flammability Tests Referenced in the NBCC 339
10.2.2.4 NBCC Fire Performance Requirements for Interior Finish Materials 345
10.2.2.5 Accredited Testing Laboratories 346
10.2.2.6 Future Developments 346

10.3 Europe 348
10.3.1 European Union 348
Björn Sundström
10.3.1.1 Introduction 348
10.3.1.2 The Euroclasses: Classification Systems and Fire Tests 350
10.3.1.3 Future Developments 376
10.3.2 National Fire Regulations and Tests 377
Edith Antonatus
10.3.2.1 Austria 378
Dieter Werner
10.3.2.2 Belgium 388
Bart Sette
10.3.2.3 France 394
Eric Guillaume
10.3.2.4 Germany 403
Edith Antonatus
10.3.2.5 Hungary 413
István Móder, Imre Juhász
10.3.2.6 Italy 417
Eleonora Anselmi
10.3.2.7 The Netherlands 428
R. J. M. (Rudolf) van Mierlo
10.3.2.8 Nordic Countries 432
Björn Sundström
10.3.2.9 Poland 435
Jadwiga Fangrat
10.3.2.10 Spain 442
Antonio Galán Penalva
10.3.2.11 Switzerland 451
Marcel Donzé
10.3.2.12 United Kingdom 465
Edith Antonatus

10.4 Asia/Pacific 479
10.4.1 China 479
Jun-Sheng Wang and Xing Jin
10.4.1.1 Statutory Regulations 479
10.4.1.2 Classification and Testing of the Fire Performance of Building Materials and Components 487
10.4.1.3 Official Approval 498
10.4.1.4 Future Developments 499
10.4.2 Japan 499
Hideki Yoshioka
10.4.2.1 Statutory Regulations 499
10.4.2.2 Classification and Testing of the Fire Performance of Building Materials and Components 500
10.4.2.3 Recent Developments of New Standard Test Methods as JIS (Japanese Industrial Standard) 504
10.4.2.4 Official Approval 509
10.4.2.5 Future Developments 510
10.4.3 Republic of Korea 512
Jungmin Choi
10.4.3.1 Statutory Regulations 512
10.4.3.2 Fire Performance Classification and Testing of Building Materials 513
10.4.3.3 Official Approval 516
10.4.3.4 Future Developments 517
10.4.4 Australia 518
Alex Webb
10.4.4.1 Statutory Regulations 518
10.4.4.2 Classification and Testing of the Fire Performance of Building Materials and Components 519
10.4.4.3 Official Approval 527
10.4.4.4 Future Developments 529
10.4.5 New Zealand 530
Peter Whiting
10.4.5.1 Statutory Regulations 530
10.4.5.2 Classification and Testing of the Fire Performance of Building Materials and Components 532
10.4.5.3 Official Approval 533

10.4.5.4 Codemark 534
10.4.5.5 Future Developments 535

11 Transportation 537

11.1 Motor Vehicles 537
Anja Hofmann and Jonas Brandt
11.1.1 Introduction 537
11.1.2 Background – Vehicle Fires 538
11.1.3 Statutory Regulations 539
11.1.4 Requirements and Tests of Interior Materials for Buses and Cars 541
11.1.5 Other Fire Safety Requirements for Vehicles 548
11.1.6 Changes in Regulations 550
11.1.7 Potential Future Changes in Regulations 552

11.2 Rail Vehicles 554
11.2.1 Introduction 554
Torben Kempers
11.2.2 Europe 556
Torben Kempers
11.2.2.1 Harmonized Railway Standard EN 45545 556
11.2.2.2 Future Developments 569
11.2.3 United States of America and Canada 570
Marc Janssens
11.2.3.1 United States of America 570
11.2.3.2 Canada 578
11.2.3.3 Future Trends 578
11.2.4 People's Republic of China 579
Shuai-Xia Tan
11.2.4.1 Statutory Regulations 579
11.2.4.2 Fire Protection Requirements, Classification, and Tests 581
11.2.4.3 Future Developments 590
11.2.5 Japan 591
Koichi Wada
11.2.5.1 Statutory Regulations 591
11.2.5.2 Classification and Testing of Materials and Components 595
11.2.5.3 Official Approval 598
11.2.5.4 Future Developments 598
11.2.6 Republic of Korea 599
Jungmin Choi

11.2.6.1 Statutory Regulations 599
11.2.6.2 Fire Hazard of Rail Vehicles 600
11.2.6.3 Fire Performance Criteria 601
11.2.6.4 Future Developments 606
11.2.7 Australia 607
Alex Webb
11.2.7.1 Statutory Regulations 607
11.2.7.2 Passenger Rolling Stock 607
11.2.7.3 Materials Fire Performance 608
11.2.7.4 Future Developments 608

11.3 Ships 609
Edith Antonatus
11.3.1 Statistics Regarding Fires on Board Passenger Ships ... 610
11.3.2 SOLAS Chapter II-2: International Regulations for Fire Safety of Ships 612
11.3.2.1 Development of SOLAS Fire Safety Regulations and Revision Activities 612
11.3.2.2 Structure of Chapter II-2 of SOLAS 613
11.3.2.3 Alternative Design and Arrangements for Fire Safety 615
11.3.3 Fire Test Procedures Code (FTP code) 616
11.3.4 Provisions for High-Speed Craft 625
11.3.5 Future Developments 626

11.4 Aircraft 627
Torben Kempers
11.4.1 Introduction 627
11.4.2 Statutory Regulations 628
11.4.3 Fire Behavior Testing 629
11.4.3.1 Bunsen Burner Flammability Testing 631
11.4.3.2 Heat Release Testing 635
11.4.3.3 Smoke Emission and Toxicity Testing 637
11.4.3.4 Additional Testing 639
11.4.3.5 Thermal/Acoustic Insulation Testing 642
11.4.4 Future Developments 644

12 Electrical Engineering and Cables 647

12.1 Electrical Engineering 647
Jun Haruhara, Mattia Ferraris, and Jürgen Troitzsch
12.1.1 International Fire Safety Requirements and Standards (IEC) 649
12.1.1.1 Objectives and Organization 649

12.1.1.2 Fire Hazard Testing, IEC TC 89 650
12.1.1.3 International Certification 675
12.1.2 Fire Safety Regulations in North America (UL, CSA) ... 676
12.1.2.1 Test and Approval Procedures of the Underwriters' Laboratories 676
12.1.2.2 CSA Test and Approval Procedures 690
12.1.3 Fire Regulations and Standards in Europe 691
12.1.3.1 European Committee for Electrotechnical Standardization (CENELEC) 691
12.1.3.2 Electrotechnical Products and EU Directives .. 692
12.1.3.3 Approval Procedures in Europe 694
12.1.4 Fire Regulations and Standards in Asia 696
12.1.5 Other Fire Safety Test Methods 698
12.1.6 Future Developments 699

12.2 Fire Hazard Assessment of Cables 705
Esther Hild
12.2.1 Small-Scale Testing 705
12.2.2 Large-Scale Testing 709
12.2.3 Side Effects of Cable Fires 719
12.2.3.1 Smoke 719
12.2.3.2 Flaming Droplets 720
12.2.3.3 Acidity 721
12.2.4 Fire Resistance Characteristics of Cables 721
12.2.5 Fire Hazard Assessment on Railway Rolling Stock Cables 725
12.2.6 Future Developments 726

13 Furniture and Furnishings 729

13.1 Introduction 729
Edith Antonatus

13.2 UK and Ireland 732
Edith Antonatus
13.2.1 Regulations and Tests for Furniture Suitable for Domestic Use 732
13.2.1.1 Test Methods Applied in the UK Regulations for Upholstered Furniture and Corresponding European Test Methods 732
13.2.1.2 Classification of Soft Upholstery in the UK ... 734
13.2.2 UK Regulations and Tests for Furniture Used in Public Areas 736

13.3 United States of America 738
Marc Janssens
13.3.1 Statutory Regulations 738
13.3.2 U.S. Smoldering Tests for Upholstered Furniture and Mattresses 740
13.3.3 U.S. Open-Flame Tests for Upholstered Furniture and Mattresses 746

13.4 Measurements and Predictions of the Burning Rate of Furniture 750
Marc Janssens
13.4.1 Measurement Methods 750
13.4.1.1 Cone Calorimeter 750
13.4.1.2 Furniture Calorimeter 751
13.4.1.3 Room Calorimeter 752
13.4.2 Model Predictions 754

13.5 Conclusions and Future Developments 755

Part III
Appendices **761**

Appendix 1 – Suppliers of Flame Retardants and Smoke Suppressants **763**

Appendix 2 – Abbreviations for Plastics and Flame Retardants ... **769**

Appendix 3 – Flame Retardants: Chemical Names and CAS Numbers **775**
Rudolf Pfaendner

Appendix 4 – Journals and Books **779**

Authors Index **781**

Standards Index **783**

Keywords Index **791**

Part I
Fundamentals

1 Introduction

1.1 Fire Prevention and Fires

Jürgen Troitzsch

The idea of fire prevention in the context of fire protection is very old. The main objectives have always been to protect life and property against fires.

Preventive fire protection has been practiced since antiquity. For example, fire safety regulations existed even in ancient Rome, and attempts were made to protect combustible materials more effectively against unwanted fires. However, fires burnt down whole cities because buildings were very close to each other, consisted mainly of wood or similar combustible materials, and open fires were used [1, 2].

Since antiquity, attempts have been made to extinguish fires by adequate fire-fighting strategies, and to prevent or delay fires by fire protection and prevention measures. Examples of early fire protection activities are given in the references [3, 4].

Today, advanced technologies in fire detection, firefighting, and passive fire protection, as well as new construction techniques, help in preventing fires, injury and loss of life, and property damage. In addition, fire safety legislation and regulations exist in developed and emerging countries all over the world. Modern fire protection and prevention plays a substantial role in reducing fire hazard. Up-to-date regulations in most developed countries ensure that high-rise buildings and premises as well as a number of especially sensitive buildings like hospitals, care homes, and buildings for public assemblies, have effective alarm systems, smoke exhaust systems, and escape ways, and are protected by fire alarms and sprinkler systems. In addition, for larger buildings including industrial buildings, compartmentation helps to limit the spread of fire and smoke. These requirements usually do not apply for single-family homes, but requirements for basic fire safety levels of building materials and products exist in a number of European countries and in China. In addition, smoke detectors have become mandatory for buildings in many countries. In general, it can be said that products used in buildings and transportation have to meet fire performance requirements regarding their contribution to ignitability, flame spread, and heat release. In addition, smoke development and toxicity requirements do apply in a few countries (such as Russia, China, and Japan) for some building products, and in many countries (for example railways in the European Union, seagoing ships, and manufacturers' requirements in aircraft) for transportation. However, there is one exception: in most countries the reaction-to-fire requirements for highway vehicles are still very low compared to other transportation means.

Parallel to the advances in fire protection, unprecedented developments in science and technologies have led to new challenges. One is scale: Today, mega-cities with high-rise buildings or towers more than 500 m high and mega-warehouses covering more than 100,000 m^2 with high fire loads have to be considered. This also applies to the extremely rapid growth of mass-transportation: in 2018, an estimated 472 billion passenger-kilometers were traveled by rail in the EU, up 1.5% from the previous year [5]. In 2017, cruise ships (some with capacity for over 5,000 passengers) carried 25.8 million passengers [6], and airplanes conveyed 4.3 billion air-passengers worldwide in 2018 [7]. On the other hand, increasing wealth and living standards are a result of the steadily growing gross national product of the industrialized and emerging countries. This is coupled to the exponential rise of combustible materials and products in use, and here most importantly of plastics, in our living environment. World plastics production increased from 1.5 million metric tons in 1950 to 359 million metric tons in 2018 [8].

Compared to the past, these global changes and developments have resulted in new fire hazards, which have led to fire disasters in high-rise buildings, entertainment centers, discotheques, road and rail tunnels, buses, ships, and aircraft. To give an idea of the diversity of fires, a comprehensive list of fires including town, building, transportation and other fires from the antiquity up to the present is compiled in the Wikipedia "List of fires" [9]. In the following sections, a few examples of more recent spectacular deadly fires compiled from newspapers and official reports give an initial, rough indication of fire causes and first items ignited. Unfortunately, in many cases, official reports on these catastrophes are not publicly available, so that fire causes and first items ignited may eventually differ from what was presented in the newspapers.

High-Rise Buildings

A number of major fires in high-rise buildings have happened in recent years. Many of these fires have led to tragic consequences because of rapid smoke and fire spread inside the building, the absence of fire alarms and sprinklers, the absence of safe and smoke-free evacuation routes, and major difficulties experienced by fire brigades in attacking the fire. While the first high-rise buildings had façades made of concrete or bricks, this has changed because of increasing requirements for energy saving and new designs. Cladding and insulation systems are increasingly applied on façades, resulting in a growing number of fires spreading along the façade and leading to high risks for the inhabitants. Therefore, in many countries the regulations for high-rise buildings are currently under review. This includes not only the improvement of fire safety requirements for façade cladding and insulation systems, but also the safety provisions inside these buildings. Early detection, sprinkler systems, and provision of safe escape routes protected against fire and smoke are key factors in avoiding fatalities in high-rise building fires. Firefighting strategies and evacuation procedures are currently being reconsidered, too.

Two key examples of fire incidents involving fire spread on façades of high-rise buildings with a large number of fatalities are the 2017 Grenfell Tower fire in London and the 2010 fire in an apartment block in Shanghai. Both happened in buildings that did not have sufficient alarm systems and sprinklers, where there were no smoke-free and safe escape routes, and that did not have an appropriate evacuation strategy. Several huge façade fires in the United Arab Emirates are good examples for the effectiveness of a comprehensive fire safety concept (see the third example below). In this country, for a number of years, the same type of rainscreen cladding used for the Grenfell tower was legally approved and frequently applied on high-rise buildings, resulting in the occurrence of a number of fast-spreading façade fires. Nevertheless, there were almost no fatalities, as the fire safety measures, escape ways and evacuation procedures allowed the occupants to be saved.

- 9 February 2009. A blaze in Beijing, China, involved the uncompleted Television Cultural Center (CCTV) building and caused one fatality. Fire cause: Unauthorized fireworks. First item ignited: The high-temperature particles from the fireworks landed on the western part of the roof, penetrated the metal panels, and ignited the insulation materials and waterproof sheets underneath the panels [10].
- 15 November 2010. A fire destroyed a 28-story high-rise apartment building in Shanghai, China, killing at least 58 people and injuring more than 70 others. Fire cause: Welding sparks. The first item ignited was PU foam on the façade installed for an external thermal insulation composite system (ETICS), but not yet covered with the outer mortar layer. Bamboo scaffoldings probably contributed to the fast fire spread [10, 11, 12].
- 31 December 2015. A fire severely damaged the 63-storey Address Downtown hotel in Dubai, United Arab Emirates. 16 persons were injured. Fire cause: Exposed wiring. The fire spread was due to the use of exterior cladding panels made of aluminum with a non-flame-retarded polyethylene core [13].
- 14 July 2017. A fire broke out in the 24-storey Grenfell Tower block of flats in North Kensington, London, UK. It caused 72 deaths, including those of two victims who later died in hospital. More than 70 others were injured and

223 people escaped. Probable fire cause: Power surge, setting faulty fridge-freezer on fire. The fire propagated and set the surrounding exterior cladding (aluminum composite panels with non-flame-retarded polyethylene core) on fire, and the air gap in the insulation system enabled a stack effect which contributed to the extent of the fire disaster [14].

Discotheques

In most of the discotheque fires listed below, which caused a high number of fatalities, basic safety rules and regulations were not followed. In many cases, decoration, internal lining and acoustic insulation were made of materials which were not flame-retarded and so ignited easily. In most cases, the ignition source was pyrotechnics, which should not have been used inside a building. Often, the major cause for the high number of fatalities was that no adequate exit ways were available, or that the exit ways were blocked.

- 29 October 1998. A discotheque fire in Gothenburg, Sweden, led to 63 fatalities and 213 injuries. Fire cause: arson. First item ignited: floor covering [15].
- 20 February 2003. 100 people were killed in the Station nightclub fire in West Warwick, Rhode Island, USA. Fire cause: pyrotechnics. First item ignited: polyurethane foam [16].
- 30 December 2004. 194 were killed and another 714 injured in the República Cromañón nightclub in Buenos Aires, Argentina. Fire cause: pyrotechnics. First item ignited: polyurethane foam [17].
- 21 September 2008. 43 people were killed and another 88 injured in the Wuwang Club fire, Shenzhen, China. Fire cause: pyrotechnics. First item ignited: flammable ceiling [18].
- 5 December 2009. 154 were killed in a fire at the Lame Horse Nightclub in Perm, Russia. Fire cause: pyrotechnics. First item ignited: plastic and dried wood twigs from the ceiling [19].
- 27 January 2013. At least 220 people were killed and hundreds injured at a fire at the Kiss nightclub in Santa Maria, Rio Grande do Sul, Brazil. Fire cause: pyrotechnics. First item ignited: acoustic insulation [20].
- 30 October 2015. 63 people were killed and 148 were injured in a fire at the Colectiv nightclub in Bucharest, Romania. Fire cause: pyrotechnics. First item ignited: non-flame-retarded foam plastic decoration column inside dance area [21].

A compilation of the deadliest public assembly and nightclub fires until 2009 is available from the NFPA [22].

Transportation

The incidents listed here mostly happened in vehicles meeting existing rules and regulations. Many of these incidents led to an improvement of the laws and standards for the respective type of vehicles, and moreover, as in the example of the Mont-Blanc Tunnel fire, to an improvement of safety measures for tunnels.

- 18 November 1987. A fire in the King's Cross London Underground Station, UK, killed 31 people. Fire cause: match flame. First item ignited: grease and detritus [23].
- 28 October 1995. Around 300 people died as a result of a subway fire in Baku, Azerbaijan. Fire cause: sparks from high voltage cables? First item ignited: not specified [24].
- 2 September 1998. Swissair Flight 111 crashed into the Atlantic Ocean near St. Margarets Bay, Nova Scotia, Canada. All 229 people on board died. Fire cause: arcing in wiring. First item ignited: metalized polyethylene terephthalate covering on a thermal acoustic insulation blanket [25].
- 24 March 1999. A truck carrying flour and margarine caught fire in the Mont-Blanc Alpine Tunnel connecting France and Italy. The fire spread involved 23 trucks and 10 cars, and lasted for 53 hours. 42 people died in this disaster. Fire cause: probably glowing cigarette. First item ignited: not specified [26].
- 11 November 2000. A fire in an ascending funicular train in the tunnel of the Gletscherbahn in Kaprun, Austria, claimed 155 lives. Fire cause: fan heater caught fire. First item ignited: oil from leaking oil pipe [27].
- 4 November 2008. A tour bus caught fire on an autobahn near Hannover, Germany, killing 20 people. Fire cause unclear: glowing cigarette or short circuit in the on-board toilet. First item ignited: not specified [28].
- 28 December 2013. A train fire killed 26 people in Andhra Pradesh state, India. Fire cause: short-circuit. First item ignited: not specified [29].

The catastrophic fires described here have been mainly initiated by small ignition sources igniting flammable items, for example, insulation foams not complying with fire safety regulations. The ignition sources identified are pyrotechnics, welding sparks, arcing in wiring, short-circuits, power surges in faulty appliances, glowing cigarettes, matches, and flammable liquids.

However, an examination of catastrophic fires only presents the fire problem from one angle. In contrast, fire statistics give a comprehensive view on fires, fatalities, and losses. These show that some ignition sources which are highlighted in catastrophic fires, such as pyrotechnics, are not relevant for fire statistics, because compared to other ignition sources, they do not contribute to a large number of deadly fires, as can be seen from the publication on fireworks [30] in the USA.

Fire statistics show that fire safety regulations and requirements contribute to public safety, resulting in less fires, fatalities, and losses. This is encouraging, but more has to be done. In 2012, the U.S. National Institute of Standards and Technology (NIST) presented a strategic roadmap aiming at reducing the nation's preventable fire burden by a third [31]. One of the main objectives is to reduce fire hazards in buildings. In the past, NIST had already developed standards for children's sleepwear, automatic sprinklers, reduced-ignition-propensity cigarettes, modernized building codes, and computer models that can predict the behavior of fire, smoke, and toxic combustion products.

Another important factor for fire safety is the need to make sure that regulations are followed. Depending on the country, the control and enforcement of fire safety regulations is different. So, it is still possible that buildings are built and used, even if the legally binding provisions for fire safety are not followed. Especially in establishments like discotheques and assembly rooms, illegal products are often used for decoration and sound insulation and, in addition, adequate escape routes are not available (or even blocked).

Furthermore, regulations on the fire performance of materials and products and their correct application are not always followed during the construction and renovation of buildings.

In building fires, another important problem in terms of fire deaths is their contents. While construction products are strictly regulated in most countries, building contents including furniture are only subject to limited fire safety requirements.

In a white paper on upholstered furniture flammability, Hall gives an overview on research and proposes development activities to reduce the flammability of upholstered furniture (by choice of covering fabrics, filling materials, flame-retardant treatment of these materials, and the use of interliners as fire barriers and fire scenarios with cigarette-ignition, small-open-flame ignition and ignition by another fire) [32].

A general review of fires and fire prevention, fire statistics, fire performance requirements and testing is presented by Troitzsch [33].

References for Section 1.1

[1] Troitzsch, J. (2004). *Plastics Flammability Handbook* (3rd edition), p. 3, Hanser Publications, Munich.

[2] Proske, U. (2008). *Catalogue of Risks: Natural, Technical, Social and Health Risks*, p. 181, Springer Verlag, Berlin.

[3] Lyons, J.W. (1970). *The Chemistry & Uses of Fire Retardants*, p. 166, Wiley, New York.

[4] Pitts, J.J. (1973). Inorganic Flame Retardants, p. 133, in *Flame Retardancy of Polymeric Materials*, Vol. 1, W.C. Kuryla and A.J. Papa (editors), Marcel Dekker, New York.

[5] *https://ec.europa.eu/eurostat/statistics-explained/index.php/Railway_passenger_transport_statistics_-_quarterly_and_annual_data.*

[6] *https://www.f-cca.com/downloads/2018-Cruise-Industry-Overview-and-Statistics.pdf.*

[7] *https://www.icao.int/annual-report-2018/Pages/the-world-of-air-transport-in-2018.aspx.*

[8] Plastics – The Facts 2019, *https://www.plasticseurope.org/application/files/9715/7129/9584/FINAL_web_version_Plastics_the_facts2019_14102019.pdf.*

[9] *https://en.wikipedia.org/wiki/List_of_fires.*

[10] Peng, L., Ni, Z. and Huang, X. (2013). Review on the fire safety of exterior wall claddings in high-rise buildings in China. *Procedia Engineering*, vol. 62, pp. 663–670.

[11] *https://www.nfpa.org/-/media/Files/News-and-Research/Fire-statistics-and-reports/Building-and-life-safety/RFFireHazardsofExteriorWallAssembliesContainingCombustibleComponents.ashx.*

[12] *http://edition.cnn.com/2010/WORLD/asiapcf/11/16/china.building.fire/index.html.*

[13] *http://www.usnews.com/news/world/articles/2016-01-20/dubai-police-exposed-wiring-likely-sparked-nye-tower-fire.*

[14] *https://en.wikipedia.org/wiki/Grenfell_Tower_fire.* Accessed May 2020.

[15] SHK (2001). The discotheque fire in Gothenburg 1998. Swedish Accident Investigation Authority. *http://www.havkom.se/en/om-shk/vaegtrafik-och-oevriga-olyckor/diskoteksbranden-i-goeteborg-1998.* Accessed July 2015.

[16] Grosshandler, W.L., Bryne, N. and Madrzykowski, D. (2005). Report of the technical investigation of the Station Nightclub fire (NIST NCSTAR 2). *http://www.nist.gov/el/disasterstudies/fire/station_nightclub_fire_2003.cfm.* Accessed July 2015.

[17] Hall, J.R. (2008). Disco fire in Buenos Aires (p. 25). *Fireworks*, NAFPA.

[18] *http://news.bbc.co.uk/2/hi/asia-pacific/7628580.stm.*

[19] *http://news.bbc.co.uk/2/hi/europe/8396587.stm.*

[20] *http://nfpatoday.blog.nfpa.org/2013/01/news-reports-nightclub-fire-in-brazil-kills-at-least-220-people.html.*

[21] *https://en.wikipedia.org/wiki/Colectiv_nightclub_fire.*

[22] *https://www.nfpa.org/Public-Education/Staying-safe/Safety-in-living-and-entertainment-spaces/Nightclubs-assembly-occupancies/Deadliest-public-assembly-and-nightclub-fires.*

[23] Fennel, D. (1988). Investigation into the King's Cross Underground Fire. Department of Transport. *http://www.railwaysarchive.co.uk/documents/DoT_KX1987.pdf.*

[24] *http://www.nytimes.com/1995/10/30/world/subway-fire-kills-300-in-caucasus-capital.html.*

[25] Transport Safety Board of Canada (2003). Aviation Investigation Report: In-flight fire leading to collision with water. Report number A98H0003. *http://www.tsb.gc.ca/eng/rapports-reports/aviation/1998/a98h0003/a98h0003.pdf.*

[26] *http://www.diva-portal.org/smash/get/diva2:962271/FULLTEXT01.pdf.*

[27] *http://www.spiegel.de/panorama/kaprun-brandursache-in-gletscherbahn-geklaert-a-155068.html.*

[28] *http://www.theguardian.com/world/2008/nov/05/germany.*

[29] *http://www.bbc.com/news/world-asia-india-25533209.*

[30] Hall, J.R. (2013). Indoor use of fireworks (p. 2). *Fireworks*, NAFPA.

[31] NIST (2012). Reducing the risk of fire in buildings and communities: A strategic roadmap to guide and prioritize research (Special Publication 1130). *http://www.nist.gov/customcf/get_pdf.cfm?pub_id=909653.*

[32] Hall, J.R. (2013). White paper on upholstered furniture flammability, NFPA. *http://www.nfpa.org/Assets/files/AboutTheCodes/277/2156%20-%20UpholsteredFurnWhitePaper.pdf.*

[33] Troitzsch, J. (2016). Fires, statistics, ignition sources, and passive fire protection measures. *Journal of Fire Sciences*, vol. 34, pp. 171–198. DOI: 10.1177/0734904116636642.

1.2 Fire Statistics

Eric Guillaume

1.2.1 Introduction

Fire is a very complex phenomenon involving many different parameters. Achieving a complete and precise understanding - including causes and circumstances of the fire, casualties, and damages, as well as firefighting issues - needs relevant, efficient, and complementary indicators [1]. These indicators may be found in fire statistics.

In many countries, provinces, municipalities, and cities, fire departments regularly collect an important number of fire-related statistics. Mostly, these statistics are published and are now increasingly available on official websites.

Fire statistics are very often used by regulators to ensure that fire regulations are based on reliable data, by firefighters to give a better response to fire incidents, and by fire experts and industries to assist in using the most relevant design fire scenarios in fire tests and engineering studies. They are used by insurers to identify and assess the risks, by authorities to define the ways of improving regulations and to justify government policies, and last (but not least) by journalists and consumer associations to highlight fire safety issues.

1.2.2 Lessons Learned from Recent ISO Work on Fire Statistics

In order to understand how fire statistics are captured and the criteria on which they are based, in 2011, the International Standardization Organization (ISO) Technical Committee ISO/TC 92 *Fire Safety* decided to collect information through a special enquiry, which was sent to all ISO/TC 92 members. The results were published in the Technical Report ISO/TR 17755-1, released in 2014 [2].

Ten countries responded to the ISO survey: Canada, the USA, France, the UK, Russia, Kenya, Australia, the People's Republic of China, the Republic of Korea, and Japan. For the authors of this document, the answers received were a total surprise: indeed, they show that fire statistics, contrary to public opinion, are mostly not comparable between countries, and, worse, in some cases, between fire departments in the same country. The study showed that this was the result of a lack of common definitions of the different terms and expressions used in fire statistics, a lack of common methodology used for collecting fire statistics, and a lack of training and qualification of fire investigators involved in the capture and the analysis of fire statistics.

Lack of Common Definitions

This lack of common definitions of the different terms and expressions used in fire statistics becomes obvious, for example, when studying the meaning of "fire fatality". Countries have different approaches in how to consider the time elapsed after a fire to record a fire fatality. Many of them, such as Australia, France, Kenya, Russia, the UK, and the USA, do not fix any time limit for recording a fire death. Without stating any reason, Canada counts a fire fatality as a result of injuries leading to death within one year and one day after the fire incident. In Japan, "death within 48 hours after fire" is regarded as fire death, whereas the Korean definition of a fire fatality is "death occurring within 24 hours after a fire".

Following this terminology issue, ISO recently published ISO/TS 17755-2 [3] as a basis of common definitions.

Lack of Common Methodology

One of the most relevant examples of the lack of common methodology for collecting fire statistics data is how fire fatalities are recorded across countries. For example, neither Canada nor Russia take into account fire fatalities from road traffic accidents, air disasters, and rail disasters. The People's Republic of China does not include fire deaths resulting from arson. France only reports fire fatalities at the location of the fire – either those discovered by rescuers, or those declared dead after unsuccessful resuscitation attempts. Therefore, the French official database does not consider fire fatalities that occurred at the hospital or during transport of casualties to the hospital.

Another example, highlighted by the recent catastrophic Grenfell Tower fire of London, is the difference in the definition of buildings. In the case of high-rise buildings (HRB), this is quite relevant: in Russia, all buildings over 50 m high (calculated from the ground to top of the building) are considered HRBs, while in the USA this category applies to buildings that are over 23 m (75 ft) high. In France, HRBs were originally those above 28 m high, except for dwellings with a limit above 50 m, but a new category has recently been introduced for all buildings between 28 m and 50 m high (all calculated from the ground to the highest floor usable by people). In Argentina, all buildings with more than 20 floors are classified as HRBs, but in Sweden HRBs must have 16 floors or more.

Lack of Training and Qualification of Firefighters

Virtually all fire reports are written by the fire officers in charge. But although these fire reports contain some sections for which training has been received (description of the situation during and after the fire, description of the fire suppression operations, description of the rescue operations, etc.), the main issue is that other sections require competencies for which the officers never received any training (origin of the fire, source of the fire, characteristics and circumstances of the fire, etc.), or for which they have insufficient time to give sound answers.

As a consequence, in Canada and the People's Republic of China, nearly all fire reports are filled in by fire officers who never received training in fire investigations. In other countries, such as Kenya, the Republic of Korea, Japan, and the USA, only a small number of fire reports are filled in by fire officers who received specially adapted training.

1.2.3 Current National Fire Statistics: Relevant Data on Specific Issues

Even if current fire statistics cannot be compared from one country to another (with a few exceptions), they can still be useful to describe the global fire safety situation and trends for a group of countries, or the specific fire safety situation of a country.

1.2.3.1 Number of Fires

In many countries, the trend is to a decreasing or (more recently) an unchanged number of fires. The number of reported fires and fires per million inhabitants for selected countries (USA [4, 5], Russia [6, 7], France [6, 8], UK [6, 9], Finland [6, 10], and Switzerland [6, 11, 12]) are illustrated in Figure 1.1 and Figure 1.2.

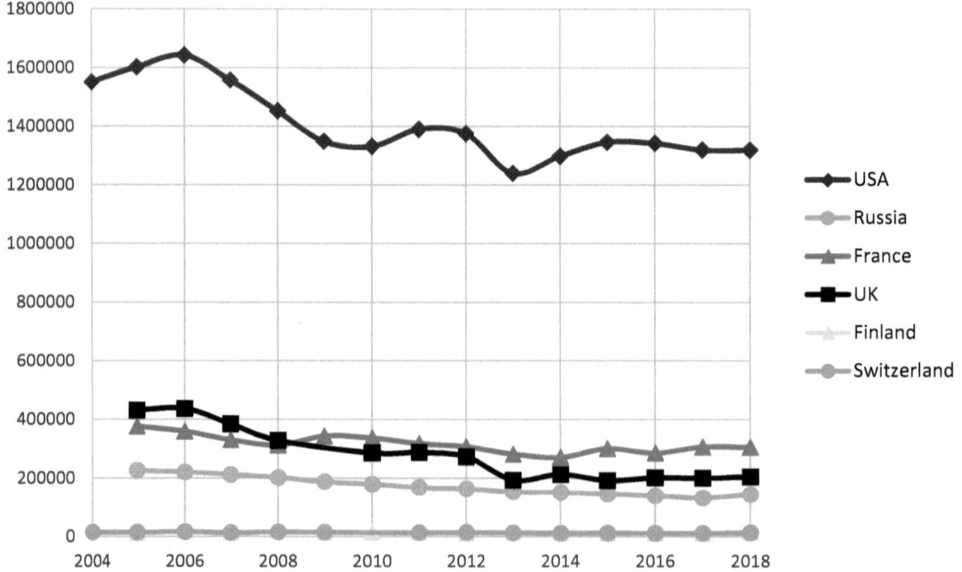

Figure 1.1 Number of fires in selected countries

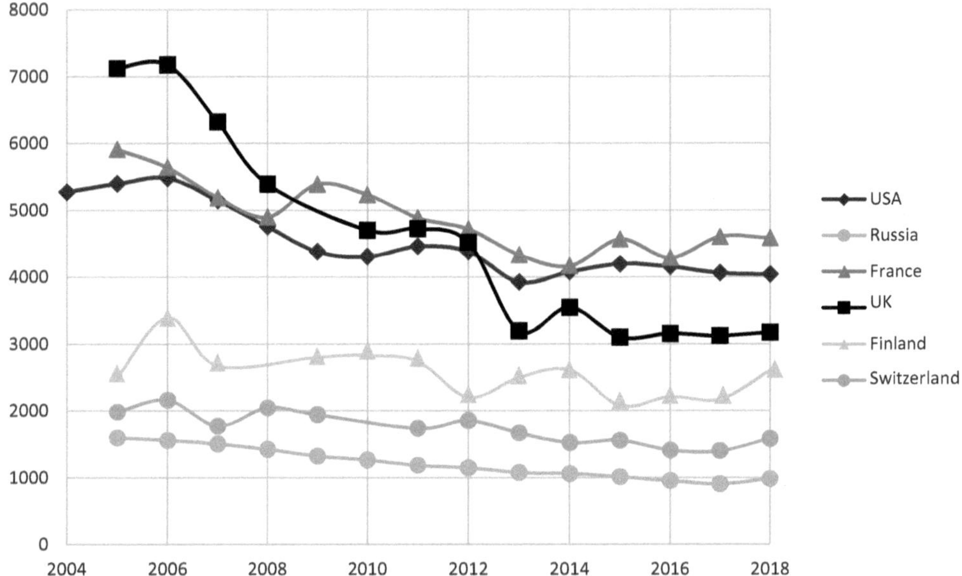

Figure 1.2 Fires per million inhabitants in selected countries

The analysis of the "World Fire Statistics" reports [6], published annually by the International Association of Fire and Rescue Service (CTIF), shows that for the 16 EU countries for which statistics were available (representing 62% of the EU population at that time), there was a decrease of 19% in the number of fires between 2006 and 2010.

Trend Uncertainties

The above statistics are called into question by recent local developments, which may reverse the global trend.

For example, in France, the last available Ministry of Interior official fire statistics [8] show an increase of 11% in the number of fires in 2015 compared to 2014. In the USA [4, 5], 1,240,000 fires were recorded in 2013, rising to 1,298,000 in 2014, representing an increase of 4.7% in one year.

In other cases, the situation seems to be unstable, with periods of increase followed by periods of decrease. An example is the number of fires in the Republic of Korea [13] from 2001 to 2009, shown in Figure 1.3.

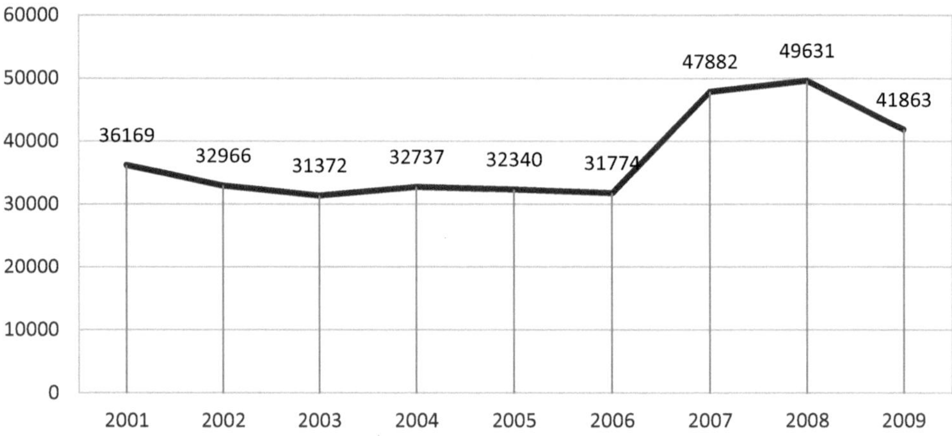

Figure 1.3 Number of fires in the Republic of Korea from 2001–2009

Local Trends to a Larger Number of Fires

As can be seen in Table 1.1, in some countries, the trend is to a larger number of fires; this may be due to an improved data collection system.

Table 1.1 Trend to a Larger Number of Fires in Selected Countries

Country	Number of reported fires					Change
	1995	2004	2006	2009	2010	
Poland [6]	72,391	–	–	159,122	–	+220%
Austria [6]	-	–	30,297	-	34,363	+13%
Ghana [6]	-	2,418	–	2,708	–	+12%

1.2.3.2 Number of Fire Fatalities

In many countries, a trend to a decreasing number of fire fatalities can be seen. However, in some countries, the situation is different and the number of fire fatalities is growing. In addition, some fire statistics show an unexpectedly high proportion of male fire fatalities, and also indicate that most fire deaths are recorded in residential building fires.

Decrease of Fire Fatalities in Many Countries

The number of fire fatalities and fire fatalities per million inhabitants for selected countries is illustrated in Figure 1.4 and Figure 1.5 (here, the curves for France are virtually covered by the UK ones).

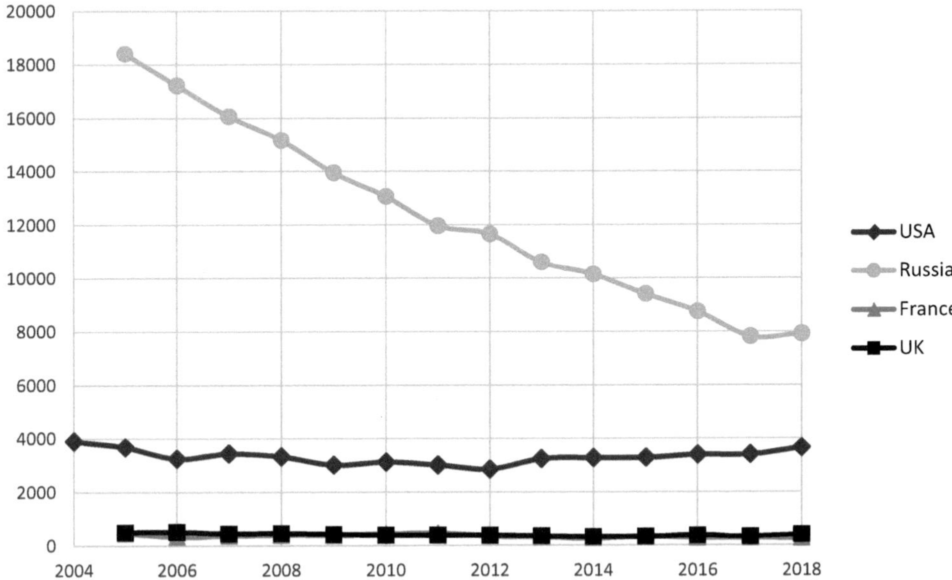

Figure 1.4 Number of fire fatalities in selected countries

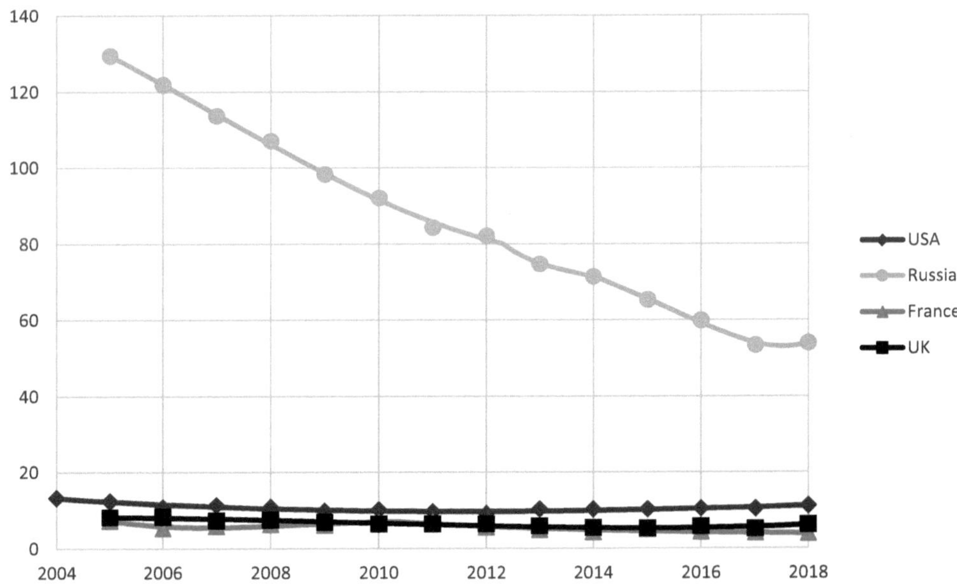

Figure 1.5 Number of fire fatalities per million inhabitants in selected countries

The “World Fire Statistics” issue no. 7 (published in 2012) covers the period 2006–2010, and shows that for the 19 EU countries for which statistical data were available (representing 74% of the EU population at that time), there was a decrease of 17% in the number of fire fatalities.

In the French official fire statistics, the number of fatalities at the location of fires (either discovered, or declared dead after unsuccessful resuscitation attempts) has decreased by 60% in the 32 years from 1982–2014, as summarized in Table 1.2.

Table 1.2 Total Number of Fire Fatalities Discovered at the Location of Fires

	Year						Difference	Change
	1982	1990	1999	2008	2012	2014		
Fatalities	702	579	460	402	362	280	422	-60%

Statistics on Japanese fire fatalities [15] show a significant proportion of suicides, and a higher-than-expected proportion of deaths of elderly people.

Trend Uncertainties

In the USA [4, 5], 2,855 fire deaths were recorded in 2012, increasing to 3,240 in 2013 and to 3,275 in 2014, representing an increase of nearly 15% in two years. In France, the Ministry of Interior’s official fire statistics [8] show an increase of 16% from the number of fire fatalities in 2014 (280) to 2015 (325).

All figures dealing with fire fatalities must be considered carefully. Indeed, the exact definition of a fire death is rarely specified in documents on fire statistics. Therefore, it is possible that a significant number of people with fire injuries died in hospital or during transport to it, and so may not have been counted in the final statement, skewing the final result.

In Ireland, the USA and Finland, the proportion of fire fatalities that were male was unexpectedly high, as shown in Table 1.3.

Table 1.3 Male Fire Fatalities in Selected Countries

Country	Year	Male fire fatalities	Female fire fatalities
Ireland [14]	2007	66%	34%
USA [4, 5]	2015	62%	38%
Finland [10]	2013	74%	26%

Fire Fatalities in Residential Buildings

Residential buildings contribute to most fire deaths because people live, cook, and sleep there.

In the official 2014 fire statistics published by the French Ministry of Interior in 2015, of the 280 fire deaths reported by the French Fire Departments, 228 fire fatalities were recorded in residential buildings, representing over 81% of all fire deaths. In a special study carried out by the Paris Police Laboratory for the three years 2012–2014, over 86% of all fire fatalities discovered at the location of fires by the Paris Fire Brigade (BSPP) were recorded in residential buildings, including one fire death in a residential high-rise building (IGH A). This is also true for some selected other countries, as shown in Table 1.4.

Table 1.4 Residential Fire Fatalities in Some Selected Countries

Country	Year	Proportion of residential fire fatalities vs total number of fire fatalities
Korea [13]	2010	65%
England [9]	2016/2017	82%
Finland [10]	2013	94%

1.2.3.3 Number of Fire Injuries

The term fire injuries is very difficult to define, and the differences between "minor injury", "moderate injury" and "severe injury" are numerous between countries – indeed, much more so than for "fire death". In some countries, there are even additional classifications – for example, in France, injuries are divided into "UR" (*Urgence Relative* = relative emergency) and "UA" (*Urgence Absolue* = absolute emergency).

Having highlighted these caveats, the current fire statistics show important differences in trends between countries. Figure 1.6 and Figure 1.7 summarize the number of fire injuries and fire injuries per million inhabitants for the USA [4, 5], Russia [6, 7], France [6, 8], and the UK [6, 9]. Here again, as for fire fatalities, we can note an unexpectedly high proportion of male fire injuries in some countries, and Table 1.5 shows the statistics for Spain [16] and the USA.

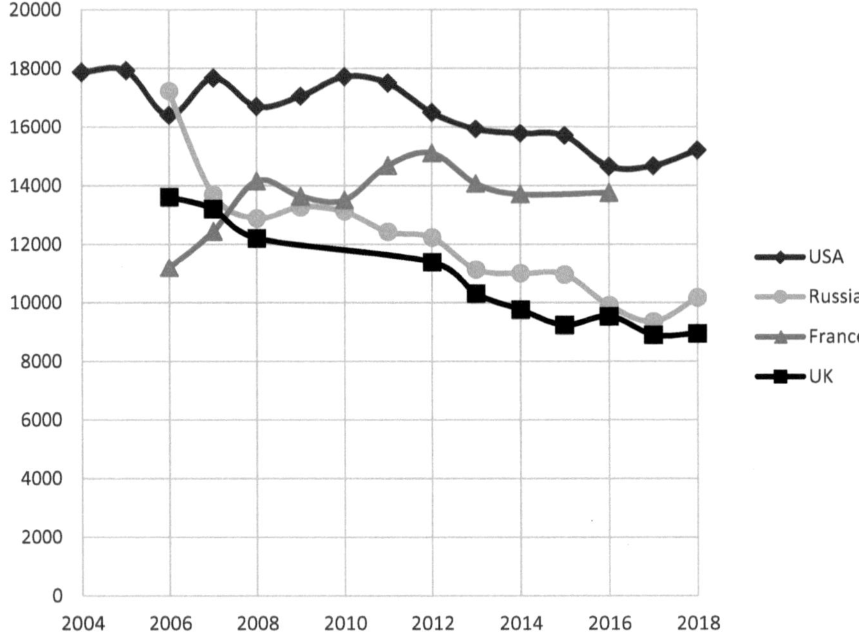

Figure 1.6 Number of fire injuries in selected countries

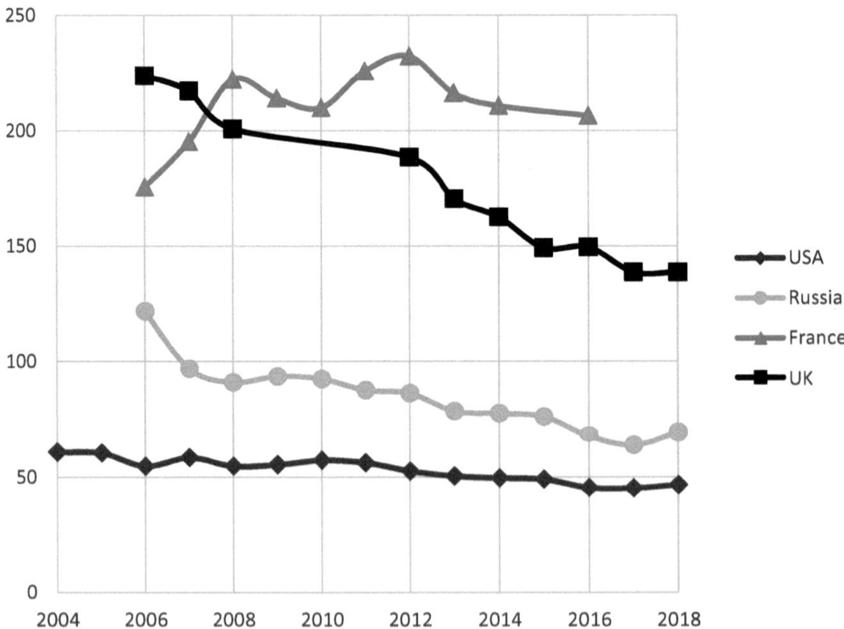

Figure 1.7 Number of fire injuries per million inhabitants in selected countries

Table 1.5 Male Fire Injuries in Spain and the USA

Country	Year	Male fire fatalities	Female fire fatalities
Spain [16]	2012	67%	33%
USA [4, 5]	2015	60%	40%

1.2.3.4 Costs of Fire Losses

Fire statistics show a surprising variation in trends for fire losses, with some countries suffering an increase, others a decrease, and others remaining stable. A small number of events - such as the major fires in the USA in 2017 and 2018 - may have an enormous impact and completely change the trend. Table 1.6 summarizes the costs for fire losses in the USA, Switzerland, Russia, and the Czech Republic, and Figure 1.8 shows year-by-year fire losses for these countries.

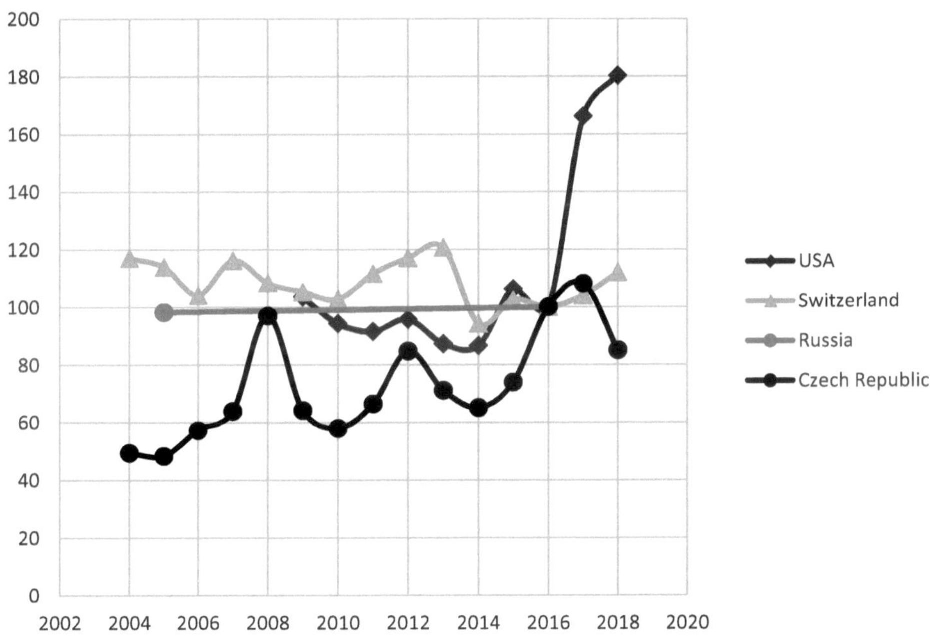

Figure 1.8 Variation of fire losses in selected countries

Table 1.6 Costs for Fire Losses in Selected Countries

Country	Year														
	2004	2005	2006	2007	2008	2009	2010	2011	2012	2013	2014	2015	2016	2017	2018
USA (billion USD, adjusted 2018) [4, 5]	–	–	–	–	–	14.7	13.4	13	13.6	12.4	12.3	15.1	14.2	23.6	25.6
Switzerland (million CHF) [12]	241	234.5	214.2	239.2	223.1	216.6	211.5	229.7	241	248.4	194	211	205.8	214.1	230.5
Russia (billion RUB) [7]	–	13.183	–	–	–	–	–	–	–	–	–	–	13.418	–	–
Czech republic (million CZK) [17]	1669.3	1634.4	1934.0	2158.5	3277.3	2169.2	1956.2	2241.8	2861.5	2402.6	2198.3	2495.9	3378.2	3653.1	2870.5

1.2.4 Future Developments

Even when fire statistics are numerous and precise, it remains difficult to derive useful conclusions from them, especially when wishing to compare the situation in different countries. To provide relevant information regarding the national fire safety situation (number of fires, fire fatalities, fire injuries, fire losses), fire statistics will have to be internationally improved through common terminology, common methodology, and common training and qualification of fire investigators. The terminology to be used is now proposed as a common basis in ISO/TS 17755-2.

Early in 2020, the European Commission decided to initiate a research program on fire statistics. The intention is to propose the harmonization of fire statistics to support national and European policies. With such initiatives, better and comparable fire statistics are expected in the near future.

References for Section 1.2

[1] Mladan, D., Subosic, D., Jakovljevic, V., *Fire and Rescue Service in the World: Status and New Challenges.* [*Medzinarodna vedecka konferencia. Riesenie krizovych situacii v specifickom prostredi*]. May 31, 2012.

[2] ISO/TR 17755:2014. Fire Safety - Overview of national fire statistics practices.

[3] ISO/TS 17755-2:2020. Fire Safety - Statistical data collection - Part 2: Vocabulary.

[4] *US Fire Statistics.* US Fire Administration (USFA), *https://www.usfa.fema.gov/data/statistics/* (accessed June 2020).

[5] *Fire in the United States (2008–2017).* US Fire Administration (USFA), National Fire Data Center. 20th Edition, November 2019. *https://www.usfa.fema.gov/data/statistics/reports/fius_2008-2017.html*

[6] Brushlinsky, N. N., Ahrens, M., Sokolov, S. V., Wagner, P., *World Fire Statistics.* Center of Fire Statistics, International Association of Fire and Rescue Services. Issue no. 23, 2018. Other issues are also analyzed.

[7] Statistics of fires and their analysis - VNIIPO, EMERCOM, VNIIPO.ru. *Fire and Fire Safety Statistics Journal*, 2016.

[8] *Statistiques Annuelles des Services d'Incendie et de Secours: Actions, Effectifs, Organisation, Moyens, Finances…*, Ministère de l'Intérieur. *https://www.interieur.gouv.fr/Publications/Statistiques/Securite-civile* (accessed June 2020).

[9] *Fire and Rescue Incident Statistics: England, April 2016 to March 2017.* Home Office. *https://www.gov.uk/government/statistics/fire-and-rescue-incident-statistics-england-april-2016-to-march-2017* (accessed June 2020).

[10] Ketola, J., Kokki, E., *Finnish Rescue Services' Pocket Statistics 2013–2017. http://info.smedu.fi/kirjasto/Sarja_D/D3_2018.pdf* (accessed June 2020).

[11] *Coordination Suisse des Sapeurs-Pompiers (CSSP). www.feukos.ch* (accessed June 2020).

[12] *Statistiques des Dommages.* AEAI (Association des établissements cantonaux d'assurance incendie) *https://www.vkg.ch/fr/assurance/r%C3%A9assurance/* (accessed June 2020).

[13] *Korean Fire Data.* KFPA. *https://www.kfpa.or.kr/eng/firedata/firedata2011.php* (accessed June 2020).

[14] *Ireland Fire Service Statistics. https://www.housing.gov.ie/community/fire-and-emergency-management/other/statistics* (accessed June 2020).

[15] *White Book on Fires 2012.* Japan Fire and Disaster Management Agency. *http://www.fdma.go.jp/html/hakusho/h24/h24/html/2-1-1b-2_1.html* (accessed June 2020).

[16] *Menos Incendios y Menos Víctimas Mortales.* Victimas de incendios en España – Fundación (MAPFRE). *https://www.fundacionmapfre.org/fundacion/es_es/descubre/menos-incendios-menos-victimas-mortales.jsp* (accessed June 2020).

[17] *Statistical Yearbooks.* Ministry of Interior – Directorate General of Fire and Rescue Service of the Czech Republic. *https://www.hzscr.cz/hasicien/article/statistical-yearbooks.aspx* (accessed June 2020).

2 The Burning of Plastics

Bernhard Schartel

Limited fire risks and hazards are crucial for materials for a variety of applications such as building construction, electrical engineering, electronics, transportation (aviation, railway vehicles, ships), and furniture and furnishings (upholstered furniture, mattresses, etc.). Thus, the fire performance of polymeric materials is a key property for several polymeric mass products, quite comparable to mechanical properties. At least 20 to 25% of all the polymeric materials used must be inherently flame resistant or flame retarded. Unfortunately, halogen-free commodity polymers such as poly(ethylene) (PE), poly(propylene) (PP), poly(styrene) (PS), poly(methyl methacrylate) (PMMA) and engineering polymers such as acrylonitrile butadiene styrene (ABS), poly(amide)s (PA 6, PA 66, etc.), bisphenol A poly(carbonate) (PC), poly(ethylene terephthalate) (PET), and poly(butylene terephthalate) (PBT) show a very high or considerable fire hazard, whereas only halogenous or high-performance polymers such as rigid poly(vinyl chloride) (PVC), poly(ether ether ketone) (PEEK), poly(ether imide) (PEI), polysulfone (PSU), poly(benzimidazole) (PBI), poly(*p*-phenylene-2,6-benzobisoxazole) (PBO), and poly(tetrafluoroethylene) (PTFE) are construed as inherently flame retarded [1]. The reason for this is that the fire hazard of organic materials is determined predominantly by their fuel production per mass during pyrolysis, or, in other words, by their remaining fire residue, their char yield (μ), and the effective heat of complete combustion of their volatile pyrolysis products (h_c) multiplied by the combustion efficiency (χ) [1–5]. Combustion efficiency is defined as heat release in an actual fire divided by the heat release for complete combustion of the volatiles. As a rule, char yield is favored only for some technical polymers and those high-performance polymers with a higher content of heteroatomic segments, and heterocyclic and aromatic groups in the main chain [6]. Quite short flexible linkages between aromatic rings tend to yield the cross-linking and aromatization prerequisite for charring. Also, polymers containing halogen exhibit rather low fire hazards, first because they release volatiles such as HBr or HCl, resulting in a reduction both in the effective heat of combustion due to fuel dilution and in combustion efficiency due to flame inhibition; and second, because the remaining unsaturated hydrocarbon structures have a strong tendency to char at the same time, which is a major reason for PVC's success as a material in building construction.

The schematic illustration of the physico-chemical processes in Figure 2.1 shows the idealized quasi-stationary flaming combustion of a horizontal polymer specimen. Idealized quasi-stationary means that even important effects, such as flame spread and dripping, the formation and growth of a char layer, and other physical and chemical processes that change over time, are disregarded. However, it becomes obvious that fire is a complex interaction of physical and chemical phenomena that involves fluid mechanics, heat and mass transport, as well as chemical reactions and kinetics in the gas and condensed phases.

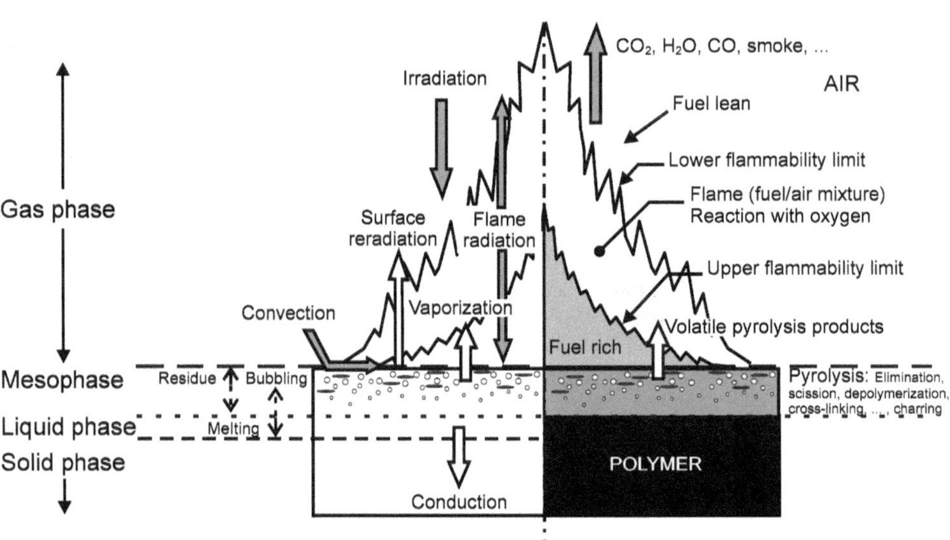

Figure 2.1 Diagram illustrating the main physical and chemical processes for the idealized quasi-stationary flaming combustion of a horizontal solid polymer

Radiation may be the dominant process in such a scenario with respect to heat transport, but conduction and convection also contribute significantly. In the condensed phase, the polymer heats up, and phase transitions such as melting, pyrolysis, and vaporization (/sublimation) occur. Here pyrolysis is generally endothermic, since the energy input must be sufficiently large to break any bond (bond dissociation energies = 200–400 kJ mol^{-1}) while leaving enough activation energy for the corresponding process. Polymer pyrolysis is determined by chemical reactions such as random scission, chain-end scission/depolymerization, the elimination of pendant groups/chain stripping, rearrangement, cross-linking, cyclization, aromatization, the fusion of aromatics, graphitization, and charring; which reactions apply is highly dependent on the chemical structure of the polymer and the interactions occurring with additives during pyrolysis [6–9]. The pyrolysis zone is, more precisely, a kind of mesophase, since solid, liquid, and volatile products occur at the same time and place. Since the boiling temperatures of some degradation products are much lower than the decomposition temperature of the polymer controlling the pyrolysis temperature, volatilization is often accompanied by intensive bubbling. Indeed, mass transport during the burning of a polymer becomes quite complex when examined in detail [10]. Mass flows are often somewhat discontinuous or turbulent. They are controlled by diffusion due to concentration and thermal gradients, as well as by phase separation and agglomeration processes [11]. The viscosity in the pyrolysis zone influences the mass transport and is controlled by the polymer, its liquid decomposition products of low molecular weight, cross-linking, charring, and increasing amounts of agglomerating inert fillers [12]. Since the burning of polymeric materials occurs in an ablative regime, the volatile evolution from a thin molten and pyrolyzing zone or decomposition front is the controlling mechanism. Quasi-stationary conditions like those illustrated in Figure 2.1, showing a pyrolysis zone running through a thermally thick polymer sample (specimen thickness is large compared to the thermal diffusion layer in the solid), are rather the exception for polymeric materials. As a rule, the final use conditions of all kinds of polymeric materials are accompanied by samples that are thermally thin or of intermediate thickness, with all charring materials showing a decreasing propagation rate of the reaction front, since the pyrolysis zones are increasingly covered and protected by a char layer.

How distinct fire behavior and flame retardancy influence fire performance in a specific fire test is not a trivial question and cannot be generalized, since an entire set of material properties such as char yield, effective heat of combustion of the volatiles, thermal conductivity, melt viscosity, and surface emissivity usually determine the fire behavior of a specimen. Furthermore, the role played by each material property differs from one fire test to the next. Each fire test simulates a distinct fire scenario in order to check whether a certain protection goal is achieved. The fire scenarios and protection goals considered vary widely with respect to different applications. Generally, fires differ crucially in their heat and mass transport, defined by characteristics such as applied heat flux, temperature, length scales, and ventilation (Table 2.1). Their behavior can be divided into three main general stages [13]:

Table 2.1 Fire Scenarios Summarized with Their Typical Parameters and Relevant Fire Properties

Fire scenario	Parameters	Ventilation	Specimen size	Fire properties
Ignition, incipient fire	Ambient temperature only, a localized ignition source with high temperatures, e.g., > 600 °C; irradiation typically negligible or small (match flames) 16–32 kW/m^{-2}	Fuel controlled fire, well ventilated	Confined to a small area, cm, small flames	Flammability, ignitability, ease of extinction
Developing fire, growth stage	Increased temperature 400–700 K, intensive irradiation 20–60 kW/m^{-2}	Well ventilated, or reduced 10–15 vol% O_2	dm–m	Flame spread, heat release rate, fire load
Fully developed fire	Temperature > 600 °C (e.g., smoke under the ceiling), often > 900 °C; high irradiation > 50 kW/m^{-2}; peak values of post flashover room fires 70–230 kW/m^{-2}, in large fires > 150–200 kW/m^{-2}, at flashover in a room fire ~20–25 kW/m^{-2} at the floor > critical heat flux for ignition of many materials	Ventilation controlled fire, lack of oxygen, O_2 < 10 vol% (high ventilation, fully developed fire) or below 5 vol% (low ventilation, fully developed fire)	> m	Fire resistance, fire stability, heat/hot gas penetration

As mentioned, properties like flammability (reaction to a small flame), flame spread, and fire penetration are not material properties, but merely describe the performance of a certain specimen in a certain fire scenario. By the way, terms such as “fire-proof”, “flash, flame or fire resistant”, “self-extinguishing”, and even “(non-)flammable” and “(in)combustible” often have different meanings depending on the standard or regulation applied; a situation that can cause confusion [14]. “Flammability” is used in this chapter in the specific sense of “showing sustained flaming, after removal of a small ignition source” and not in the general sense of “fire behavior”. Fire tests simulating similar fire scenarios do not necessarily deliver equal results for specific materials nor for a ranking of materials. Even rather minor differences in the fire scenario can play a crucial role and highlight different material properties. For instance, the ignition and flammability hazards are often crucially different if a pilot flame, irradiation, a glow wire, or a cigarette are used [15]. Further, some material properties are important for the performance of the specimen in some tests, but not in others. The melt viscosity of the material, for instance, and the resulting heat removal through dripping or melt flow, influence some flammability tests crucially [16–18], whereas other fire tests are not influenced at all by these phenomena. It should be noted that, in practice, developing flame retarded polymeric materials basically means that a material is developed in order to pass a specific fire test for a defined specimen. Sometimes this is achieved with little or no performance improvement in another fire test. In most cases, this aspect also favors screening and empirical approaches rather than trying to understand the structure-property relationships of burning polymers in detail.

2.1 Production of Volatile Fuel: Heating, Pyrolysis, Charring, Heat of Combustion

Fire behavior is defined as the response of a defined specimen in a defined fire scenario. Thus, the temperature profile of a specimen as a function of place and time is the main, probably the most important, response controlling the burning of a polymer specimen and consequently its fire properties. The transfer of thermal energy is limited to two fundamental physical modes: radiation and conduction. Whereas in radiation the energy is transmitted by electromagnetic waves, conduction requires a medium. Further, convection plays a key role in fires, defined as the combination of conduction (and/or radiation) and flow of the transmitting medium [19]. Radiation is generated by different sources, such as flames or hot surfaces, which differ in emitted wavelengths. The main thermal radiation spectrum is between 1 and 100 μm and therefore classified as infrared radiation. In fires, thermal radiation occurs in heat transfer between surfaces like ceilings and walls as well as through absorption and emission by soot and gases [20]. The radiant heat flux per unit area is given by Planck's law:

$$\dot{q}'' = \varepsilon \sigma T^4 \tag{2.1}$$

where ε is the emissivity and σ is the Stefan-Boltzmann constant [2]. The emissivity for solids and liquids varies between 0 and 1 and depends on the wavelength but can often be assumed as a constant close to 1 under fire heating conditions. Eq. (2.1) also describes the surface re-radiation in Figure 2.1, efficiently reducing the effective heat flux feeding the pyrolysis when higher surface temperatures are reached. For soot, an allocated component in combustion gases, the emissivity is constant over all wavelengths. Most of the radiation in fires is emitted by soot.

Radiative heat transfer is the main energy input in large-scale fires. Absorptivity, defined as the fraction of the occurring radiation that is absorbed by the polymer, has a significant influence on the material's response [21]. This absorption can take place on the material's surface or deep within more transparent polymeric materials. Both surface and in-depth absorption of radiation plays a major role in fire science since it affects the ignition and flame spread. Indeed, even the color of the polymer has an impact on heat absorption. The absorption of radiation for black PMMA, for instance, is significantly higher than for a transparent or white polymer [22]. Carbon nanoparticles were reported to focus all heat absorption at the surface layer, reducing the t_{ig} [23]. Depending on polymer and wavelength, radiation is not only absorbed but also reflected. Back in the 1970s, the important role of the reflectance-absorptance characteristic of the polymer surface and its dependence on the radiation sources was discussed [24]. Increasing the reflectivity of polymers in the range of infrared radiation, e.g., by adding an infrared mirror coating, can reduce heat absorption by more than an order of magnitude. As time to ignition can be shifted by a factor greater than 10, this represents an important approach to flame retardancy [25]. For the following discussions, e.g., the equations for ignition, heat release, and flame spread, heat absorption in depth is neglected, with the absorption at the surface always assumed to be complete without any reflection.

Conductive heat transfer is characterized by diffusion of energy through a medium, such as the solid polymer. This diffusion takes place from a point of higher temperature to a point of lower temperature. Thus, heat conduction into a polymer specimen functions as cooling for the surface when exposed to irradiation. The conductive heat flux is given by Fourier's law and can be written for a one-dimensional homogeneous and isotropic object as:

$$\dot{q}'' = -\kappa \frac{dT}{dy} \tag{2.2}$$

where κ is the thermal conductivity of the medium [26]. Polymers typically exhibit low thermal conductivity, which is responsible for their short times to ignition. The product of thermal conductivity, density (ρ), and heat capacity (c) is called thermal inertia and characterizes the rate of the surface temperature rise of the polymer when exposed to an external heat flux. The surface of materials with a lower thermal inertia heats up more quickly, which results in a shorter time to ignition. This phenomenon increases when it comes to heating cellular polymers, as these not only have an even lower κ, but typically also a low ρ, so that the surface temperature of foams rapidly rises to the fire point [27, 28].

Convection is the conductive heat flux in a turbulent flow between a fluid and a static surface averaged over time. Heat transfer by convection always depends on the motion of the medium, which influences the temperature gradient at the surface. The convection heat transfer coefficient (h_{conv}) describes this phenomenon. Therefore, the convective heat flux is given by [2]:

$$\dot{q}'' = h_{conv}\left(T - T_s\right) \tag{2.3}$$

Radiation and convection are the two ways in which polymer specimens heat up in a fire (Figure 2.1). What is more, convection is believed to replace conduction as the main heat transfer mechanism when the polymer has melted. Since convection depends on the flow field of the fluids, the viscosity of polymeric melts becomes important. Several fillers, in particular anisotropic nanoparticles in nanocomposites, increase the melt viscosity of polymers, an approach that is believed to hinder convective flows in the molten polymer. This may result in decreased times to ignition for nanocomposites compared to the pristine polymers [29].

However, not only the heat sources, heat transfer, and material properties like absorption, density, and heat capacity must be considered in order to understand the temperature profile of a polymer specimen in fire, but also, of course, the dimensions of the specimen itself. This becomes most obvious when thermally thin specimens are compared with thermally thick ones. In thermally thin specimens, the whole specimen heats up at once and contributes to the fuel support of the flame, whereas in thermally thick specimens, a pyrolysis zone or front runs through the specimen.

Polymer plates and slabs usually burn with anaerobic thermal decomposition (= pyrolysis in the sense of its chemical definition) [14], because the oxygen concentration below a stable flame in contact with the specimen surface is crucially reduced. Oxygen absorption and the diffusion rate are proposed to be too low to cause appreciable thermo-oxidative decomposition. Thus, fire residues taken directly after flameout match the residues obtained in thermogravimetry under N_2, but not those obtained under air [1, 30–32]. The surface layer shows thermo-oxidative decomposition before ignition; the fire residue after flameout resulting in an afterglow, e.g., PA 66 is reported to form a black skin before ignition [33]; carbon fillers are thermally stable during burning with a stable flame but are consumed in a pronounced afterglow [34]. Well flame retarded polymeric materials, in particular effective intumescent systems, often tend to burn without a stable flame, such that a thermo-oxidative contribution must be considered. Figure 2.2 shows the remarkable burning behavior of an epoxy resin in the cone calorimeter using a frame. Four different stages can be monitored. First, before ignition, an increase in the temperature of the surface layer in contact with the air atmosphere; second, after ignition, burning with a stable flame based on pure anaerobe pyrolysis – also characterized by the char yield of pyrolysis; third, a rapid and strong decrease in HRR attributed to a switch from a stable flame to instable surface flames, also consuming the char formed during pyrolysis; and finally, an afterglow: exothermic thermo-oxidative decomposition of the remaining char.

In a somewhat similar manner, thermo-oxidation may play a role during the burning of materials that show a high surface-to-volume ratio, such as fibers, textiles, and foams. However, in several important fire tests for polymeric materials, such as the UL 94 burning chamber and the oxygen index (OI), a burner flame is applied to the specimen, removed, and the subsequent extinction observed. Thus, this procedure is accompanied by anaerobic pyrolysis. In contrast, tests of smoldering are based on exothermic thermo-oxidative decomposition.

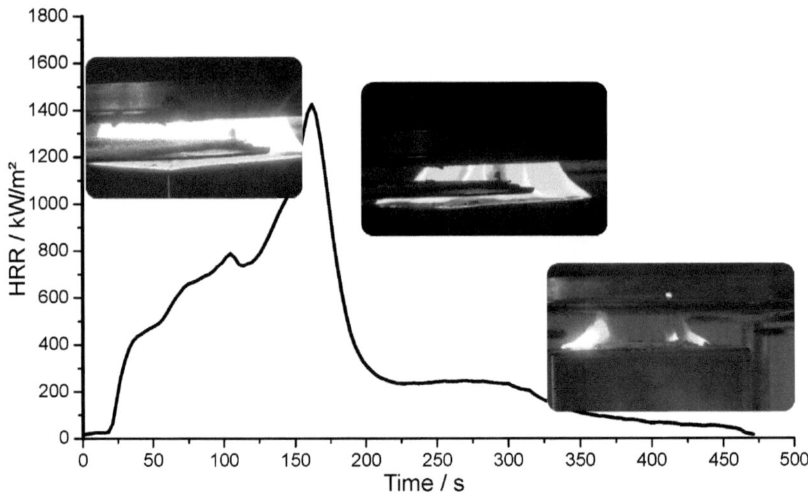

Figure 2.2 Burning of an epoxy resin in the cone calorimeter using a frame. Before ignition, the surface layer in contact with the air atmosphere heats up; after ignition, HRR increases rapidly, caused by burning with a stable flame based on pure anaerobic pyrolysis - the rapid and strong decrease in HRR around 170 s is attributed to a switch from a stable flame to instable surface flames consuming material under the frame and char formed in pyrolysis; and finally an afterglow, further exothermic thermo-oxidative decomposition of the remaining char

In a good approximation, the pyrolysis temperature is controlled by the bond dissociation energies of the weakest links in the polymer backbone [35]. The dissociation energies depend not only on the bond, such as C–C, C=C, or C–O, but also on the side groups, such as CH_3, 2 × CH_3, CH_2–CH_2–C_6H_5, or 2 × F, which have a major impact on the stability of polyolefins. Further, structural defects, branching, and end groups play an important role in polymer decomposition and thus influence the decomposition temperature. In general, chain branching, double bonds, and introducing O weaken linear carbon hydrogen structures, and strengthen aromatic rings and cross-linking. How and how much the thermo-oxidation differs from thermal decomposition obviously depends on the sensitivity of the polymeric structure to thermo-oxidation. Thermogravimetry is an extremely powerful tool to investigate thermal and thermo-oxidative decomposition. Figure 2.3 sketches the typical thermal and thermo-oxidative decomposition of several polymers monitored by thermogravimetry using a constant heating rate of 10 or 20 K/min under nitrogen and oxygen, respectively.

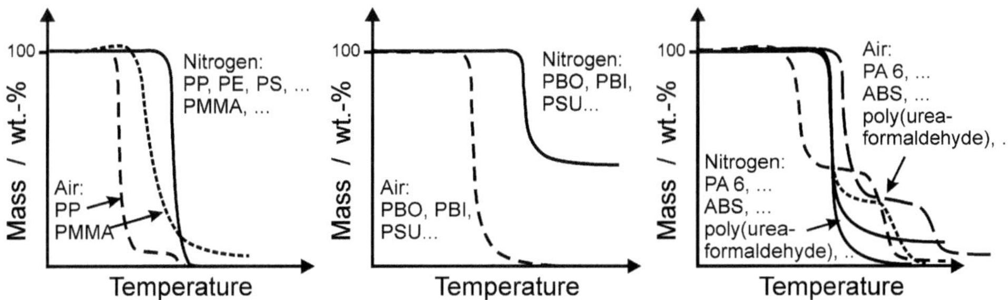

Figure 2.3 Typical types of thermal and thermo-oxidative decomposition of different polymers as monitored by thermogravimetry. Decomposition temperature, number of decomposition steps, and char yield depend strongly on the sensitivity of the different polymers to oxidation

During thermal decomposition, charring organic materials leave a large share of the original carbon content, or in other words fuel, as carbonaceous residue, which is often multicellular char. Inorganic residues are called ash. Many flame retarded polymers form thermally stable or intumescent inorganic-carbonaceous fire residues. Therefore, the formation of voluminous char must occur at the expense of other reactions that can form volatiles. Char is formed in the pyrolysis zone at temperatures between 300 and 600 °C. The upper surface of the fire residue, which is exposed to the irradiation, reaches up to 600–900 °C, while the part exposed to the flame can reach up to 1500 °C. Thus, the material produces fire retardant char in the range from 300 to 1500 °C [6]. After decomposition of the top layer of the charring materials, the volatile products in the deeper layers must pass through the char to reach the surface and flame. This limits the gas transport rate and can cause the volatiles to undergo secondary reactions [36]. However, the increase in volume is usually limited, indicating that negligible amounts of the volatiles are captured within the fire residue. Fire residues, in particular cellular chars with low density, effectively reduce the heat flux to the pyrolysis front through thermal insulation and heat shielding. Radiation of the hot surface compensates for the external irradiation and the radiation from the flame. Whereas the surface temperature of a non-charring polymer is characterized by the pyrolysis temperature, the much hotter residual top layer acts as an efficient heat shield, as shown by Eq. (2.3) and Figure 2.1, offsetting much of the external heat flux and feedback from the flame. This reduction in the effective heat flux in pyrolysis due to a heat shield exerts the greatest impact on fire residue formation, particularly in terms of the mass loss rate and HRR [37]. Furthermore, the decreased mass transfer results in the production of less heat in the flame, which can be seen as a self-reinforcing fire-retardant mechanism.

The char yield affects the amount of fuel release (fire load) as well as the fuel release rate; the physical properties of the residual layer strongly affect the rate of pyrolysis. Important characteristics of the char are not only the amount, but also its density, permeability, oxidation resistance, and thermal insulation properties [6]. The thickness of the virgin material and of the char layer, as well as the location of the pyrolysis zone, are all functions of time. With a growing char layer, heat transport to the virgin polymer decreases as a result of decomposition and the heat release rate. Char from different technical polymers such as PC was characterized via FTIR, Raman spectroscopy, and solid-state NMR. Char consisted of small micro graphene-like or graphitic areas loosely bonded with various oxygen functionalities via aromatic linkages [6, 38]. Thus the typical char layer consists of amorphous aromatic carbon, a precursor of graphite.

The important role of char yield (μ) was pointed out back in the 1970s by van Krevelen, who reported on the linear relationship between μ/wt% of pure halogen-free polymers measured with thermogravimetry under nitrogen and their fire property oxygen index (OI) [30, 31]:

$$\mathrm{OI} = 17.5 + 0.4\mu \tag{2.4}$$

This approach was refined to take char yield and the effective heat of combustion of the pyrolysis products into account [1, 3, 39, 40]. Indeed, the effective heat of combustion of the volatile pyrolysis products of polymers is observed to lie between 12 and 44 kJ g^{-1} [1, 2, 39]. When the fire behavior of polymeric materials is considered, especially if they contain fillers and flame retardants, it becomes clear that the char yield and type of fuel are the two most important factors, although not the only ones. The char yields of polymers are close to 0 wt% for most of the commodity polymers and around 15–30 wt% for most technical polymers, while some high-performance polymers achieve char yields >70 wt%. Frequently, the effective heat of combustion of the volatile pyrolysis products is ranked in the opposite order, or, much more importantly, in the order of the corresponding mass loss. Thus, polyolefins decompose close to 100 wt% and show effective heats of combustion (HOC) and effective heats of complete combustion (h_c) of the volatiles measured between HOC - h_c = 39–44 MJ/kg. Several technical polymers show mass losses of around 70–90 wt%, with the effective heat of combustion of their volatile pyrolysis product around h_c = 20 to 30 MJ/kg. Natural polymers such as cellulose or lignin are characterized by h_c < 20 MJ/kg. Table 2.2 lists these values for some polymers.

Table 2.2 Char Yield (μ), Measured Heat of Combustion (HOC) - Effective Heat of Complete Combustion (h_c) of the Volatile Pyrolysis Products, and Oxygen Index (OI) of Some Polymers

	μ/wt%	Measured HOC - h_c (complete combustion) of the volatiles/MJ/kg	OI/vol%
PP	0	41.9–44.0	17
PE	0	40.0–43.5	18
PS	0	27.9–42.0	18
ABS	0	29.0–37.0	18

	μ/wt%	Measured HOC – h_c (complete combustion) of the volatiles/MJ/kg	OI/vol%
PA 6	1-2	25.8-29.8	21-24
PA 66	4	25.2-27.4	24
PMMA	2	24.2-24.9	17
Polyurethane (PU)	3-13	16.3-23.7	17
PVC	11	7.7-11.3	50
PBT		20.9-23.8	23
PET	13	14.3-24.1	20-21
Epoxy resin	4-10	20.0-27.0	20-26
PC	19-25	21.2-23.3	26
PSU	28-30	20.4-22.4	31-32
PEEK	50-54	21.3-23.2	35
PEI	52-53	16.7-23.2	47
PBI	70-75	16.2-22	41-42
PBO	75	28.6	56
Cellulose	4-20	11.3-20.0	18.5

Data taken from [1, 39] to give an overview. Note that the different polymer products and different methods used cause significant variations, particularly in HOC and h_c when HOC is measured with combustion efficiencies as low as 0.75, but also in char yields when determined at different temperatures.

Natural fires are not characterized by complete combustion. This means that the carbonaceous char of the residue contains combustion energy, but incomplete oxidation also occurs in the flame, producing products like CO, smoke, and soot. Indeed, the smoke layer below the ceiling sometimes ignites again in flashover scenarios. The combustion efficiency χ is HOC/h_c and usually lies between 0.75 and 1 in a diffusion flame; combustion efficiency is reduced by flame retardants functioning via flame inhibition as a mode of action. The net heat of combustion of the polymer $h_{c,polym}$ is the calorific energy generated by the chemical reactions leading to complete combustion per unit mass of fuel, assuming that the water produced is in the gaseous state. This value is measured in the oxygen bomb calorimeter or can be calculated from the standard heats of formation of the materials [39]. The net heat of combustion of the polymer $h_{c,polym}$ equals the sum of $(1 - \mu) \cdot h_c + \mu \cdot h_{c,char} = h_{c,polym}$, with $h_{c,char}$ the heat of combustion of the char.

As the amount and the quality of the fuel are two of the most important parameters affecting the fire hazard of polymers, at first glance char formation would appear to reduce the amount of fuel released. Although this is correct, charring consists of much more than merely storing fuel, or simply replacing fuel with inert filler. The effective heat of combustion of the volatiles differs from the effective heat of combustion for complete combustion as measured in the bomb calorimeter, for instance. The effective heat of complete combustion of the char equals, in a good approximation, the effective heat of complete combustion of graphite: 32.9 MJ/kg, or perhaps somewhat less, 30.0–32.9 MJ/kg, if it contains more heteroatoms. Thus, the effective heat of the complete combustion of char differs from the effective heat of combustion of the volatile pyrolysis products and from the effective heat of complete combustion of the polymers. Considering that pyrolysis in the condensed phase and oxidation in the flame consist of rather ordinary chemical reactions, the conservation of atoms during decomposition and oxidation equals, in a good approximation, a conservation of the heat of combustion when the polymeric educts are compared with the pyrolysis and combustion products. Although all of the data are available, this idea is seldom used to discuss the fire behavior and flame retardancy of polymers. The charring of polyolefins is less efficient than charring in technical polymers. Because the heat stored in carbonaceous residue per mass is lower than the effective heat of combustion of polyolefins, the effective heat of combustion of the volatiles increases. Charring in polyolefins is less efficient than replacing an equivalent vol% of the polymer. In contrast, charring in technical polymers, high-performance polymers, and natural polymers is very efficient, since the effective heat of complete combustion of the char is greater than the effective heat of combustion of the polymers. Increasing the char yield goes along with decreasing the effective heat of combustion of the volatile pyrolysis product. Introducing charring in such polymers is more efficient than replacing the equivalent amount of vol% fuel.

2.2 Decomposition of Polymers

The thermal decomposition of polyolefins, such as PP, PE, of PS, and of PMMA, can be understood as quasi the opposite (Figure 2.4) of their free-radical synthesis, resulting in the deterioration of molecular weight. Rather well-defined radical chain reactions characterize the depolymerization into monomers, and chain transfer induces random chain scission into oligomeric fragments.

The first step of decomposition entails the generation of radicals due to thermally activated bond scission. In the case of absent chain transfer, depolymerization produces one monomer at a time in the propagation step [41]. The most important example for nearly perfect depolymerization is PMMA, for which the monomer yield amounts to nearly 100% [42, 43], and where steric hindrance was identified, preventing chain transfer. The radical is stabilized by the aromatic side group in PS, resulting in monomer yields of around 40% [43], whereas for PP the chain transfer becomes the controlling characteristic. PP shows monomer yields of only around 2%, with the main pyrolysis products being oligomeric chain fragments. In this case, intramolecular transfer results in oligomeric fragments and the intermolecular transfer in statistical chain scission. Seemingly random chain scission occurs at any position of the polymer backbone, resulting in a rapid decrease in molecular weight. Decomposition of PS at lower temperatures is characterized by larger amounts of dimers, trimers, etc., due to a lower amount of initiation and chain transfer, and the parallel occurrence of depolymerization [44]. Decomposing PS at higher temperatures delivers more monomers and fragments of monomers due to a larger amount of initial initiation followed by depolymerization. Increasing the residence time of PS in the pyrolysis zone, e.g., in PS/layered silicate nanocomposites, also shifts the pyrolysis products distribution toward monomers and fragments of monomers [45]. However, such large and significant changes in pyrolysis products must not be misinterpreted as always relevant for fire behavior. Often, they result in only tiny changes in the overall effective heat of combustion of the volatiles and mass loss, so that the ablative phenomena of sustained burning or flammability are hardly influenced. Nevertheless, radical generators are used successfully as adjuvants, enhancing the initial molecular weight reduction in polyolefins to increase melt flow and dripping in order to achieve V-2 (extinguishing by flaming dripping) or dripping V-0 (rapid extinguishing by non-flaming dripping) classifications in UL 94, and in PS foams to pass the reaction to fire test [46].

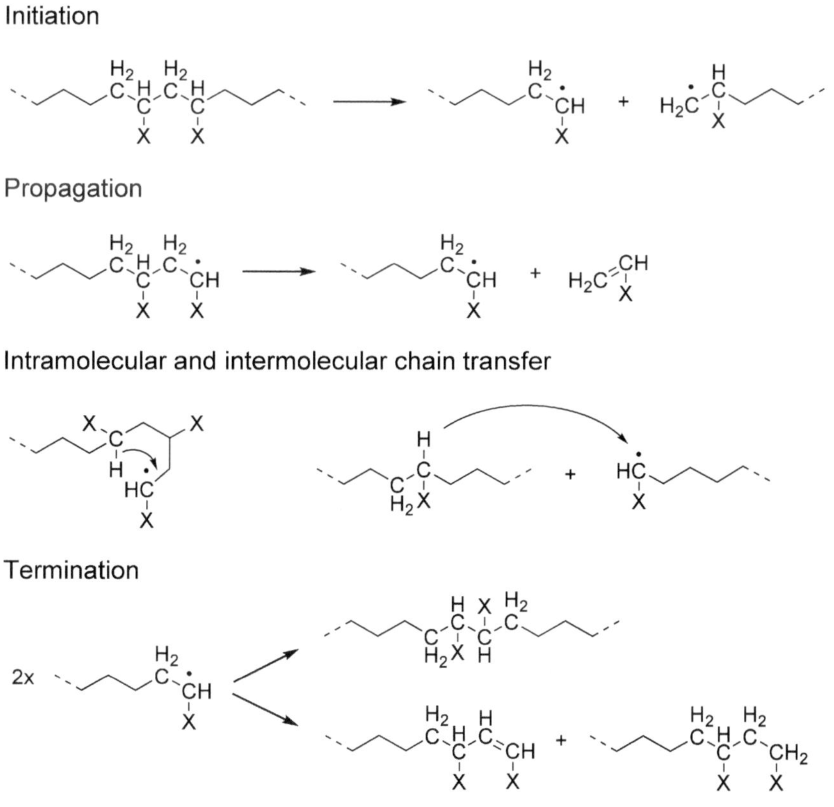

Initiation

Propagation

Intramolecular and intermolecular chain transfer

Termination

Figure 2.4 General decomposition scheme of polyolefins

For several polymers, such as polyvinyl chloride (PVC), polyvinyl fluoride (PVF), polyvinyl acetate (PVA), and polyacrylonitrile (PAN), the thermal stability of side groups is much weaker than the backbones, so that the reactions of the side groups control thermal decomposition. The thermal decomposition of PVC is due to side groups (HCl) stripped off from the chain (Figure 2.5) at relatively low temperatures (about 100–120 °C) [47]. The remaining dehydrogenated backbone tends to cross-link and aromatize, functioning as a precursor for charring.

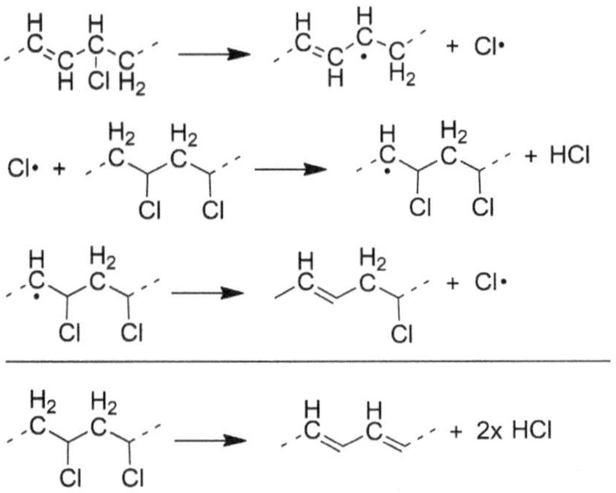

Figure 2.5 Side-group elimination (chain stripping) of PVC

Polymers with heteroatoms in their backbone show rather specific decomposition pathways, starting with the scission of the weakest bond. Figure 2.6 shows one possible decomposition route for PET [48–50]. Molecule dynamics cause a six-membered cyclic transition state, which, in random chain scissions, leads to a cut of the alkyl-oxygen bond, producing carboxylic acids and short chain hydrocarbons. Polymers with heteroatoms often decompose via competing decomposition pathways, since several different bond scissions can occur, with hydrolytic chain scission taking place at the same time, or alternative decomposition pathways starting with rearrangements [51]. A few polymers show quite dominant major decomposition pathways, such as polyamide 6 [52], for which the amide bond scission cyclization leads to the formation of caprolactam (Figure 2.6).

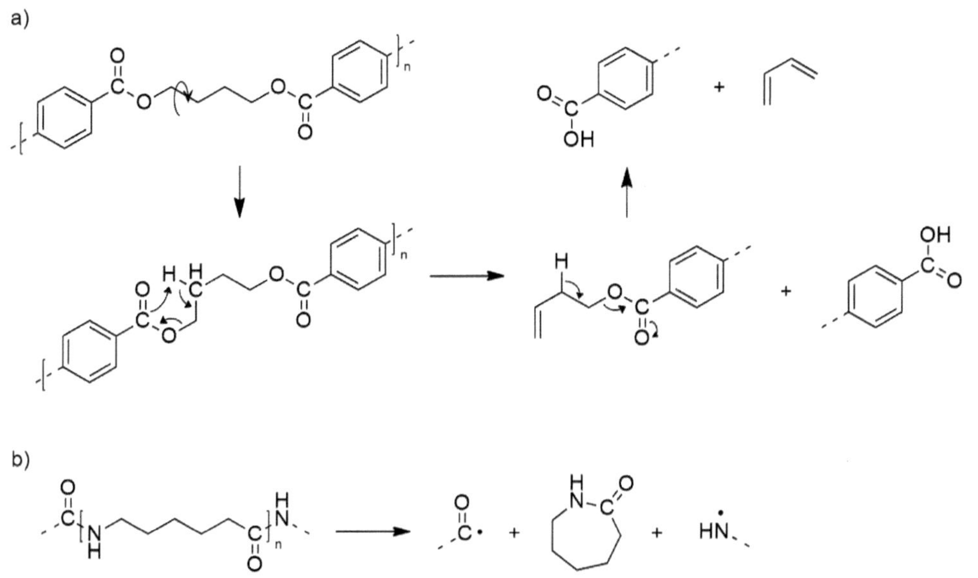

a)

b)

Figure 2.6 Thermal decomposition of a) PET and b) PA 6

Decomposition pathways that are characterized by dehydrogenization, cross-linking, and aromatization open the door to graphitization and the formation of nonvolatile char. Usually, the formation of volatiles goes along with the dehydrogenization of the remaining compound, resulting in cross-linking or aromatization. Cross-linking can be used to increase charring [53], and some groups function as precursors for charring. In the case of PAN (Figure 2.7), the whole polymer functions as a precursor for char at low heating rates [54]. The first step of its decomposition is a cyclization, at which nearby functional CN groups build linkages between carbon and nitrogen, while some ammonia and hydrogen cyanide evolve. At temperatures between 625 K and 975 K, primarily hydrogen is split off, whereby cross-linking can take place and the structure may start to carbonize. At higher temperatures, nitrogen is eliminated, leading to pure carbon structures.

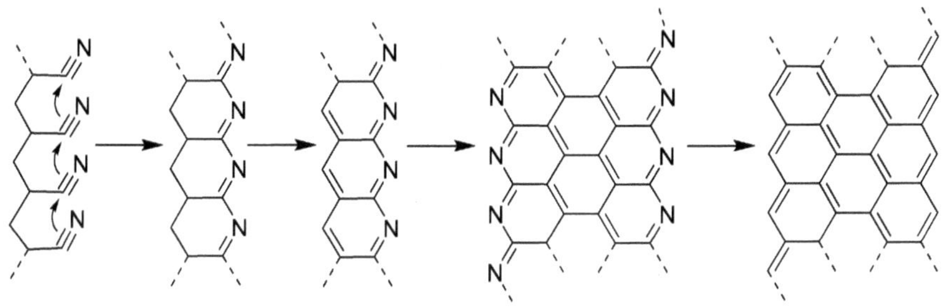

Figure 2.7 Charring mechanism of PAN

Polymer pyrolysis is controlled by various mechanisms, including depolymerization, random scission, chain stripping, rearrangements, cyclization, and cross-linking. Within the subsequent reactions, oligomers and monomers can be formed, as well as all kinds of volatile decomposition products. For some polymers, a portion of the pyrolysis products remains as carbonaceous char in the condensed phase via dehydrogenization, cross-linking, aromatization, and graphitization. The decomposition pathways of a polymer and its char yield are quite specific to its chemical structure.

2.3 Ignition and Smoldering

Preventing sustained ignition, in particular of components likely to be the first to start a fire, is the most valuable fire protection goal, especially for materials used in electrical engineering and electronics. Fire tests are used to characterize ignitability, flammability, and reaction to a realistic ignition source. Several fire tests use contact with a small flame, while others are more specific, e.g., using a glowing wire to simulate an overheated wire due to a short circuit. The ignition sources determine the small length scale of tests, such as the OI (or limiting oxygen index LOI), UL 94, and glow wire test. Apart from irradiation from the ignition source itself, there is no external heat flux impact on the tested material. However, most of the tests operating in the ignition fire scenario assess the sustained ignition of a specimen by assessing the flammability of the material. These do not monitor the critical values for ignition, such as time to ignition, critical heat flux for ignition, or critical mass loss rate for ignition, but measure the time to extinction and speed of the flame front.

Ignition generally occurs under a set of critical conditions in a gas volume: oxygen concentration, fuel concentration, and the fuel/oxygen ratio, pressure, temperature, and the energy of the ignition source, such as a spark, hot surface, or a pilot flame. In the representative setup sketched in Figure 2.1 and Figure 2.8, representing the cone calorimeter delivering irradiation and a spark igniter, respectively, several parameters are determined by the ambient conditions, with ignition identified as occurring at a critical minimum fuel release rate per unit area. The critical energy density determines a minimum critical HRR at ignition. This critical HRR for ignition equals the minimum critical mass flux multiplied by the heat of combustion of fuel gases. This critical HRR was found to be around 20 kW/m^2 per unit area [3]; the critical mass loss rate per unit area was around 1 g/sm^2 at the flash point. Similar critical values in HRR and mass loss rate can be expressed for the occurrence of extinction [3]. When the mass flow of fuel gases falls below around 2 to 4 g/sm^2 or when the HRR is lower than 50 to 100 kW/m^2, extinction of the flame takes place. Sustained ignition occurs only when a quasi-stationary stage follows the ignition event, with mass loss rates and HRR greater than the critical values for extinguishing. The additional contribution of the heat flux of the flame $\dot{q}''_{flame}$ (Figure 2.8) to the effective heat flux transformed into the fuel production rate must be large enough to prevent extinction; otherwise only flashing occurs.

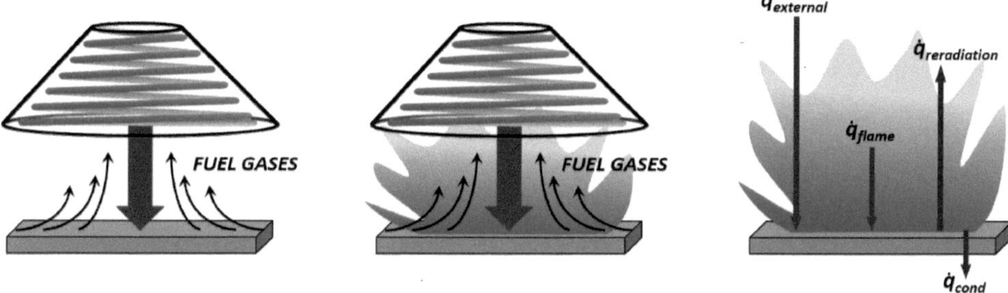

Figure 2.8 Illustration showing heating and fuel release, ignition, and sustained ignition

The critical condition for igniting a surface under irradiation can be expressed in various variables, such as the critical HRR or mass loss rate discussed earlier, but also in terms of the heat of gasification per unit mass of volatile fuel L_g or ignition temperature T_{ig}. The heat of gasification per unit mass of polymer h_g is the sum of the heat of vaporization Δh_v, the heat Δh_s needed to heat up the material to T_{ig}, the heat needed to reach phase transition Δh_f and to achieve decomposition Δh_d (Eq. (2.5)). For most of the polymers, $L_g/2$ equals, in a satisfactory approximation, the integral of the specific heat capacity at the starting temperature $c_p(T) = c_p(T_0) = c_0$ from the starting temperature T_0 to the temperature of ignition T_{ig} (Eq. (2.6)) [3]. The ignition temperature T_{ig} observed at the surface of the specimen is found to be specific to the material's characteristics in setups as sketched in Figure 2.1 and Figure 2.8. T_{ig} was reported to be characteristic for each polymer and rather independent from the external heat flux; both are confirmed experimentally to be correct for a wide range of irradiation levels [1]. It was proposed that

$$L_g = \frac{h_g}{(1-\mu)} = \frac{1}{(1-\mu)}\left(\Delta h_s + \Delta h_f + \Delta h_d + \Delta h_v\right) \tag{2.5}$$

$$L_g \approx \frac{c_0 T_{ig}^2}{T_0} \text{ and } T_{ig} \approx \left[\frac{T_0 L_g}{c_0}\right]^{\frac{1}{2}} \tag{2.6}$$

The time to ignition t_{ig} can be understood as the time to reach T_{ig}. It decreases with increasing irradiation and is dependent mainly on the thermal conductivity κ, the density ρ, and the heat capacity c_p of the material (Eq. (2.7)). For thermally thin samples, all of the material heats up at once, so that the thickness d controls the result rather than the thermal conductivity (Eq. (2.7)). Figure 2.9 shows the dependency of the time to ignition on the external heat flux q''_{ext}. In this way, the critical heat flux needed for ignition may be determined.

$$t_{ig(thick)} = \frac{\pi}{4}\kappa\rho c_p\left(\frac{T_{ig}-T_0}{\dot{q}''_{ext}-\dot{q}''_{crit}}\right)^2 \qquad \frac{1}{\sqrt{t_{ig(thick)}}} = \frac{2\left(\dot{q}''_{ext}-\dot{q}''_{crit}\right)}{\left(T_{ig}-T_0\right)\sqrt{\pi\kappa\rho c_p}}$$

$$t_{ig(thin)} = \rho c_p d\frac{T_{ig}-T_0}{\dot{q}''_{ext}-\dot{q}''_{crit}} \qquad (2.7)$$

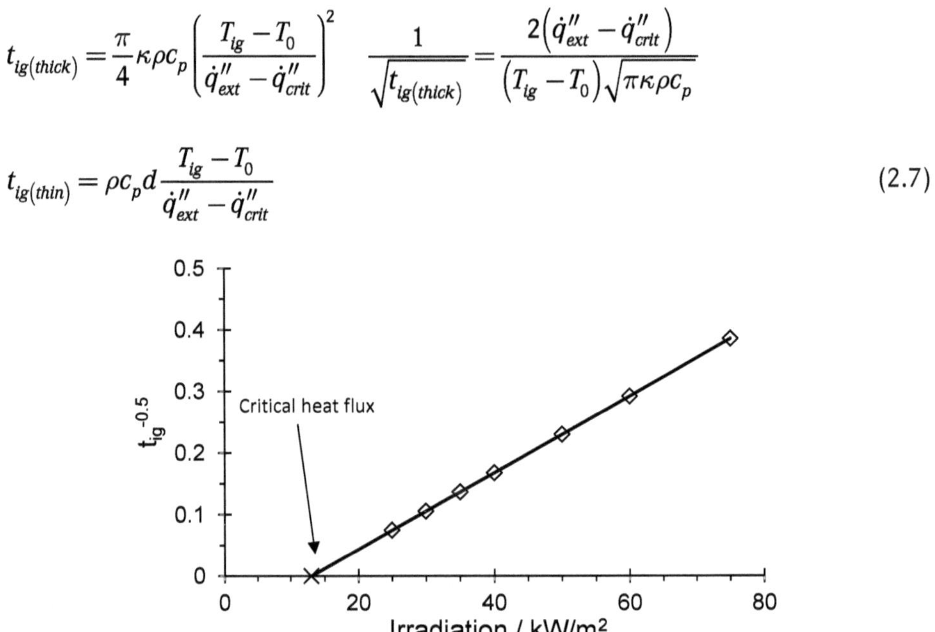

Figure 2.9 $t_{ig}^{-0.5}$ plotted versus irradiation, showing the linear relationship according to Eq. (2.7) for thermally thick specimens as a straight line and the critical heat flux for ignition as the *x*-axis intercept

In contrast to the ignition of a flaming fire, smoldering is defined as flameless combustion of a material. It is a rather low-temperature, heterogeneous, exothermic thermo-oxidative decomposition, in which the decomposition products are partially oxidized. Smoldering is a significant fire hazard, because it is easily initiated, i.e., it requires less energetic ignition sources, such as cigarettes, hot materials, or glowing wires, and may be difficult to detect. The critical heat fluxes needed to start smoldering are between ¼ and ½ of the critical heat fluxes for sustained flaming ignition [55, 56]. The initiation of the smoldering process is determined by the oxidation kinetics of the fuel, and the propagation is regulated by the rate of oxygen transport to the reaction zone. Materials with a large surface area per unit of mass, like many foams, and all kinds of fibers, cotton, wood, and cellulose, are predisposed to sustain smoldering because of enhanced oxygen diffusion, and act as insulation barriers to reduce the heat loss, which sustains smoldering. The typical propagation rates are 10^{-3}–10^{-2} cm/s and thus orders of magnitude smaller than the typical flame spread rates of flaming burning, around 10^{-1}–10^{2} cm/s [56].

Furthermore, smoldering, even after minutes or hours, can act as an ignition source for flaming combustion. Smoldering can be seen as a pathway to flaming fires for heat sources too weak to produce a flame directly, and many domestic fires are believed to begin with smoldering. The classic example is the burning cigarette in contact with upholstered furniture, where smoldering combustion can be sustained to the point where the evolved gases reach their flammability limits and flames rapidly grow. During the smoldering process, high yields of toxic gases evolve, e.g., carbon monoxide. Because smoldering is often underestimated, there is a lack of extensive studies and models in the literature, in contrast to the wealth of material on flaming combustion. A comprehensive debate on smoldering combustion was presented elsewhere [57]. Incidentally, afterglow, the thermo-oxidative decomposition of the char after flameout, can be understood as the smoldering of porous carbonaceous char.

Melt flow and dripping influence fire properties, e.g., the reaction to an ignition source. Melt flow or dripping can be both detrimental and beneficial. Flaming dripping provides additional ignition sources, may increase flame spread, and may build up liquid pool fires. On the other hand, melt flow and dripping remove mass and heat from the actual pyrolysis zone, thus supporting or causing extinction. In effect, well-adjusted dripping behavior is crucial for passing many fire tests, such as the burner test for building products, the German Brandschacht, OI, glow wire tests, and the vertical UL 94 classification. Commercial flame retardant thermoplastics often contain special polytetrafluoroethylene (PTFE) fibers that cause flow limits at low shear rates in order to achieve V-0 non-dripping classification in the UL 94 test [46]. An enlightening example is PC/ABS flame retarded with bisphenol A-bis(diphenyl phosphate) BDP and less than 1 wt% PTFE fibers [58]. The shear stress versus shear rate master curves of PC/ABS, PC/ABS/BDP, and PC/ABS/BDP/PTFE are shown in Figure 2.10(a). The flame retardant BDP is a well-known plasticizer that crucially reduces viscosity, which is usually favored to improve processability, but also considerably enhances dripping. Adding PTFE fibers reduces dripping, inducing a flow limit at low shear rates, but preserves the improved processing properties at high shear rates. The desired V-0 classification in the UL 94 test is reached only when the two additives are combined in PC/ABS+BDP+PTFE. Other inorganic fillers and nanocomposites also function as anti-dripping agents [59, 60]. On the other hand, radical generators in combination with flame inhibitors are the typical example for exploiting enhanced melt flow or dripping in PS foams to pass fire tests so that building products and thin PP products can obtain a V-2 (flaming dripping extinction) or non-flaming dripping V-0 classification [61, 62]. Generating additional radicals during the early stages of decomposition in the condensed phase enhances the statistical scission of the long polyolefinic chains. The reduced viscosity of the pyrolyzing melt is accompanied by enhanced or earlier dripping, aimed at the rapid extinction of flaming. Figure

2.10(b) shows a computer simulation of such a non-flaming dripping V-0 behavior in a UL 94 test [63]. The simulated temperature demonstrates that in the UL 94 setup, the flame-retardant mode of action can be understood as an efficient cooling effect. The hot dripping removes enough energy so that the remaining slab extinguishes. In other tests, the effect is described as retreating from the ignition source. It is not only thermoplastics that show a large influence by melt flow and dripping at temperatures above their glass, melting, and decomposition temperatures, but also thermosets. For these, after a small delay, the pyrolysis products reach equilibrium above the decomposition temperature, yielding molten materials that behave quite similarly to thermoplastics.

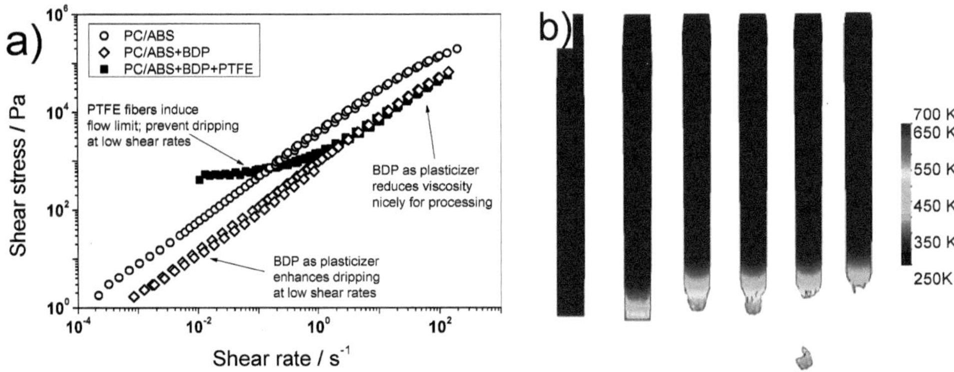

Figure 2.10 (a) Shear stress versus shear rate master curves (523 K) of PC/ABS (open circles), PC/ABS+BDP (open diamonds), and PC/ABS+BDP+PTFE (filled squares); (b) images of the PFEM simulation for extinction (V-0) of PP through non-flaming dripping

2.4 Steady Burning and Flame Spread

Combustion is the exothermic process in which a fuel is rapidly oxidized, thereby consuming oxygen and producing heat and light. If the fuel is thoroughly mixed with air, and the amount of oxygen is constant, the ensuing combustion is complete; this is referred to as a "premixed flame". In common fire scenarios, however, the amount of oxygen that can reach the fuel is limited by the laws of diffusion, hence the term "diffusion flame" [64]. A diffusion flame is characterized by its incandescent soot of a yellow-orange color, a result of incomplete combustion, as well as the lack of a clear flame front. Due to the dominating diffusion laws, the oxygen concentration from the periphery of the flame to its center is described by a gradient: the interfacial region at the material surface decomposes under anaerobic, pyrolytic conditions, while the flame zone above is characterized by thermo-oxidative reactions. In all cases, the high amount of thermal energy causes the bulk material to decompose and volatize, causing a flux of mass $\dot{m}''$ into the gas phase. The material flux is related to the combustion in the flame zone by the resulting effective heat flux $\dot{q}''_{eff}$ used for pyrolysis [3]:

$$\dot{m}'' \cdot L_g = \dot{q}''_{eff} \tag{2.8}$$

The conditions needed to produce a flux of fuel into the flame are dependent on the underlying decomposition reactions, which are inherently specific to the material's composition and its properties (heat capacity, heat conductivity, ratio of mass to surface area, etc.). The pyrolysis of the polymer results in the release of volatile fuel into the gas phase; covalent bonds of these fuel molecules are immediately broken further (bond scission) in the flame zone. This homolytic cleavage leads to the formation of two free-radical moieties, which is the first step in a vast series of chain reactions and is referred to as the initiation reaction (Figure 2.11). The reaction of a hydrogen radical with molecular oxygen is further known as the branching reaction, as both the resulting hydroxyl and oxygen radicals make a crucial contribution to the resulting radical chain reactions. The hydroxyl radical may interact with various combustion products, such as carbon monoxide, hydrogen, or hydrocarbons to form new free radicals, thus propagating the decomposition mechanism (Figure 2.11). Termination of the free-radical chain reaction occurs when two free radicals react with one another, as is the case when a hydroxyl and a hydrogen radical, or two hydrogen free radicals, interact to form water or molecular hydrogen, respectively (Figure 2.11).

Initiation:	$RH \rightarrow R\bullet + H\bullet$
	$C_nH_{2n+2} \rightarrow C_{n-1}H_{2n-1}\bullet + CH_3\bullet$
Branching:	$O + H_2 \rightarrow OH + H$
	$H\bullet + O_2 \rightarrow HO\bullet + O\bullet$
Propagation:	$HO\bullet + CO \rightarrow CO_2 + H\bullet$
	$HO\bullet + H_2 \rightarrow H_2O + H\bullet$
	(C_xH_n consumption)
	$RH + HO\bullet \rightarrow RH\bullet + H_2O$
	$RH + O \rightarrow RH\bullet + HO\bullet$
	$RH_2 + H\bullet \rightarrow RH\bullet + H_2$
	$RH\bullet + O_2 \rightarrow RO + HO\bullet$
	...
	$HCO\bullet \rightarrow CO + H\bullet$
Most exothermic:	$CO + HO\bullet \rightarrow CO_2 + H\bullet$
Termination:	$H\bullet + HO\bullet \rightarrow H_2O$
	$H\bullet + H\bullet \rightarrow H_2$

Figure 2.11
Scheme of oxidative chain reaction in the flame

Of the abundant and complex free-radical reactions taking place in a fire scenario, the formation of carbon dioxide (and a hydrogen radical) from the reaction of hydroxy free radicals with carbon monoxide is the most exothermic reaction and is the largest contributor to the heat released in a fire scenario.

The sum of all exothermic reactions in a fire scenario is known as the total heat released (THR); it is the integral of the heat release rate (HRR) curve over time. The HRR is most commonly determined by measuring oxygen consumption, and the equation used to calculate the HRR considers the material's behavior under fire conditions and the heat fluxes working upon it. When ignition is continuous and the burning rate is constant, the HRR of the material is equal to the product of the combustion efficiency χ, the heat of complete combustion of fuel gases h_c and the mass flux $\dot{m}''$, which, according to Eq. (2.9), is related to the net heat flux $\dot{q}''_{net}$ via the enthalpy of gasification L_g [1]:

$$\mathrm{HRR} = \chi \cdot h_c \cdot \dot{m}'' = \chi \frac{h_c}{L_g} \dot{q}''_{net}$$

$$\dot{q}''_{net} = \dot{q}''_{ext} + \dot{q}''_{flame} - \dot{q}''_{rerad} - \dot{q}''_{loss} \tag{2.9}$$

The combustion efficiency χ is the ratio between the measured effective heat of combustion (HOC) and the heat of complete combustion h_c; its values range from $\chi = 1$ for complete and $\chi < 1$ for incomplete oxidization. It depends somewhat on the material burning, but also on the ventilation of the fire scenario. Flame inhibition, one of the most important modes of action used in flame-retarded polymers, reduces the combustion efficiency considerably. The net heat flux used for pyrolysis can be written as the sum of the heat fluxes in steady flaming combustion (Eq. (2.5), Eq. (2.9), Figure 2.1, Figure 2.8), where $\dot{q}''_{ext}$ is the external heat flux working upon the material per unit area, $\dot{q}''_{flame}$ is the heat flux of the flame, $\dot{q}''_{rerad}$ is the heat flux caused by reradiation of the hot surface to the environment, and $\dot{q}''_{loss}$ is the loss of heat into the specimen and surrounding environment, such as the specimen holder. The product of combustion efficiency and the ratio of heat of complete combustion of the fuel gases to enthalpy of gasification is known as the heat release parameter HRP [1]:

$$\mathrm{HRP} = \chi \cdot \frac{h_c}{L_g} = \chi \cdot (1 - \mu) \cdot \frac{h_c}{h_g} \tag{2.10}$$

A time dependent factor ($\theta(t)$) between 0 and 1 may be introduced for the influence of the protective residual layer formed during burning. When these factors are combined, the resulting equation describes the dependency of the HRR on the heat release parameter, the protective layer, and the net heat flux (Eq. (2.11)).

$$\mathrm{HRR}(t) = \chi \cdot \theta(t) \cdot (1 - \mu) \cdot \frac{h_c}{h_g} \cdot \left(\dot{q}''_{ext} + \dot{q}''_{flame} - \dot{q}''_{rerad} - \dot{q}''_{loss}\right) \tag{2.11}$$

HRR is believed to be the most important fire property to assess the fire hazard of materials [65]. Therefore, Eq. (2.11) is highlighted as the most important statement of this chapter. Apart from the parameter of heat absorption, which is set to 1 and thus neglected, it combines all of the important characteristics controlling fire behavior, including char yield, combustion efficiency, heat of combustion of the volatiles, and heat of gasification. The empirical factor $\theta(t)$ is added arbitrarily to account for the changing protective layer properties of the fire residue during burning, which results in HRR changing as a function of t. This may be superfluous if h_g is used as $h_g(t)$ to account for the impact of the residual protective layer, for instance, or if Eq. (2.11) is used to describe a constant, steady state HRR. Thus, in all relevant sources [1–3], Eq. (2.11) is always given without $\theta(t)$; thus, $\theta(t)$ may be ignored or added as an empirical contribution after the fact.

However, it must not escape our notice that without $\theta(t)$, Eq. (2.11) also covers the greatest effect of residual protective layers: the heat shielding effect. In contrast to the effects of absorption in depth and heat reflection mentioned at the beginning of this chapter, heat shielding is not connected to reduced heat absorption or increased heat reflection but is based on increased reradiation of the hot surface $\dot{q}''_{rerad}$. The heat shielding effect has been reported as the most important flame-retardant mode of action in layered silicate nanocomposites of non-charring polymers, for instance; in some systems, it may be the only relevant one [38]. The key to understanding this is to apply Eq. (2.1) and quantify the role (to the fourth power) of the surface temperature T in $\dot{q}''_{rerad}$ of Eq. (2.11). The surface temperature of non-charring polymers equals their pyrolysis temperature, whereas the carbonaceous, inorganic-carbonaceous, or inorganic residual layer can heat up to much higher temperatures. Thus, jumping from a pyrolysis temperature of 420 °C to a residue surface temperature of 720 °C entails an increase in $\dot{q}''_{rerad}$ from 10 kW/m^2 to 50 kW/m^2, compensating for the feedback from a flame or the external heat flux in a cone calorimeter experiment.

Once a material is successfully ignited, the flame is stable and starts expanding, known as flame spread. This can happen in air, along surfaces, or through porous solids [66]. It means that the affected area always offers enough combustible material to supply the flame. With respect to the quality of the fuel, an adequate amount of fuel must be provided over time. The phenomenon can be regarded as many small fuel elements igniting one after the other. The velocity of the flame spread depends on how fast the nearby fuel elements are heated up to the ignition temperature (T_{ig}). The heat flux can be provided by the burning material or other external sources. If the current element is not burning intensively or long enough to ignite the next one, the flame extinguishes [67]. Flame spread is the most important fire hazard in developing fire, but must not be confused with fire spread. Fire spread means the advancing of a fire front and thus can be flaming or smoldering [66].

The speed of the flame spread depends on its mode. For wind-aided flame spread, the air flows in the same direction as that in which the front of the flame is propagating. Another expression for this is concurrent flow. For opposed flow flame spread, the air flows in the opposite direction from the flames. Natural flow is caused by the buoyancy of the flame. Thus, if the specimen's surface is horizontally aligned, there is an opposed flow in all directions of the plane. If the specimen is aligned vertically, there is a strong wind-aided flame spread upwards and a much smaller opposed flow flame spread downwards (Figure 2.12). In contrast to natural flow, forced flow is produced by meteorological flow or a fan. The principle remains the same for wind-aided or opposed flame spread. The flow conditions affect the distance affected directly by the flame δ*f* and thus the velocity of the flame spread [66]. By the way, this very different area of material in direct contact with the flame is also the main difference between vertical UL 94 and OI, which assess wind-aided and opposed flow flame speed, respectively. Furthermore, the flame temperature of many solids decreases with a lower oxygen concentration in the air. This often limits the flame spread on room ceilings [68], which is a good example of a natural wind-aided flame spread [66].

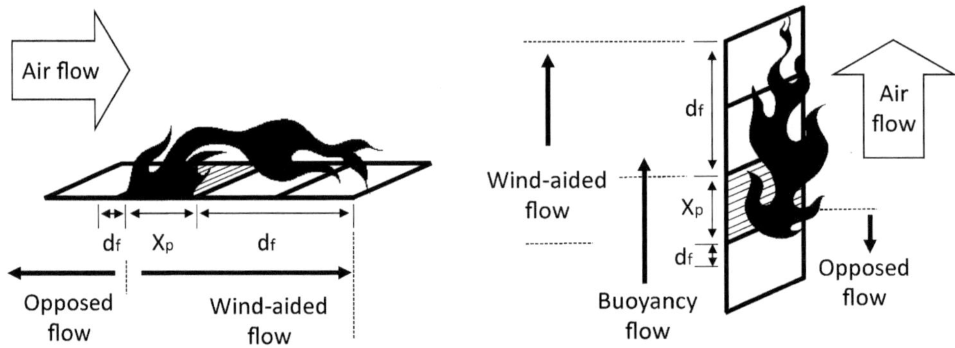

Figure 2.12 Illustration depicting wind-aided flame spread and opposed flow flame spread

The velocity *v* of flame spread is given by the difference between the energy provided by the heat flux from the flame to the fuel length δ*f* and the energy required to heat up to the ignition temperature. The heat flux depends on the heat release rate of the material. For a thermally thin material with a thickness *d*, using Eq. (2.7), this balance can be expressed as [66, 68]:

$$v \cdot \rho \cdot d \cdot c_p \cdot \left(T_{ig} - T_0\right) = \dot{q}''_{flame} \cdot \delta_f$$

$$v = \frac{\dot{q}''_{flame} \cdot \delta_f}{\rho \cdot d \cdot c_p \cdot \left(T_{ig} - T_0\right)} = \frac{\delta_f}{t_{ig}} \tag{2.12}$$

For a thermally thick specimen:

$$v = \frac{\delta_f}{t_{ig}} = \frac{(\dot{2q}''_{flame})^2 \cdot \delta_f}{\pi \cdot \rho \cdot k \cdot c_p \cdot \left(T_{ig} - T_0\right)^2} \tag{2.13}$$

2.5 Fire Load and Fire Resistance

In a fully developed fire, all combustible materials are believed to be on fire, and therefore this state is characterized by extreme high heat fluxes and temperatures. The fire safety objectives have changed. Ignitability, flammability, reaction to fire, and burning behavior are no longer substantial; the point is to gain time for evacuation and firefighting before the collapse of the construction and to prevent the fire from spreading further. The crucial fire properties in a fully developed fire scenario are fire load and fire resistance. Fire load is the quantity of heat that can be released during the complete combustion of all the combustible materials involved (ISO 13943). To assess the fire load of certain materials, measurements in the bomb calorimeter, cone calorimeter, or micro combustion calorimeter (MCC or pyrolysis combustion flow calorimeter, PCFC) are used [1]. In the bomb calorimeter, the specimen is combusted completely under a pure oxygen atmosphere. The result is the net heat of combustion per mass of the specimen, which can be used to calculate the worst-case fire load. More realistically, the total heat evolved (THE) in the pyrolysis of polymeric materials, along with their distinct char yields, is determined by integrating the heat release rate measured in the cone calorimeter or the MCC. Thus, the cone calorimeter delivers the HOC for a real specimen in a well-ventilated fire, and the MCC the h_c for the complete combustion of the volatiles as discussed earlier. Along with the mass of the products or components, the fire load can be assessed. Apart from the calorimetric assessment, non-combustibility tests are performed, such as ISO 1182 on construction products of the highest classification in Europe. The cylindrical specimen is placed in a preheated electric furnace, and the mass loss and temperature rise caused by the specimen are monitored within the high-temperature furnace. Since these tests also focus on complete combustion, all polymeric materials, which are generally combustible, usually fail such tests.

Fire resistance is the ability of a test specimen to withstand fire or give protection from it for a period of time (ISO 13943). Fire resistance is usually a property of a component rather than a material. Fire resistance properties are relevant only for polymeric materials used in fire-resistant components, systems, and structures. Common criteria are fire stability, thermal insulation, or functioning as a barrier for heat, fire, and smoke for a certain time. Fire resistance is often tested in nearly full-scale tests, e.g., the column furnace for steel beams or the wall furnace for fireproof doors. A key technology in this field – intumescent coatings – is based on polymer binders. Intumescent coatings are used to increase the fire resistance of various structures like steel constructions, wooden constructions, or composite structures. An intumescent coating reduces the heat transfer through the formation of a multicellular foam residue when exposed to heat. To achieve massive expansion, three major ingredients are essential [69–73]: an acid source (often ammonium polyphosphate), a char builder (any kind of polyol, for example), and a blowing agent (often melamine). During the heating of the coating, the polymeric binder melts, and the binder and char builder start to decompose. The acid formed from the acid source cross-links the char to a viscous melt. At this moment, the blowing agent is activated and releases gas, which is trapped in the melt. The increasing amount of gas bubbles and the increasing bubble size leads to a massive expansion of the coating and to the formation of a low heat conducting foam. Additionally, some of the chemical reactions leading to the foaming are endothermic and thus also consume heat.

Due to the trend toward lightweight construction in transport, especially in aviation and shipping, more and more fiber reinforced polymer composites are used in applications bearing heavy loads. Fiber reinforced polymers are replacing load-bearing steel or aluminum structures. These composite structures have two major disadvantages compared to aluminum or steel structures when it comes to fire safety [74]. First, due to their polymeric ingredients, they increase the flammability and fire load of the construction. Second, the fire stability of composites may be a problem due to their specific fire behavior. Reaching a material temperature around the glass transition temperature of the used epoxy matrix ($T > 100–200\ °C$) can result in a massive loss of mechanical stability and ultimately in the collapse of the composite structure.

■ 2.6 Conclusions

The burning of a polymer is a physico-chemical process strongly influenced by the coupling of a chemical reaction – oxidation of fuel – in the gas phase with a chemical decomposition reaction – pyrolysis – in the condensed phase via heat and mass transfer. The heat and mass flux control the intensity of fire and the ablation of fuel. Indeed, the temperature profile as a function of time may be one of the most important responses of a specimen to understand its burning behavior. Further, several physical phenomena, such as the heat absorption of the materials, thermal conductivity, and also melt flow and dripping, play a major role in determining ignition, flammability, and fire behavior. The burning of a polymer is very complex.

The various phenomena interact with each other, e.g., pyrolysis also influences the viscosity of the melt, and, thus, whether dripping or charring results in a protective layer, increasing the shielding effect of the residual protective layer. Only a detailed and comprehensive description opens the door to a well-founded understanding of the burning behavior of polymeric materials.

References for Chapter 2

[1] Lyon, R.E., Plastics and rubber, in *Handbook of Building Materials for Fire Protection*, Harper, C.A. (Ed.) (2004) McGraw-Hill, pp. 3.1–3.51.

[2] Quintiere, J.G., *Fundamentals of Fire Phenomena* (2006) Wiley, Chichester.

[3] Lyon, R.E., Janssens, M.L., Polymer flammability and fire behavior, in *Encyclopedia of Polymer Science & Technology*, 3rd Edition, Mark, H.F., Bikales, N.M. (Ed.) (2005) Wiley, New York. pp. 1–73.

[4] Drysdale, D.D., Fire safety design requirements of flame-retarded materials, in *Fire Retardant Materials*, Horrocks, A.R., Price, D. (Eds.) (2001) Woodhead Publishing, Cambridge, pp. 378–397.

[5] Lyon, R.E., Quintiere, J.G., Criteria for piloted ignition of combustible solids. *Combust. Flame* (2007) 151, pp. 551–559.

[6] Levchik, S.V., Wilkie, C.A., Char formation, in *Fire Retardancy of Polymeric Materials*, Grand, A.F., Wilkie, C.A. (Ed.) (2000) Marcel Dekker, New York, pp. 171–215.

[7] Hirschler, M.M., Chemical aspects of thermal decomposition of polymeric materials, in *Fire Retardancy of Polymeric Materials*, Grand, A.F., Wilkie, C.A. (Ed.) (2000) Marcel Dekker, New York, pp. 27–79.

[8] Schartel, B., Perret, B., Dittrich, B., Ciesielski, M., Krämer, J., Müller, P., Altstädt, V., Zang, L., Döring, M., Flame Retardancy of Polymers: The role of specific reactions in the condensed phase. *Macromol. Mater. Eng.* (2016) 301, pp. 9–35.

[9] Pearce, E.A., Weil, E.D., Barinov, V.Y., Fire smart polymers, in *Fire and Polymers, Materials and Solutions for Hazard Prevention*, ACS Symposium Series 797, Nelson, G.L., Wilkie, C.A. (Ed.) (2001) American Chemical Society, Washington, pp. 37–48.

[10] Thomson, H.E., Drysdale, D.D., Flammability of plastics. 1. Ignition temperatures. *Fire Mater.* (1987) 11, pp. 163–172.

[11] Schartel, B., Weiß, A., Sturm, H., Kleemeier, M., Hartwig, A., Vogt, C., Fischer, R.X., Layered silicate epoxy nanocomposites: Formation of the inorganic-carbonaceous fire protection layer. *Polym. Adv. Technol.* (2011) 22, pp. 1581–1592.

[12] Kashiwagi, T., in *Twenty-fifth Symposium on Combustion*, Pittsburgh (1994) The Combustion Institute Inc., Pittsburgh, pp. 1423–1437.

[13] Schartel, B., Uses of fire tests in materials flammabilty development, in *Fire retardancy of polymeric materials*, Wilkie, C.A., Morgan A.B. (Ed.) (2010) CRC Press, Boca Raton, pp. 387–420.

[14] Schartel, B., Wilkie, C.A., Camino, G., Recommendations on the scientific approach to polymer flame retardancy: Part 1 – scientific terms and methods. *J. Fire Sci.* (2016) 34, pp. 447–467.

[15] Babrauskas, V., *Ignition Handbook* (2003) Fire Science Publishers, Issaquah.

[16] Reimschuessel, H.K., Shalaby, S.W., Pearce, E.M., On the oxygen index of nylon 6. *J. Fire Flammability* (1973) 4, pp. 299–308.

[17] Wang, Y., Zhang, F., Chen, X., Jin, Y., Zhang J., Burning and dripping behaviors of polymers under UL 94 vertical burning test conditions. *Fire Mater.* (2010) 34, pp. 203–215.

[18] Dupretz, R., Fontaine, G., Duquesne, S., Bourbigot, S., Instrumentation of UL-94 test: understanding of mechanisms involved in fire retardancy of polymers. *Polym. Adv. Technol.* (2015) 26, pp. 865–873.

[19] Atreya, A., Convection Heat Transfer, in *SFPE Handbook of Fire Protection Engineering*, 3rd Edition, DiNenno, P.J., Drysdale, D.D., Beyler, C.L., Walton, W.D., Custor, R.L.P., Hall Jr., J.R., Watts Jr., J.M. (Ed.) (2002) SPFE, Quincy, pp. 1.44–1.72.

[20] Tien, C.L., Lee, K.Y., Stretton A.J., Radiation Heat Transfer, in *SFPE Handbook of Fire Protection Engineering*, 3rd Edition, DiNenno, P.J., Drysdale, D.D., Beyler, C.L., Walton, W.D., Custor, R.L.P., Hall Jr., J.R., Watts Jr., J.M. (Ed.) (2002) SPFE, Quincy, pp. 1.73–1.89.

[21] Lautenberger, C., Radiation Heat Transfer, in *SFPE Handbook of Fire Protection Engineering*, Hurley, M.J. (Ed.) (2015) Springer, New York, pp. 102–137.

[22] Linteris, G., Wilthan, B., Zammarano, M., Hanssen, L., Absorption and reflection of infrared radiation by polymers in fire-like environments. *Fire Mater.* (2012) 36, pp. 537–553.

[23] Dittrich, B., Wartig, K.-A., Hofmann, D., Mülhaupt, R., Schartel, B., Carbon black, multiwall carbon nanotubes, expanded graphite and functionalized graphene flame retarded polypropylene nanocomposites. *Polym. Adv. Technol.* (2013) 24, pp. 916–926.

[24] Hallman, J.R., Welker, J.R., Sliepcevich, C.M., Polymer surface reflectance-absorptance characteristics. *Polym. Eng. Sci.* (1974) 14, pp. 717–723.

[25] Schartel, B., Beck, U., Bahr, H., Hertwig, A., Knoll, U., Weise, M., Sub-micrometre coatings as an infrared mirror: A new route to flame retardancy. *Fire Mater.* (2012) 36, pp. 671–677.

[26] Rockett, J.A., Mike J.A., Conduction of Heat in Solids, in *SFPE Handbook of Fire Protection Engineering*, 3rd Edition, DiNenno, P.J., Drysdale, D.D., Beyler, C.L., Walton, W.D., Custor, R.L.P., Hall Jr., J.R., Watts Jr., J.M. (Ed.) (2002), SPFE, Quincy, pp. 1.27–1.43.

[27] Madrzykowski, D., Stroup, D.W., in *Fire Protection Handbook*, 20th Edition (2008) National Fire Protection Assoc, Quincy, pp. 2-31–2-48.

[28] Drysdale, D.D., Fundamentals of the Fire Behaviour of Cellular Polymers, in *Fire and Cellular Polymers*, Buist, J.M., Grayson, S.J., Woolley, W.D. (Ed.) (1986) Elsevier, Barking, pp. 61–75.

[29] Fina, A., Camino, G., Ignition mechanisms in polymers and polymer nanocomposites. *Polym. Adv. Technol.* (2011) 22, pp. 1147–1155.

[30] van Krevelen, D.W., Some basic aspects of flame resistance of polymeric materials. *Polymer* (1975) 16, pp. 615–620.

[31] van Krevelen, D.W., *Properties of Polymers: Their Correlation with Chemical Structure* (1990) Elsevier, Amsterdam.

[32] Schartel, B., Wilkie, C.A., Camino, G., Recommendations on the scientific approach to polymer flame retardancy: Part 2 – concepts. *J. Fire Sci.* (2017) 35, pp. 3–20.

[33] Schartel, B., Kunze, R., Neubert, D., Red phosphorus-controlled decomposition for fire retardant PA 66. *J Appl. Polym. Sci.* (2002) 83, pp. 2060–2071.

[34] Dittrich, B., Wartig, K.-A., Hofmann, D., Mülhaupt, R., Schartel, B., Flame retardancy through carbon nanomaterials: Carbon black, multiwall carbon nanotubes, expanded graphite, multi-layer graphene and graphene in polypropylene. *Polym. Degrad. Stab.* (2013) 98, pp. 1495–1505.

[35] Mita, I., Effects of structure on degradation and stability of polymers, in *Aspects of Degradation and Stabilization of Polymers*, Jellinek, H.H.G. (Ed.) (1978) Elsevier, Amsterdam, pp. 247–294.

[36] Weil, E.D., Hansen, R.N., Patel, N., Prospective Approaches to More Efficient Flame-Retardant Systems, in *Fire and Polymers. Hazards Identification and Prevention*, Nelson, G.L. (Ed.) (1990) ACS Symposium Series 425, ACS Washington DC, pp. 97–108.

[37] Wu, G.M., Schartel, B., Bahr, H., Kleemeier, M., Yu, D., Hartwig A., Experimental and quantitative assessment of flame retardancy by the shielding effect in layered silicate epoxy nanocomposites. *Combust. Flame* (2012) 159, pp. 3616–3623.

[38] Factor, A., Char formation in aromatic engineering polymers, in *Fire and Polymers. Hazards Identification and Prevention*, ACS Symposium Series 425, Nelson, G.L. (Ed.) (1989) ACS, Washington, pp. 275–287.

[39] Tewarson, A., Generation of heat and chemical compounds in fire, in *SFPE Handbook of Fire Protection Engineering*, 3rd Edition, DiNenno, P.J., Drysdale, D.D., Beyler, C.L., Walton, W.D., Custor, R.L.P., Hall Jr., J.R., Watts Jr., J.M. (Ed.) (2002) SPFE, Quincy, pp. 3.4:3-82–3-161.

[40] Walters, R.N., Lyon, R.E., Molar group contributions to polymer flammability. *J. Appl. Polym. Sci.* (2003) 87, pp. 548–563.

[41] Grassie, N., Scott, G., *Polymer Degradation and Stabilisation*. (1985) Cambridge University Press, Cambridge.

[42] Achilias, D.S., Chemical recycling of poly(methyl methacrylate) by pyrolysis. Potential use of the liquid fraction as a raw material for the reproduction of the polymer. *Eur. Polym. J.* (2007) 43, pp. 2564–2575.

[43] Kelen, T., *Polymer Degradation* (1983) Van Nostrand Reinhold Company, New York.

[44] Faravelli, T., Pinciroli, M., Pisano, F., Bozzano, G., Dente, M., Ranzi. E., Thermal degradation of polystyrene. *J. Anal. Appl. Pyrolysis* (2001) 60, pp. 103–121.

[45] Jang, B.N., Wilkie, C.A., The thermal degradation of polystyrene nanocomposite. *Polymer* (2005) 46, pp. 2933–2942.

[46] Weil, E.D., Levchik, S.V., *Flame Retardants for Plastics and Textiles*, 2nd Edition (2016) Hanser, Munich.

[47] Yu, J., Sun, L.S., Ma, C., Qiao, Y., Yao, H., Thermal degradation of PVC: A review. *Waste Manag.* (2016) 48, pp. 300–314.

[48] Holland, B.J., Hay, J.N., The thermal degradation of PET and analogous polyesters measured by thermal analysis-fourier transform infrared spectroscopy. *Polymer* (2002) 43, pp. 1835–1847.

[49] Bounekhel, B., McNeill, I.C., Thermal-degradation studies of terephthalate polyesters. 2. Poly(ether-esters). *Polym. Degrad. Stab.* (1995) 49, pp. 347–352.

[50] Montaudo, G., Puglisi, C., Samperi, F., Primary thermal-degradation mechanisms of PET and PBT. *Polym. Degrad. Stab.* (1993) 42, pp. 13–28.

[51] Jang, B.N., Wilkie, C.A., A TGA/FTIR and mass spectral study on the thermal degradation of bisphenol A polycarbonate. *Polym. Degrad. Stab.* (2004) 86, pp. 419–430.

[52] Levchik, S.V., Weil, E.D., Lewin, M., Review thermal decomposition of aliphatic nylons. *Polym. Int.* (1999) 48, pp. 532–557.

[53] Levchik, G.F., Si, K., Levchik, S.V., Camino, G., Wilkie C.A., The correlation between cross-linking and thermal stability: Cross-linked polystyrenes and polymethacrylates. *Polym. Degrad. Stab.* (1999) 65, pp. 395–403.

[54] Grassie, N., The pyrolysis of acrylonitrile homopolymers and copolymers, in *Developments in Polymer Degradation*, vol. 1, Grassie, N. (Ed.) (1977) Applied Science publishers, London, pp. 137–169.

[55] Hadden, R., Alkatib, A., Rein, G., Torero, J.L., Radiant Ignition of Polyurethane Foam: The Effect of Sample Size. *Fire Technol.* (2014) 50, pp. 673–691.

[56] Ohlemiller, T.J., Smoldering Combustion, in *SFPE Handbook of Fire Protection Engineering*, 3rd Edition, DiNenno, P.J., Drysdale, D.D., Beyler, C.L., Walton, W.D., Custor, R.L.P., Hall Jr., J.R., Watts Jr., J.M. (Ed.) (2002), SPFE, Quincy, pp. 2-200–2-210.

[57] Rein, G., Smoldering Combustion, in *SFPE Handbook of Fire Protection Engineering*, Hurley, M.J. (Ed.) (2015) Springer, New York, ch. 19, pp. 581–603.

[58] Kempel, F., Schartel, B., Marti, J.M., Butler, K.M., Rossi, R., Idelsohn, S.R., Oñate, E., Hofmann, A., Modelling the vertical UL 94 test: Competition and collaboration between melt dripping, gasification and combustion. *Fire Mater.* (2015) 39, pp. 570–584.

[59] Wawrzyn, E., Schartel, B., Karrasch, A., Jäger, C., Flame-retarded bisphenol A polycarbonate/silicon rubber/bisphenol A bis(diphenyl phosphate): Adding inorganic additives. *Polym. Degrad. Stab.* (2014) 106, pp. 74–87.

[60] Schartel, B., Braun, U., Knoll, U., Bartholmai, M., Goering, H., Neubert, D., Pötschke, P., Mechanical, Thermal and Fire Behavior of Bisphenol A Polycarbonate/Multiwall Carbon Nanotube Nanocomposites. *Polym. Eng. Sci.* (2008) 48, pp. 149–158.

[61] Eichhorn, J., Synergism of free radical initiators with self-extinguishing additives in vinyl aromatic polymers. *J. Appl. Polym. Sci.* (1964) 8, pp. 2497–2524.

[62] Pawelec, W., Aubert, M., Pfaendner, R., Hoppe, H., Wilen, C. E., Triazene compounds as a novel and effective class of flame retardants for polypropylene. *Polym. Degrad. Stab.* (2012) 97, pp. 948–954.

[63] Matzen, M., Marti, J. M., Oñate, E., Idelsohn, S., Schartel, B., Understanding dripping V-0 polypropylene through advanced UL 94 investigations and particle finite element method simulations, *in preparation.*

[64] Glassman, I., Yetter, R. A., *Combustion*, 4th Edition (2008) Academic Press, Burlington, pp. 311–377.

[65] Babrauskas, V., Peacock, R., Heat release rate – the single most important variable in fire hazard. *Fire Safety J.* (1992) 18, pp. 255–272.

[66] Quintiere, J. G., *Principles of Fire Behaviour* (1998) Delmar Publishers, Albany.

[67] Mowrer, F. W., Fundamentals of the fire hazards of materials, in *Handbook of building materials for fire protection*, Harper, C. A. (Ed.) (2004) McGraw-Hill, New York, pp. 1.1–1.50.

[68] Quintiere, J. G., Surface flame spread, in *SFPE Handbook of Fire Protection Engineering*, 3rd Edition, DiNenno, P. J., Drysdale, D. D., Beyler, C. L., Walton, W. D., Custor, R. L. P., Hall Jr., J. R., Watts Jr., J. M. (Ed.) (2002) SPFE, Quincy, pp. 2-246–2-257.

[69] Vandersall, H., Intumescent coating systems, their development and chemistry. *J. Fire Flammability* (1971) 2, pp. 97–140.

[70] Bourbigot, S., Le Bras, M., Duquesne, S., Rochery, M., Recent advances for intumescent polymers. *Macromol. Mater. Eng.* (2004) 289, pp. 499–511.

[71] Alongi, J., Han, Z. D., Bourbigot, S., Intumescence: Tradition versus novelty. A comprehensive review. *Prog. Polym. Sci.* (2015) 51, pp. 28–73.

[72] Duquesne, S., Magnet, S., Jama, C., Delobel, R., Intumescent paints: fire protective coatings for metallic substrates. *Surf. Coat. Technol.* (2004) 180–181, pp. 302–307.

[73] Weil, E. D., Fire-protective and flame-retardant coatings – A state-of-the-art review. *J. Fire Sci.* (2011) 29, pp. 259–296.

[74] Mouritz, A. P., Gibson, A. G., *Fire Properties of Polymer Composite Materials* (2006) Springer, Dordrecht.

3 Flame Retardants and Flame-Retarded Plastics

3.1 Flame Retardants

Manfred Döring, Lara Greiner, and Daniela Goedderz

3.1.1 Importance, Development, and Market

Rapid progress in the field of flame-retarded polymers has taken place in the last 50 years. New flame retardant formulations have entered the market due to an increasing diversity of products, which have required new regulations and standards concerning flame retardancy. In response to legal requirements, the development of non-toxic and preferably non-volatile flame retardants will increasingly become of interest for the industry due to bans of certain halogen-containing flame retardants, as well as other toxic and environmentally hazardous flame retardants [1–4]. In this context, phosphorus-based compounds in particular are an important aspect of fire retardancy investigations.

A trend towards polymeric and reactive flame retardants is taking place to avoid the undesirable migration of flame retardant substances from finished plastic materials. These migration effects can result in human exposure to potentially toxic chemicals and the deterioration of the flame retardant efficiency of the plastic material [5–6]. Halogen-containing flame retardants still have a significant share of the market due to their high flame retardant efficiency (Figure 3.1). Aluminum trihydroxide (ATH) has a very large market share (34%) as well. High loadings of ATH are required for achieving sufficient flame retardancy, but they diminish the mechanical properties of the resulting finished plastic. However, environmental aspects and low prices are still decisive factors for the usage of such inorganic hydroxides. In general, the market acceptance of flame retardant formulations is dependent on the fire behavior of the finished plastic in standard burning test setups and the price-performance ratio. Non-halogenated flame retardants (including those based on phosphorus and nitrogen) hold a share of 29% of the global market [7–9].

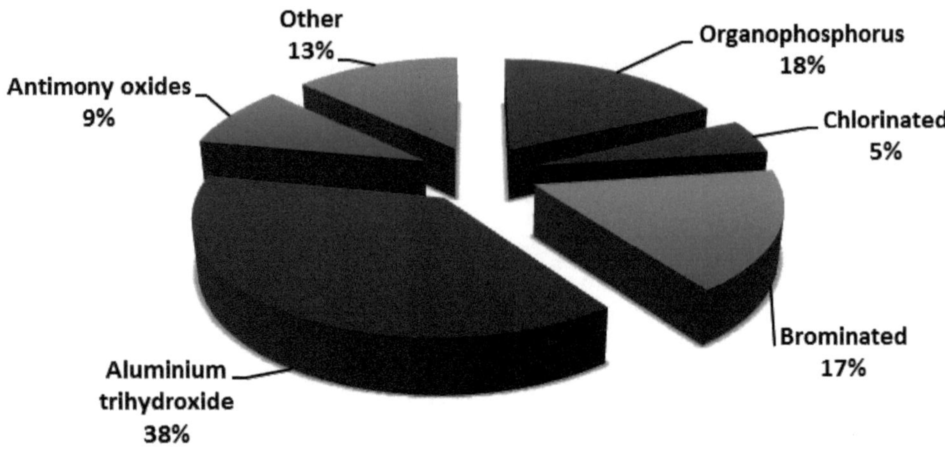

Figure 3.1 Global flame retardant market in 2016 according to chemical class [9]

A major problem for flame-retarded plastics is their limited recycling rate. Plastic waste is a major environmental challenge, and recycling this waste is one of the most important tasks. For this reason, the waste management plan of the European Union (EU) sets out an increase in the recycling rate of plastic waste of up to 70% by 2020 [10–11]. Efforts are being made to improve the recyclability of halogen-free flame-retarded plastics. The recycling process becomes more complicated when fibers or additives are added to plastics [8, 10, 12–13]. Ongoing projects demonstrate that multi-processing extrusion can result in a better distribution of flame retardants in the polymer matrix, while maintaining their flame retardant efficiency [10, 14–15]. Furthermore, there is a growing interest in making fire retardants more sustainable by using renewable raw materials [16].

3.1.2 Mode of Action

While a few polymers have inherent flame-retardant properties (i. e., teflon, polyimides, polybenzimidazoles, and phenolic resins), others need the incorporation of flame retardants to meet specific fire safety requirements. Flame retardants inhibit or even suppress the combustion process by acting chemically and/or physically in the solid, liquid, or gas phase. They act during the various stages of the combustion process, including heating, decomposition, ignition, or flame spread. Different modes of action are described in the following subsections.

It should be noted that combustion is a complex process in which individual stages may occur simultaneously, albeit with one dominating. The choice of flame retardants depends on whether gas-phase or condensed-phase activity is desired. For example, halogenated compounds act in the gas phase, while phosphates (ammonium polyphosphate (APP), melamine polyphosphate (MPP), and phosphoric esters) act as polyphosphoric acid producers in the condensed phase via char formation. Other parameters that need to be considered for flame retardants are their stability at polymer-processing temperatures, their compatibility with the polymer, as well as toxicity and cost.

3.1.2.1 Physical Action

Formation of a protective layer, cooling and dilution are possibilities to retard combustion by physical action.

The formation of a protective layer with low thermal conductivity by additives under an external heat flux leads to a reduction of oxygen flow and heat transfer between heat source and material. It therefore reduces the degradation rate of the polymer and the fuel flow in terms of pyrolysis gases that feed the flame. Pyrolysis of particular phosphorus additives leads to the formation of pyrophosphoric or polyphosphoric compounds that form a protective, vitreous, thermally stable barrier. Boric acid-based additives, inorganic borates, and low-melting glasses act similarly.

Endothermic degradation reactions of additives like ATH can also have an effect on the energy balance of combustion. Cooling of the substrate occurs and the temperature required for sustaining the combustion process is no longer reached. The efficiency of ATH depends on the amount incorporated into the polymer.

Finally, fillers (e.g., talc and chalk), other inert substances, and additives evolving inert gases (e.g., water, CO_2) on decomposition dilute the fuel in the solid and gaseous phases. This helps the gas mixture to stay below the ignition limit.

3.1.2.2 Chemical Action

Chemical action can be divided into two main categories, according to whether the action takes place in the gas phase or the condensed phase.

Gas Phase

If a flame retardant or its degradation products act in the gas phase, the radical mechanism of the combustion process in the gas phase is interrupted. The reactive radical concentration (H·, ·OH) decreases, the exothermic processes that occur in the flame are stopped, the system cools down and the supply of flammable gases is reduced until the flame eventually extinguishes. The flame retardant, while inhibiting the flame, may increase the toxicity of the fire gases because through incomplete combustion, the reactions in the flame zone will lead to highly toxic CO, other partially burnt products, and irritants.

Typically, halogenated flame retardants show gas-phase activity. A variety of chlorinated and brominated flame retardants are commercially available but the use of non-polymeric halogenated flame retardants is increasingly restricted by legislation. The flame retardancy is caused by the formation of hydrogen halides that react with H· or ·OH radicals formed in the gas phase. These two reactive radicals are known to catalyze the degradation of polymers in the combustion process [17].

Condensed Phase

Different mechanisms are observed in the condensed phase. The flame retardant can cause accelerated breakdown of the polymer, so that it melts, becomes liquid, and the polymer flows away from the sphere of influence of the flame (retreat effect). Based on this mechanism, flame retardant materials with high surface areas (foams and films) pass specific flame propagation tests [4, 18–19].

One of the most important chemical actions is the formation of a carbonaceous layer on the polymer surface (barrier effect). As an example, flame retardants may generate double bonds and linkage in polymers by condensation of oxygen-rich organic compounds. Cyclizing and cross-linking processes lead to a carbonaceous layer acting as a barrier against oxygen, pyrolysis gases, thermal feedback, smoke, and products of incomplete combustion. Polyphosphates (APP, MPP, and phosphoric esters) act in this way by forming non-volatile polyphosphoric acid when heated. A special case of this action is intumescence, mainly used in coatings. The additional incorporation of blowing agents into the flame retardant formulation leads at higher temperatures to a voluminous, insulating protective layer. The intumescent formulations consist of an acid source (e.g., APP), a carbon supplying agent (e.g., polyalcohol), and a blowing agent (e.g., melamine). The mixture has to be adjusted for the specific applications needed. Intumescence leads to much better insulation of the polymer and other materials under the protective layer.

Flame retardants may be used as such or as mixtures causing a synergistic effect in polymers. A synergistic effect is defined as being one that is greater than the sum of the individual actions, and helps to reduce the total amount of flame retardants used. Special synergists are discussed later in this chapter.

A distinction is made between reactive and additive flame retardants. Reactive flame retardants are incorporated into the polymer network (for example in thermosets like unsaturated polyester resins, epoxy resins, and polyurethanes, or by copolymerization in fiber-grade PET). They do not migrate or volatize, they have no plasticizing effect, and do not result in a negative effect on the thermal stability of the polymer. Additive flame retardants are mainly used in thermoplastics, but also in thermosets and elastomers. They act like additives or fillers if they are non-compatible, or as plasticizers if they are compatible with the polymer.

Combinations of flame retardants with synergistic effects are often used, and current developments are aimed at polymeric flame retardants. They do not migrate, show better compatibility with polymer matrices (blending), and have no negative effect on important properties like the glass transition temperature of the polymer matrix. For example, poly(pentabromobenzylacrylate) (see Section 3.1.3.2) is a suitable polymeric flame retardant for ABS or PBT applications and an alternative to legacy flame retardants with toxicological concerns. The use of hexabromocyclododecane (HBCD), the flame retardant of choice in PS foams, has been banned because of its persistence, bioaccumulation, and toxicity in the environment. An alternative for thermal insulation polystyrene foams is a new polymeric flame retardant based on brominated butadiene-styrene block copolymers [20].

3.1.3 Important Flame Retardant Classes

The following sections give an overview of the categories of flame retardants. Inorganic flame retardants like mineral fillers and metal hydroxides, nitrogen-containing flame retardants like triazine-based substances, halogenated flame retardants, and phosphorus-containing flame retardants are the major players for achieving flame retardancy of polymers. Radical-forming agents are often powerful synergists in flame retardant formulations by enhancing the flame retardant efficiency and thus decreasing the total amount of flame retardants needed. Some radical-forming agents combine flame retardancy with light stability or an antioxidant effect. Flame retardant efficiency and cost-effectiveness play an important role regarding the choice of flame retardant formulations for specific polymer applications.

3.1.3.1 Metal Hydroxides and Mineral Fillers

Inorganic flame retardants are an important group of flame retardants due to their low prices and low toxicity. However, relatively high concentrations are required for an effective flame retardant effect, and they can influence the mechanical polymer properties negatively [21, 22]. In most cases, inorganic flame retardants are based on carbonates or hydroxides of group II or III elements [7].

The interaction between hydrophobic polymers and hydrophilic fillers is very weak. High loadings of inorganic fillers have a negative impact on the material's properties and its processability. To overcome these problems, inorganic fillers can be surface-modified to increase their compatibility with the polymer. Organosilanes, silicones, and polymers modified with maleic anhydride are suitable surface modifiers [23–25]. The polymer matrix itself can also be modified. Nachtigall et al. used either vinyltriethoxysilane (VTES) or maleic anhydride (MA) as coupling agents to synthesize polypropylene (PP)-ATH composites. The best results were obtained with VTES, which resulted in a higher degree of functionalization and better mechanical properties [26].

The mode of action of certain inorganic flame retardants is based on endothermic decomposition of these filler-type materials (cooling), leading to the release of water or inert gases at pyrolysis temperatures, which decrease the amount of fuel and oxygen in the flame. Inorganic flame retardants do not evaporate under heat exposure like organic flame retardants. Instead, they reduce the total amount of fuel and increase the thermal conductivity and heat capacity by shielding the surface and the oxides remaining in the char. This results in an insulating effect that prevents outbound radiation, and stops flows of oxygen, and flammable pyrolysis products [7, 21, 27–30]. The endothermic dehydration of metal hydroxides leads to the formation of a non-cohesive residue consisting of the corresponding metal oxide [31, 32]. Inert fillers also dilute the polymer in the condensed phase and increase the amount of thermal energy needed to raise the temperature of the composition to the pyrolysis level, due to the high heat capacity of the fillers. The released water and the decomposition products can undergo endothermic interactions in the flame (reactions 3.1 and 3.2) [33].

The endothermic decomposition reaction takes place according to the following reaction schemes [28]:

$$2\ Al(OH)_{3(s)} \rightarrow Al_2O_{3(s)} + 3\ H_2O_{(g)} + 1300\ kJ\ g^{-1} \tag{3.1}$$

$$Mg(OH)_{2(s)} \rightarrow MgO_{(s)} + H_2O_{(g)} + 1450\ kJ\ g^{-1} \tag{3.2}$$

To ensure the flame retardant efficiency of such non-combustible fillers, the decomposition temperature must be in a narrow temperature range – above the processing temperature of the polymer, but before the decomposition of the polymer begins. Inorganic fillers like aluminum hydroxides (ATH, boehmite), magnesium dihydroxide (MDH) (commonly referred to as magnesium hydroxide), and magnesium carbonate hydroxide are the most important commercially available representatives of this flame retardant category, and are used mainly in polyethylene (PE), polypropylene (PP), and ethylene vinyl acetate copolymer (EVA). Magnesium hydroxide is also used for polyamides (PAs) [21, 34].

ATH can be used in synergistic formulations with phosphorus compounds in which the phosphorus atom has a carbon or hydrogen rich environment, whereas no synergistic effect results from formulations containing ATH and phosphates rich in oxygen [35].

Table 3.1 summarizes the most important physical properties of some selected mineral fillers as flame retardants [7, 28, 36–38].

Table 3.1 Physical Properties of Selected Mineral Fillers

Filler	Formula	T_{decomp}/°C	ΔH_{decomp}/kJ g^{-1}
Aluminum trihydroxide	$Al(OH)_3$	180-200	1300
Magnesium dihydroxide	$Mg(OH)_2$	300-320	1450
Calcium hydroxide	$Ca(OH)_2$	430-450	1150
Nesquehonite	$MgCO_3 \cdot 3\ H_2O$	70-100	1750
Hydromagnesite	$Mg_5(CO_3)_4(OH)_2 \cdot 4\ H_2O$	220-240	1300
Huntite	$Mg_3Ca(CO_3)_4$	400	980
Ultracarb	Hydromagnesite-Huntite (60:40)	220-400	1172
Boehmite	AlO(OH)	340-350	560
Kaolinite	$Al_2Si_2O_5(OH)_4$	550-600	650

Aluminum trihydroxide (ATH) is one of the most used flame retardants and decomposes in a temperature range of 180–200 °C [36] endothermically, with water loss between 1170 J g^{-1} and 1300 J g^{-1} [36, 39–40]. The decomposition parameters depend on the physical form and the particle size of the aluminum hydroxide [28].

The synergism between magnesium hydroxide and phosphoric compounds is already known. For example, a combination of 10 wt% of a cyclic phosphonate and magnesium hydroxide provides a UL 94 V-0 rating, whereas 10 wt% of the phosphonate or 10 wt% of magnesium hydroxide alone in glass-fiber-reinforced PBT only achieves UL 94 V-2 [35, 41, 42].

Ethylene–vinyl acetate copolymer (EVA) can be loaded with relatively high amounts of mineral fillers without strongly affecting the characteristics of the polymer. The main reasons for this phenomenon are the EVA elastomeric properties and the polarity of the acetate side chains. EVA–metal hydroxide composites (EVA–MH) lead to the release of acetone as a pyrolysis product due to the catalytic action of metal oxides (formed by the thermal decomposition of metal hydroxides) upon acetic acid [43].

ATH and magnesium hydroxide are used as flame retardants for wires and cables. High loadings are required for meeting flame retardancy requirements, such as preventing flame propagation. However, compounding and extrusion are challenging because of their high densities, while the lack of flexibility of the finished products can be problematic. Solutions to this problem are nanometer-scale nanocomposites consisting of organoclays or carbon nanotubes dispersed in the polymer matrix [44].

Figure 3.2 shows HRR curves for EVA composites with calcium- and magnesium-containing fillers [31].

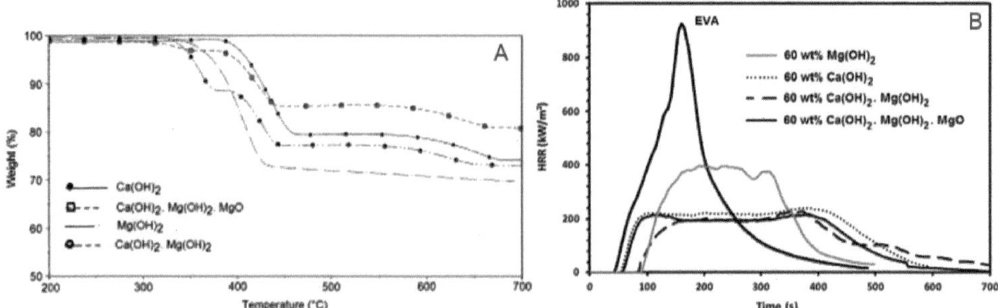

Figure 3.2 TGA curves for $Ca(OH)_2$, $Ca(OH)_2 \cdot Mg(OH)_2 \cdot MgO$, $Mg(OH)_2$ and $Ca(OH)_2 \cdot Mg(OH)_2$ under air (20 °C/min) (A) and heat release rate (HRR) curves for EVA and EVA composites containing calcium- and magnesium-based fillers (50 kW/m²) (B)

Polymer-clay composite structures consist of modified layered silicates, which are dispersed in a polymer matrix and possess exceptional properties in comparison to the neat polymer. Nanocomposite structures are formed by the intercalation of the polymer between the silicate layers, whereby intercalated and exfoliated structures are formed during dispersion. A microcomposite structure is obtained when no intercalation of the polymer between the silicate layers is possible and a phase separation has occurred.

Nanocomposites can be obtained by in situ intercalative polymerization, intercalation of polymer from solution, template synthesis or melt intercalation [45–47]. Camino et al. [48] investigated the microstructures and fire behavior of EVA-organoclay microcomposites and nanocomposites. A nanocomposite was obtained by using a montmorillonite clay, while aminododecanoic acid-exchanged fluorohectorite led to a microcomposite. They demonstrated that nanodispersed morphology leads to a decrease of the rate of combustion, while the microcomposite shows the same fire behavior as neat EVA. The clay layers reassemble and promote the charring of the polymer, resulting in the formation of a protective charred ceramic surface layer [48, 49]. The degradation pathway of a polymer is influenced by the incorporation of a clay, which results in a different fire behavior. The protective surface layer formed during combustion leads to excessive heating in the condensed phase, resulting in extensive random scission. This effect is intensified by a significant reduction of HRR. In this context, Wilkie et al. correlated the degradation pathway of a polymer with the peak heat release rate [49, 50].

To reduce HRR significantly, commonly used inorganic fillers like $Mg(OH)_2$ or $Al(OH)_3$ require very high loadings between 40–65 wt%, while nanofillers only require very low loadings between 0.5–7 wt% [51, 52]. Layered double hydroxides (LDH) can be used as nanofillers and consist of lamellar inorganic crystalline materials, with the anions in the interlayer space being exchangeable. Anionic surfactants can be used to increase the surface hydrophobicity and to enhance the space between the layers. The decomposition of those LDHs leads to the release of interlayer water and intercalated anions, resulting in mixed metal oxides [53]. The flame retardant efficiency is based on the dilution of oxygen, the formation of char, smoke suppression, and heat adsorption during endothermic degradation [53–56].

Other nanocomposite materials are flame retardant systems with carbon nanotubes (CNTs) as nanofillers. CNTs possess a cylindrical nanostructure and can be grouped into small-diameter (1–2 nm) single-walled nanotubes (SWNTs) or larger-diameter (10–100 nm) multi-walled nanotubes (MWNTs). Kashiwagi et al. [57] demonstrated that 0.5 wt% of SWNTs in PMMA result in a decrease in HRR due to the formation of a protective layer acting as a thermal shield [49, 58, 59]. Graphene oxide, expanded graphene, multilayered graphene (MLG), and graphene itself can also improve the thermal stability of polymers like PVA and reduce the peak HRR. The flame retardant efficiency is achieved by the formation of a compact char layer during the combustion process that prevents heat and mass transfer [51, 60–62].

3.1.3.2 Halogenated Flame Retardants

Halogenated flame retardants based on brominated or chlorinated substances are commercially available. However, the use of some halogenated flame retardants is restricted because of concerns regarding their persistence, bioaccumulation, and toxicity (PBT) [63]. Typically, these flame retardants are added to the polymer as additives, or halogenated structures are introduced as reactive components. Their effectiveness is related to the ease of scission of the carbon–halogen bond, which increases from fluorides to chlorides, bromides, and iodides [64]. The bond between carbon and fluorine is too strong to interfere with the combustion process, whereas the bond between carbon and iodine is too weak. However, polytetrafluoroethylene is used as an anti-dripping agent [65]. Bromine-containing flame retardants show the highest efficiency because the bonding energy between carbon and bromine enables the flame retardant to interfere at a more favorable point in the combustion process. Additionally, the effective agent HBr is liberated over a narrow temperature range so that it is available at a high concentration in the flame zone [30]. Chlorine-containing flame retardants are less effective, because they release HCl over a wider temperature range, resulting in a lower concentration in the flame zone. The flame retardant has to be adjusted to the polymer and the method of incorporation.

Halogenated flame retardants operate in the gas phase (Section 3.1.2) [30], by interfering with the radical chain mechanism [66, 67]. First, in the halogenated compound, the bond between the carbon and halogen atom breaks. A halogen radical X· and an R· radical (the rest of the organic compound) are generated. X· reacts with a hydrogen atom of the polymer chain (P–H) to form the hydrogen halide H–X [68]. The hydrogen halides react with the high-energy ·OH [69–72] and H· [73] radicals so that these radicals are removed from the burning zone and are replaced by low-energy halogen radicals (Figure 3.3). The hydrogen halide consumed is regenerated by reaction with hydrocarbon, so overall it acts as a catalyst [74].

The basic steps [40] are shown in Figure 3.3:

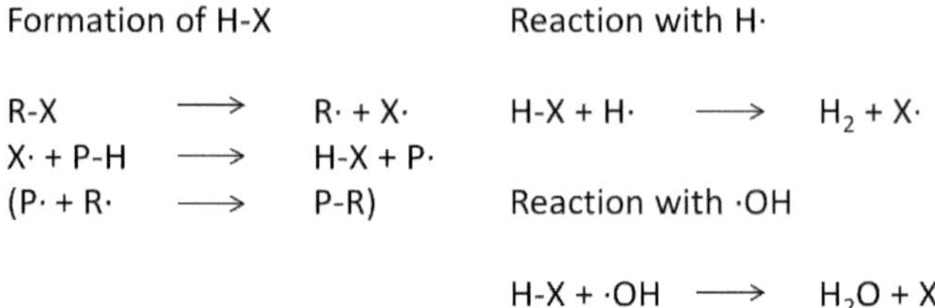

Figure 3.3 Basic steps of the gas-phase activity of halogenated flame retardants

Chlorine-containing flame retardants are commercially mostly used as chlorinated hydrocarbons or chlorinated cycloaliphatics, which are cheap and have good light stability. Compared to bromine-containing flame retardants, their lower efficiency requires high loadings to provide a sufficient level of flame retardancy, which may affect the properties of the polymer materials. Commercially available chlorinated hydrocarbons are chloroparaffins, which depending on the chlorine content (40–70%) can be liquid or solid. Their poor thermal stability (dehydrochlorination starts at 180 °C) restricts their use to PVC, LDPE and polyurethanes. Cycloaliphatic chlorine compounds show better thermal stability and have found widespread applications as flame retardants in polymers. Hexachloro-*endo*-methylenetetrahydrophthalic acid (HET acid) is used in unsaturated polyester resins. Dodecachloropentacyclooctadeca-7,15-diene, which is known as Dechlorane Plus® and stable up to is melting point of about 350 °C, is used in polyamides and polybutylene terephthalates. In polyurethane (PU) applications, the liquid tris(chloroisopropyl) phosphate (TCPP) is used. In contrast to other chlorophosphates such as TCEP (tris(2-chloroethyl)phosphate) and TDCP (tris(1,3-dichloro-2-propyl)phosphate), this substance is not banned.

Because of the volatility and consequent "fogging" contribution of chlorinated flame retardants, the search for alternatives is important. In particular, phosphorus-containing flame retardants may be used in PU foams. For example, commercially available propionates and reactive phosphonate polyols functionalized with 9,10-dihydro-9-oxa-10-phosphaphenanthrene-10-oxide (DOPO) have low volatility because they become part of the PU network.

Bromine-containing flame retardants are widely used and commercialized. Versatile applications in nearly all polymers are possible, and they have a low tendency to bleed out. In many cases, antimony trioxide (ATO) is added as a synergist, and the halogen–antimony synergism has been described for a variety of polymers [75]. On heating, these flame retardant systems evolve volatile metal halides, which are well-known synergistic flame inhibitors of much greater effectiveness than hydrogen halides evolved in the absence of ATO [76, 77]. Various phosphorus-containing compounds have been described as effective alternatives to ATO [78]. Compared to chlorinated flame retardants, brominated compounds exhibit a poorer light stability and are more expensive. They are available as both reactive and additive flame retardants.

Depending on the polymer and the application, aliphatic, cycloaliphatic, or aromatic brominated flame retardants may be used. A versatile and commonly known brominated phenol derivative is tetrabromobisphenol A (TBBA, Figure 3.4). It is primarily used for flame retardant epoxy resins, in particular as a flame retardant matrix resin for printed circuit boards. The incorporation takes place by preformulation with the epoxy resin, and a variety of oligomeric and polymeric derivatives are also used based on that reaction. Low molecular weight TBBA-based epoxy resins can be used in unsaturated polyester (UP), phenolic (PF) and epoxy (EP) resins. High molecular weight grades are suitable for engineering thermoplastics like polyamide 6, PBT, polycarbonate (PC), or high-impact polystyrene (HIPS) and ABS. TBBA-bis(2,3-dibromopropylether) is mainly used in PP and HIPS, provided that the V-2 classification in UL 94 vertical tests is sufficient. The phenoxy-terminated carbonate oligomer is used in polybutylene terephthalate. Other brominated phenol derivatives are bis(tribromophenoxy)ethane, which is mainly used in ABS and HIPS, as well as polydibromophenylene ether for applications in polyamides [39, 80–81].

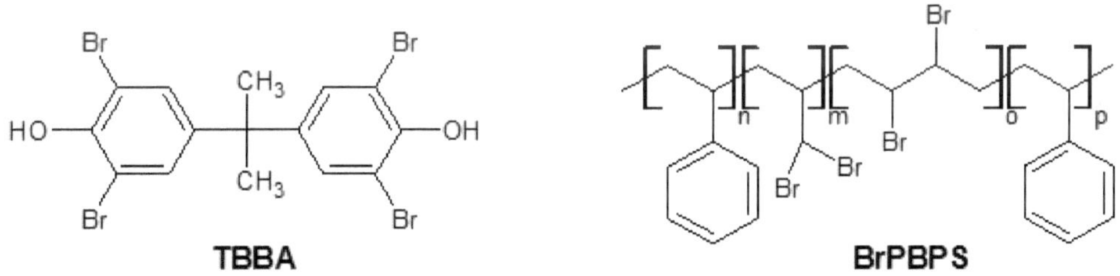

TBBA

BrPBPS

Figure 3.4 Molecular structure of tetrabromobisphenol A and of polymeric brominated polybutadiene-polystyrene (BrPBPS) [79]

Since polybrominated diphenyl ethers are banned, flame retardants not belonging to the biphenyl or diphenyl ether classes have been developed for various applications [82–83]. Among these compounds, bis(pentabromophenyl)ethane is mainly used in HIPS, polyolefins, and engineering plastics. Other reactive brominated flame retardants are tetrabromophthalic anhydrides (used in unsaturated polyesters) and their derivatives, such as 1,2-bis(tetrabromophthalimide)ethane, which are used in HIPS, ABS, PP and PBT. These compounds have high UV and thermal stability. Tetrabromophthalate diol is used as polyol component in rigid PU foam, and plasticising tetrabromophthalate ester (which is a liquid) is applied in PVC, elastomers, adhesives, and coatings. As well as tetrabromobisphenol A polycarbonate oligomers, brominated oligomers and polymers such as polydibromophenylene oxide, brominated polystyrene, and poly(pentabromobenzyl acrylate) are used as non-volatile and non-toxic flame retardants [39, 81].

Exceptions to the gas-phase mechanism of chlorinated flame retardants have been observed [81]. Fenimore et al. [84, 85] showed by measurements of the oxygen and nitrous oxide index of chlorinated polyethylene that here they act in the condensed phase. Studies into the fire-retardant action of halogenated compounds in the condensed phase have mostly been performed on mixtures of chloroparaffin with vinyl polymers such as polyethylene.

3.1.3.3 Phosphorus-Containing Flame Retardants

The patent and other literature on halogen-free flame retardants predominantly focuses on phosphorus-based products, which primarily include phosphorus in the form of phosphines, phosphine oxides, phosphinates, phosphonates, red phosphorus, or phosphates.

The mode of action depends on the phosphorus species and the polymer in which they are incorporated [86–91], as well as on other additives. The phosphorus species most probably act in the condensed phase by char promotion [52, 92–94], leading to intumescence in the case of an added blowing agent [39, 40, 95, 96] or through inorganic glass formation [86, 97]; it may also act in the gas phase by flame inhibition [39, 40, 70, 98]. In fact, for many systems, there is significant action in both the gas and the condensed phases [86, 87, 99–101]. The mode of action depends on the chemical environment of the phosphorus (carbon/hydrogen-rich leads to action in the gas phase, oxygen-rich leads to action in the condensed phase). Without any further additives, phosphorus-containing flame retardants are particularly effective in polymers with high oxygen content, and show weaker efficacy in polyolefins and styrenics [102]. Polymeric systems are a current focus of research into phosphorus-containing flame retardants, due to their lower volatility and smaller influence on the material properties of the flame-retarded polymeric material.

The gas-phase activity of phosphorus-containing flame retardants can be described in a similar way as for halogen-containing flame retardants [70, 103]. The efficiency of phosphorus in the gas phase is nearly the same as that of HBr [102]. Here, the PO· radical plays a major role [34, 70, 104, 105]. Important reactions [106] are shown in Figure 3.5.

$$PO\cdot + H\cdot \longrightarrow HPO$$
$$PO\cdot + \cdot OH \longrightarrow HPO_2$$
$$HPO + H\cdot \longrightarrow H_2 + PO\cdot$$
$$PO\cdot + H_2 + \cdot OH \longrightarrow H_2O + HPO$$
$$HPO_2\cdot + H\cdot \longrightarrow H_2O + PO$$
$$HPO_2\cdot + H\cdot \longrightarrow H_2 + PO_2$$
$$HPO_2\cdot + \cdot OH \longrightarrow H_2O + PO_2$$

Figure 3.5 Key reactions of flame inhibition for phosphorus-containing flame retardants

DOPO (9,10-dihydro-9-oxa-10-phosphaphenanthrene-10-oxide) and its derivatives are well-known to react primarily in the gas phase [107]. The driving force for the formation of PO· radicals is assumed to be the thermodynamically stable dibenzofuran (DBF). The most important reactions [108] are shown in Figure 3.6 [21, 107–109].

$$HPO \longrightarrow H\cdot + PO\cdot \quad 278.8\ kJ\cdot mol^{-1}$$
$$DOPO \longrightarrow H\cdot + \cdot DOPO\text{-}H \quad 354.2\ kJ\cdot mol^{-1}$$
$$\cdot DOPO\text{-}H \longrightarrow PO\cdot + DBF \quad 88.6\ kJ\cdot mol^{-1}$$
$$DOPO \longrightarrow HPO + DBF \quad 163.9\ kJ\cdot mol^{-1}$$
$$DOPO + H\cdot \longrightarrow H_2 + \cdot DOPO\text{-}H \quad -105.1\ kJ\cdot mol^{-1}$$
$$DOPO + \cdot OH \longrightarrow H_2O + \cdot DOPO\text{-}H \quad -148.8\ kJ\cdot mol^{-1}$$
$$DOPO + H\cdot \longrightarrow PO\cdot + H_2 + DBF \quad -16.5\ kJ\cdot mol^{-1}$$
$$DOPO + \cdot OH \longrightarrow PO\cdot + H_2O + DBF \quad -60.3\ kJ\cdot mol^{-1}$$

DOPO

DBF

·DOPO-H

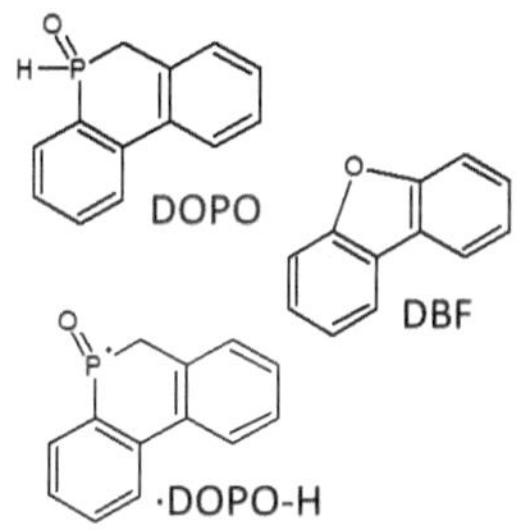

Figure 3.6 Thermal behavior of DOPO and formation of PO· radicals and dibenzofuran, with calculated reaction energies (DFT)

The action in the condensed phase is based on the chemical interaction between the flame retardant and the polymer. The diminishing activity with decreasing hydroxyl group content of the polymer or additive compound can be related to the susceptibility of the polymers to dehydration and char formation [110, 111]. Whereas celluloses are adequately flame retarded with around 2% phosphorus, 5–15% is needed for polyolefins [64].

Typical observations for an occurred activity in the condensed phase is the formation of phosphate glasses and polyphosphoric acid [40, 93, 112]. During the combustion process, phosphoric acid is formed, which condenses to pyrophosphoric structures. While the released water dilutes the gas phase, phosphoric acid and polyphosphoric acid force the dehydration reaction of terminal hydroxyl functions. This dehydration of the polymeric structure induces the formation of carbocations and double-bonds and therefore, at high temperatures, cyclization, cross-linking, and aromatization/graphitization occurs and a carbonized layer is formed.

Phosphoric and polyphosphoric acids can, together with the carbonized residues, additionally form a molten viscous surface layer protecting the polymer substrate against flame and oxygen. This carbonized layer (char) limits the volatilization of fuel, prevents the formation of new free radicals, and limits oxygen diffusion, which reduces combustion and insulates the polymer underneath from heat [49].

Commercially available phosphorus-containing flame retardants (and those in development) can be divided into several categories. Coated red phosphorus is the most concentrated source of phosphorus for flame retardancy. It is very effective in polymers such as polyesters, polyamides, and polyurethanes. It is also used for coating textiles [49]. For example, Class V-0 to UL 94 is achieved if 6 to 8 wt% of red phosphorus is added to glass-filled PA6.6 [113].

The group of phosphates includes the inorganic phosphates ammonium polyphosphate (APP, Figure 3.7) and melamine polyphosphate (MPP), which are inorganic salts of polyphosphoric acid. The chain length of polyphosphates is variable, and affects crystallinity. Long-chain APPs (type II, chain length: 700–1000) decompose to ammonia and polyphosphoric acid at temperatures higher than 270 °C, whereas short-chain APPs (type I, chain length: 30–150) decompose at temperatures above 230 °C [114]. Therefore, the APP used can be adapted to the decomposition temperature of the polymer. Low molecular weight APP shows better hydrolytic stability. Thermoset coatings [115] of APP granules further improve the hydrolysis resistance. APP can be used as an acid-forming agent in oxygen- and/or nitrogen-containing polymers like polyurethanes or epoxy resins [116]. The polyphosphoric acid reacts with the oxygen- or nitrogen-containing polymers and catalyzes their dehydration reaction and char formation [117]. Together with char-forming nitrogenous resins, APP is used in polyolefins and ethylene-vinyl acetate (EVA) as well as in intumescent coatings [8]. Due to its higher thermal stability, MPP is used as flame retardant in engineering plastics, usually in combination with gas-phase-active phosphorus compounds.

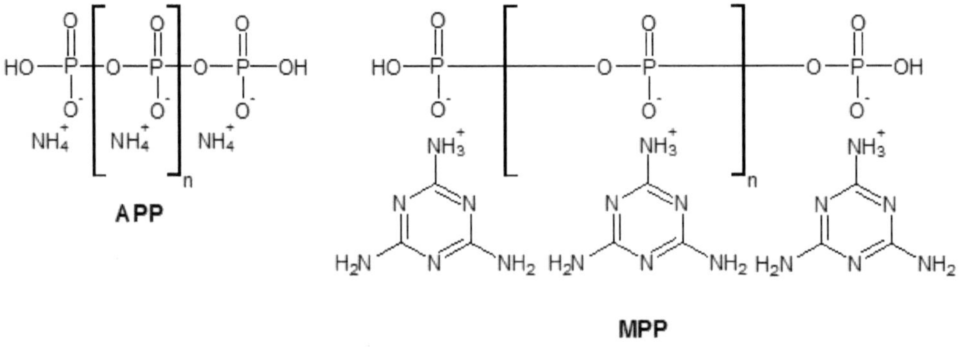

APP

MPP

Figure 3.7 Structure of ammonium polyphosphate and melamine polyphosphate

Organic phosphorus-based compounds can either act as additives or be incorporated into the polymer network. The main groups are phosphates, phosphonates, and phosphinated salts and esters [42]. Phosphate esters are mainly used as flame retardant plasticizers in polyvinylchloride and engineering plastics (polyphenylene oxide–high-impact polystyrene, polycarbonate–acrylonitrile butadiene styrene (PC–ABS) blends, polycarbonate), or phenolic resins and coatings. Phosphates, phosphonates, and phosphinates are used for different applications. For example, reactive phosphorus-containing compounds are used for polyurethane foams and epoxy resins.

Phosphate esters like triphenyl phosphate (TPP), or the alternative non-toxic compounds *tert*-butylphenyl diphenyl phosphate and cresyl diphenyl phosphate, have a high volatility resulting in a relatively low fire retardant efficiency [87]. Alternatives to these phosphate esters are oligomeric derivatives of resorcinol bis(diphenyl phosphate) (RDP), bisphenol A bis(diphenyl phosphate) (BDP), and the biphosphate resorcinol bis(dixylenyl phosphate) (RXP) (Figure 3.8). RDP, BDP, and RXP are less volatile and have a higher thermal stability than TPP. In PC/ABS (5:1) formulations, BDP and RDP achieve UL 94 V-0 ratings when 1 wt% of phosphorus is incorporated. BDP shows better fire-retardant efficiency, hydrolytic stability and thermal stability than RDP. RXP has a higher hydrolytic stability than BDP, because of steric hindrance provided by the 2,6-xylyl groups; its flame retardant efficiency is similar to that of BDP [118]. The primary fire retardant action is likely to occur in the condensed phase by the formation of polyphosphoric acid [119, 120].

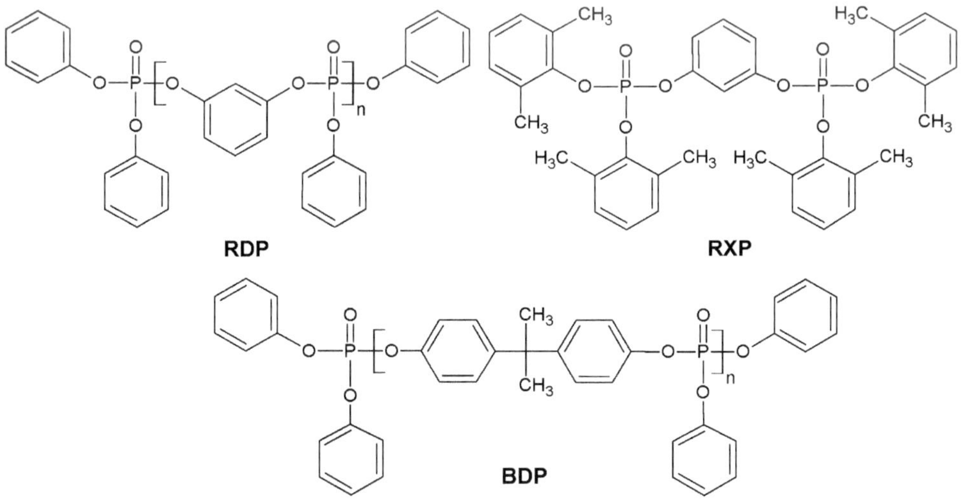

RDP

RXP

BDP

Figure 3.8 Molecular structures of polymeric derivatives of resorcinol bis(diphenyl phosphate), bisphenol A bis(diphenyl phosphate) and resorcinol bis(dixylenyl phosphate)

Phosphonates, especially polymeric methylphosphonates, have been commercialized. Used as additives, such polymeric flame retardants are less volatile, less prone to migrate during compounding and processing, and their plasticizing effect is reduced. Polymeric methylphosphonates (Figure 3.9) were developed for copolymerization with polycarbonate to obtain a transparent flame-retardant polymer. The company FRX Polymers produces aliphatic polyphosphonate homopolymers [121], copolymers [122, 123], and branched polymers [124] as polymeric flame retardant additives for use in PET fibers and foams, PC, epoxy resins, and for use on their own as transparent polymers [125]. Typically, phosphonates show condensed-phase flame-retardant activity [126]. A molecular bridged methylphosphonate (Figure 3.9) has been designed, mainly for polyolefin applications in textiles and stretched films.

Products from diethylphosphinic acid salts like the aluminum salt (DEPAL, Figure 3.9) or the zinc salt (DEPZn) have become established for polyamides, thermoplastic polyesters, and certain thermosets, amongst others [127]. For PA6, PA66, and PBT [128], synergistic blends with melamine polyphosphate are used. While loadings of 25 wt% in glass-fiber-reinforced PA66 are needed to achieve effectiveness [129], the phosphinate content can be reduced to less than 20 wt% in the presence of MPP [130].

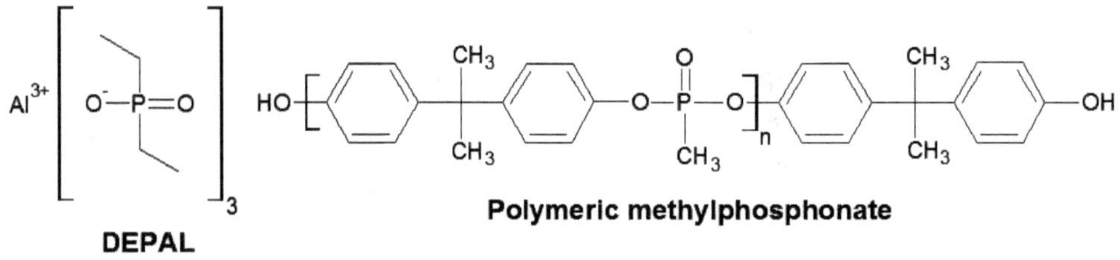

Al^{3+} $[O^{-}–P=O]_3$ DEPAL

HO–[C₆H₄–C(CH₃)₂–C₆H₄–O–P(=O)(CH₃)–O]ₙ–C₆H₄–C(CH₃)₂–C₆H₄–OH Polymeric methylphosphonate

Figure 3.9 Molecular structures of DEPAL and polymeric methylphosphonate

Melamine derivatives like MPP alone require relatively high loadings of about 30 wt% for a UL 94 V-0 rating in glass fiber-reinforced PA66 or PBT without synergists [42, 131]. In addition, flame retardants containing melamine poly(metal phosphates) have been introduced for similar applications [132].

Phenoxyphosphazene compounds (mixtures of poly- and cyclobisphenoxyphosphazenes) are mainly used for ABS and glass-fiber-reinforced epoxy resin applications like printed circuit boards. Phenoxyphosphazenes are thermally and hydrolytically stable compounds containing phosphorus and nitrogen. They are effective in PC/ABS at 12 to 15 wt% [133]. In general, the flame retardant mechanism of the phosphazene compound depends on the phosphazene substituent and the polymer material as matrix, and occurs in the gas phase as well as in the condensed phase [134]. It is claimed that phosphates like RDP or TPP are synergistic with phosphazenes [135].

DOPO (see above) and its derivatives are the best-known gas-phase active flame retardants commonly used in polymers. Reactive compounds are especially used to flame-retard high-performance epoxy resins [136], and may be preformulated with the epoxy resin (fusion process, where the phosphorus component reacts with the epoxy groups). Higher-functional flame retardants are more suitable where the effect on the glass transition temperatures and the mechanical properties has to be as small as possible. Examples are the commercially available bifunctional DOPO-HQ, the polymeric HFC-X [137], and tetrakis(DOPO-methyl)bisphenol A [138, 139]. The structures of these flame retardants are shown in Figure 3.10. DOPO-containing compounds that act as flame retardants as well as co-curing agents have been developed, and amines [140–142], phenols [143, 144], and anhydrides [145] can be used as curing functionalities. DOPO-EDA (Figure 3.10), for example, is used in epoxy resins [146] or in PU foams [147].

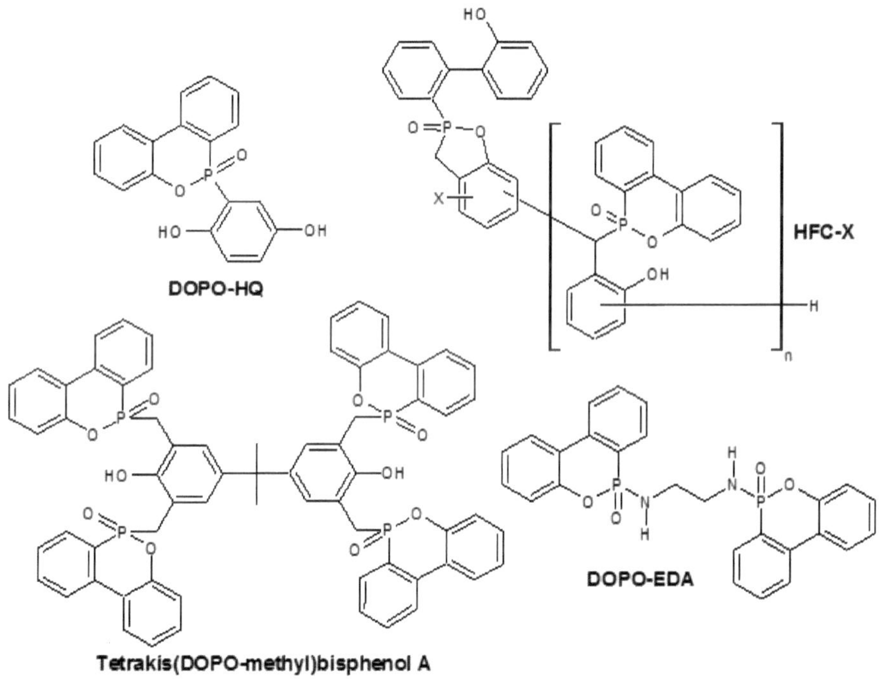

DOPO-HQ

HFC-X

Tetrakis(DOPO-methyl)bisphenol A

DOPO-EDA

Figure 3.10 Reactive DOPO derivatives for use in epoxy resins

The reactive adduct of itaconic acid and DOPO [148, 149] is used in PET fibers and PU lacquers. For thermoplastics, additive flame retardants based on DOPO are mostly used. For example, the bridged DOPO derivatives (DiDOPO) shown in Figure 3.11 were synthesized and showed high thermal stabilities (> 340 °C) that make them suitable for applications in high-temperature polyamides and epoxy resins [150–154]. Especially for the derivative with R = C_6H_5, the synthesis is atom-economic.

As depicted above, current research focusses on polymeric flame retardants. For example, PolyDOPAc [155, 156], shown in Figure 3.11, can be used as polymeric flame retardant in PA66 and epoxy resins. This structure is based on the low-molecular-weight compound DOPAc-3-THIC that was tested in high-performance epoxy resins and composites thereof [157]. Other polymeric flame retardants are, for example, DOPO-containing polyacrylate compounds [156] for application in PA.

Besides polymeric flame retardants, efficient synergistic flame retardants containing phosphorus in different chemical environments are investigated. Low molecular weight compounds like the Atherton–Todd reaction product of DOPO and trihydroxymethylphosphine oxide (THPO) [158] were synthesized. DOPO-THPO was tested in epoxy resins. and low phosphorus loadings of 0.33 wt% resulted in a V-0 classification in UL 94 tests. A polymeric phosphorus-containing flame retardant containing different phosphorus species is shown in Figure 3.11 [159]. This flame retardant was also successfully tested in epoxy resins and composites.

DiDOPO
R = H / C_5H_5

DOPAc-3-THIC

PolyDOPAc

PFR

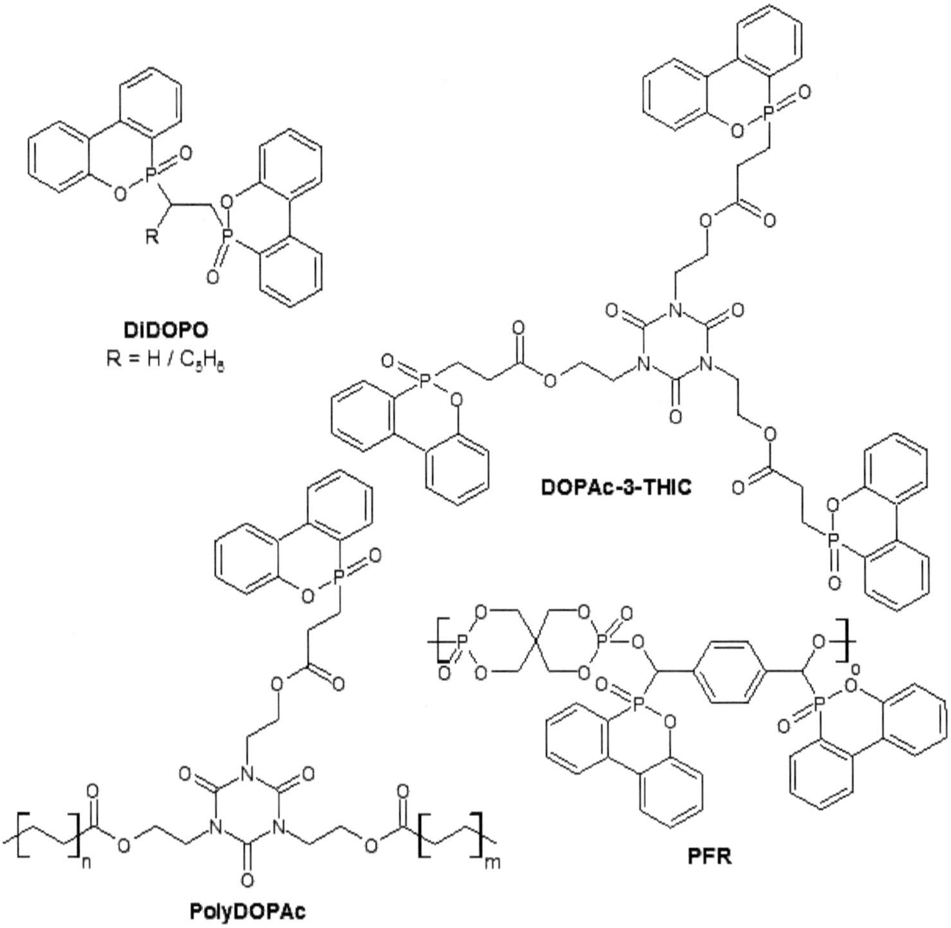

Figure 3.11 Additive DOPO derivatives

The versatility of DOPO is shown by zinc salts with the ring-opened DOPO–water adduct or oxidized DOPO (ZnDOPOx), which have a particularly high thermal stability and can be used as flame retardants for high-performance polyamides [160, 161]. Figure 3.12 shows thermogravimetric measurements of five DOPO-containing flame retardant additives. Dependent on the environment of DOPO, the thermal stability of the flame retardant can be adjusted to the desired application. Butyl *6H*-dibenz[*c,e*][1,2]oxaphosphorin-6-propanoate 6-oxide (DOPAc-bu) is a liquid flame retardant developed for PU foams and is volatile, as is shown by a residue of 0% [162]. Polymeric compounds like PolyDOPAc show higher thermal stability, as described above.

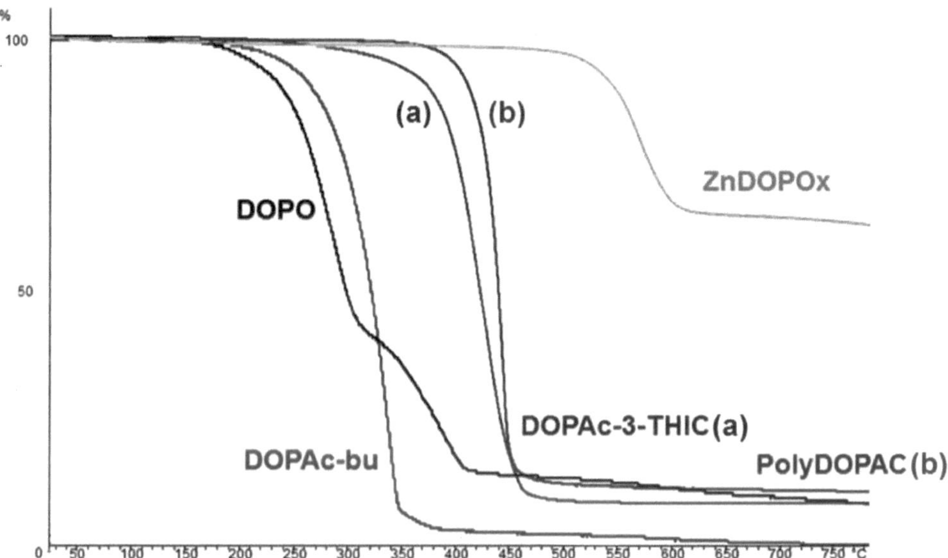

Figure 3.12 Thermogravimetry curves of DOPO-containing flame retardants

Intumescent flame retardants mostly contain a phosphorus compound as the acid-forming component. Intumescence is essentially a special case of a condensed-phase mechanism. In general, the intumescence effect is based upon an expanded carbonized layer on the surface of the polymer during thermal carbonification. It acts like a foamed char by being an insulating barrier, reducing heat transfer from the polymer towards the polymer surface, and limiting fuel transfer from the polymer towards the flame, as well as the diffusion of oxygen into the material.

There are three components used to gain intumescence: a charring agent, a strong acid to promote charring, and a blowing agent. The acid source is needed to promote the dehydration of the carbonizing agent. Due to the formation of non-volatile polyphosphoric acid, phosphates are recommended acid sources. Additional carbonizing agents (OH-rich compounds like pentaerythritol or carbohydrates) are dehydrated by the acid to form a char, while the blowing agent decomposes and releases gas. This leads to the expansion of the polymer and the formation of a swollen multicellular layer. The liberation of the acid has to occur below the decomposition temperature of the carbonizing agent, while the dehydration should take place within the decomposition temperature range of the polymer. Melamine, melamine salts, dicyandiamide, or urea are used as blowing agents [163]. The gas must be released during the thermal decomposition of the carbonizing agent to trigger the expansion of the mechanically stable carbonized layer.

APP or MPP [164] may be used as carbonization agents. APP and particularly melamine additionally work as blowing agents. Oxygen-rich polymers like epoxy resins or polyamide-6 [165–167] and polyurethanes [168] can be used as carbonization agents. Synergistic agents may be zeolites [169] and clay materials [170]. A fourth ingredient may be a compatibilizer of the additives such as ethylene–vinyl acetate copolymers or other functional polymers [171, 172].

One of the first developed charring systems was PP–APP–pentaerythritol [173, 174]. The first stage (reaction of the acidic species with the carbonization agent and formation of esters mixtures) takes place below 280 °C. At about 280 °C the carbonization process (Friedel–Crafts reactions and free radical processes) takes place. Gaseous products are produced within a temperature range of 280 °C to 350 °C. The intumescent material decomposes at higher temperatures and loses its foamed character at about 430 °C [174].

3.1.3.4 Organic Radical-Forming Agents and N-, O- and S-Containing Synergists

The mode of action of flame retardants producing free radicals that interact with the combustion process by terminating reactive radicals like ·OH and H· has already been discussed in this chapter (see Section 3.1.2.2). Besides organophosphorus radical sources, various organic radical-forming agents (whose effectiveness in "thin-wall" applications like films or foams is well investigated) have been developed. Typically, organic radical-forming agents are used in synergistic flame retardant systems, mainly in combination with phosphorus-containing flame retardants. The most important classes of radical-forming agents [175, 176] are shown in Figure 3.13. Nitrogen-based compounds like *N*-alkoxy hindered amines (NOR), azoalkanes and oxyimides, as well as sulfur-based compounds like disulfides and sulfenamides, are the most important radical-forming agents [4, 19, 175, 177–180]. Some radical-forming agents not only provide flame retardancy, but also work as light stabilizers or antioxidants in some polymer systems [177, 181].

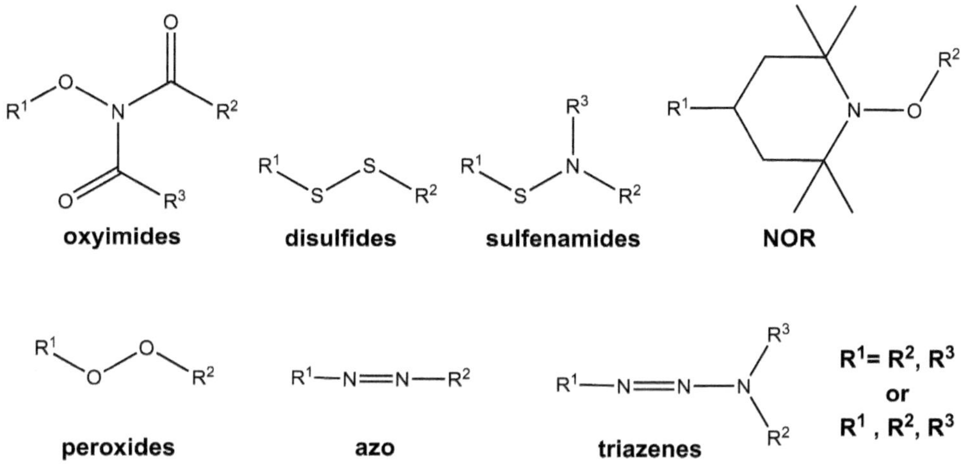

Figure 3.13 Most important organic radical-forming agents for flame retardant formulations

Peroxide compounds are also effective radical-forming agents, but show relatively low thermostability compared to other radical-forming agents. Temperatures of 80–150 °C are necessary for the thermal decomposition of the O–O bond. Some peroxy compounds like dicumyl peroxide, 1,3-bis[2-*tert*-butylperoxy)isopropyl]benzene, and 2,5-dimethyl-2,5-di(*tert*-butylperoxy)hexane are used to accelerate the action of brominated flame retardants, e.g., in polystyrene foams [180, 182–185].

N-Alkoxy-hindered amines (NOR) generate free radicals that are able to affect the combustion process and the related radical mechanisms. A synergy of NORs and brominated flame retardants has been described by Kaprinidis et al. [186]. 13 phr of tris(3-bromo-2,2-bis(bromomethyl)propyl)phosphate and 1 phr of a substance containing a HALS (hindered amine light stabilizer) met UL 94 V-0, whereas 13 phr of the halogenated flame retardant itself only achieved a UL 94 V-2 classification [186–189].

Wilén et al. have investigated azoalkanes as a new substance class regarding flame retardancy. For the first time, they obtained evidence of flame retardancy for azoalkanes alone at very low loadings (max. 1 wt% in polypropylene) [188, 190]. Azoalkanes and NOR moieties can be unified in one molecule called bis(1-propyloxy-2,2,6,6-tetramethylpiperidyl)-4-diazene (AZONOR, Figure 3.14), which provides self-extinguishing behavior and flame retardancy for polypropylene [177, 188, 190–192]. The thermal decomposition mechanism of related azoalkanes and triazene compounds lead to dissociation into free radicals and nitrogen. NORs are able to influence the release of bromine radicals from brominated flame retardants, so improving the flame retardant efficiency in polypropylene "thin-wall" applications like fibers, foams or films, as well as providing light stability [190, 193, 194].

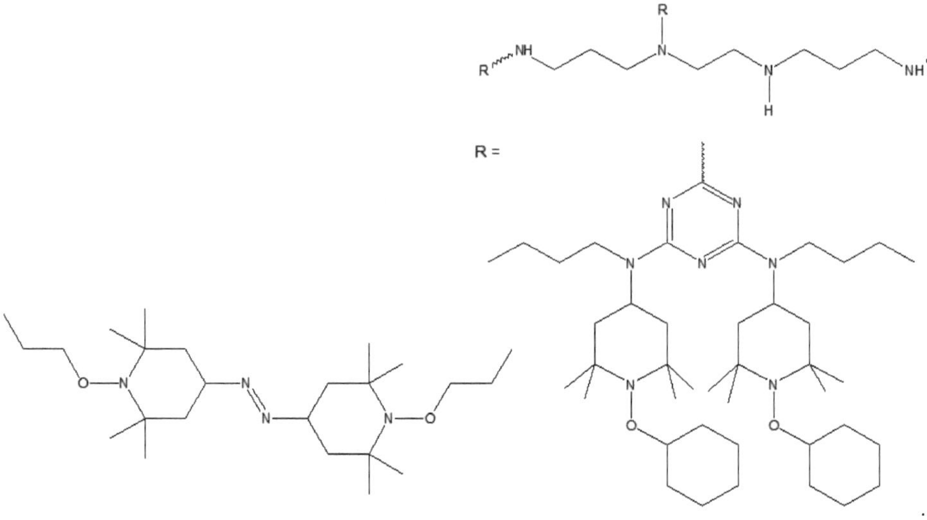

AZONOR

Flamestab NOR 116

Figure 3.14 Structures of Flamestab NOR 116 and AZONOR

The radical-scavenging activity of *N*-hydroxyimides and their corresponding oxyimides (Figure 3.14) is well known [195, 196]. The flame retardant efficiency of the oxyimides can be improved by adding phosphorus-containing compounds like phosphinates or phosphonates. For example, 6 wt% of pentaerythritol spirobis(methylphosphonate) and 2 wt% of an organic oxyimide (phthalimide ester of trimesic acid) in polypropylene result in a UL 94 classification of V-0 (sample thickness: 1.5 mm) [175, 179].

The sulfur analogs of oxyimides are called sulfenamides, and also represent a class of radical-forming agents (Figure 3.13). Wilén et al. investigated the flame retardant efficiency of this new substance class in polypropylene, polystyrene, and low-density polyethylene. Thermolysis of sulfenamides results in the formation of aminyl and thiyl radicals [176, 197, 198]. 1 wt% of 9-[(4-methoxyphenyl)thio]-9*H*-carbazole in polypropylene films (200 μm thickness) has an average burning time of 7 s and passes the German DIN 4102-1 B2 classification. Other sulfenamide structures with a maximum amount of 1 wt% in polypropylene films meet DIN 4102-1 B2, such as 1 mm thick polypropylene panels [176].

Disulfides are also effective radical-forming agents. and were investigated in detail concerning flame retardant formulations for various polymer systems [4, 19, 178, 180]. The disulfide bond cleaves at temperatures of 200–300 °C, which is higher than the peroxide bond cleavage at 80–150 °C [180, 183]. Examples of this substance class are aromatic and aliphatic disulfides, heteroatom- and heterocycle-substituted disulfides, and thiuram disulfides [180]. The possible reaction mechanisms for thermolysis of different disulfide compounds have been investigated in detail by Reyniers et al. [199] and Nakhmanovich et al. [200]. The efficiency of radical forming agents like disulfides and oxyimides has been recently investigated in PET [19].

Polystyrene foams with decreased flammability were developed in 1970 by mixing pre-expanded polystyrene beads with elemental sulfur [201]. Over 20 years later an effective synergistic flame retardant formulation consisting of organophosphorus compounds and elemental sulfur was developed [202]. Döring et al. investigated possible synergistic flame retardant formulations of organic disulfides and phosphorus-containing compounds for bulk and foamed polystyrene samples. As a phosphorus-free flame retardant, 2,2'-dithiobis(benzothiazole) provides self-extinguishing behavior for bulk polystyrene samples. The fragments of the flame retardants are intercepted by the polymer matrix and released together with fragments of the polymer matrix at lower temperature. The accelerated decomposition of the polystyrene matrix is an important aspect of the fire retardancy of polystyrene, because the lowered molecular weight results in a higher melt flow so that combustible material is abstracted from the flame. This retreat effect is caused by a chemical interaction between the polymer matrix and the released fragments of the added flame retardants/synergists [4, 19, 178].

Other radical generators like polydopamine (PDA) were investigated for flame retardant efficiency. PDA is an environmentally friendly bio-based radical scavenger obtained by oxidative polymerization of dopamine [203]. In 2015, Ellison et al. developed a flame retardant nanocoating based on an aqueous PDA solution for flexible polyurethane (PU) foams; it decreased the peak heat release rate (PHRR) by 67% compared to neat PU foam. The radical scavenging activity of PDA has been investigated many times. PDA is difficult to ignite, showing no detectable PHRR in micro-combustion calorimetry experiments [203–205].

3.1.3.5 Synergistic Flame Retardant Systems

A flame retardant formulation consisting of at least two components can result in an synergistic [4, 206], antagonistic [170], or additional [207] flame retardant effect. A synergistic flame retardant effect means that less flame retardant is needed in the polymer, and so the impact on the material properties of the polymer is smaller. A synergistic flame retardant combination aims at optimizing the efficiency of the flame retardant used.

A synergistic effect can be achieved by combining flame retardants with different flame retardant mechanisms, or by combining flame retardants to intensify the same mechanism [208].

The most synergistic systems are listed in Table 3.2.

Table 3.2 Synergistic Flame Retardant Formulations

Type of flame retardants	Examples	Synergists	Usual flame retardant loading/%	Polymers	References
Inorganic flame retardants	$Mg(OH)_2$, AlO(OH)	Nanoclays, EG, phosphorus compounds	> 60	EVA	[35, 209–211]
Halogenated flame retardants	Brominated compounds	Sb_2O_3, borates, radical generators, phosphorus compounds	25–35	Polyolefins	[186, 210, 212–218]
Intumescent flame retardants	APP	Polyols, metal salts, triazine derivatives	20–30	UP resin, polyolefins, coatings	[219, 220]
Radical generators	Disulfides, Sulfenamides, NOR, azo- and triazene compounds	Halogen and phosphorus compounds	1–15	Polyolefins	[4, 210, 212, 221]

Synergistic formulations containing phosphorus compounds with different chemical environments are known [222, 223]. Substances containing phosphorus and either silicone [224–226] or sulfur [4, 178] within one molecule are also described in the literature.

A flame retardant formulation consisting of a halogen-containing flame retardant and an antimony oxide is a longstanding and very effective synergistic mixture. Antimony oxides alone have no flame retardant effect. The enhanced efficiency of the halogen-containing flame retardant is based on the facilitated transfer of the active agents into the gas phase by the halogenated antimony compound [76, 214, 227].

An example of combining different flame retardant mechanisms was investigated by Schartel et al. The investigation focused on the efficiency of flame retardant formulations consisting of nanometric metal oxides like TiO_2, Al_2O_3, or Fe_2O_3 and aluminum (diethylphosphinate) (DEPAL) in poly(butylene terephthalate) (PBT). The combination of 1–2 wt% of metal oxides and 5–10 wt% of DEPAL in PBT improves the UL 94 rating from V-1/V-2 to V-0, and showed a synergistic effect concerning the total heat evolved and an antagonistic effect concerning LOI results. Although most of the DEPAL is converted into diethylphosphinic acid in the gas phase, the addition of DEPAL to PBT leads to the formation of a phosphinate-terephthalate salt in the solid phase. Metal oxides added to PBT sustain the generation of a stable carbonaceous char due to the Lewis acid activity of the metal ion and the resulting interaction with the polymer. The combination of the gas and condensed phase mechanism leads to a synergistic effect in PBT and thus influences the decomposition pathway of the polymer [228, 229].

A well-known synergistic formulation is melamine polyphosphate (MPP) and aluminum diethylphosphinate (DEPAL) in polyesters and polyamides. MPP alone promotes the dilution of fuel and the formation of a protective barrier, whereas the flame retardant effect of DEPAL mainly takes place in the gas phase. If MPP and DEPAL are combined, a protective layer of aluminum phosphate is formed, representing the main flame retardant mechanism [230, 231].

MDH can be added to APP in a synergistic flame-retarded formulation of ethylene-vinyl acetate copolymer. APP acts as synergist in EVA–MDH composites. The combination of MDH and APP results in the formation of a thermal barrier as a protecting layer consisting of a magnesium phosphate glass and a magnesium oxide ceramic [232]. A flame-retardant formulation consisting of crystalline ATH (4.5 m^2 g^{-1}) and aluminum hypophosphite (AHP) or DEPZn shows a synergistic effect in unsaturated polyester (UP) resins compared to amorphous ATH (300 m^2 g^{-1}). The residue of the flame retardant formulations in UP resins is more fragile using amorphous ATH, whereas a dense and tough residue is formed by the use of crystalline ATH instead of amorphous ATH. An additional barrier formation takes place at higher temperatures [35].

3.1.3.6 Other Flame Retardants

Low-melting glasses can be used as flame retardants due to different modes of action. They improve the thermal stability of the polymer by generating a char barrier to shield the undecomposed polymer. The oxidation of the polymer is also prevented. Low-melting glasses generate a more continuous char surface with a lower surface area. In addition, the combustion char is held together by a greater degree of structural integrity [233, 234].

Expandable graphite (EG) is an intercalation compound consisting of parallel planes of carbon atoms, in which the planes are not connected by covalent bonds. So, molecules like sulfuric acid can be inserted between them. Under heat exposure, the inserted molecules start to decompose, generating gas, which forces the graphite layer planes apart and leads to expansion of the material. The EG generates a low-density, non-combustible material with a twisted shape or worm-like structure. EG flakes may form an insulating layer similar to intumescence, with an up to 100-fold increase in volume. The EG layer also reflects up to 50% of radiant heat [235–239]. The main application of EG is in rigid polyurethane foams, where its insulative and expansion properties make it suitable as sealants or firestops, for which condensed-phase activity is required [240–242].

In general, nitrogen-containing flame retardants show a lower efficiency than other flame retardant additives, but their efficiency can be enhanced by adding an appropriate synergist. They also can be part of intumescent flame-retarded formulations. Commercial nitrogen compounds are melamine and melamine derivatives, such as the thermally stable salts of melamine cyanurate with strong acids. Their mode of action is based on diluting effects due to releasing inert gases, increasing char yield and acting as a heat sink. They act in the gas and condensed phases and also as smoke suppressants. Melamine and its derivatives promote non-flaming dripping in polyamides, in that the polymer moves away from the pyrolysis zone. Their main field of application is in polyamide (PA) and poly(butylene terephthalate) (PBT) [21, 42, 49, 132, 164, 243–248].

Melamine salts display higher flame retardant activity in the condensed phase than neat melamine [49]. The decomposition of melamine phosphate at a temperature of more than 250 °C leads to the formation of melamine pyrophosphate with release of water, melamine, and phosphoric acid, which can phosphorylate polymers [49, 249].

Melamine and related derivatives decompose with release of ammonia, which dilutes combustible gases and leads to the endothermal formation of melamine condensates like melam, melem, and melon [49, 250] (Figure 3.15), which also show flame retardant effects [49, 249].

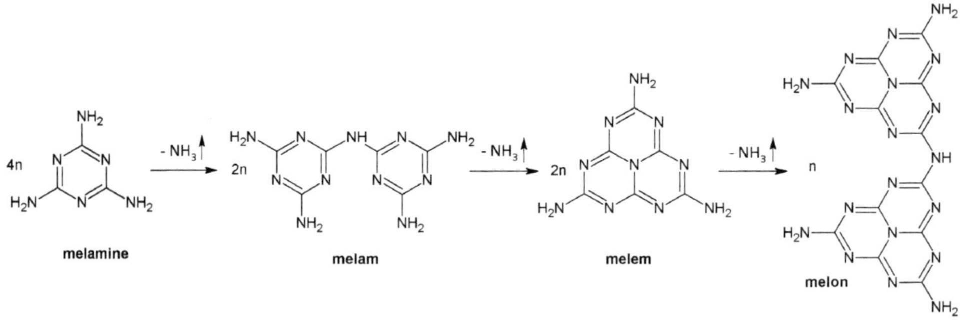

Figure 3.15 Thermal decomposition of melamine and melamine derivatives to the corresponding condensation products melam, melem, and melon

A melamine-containing polymer in combination with APP can be used as an intumescent and synergistic flame retardant formulation in polyolefins. The polymer consists of a triazine bridged system containing piperazine and morpholine, and can be synthesized without organic solvents. Commonly, the triazine-based polymer is blended with APP in a 3 : 1 ratio and can be used for polyurethane foams, PE, PP, and PP copolymers [251–255].

There are many boron-based flame retardants commercially available for wood/cellulose coatings, polyamide, PVC, or epoxy applications. The zinc borate $2\ ZnO \cdot 3\ B_2O_3 \cdot 3.5\ H_2O$ has a thermostability of about 290 °C and is used for PVC, polyolefins, polyamides and epoxy applications. The anhydrous type of zinc borate ($2\ ZnO \cdot 3\ B_2O_3$) has a thermostability of up to 500 °C and is suitable for engineering plastics thermostable above 300 °C. Depending on the polymer matrix and the halogen source, zinc borates can take effect as flame retardants, afterglow suppressants [256, 257], smoke suppressants [258–260], or anti-arcing agents [261]. They act in the condensed phase, promoting the formation of a cellular char while releasing water. Ning et al. studied the flame retardant and smoke suppression properties of aluminum trihydroxide and zinc borate in PVC and observed a synergistic effect for flame retardancy and smoke suppression. A small amount of both additives decreases the released amount of aromatics during the combustion process and promotes the char formation. Zinc-containing organic compounds are good smoke suppressants for chlorine-containing polymers due to the formation of $ZnCl_2$ during decomposition. $ZnCl_2$ is a strong Lewis acid and a catalyst for the ionic dehydrochlorination of chlorine-containing polymers [227, 258, 262–270].

Zinc stannates and hydroxystannates are synergists for flame retardant formulations containing brominated flame retardants. The application of zinc hydroxystannate is limited to processing temperatures below 180 °C, while zinc stannate has a thermostability of 400 °C. Like zinc borate, they always act as smoke suppressants. Zinc stannate is a well-known synergist for bromine-containing flame retardants in polyamide 6 and 66. Zinc stannate and phosphorus-containing flame retardants act as synergists in PC/ABS blends. The mode of action of zinc stannates as additional synergists for halogen-containing flame retardants is comparable to that of antimony oxide, with char promotion and the decrease of carbon monoxide formation as supplementary effects [271–276].

Metal molybdates and molybdenum oxides promote char formation and can be used to control the smoke emission during the combustion process. Commercially available smoke suppressants and flame retardants contain calcium, zinc molybdates, zinc molybdate–magnesium silicate complexes, and zinc molybdate–magnesium hydroxide complexes. The mode of action of molybdenum compounds is limited to the condensed phase, traceable in the char residue. The smoke-suppressant characteristics of molybdenum compounds are based on the increased char level. The emission of aromatic compounds during the combustion of PVC can also be decreased. A synergistic flame retardant effect occurs when halogen-containing compounds and molybdenum compounds are combined. Further studies have disclosed a synergism with metal hydroxides like magnesium hydroxide in flexible PVC [277–285].

Silicone-containing compounds like silicas, silicones, silicates, silanes, or silsesquioxanes can be applied as flame retardants for PS, PP, PC, and EVA. Silicone-containing compounds can be used as fillers, copolymers, or as polymer matrix. Silicone-based flame retardants can also be applied as coatings consisting of fillers or silicone copolymers, organosilanes, silsesquioxanes, or silicates, and are suitable substitutes for halogen- or phosphorus-containing flame retardants. Silicone-based materials as flame retardants produce protective surface coatings during the combustion process and decrease the rate of heat release. Silicone polymers possess the overall formula $(R_mSi(O)_{4-m/2})_n$ with m = 1–3 and n ≥ 2.

Iji et al. studied the LOI (Limiting Oxygen Index) values of silicone-containing compounds with different silicone polymer structures in polycarbonate (PC) and PC/ABS resins. They obtained flame retardancy and good mechanical properties (strength, moulding) in polycarbonate. They varied the chain topology (branched or linear chain structure), the side groups (mixtures of phenyl and methyl with different ratios) and the terminal functional groups (methyl, hydroxyl, vinyl, methoxy). The best LOI results were obtained with a silicone polymer consisting of a branched chain type and methyl groups as terminal functional groups and a methyl/phenyl ratio of 40/60 along the polycarbonate chain. The silicone concentration in polycarbonate was 5 phr.

Silsesquioxanes as inorganic-organic hybrid materials have the empirical formula $RSiO_{3/2}$, where R represents a hydrogen atom or a carbon moiety. Polyhedral oligomeric silsesquioxanes (POSS) are classified as an intermediate structure of silicones and silicas [107, 286] (Figure 3.16) and can be incorporated into polymer matrices by grafting, copolymerization or blending. They possess very good thermoxidative stability, good heat resistance and environmental neutrality, which makes them promising as flame retardants. POSS is used in polystyrene, polycarbonate, polyester, vinyl ester, PU, and epoxy resin applications [49, 227, 287, 288].

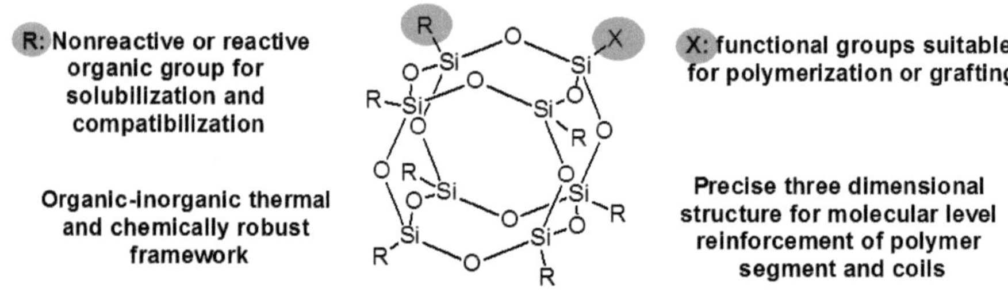

Figure 3.16 General structure of polyhedral oligomeric silsesquioxanes (POSS)

3.1.4 Greener Alternatives to Common Flame Retardants

Although aspects such as fire performance, manufacturing, cost-benefit considerations, and recycling are key points in choosing a flame retardant, toxicity is also important [6, 40, 289, 290].

Certain flame retardant systems like halogenated substances-antimony trioxide, phosphorus-halogen and organo-phosphorus compounds (i.e., TPP), as well as fluoropolymers as antidrips, may release environmentally harmful combustion products, whereas phosphorus, phosphorus-nitrogen, and mineral flame retardants may have a lower tendency to release toxic combustion products. Weathering and mechanical deterioration may also promote the release of possible toxic products during use of the flame-retarded product.

Greener approaches to achieve fire retardancy involve the use of renewable raw materials as basic materials for the synthesis of flame retardants. Efforts have been made in using bio-based platform chemicals for the synthesis of flame retardants, which require a biomass origin or a bioprocessing route. Certain bio-based chemicals from renewable resources like acrylic acid, itaconic acid, pentaerythritol, or glycerine can be used for producing flame retardants. Likewise, certain polymers can be manufactured from bio-based chemicals. Lignin as bio-polymeric polyphenol promotes char formation during the combustion process due to its aromatic structure. Bourbigot et al. phosphorylated lignin through a grafting process and flame retarded acrylonitrile butadiene styrene (ABS). A concentration of 30 wt% phosphorylated lignin in ABS resulted in a decrease of the peak heat release rate pHRR (cone calorimetry) of 58% in comparison to neat ABS. 30 wt% of non-phosphorylated lignin promotes faster ignition and shows a decrease of 43% pHRR compared to neat ABS [291–296].

References for Section 3.1

[1] Directive 2006/1907/EC: REACH. Official Journal European Union: 2006; Vol. 396, pp. 1–849.

[2] Directive 2011/65/EU: RoHS. Official Journal European Union: 2011; Vol. 174, pp. 88–110.

[3] Directive 2012/19/EU:WEEE. Official Journal European Union: 2012; Vol. 197, pp. 38–71.

[4] Wagner, J.; Deglmann, P.; Fuchs, S.; Ciesielski, M.; Fleckenstein, C. A.; Döring, M., A flame retardant synergism of organic disulfides and phosphorous compounds. *Polymer Degradation and Stability* 2016, 129, 63–76.

[5] PINFA, *http://www.pinfa.eu/index.php/en/about-pinfa/mission.*

[6] Papaspyrides, C. D.; Kiliaris, P., *Polymer Green Flame Retardants.* Elsevier Science: 2014.

[7] Hull, T. R.; Witkowski, A.; Hollingbery, L., Fire retardant action of mineral fillers. *Polymer Degradation and Stability* 2011, *96* (8), 1462–1469.

[8] Horrocks, A. R.; Price, D., *Advances in Fire Retardant Materials.* Elsevier Science: 2008.

[9] The flame retardants market, *https://www.flameretardants-online.com/flame-retardants/market* (accessed 15.01.2019), 2017.

[10] Schultheis, C., Halogen-free flame retarded plastics: A demanding challenge in recyclability? In *FRPM 2017*, Manchester, 2017.

[11] Directive 2008/98/EC of the European Parliament and of the Council. In *13*, 2008.

[12] Luda, M. P.; Euringer, N.; Moratti, U.; Zanetti, M., WEEE recycling: Pyrolysis of fire retardant model polymers. *Waste Management* 2005, 25 (2), 203–208.

[13] Pivnenko, K.; Granby, K.; Eriksson, E.; Astrup, T. F., Recycling of plastic waste: Screening for brominated flame retardants (BFRs). *Waste Management* 2017, 69, 101–109.

[14] Peeters, J. R.; Vanegas, P.; Tange, L.; Van Houwelingen, J.; Duflou, J. R., Closed loop recycling of plastics containing flame retardants. *Resources, Conservation and Recycling* 2014, 84, 35–43.

[15] Metzsch-Zillingen, E., Pilot project for recycling halogen-free flame retardant plastics – examining the challenges and on-going work. In *Fire Resistance in Plastics*, Köln, 2017.

[16] Camino, G., In *Biobased Fire Retardant Materials*, FRPM 2017, Manchester, Manchester, 2017.

[17] Minkoff, G.J., *Chemistry of combustion reactions*. Butterworths: 1962.

[18] Wagner, J., Halogenfreie Flammschutzmittelmischungen für Polystyrol-Schäume. Ruprecht-Karls-Universität Heidelberg, 2012.

[19] Goedderz, D., Halogen-free flame retardant formulations containing disulfide compounds for polystyrene and poly(ethylene terephthalate) applications. In *FRPM 2017*, Manchester, 2017.

[20] King, B.A.; Stobby, W.G.; Murray, D.J.; Worku, A.Z.; Beulich, I.; Tinetti, S.M.; Hahn, S.F.; Drumright, R.E., Brominated butadiene/vinyl aromatic copolymers, blends of such copolymers with a vinyl aromatic polymer and polymeric foams formed from such blends. EP1957544B1, 2010.

[21] Klinkowski, C.; Burk, B.; Bärmann, F.; Döring, M., Moderne Flammschutzmittel für Kunststoffe. *Chemie in Unserer Zeit* 2015, *49* (2), 96–105.

[22] Martins, M.S.S.; Schartel, B.; Magalhães, F.D.; Pereira, C.M.C., The effect of traditional flame retardants, nanoclays and carbon nanotubes in the fire performance of epoxy resin composites. *Fire and Materials* 2017, *41* (2), 111–130.

[23] Schaeling, J.; Herbiet, R.; Hillekamps, H. Surface-coated magnesium hydroxide. US20040127602A1, 2004.

[24] Ramdatt, P.E., Surface-modified non-halogenated mineral fillers. US8378008, 2013.

[25] Hornsby, P.; Watson, C., Interfacial modification of polypropylene composites filled with magnesium hydroxide. *Journal of Materials Science* 1995, *30* (21), 5347–5355.

[26] Nachtigall, S.M.B.; Miotto, M.; Schneider, E.E.; Mauler, R.S.; Camargo Forte, M.M., Macromolecular coupling agents for flame retardant materials. *European Polymer Journal* 2006, *42* (5), 990–999.

[27] Döring, M.Y.B.; Held, I., *Innovative Flame retardants in E&E* Applications. pinfa, Brüssel: 2010; Vol. 3. Aufl.

[28] Hollingbery, L.A.; Hull, T.R., The thermal decomposition of huntite and hydromagnesite—A review. *Thermochimica Acta* 2010, *509* (1), 1–11.

[29] Hornsby, P.R., The application of magnesium hydroxide as a fire retardant and smoke-suppressing additive for polymers. *Fire and Materials* 1994, 18 (5), 269–276.

[30] Troitzsch, J.H., Overview of flame retardants. *Chemistry Today* 1998, *16*.

[31] Laoutid, F.; Lorgouilloux, M.; Lesueur, D.; Bonnaud, L.; Dubois, P., Calcium-based hydrated minerals: Promising halogen-free flame retardant and fire resistant additives for polyethylene and ethylene vinyl acetate copolymers. *Polymer Degradation and Stability* 2013, *98* (9), 1617–1625.

[32] Camino, G.; Maffezzoli, A.; Braglia, M.; De Lazzaro, M.; Zammarano, M., Effect of hydroxides and hydroxycarbonate structure on fire retardant effectiveness and mechanical properties in ethylene-vinyl acetate copolymer. *Polymer Degradation and Stability* 2001, *74* (3), 457–464.

[33] Hornsby, P.R.; Watson, C.L., A study of the mechanism of flame retardance and smoke suppression in polymers filled with magnesium hydroxide. *Polymer Degradation and Stability* 1990, *30* (1), 73–87.

[34] Green, J., Mechanisms for flame retardancy and smoke suppression – A review. *Journal of Fire & Flammability* 1996, *14* (6), 426–442.

[35] Reuter, J.; Greiner, L.; Schönberger, F.; Döring, M.J.J.o.A.P.S., Synergistic flame retardant interplay of phosphorus containing flame retardants with aluminum trihydrate depending on the specific surface area in unsaturated polyester resin. *Journal of Applied Polymer Science* 2019, *136* (13), 47270.

[36] Rothon, R., *Particulate-filled Polymer Composites*. Rapra Technology: 2003.

[37] Batistella, M.; Otazaghine, B.; Sonnier, R.; Caro-Bretelle, A.-S.; Petter, C.; Lopez-Cuesta, J.-M., Fire retardancy of ethylene vinyl acetate/ultrafine kaolinite composites. *Polymer Degradation and Stability* 2014, *100*, 54–62.

[38] Vaughan, F., Energy changes when kaolin minerals are heated. *Clay Miner. Bull.* **1955,** *2* (13), 265–274.

[39] Grand, A.F.; Wilkie, C.A., *Fire Retardancy of Polymeric Materials.* CRC Press: 2000.

[40] Horrocks, A.R.; Price, D., *Fire Retardant Materials.* CRC Press: 2001.

[41] Klatt, M.; Heitz, T.; Gareiss, B., Flame-proof thermoplastic moulding materials. US6306941 B1, 2001.

[42] Levchik, S.V.; Weil, E.D., Flame retardancy of thermoplastic polyesters–a review of the recent literature. *Polymer International* 2005, *54* (1), 11–35.

[43] Hewitt, F.; Rhebat, D.E.; Witkowski, A.; Hull, T.R., An experimental and numerical model for the release of acetone from decomposing EVA containing aluminium, magnesium or calcium hydroxide fire retardants. *Polymer Degradation and Stability* 2016, *127* (Supplement C), 65–78.

[44] Beyer, G., Flame retardancy of nanocomposites based on organoclays and carbon nanotubes with aluminium trihydrate. *Polymers for Advanced Technologies* 2006, *17* (4), 218–225.

[45] Beyer, G., Nanocomposites: a new class of flame retardants for polymers. *Plastics, Additives and Compounding* 2002, *4* (10), 22–28.

[46] Morgan, A.B.; Wilkie, C.A., *The Non-halogenated Flame Retardant Handbook.* Wiley: 2014.

[47] Kiliaris, P.; Papaspyrides, C.D., Polymer/layered silicate (clay) nanocomposites: An overview of flame retardancy. *Progress in Polymer Science* 2010, *35* (7), 902–958.

[48] Zanetti, M.; Kashiwagi, T.; Falqui, L.; Camino, G., Cone calorimeter combustion and gasification studies of polymer layered silicate nanocomposites. *Chemistry of Materials* 2002, *14* (2), 881–887.

[49] Laoutid, F.; Bonnaud, L.; Alexandre, M.; Lopez-Cuesta, J.M.; Dubois, P., New prospects in flame retardant polymer materials: From fundamentals to nanocomposites. *Materials Science and Engineering: R: Reports* 2009, *63* (3), 100–125.

[50] Jang, B.N.; Costache, M.; Wilkie, C.A., The relationship between thermal degradation behavior of polymer and the fire retardancy of polymer/clay nanocomposites. *Polymer* 2005, *46* (24), 10678–10687.

[51] Dittrich, B.; Wartig, K.-A.; Hofmann, D.; Mülhaupt, R.; Schartel, B., Flame retardancy through carbon nanomaterials: Carbon black, multiwall nanotubes, expanded graphite, multi-layer graphene and graphene in polypropylene. *Polymer Degradation and Stability* 2013, *98* (8), 1495–1505.

[52] Morgan, A.B.; Wilkie, C.A., *Flame Retardant Polymer Nanocomposites.* Wiley: 2007.

[53] Wang, Q.; Undrell, J.P.; Gao, Y.; Cai, G.; Buffet, J.-C.; Wilkie, C.A.; O'Hare, D., Synthesis of flame-retardant polypropylene/LDH-borate nanocomposites. *Macromolecules* 2013, *46* (15), 6145–6150.

[54] Costa, F.R.; Wagenknecht, U.; Heinrich, G., LDPE/Mg-Al layered double hydroxide nanocomposite: Thermal and flammability properties. *Polymer Degradation and Stability* 2007, *92* (10), 1813–1823.

[55] Vaccari, A., Preparation and catalytic properties of cationic and anionic clays. *Catalysis Today* 1998, *41* (1), 53–71.

[56] Matusinovic, Z.; Feng, J.; Wilkie, C.A., The role of dispersion of LDH in fire retardancy: The effect of different divalent metals in benzoic acid modified LDH on dispersion and fire retardant properties of polystyrene- and poly(methyl-methacrylate)-LDH-B nanocomposites. *Polymer Degradation and Stability* 2013, *98* (8), 1515–1525.

[57] Kashiwagi, T.; Du, F.; Winey, K.I.; Groth, K.M.; Shields, J.R.; Bellayer, S.P.; Kim, H.; Douglas, J.F., Flammability properties of polymer nanocomposites with single-walled carbon nanotubes: effects of nanotube dispersion and concentration. *Polymer* 2005, *46* (2), 471–481.

[58] McNally, T.; Pötschke, P., *Polymer-Carbon Nanotube Composites: Preparation, Properties and Applications.* Elsevier Science: 2011.

[59] Kashiwagi, T.; Grulke, E.; Hilding, J.; Groth, K.; Harris, R.; Butler, K.; Shields, J.; Kharchenko, S.; Douglas, J., Thermal and flammability properties of polypropylene/carbon nanotube nanocomposites. *Polymer* 2004, *45* (12), 4227–4239.

[60] Sang, B.; Li, Z.-W.; Li, X.-H.; Yu, L.-G.; Zhang, Z.-J., Graphene-based flame retardants: a review. *Journal of Materials Science* 2016, *51* (18), 8271–8295.

[61] Huang, G.; Gao, J.; Wang, X.; Liang, H.; Ge, C., How can graphene reduce the flammability of polymer nanocomposites? *Materials Letters* 2012, *66* (1), 187–189.

[62] Potts, J. R.; Dreyer, D. R.; Bielawski, C. W.; Ruoff, R. S., Graphene-based polymer nanocomposites. *Polymer* 2011, *52* (1), 5–25.

[63] Directive 2002/95/EC: RoHS. Official Journal European Union: 2003.

[64] Lyons, J. W., Chemistry and uses of fire retardants. 1970.

[65] Roma, P.; Camino, G.; Luda, M., Mechanistic studies on fire retardant action of fluorinated additives in ABS. *Fire and materials* 1997, *21* (5), 199–204.

[66] Avondo, G.; Vovelle, C.; Delbourgo, R., The role of phosphorus and bromine in flame retardancy. *Combustion and Flame* 1978, *31*, 7–16.

[67] Williams, F. A., Combustion A2 – Meyers, R. A., In *Encyclopedia of Physical Science and Technology (Third Edition)*, Academic Press: New York, 2003; pp. 315–338.

[68] Wilson, W.; O'Donovan, J.; Fristrom, R., In *Flame inhibition by halogen compounds*, Symposium (International) on combustion, Elsevier: 1969; pp. 929–942.

[69] Rosser, W.; Wise, H.; Miller, J., In *Mechanism of combustion inhibition by compounds containing halogen*, Symposium (International) on Combustion, Elsevier: 1958; pp. 175–182.

[70] Hastie, J., Molecular basis of flame inhibition. *Journal of Research* 1973, *77*, 733–754.

[71] Levy, A.; Droege, J.; Tighe, J.; Foster, J. In *The inhibition of lean methane flames*, Symposium (International) on Combustion, Elsevier: 1961; pp. 524–533.

[72] Petrella, R., Factors affecting the combustion of polystyrene and styrene. In *Flame-retardant Polymeric Materials*, Springer: 1978; pp. 159–201.

[73] Sawyer, R. M. F. R. F., In *Flame Inhibition Chemistry Advisory Group for Aerospace Research and Development Conference Proceedings*, No. 84 on Aircraft Fuels, Lubricants and Fire Safety, North Atlantic Treatly Organization: 1971.

[74] Day, M.; Stamp, D.; Thompson, K.; Dixon-Lewis, G., In *Inhibition of hydrogen-air and hydrogen-nitrous oxide flames by halogen compounds*, Symposium (International) on Combustion, Elsevier: 1971; pp. 705–712.

[75] Khanna, Y.; Pearce, E. M., Synergism and flame retardancy. In *Flame-retardant polymeric materials*, Springer: 1978; pp. 43–61.

[76] Pitts, J. J.; Scott, P. H.; Powell, D. G., Thermal decomposition of antimony oxychloride and mode in flame retardancy. *Journal of Cellular Plastics* 1970, *6* (1), 35–37.

[77] Brauman, S. K.; Brolly, A. S., Sb//2-O//3-halogen fire retardance in polymers – 1. General mode of action. *Journal of Fire Retardant Chemistry* 1976, *3* (2), 66–83.

[78] Shtekler, R.; Yaakov, Y. B.; Georlette, P.; Levchik, S. V., Antimony trioxide free flame retardant thermoplastic composition. US9475933B2, 2016.

[79] Beach, M. W.; Hull, J. W.; King, B. A.; Beulich, I. I.; Stobby, B. G.; Kram, S. L.; Gorman, D. B., Development of a new class of brominated polymeric flame retardants based on copolymers of styrene and polybutadiene. *Polymer Degradation and Stability* 2017, *135*, 99–110.

[80] Eckel, T., The most important flame retardant plastics. *Plastics Flammability Handbook*. Munich: Hanser 2004, 158–72.

[81] Troitzsch, J., *International Plastics Flammability Handbook: Principles, Regulations, Testing and Approval*. Hanser: 1983.

[82] Smith, R.; Georlette, P.; Finberg, I.; Reznick, G., Development of environmentally friendly multifunctional flame retardants for commodity and engineering plastics. *Polymer Degradation and Stability* 1996, *54* (2-3), 167-173.

[83] Finberg, I.; Utevski, L.; Kallos, M., Styrenics plastics/brominated indan as highly compatible systems. *IMEC* 1997, *8*, 19.

[84] Fenimore, C.P.; Martin, F.J., Flammability of polymers. *Combustion and Flame* 1966, *10* (2), 135-139.

[85] Fenimore, C.; Jones, G., Modes of inhibiting polymer flammability. *Combustion and Flame* 1966, *10* (3), 295-301.

[86] Brauman, S.K., Phosphorus fire retardance in polymers. I - General mode of action. II - Retardant-polymer substrate interactions. *Journal of Fire Retardant Chemistry* 1977, *4*, 18-58.

[87] Pawlowski, K.H.; Schartel, B., Flame retardancy mechanisms of triphenyl phosphate, resorcinol bis (diphenyl phosphate) and bisphenol A bis (diphenyl phosphate) in polycarbonate/acrylonitrile-butadiene-styrene blends. *Polymer International* 2007, *56* (11), 1404-1414.

[88] Pearce, E.M.; Weil, E.D.; Barinov, V.Y., *Fire smart polymers.* ACS Publications: 2001.

[89] Nelson, G.L.; Wilkie, C.A., *Fire retardancy in 2001.* ACS Publications: 2001.

[90] Braun, U.; Balabanovich, A.I.; Schartel, B.; Knoll, U.; Artner, J.; Ciesielski, M.; Döring, M.; Perez, R.; Sandler, J.K.; Altstädt, V., Influence of the oxidation state of phosphorus on the decomposition and fire behaviour of flame-retarded epoxy resin composites. *Polymer* 2006, *47* (26), 8495-8508.

[91] Pawlowski, K.H.; Schartel, B.; Fichera, M.A.; Jäger, C., Flame retardancy mechanisms of bisphenol A bis(diphenyl phosphate) in combination with zinc borate in bisphenol A polycarbonate/acrylonitrile-butadiene-styrene blends. *Thermochimica Acta* 2010, *498* (1), 92-99.

[92] Lewin, M.; Weil, E., Mechanisms and modes of action in flame retardancy of polymers. *Fire Retardant Materials* 2001, 31-68.

[93] Troitzsch, J., *Plastics Flammability Handbook: Principles, Regulations, Testing, And Approval.* Hanser Verlag: 2004.

[94] Schartel, B.; Kunze, R.; Neubert, D., Red phosphorus-controlled decomposition for fire retardant PA 66. *Journal of Applied Polymer Science* 2002, *83* (10), 2060-2071.

[95] Bourbigot, S.; Le Bras, M.; Duquesne, S.; Rochery, M., Recent advances for intumescent polymers. *Macromolecular Materials and Engineering* 2004, *289* (6), 499-511.

[96] Kunze, R.; Schartel, B.; Bartholmai, M.; Neubert, D.; Schriever, R., TG-MS and TG-FTIR applied for an unambiguous thermal analysis of intumescent coatings. *Journal of Thermal Analysis and Calorimetry* 2002, *70* (3), 897-909.

[97] Levchik, S.V.; Levchik, G.F.; Balabanovich, A.I.; Weil, E.D.; Klatt, M., Phosphorus oxynitride: a thermally stable fire retardant additive for polyamide 6 and poly (butylene terephthalate). *Angewandte Makromolekulare Chemie* 1999, *264*, 48-55.

[98] Braun, U.; Schartel, B., Effect of red phosphorus and melamine polyphosphate on the fire behavior of HIPS. *Journal of fire sciences* 2005, *23* (1), 5-30.

[99] Brauman, S.K.; Fishman, N., Phosphorus fire retardance in polymers. 3. Some aspects of combustion performance. *Journal of Fire Retardant Chem* 1977, *4*, 93-111.

[100] Brauman, S., Phosphorus fire retardance in polymers. 4. Poly (ethylene terephthalate)-ammonium polyphosphate, a Model System. *Journal of Fire Retardant Chemistry* 1980, *7* (2), 61-68.

[101] Schartel, B.; Balabanovich, A.I.; Braun, U.; Knoll, U.; Artner, J.; Ciesielski, M.; Döring, M.; Perez, R.; Sandler, J.K.W.; Altstädt, V.; Hoffmann, T.; Pospiech, D., Pyrolysis of epoxy resins and fire behavior of epoxy resin composites flame-retarded with 9,10-dihydro-9-oxa-10-phosphaphenanthrene-10-oxide additives. *Journal of Applied Polymer Science* 2007, *104* (4), 2260-2269.

[102] Horrocks, A.R.; Price, D.; Price, D., *Fire retardant materials.* Woodhead Publishing: 2001.

[103] Hastie, J.W., Mass spectrometric studies of flame inhibition: Analysis of antimony trihalides in flames. *Combustion and Flame* 1973, *21* (1), 49–54.

[104] Fenimore, C.; Jones, G., Phosphorus in the burnt gas from fuel-rich hydrogen-oxygen flames. *Combustion and Flame* 1964, *8* (2), 133–137.

[105] Jayaweera, T.; Melius, C.; Pitz, W.; Westbrook, C.; Korobeinichev, O.; Shvartsberg, V.; Shmakov, A.; Rybitskaya, I.; Curran, H., Flame inhibition by phosphorus-containing compounds over a range of equivalence ratios. *Combustion and flame* 2005, *140* (1), 103–115.

[106] Schartel, B., Phosphorus-based flame retardancy mechanisms—old hat or a starting point for future development? *Materials* 2010, *3* (10), 4710–4745.

[107] Wilkie, C.A.; Morgan, A.B.; Nelson, G.L., *Fire and Polymers V*. American Chemical Society: 2009; Vol. 1013, p. 456.

[108] Döring, M., Comparison of phosphorous-containing flame retardants in epoxy resins. In *BCC Conference on Flame Retardancy*, Stamford, 2007.

[109] Schäfer, A., Gasphasenaktive phosphacyclische Flammschutzmittel und deren Wirkmechanismen. 2008.

[110] Weil, E., *Encyclopedia of Polymer Science and Technology*. Wiley Interscience, New York: 1986.

[111] Aronson, A., In *Phosphorous Chemistry*, ACS Symposium, 1992; p. 218.

[112] Levchik, S.V., Introduction to flame retardancy and polymer flammability. *Flame Retardant Polymer Nanocomposites* 2007, 1–29.

[113] Davis, J.; Huggard, M., The technology of halogen-free flame retardant phosphorus additives for polymeric systems. *Journal of Vinyl and Additive Technology* 1996, *2* (1), 69–75.

[114] Liu, G.; Chen, W.; Liu, X.; Yu, J., Controllable synthesis and characterization of ammonium polyphosphate with crystalline form V by phosphoric acid process. *Polymer Degradation and Stability* 2010, *95* (9), 1834–1841.

[115] Futterer, T., In A review about the use of novel intumescent flame retardants in thermoplastic materials, Fall FRCA Conference, Cleveland, Cleveland, 2002.

[116] Duquesne, S.; Le Bras, M.; Bourbigot, S.; Delobel, R.; Camino, G.; Eling, B.; Lindsay, C.; Roels, T.; Vezin, H., Mechanism of fire retardancy of polyurethanes using ammonium polyphosphate. *Journal of Applied Polymer Science* 2001, *82* (13), 3262–3274.

[117] Camino, G.; Grassie, N.; McNeill, I., Influence of the fire retardant, ammonium polyphosphate, on the thermal degradation of poly (methyl methacrylate). *Journal of Polymer Science Part A: Polymer Chemistry* 1978, *16* (1), 95–106.

[118] Katayama, M.; Ito, M.; Otsuka, Y., Polycarbonate resin composition containing block copolymer. US6316579, 2001.

[119] Levchik, S.V.; Bright, D.A.; Alessio, G.R.; Dashevsky, S., New halogen-free fire retardant for engineering plastic applications. *Journal of Vinyl and Additive Technology* 2001, *7* (2), 98–103.

[120] Levchik, S.V.; Bright, D.A.; Moy, P.; Dashevsky, S., New developments in fire retardant non-halogen aromatic phosphates. *Journal of Vinyl and Additive Technology* 2000, *6* (3), 123–128.

[121] Freitag, D.; Go, P.; Stahl, G., Compositions comprising polyphosphonates and additives that exhibit an advantageous combination of properties, and methods related thereto. US2007203269A1, 2007.

[122] Freitag, D., Method for the production of block copolycarbonate/phosphonates and compositions therefrom. WO2007065094A2, 2007.

[123] Freitag, D., Poly (block-phosphonato-ester) and poly (block-phosphonato-carbonate) and methods of making same. WO2007022008A2, 2007.

[124] Freitag, D.; Go, P., Insoluble and branched polyphosphonates and methods related thereto. WO2009018336A3, 2009.

[125] Lens, J.-P., In Novel polymeric, non-halogenated flame retardants with broad applicability in multiple industries, *Abstracts of Papers of the American Chemical Society*, 2012.

[126] Chen, L.; Wang, Y.-Z., Aryl polyphosphonates: Useful halogen-free flame retardants for polymers. *Materials* 2010, *3* (10), 4746.

[127] Nass, B.D.I.; Schlosser, E.D.I.; Wanzke, W.D., Flammschutzmittel-Kombination für thermoplastische Polymere I. EP1024167B1, 2005.

[128] Costanzi, S.; Leonardi, M., Polyamide compositions flame retarded with aluminium hypophosphite. WO2005075566A1, 2005.

[129] Kleiner, H.-J.; Budzinsky, W.; Kirsch, G., Flameproofed polyester molding composition. US5780534A, 1998.

[130] Nass, B.D.I.; Wanzke, W.D., Synergistische Flammschutzmittel-Kombination für Polymere. 1999.

[131] Penn, R.E., Flame retarded poly(butylene terephthalate) composition. US5814690A, 1998.

[132] Müller, P.; Schartel, B., Melamine poly(metal phosphates) as flame retardant in epoxy resin: Performance, modes of action, and synergy. *Journal of Applied Polymer Science* 2016, *133* (24), 43549.

[133] Maruyama, K; Motoshige, R., Flame retardant thermoplastic resin composition. EP0728811B1, 1996.

[134] Allen, C.W., The use of phosphazenes as fire resistant materials. *Journal of Fire Sciences* 1993, *11* (4), 320–328.

[135] Lim, J.C.; Lee, J.H.; Kwon, I.H. Flame retardant thermoplastic resin composition. US6630524B1, 2003.

[136] Wang, D.-Y., *Novel Fire Retardant Polymers and Composite Materials*. Woodhead Publishing: 2016.

[137] Hayashi, K.; Satou, Y.; Nakamura, T. Phosphorus-atom-containing oligomer composition, curable resin composition, cured product thereof, and printed circuit board. WO2012124689A1, 2012.

[138] Kim, S.N.; Kim, H.S.J.S.C.; Choi, W.H.; Lim, B.G., Organic phosphorous flame retardant and manufacturing method thereof. KR101084507B1, 2009.

[139] Gan, J., Phosphorus-containing compounds useful for making halogen-free ignition-resistant polymers. US8202948B2, 2012.

[140] Chen, T.; Chen, X.; Wang, M.; Hou, P.; Jie, C.; Li, J.; Xu, Y.; Zeng, B.; Dai, L., A novel halogen-free co-curing agent with linear multi-aromatic rigid structure as flame-retardant modifier in epoxy resin. *Polymers for Advanced Technologies* 2018, *29* (1), 603–611.

[141] Artner, J.; Ciesielski, M.; Walter, O.; Döring, M.; Perez, R.M.; Sandler, J.K.W.; Altstädt, V.; Schartel, B., A novel DOPO-Based diamine as hardener and flame retardant for epoxy resin systems. *Macromolecular Materials and Engineering* 2008, *293* (6), 503–514.

[142] Lin, C.H.; Cai, S.X.; Lin, C.H., Flame-retardant epoxy resins with high glass-transition temperatures. II. Using a novel hexafunctional curing agent: 9,10-dihydro-9-oxa-10-phosphaphenanthrene 10-yl-tris(4-aminophenyl) methane. *Journal of Polymer Science Part A: Polymer Chemistry* 2005, *43* (23), 5971–5986.

[143] Xiao, L.; Sun, D.-C.; Niu, T.-l.; Yao, Y.-w., Syntheses of two dopo-based reactive additives as flame retardants and co-curing agents for epoxy resins. *Phosphorus, Sulfur, and Silicon and the Related Elements* 2014, *189* (10), 1564–1571.

[144] Liu, Y.L., Flame-retardant epoxy resins from novel phosphorus-containing novolac. *Polymer* 2001, *42* (8), 3445–3454.

[145] Wirasaputra, A.; Yao, X.; Zhu, Y.; Liu, S.; Yuan, Y.; Zhao, J.; Fu, Y., Flame-retarded epoxy resins with a curing agent of DOPO-triazine based anhydride. *Macromolecular Materials and Engineering* 2016, *301* (8), 982–991.

[146] Glauner, U.; Storzer, U.; Keller, H.; Rieckert, H. Halogen-free flame retardants. EP2144950B1, 2013.

[147] Chang, S. J.; Chang, F. C., Synthesis and characterization of copolyesters containing the phosphorus linking pendent groups. *Journal of Applied Polymer Science* 1999, *72* (1), 109–122.

[148] Long, L.; Chang, Q.; He, W.; Xiang, Y.; Qin, S.; Yin, J.; Yu, J., Effects of bridged DOPO derivatives on the thermal stability and flame retardant properties of poly (lactic acid). *Polymer Degradation and Stability* 2017, *139*, 55–66.

[149] Huang, W.; He, W.; Long, L.; Yan, W.; He, M.; Qin, S.; Yu, J., Highly efficient flame-retardant glass-fiber-reinforced polyamide 6T system based on a novel DOPO-based derivative: Flame retardancy, thermal decomposition, and pyrolysis behavior. *Polymer Degradation and Stability* 2018, *148*, 26–41.

[150] Scholz, W. F.; Paul, S. Circuit materials with improved fire retardant system and articles formed therefrom. US20150351237A1, 2015.

[151] Uwe Dittrich, B. J., Manfred Döring, Michael Ciesielski Derivatives of 9,10-dihydro-9-oxa-10-phosphaphenanthrene-10-oxide. US20050020739A1, 2003.

[152] Yao, Q.; Mack, A. G.; Junzuo, W., Process for preparation of DOPO-derived compounds and compositions thereof. WO2011123389, 2011.

[153] Yu, B.; Lo, S. M.; Hu, Y. In *Synthesis, structure-property relationships of DOPO-phosphonamidates and their flame retardant application*, Key Engineering Materials, Trans Tech Publ: 2016; pp 102–113.

[154] Neisius, N. M.; Lutz, M.; Rentsch, D.; Hemberger, P.; Gaan, S., Synthesis of DOPO-based phosphonamidates and their thermal properties. *Industrial & Engineering Chemistry Research* **2014,** *53* (8), 2889–2896.

[155] Schmidt, C.; Ciesielski, M.; Greiner, L.; Döring, M., Novel organophosphorus flame retardants and their synergistic application in novolac epoxy resin. *Polymer degradation and stability* **2018,** *158*, 190–201.

[156] Wermter, H.; Futterer, T.; Vogt, R.; Doering, M.; Ciesielski, M., Duromer, production method, use and compositions. EP2956491B1, 2014.

[157] Perret, B.; Schartel, B.; Stöß, K.; Ciesielski, M.; Diederichs, J.; Döring, M.; Krämer, J.; Altstädt, V., A new halogen-free flame retardant based on 9,10-dihydro-9-oxa-10-phosphaphenanthrene-10-oxide for epoxy resins and their carbon fiber composites for the automotive and aviation industries. *Macromolecular Materials And Engineering* 2011, *296* (1), 14–30.

[158] Wang, J.; Ma, C.; Wang, P.; Qiu, S.; Cai, W.; Hu, Y., Ultra-low phosphorus loading to achieve the superior flame retardancy of epoxy resin. *Polymer Degradation and Stability* 2018, *149*, 119–128.

[159] Wang, X.; Hu, Y.; Song, L.; Yang, H.; Xing, W.; Lu, H., Synthesis and characterization of a DOPO-substitued organophosphorus oligomer and its application in flame retardant epoxy resins. *Progress in Organic Coatings* 2011, *71*, 72–82.

[160] Ahlmann, M. D.; Dittrich, U. D.; Döring, M. P. D.; Just, B. D.; Keller, H. D.; Storzer, U. D., New phosphinate complex useful as flame protective agent for polyester, polyamide and epoxide resin and their manufactured products. DE102004049614A1, 2006.

[161] Burk, B., Entwicklung neuer flammschutzmittel basierend auf derivaten des 9,10-dihydro-10-oxaphosphaphenanthren-10-oxids. Ruprechts-Karl-Universität Heidelberg, 2016.

[162] Endo, S.; Kashihara, T.; Osako, A.; Shizuki, T.; Ikegami, T., Neue phosphor-enthaltende Verbindungen. DE 2646218 A1, 1977.

[163] Alongi, J.; Han, Z.; Bourbigot, S., Intumescence: Tradition versus novelty. A comprehensive review. *Progress in Polymer Science* 2015, *51*, 28–73.

[164] Levchik, S. V.; Balabanovich, A. I.; Levchik, G. F.; Costa, L., Effect of melamine and its salts on combustion and thermal decomposition of polyamide 6. *Fire and Materials* 1997, *21* (2), 75–83.

[165] Almeras, X.; Dabrowski, F.; Le Bras, M.; Poutch, F.; Bourbigot, S.; Marosi, G.; Anna, P., Using polyamide-6 as charring agent in intumescent polypropylene formulations. I. Effect of the

compatibilising agent on the fire retardancy performance. *Polymer Degradation and Stability* 2002, *77* (2), 305-313.

[166] Le Bras, M.; Bourbigot, S.; Felix, E.; Pouille, F.; Siat, C.; Traisnel, M., Characterization of a polyamide-6-based intumescent additive for thermoplastic formulations. *Polymer* 2000, *41* (14), 5283-5296.

[167] Bourbigot, S.; Duquesne, S.; Leroy, J.-M., Modeling of heat transfer of a polypropylene-based intumescent system during combustion. *Journal of Fire Sciences* 1999, *17* (1), 42-56.

[168] Bras, M.L.; Bugajny, M.; Lefebvre, J.M.; Bourbigot, S., Use of polyurethanes as char-forming agents in polypropylene intumescent formulations. *Polymer International* 2000, *49* (10), 1115-1124.

[169] Bourbigot, S.; Le Bras, M.; Delobel, R.; Decressain, R.; Amoureux, J.-P., Synergistic effect of zeolite in an intumescence process: study of the carbonaceous structures using solid-state NMR. *Journal of the Chemical Society, Faraday Transactions* 1996, *92* (1), 149-158.

[170] Le Bras, M.; Bourbigot, S., Mineral fillers in intumescent fire retardant formulations–criteria for the choice of a natural clay filler for the ammonium polyphosphate/pentaerythritol/polypropylene system. *Fire and Materials* 1996, *20* (1), 39-49.

[171] Davis, R.; Jarrett, W.L.; Mathias, L.J., Solution 13C NMR spectroscopy of polyamide homopolymers (nylons 6, 11, 12, 66, 69, 610 and 612) and several commercial copolymers. *Polymer* 2001, *42* (6), 2621-2626.

[172] Liu, N.; Baker, W., Reactive polymers for blend compatibilization. *Advances in Polymer Technology* 1992, *11* (4), 249-262.

[173] Camino, G.; Costa, L.; Trossarelli, L., Study of the mechanism of intumescence in fire retardant polymers: Part II–Mechanism of action in polypropylene-ammonium polyphosphate-pentaerythritol mixtures. *Polymer Degradation and Stability* 1984, *7* (1), 25-31.

[174] Delobel, R.; Bras, M.L.; Ouassou, N.; Alistiqsa, F., Thermal behaviours of ammonium polyphosphate-pentaerythritol and ammonium pyrophosphate-pentaerythritol intumescent additives in polypropylene formulations. *Journal of Fire Sciences* 1990, *8* (2), 85-108.

[175] Pfaendner, R.; Metzsch-Zilligen, E.; STEC, M. Verwendung von organischen oxyimiden als flammschutzmittel für kunststoffe sowie flammgeschützte kunststoffzusammensetzung und hieraus hergestellte formteile. WO 2014154636 A1, 2014.

[176] Tirri, T.; Aubert, M.; Pawelec, W.; Holappa, A.; Wilén, C.-E., Structure-property studies on a new family of halogen free flame retardants based on sulfenamide and related structures. *Polymers* 2016, *8* (10), 360.

[177] Pawelec, W.; Aubert, M.; Pfaendner, R.; Hoppe, H.; Wilén, C.-E., Triazene compounds as a novel and effective class of flame retardants for polypropylene. *Polymer degradation and stability* 2012, *97* (6), 948-954.

[178] Wagner, S., Neue phosphorbasierte flammschutzmittel für technische Kunststoffe und Epoxidharze. Ruprecht-Karls-Universität Heidelberg, 2012.

[179] Pfaendner, R.; Mazurowski, M., Verwendung von phosphorhaltigen organischen oxyimiden als flammschutzmittel und/oder als Stabilisatoren für Kunststoffe, flammgeschützte und/oder stabilisierte Kunststoffzusammensetzung, Verfahren zu deren Herstellung sowie Formteile, Lacke und Beschichtungen. WO2016042040 A1, 2016.

[180] Pawelec, W.; Holappa, A.; Tirri, T.; Aubert, M.; Hoppe, H.; Pfaendner, R.; Wilén, C.-E., Disulfides - Effective radical generators for flame retardancy of polypropylene. *Polymer Degradation and Stability* 2014, *110*, 447-456.

[181] Wilén, C.-E., In *Materiaalien sovellusket yrityksille Teemana muovit*, Muovien lisäaineet, Turku, Finland, Turku, Finland, 2016.

[182] Weil, E.D.; Levchik, S.V., Flame retardants for polystyrenes in commercial use or development. *Journal of fire sciences* 2007, *25* (3), 241-265.

[183] Chatgilialoglu, C.; Asmus, K.-D., *Sulfur-Centered Reactive Intermediates in Chemistry and Biology*. Springer Science & Business Media: 2013; Vol. 197.

[184] Jacob, E., Method for making self-extinguishing alkenyl aromatic resin compositions comprising incorporating therein an organic bromine-containing compound and an organic peroxide. US3058926A, 1962.

[185] Eichhorn, J., Flame-proof alkenyl aromatic polymers. US3124557A, 1964.

[186] Kaprinidis, N.; Shields, P.; Leslie, G., Antimony free flame retardant systems containing Flamestab NOR 116 for polypropylene molding. *Interscience Communication* 2002, pp. 95–106.

[187] Benoit, D.; Chaplinski, V.; Braslau, R.; Hawker, C.J., Development of a universal alkoxyamine for "living" free radical polymerizations. *Journal of the American Chemical Society* 1999, *121* (16), 3904–3920.

[188] Nicolas, R.C.; Wilén, C.-E.; Roth, M.; Pfaendner, R.; King, R.E., Azoalkanes: A novel class of flame retardants. *Macromolecular Rapid Communications* 2006, *27* (12), 976–981.

[189] Marney, D.C.O.; Russell, L.J.; Stark, T.M., The influence of an N-alkoxy HALS on the decomposition of a brominated fire retardant in polypropylene. *Polymer Degradation and Stability* 2008, *93* (3), 714–722.

[190] Aubert, M.; Wilén, C.-E.; Pfaendner, R.; Kniesel, S.; Hoppe, H.; Roth, M., Bis(1-propyloxy-2,2,6,6-tetramethylpiperidin-4-yl)-diazene – An innovative multifunctional radical generator providing flame retardancy to polypropylene even after extended artificial weathering. *Polymer Degradation and Stability* 2011, *96* (3), 328–333.

[191] Aubert, M.; Roth, M.; Pfaendner, R.; Wilén, C.-E., Azoalkanes: A novel class of additives for cross-linking and controlled degradation of polyolefins. *Macromolecular Materials and Engineering* 2007, *292* (6), 707–714.

[192] Aubert, M.; Nicolas, R.C.; Pawelec, W.; Wilén, C.-E.; Roth, M.; Pfaendner, R., Azoalkanes—novel flame retardants and their structure-property relationship. *Polymers for Advanced Technologies* 2011, *22* (11), 1529–1538.

[193] Pfaendner, R., How will additives shape the future of plastics? *Polymer Degradation and Stability* 2006, *91* (9), 2249–2256.

[194] Zhang, S.; Horrocks, A.R.; Hull, R.; Kandola, B.K., Flammability, degradation and structural characterization of fibre-forming polypropylene containing nanoclay-flame retardant combinations. *Polymer Degradation and Stability* 2006, *91* (4), 719–725.

[195] Hermans, I.; Jacobs, P.; Peeters, J., Autoxidation catalysis with N-hydroxyimides: more-reactive radicals or just more radicals? *Physical Chemistry Chemical Physics* 2007, *9* (6), 686–690.

[196] Amorati, R.; Lucarini, M.; Mugnaini, V.; Pedulli, G.F.; Minisci, F.; Recupero, F.; Fontana, F.; Astolfi, P.; Greci, L., Hydroxylamines as oxidation catalysts: Thermochemical and kinetic studies. *The Journal of Organic Chemistry* 2003, *68* (5), 1747–1754.

[197] Wilén, E.-C.; Aubert, M.; Tirri, T.; Pawelec, W. Sulfenamides as flame retardants. US2016289566A1, 2016.

[198] Jacob, E. Self-extinguishing alkenyl aromatic resins containing sulfenamides. US3296340A, 1967.

[199] Vandeputte, A.G.; Reyniers, M.-F.; Marin, G.B., Theoretical study of the thermal decomposition of dimethyl disulfide. *The Journal of Physical Chemistry A* 2010, *114* (39), 10531–10549.

[200] Glotova, T.E.; Nakhmanovich, A.S.; Lopyrev, V.A., Thermal decomposition and isomerization of 2,2'-di(benzoxazolyl) disulfide. *Russian Chemical Bulletin* 1993, *42* (10), 1753–1755.

[201] Van, R.G.W., Elementary sulfur as flame-retardant in plastic foams. US 3542701 A, 1970.

[202] Prindle, J.C.; Nalepa, C.J.; Kumar, G. Flame retardant compositions for use in styrenic polymers. WO 9910429 A1, 1999.

[203] Cho, J.H.; Vasagar, V.; Shanmuganathan, K.; Jones, A.R.; Nazarenko, S.; Ellison, C.J., Bioinspired catecholic flame retardant nanocoating for flexible polyurethane foams. *Chemistry of Materials* 2015, *27* (19), 6784–6790.

[204] Feng, K.; Hou, L.; Tang, B.; Wu, P., A self-protected self-cleaning ultrafiltration membrane by using polydopamine as a free-radical scavenger. *Journal of Membrane Science* 2015, *490*, 120–128.

[205] Roberts, B.C.; Jones, A.R.; Ezekoye, O.A.; Ellison, C.J.; Webber, M.E., Development of kinetic parameters for polyurethane thermal degradation modeling featuring a bioinspired catecholic flame retardant. *Combustion and Flame* 2017, *177* (Supplement C), 184–192.

[206] Lewin, M., Synergism and catalysis in flame retardancy of polymers. *Polymers for Advanced Technologies* 2001, *12* (3–4), 215–222.

[270] Le Bras, M.; Bourbigot, S.; Le Tallec, Y.; Laureyns, J., Synergy in intumescence–application to β-cyclodextrin carbonisation agent in intumescent additives for fire retardant polyethylene formulations. *Polymer Degradation and Stability* 1997, *56* (1), 11–21.

[208] Visakh, P.M.; Arao, Y., *Flame Retardants: Polymer Blends, Composites and Nanocomposites*. Springer International Publishing: 2015.

[209] Zhao, C.-X.; Liu, Y.; Wang, D.-Y.; Wang, D.-L.; Wang, Y.-Z., Synergistic effect of ammonium polyphosphate and layered double hydroxide on flame retardant properties of poly(vinyl alcohol). *Polymer Degradation and Stability* 2008, *93* (7), 1323–1331.

[210] Weil, E.D., Synergists, adjuvants and antagonists in flame-retardant systems. *Fire retardancy of polymeric materials* 2000, 115–145.

[211] Yu, L.; Chen, L.; Dong, L.-P.; Li, L.-J.; Wang, Y.-Z., Organic–inorganic hybrid flame retardant: preparation, characterization and application in EVA. *RSC Advances* 2014, *4* (34), 17812–17821.

[212] Pearce, E., *Flame-Retardant Polymeric Materials*. Springer US: 2012.

[213] Zanetti, M.; Camino, G.; Canavese, D.; Morgan, A.B.; Lamelas, F.J.; Wilkie, C.A., Fire retardant halogen–antimony–clay synergism in polypropylene layered silicate nanocomposites. *Chemistry of Materials* 2002, *14* (1), 189–193.

[214] Wang, M.Y.; Horrocks, A.R.; Horrocks, S.; Hall, M.E.; Pearson, J.S.; Clegg, S., Flame retardant textile back-coatings. Part 1: Antimony-halogen system interactions and the effect of replacement by phosphorus-containing agents. *Journal of Fire Sciences* 2000, *18* (4), 265–294.

[215] Handa, T.; Nagashima, T.; Ebihara, N., Synergistic action of Sb_2O_3 with bromine-containing flame retardant in polyolefins. 1—The variance in their performance versus Sb/Br ratio. *Fire and Materials* 1979, *3* (2), 61–67.

[216] Costa, L.; Goberti, P.; Paganetto, G.; Camino, G.; Sgarzi, P., Thermal behaviour of chlorine-antimony fire-retardant systems. *Polymer Degradation and Stability* 1990, *30* (1), 13–28.

[217] Davies, P.J.; Horrocks, A.R.; Alderson, A., Possible phosphorus/halogen synergism in flame retardant textile backcoatings. *Fire and Materials* 2002, *26* (4–5), 235–242.

[218] Green, J., Phosphorus-bromine flame retardant synergy in polycarbonate/ABS blends. *Polymer Degradation and Stability* 1996, *54* (2), 189–193.

[219] Bourbigot, S.; Duquesne, S., Intumescence and nanocomposites: a novel route for flame-retarding polymeric materials. *Flame Retardant Polymer Nanocomposites* 2007, 131–62.

[220] Vannier, A.; Duquesne, S.; Bourbigot, S.; Castrovinci, A.; Camino, G.; Delobel, R., The use of POSS as synergist in intumescent recycled poly(ethylene terephthalate). *Polymer Degradation and Stability* 2008, *93* (4), 818–826.

[221] Leu, T.-S.; Wang, C.-S., Synergistic effect of a phosphorus–nitrogen flame retardant on engineering plastics. *Journal of Applied Polymer Science* 2004, *92* (1), 410–417.

[222] Salmeia, K.A.; Fage, J.; Liang, S.; Gaan, S., An overview of mode of action and analytical methods for evaluation of gas phase activities of flame retardants. *Polymers* 2015, *7* (3), 504–526.

[223] Stelzig, T.; Bommer, L.; Gaan, S.; Buczko, A. Dopo-based hybrid flame retardants. WO 2015140105 A1, 2015.

[224] Song, T.; Li, Z.S.; Liu, J.G.; Yang, S.Y., Novel phosphorus–silicon synergistic flame retardants: Synthesis and characterization. *Chinese Chemical Letters* 2012, *23* (7), 793–796.

[225] Song, S.; Ma, J.; Cao, K.; Chang, G.; Huang, Y.; Yang, J., Synthesis of a novel dicyclic silicon-/phosphorus hybrid and its performance on flame retardancy of epoxy resin. *Polymer Degradation and Stability* 2014, *99*, 43–52.

[226] Dow Corning develops advanced flame retardant synergist for polyamide compounds. *Additives for Polymers* 2017, *2017* (7), 2.

[227] Wilkie, C.A.; Morgan, A.B., *Fire Retardancy of Polymeric Materials*, Second Edition. CRC Press: 2009.

[228] Gallo, E.; Braun, U.; Schartel, B.; Russo, P.; Acierno, D., Halogen-free flame retarded poly(butylene terephthalate) (PBT) using metal oxides/PBT nanocomposites in combination with aluminium phosphinate. *Polymer Degradation and Stability* 2009, *94* (8), 1245–1253.

[229] Gallo, E.; Schartel, B.; Braun, U.; Russo, P.; Acierno, D., Fire retardant synergisms between nanometric Fe2O3 and aluminum phosphinate in poly(butylene terephthalate). *Polymers for Advanced Technologies* 2011, *22* (12), 2382–2391.

[230] Braun, U.; Bahr, H.; Schartel, B., Fire retardancy effect of aluminium phosphinate and melamine polyphosphate in glass fibre reinforced polyamide 6. In *e-Polymers*, 2010; Vol. 10.

[231] Braun, U.; Schartel, B.; Fichera, M.A.; Jäger, C., Flame retardancy mechanisms of aluminium phosphinate in combination with melamine polyphosphate and zinc borate in glass-fibre reinforced polyamide 6,6. *Polymer Degradation and Stability* 2007, *92* (8), 1528–1545.

[232] Bourbigot, S.; Duquesne, S.; Sébih, Z.; Ségura, S.; Delobel, R., Synergistic aspects of the combination of magnesium hydroxide and ammonium polyphosphate in flame retardancy of ethylene-vinyl acetate copolymer. ACS Publications: 2006.

[233] Kroenke, W.J., Low-melting sulphate glasses and glass-ceramics, and their utility as fire and smoke retarder additives for poly(vinyl chloride). *Journal of Materials Science* 1986, *21* (4), 1123–1133.

[234] Bourbigot, S.; Duquesne, S., Fire retardant polymers: recent developments and opportunities. *Journal of Materials Chemistry* 2007, *17* (22), 2283–2300.

[235] Krassowski, D.; Hutchings, D.; Qureshi, S., Expandable graphite flake as an additive for a new flame retardant resin. *GrafTech International Ltd* 2012, 1–9.

[236] Marosi, G.; Tohl, A.; Bertalan, G.; Anna, P.; Maatoug, M.A.; Ravadits, I.; Bertóti, I.; Tóth, A., Modified interfaces in multicomponent polypropylene fibers. *Composites Part A: Applied Science and Manufacturing* 1998, *29* (9), 1305–1311.

[237] Okisaki, F., In *Flamecut GREP series – New non-halogenated flame-retardant systems*, Fire Retardant Chemicals Association, Spring Meeting, San Francisco, 1997; pp. 11–24.

[238] Duquesne, S.; Le Bras, M.; Bourbigot, S.; Delobel, R.; Camino, G.; Eling, B.; Lindsay, C.; Roels, T., Thermal degradation of polyurethane and polyurethane/expandable graphite coatings. *Polymer Degradation and Stability* 2001, *74* (3), 493–499.

[239] Duquesne, S.; Bras, M.L.; Bourbigot, S.; Delobel, R.; Vezin, H.; Camino, G.; Eling, B.; Lindsay, C.; Roels, T., Expandable graphite: A fire retardant additive for polyurethane coatings. *Fire and Materials* 2003, *27* (3), 103–117.

[24] Modesti, M.; Lorenzetti, A.; Simioni, F.; Camino, G., Expandable graphite as an intumescent flame retardant in polyisocyanurate-polyurethane foams. *Polymer Degradation and Stability* 2002, *77* (2), 195–202.

[241] Lorenzetti, A.; Dittrich, B.; Schartel, B.; Roso, M.; Modesti, M., Expandable graphite in polyurethane foams: The effect of expansion volume and intercalants on flame retardancy. *Journal of Applied Polymer Science* 2017, *134* (31), 45173.

[242] Modesti, M.; Lorenzetti, A., Flame retardancy of polyisocyanurate-polyurethane foams: use of different charring agents. *Polymer Degradation and Stability* 2002, *78* (2), 341–347.

[243] Naik, A.D.; Fontaine, G.; Samyn, F.; Delva, X.; Bourgeois, Y.; Bourbigot, S., Melamine integrated metal phosphates as non-halogenated flame retardants: Synergism with aluminium phosphinate

for flame retardancy in glass fiber reinforced polyamide 66. *Polymer Degradation and Stability* 2013, *98* (12), 2653-2662.

[244] Muthu, S. S.; Gardetti, M. A., *Green Fashion*. Springer Singapore: 2016.

[245] Leistner, M.; Pfaendner, R.; Trupti, D.; Köstler, H. G.; Wehner, W. Flammschutzmittelzusammensetzung Flame retardant composition. DE102013210902A1, 2014.

[246] Matzen, M.; Kandola, B.; Huth, C.; Schartel, B., Influence of flame retardants on the melt dripping behaviour of thermoplastic polymers. *Materials* 2015, *8* (9), 5621-5646.

[247] Kiliaris, P.; Papaspyrides, C. D.; Pfaendner, R., Polyamide 6 filled with melamine cyanurate and layered silicates: Evaluation of flame retardancy and physical properties. *Macromolecular Materials and Engineering* 2008, *293* (9), 740-751.

[248] Price, D.; Liu, Y.; Milnes, G. J.; Hull, R.; Kandola, B. K.; Horrocks, A. R., An investigation into the mechanism of flame retardancy and smoke suppression by melamine in flexible polyurethane foam. *Fire and Materials* 2002, *26* (4-5), 201-206.

[249] Costa, L.; Camino, G.; Luda di Cortemiglia, M. P., Mechanism of thermal degradation of fire-retardant melamine salts. In *Fire and Polymers*, American Chemical Society: 1990; Vol. 425, 211-238.

[250] Costa, L.; Camino, G., Thermal behaviour of melamine. *Journal of thermal analysis* 1988, *34* (2), 423-429.

[251] Cipolli, R.; Oriani, R.; Masarati, E.; Nucida, G. Melaminic polycondensates. EP0542374A1, 1993.

[252] Kaul, B. L., Polytriazinyl compounds as flame retardants and light stabilizers. US8202924B2, 2012.

[253] Enescu, D.; Frache, A.; Lavaselli, M.; Monticelli, O.; Marino, F., Novel phosphorous-nitrogen intumescent flame retardant system. Its effects on flame retardancy and thermal properties of polypropylene. *Polymer Degradation and Stability* 2013, *98* (1), 297-305.

[254] Kaul, B. L., Polytriazinyl compounds as flame retardants and light stabilizers. EP2130854A3, 2010.

[255] Li, B.; Xu, M., Effect of a novel charring-foaming agent on flame retardancy and thermal degradation of intumescent flame retardant polypropylene. *Polymer Degradation and Stability* 2006, *91* (6), 1380-1386.

[256] Weil, E. D.; Levchik, S.; Moy, P., Flame and smoke retardants in vinyl chloride polymers - Commercial usage and current developments. *Journal of Fire Sciences* 2006, *24* (3), 211-236.

[257] Shen, K. K.; Griffin, T. S., Zinc Borate as a flame retardant, smoke suppressant, and afterglow suppressant in polymers. In *Fire and Polymers*, American Chemical Society: 1990; Vol. 425, pp. 157-177.

[258] Ning, Y.; Guo, S., Flame-retardant and smoke-suppressant properties of zinc borate and aluminum trihydrate-filled rigid PVC. *Journal of Applied Polymer Science* 2000, *77* (14), 3119-3127.

[259] Ferm, D. J.; Shen, K. K., The effect of zinc borate in combination with ammonium octamolybdate or zinc stannate on smoke suppression in flexible PVC. *Journal of Vinyl and Additive Technology* 1997, *3* (1), 33-41.

[260] Kim, S., Flame retardancy and smoke suppression of magnesium hydroxide filled polyethylene. *Journal of Polymer Science Part B: Polymer Physics* 2003, *41* (9), 936-944.

[261] *The Electronic Information Age and Its Demands on Fire Safety*. Taylor & Francis: 1995.

[262] *FRCA: Fire Safety Development Spring 2000 Conference*. Taylor & Francis: 2000.

[263] Hulbert, R. W.; Nies, N. P., Zinc borate of low hydration and method for preparing same. US3649172A1, 1972.

[264] Council, N. R., *Toxicological risks of selected flame-retardant chemicals*. National Academies Press: 2000.

[265] Lehmann, H. A.; Sperschneider, K.; Kessler, G., Zur Chemie und Konstitution borsaurer Salze. XVIII. Über wasserhaltige Zinkborate. *Zeitschrift für Anorganische und Allgemeine Chemie* 1967, *354* (1-2), 37-43.

[266] Sprague, R.W.; Shen, K.K., The use of boron products in cellulose insulation. *Journal of Thermal Insulation* 1979, *2* (4), 161–174.

[267] Hulbert, R.W.; Nies, N.P., Zinc borate of low hydration and method for preparing same. US3549316A, 1970.

[268] Alongi, J.; Malucelli, G., Cotton fabrics treated with novel oxidic phases acting as effective smoke suppressants. *Carbohydrate Polymers* 2012, *90* (1), 251–260.

[269] Stoeva, S.; Karaivanova, M.; Benev, D., Poly(vinyl chloride) composition. II. Study of the flammability and smoke-evolution of unplasticized poly(vinyl chloride) and fire-retardant additives. *Journal of Applied Polymer Science* 1992, *46* (1), 119–127.

[270] Directive 98/8/EC of the European Parliament and of the Council. 2010.

[271] Horrocks, A.R.; Smart, G.; Hörold, S.; Wanzke, W.; Schlosser, E.; Williams, J., The combined effects of zinc stannate and aluminium diethyl phosphinate on the burning behaviour of glass fibre-reinforced, high temperature polyamide (HTPA). *Polymer Degradation and Stability* 2014, *104*, 95–103.

[272] Jung, H.C.; Kim, W.N.; Lee, C.R.; Suh, K.S.; Kim, S.-R., Properties of flame-retarding blends of polycarbonate and poly (acrylonitrile-butadiene-styrene). *Journal of polymer engineering* 1998, *18* (1–2), 115–130.

[273] Horrocks, A.R.; Smart, G.; Kandola, B.; Holdsworth, A.; Price, D., Zinc stannate interactions with flame retardants in polyamides; Part 1: Synergies with organobromine-containing flame retardants in polyamides 6 (PA6) and 6.6 (PA6.6). *Polymer Degradation and Stability* 2012, *97* (12), 2503–2510.

[274] Horrocks, A.R.; Smart, G.; Price, D.; Kandola, B., Zinc stannates as alternative synergists in selected flame retardant systems. *Journal of Fire Sciences* 2009, *27* (5), 495–521.

[275] Daniels, C.; Herbert, M.J.; Rai, M., Halogenated polymeric formulation containing divalent (hydroxy) stannate and antimony compounds as synergistic flame-retardants. US6245846B1, 2001.

[276] Cusack, P.; Heer, M.; Monk, A., Zinc hydroxystannate as an alternative synergist to antimony trioxide in polyester resins containing halogenated flame retardants. *Polymer Degradation and Stability* 1997, *58* (1–2), 229–237.

[277] Walker, J.K.; Luke, E.; Chen, T.; Ivarov, A., In *Synergies of metal hydroxides and metal molybdates in low-smoke flexible PVC*, Proceedings of the 57th IWCS Conference, Providence, RI, 2008.

[278] Saischek, G.D.I.; Stern, G.D.I.D.; Weinrotter, K.D.I.D.; Schmidt, H.D.; Kraessig, H.D.; Baumann, S.; Verwanger, A., Trockengesponnene Polyacrylnitrilfaser und Verfahren zu deren Herstellung. DE33181005A1, 1984.

[279] Coaker, A.W., Fire and flame retardants for PVC. *Journal of Vinyl and Additive Technology* 2003, *9* (3), 108–115.

[280] Lewin, M.; Atlas, S.M.; Pearce, E.M., *Flame - Retardant Polymeric Materials*. Springer US: 2012.

[281] Starns, W.H.; Edelson, D., Mechanistic aspects of the behavior of molybdenum(VI) oxide as a fire-retardant additive for poly(vinyl chloride). An interpretive review. *Macromolecules* 1979, *12* (5), 797–802.

[282] Lattimer, R.P.; Kroenke, W.J., The functional role of molybdenum trioxide as a smoke retarder additive in rigid poly(vinyl chloride). *Journal of Applied Polymer Science* 1981, *26* (4), 1191–1210.

[283] Skinner, G.A.; Haines, P.J., Molybdenum compounds as flame-retardants and smoke-suppressants in halogenated polymers. *Fire and Materials* 1986, *10* (2), 63–69.

[284] Lum, R.M., MoO_3 additives for PVC: A study of the molecular interactions. *Journal of Applied Polymer Science* 1979, *23* (4), 1247–1263.

[285] Edelson, D.; Kuck, V.J.; Lum, R.M.; Scalco, E.; Starnes, W.H.; Kaufman, S., Anomalous behavior of molybdenum oxide as a fire retardant for polyvinyl chloride. *Combustion and Flame* 1980, *38*, 271–283.

[286] Zhang, W.; Camino, G.; Yang, R., Polymer/polyhedral oligomeric silsesquioxane (POSS) nanocomposites: An overview of fire retardance. *Progress in Polymer Science* 2017, *67*, 77–125.

[287] Iji, M.; Serizawa, S., Silicone derivatives as new flame retardants for aromatic thermoplastics used in electronic devices. *Polymers for Advanced Technologies* 1998, *9* (10–11), 593–600.

[288] Liang, S.; Neisius, N. M.; Gaan, S., Recent developments in flame retardant polymeric coatings. *Progress in Organic Coatings* 2013, *76* (11), 1642–1665.

[289] Darnerud, P. O., Toxic effects of brominated flame retardants in man and in wildlife. *Environment International* 2003, *29* (6), 841–853.

[290] Zhang, M.; Buekens, A.; Li, X., Brominated flame retardants and the formation of dioxins and furans in fires and combustion. *Journal of Hazardous Materials* 2016, *304*, 26–39.

[291] Sonnier, R.; Taguet, A.; Ferry, L.; Lopez-Cuesta, J. M., *Towards Bio-based Flame Retardant Polymers*. Springer International Publishing: 2017.

[292] Fink, J. K., *The Chemistry of Bio-based Polymers*. Wiley: 2014.

[293] Costes, L.; Laoutid, F.; Brohez, S.; Dubois, P., Bio-based flame retardants: When nature meets fire protection. *Materials Science and Engineering: R: Reports* 2017, *117*, 1–25.

[294] Philp, J. C.; Ritchie, R. J.; Allan, J. E. M., Biobased chemicals: the convergence of green chemistry with industrial biotechnology. *Trends in Biotechnology* 2013, *31* (4), 219–222.

[295] Willke, T.; Vorlop, K.-D., Biotechnological production of itaconic acid. *Applied Microbiology and Biotechnology* 2001, *56* (3), 289–295.

[296] Prieur, B.; Meub, M.; Wittemann, M.; Klein, R.; Bellayer, S.; Fontaine, G.; Bourbigot, S., Phosphorylation of lignin to flame retard acrylonitrile butadiene styrene (ABS). *Polymer Degradation and Stability* 2016, *127*, 32–43.

3.2 Flame-Retarded Plastics

Rudolf Pfaendner

3.2.1 Introduction

This section focuses on formulations to provide plastics with flame-retarded or fire-retarded properties from the perspective of the polymer structure. Therefore, the following sub-chapters are grouped according to industrially important polymer classes and not according to flame retardants or markets. The flame retardants considered are mainly to be used as additives incorporated into the polymer, only exceptionally in the polymerization or polycondensation process, and not as a coating or through impregnation processes upon a shaped product. Each sub-chapter reviews commercially available solutions for the individual polymers and technological trends if applicable. As the number of publications on flame-retarded polymers is enormous for most polymer classes and is continuously increasing, recent findings will be selected on the basis of innovative approaches. Moreover, areas where dynamic developments are seen will be described in more detail than fields where traditional solutions are preferred.

The flame retardants mentioned in the following are described by their chemical name. In addition, their CAS numbers are listed in Appendix 3 at the end of the book. Details about the mode of action are shown in the previous section (Section 3.1). The flame retardancy of formulations and average loadings of flame retardants achieved are exemplified in standard tests, which are described in detail in Part II of this book. Although many flammability standards are in use and, in addition, many other criteria are of importance for selected applications besides flammability, Underwriter Laboratories UL 94 V was selected to describe comparable data for the different polymer classes, as this standard is most common in the literature and very well known in the polymer world.

As an introduction step, a simplified overview of important combinations of polymer/flame retardants for orientation purposes is shown in Table 3.3. The polymer list in Table 3.3 follows the sub-chapters. The selected flame retardant classes cover halogens (i.e., mainly brominated flame retardants), inorganics (such as aluminum trihydroxide and magnesium dihydroxide), intumescents, including barrier forming systems based on phosphorus and/or nitrogen (e.g., ammonium polyphosphate, melamine derivatives, phosphinates, among others), organic phosphorus compounds (e.g., phosphate esters, phosphaphenanthrene oxide), and radical generators (e.g., alkoxyamines, oxyimides). As many combinations are used in reality and many more have been experimentally evaluated, Table 3.3 can only serve as a rough indication; moreover, different applications/markets have different requirements. Further details are found later or in more detail in the cited references. In the following, the polymers consuming the majority of flame retardants (PP, PE, PS, PA, PBT, epoxy) are described in more detail.

Table 3.3 Overview of Flame-Retarded Polymers

Polymer (thermoplastics, thermosets, elastomers)	Important FR markets	Halogens	Inorganics	P/N-intumescent	Organic P	Radical generators	Others
Polypropylene	E+E, textiles, building and construction	✓✓	✓✓✓	✓✓		✓✓✓	
Polyethylene	Wire and cable	✓✓	✓✓✓	✓✓		✓✓✓	
Other polyolefins (EPDM, EVA, COC, poly(methyl-4-pentene))	E+E	✓✓	✓✓✓	✓✓		✓✓	
Polystyrenes	Building and construction	✓✓✓		✓	✓	✓	

Polymer (thermoplastics, thermosets, elastomers)	Important FR markets	Halogens	Inorganics	P/N-intumescent	Organic P	Radical generators	Others
Styrene copolymers (ABS, ASA)	E+E, transport	✓✓		✓✓	✓✓✓		
Polyacrylates and polymethacrylates	Building and construction		✓	✓	✓✓		
Polyoxymethylene and polyacetals	E+E		✓	✓	✓		
Polyamides	E+E, textiles	✓✓✓	✓	✓✓✓			
Polyesters (PBT, PET, PLA)	E+E, textiles	✓✓✓	✓	✓✓✓	✓✓		
Polycarbonate and polycarbonate blends	Building and construction, E+E	✓✓			✓		
Polyurethanes (TPU)	Transport, wire and cable	✓	✓✓	✓✓	✓✓	✓✓	
Polyvinylchloride (PVC)	Wire and cable, building and construction	✓✓	✓✓✓		✓✓✓		
Other halogenated polymers (PTFE)							No need
High-performance polymers (PEI, PEEK, PPS, PES, PSU, PI, LCP)	Transport, E+E						No need
Epoxy resins	E+E, transport	✓✓✓	✓	✓✓	✓✓✓	✓	
Polyurethanes (thermoset)	Building and construction, furniture	✓✓✓		✓✓	✓✓✓		
Phenolic/formaldehyde resins	E+E, transport				✓✓		
Unsaturated polyesters	Building and construction, transport	✓✓	✓✓✓	✓✓	✓		
Elastomers	Transport	✓✓	✓✓✓	✓	✓		

3.2.2 Thermoplastics

The common way to provide flame-retarded thermoplastics is to introduce suitable flame retardants as additives through a processing or compounding step or sometimes during polymer production. An example for the latter process is expanded polystyrene (EPS). Polyester fibers are the only large-scale exception as the flame-retarded component, e.g., 2-carboxyethyl phenyl phosphinic acid is introduced already as a comonomer in the polycondensation process.

3.2.2.1 Polypropylene

The flammability of polypropylene (PP) is characterized by a high heat of combustion, high heat flux from flame, and medium smoke generation [1]. Thermal degradation of PP takes place via random chain scission. Pyrolysis and combustion experiment results show that aliphatic, unsaturated, and aromatic hydrocarbons are the major breakdown molecules. In the presence of oxygen low molecular weight alcohols, aldehydes and ketones are detected additionally [2]. As one of the most universal polymers, PP flame-retarded formulations have been developed with a broad range of flame retardants, mainly brominated compounds, magnesium dihydroxide, intumescent systems based on ammonium polyphosphate, and, more recently, radical generators using preferably P/N synergism (Table 3.4).

Table 3.4 Representative Flame Retardant Formulations for Polypropylene

Type of flame retardants	Examples	Synergists	Usual flame retardant loading range for PP (%)
Inorganics	$Mg(OH)_2$, AlO(OH)	Layered silicates (nanoclays), zinc borates, carbon, phosphorus derivatives	60–80
Halogens	Brominated compounds	Sb_2O_3, borates, radical generators	25–35
Intumescent	Ammonium polyphosphate (APP)	Polyols, metal oxides, metal salts, triazine derivatives	20–30
Radical generators	Alkoxyamines, oxyimides, sulfenamides	Phosphonates, phosphazenes	1–15

Magnesium dihydroxide (MDH) is the inorganic flame retardant of choice for polypropylene, whereas different synthetic and natural (brucite) materials in different particle sizes and coatings are available. For PP (and other polyolefins), coatings based on stearic acid are preferred cost performance wise. Required loadings of inorganic flame retardants to achieve UL 94 V-0 are around 60 weight%. An inorganic alternative to MDH is boehmite (AlO(OH)), which decomposes like MDH above 300 °C. As the energy of decomposition and the energy absorbed is less (40% resp. 50%, [3]) compared to MDH, an even higher loading is necessary.

The traditional area of brominated flame retardants is today represented through decabromodiphenyl ethane and the bis(2,3-dibromopropylether) of tetrabromobisphenol-A in combination with antimony oxide (ATO) in concentrations between 15 and 25% of brominated additive and 4–9% of ATO to pass UL 94 V-0. V-2 formulations can be achieved through 2–5% brominated flame retardant, e.g., tris(tribromoneopentyl)phosphate and 1–2% ATO. In ATO-free formulations, ATO is replaced by radical generators such as alkoxyamines [4–6], calcium borate on a silicate carrier [7] or polyphosphonates [8].

Intumescent formulations contain ammonium polyphosphate (APP) as the main component and are available as fully formulated systems whereas a loading of 20–30% is usually needed to achieve UL 94 V-0. PP formulations containing modifications of APP and combinations of APP with many available and potential flame retardant synergists have been published with the main target of stabilizing the intumescent barrier layer, to reduce the overall loading, burning times, and smoke densities. Among APP modifications, encapsulated APP is most popular [9, 10]. Classic mixtures of APP with pentaerythritol (PER), dipentaerythritol or tris(hydroxyethyl)isocyanurate [11, 12] have been combined additionally with minerals such as layered double hydroxides [13, 14], montmorillonite nanocomposites [15, 16], sepiolite [17, 18], vermiculite [19], but as well with classical synergists such as ATO [20], zinc borate [21], zinc hydroxystannate [22], metal oxides, and metal salts [23, 24]. Another option to increase the performance of APP in PP is a synthetic modification, e.g., with ethylenediamine [25] or ethanolamine [26] or the combination of APP with other phosphorus- or nitrogen-based flame retardants, e.g., phosphazene [27], phosphorus diamidates [28], aluminum hypophosphite [29], hyperbranched polyurea [30], or alkoxyamines [31]. Polymeric co-components of APP are, for example, poly(triazinyl, morpholonyl)triazine [32], a condensate of *N,N'*-bis(2-aminoethyl)phenyl phosphorodiamidate with a piperidine substituted triazine [33] or a condensate from tris(2-hydroxyethyl)isocyanurate with a condensate from tris(2-hydroxyethyl)isocyanurate and spirocyclic pentaerythritoldiphosphoryl dichloride [34]. Many combinations show some improvement, i.e., reduced loading, for UL 94 V-0, increased Limiting Oxygen Index LOI, reduced peak heat release rate in cone experiments, or reduced smoke density versus the standard APP/PER formulation. For example, the addition of 0.1 to 0.5% of alkoxyamine such as a cyclohexyl derivative to an APP/PER system with a total concentration of 25% results in UL 94 V-0 (3.2 mm) rating versus V-1 without alkoxyamine and, in parallel, an increase of LOI from 31% to 35% [31].

Other flame-retarded PP formulations are based on phosphate esters and salts, where improved hydrolytic stability and light stability in comparison to APP is claimed, for example, ethylenediamine phosphate salt or piperazine pyrophosphate [35] or even phosphonates such as dimethylspirophosphonate [36].

Radical generators are the most recent developments to provide flame retardant PP originally in the form of films and fibers. Efficient chemical structures comprise alkoxyamines [37], sulfenamides [38], or oxyimides [39–42]. In combination with, e.g., phosphorus synergists, flame-retarded PP in thick sections is possible as well. The oxyimides are exceptional through their variability and high thermal stability. In combination with phosphorus synergists such as cyclic phosphonates, UL 94 V-0 (1.6 mm) is achieved at loadings below 10%.

3.2.2.2 Polyethylene

Like other polyolefins, and, in comparison to other standard plastics (PS, PMMA, PC), polyethylene (PE) is a polymer with a high heat of combustion, high rate of heat release, medium time to ignition, and a low smoke release [1]. When taking the LOI (Limiting Oxygen Index, ISO 4589) values as a measure of the flammability of polymers, with a value of 18%, PE ranks among the most flammable commodity and engineering plastics [43]. Thermal degradation of PE takes place via random and chain-end scission and results in the formation of diolefin, olefin, and paraffin fragments and lower molecular weight products at a higher temperature of pyrolysis [44]. Combustion of PE yields mainly CO_2, CO, and hydrocarbons; however, aromatic compounds are detected with increasing temperature through pyro synthesis reactions [45].

Flame-retarded polyethylene formulations are dominated through inorganic or mineral fillers and mainly aluminum hydroxide (ATH, $Al(OH)_3$), boehmite (AlO(OH)) and magnesium dihydroxide (MDH, $Mg(OH)_2$) (Table 3.5).

Further mineral flame retardants are huntite ($Mg_3Ca(CO_3)_4$) and hydromagnesite ($Mg_5\,(CO_3)_4(OH)_2{\cdot}4H_2O$) or a natural mixture of both. The main application area of PE flame retarded with inorganics is wire and cable, often based on blends of LLDPE or LDPE and EVA. 60–65% of ATH or up to 80% of MDH are required to provide flame retardancy and to pass standards such as UL-1072 and UL-1277 [46, 47]. Flame retardant synergists such as calcium borate on a silica carrier [48], zinc borate, zinc (hydroxy) stannate [49], or nanoclays are used to stabilize the oxide layer formed. Moreover, coupling agents/compatibilizers are essential to achieve a reasonable level of mechanical performance.

Table 3.5 Representative Flame Retardant Formulations for Polyethylene

Type of flame retardants	Examples	Synergists	Usual flame retardant loading range for PE (%)
Inorganics	$Al(OH)_3$, $Mg(OH)_2$, $AlO(OH)$	Layered silicates (nanoclays), zinc borates, carbon, phosphorus derivatives	60–80
Halogens	Chloroparaffins, (polymeric) brominated compounds	Sb_2O_3, zinc borates, radical generators	25–35
Phosphorus P/N Intumescent	Ammonium polyphosphate (APP) Phosphates Expandable graphite	Polyols, melamine (derivatives), triazine derivatives, metal oxides, phosphorus derivatives	30–40
Radical generators	Alkoxyamines, oxyimides	Phosphonates	1–15

A representative example of synergistic performance to lower the high loadings of inorganic flame retardants is the combination of ATH and nanoclays. By adding 2.5% of organically modified clay (montmorillonite), the peak heat release rate in the cone calorimeter test is substantially reduced [50] and with 5% of the clay, the loading of ATH in EVA cable materials can be decreased from 65% to 45% without sacrificing the low heat release rate [51]. Even very severe cable standards such as UL 1666 can only be passed with a combination of EVA, ATH, and clay. The reason for the improved performance is the formation of an oxide layer with improved physical stability. As potential alternatives to clay as a co-component in ATH/MDH-based flame retardants, halloysites [52, 53], layered double hydroxides [54], carbon nanotubes [55], silica, carbon black [56], and graphene [57] have been reported.

Another improvement of inorganic flame retardants was shown through the combination with phosphorus compounds, e.g., hypophosphites [58], phosphonites or phosphonates [59], or red phosphorus [60].

Preferred brominated flame retardants for PE are aromatic bromine derivatives such as decabromodiphenyl ethane, ethylene bis(tetrabromophthalimide), aliphatic bromine derivatives such as tris(tribromoneopentyl) phosphate, 1,3,5-tris(2,3-dibromopropyl) isocyanurate, or mixed aliphatic-aromatic derivatives such as tetrabromobisphenol-A-bis(2,3-dibromopropylether). More recently, polymeric structures with aromatic and or aliphatic bromine groups were introduced such as poly(pentabromobenzyl) acrylate, polystyrene-brominated butadiene-polystyrene block copolymers, or brominated polyphenylene ether.

The loadings of brominated flame retardants necessary to achieve a UL 94 V-0 classification of, e.g., HDPE, are 19 to 22% of brominated flame retardant (19% of decabromodiphenyl ethane, 21% of poly(pentabromobenzyl) acrylate or 22.3% of ethylene bis(tetrabromophthalimide)) together with antimony trioxide (ATO) in a preferred ratio of 1:3, corresponding to 6.3% [61]. In cross-linked PE (20% talc-filled, dicumyl peroxide cross-linked), e.g., 21% of brominated flame retardant (decabromodiphenyl ethane or ethylene bis(tetrabromophthalimide)) and 7% of ATO are used together with 1% nanoclay [61].

In general, it seems that the overall flame retardant concentration necessary increases from LDPE (low density polyethylene) over LLDPE (linear low density polyethylene) to HDPE (high density polyethylene) as shown for the UL 94 V-2 classification. There, as a formulation example for the no longer used decabromodiphenyl ether (Deca), LDPE needs 6% Deca and 2% ATO; LLDPE needs 6% and 3% of ATO; and HDPE needs 8% and 3% of ATO [62].

Potential synergistic flame retardants in addition to ATO may be zinc borate, calcium borate, or radical initiators such as 2,3-dimethyl-2,3-diphenylbutane [63] or sterically hindered alkoxy amines, e.g., the N-cyclohexyl derivative.

In addition to brominated flame retardants, chlorinated compounds are used in polyethylene formulations, e.g., polychlorinated paraffins. PE formulations to pass the standard UL 94 V-0 contain 24% of chlorinated paraffins and 10% of ATO [64].

Typical phosphorus-based compounds such as phosphate esters or phosphinates are not very efficient in PE. The most used intumescent approach with PE is based on ammonium polyphosphate (APP). Many synergistic combinations have been described based on APP [65], such as alternative carbon sources (e.g., polysaccharides, cyclodextrins [66], red phosphorus, different metal oxides, talc, zeolites, zinc borate, halloysites [67], and others). A more recent synergistic nitrogen compound is a triazine derivative with morpholine groups [32]. In general, a flame retardant loading of a minimum of 30% based on APP is needed to achieve a UL 94 V-0 classification in a PE formulation. For example, APP (22%) in combination with a triazine charring agent (8%) can reach the V-0 classification [68].

Alternative intumescent phosphorus-based flame retardants for PE are piperazine pyrophosphates [69] or ethylene diamine phosphate in combination with melamine or melamine polyphosphate [70]. The concentration to meet UL 94 V-0 is above 30%.

The performance of phosphinates, such as aluminum diethyl phosphinate or aluminum hypophosphite as well as DOPO derivatives, which are often efficient flame retardants in engineering plastics, is pretty low. For example, to achieve UL 94 V-0 (4 mm) in LDPE, 33% of aluminum hypophosphite is needed; 29% of the additive results only in UL 94 V-2, and below 23%, it is classified as "not rated" [71]. However, there is a beneficial synergistic behavior when the phosphinates are combined with radical generators such as alkoxyamines [72] or oxyimides [39], where total concentrations of 10% flame retardant achieve at least a UL 94 V-2 classification.

Besides polypropylene, radical generators are a novel solution to provide flame-retarded polyethylene mainly in the form of films, fibers, and foams, where comparatively low concentrations can be sufficient to pass the required standards. Concentrations of alkoxy amine flame retardants are in the range of 1–3%, e.g., a HDPE film (200 μm) containing 1% of alkoxy amine passes German DIN 4102 B2 and a LLDPE film passes UL 94 VTM-2 [73]. Improved versions of alkoxy amines in the form of azo derivatives have been developed, where only 0.5% of additive in LDPE is sufficient to pass the DIN 4102 B2 standard [74]. A flame retardant performance similar to alkoxy amines is shown by azo compounds [75], triazenes [76], disulfides [77], sulfenamides [38], and oxyimides [39–42].

Another approach to increase the performance of radical generators, to reduce the loading, or to pass more stringent standards is a synergistic combination with selected phosphorus compounds such as phosphonates or phosphazenes [39], e.g., the combination of an oxyimide and dimethylspirophosphonate in a ratio of 1:4 is capable of reaching UL 94 V-0 (1.6 mm) at an overall loading of 10% in HDPE.

3.2.2.3 Other Polyolefins

EPDM rubber (ethylene-propylene-diene rubber) or its PP blends can be flame retarded in a similar way with brominated flame retardants as described earlier. Apart from halogenated flame retardants, intumescent systems based on ammonium polyphosphate (APP), e.g., in microencapsulated form or in combination with pentaerythritol (PER), are proposed; loadings of 40% are necessary for UL 94 V-0 (3 mm) [78, 79]. When using aluminum trihydroxide (ATH) as the main flame retardant, combinations with a small amount of nanoclay (2–3%) are beneficial and UL 94 V-0 (3 mm) is reached in the range of 50% of flame retardant [80].

Ethylene-vinyl acetate copolymer is somewhat less flammable than polyethylene and the LOI increases with increasing vinyl acetate (VA) content (60% VA LOI = 19% [81]). ATH flame retardant is used in the range of 55–60%. Combinations with melamine derivatives, such as melamine borate or melamine polyphosphate, support an additional condensed phase mechanism [81]. Magnesium dihydroxide (MDH) at a concentration of 50% results in UL 94 V-1 (2 mm), addition of montmorillonite (about 3%) and organically modified montmorillonite (about 6%) allows V-0 [82]. Alternatively, MDH flame retardancy can be improved through the addition of expandable or expanded graphite [83].

Flame-retarded polycycloolefins (COC) have not been an important area of research so far. Brominated phosphates, oligomeric brominated polycarbonates combined with phosphates, and 4,6-tris(2,4,6-tribromophenoxy)-1,3,5-triazine have been reported as potential flame retardants for COC [84].

Poly-4-methyl-1-pentene can be flame retarded through the usual polyolefin additives, e.g., brominated flame retardants in combination with antimony trioxide (ATO) and melamine cyanurate [85], phosphaphenanthrene oxide derivatives [86], or aluminum diethyl phosphinate in combination with PTFE [87], but only UL 94 V-2 is achieved. Poly-1-butene can be flame retarded through MDH at concentrations around 60% [88].

3.2.2.4 Polystyrenes

Polystyrene degrades thermally by a radical depolymerization reaction through random chain scission yielding mainly the monomeric styrene and some oligomers via intramolecular radical transfer reactions [89]. As there is no char formation, the flammability of styrene with a LOI of 19 is quite high [90]. Besides the classical brominated systems/antimony oxide combinations, concepts to improve the flame retardancy of crystal PS and copolymers such as HIPS (high impact polystyrene) and ABS (acrylonitrile butadiene styrene) are aimed at increasing char formation via blends (e.g., PPE/PS or ABS/PC; PPE: polyphenylene ether, PC: polycarbonate) with less flammable polymers in combinations with halogen-free systems and/or via providing intumescent systems (Table 3.6).

Table 3.6 Representative Flame Retardant Formulations for Styrenics

Type of polystyrene	Type of flame retardants	Examples	Synergists	Usual flame retardant loading range for PS (%)
HIPS	Halogens	(Oligomeric) brominated compounds	Sb_2O_3, zinc borates, radical generators	15-25
EPS foam	Halogens	Polymeric brominated compounds	Radical generators	2-4
ABS	Halogens	Brominated compounds	Sb_2O_3, PTFE	15-25
ABS	Phosphorus	Phosphates, phosphonates	Lewis acids, nanocomposites	15-35

Formulations to achieve UL 94 V-0 (1.6 mm) comprise mainly aromatic brominated flame retardants such as 4,6-tris(2,4,6-tribromophenoxy)-1,3,5-triazine, brominated epoxy polymers, or end-capped brominated epoxy oligomers in combinations with antimony trioxide (ATO) and PTFE. The concentrations are in the range of 15–20% brominated FR and 1.5 to 4% ATO. For UL 94 V-2, around 6% aliphatic brominated FR, such as bis(2,3-dibromopropyl)ether of tetrabromobisphenol-A or tris(tribromoneopentyl) phosphate together with around 1.5% of ATO, are sufficient.

Phosphate flame retardants have a limited performance in polystyrene. Thus, only a combination with char-forming polymers like HIPS/PPE (70:30) meets UL 94 V-0 when 20% of resorcinol bis(di-2,6-xylylphosphate) is used [90]. Many other combinations of halogen-free flame retardants have been tested in polystyrene and HIPS, for example, magnesium dihydroxide (MDH) in combination with microencapsulated red phosphorus [91] (40–50% flame retardant to achieve UL 94 V-0 (3 mm)) or expandable graphite [92], aluminum hypophosphite [93] (25% flame retardant for UL 94 V-0), or ammonium polyphosphate (APP) combinations [94] with poly(1,3,5-triazine-2-aminoethanol)diethylenetriamine as a novel charring agent (30% flame retardant for UL 94 V-0 (3.2 mm)). The addition of nanoclays to any intumescent system is a method that has been investigated further to improve the flame retardancy of polystyrenes [95].

A specific sub-group of polystyrene applications is foam in the form of EPS or XPS, which is mainly used in construction applications. With the phase-out of hexabromocyclododecane (HBCD), the replacement of choice is a polystyrene-brominated polybutadiene-polystyrene block copolymer [96, 96a]. In a concentration of 2–3%, e.g., German DIN 4102 B2 [97] is passed. Furthermore, EPS can be manufactured in the presence of radical generators such as dicumylperoxide or dicumyl(2,3-dimethyl-2,3-diphenyl butane) to reduce the overall FR loading.

In other investigations, the flame retardancy of polystyrene foam applications could be achieved through combinations of sulfur or disulfides with phosphorus organic compounds [98] or through the use of a macromolecular nitrogen-phosphorus intumescent system [99]. A further approach based on radical generators to improve the flame retardancy of polystyrene is the addition of azo-alkoxyamines to brominated systems as demonstrated in HIPS films [74].

3.2.2.5 Styrene Copolymers

The family of styrene copolymers comprises SAN (styrene acrylonitrile), ABS (acrylonitrile butadiene styrene), ASA (acrylonitrile styrene acrylate), MBS (methacrylate butadiene styrene), MABS (methacrylate acrylonitrile butadiene styrene), and others, the most important polymer in this class being ABS. Polyacrylonitrile (PAN) is a polymer of high flammability despite the cyclization reaction, which takes place at temperatures above 250 °C resulting in char formation. Incorporation of styrene as in SAN or ABS increases the flammability of the resulting copolymers, whereas some acrylates and vinyl acetates reduce it [100, 101]. With a combination of acrylonitrile, butadiene, and styrene in ABS polymers (LOI = 18), comparatively high loadings of flame retardants are needed. Brominated flame retardants achieving UL 94 V-0 comprise tris(tribromophenyl) cyanurate, brominated epoxy oligomers, or decabromodiphenyl ethane (concentration: 15–20%) in combination with antimony oxide (ATO) (2–4%) and PTFE as an anti-dripping agent. Standard UL 94 V-2 can be achieved with brominated flame retardants at loadings below 10% and without ATO.

Halogen-free flame retardant approaches for ABS mainly follow the strategy to increase char formation in the combustion process, e.g., by using phosphorus compounds, intumescent systems, or nanocomposites [102]. Phosphonates are more efficient than phosphates; 15–35% are required to achieve UL 94 V-0. Furthermore, cyclic alkyl phosphonates were claimed to be the most efficient [103]. Another possibility to increase char formation is via the addition of Lewis acids, such as $ZnCl_2$ [102], to phosphorus flame retarded ABS.

ASA and other styrene copolymers may be flame retarded like ABS. For example, brominated epoxides were used in a concentration of 21% to achieve UL 94 V-0 [104].

3.2.2.6 Polymethacrylates and Polyacrylates

Thermal degradation of polymethylmethacrylate (PMMA) results in depolymerization and monomer formation without any char residues [105]. This radical decomposition process is accelerated in the presence of oxygen. Even under combustion conditions, monomer is found besides mainly CO_2 [106]. The heat release rate of PMMA in combustion is in the range of other engineering plastics (26.75 MJ/kg [107]) and lower than that of polyolefins; smoke formation is significant. Therefore, the strategy to flame retard PMMA is to increase the char by combining phosphorus species, e.g., ammonium polyphosphate (APP), melamine polyphosphate, or aluminum diethyl phosphinate with inorganic oxides such as TiO_2 or nanomaterials such as montmorillonite, sepiolite, or halloysite [108–110]. A loading of 25% is needed to prove a substantial decrease in flammability in cone calorimetry experiments. A further challenge is to keep the transparency of PMMA, where nanosized fillers or soluble phosphates such as triphenyl phosphate [111] are beneficial. Furthermore 9,10-dihydro-9-oxa-10-phosphaphenanthrene-10-oxide (DOPO) derivatives contribute to char formation in PMMA [112, 113] and DOPO derivatives of acrylates can act as flame retardants by themselves [114]. Flammability of polyacrylates may be reduced in a similar way as that of PMMA; however, it is still difficult to achieve standards such as UL 94 V-0 and to keep transparency for all acrylates and methacrylates.

3.2.2.7 Polyoxymethylene and Polyacetals

Polyoxymethylene (POM) and polyacetal copolymers are challenging to flame retard, as it is a polymer class of significant commercial importance with the lowest LOI in the range of 15–16%. At high temperatures, the polymers are easily depolymerized into formaldehyde. The degradation is furthermore catalyzed through Lewis acids. Therefore, flame retardant systems of brominated flame retardants and antimony oxide (ATO) are not efficient [115]. Combinations based on phosphorus nitrogen synergism in intumescent systems, e.g., ammonium polyphosphate (APP) and melamine or melamine cyanurate, require more than 35% loading to achieve V-1 classification [116, 117]. To use novolac as an additional component seems to be beneficial to achieve UL 94 V-0 classifications; however, a loading around 40% is mandatory [115, 117]. More complex flame retarded POM formulations were composed of magnesium dihydroxide (MDH), melamine, triphenyl phosphate, and novolac; however, 55% flame retardant mixture was needed for UL 94 V-0 (1.6 mm) [118].

3.2.2.8 Polyamides

Flame-retarded polyamides are mainly based on polyamide 6 and polyamide 6.6 and blends thereof, often glass-fiber-reinforced for E&E (electro and electronics) applications. Even at high temperatures, the degradation of polyamide 6 leads to the formation of the monomer caprolactam [119, 120]. The main degradation product of polyamide 6.6 is cyclopentanone formed from adipic acid in an intramolecular cyclization process [119, 120].

Besides classical brominated systems to flame retard polyamides, the choice of efficient halogen-free solutions is constantly growing (Table 3.7). Today, brominated flame retardants for polyamides preferably comprise polymers such as brominated polystyrene, brominated epoxy polymers, brominated polyacrylates, mainly in combination with antimony oxide (ATO). As a low molecular weight alternative, decabromodiphenyl ethane may be used. Loadings to achieve UL 94 V-0 are in the range of 20–25%. The ratio of brominated flame retardant to ATO is around 4:1; PA 6.6 usually needs a somewhat higher loading than PA 6, regardless of whether it is reinforced or not. For example, a formulation with18% brominated polystyrene and 5% ATO in 30% glass-fiber-reinforced polyamide 6.6 meets UL 94 V-0 (0.8 mm) [121].

Table 3.7 Representative Flame Retardant Formulations for Polyamides

Type of flame retardants	Examples	Synergists	Usual flame retardant loading range for PA (%)
Halogens	Polymeric brominated compounds (e.g., brominated polystyrenes, epoxy polymers, polyacrylates)	Sb_2O_3, zinc stannate	20–25
Phosphorus	Phosphinates (e.g., aluminum diethyl phosphinate)	Melamine cyanurate, melamine polyphosphate	15–20
P/N	Melamine polyphosphates	Metal salts	
Phosphorus	Red phosphorus		5–10
P/N	Melamine cyanurate (PA 6)		5–15

In unreinforced polyamide 6, melamine cyanurate is the first choice in terms of cost, and between 6% and 12% are proposed to pass UL 94 V-0; a further reduction of the concentration is claimed through addition of a specific glass [122].

Apart from red phosphorus, which is active due to its "undiluted" phosphorus concentration with loadings below 10% (e.g., in PA 6.6, 7.2% for UL 94 V-0 (1.6 mm)) [123]), the commercial phosphorus-based alternatives for polyamides are organic phosphinates, mainly aluminum diethyl phosphinate. It is frequently applied in combinations with melamine polyphosphate or other metal phosphorus components [124]. Many other synergists have been investigated for aluminum diethyl phosphinate [35, 125], where melamine cyanurate is a suitable synergist for polyamide 6. The addition of zinc stannate to phosphinates can decrease smoke density [126] and helps to reduce corrosion in processing machinery [127]. The overall flame retardant concentration of metal phosphinates and their combinations is usually between15 and 20%.

Another phosphorus-based system for polyamides is the combination of a polymeric metal polymelaminephosphate, e.g., aluminum polymelaminephosphate or magnesium polymelaminephosphate, with a polymeric phosphaphenanthrene oxide (DOPO) derivative, i.e., poly-zinc DOPO [128], where little or no corrosion of processing machinery is claimed to be the main advantage [129]. DOPO derivatives alone were shown to be active flame retardants in polyamides, provided the heat stability was high enough [130, 131]. As a recent example, glycerol tri(DOPO-acrylate) can be mentioned, where a 25% loading was used for UL 94 V-0 (1.6 mm) in PA-6 and PA-6.6 [132].

Inorganic flame retardants are of less importance for polyamides; however, magnesium dihydroxide (MDH) is suitable with a loading of around 60%. Boehmite can be used in combination with phosphinates [133].

Today, flame-retarded high temperature-resistant polyamides (polyphthalamides) including polyamide 4.6 are a small but challenging part for flame-retarded formulations due to high processing temperatures. However, in principle, similar formulations like for polyamide 6.6 can be used when selecting products with a high temperature stability of up to 350 °C. Other polyamides such as PA 11, PA 12, PA 1010, or PA 1212 are of minor importance with regard to flame retardant formulations. For example, 20% aluminum diethyl phosphinate in combination with 7.5% of nanoclays passes UL 94 V-0 (1.6 mm) for polyamide 11 and 12 [134]. Due to the lower processing temperatures compared to PA 6 and PA 6.6, even ammonium polyphosphate (APP) is an acceptable flame retardant as demonstrated for polyamide 1010 [135].

3.2.2.9 Polyesters

The main flame-retarded polyester substrate of interest is polybutylene terephthalate (PBT) in E & E applications. Flame-retarded polyethylene terephthalate (PET) is less important (with the exception of the fibers discussed in Chapter 4) with flame retardant formulations similar to PBT; however, the higher processing temperature has to be considered. Polylactic acid (polylactide, PLA), an aliphatic polyester, is an example of polymers from natural resources, with which numerous research groups have been working on flame-retarded solutions.

The first step of thermal degradation of polyesters (PBT, PET) is the formation of acids and olefins, likely via a cyclic transition step. Under combustion conditions, aldehydes, ethylene, butadiene, and tetrahydrofuran (THF) are formed as flammable, low molecular degradation products [136].

Flame-retarded PBT, often glass-fiber-reinforced and mainly for E & E applications, has already been formulated for many years with polymeric brominated flame retardants, e.g., brominated polystyrene in combination with antimony oxide (ATO) as a synergist. Halogen-free alternatives for PBT are growing; however, the commercial solutions are still less successful compared to polyamides (Table 3.8). Polymeric alternatives to brominated styrene, which is sometimes difficult to process and shows an impact on mechanical properties, are brominated epoxy polymers, brominated polyacrylates, or brominated oligo-carbonates. Low-molecular-weight brominated alternatives for PBT include decabromodiphenyl ethane. Addition of an anti-drip agent like PTFE can reduce the overall loading. Total concentrations of brominated flame retardant and ATO are between 10 and 20% with the ratio of brominated flame retardant to antimony oxide in the range of 3:1. For example, 30% glass-fiber-reinforced PBT meets UL 94 V-0 (3.2 mm) with 10% of brominated styrene or other brominated flame retardants [137, 138]. To pass UL 94 tests at smaller specimen thicknesses, an increased concentration together with ATO is needed; the loading of reinforced and unreinforced PBT remains at a similar level. Antimony-free flame-retarded PBT with brominated flame retardants can be achieved through combinations with phosphonates (e.g., aluminum methylmethyl phosphonate), phosphinates (aluminum diethyl phosphinate) [139], or pyrophosphates [140]. Moreover, radical generators have been mentioned as synergists in antimony-free systems in the form of 2,3-diphenyl-2,3-dimethylbutane [141, 142].

Table 3.8 Representative Flame Retardant Formulations for Polybutylene Terephthalate (PBT)

Type of flame retardants	Examples	Synergists	Usual flame retardant loading range for PBT (%)
Halogens	Polymeric brominated compounds (brominated polystyrenes, epoxy polymers, polyacrylates)	Sb_2O_3, anti-dripping agent (PTFE), phosphinates, phosphonates	10–25
Phosphorus P/N Intumescent	Phosphinates (e. g., aluminum diethyl phosphinate)	Triazine derivatives (melamine cyanurate melamine polyphosphate) nanocomposites, metal oxides	15–20

For halogen-free flame-retarded PBT, organic phosphinates in combination with nitrogen compounds are first choice, e. g., aluminum diethyl phosphinate (approximately 15% for UL 94 V-0) in combination with various synergists such as melamine polyphosphate [125, 143], melamine cyanurate [144], or mixed-metal melamine polyphosphates [145]. Further synergists for flame-retarded PBT, e. g., nano metal oxides [146], polyphosphonates [147], boehmite [125, 148], porous boron silicate glass [149], carbon nanotubes [150], sepiolite [151], and, more recently, 4,4'-bishydroxydeoxybenzoin-polyphosphonate [152] have been described (ca. 15–20% for UL 94 V-0). The corrosivity on the processing equipment by the phosphinate is reduced, e. g., through epoxides, metal oxides, or metal hydroxides [153]. Alternatives to aluminum diethyl phosphinate can be aluminum hypophosphite [154–156], eventually combined with melamine cyanurate or melamine polyphosphate at around 20% to achieve UL 94 V-0 (1.6 mm). Red phosphorus is of less importance in PBT, where mainly the V-2 classification is achieved [157]; some phosphate esters, such as bisphenol-A-biscresyl phosphate, showed reduced flammability in concentrations below 10% as well [158].

Naturally occurring metal hydroxides are not suitable for polyesters; however, inert inorganic synergists have been evaluated in some examples, e. g., aluminum diethyl phosphinate and metal oxides [146], montmorillonite in combination with ammonium polyphosphate (APP) and melamine cyanurate [159], surface modified carbon nanotubes [160], or modified silica [161].

Polyethylene terephthalate (PET) for engineering applications can be flame retarded in a similar way to PBT. In addition to the PBT formulations proposed, more recently, DOPO derivatives were described, e. g., cyclotriphosphazene derivatives [162], where the UL 94 V-0 (3.2 mm) classification for unreinforced PET is met with 5% loading. For PET fibers, compatible/soluble flame retardants as additives are preferred, e. g., DOPO containing polyesters are proposed [163, 164], which may be used as well in PBT [165]. Moreover, the state of the art in PET fibers is mainly the incorporation of flame-retardant monomers active during synthesis, e.g., 2-carboxyethyl methyl phosphinic acid, 2-carboxyethyl phenyl phosphinic acid, or the adduct of DOPO with itaconic acid [164]. A further group of efficient polymeric additives in PET is the class of aryl polyphosphonates [166]; UL 94 V-0 can be achieved with 5% loading.

Polylactic acid, an upcoming aliphatic polyester from natural resources, is an area of research including flame-retarded formulations. Contrary to aliphatic-aromatic polyesters, the processing temperatures are much lower, around 200 °C, where the hydrolytic and processing instability has to be considered when choosing appropriate flame retardants. Combustion energy and total heat release rate of PLA is much higher compared to PET and PBT [167]. Classic intumescent systems such as APP meet UL 94 V-2 at 30% loading [168]. Therefore, many combinations have been tested to improve intumescent systems to UL 94 V-0, for example, combinations of APP and chitosan [169], of APP, melamine and montmorillonite [170], of APP and expandable graphite [168], layered double hydroxides [171], or lignin [172]. As a PLA specific topic, flame retardants from natural resources have been incorporated such as phytic acid [172, 173]; however, with 20% phytic acid only UL 94 V-2 is achieved [172].

For other polyesters from natural resources, such as polybutylene succinate (PBS), polyhydroxybutyrate, and polyhydroxyvalerate, only very scarce literature exists: melamine phosphate, melamine phosphite, and melamine hypophosphite were studied in PBS, where with 30% loading, UL 94 V-0 (3 mm) was obtained [174].

3.2.2.10 Polycarbonate and Polycarbonate Blends

Primary degradation products of the thermal degradation of polycarbonate (PC) are carbon dioxide and bisphenol-A accompanied by molecular rearrangement processes resulting in extensive cross-linking and char formation [175]. Thus, PC is usually self-extinguishing in UL 94 testing (V-2). The addition of flame retardants is only mandatory for severe requirements.

Some of the most efficient flame retardants in polycarbonate are organic metal salts, namely potassium perfluorobutane sulfonate, which is used in the range of 0.05–0.5% [176], often in combination with an anti-dripping agent, e.g., PTFE for smaller thicknesses of test pieces. Transparency of PC is almost not influenced through the low concentrations to be used. The sulfonate acts in a catalytic way by accelerating the char formation of polycarbonate and through enhanced cross-linking [176]. Therefore, it is not surprising that other alkali metal salts such as lithium or the sodium salt of perfluorobutane sulfonate act similarly as well as fluorine-free sulfonates, e.g., potassium-*p*-toluene sulfonate or potassium diphenylsulfone sulfonate [177]. New organic metal salts are still a research topic, e.g., 1,3,5,7-terakis(phenyl-4-sodium sulfonate) adamantane [178] or phosphonium sulfonates [179].

Brominated flame retardants are used in the range of 10% to achieve UL 94 V-0, e.g., phenoxy-terminated carbonate oligomers from tetrabromobisphenol-A or brominated polystyrene.

Phosphorus-based flame retardants may be too volatile and, due to transesterification reactions, too reactive to be used in polycarbonates, whereas oligomeric aromatic phosphate esters can reduce the negative impact. On the other hand, due to a plasticizing effect, polycarbonate processing can be improved. Combinations of phosphazenes with phosphate esters have been investigated as well; however, 14.5% were necessary to pass UL 94 V-0 [180].

Phosphates such as bisphenol-A bis(diphenyl phosphate) or resorcinol bis(diphenyl phosphate) are the domain of flame-retarded PC/ABS blends. Around 10% of phosphate esters are used at 20% ABS blend content to achieve UL 94 V-0 in the presence of 0.5% PTFE as an anti-dripping agent. Resorcinol bis(diphenyl phosphate) with 8% loading meets UL 94 V-0 (1.6 and 3.2 mm) [181]. Increasing the ABS component needs a higher FR concentration. The phosphate structure influences the decomposition, the activity in the gas and condensed phase, and the ability of transesterification and cross-linking reactions. Although V-0 is achieved with the same phosphorus concentration (corresponding to 10.3–13.8% of flame retardant), it seems that steric hindrance like in resorcinol bis(dixylyl phosphate) leads to a somewhat reduced performance [182]. Alternative phosphorus-containing flame retardants are cyclic, polymeric, cross-linked [183], or substituted phosphazenes [184].

Today, brominated flame retardants for PC/ABS blends can be polymeric, such as polybrominated polyphenylene ethers, but also of low molecular weight, such as ethylene bis(tetrabromophthalimide), in concentrations around or just below 10% for UL 94 V-0.

Besides PC/ABS, flame-retarded PC/PBT and PC/PET blends have industrial relevance and can be flame retarded more easily compared to PC/ABS blends. For example, PC/PET is flame retarded with 4% resorcinol bis(diphenyl phosphate) [185] to pass UL 94 V-0. PC/PBT blends require only a minor loading increase compared to PC/PET, e.g., 5% hydroquinone bis(diphenyl phosphate) in combination with 0.5% PTFE anti-dripping agent [186]. However, control of the morphology of PC/polyester blends plays an additional role in flame retardancy [187].

3.2.2.11 Polyurethanes

Thermoplastic polyurethanes (TPU) degrade thermally at temperatures between 200 and 300 °C into their monomers, i.e., isocyanates and alcohols, or by the formation of amines with the elimination of CO_2, and, moreover, of heterocyclic nitrogen-containing molecules. At higher pyrolysis temperatures, polycyclic aromatics are formed [188]. The individual products depend on the monomer composition

(e. g., diisocyanate type, polyetherpolyol, or polyesterpolyol). In consideration of the low degradation temperature with a relatively low LOI of 23% [189], the flame retardancy of TPU is quite challenging and many potential flame retardants have been tested [190, 191].

Inorganic flame retardants, i. e., aluminum trihydroxide (ATH) or magnesium dihydroxide (MDH) are used in TPU cable formulations; however, they require a loading between 70 and 80% for UL 94 V-2 [192]. Zinc borate is often used as a co-component to improve the electrical properties [193]. A possibility to reduce the high loading of ATH or MDH is the combination with phosphates, which act in parallel as plasticizers [194].

Ammonium polyphosphate (APP) is another suitable flame retardant and has been used in many combinations, e. g., together with resorcinol bis(diphenyl phosphate), borates, or montmorillonite [191]; around 40% of flame retardant combinations are needed to pass UL 94 V-0 (1 mm).

Further phosphorus-based flame retardants for thermoplastic polyurethanes are red phosphorus [195], phosphaphenanthrene oxide (DOPO) [196], and phosphinates such as aluminum diethyl phosphinate [191, 197–199]; around 20% phosphinates and 15% synergistic co-components are applied.

Out of the class of nitrogen flame retardants, melamine cyanurate is an option, often utilized in conjunction with phosphates [200], with melamine polyphosphate [201], or aluminum diethyl phosphinate [202]; loadings are above 30%.

A potential new approach for thermoplastic polyurethanes is the use of radical generators or combinations with phosphorus-containing flame retardants, where concentrations below 20% allow them to pass UL 94 V-0 (1.6 mm) [39, 203].

3.2.2.12 Polyvinylchloride

Due to its high chlorine content, rigid PVC is self-extinguishing when the ignition source is removed. Furthermore, in case of fire, the heat release rate is lower than that of most polymers [204]. During pyrolysis and thermooxidation experiments, HCl and aromatics such as benzene, toluene, chlorobenzene, and divinylbenzene are found [205], molecules which contribute extensively to smoke formation. Therefore, smoke suppressants based on molybdates, borates, stannates, and antimony oxide (ATO) have commercial relevance for PVC applications in addition to flame retardants (see Section 5.2).

The flammability of plasticized PVC depends strongly on the type, volatility, and concentration of the plasticizer. Plasticizers of low flammability such as phosphates, brominated phthalates, or chlorinated or brominated paraffins allow for the same low flammability as rigid PVC [206, 207].

As inorganic flame retardants, aluminum trihydroxide (ATH), magnesium dihydroxide (MDH), and combinations thereof are used, for example, in cable formulations, where loadings around 20% in combination with smoke suppressants are applied [206, 207]. Chlorinated/brominated paraffins as plasticizers and flame retardants are in the same range. Aryl phosphates such as diphenylcresyl phosphate are sometimes formulated in combination with other plasticizers up to 20%. More recently, phosphate or even phosphaphenanthrene oxide (DOPO)-modified plasticizers based on natural resources like soybean oil [208, 209] or castor oil [210] were proposed.

3.2.2.13 Other Halogenated Polymers

Highly chlorinated polymers such as polyvinylidene chloride (PVDC) or fluorinated polymers such as polyvinylidene fluoride (PVDF) or polytetrafluoroethylene (PTFE) are non-flammable and flame retardants by themselves. A rating of UL 94 V-0 is achieved through all major fluorinated polymers such as PTFE, FEP, PFA, ETFE, ECTFE, or PVDF [211]. Fluorinated polymers with high fluorine concentrations show a LOI > 95, i.e., these burn only in pure oxygen. Finely dispersed PTFE powder is often used as an anti-dripping agent in concentrations around 0.5% (0.01–1%), mainly in combination with brominated flame retardants. The anti-dripping behavior of PTFE is explained through fibrillation and enhanced shear viscosity and elasticity of the polymer melt [212, 213]. Moreover, PVDF can be used as a blend component to improve flame retardancy as demonstrated for ABS, preferably in combination with halloysite nanotubes [214].

3.2.2.14 High-Performance Polymers

High-performance polymers of a larger scale include polyetheretherketones (PEEK), polysulfones (PSU), polyethersulfones (PES), polyphenylene sulfides (PPS), polyetherimides (PEI), polyamideimides (PAI), polyimides (PI), liquid crystal polymers (LCP), and polyphenylene oxide (PPO). In comparison to most standard polymers, these mainly aromatic polymers are usually seen as self-extinguishing and ranked UL 94 V-0 [215, 216] without any added flame retardants with the exception of PSU and PPO blends. Also, due to high processing temperatures between 300 and 400 °C, many flame retardants are excluded due to insufficient thermal stability.

PEI shows, depending on its structure, a LOI value between 47 and 54, and achieves, for example, all ratings like UL 94 V-0 at 0.64 mm, 5V at 1.5 mm, and German DIN 4102 B1 [215]. Molybdic oxide is active as a smoke suppressant [217]. PAI with a LOI of 45% or above is ranked UL 94 V-0 (1.2 mm) [218].

The flame retardancy of PEEK with a LOI of 37.3% is influenced in the presence of fillers; however, UL 94 V-0 (3 mm) is always reached [219]. The addition of 0.5% of carbon nanotubes shows a minor improvement in flame retardancy [220]. Another way to reduce flammability is the addition of perfluorinated sulfonate salts [221].

PPS with a LOI of 44 is classified as V-0 at 0.8 mm [215].

Polyimides and LCPs cover a range of materials with different structures; thus, the flammability is structure dependent. Inorganic additives such as silica, graphite, and aluminum oxide increase the char through a ceramifying strategy [222, 223]. In a similar way, char formation is increased in the presence of halloysite nanotubes in polyethersulfones [224].

The classification of polysulfones (PSU) depends heavily on the test sample size: 1.6 mm thickness is not classified, 3.2 mm results in UL 94 V-2, and 4.5 mm in V-0. Reduced flammability is claimed to be achieved through the addition of perfluorinated sulfonate salts [221].

With a LOI of 29, PPO as such is ranked as self-extinguishing. The commercially important applications are blends with HIPS, which require the addition of flame retardants. Phosphates such as resorcinol bis(diphenyl phosphate) are efficient at 20% concentration for UL 94 V-0 at 1.6/3.2 mm [181]. Combinations with mica were shown to give some positive additional effects [225]. In a PPO/HIPS 40:60 blend, brominated polycarbonate or brominated phosphate achieved UL 94 V-0 (1.6 mm) either with 12% end-capped brominated polycarbonate plus 2.5% antimony trioxide (ATO), with 10.7% brominated phosphate, or with 13% triphenyl phosphate [226].

3.2.3 Thermosets

Most thermoset materials can be flame retarded by additives such as fillers or by reactive flame-retarded molecules, which are incorporated in the polymer network during curing or through combinations of both strategies, which offer a larger choice of potential formulations.

3.2.3.1 Epoxy Resins

The degradation and combustion processes of epoxides depend on the structure of the epoxide (e. g., aliphatic or aromatic), the hardener (e. g., amine, anhydride), and the cross-linking density [227]. In general, flammability is reduced with increasing aromatic content and increasing cross-linking density. Both effects lead to enhanced charring. Because of the importance of flame-retarded epoxy resins in electronic applications, e. g., as printed circuit boards, and in transport applications, e. g., as composite structures, epoxy resins are one of the primary fields, where novel flame retardants and novel flame-retardant classes are evaluated. Thus, the literature on flame-retarded epoxy resins is extensive [130, 227–230]. Preferred commercial solutions are brominated systems and phosphaphenanthrene oxide (DOPO)-based alternatives (Table 3.9).

Table 3.9 Representative Flame Retardant Formulations for Epoxy Resins

Type of flame retardants	Examples	Synergists	Usual flame retardant loading range for epoxy resins (%)
Inorganics	Mainly as a synergist: $Al(OH)_3$, AlO(OH)		
Halogens	Reactive brominated compounds: TBBA	Sb_2O_3, zinc stannate	20–30
Phosphorus P/N Intumescent	Phosphinates (e.g., aluminum diethyl phosphinate) Phosphonates Melamine polymetal phosphates	Melamine polyphosphate, hydroxides	15–25
Phosphorus	Phosphaphenanthrene oxide derivatives (DOPO, DOPO-HQ)		10–15

The dominant reactive flame retardant in epoxy resins is tetrabromobisphenol-A. In addition, the diglycidylether of tetrabromobisphenol-A may be used. The brominated non-reactive additive of choice is decabromodiphenyl ethane. The bromine content of an epoxy resin to pass UL 94 V-0 is about 20%.

The preferred halogen-free flame retardant alternative for electronic applications is 9,10-dihydro-9-oxa-10-phosphaphenanthrene-10-oxide (DOPO), which is incorporated in the epoxy resin network through reaction of the P-H group with the epoxide ring. In order to keep the reactive functionality, 2-(6-oxido-6*H*-dibenz[c,e][1,2]oxaphophorin-6-yl)-1,4-benzenediol (DOPO-HQ) is proposed. Roughly 10–15% of DOPO or DOPO derivatives, corresponding to a 1.5–2% phosphorus content, are applied to meet UL 94 V-0, depending on the epoxy resin and the hardener. Apart from these well-established commercial DOPO-based flame retardants, a large number of DOPO derivatives have been synthesized benefitting from the reactivity of the P-H group and, therefore, accessibility to complex structures. For example, it is possible to introduce other flame-retardant active molecules via reaction with Schiff bases. Through this process and the introduction of piperidine groups [231] or phenolic groups [232], which act additionally as co-curing agents, LOI (from 23.8 to 32.8%) and flame retardancy (UL 94 from not classified to V-0 (3 mm)) can be enhanced and the overall phosphorus content reduced.

Further non-reactive flame retardant additives in epoxy resins are metal phosphinates such as aluminum diethyl phosphinate or zinc diethyl phosphinate, melamine polyphosphate, or melamine poly(metal phosphates), where a loading of around 20% is used [233]. Moreover, synergistic effects are reported with various combinations of phosphorus derivatives and inorganic compounds such as boehmite [229] or silicium derivatives [230, 234]. As an exception, a strong antagonistic effect is found for DOPO in combination with magnesium dihydroxide (MDH) [235]. Furthermore, red phosphorus and even ammonium polyphosphate (APP) in combination with montmorillonite can be used with 10% of the combination to achieve V-0 at 3.2 mm [236].

Finally, as an example of a reactive oligomeric phosphonate, poly(1,3-phenylene methylphosphonate) is mentioned. There, 20–30% is needed to pass V-0 at 1.6 mm [237].

3.2.3.2 Polyurethanes

The main application for thermoset polyurethanes is rigid or flexible foam, e.g., for furniture, building, or transport applications. Thermal decomposition of polyurethane foams results in the dissociation of the urethane linkage and formation of amines through CO_2 elimination; combustion in air leads, in addition, to CO_2, H_2O, and to the formation of various CN-containing structures such as HCN with nearly no char formation. Smoke density is quite high [238, 239]. The decomposition process depends as well on the monomers (hard, flexible segments, aromatic, aliphatic structures), the cross-linking density, and the foam structure (cell dimensions). Flame retardants in use are either additives or reactive co-components mainly in the form of alcohols.

Classic flame retardants for rigid and flexible foams are tris(2-chloroethyl) phosphate or tris(1-chloro-2-propyl) phosphate; around 15% of the flame retardants are applied in formulations [240]. Nearly all readily available flame retardants have been investigated and are used in selected polyurethane foam applications in the form of additives including phosphorus (red phosphorus, phosphates such as triphenyl phosphate, phosphonates such as dimethyl propyl phosphonate, ammonium polyphosphate (APP), phosphaphenanthrene oxide (DOPO derivatives)), nitrogen derivatives (melamine, melamine cyanurate, melamine polyphosphate), polymeric silicone derivatives, and inorganic flame retardants (e.g., aluminum trihydroxide (ATH)). For example, melamine polyphosphate is a somewhat more efficient flame retardant than melamine cyanurate; a significant influence on burning behavior is found at a loading of 10% or above [241]. With microencapsulated melamine polyphosphate, UL 94 V-0 (3 mm) is achieved at a loading of 30% [242]. Ammonium polyphosphate (APP) and microencapsulated APP to reduce hydrolytic cleavage and water leaching achieve UL 94 V-0 (8 mm) at loadings between 14 and 17%. DOPO phosphonamidates show a UL 94 HBF rating of HF-1 at concentrations as low as 5% [244].

Reactive alcohols, which are incorporated into the polyurethane network, are, for example, tris(tribromoneopentyl) alcohol, dibromoneopentyl glycol, tetrabromophthalic acid derivatives, or diethyl-*N,N'*bis(2-hydroxyethyl) aminomethylphosphonate. In addition, further phosphorus-containing reactive flame retardant polycondensation products with ethylene glycol and phosphoric acid ester units have to be mentioned; depending on the application, loadings between 5 and 30% are used.

3.2.3.3 Phenolic Formaldehyde Resins

Phenolic formaldehyde resins are, in general, less flammable than most standard polymers due to a high content of aromatic structures, leading to substantial char formation during combustion despite a comparatively low thermal stability. Moreover, smoke formation is very low at least in the presence of glass cloth [245]. Due to this inherent flame retardancy, phenolic formaldehyde resins have been used as flame retardant components in other polymers, e.g., in ABS [246], unsaturated polyesters [247], or epoxy resins [248]. To improve flame retardancy (and thermal stability), active molecules such as phosphoric acid or boric acid can be introduced during the polycondensation stage. Alternatively, brominated phenol and brominated phenol phosphoric acids were shown to increase the LOI from 25/35 (linear/cured phenolic resin) up to 48/56% [249]. Introduction of heterocyclic structures such as melamine improve flame retardancy too [250]. As non-reactive co-additives, phosphorus compounds such as resorcinol bis(diphenyl) phosphate, bisphenol-A-bis(diphenyl) phosphate or phosphaphenanthrene oxide (DOPO) were claimed [247].

3.2.3.4 Unsaturated Polyesters

Resins known as unsaturated polyesters (UPE) usually contain a large amount of styrene as a diluent and cross-linking agent; therefore, the combustion process results in the formation of monomeric styrene [251]. For this reason, UPE resins burn easily with a LOI around 19. As one of the important aspects of UPE resins is their low cost, suitable flame retardants have to be low-cost, too. Therefore, aluminum hydroxide (ATH) is often the additive of choice, where about 55% is used to achieve UL 94 V-0 (3.2 mm) [252]. The same concentration range is needed for APP in combination with triphenyl phosphate [253]; loading can be reduced by adding expandable graphite [254]. Brominated flame retardants used are decabromodiphenyl ethane or brominated epoxy polymer in combination with antimony trioxide (ATO); UL 94 V-0 is met with loadings of around 20% of brominated additive and 5% ATO. Alternatively, dibromoneopentyl glycol, any halogen-substituted anhydrides, or diols can be used as flame retardant reactive components in the synthesis of unsaturated polyesters. Furthermore, flame retardants containing unsaturated groups are added in the curing/polymerization stage of UPEs to be incorporated in the formed network. An example of the latter approach is 1-oxo-2,6,7-trioxa-1-phosphabicyclo-[2.2.2]-octane-methyl diallyl phosphate [255].

3.2.4 Elastomers

Flame-retarded elastomers are partially covered in Section 3.2.2.3: Other Polyolefins (EPDM and EVA) and in Section 3.2.2.11: Thermoplastic Polyurethanes. Elastomers covered here comprise natural rubber (NR), chlorinated rubber (CR), acrylonitrile-butadiene (NBR), styrene-butadiene copolymers (SBR), and silicone rubber (MQ). To draw general trends and to elaborate the performance of flame retardants in elastomers is much more challenging than with thermoplastics and thermosets, as elastomers contain many different ingredients sometimes in high concentrations and the vulcanization process influences the overall properties.

It seems that research on improving the flame retardancy of elastomers is focused on the use of mineral fillers such as aluminum trihydroxide (ATH), layered double hydroxides (LDH), and nanocomposite structures such as carbon nanotubes, montmorillonite, halloysites, and others [256–258]. For example, the LOI of natural rubber in a model formulation is increased from 18.2% to 26.2% in the presence of 70% ATH (relative to NR rubber) and UL 94 V-1 (3 mm) is achieved. To reach UL 94 V-0, a combination of ATH (50%) and chlorinated paraffin (20%) on top is necessary [259]. A halogen-free alternative is the combination of 65% ATH and 5% organoclay, where a LOI of 28% and UL 94 V-0 can be realized [259]. Brominated flame retardants such as decabromodiphenyl ethane are used in combination with antimony oxide (ATO) (ratio 3:1) in total concentrations above 25% in NR.

Through the presence of chlorine in its molecule, polychloroprene (CR) is inherently less flammable (LOI 32%) [261] than NR. Thus, CR rubber achieves UL 94 V-0 (1.75 mm) with 5% magnesium dihydroxide (MDH) in the presence of zinc borate and ATO [260]. In addition, a combination of 8–10% ammonium sulfamate and 3–5% ATO increases the LOI up to 44% [261].

A number of investigations have been carried out with NBR rubber in the presence of nanomaterials [256, 257], where mainly LOI values and cone data have been published. For example, the addition of organically modified montmorillonite to NBR (22% acrylonitrile) may increase the LOI from 21% up to 37% at 5% loading depending on the cross-linking either through peroxide or sulfur [262].

Like in other elastomers, the flame retardancy of styrene butadiene rubber (SBR) is improved through the addition of minerals such as ATH or MDH in combination with chlorinated paraffins [257] or through the introduction of nanoclays. An improved performance is found for combinations of MDH with organically modified montmorillonite proven through an increase of the LOI from 19% to 25% (30% MDH and 20% montmorillonite). Intumescent systems may be an alternative approach as shown through the combination of encapsulated epoxy or melamine-formaldehyde-encapsulated ammonium polyphosphate (APP) in combination with pentaerythritol (PER). UL 94 V-0 is found for a combination of 30% APP and 10% PER in SBS rubber [264].

Silicone rubbers possess better flame retardant properties than carbon-based polymers as the formation of SiO_2, which acts as a barrier layer, can take place during combustion. The overall heat release rate of silicone rubbers is comparatively low [265]. Thus, silicones are introduced as flame retardants in other polymers. The strategy to improve the flame retardancy of silicone is to add fillers to enhance the barrier formation with components such as silica, calcium carbonate, wollastonite, mica, and others [265]. In addition, organic flame retardants have been proposed, such as phosphaphenanthrene oxide (DOPO)-modified silicones [266], or functionalized zirconium phosphate as an additive. With only 4% used, the latter shifts a silicone rubber from UL 94 not classified to V-0 [267].

References for Section 3.2

[1] Irvine, D.J., McCluskey, J.A., Robinson, I.M., *Polym. Degrad. Stab.* (2000) 67, pp. 383–396.

[2] Woolley, W.D., Ames, S.A., Fardell, P.J., *Fire Mater.* (1979) 3, pp. 110–120.

[3] Hull, T.R., Witkowski, A., Hollingbery, L., *Polym. Degrad. Stab.* (2011) 96, pp. 1462–1469.

[4] EP 0792911, inv.: Tennesen, C.A., assigned to FMC Corporation, pri. 27.02.1996.

[5] WO 2004035673 inv.: Kaprinidis, N., Lelli, N. (2004) assigned to Ciba Specialty Chemicals Holding Inc., pri. 17.10.2002/13.01.2003.

[6] WO 2004035671, inv.: Kaprinidis, N., Lelli, N. (2004) assigned to Ciba Specialty Chemicals Holding Inc., pri. 17.10.2002.

[7] WO 2015123042, Inv.: Assor, Y., Melamed, L., Eden, E., Leifer, M., Levchik, S.V. (2015) assigned to ICL-IP America Inc., pri. 12.02.2014.

[8] N.N., FRX Polymers' Nofia flame retardants find new application as antimony synergist replacement, *Additives for Polymers* (2017) 10, 2.

[9] Lim, K.S., Bee, S.T., Sin, L.T., Tee, T.T., Ratnam, C.T., Hui, D., Rahmat, A.R., *Compos., Part B* (2016) 84, pp. 155–174.

[10] Wu, K., Zhang, Y., Hu, W., Lian, J., Hu, Y., *Compos. Sci. Technol.* (2013) 81, 17–23.

[11] Gao, S., Zhao, X., Liu, G., *J. Appl. Polym. Sci.* (2017), p. 134.

[12] Chen, W., Yuan, S., Sheng, Y., Liu, G., *J. Appl. Polym. Sci.* (2015) 132, pp. 41214/1-41214/8.

[13] Zhang, M., Ding, P., Qu, B., *Polym. Compos.* (2009) 30, pp. 1000–1006.

[14] Ding, P., Tang, S., Yang, H., Shi, L., *Adv. Mater. Res. (Zurich, Switzerland)* (2010) 87–88, pp. 427–432.

[15] Yi, D., Yong, R., *J. Appl. Polym. Sci.* (2010) 118, pp. 834–840.

[16] Wen, P., Wang, D., Liu, J.J., Zhan, J., Hu, Y., Yuen, R.K.K., *Polym. Adv. Technol.* (2017) 28, pp. 679–685.

[17] Huang, N.H., Chen, Z.J., Wang, J.Q., Wei, P., *eXPRESS Polym. Lett.* (2010) 4, pp. 743–752.

[18] Kaynak, E., Ureyen, M.E., Koparal, A.S., *e-Polym.* (2017) 17, pp. 341–348.

[19] Ren, Q., Zhang, Y., Li, J., Li, J.C., *J. Appl. Polym. Sci.* (2011) 120, pp. 1225–1233.

[20] Li, N., Xia, Y., Mao, Z., Wang, L., Guan, Y., Zheng, A., *J. Polym. Sci.* (2012) 97, pp. 1737–1744.

[21] Dogan, M., Yilmaz, A., Bayramli, E., *Polym. Degrad. Stab.* (2010) 95, pp. 2584–2588.

[22] Su, X., Yi, Y., Tao, J., Qi, H., *Polym. Degrad. Stab.* (2012) 97, pp. 2128–2135.

[23] Zhang, Y., Li, X., Fang, Z., Hull, T.R., Kelarakis, A., Stec, A.A., *Polym. Degrad. Stab.* (2017) 136, pp. 139–145.

[24] Wu, N., Yang, R., *Polym. Adv. Technol.* (2011) 22, pp. 495–501.

[25] Shao, Z.B., Deng, C., Tan, Y., Chen, M.J., Chen, L., Wang, Y.Z., *Polym. Degrad. Stab.* (2014) 106, pp. 88–96.

[26] Shao, Z.B., Deng, C., Tan, Y., Yu, L., Chen, M.J., Chen, L., Wang, Y.Z., *J. Mater. Chem A* (2014) 2, pp. 13955–13965.

[27] Wen, P., Tai, Q., Hu, Y., Yuen, R.K.K., *Ind. Eng. Chem. Res.* (2016) 55, pp. 8018–8024.

[28] Ye, X., Wang, Y., Zhao, Z., Yan, H., *Polym. Degrad. Stab.* (2017) 142, pp. 29–41.

[29] Xu, M., Wang, J., Ding, Y., Li, B., *Chin. J. Polym. Sci.* (2015) 33, pp. 318–328.

[30] Yan, H., Zhao, Z., Ge, W., Zhang, N., Jin, Q., *Ind. Eng. Chem. Res.* (2017) 56, pp. 8408–8415.

[31] Xie, H., Lai, X., Zhou, R., Li, H., Zhang, Y., Zeng, X., Guo, J., *Polym. Degrad. Stab.* (2015) 118, pp. 167–177.

[32] Kaul, B.L., *Rubber Fibres Plastics International* (2016) 11, pp. 190–196.

[33] Xie, H., Lai, X., Li, H., Zeng, X., *Polym. Degrad. Stab.* (2016) 130, pp. 68–77.

[34] Li, H., Liu, J., Zhao, W., Wang, X., Wang, D., *J. Fire Sci.* (2016) 34, pp. 104–119.

[35] Hoerold, S., Phosphorus-based and intumescent flame retardants in *Polymer Green Flame Retardants,* Papaspyrides, C.D., Kiliaris, P. (Ed.) (2014), pp. 221–254.

[36] WO 2010146033 inv.: Butz, V. (2010) assigned to Thor GmbH, pri. 16.06.2009.

[37] Wilen, C.E., Pfaendner, R., Design and utilization of nitrogen containing flame retardants based on N-alkoxyamines, azoalkanes and related compounds in *Polymer Green Flame Retardants,* Papaspyrides, C.D., Kiliaris, P. (Ed.) (2014), pp. 267–288.

[38] WO 2015067736, inv.: Wilen, C.E., Aubert, M., Tierri, T., Pawelec, W. (2015) pri. 14.05.2015.

[39] WO 2014154636, inv.: Pfaendner, R., Metzsch-Zilligen, E., Stec, M. (2014) assigned to Fraunhofer Gesellschaft, pri. 25.03.2013.

[40] WO 2015180888, inv.: Pfaendner, R., Mazurowski, M. (2015) assigned to Fraunhofer Gesellschaft, pri. 28.05.2014.

[41] WO 2016042043, inv.: Pfaendner, R., Mazurowski, M. (2016) assigned to Fraunhofer Gesellschaft, pri. 18.09.2014.

[42] WO 2016042040, inv.: Pfaendner, R., Mazurowski, M. (2016) assigned to Fraunhofer Gesellschaft, pri. 18.09.2014.

[43] Schartel, B., Uses of fire tests in materials flammability development in *Fire retardancy of polymeric materials,* 2nd edition, Wilkie, C.A., Morgan, A.B. (Ed.) (2010) Boca Raton, pp. 387–420.

[44] Ueno, T., Nakashima, E., Takeda, K., *Polym. Degrad. Stab.* (2010) 95, pp. 1862–1869.

[45] Font, R., Aracil, I., Fullana, A., Conesa, J.A., *Chemosphere* (2004) 57, pp. 615–627.

[46] Sauerwein, R., Mineral filler flame retardants in *Non-halogenated flame retardant handbook,* Morgan, A.B., Wilkie, C.A. (Ed.) (2014) Hoboken, NJ, pp. 75–141.

[47] Weil, E.D., Levchik, S.V., *J. Fire Sci.* (2008) 26, pp. 5–43.

[48] WO 2013085788, inv.: Levchik, S.V., Moy, P., Alessio, G.R., Shawhan, G., Innes, J. (2013) assigned to ICL-IP America Inc., pri. 09.12.2011/12.06.2012.

[49] Cross, M.S., Cusack, P.A., Hornsby, P.R., *Polym. Degrad. Stab.* (2003) 79, pp. 309–318.

[50] Wilkie, C.A., Zhang, J., *Polym. Adv. Technol.* (2005) 16, pp. 549–553.

[51] Beyer, G., Flame retardancy from polymer nanocomposites – from research to technical products in *Industry Guide to polymer nanocomposites,* Beyer, G. (Ed.) (2009) Bristol, pp. 159–199.

[52] Beyer, G., *Proceedings of the Annual Conference on Recent Advances in Flame Retardancy of Polymeric Materials 22nd* (2012) pp. 293–301.

[53] Du, M., Guo, B., Jia, D., *Polym. Int.* (2010) 59, pp. 574–582.

[54] Zhang, G., Ding, P., Zhang, M., Qu, B., *Polym. Degrad. Stab.* (2007) 92, pp. 1715–1720.

[55] Beyer, G., *Fire Mater.* (2002) 26, pp. 291–293.

[56] Gong, J., Niu, R., Tian, N., Chen, X., Wen, X., Liu, J., Sun, Z., Mijowska, E., Tang, T., *Polymer* (2014) 55, pp. 2998–3007.

[57] Han, Z., Wang, Y., Dong, W., Wang, P., *Mat. Lett.* (2014) 128, pp. 275–278.

[58] WO 2013045965, inv.: Zucchelli, U. (2013) assigned to Italmatch Chemicals S.p.A., pri. 28.09.2011.

[59] WO 2013116283, inv.: Levchik, S.V., Zilberman, J., Desikan, A.N., Alessio, G.R. (2013) assigned to ICL-IP America Inc., pri. 01.02.2012.

[60] Wang, Z., Qu, B., Fan, W., Huang, P., *J. Appl. Polym. Sci.* (2001) 81, pp. 206–214.

[61] Moore, M., New solutions for flame -retarded polyolefins, presentation *Fire resistance in plastics* (2014) Cologne, December 9–12.

[62] Weil, E.D., Levchik, S.V., *J. Fire Sci.* (2008) 26, pp. 5–43.

[63] US 6737456 inv.: Bar-Yakov, Y., Hini, S. (2004) assigned to Bromine Compounds Ltd., pri. 27.02.2001.

[64] Flame retardants for superior, cost-effective performance, product brochure, Dover Chemical Corporation, June 2013, accessible at *www.doverchem.com/Portals/0/FR.pdf*, accessed on 12.03.2018.

[65] Bourbigot, S., Le Bras, M., Ducquesne, S., Rochery, M., *Macromol. Mat. Eng.* (2014) 289, pp. 499–511.

[66] Alongi, J., Poskovic, M., Visakh, P.M., Frache, A., Malucelli, G., *Carbohydr. Polym.* (2012) 88, pp. 1387–1394.

[67] Zhao, J., Deng, C.L., Du, S.L., Chen, L., Debg, C., Wang, Y.Z., *J. Appl. Polym. Sci.* (2014) 131, p. 40065.

[68] Hu, X., Li, Y.L., Wang, Y.Z., *Macromol. Mat. Eng.* (2004) 289, pp. 208–112.

[69] ADK STAB FP-2100J, Product data sheet, Adeka Palmarole, 24/07/2007, accessible at *http://www.adeka-pa.eu/applications/building-construction-furniture/wood-plastic-composites/polyolefins/flame-retardantswww.adeka-palmarole.com/* accessed on 12.03.2018.

[70] US 6733697 inv.: Rhodes, M.S., Izrailev, L., Tuareck, J., Rhodes, P.S. (2004) pri. 22.07.2002.

[71] Tian, S., He, H., Wang, D., Yu, P., Jia, Y., Luo, Y., *Fire Mater.* (2017) 0, pp. 1-10.

[72] WO 2011117266, inv.: Xalter, R. (2011) assigned to BASF SE, pri. 25.03.2010.

[73] WO 2013136285, inv.: Lips, G., Huber, G., Hoppe, H., Le Gal, A. (2013) assigned to BASF SE, pri. 16.03.2012.

[74] Aubert, M., Tirri, T., Wilen, C.E., Francois-Heude, A., Pfaendner, R., Hoppe, H., Roth, M., *Polym. Degrad. Stab.* (2012) 97, pp. 1438–1446.

[75] Aubert, M., Nicolas, R.C., Pawelec, W., Wilen, C.E., Roth, M., Pfaendner, R., *Polym. Adv. Technol.* (2011) 22, pp. 1529–1538.

[76] Tirri, T., Aubert, M., Pawelec, W., Wilen, C.E., Pfaendner, R., Hoppe, H., Roth, M., Sinkkonen, J., *J. Appl. Polym. Sci.* (2014) 131, p. 40413.

[77] Pawelec, W., Holappa, A., Tirri, T., Aubert, M., Hoppe, H., Pfaendner, R., Wilen, C.E., *Polym. Degrad. Stab.* (2014) 110, pp. 447–456.

[78] Wang, Z., Gu, Z., Jiang, P., *Adv. Mater. Res.* (2012) 415–417, pp. 399–402.

[79] Tang, G., Hu, Y., Song, L., *Procedia Eng.* (2013) 62, pp. 371–376.

[80] Yen, Y.Y., Wang, H.T., Guo, W.J., *J. Appl. Polym. Sci.* (2013) pp. 2042–2048.

[81] Hoffendahl, C., Fontaine, G., Duquesne, S., Taschner, F., Mezger, M., Bourbigot, S., *Polym. Degrad. Stab.* (2015) 115, pp. 77–88.

[82] Szep, A., Szabo, A., Toth, N., Anna, P., Marosi, G., *Polym. Degrad. Stab.* (2006) 91, pp. 593–599.

[83] Vahabi, H., Ravesthian, A., Fasihi, M., Sonnier, R., Saeb, M.R., Dumarzet, L., Kandola, B.K., *Polym. Degrad. Stab.* (2017) 143, pp. 164–175.

[84] JP 2010037349 inv.: Matsumura, T., Okaniwa, M., assigned to JSR Corp. pri. 31.07.2008.

[85] JP 55082142, assigned to Matsushita Electric Industrial Co., pri. 18.12.1978.

[86] JP 2006328100, inv.: Saito, T., Kim, S.H., Kim, J.C., Park, J.H., assigned to Songwon Industrial Co., pri. 23.05.2005.

[87] JP 2012237019, inv.: Fujii, K., Fujiwara, K., Yoshida, K., Endo, K., assigned to Mitsui Chemicals Inc., pri. 10.09.2012.

[88] EP 131358, inv.: Yusawa, M., Igarash, C., assigned to Mitsubishi Petrochemical Co. Ltd., prio. 17.05.1983.

[89] Bourbigot, S., Gilman, J.W., Wilkie, C.A., *Polym. Degrad. Stab.* (2004) 84, pp. 483–492.

[90] Jang, B.N., Jung, I., Choi, J., *J. Appl. Polym. Sci.* (2009) 112, pp. 2669–2675.

[91] Liu, J., Yu, Z., Chang, H., Zhang, Y., Shi, Y., Luo, J., Pan, B., Lu, C., *Polym. Degrad. Stab.* (2014) 103, pp. 83–95.

[92] Liu, J., Zhang, Y., Peng, S., Pan, B., Lu, C., Liu, H., Ma, J., Niu, Q., *Polym. Degrad. Stab.* (2015) 121, pp. 261–270.

[93] Yan, Y.W., Huang, J.Q., Guan, Y.H., Shang, K., Jian, R.K., Wang, Y.Z., *Polym. Degrad. Stab.* (2014) 99, pp. 35–42.

[94] Yan, Y.W., Chen, L., Jian, R.K., Kong, S., Wang, Y.Z., *Polym. Degrad. Stab.* (2012) 97, pp. 1423–1431.

[95] Kiliaris, P., Papaspyrides, C.D., *Progr. Polym. Sci.* (2010) 35, pp. 902–958.

[96] Beach, M.W., Rondan, N.G., Froese, R.D., Gerhart, B.B., Green, J.G., Stobby, B.G., Shmakov, A.G., Shvartsberg, V.M., Korobeinichev, O.P., *Polym. Degrad. Stab.* (2008) 93, pp. 1664–1673.

[96a] Beach, M.W., Beaudoin, D.A., Beulich, I., Bloom, J.C., Davis, J.W., Hollnagel, H.M., Hull, J.W., King, B., Kram, S., Lukas, C., Matteucci, M., Morgan, T., Stobby, B., *Cell. Polym.* (2013) 32, pp. 229–236.

[97] WO 2007058736 inv.: King, B., Stobby, W.G., Murray, D.J., Worku, A.Z., Beulich, I., Tinetti, S.M., Hahn, S.F., Drumright, R.E. (2007) assigned to Dow Global Technologies Inc., pri. 12.11.2005.

[98] Wagner, J., Deglmann, P., Fuchs, S., Ciesielski, M., Fleckenstein, C.A., Döring, M., *Polym. Degrad. Stab.* (2016) 129, pp. 63–76.

[99] Wang, G., Chen, X., Liu, P., Bai, S., *J. Appl. Polym. Sci.* (2017), p. 44356.

[100] Crook, V., Ebdon, J., Hunt, B., Joseph, P., Wyman, P., *Polym. Degrad. Stab.* (2010) 95, pp. 2260–2268.

[101] Hall, M.E., Zhang, J., Horrocks, A.R., *Fire Mater.* (1994) 18, pp. 231–241.

[102] Feng, J., Carpanese, C., Fina, A., *Polym. Degrad. Stab.* (2016) 129, pp. 319–327.

[103] Hoang, D.Q., Kim, J., Jang, B.N., *Polym. Degrad. Stab.* (2008) 93, pp. 2042–2047.

[104] WO 2008149260, inv.: Huijs, F., Arnould, D. (2008) assigned to Sabic Innovative Plastics IP B.V., pri. 08.06.2007.

[105] Chang, T.C., Yu, P.Y., Hong, Y.S., Wu, T.R., Chiu, Y.S., *Polym. Degrad. Stab.* (2002) 77, pp. 29–34.

[106] Tewarson, A., *Fire Mater.* (1976) 1, pp. 90–96.

[107] Hsieh, F.Y., Hirsch, D.B., Beeson, H.D., *Fire Mater.* (2003) 27, pp. 9–17.

[108] Friederich, B., Laachachi, A., Ferriol, M., Cochez, M., Sonnier, R., Toniazzo, V., Ruch, D., *Polym. Degrad. Stab.* (2012) 97, pp. 2154–2161.

[109] Vahabi, H., Lin, Q., Vagner, C., Cochez, M., Ferriol, M., Laheurte, P., *Polym. Degrad. Stab.* (2016) 124, pp. 60–67.

[110] Laachachi, A., Cochez, M., Leroy, E., Ferriol, M., Lopez-Cuesta, J.M., *Polym. Degrad. Stab.* (2007) 92, pp. 61–69.

[111] Kim, S., Wilkie, C.A., *Polym. Adv. Technol.* (2008) 19, pp. 496–506.

[112] Qian, X., Jin, J., Lu, L., Shao, G., Jiang, S., *J. Polym. Res.* (2017) 24, p. 45.

[113] Krala, G., Ubowska, A., Kowalczyk, K., *Polym. Eng. Sci.* (2014) DOI 10.1002/pen.23644.

[114] US 20150368405 inv.: Wermter, H., Futterer, T., Vogt, R., Doering, M., Ciesielski, M. (2015) assigned to Chemische Fabrik Budenheim KG, pri. 14.02.2013.

[115] Harashina, H., Tajima, Y., Itoh, T., *Polym. Degrad. Stab.* (2006) 91, pp. 1996–2002.

[116] Sun, S., He, Y., Wang, X., Wu, D., *J. Appl. Polym. Sci.* (2010) 118, pp. 611–620.

[117] Zhang, Q., Chen, Y., *J. Polym. Res.* (2011) 18, pp. 293–303.

[118] Liu, Y., Wang, Z., Wang, Q., *J. Appl. Polym. Sci.* (2012) 125, pp. 968–974.

[119] Levchik, S. V., Weil, E. D., Lewin, M., *Polym. Int.* (1999) 48, pp. 523–557.

[120] Gijsman, P., Steenbakkers, R., Fürst, C., Kersjes, J., *Polym. Degrad. Stab.* (2002) 78, pp. 219–224.

[121] De Schryver, D., Landry, S. D., Reed, J. S., *Polym. Degrad. Stab.* (1999) 64, pp. 471–477.

[122] WO 2013131545 inv.: Voss, H. J., Ferner, U. (2013) assigned to Trovotech GmbH, pri. 07.03.2012.

[123] Isitman, N. A., Gunduz, H. O., Kaynak, C., *J. Fire Sci.* (2010) 28, pp. 87–100.

[124] WO 2015113740, inv.: Bauer, H., Hoerold, S., Sicken, M. (2015) assigned to Clariant International Ltd., pri. 29.01.2014.

[125] Rakotomalala, M., presentation, Fire resistance in plastics (2012) Köln, Nov. 27–29.

[126] Horrocks, A. R., Smart, G., Hörold, S., Wanzke, W., Schlosser, E., Williams, J., *Polym. Degrad. Stab.* (2014) 104, pp. 95–103.

[127] US 8541489, inv.: Yin, Y. (2013) assigned to E. I. du Pont de Nemours and Company, pri. 01.07.2008.

[128] WO 2014124990, inv.: Leistner, M., Pfaendner, R., Trupti, D., Koestler, H. G., Wehner, W. (2014) assigned to Floridienne S. A., pri. 13.02.2013/11.06.2013.

[129] Pfaendner, R., Doering, M., *Kunststoffe* (2014), 104(8), pp. 71–76.

[130] Wendels, S., Chavez, T., Bonnet, M., Salmeia, K. A., Gaan, S., *Materials* (2017) 10, p. 784.

[131] Wagner, S., Doering, M., *Materials* (2010) 3, p. 4300.

[132] Xie, M., Zhang, S., Ding, Y., Wang, F., Liu, P., Tang, H., Wang, Y., Yang, M., *J. Appl. Polym. Sci.* (2017), p. 44892.

[133] *http://www.chemanager-online.com/en/topics/chemicals-distribution/mineral-based-flame-retardants-engineering-plastics*, accessed on 12.03.2018.

[134] Lao, S. C., Wu, C., Moon, T. J., Koo, J. H., Morgan, A., Pilato, L., Wissler, G., *J. Comput. Mater.* (2009) 43, pp. 1803–1818.

[135] Battegazzore, D., Alongi, J., Fontaine, G., Frache, A., Bourbigot, S., Malucelli, G., *RSC Adv.* (2015) 5, pp. 39424–39432.

[136] Levchik, S. V., Weil, E. D., *Polym. Adv. Technol.* (2004) 15, pp. 691–700.

[137] Green, J., Chung, J., *J. Fire Sci.* (1990) 8, pp. 254–265.

[138] Suzanne, M., Ramani, A., Ukleja, S., McKee, M., Zhang, J., Delichatsios, M. A., Patel, P., Clarke, P., Cusack, P., *Fire Mater.* (2017) 1–10 DOI 10:1002/fam.2453.

[139] WO 2012088080, inv.: Shtekler, R., Yaakov, Y. B., Georlette, P., Levchik, S. V. (2012) assigned to ICL-IP America Inc., prio. 22.12.2010.

[140] US 5554674, inv.: Hamilton, D. G. (1996) assigned to General Electric Company, pri. 07.04.1995.

[141] JP11199784, inv.: Hironaka, K., Nishijima, K., assigned to Teijin Ltd, pri. 17.10.1989.

[142] JP 03131655, inv.: Yamauchi, K., Matsumoto, H., Inoue, S., assigned to Toray Ind., pri. 14.10.1998.

[143] Sullalti, S., Colonna, M., Berti, C., Fiorini, M., Karanam, S., *Polym. Degrad. Stab.* (2012) 97, pp. 566–572.

[144] Braun, U., Bahr, H., Sturm, H., Schartel, B., *Polym. Adv. Technol.* (2008) 19, pp. 680–692.

[145] Louisy, J., Vortrag Fachtagung "Trends im Brandschutz und innovative Flammschutzmittel für Kunststoffe", Würzburg 11./12.09.2013.

[146] Gallo, E., Braun, U., Schartel, B., Russo, P., Acierno, D., *Polym. Degrad. Stab.* (2009) 94, pp. 1245–1253.

[147] WO 2010080491, inv.: Wanzke, W., Hoerold, S., Lebel, M., Freitag, K.D. (2010) assigned to Clariant International Ltd, FRX Polymers Inc., pri. 08.01.2009.

[148] EP 2025710, inv.: Hoerold, S., Wanzke, W., assigned to Clariant International Ltd., pri. 06.08.2007.

[149] WO 2017097385, inv.: Voss, H.J. (2017) assigned to Trovotech GmbH, pri. 07.12.2015.

[150] Yang, W., Yang, B., Lu, H., Song, L., Hu, Y., *Ind. Eng. Chem. Res.* (2014) 53, pp. 18489–18496.

[151] Nazare, S., Hull, T.R., Biswas, B., Samyn, F., Bourbigot, S., Jama, C., Castrovinci, A., Fina, A., Camino, G., *Fire retardancy of polymers: new strategies and mechanisms*, Hull, T.R., Kandola, K. (Ed.) (2009), The Royal Society of Chemistry, Cambridge, UK, July 4–6, pp. 168–183.

[152] Hu, X., Wang, Y., Yu, J., Zhu, J., Hu, Z., *J. Appl. Polym. Sci.* (2017) DOI: 10.1002/APP.45537.

[153] WO 2012072739, inv.: Minder, E., Herbst, H. (2012) assigned to BASF SE, pri. 02.12.2010.

[154] WO 2005121232, inv.: Costanzi, S., Leonardi, M. (2005) assigned to Italmatch Chemicals S.P.A., pri. 10.06.2004.

[155] Yang, W., Song, L., Hu, Y., Lu, H., Yuen, R.K.K., *J. Appl. Polym. Sci.* (2011) 122, pp. 1480–1488.

[156] Chen, L., Luo, Y., Hu, Z., Lin, G.P., Zhao, B., Wang, Y.Z., *Polym. Degrad. Stab.* (2012) 97, pp. 158–165.

[157] Balabaovich, A.I., Zevaco, T.A., Schnabel, W., *Macromol. Mater. Eng.* (2004) 289, pp. 181–190.

[158] Ishikawa, T., Maki, I., Takeda, K., *J. Appl. Polym. Sci.* (2004) 92, pp. 2326–2333.

[159] Yang, W., Kan, Y., Song, L., Hu, Y., Lu, H., Yuen, R.K.K., *Polym. Adv. Technol.* (2011) 22, pp. 2564–2570.

[160] Yang, W., Zhou, H., Yang, B., Lu, H., Song, L., Hu, Y., *Polym. Compos.* (2016) 37, pp. 1812–1820.

[161] Courtat, J., Melis, F., Taulemesse, J.M., Bounor-Legare, V., Sonnier, R., Ferry, L., Cassagnau, P., *Polym. Degrad. Stab.* (2017) 143, pp. 74–84.

[162] Fang, Y., Zhou, X., Xing, Z., Wu, Y., *J. Appl. Polym. Sci.* (2017) DOI: 10.1002/APP.45246.

[163] Chang, S.J., Chang, F.C., *J. Appl. Polym. Sci.* (1999) 72, pp. 109–122.

[164] Levchik, S.V., Weil, E.D., *Polym. Int.* (2005) 54, pp. 11–35.

[165] Brehme, S., Schartel, B., Goebbels, J., Fischer, O., Pospiech, D., Bykov, Y., Döring, M., *Polym. Degrad. Stab.* (2011) 96, pp. 875–884.

[166] Chen, L., Wang, Y.Z., *Materials* (2010) 3, pp. 4746–4760.

[167] Okoshi, M., Thumsorn, S., Hamada, H., *Energy Procedia* (2016) 89, pp. 38–44.

[168] Zhu, H., Zhu, Q., Li, J., Tao, K., Xue, L., Yan, Q., *Polym. Degrad. Stab.* (2011) 96, pp. 183–189.

[169] Chen, C., Gu, X., Jin, X., Sun, J., Zhang, S., *Carbohydr. Polym.* (2017) 157, pp. 1586–1593.

[170] Lesaffre, N., Bellayer, S., Vezin, H., Fontaine, G., Jiminez, M., Bourbigot, S., *Polym. Degrad. Stab.* (2017) 139, pp. 143–164.

[171] Jin, X., Gu, X., Chen, C., Tang, W., Li, H., Liu, X., Bourbigot, S., Zhang, Z., Sun, J., Zhang, S., *J. Mater. Sci.* (2017) 52, pp. 12235–12250.

[172] Costes, L., Laoutid, F., Brhez, S., Delvosalle, C., Dubois, P., *Eur. Polym. J.* (2017) 94, pp. 270–285.

[173] Cheng, X.W., Guan, J.P., Tang, R.C., Liu, K.Q., *J. Cleaner Prod.* (2016) 124, pp. 114–119.

[174] Yang, H., Song, L., Tai, Q., Wang, X., Yu, B., Yuan, Y., Hu, Y., Yuen, R.K.K., *Polym. Degrad. Stab.* (2014) 105, pp. 248–256.

[175] Levchik, S.V., Weil, E.D., *Polym. Int.* (2005) 54, pp. 981–998.

[176] Huang, X., Ouyang, X., Ning, F., Wang, J., *Polym. Degrad. Stab.* (2006) 91, pp. 600–613.

[177] Nodera, A., Kanai, T., *J. Appl. Polym. Sci.* (2004) 94, pp. 2131–2139.

[178] Guo, J., Wang, Y., Feng, L., Zhong, X., Yang, C., Liu, S., Cui, Y., *Polymer* (Korea) (2013) 37, pp. 437–441.

[179] Hou, S., Zhang, Y. J., Jiang, P., *Polym. Degrad. Stab.* (2016) 130, pp. 165–172.

[180] Yang, Y., Kong, W., Cai, X., *Polym. Degrad. Stab.* (2016) 134, pp. 136–143.

[181] Bright, D. A., Dashevsky, S., Moy, P. Y., Williams, B., *J. Vinyl Addit. Technol.* (1997) 3, pp. 170–174.

[182] Despinasse, M. C., Schartel, B., *Polym. Degrad. Stab.* (2012) 97, pp. 2571–2580.

[183] EP 0945478, inv.: Nakacho, Y., Yabuhara, T., Tada, Y., Nishioka, Y., assigned to Otsuka Chemical Company Ltd., pri. 15.10.1997/16.02.1998.

[184] WO 2009055993, inv.: Su, K. C. H., Jow, J., Zhu, J. L., Tang, X., Bai, P., Liu, S., Lai, S. (2009) assigned to Dow Global Technologies Inc., pri. 02.11.2007.

[185] Levchik, S. V., Weil, E. D., *J. Fire Sci.* (2006) 24, pp. 137–151.

[186] WO 2014168869 inv.: Levchik, S. V., Alessio, G. R., Shtekler, R., Eden, E., Georlette, P. (2014) assigned to ICL-IP America Inc., pri. 10.04.2013.

[187] Sonnier, R., Viretto, A., Taguet, A., Lopez-Cuesta, J. M., *J. Appl. Polym. Sci.* (2012) 125, pp. 3148–3158.

[188] Herrera, M., Matuschek, G., Kettrup, A., *Polym. Degrad. Stab.* (2002) 78, pp. 323–331.

[189] Tabuani, D., Belluccci, F., Terenzi, A., Camino, G., *Polym. Degrad. Stab.* (2012) 97, pp. 2594–2601.

[190] Levchik, S. V., Weil E. D., *Polym. Int.* (2004) 53, pp. 1585–1610.

[191] Told, A., Harakaly, G., Szolnoki, B., Zimonyi, E., Marosi, G., *Polym. Degrad. Stab.* (2012) 97, pp. 2524–2530.

[192] Pinto, U. A., Visconte, L. L. Y., Gallo, J., Nunes, R. C. R., *Polym. Degrad. Stab.* (2000) 69, pp. 257–260.

[193] Shen, K. K., Ferm, D. J., in *Proceedings of the conference on recent advances in flame retardancy of polymeric materials*, Lewin, M. (Ed.) (1995) Stamford, USA, pp. 239–247.

[194] EP 1167429, inv.: Mehl, A. (2002) assigned to Nexans, pri. 14.06.2000.

[195] WO 2008151892, inv.: Muessig, B. (2008) assigned to TESA AG, pri. 13.06.2007.

[196] WO 2008151893, inv.: Muessig, B. (2008) assigned to TESA AG, pri. 13.06.2007/19.06.2007.

[197] WO 2008151897, inv.: Muessig, B. (2008) assigned to TESA AG, pri. 13.06.2007.

[198] EP 1391472, inv.: Braeuer, W., Heidingsfeld, H., Peerlings, H., Trabert, L. (2004) assigned to Bayer AG, pri. 21.08.2002.

[199] WO2006121549, inv.: Siddhamalli, S. K., Brown, C. A. (2006) assigned to Noveon Inc., pri. 13.04.2005.

[200] EP 617079, inv.: Hackl, C., Lehrich, F., Bittner, G. (1994) assigned to Elastogran GmbH, pri. 22.03.1993.

[201] WO 2013087733, inv.: Wermter, H., Futterer, T., Fibla, V. M., Goldammer, M. (2013) assigned to Chemische Fabrik Budenheim KG, pri. 13.12.2011.

[202] DE 102012008957, inv.: Seong, C. Y., Kim, D. H., Kim, K. J., Kim, Y. S., Bae, H. C. N. (2013) assigned to TSC Co. Ltd., pri. 29.12.2011.

[203] Tirri, T., Aubert, M., Wilen, C. E., Pfaendner, R., Hoppe, H., *Polym. Degrad. Stab.* (2012) 97, pp. 375–382.

[204] Hirschler, M. M., *Fire Mater.* (2017), pp. 1–14.

[205] Huggett, C., Levin, B. C., *Fire Mater.* (1987) 11, pp. 131–142.

[206] Weil, E. D., Levchik, S., Moy, P., *J. Fire Sci.* (2006) 24, pp. 211–236.

[207] Coaker, A. W., *J. Vinyl Addit. Technol.* (2003) 9, pp. 108–115.

[208] Jia, P., Zhang, M., Hu, L., Bo, C., Zhou, Y., *J. Therm. Anal. Calorim.* (2015) 120, pp. 1731–1740.

[209] Jia, P., Zhang, M., Hu, L., Bo, C., Zhou, Y., *Thermochim. Acta* (2015) 613, pp. 113–120.

[210] Jia, P., Zhang, M., Hu, L., Bo, C., Zhou, J., Feng, G., Zhou, Y., *Polym. Degrad. Stab.* (2015) 121, pp. 292–302.

[211] *http://www.zeusinc.com*, accessed on 14.03.2018.

[212] Zhu, X., Xu, H., Lu, J., Wang, J., Zhou, S., *J. Polym. Res.* (2008) 15, pp. 295–300.

[213] Obr, A., Zatloukal, M., AIP Conf. Proc. *Novel Trends in Rheology IV* (2011) 1375, pp. 248–252.

[214] Remanan, S., Sharma, M., Jayashree, P., Prameswaranpillai, J., Fabian, T., Shih, J., Shankarappa, P., Nuggehalli, B., Bose, S., *Mater. Res. Express* (2017) 4, p. 065301.

[215] *Ullmann's Encyclopedia of Industrial Chemistry: Polymers, High-Temperature* (2012) DOI: 10.1002/14356007.a21_449.pub3.

[216] Chen, J., Zhun, L., Wang, J., Qian, J., *Appl. Mech. Mater.* (2014) 584–586, pp. 1523–1526.

[217] EP 0637607 inv.: Glaser, R.H., assigned to General Electric Company, pri. 05.08.1993.

[218] Torlon PAI, Design Guide, Solvay Specialty Polymers, accessible at *https://www.solvayultrapolymers.com/en/binaries/Torlon-PAI-Design-Guide_EN-227547.pdf*, accessed on 14.03.2018.

[219] Patel, P., Hull, T.R., Lyon, R.E., Stoliarov, S.I., Walters, R.N., Crowley, S., Safronava, N., *Polym. Degrad. Stab.* (2011) 96, pp. 12–22.

[220] Patel, P., Stec, A.A., Hull, T.R., Naffakh, M., Diez-Pascual, A.M., Ellis, G., Safronova, N., Lyon, R.E., *Polym. Degrad. Stab.* (2012) 97, pp. 2492–2502.

[221] WO 2013079383, inv.: Sriram, S.R., Tortelli, V., Grover, G., Monzani, C., Johnson, S.P. (2013) assigned to Solvay Specialty Polymers USA LLC, pri. 28.11.2011.

[222] Liu, J., Gao, Y., Wang, F., Wu, M., *J. Appl. Polym. Sci.* (2000) 75, pp. 384–389.

[223] Morgan, A.B., Putthanarat, S., *Polym. Degrad. Stab.* (2011) 96, pp. 23–32.

[224] Lecouvet, B., Sclavons, M., Bourbigot, S., Bailly, C., *Polym. Degrad. Stab.* (2013) 98, pp. 1993–2004.

[225] Jiang, S., Liu, S., Zhao, J., Chen, X., *Polym. Degrad. Stab.* (2013) 98, pp. 2765–2773.

[226] Green, J., *J. Fire Sci.* (1994) 12, pp. 388–408.

[227] Levchik, S.V., Weil, E.D., *Polym. Int.* (2004) 53, pp. 1901–1929.

[228] Weil, E.D., Levchik, S., *J. Fire Sci.* (2004) 22, pp. 25–40.

[229] Rakotomalala, M., Wagner, S., Doering, M., *Materials* (2010) 3, pp. 4300–4327.

[230] Gerard, C., Fontaine, G., Bourbigot, S., *Materials* (2010) 3, pp. 4476–4499.

[231] Huo, S., Wang, J., Yang, S., Zhang, B., Chen, X., Wu, Q., Yang, L., *Polym. Degrad. Stab.* (2017) 146, pp. 250–259.

[232] Huang, S., Hou, X., Li, J., Tian, X., Yu, Q., Wang, Z., *High Perform. Polym.* (2017) 1–11, DOI 10.1177/0954008317745957.

[233] Mueller, P., Schartel, B., *J. Appl. Polym. Sci.* (2016), p. 43549.

[234] Chrusciel, J.J., Lesniak, E., *Progr. Polym. Sci.* (2015) 41, pp. 67–121.

[235] Liao, H., liu, Y., Jiang, J., Li, J., Liu, Y., *J. Fire Sci.* (2016) 34, pp. 3–12.

[236] He, X., Zhang, W., Yi, D., Yang, R., *J. Fire Sci.* (2016) 34, pp. 212–225.

[237] Levchik, S., Pietrowski, A., Weil, E., Yao, Q., *Polym. Degrad. Stab.* (2005) 88, pp. 57–62.

[238] Singh, H., Jain, A.K., *J. Appl. Polym. Sci.* (2009) 111, pp. 1115–1143.

[239] Chattopadhyay, D.K., Webster, D.C., *Progr. Polym. Sci.* (2009) 34, pp. 1068–1133.

[240] Weil, E.D., Levchik, S.V., *J. Fire Sci.* (2004) 22, pp. 183–210.

[241] Thirumal, M., Khastgir, D., Nando, G.B., Naik, Y.P., Singha, N.K., *Polym. Degrad. Stab.* (2010) 95, pp. 1138–1145.

[242] Liu, S.H., Kuan, C.F., Kuan, H.C., Shen, M.Y., Yang, J.M., Chiang, C.L., *Polymers* (2017) 9, 407 DOI:10.3390/polym9090407.

[243] Luo, F., Wu, K., Lu, M., Nie, S., Li, X., Guan, X., *J. Therm. Anal. Calorim.* (2015) 120, pp. 1327–1335.

[244] Gaan, S., Liang, S., Mispreuve, H., Perler, H., Naescher, R., Neisius, M., *Polym. Degrad. Stab.* (2015) 113, pp. 180–188.

[245] Kourtides, D.A., Gilwee, W.J., Parker, J.A., *Polym. Eng. Sci.* (1979) 19, pp. 24–29.

[246] Costa, L., Rossi di Montelera, L., Camino, G., Weil, E.D., Pearce, E.M., *J. Appl. Polym. Sci.* (1998) 68, pp. 1067–1076.

[247] Kandola, B.K., Krishnan, L., Ebdon, J.R., *Polym. Degrad. Stab.* (2014) 106, pp. 129–137.

[248] Tyberg, C.S., Bergeron, K., Sankarapandian, M., Shih, P., Loos, A.C., Dillard, D.A., McGrath, J.E., Riffle, J.S., Soranthia, U., *Polymer* (2000) 41, pp. 5053–5062.

[249] Antony, R., Pillai, C.K.S., *J. Appl. Polym. Sci.* (1994) 54, pp. 429–438.

[250] EP 1013684, inv.: Kiuchi, Y., Iji, M., Soyama, M. (2000) assigned to NEC Corporation, pri. 15.12.1998/07.04.1999/19.11.1999.

[251] Kandare, E., Kandola, B.K., Price, D., Nazare, S., Horrocks, R.A., *Polym. Degrad. Stab.* (2008) 93, pp. 1996–2006.

[252] US 20170130030, inv.: Levchik, S., Pack, S. (2017) assigned to ICL-IP America Inc., pri. 03.08.2016.

[253] Pan, L.L., Li, G.Y., Su, Y.C., Lian, J.S., *Polym. Degrad. Stab.* (2012) 97, pp. 1801–1806.

[254] Shih, Y.F., Wang, Y.T., Jeng, R.J., Wie, K.M., *Polym. Degrad. Stab.* (2004) 86, pp. 339–348.

[255] Dai, K., Song, L., Jiang, S., Yu, B., Yang, W., Yuen, R.K.K., Hu, Y., *Polym. Degrad. Stab.* (2013) 98, pp. 2033–2040.

[256] Khobragade, P.S., Hansora, D.P., Naik, J.B., Chatterje, A., *Polym. Degrad. Stab.* (2016) 130, pp. 194–244.

[257] Srivastava, S.K., Kuila, T., Fire retardancy of elastomers and elastomer nanocomposites in *Polymer Green Flame Retardants*, Papaspyrides, C.D., Kiliaris, P. (Ed.) (2014), pp. 597–651.

[258] Kind, D.J., Hull, T.R., *Polym. Degrad. Stab.* (2012) 97, pp. 201–213.

[259] Ismawi, D.H.A., Harper, J.F., Ansarifar, A., *J. Rubber Res.* (2008) 11, pp. 223–236.

[260] Fernando, N.T., Liyanage, S.S., *J. Natl. Sci. Found. Sri Lanka* (2009) 37, pp. 33–39.

[261] Karaivanova, M., Mihailova, N., Gjurova, K., *J. Appl. Polym. Sci.* (1990) 40, pp. 1939–1949.

[262] Janowska, G., Kucharska-Jastrzabek, A., Rybinski, P., *J. Therm. Anal. Calorim.* (2011) 103, pp. 1039–1046.

[263] Zhang, H., Wang, Y., Wu, Y., Zhang, L., Yang, J., *J. Appl. Polym. Sci.* (2005) 97, pp. 844–849.

[264] Huang, Y., Yang, J., Wang, Z., *J. Therm. Compos. Mater.* (2017) 30, pp. 816–826.

[265] Hamdani, S., Longuet, C., Perrin, D., Lopez-Cuesta, J.M., Ganachaud, F., *Polym. Degrad. Stab.* (2009) 94, pp. 465–495.

[266] Chen, W., Liu, Y., Xu, C., Liu, Y., Wang, Q., *RSC Adv.* (2017) 7, pp. 39786–39795.

[267] Zhang, Y., Zeng, X., Lai, X., Li, H., *RSC Adv.* (2018) 8, pp. 111–121.

4 Textiles

A. Richard Horrocks and Baljinder K. Kandola

4.1 Introduction

The high fire hazard posed by textiles, both in historical times and to the present day, is a consequence of the high surface area of the fibers present and the ease of access to atmospheric oxygen. However, in spite of their combustibility, across the world very few comprehensive fire statistics exist, especially those which attempt to relate deaths and injuries to cause, such as ignition and burning propagation properties of textile materials.

The annual UK Fire Statistics [1] are some of the most comprehensive available, and could be considered representative of a typical European country (the UK population is about 66 million). For instance, up to 2015/16, these statistics have demonstrated that while about 20% of fires in dwellings are caused by textiles being the first ignited material, over 50% of the fatalities are caused by these fires. Figure 4.1 presents typical data during the last 35 years, but it should be noted that since 1993, such detailed data has not been so freely available.

This data shows that, in general, deaths from fires in UK dwellings fluctuated around 700 per annum between 1982 and 1988; since then, they have fallen significantly, reaching 260–300 per annum during the 2014–2016 period. Fatalities from textile-related fires show a similar pattern, and it may be concluded that legislation associated with the mandatory sale of flame-retarded upholstered furnishing fabrics into the domestic UK market since 1989 has played a significant role in these reductions [2]. If these figures are representative of a typical EU community member, then it is likely that deaths from dwelling fires associated with textiles are of the order of 2000 per annum within the European Community.

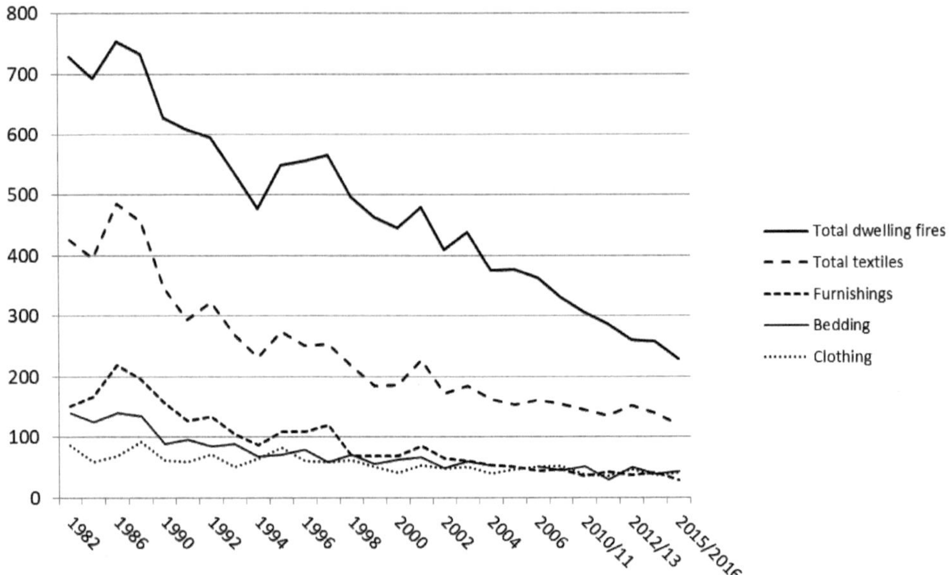

Figure 4.1 Annual UK fatalities in dwelling fires for different textile materials

It is rarely the direct causes of the fire, such as burn severity, which are the prime causes of death, but the effects of the smoke and emitted fire gases, which initially cause disorientation and impede escape, followed by incapacitation, asphyxiation and death [1]. Only in clothing-related fires are injury and death caused primarily by burns, such as when loose-fitting clothes (e.g., nightwear and summer dresses) are worn directly over the body.

4.1.1 Burning Behavior of Textile Fibers and Fabrics

The burning behavior of fibers is influenced and often determined by a number of thermal transition temperatures and thermodynamic parameters. Table 4.1 [3] lists the commonly available fibers with their physical glass (T_g) and melting (T_m) transitions, if appropriate, which may be compared with their chemically related transitions of pyrolysis (T_p) and ignition and the onset of flaming combustion (T_c). In addition, typical values of flame temperature and heats of combustion are given. Generally, the lower the respective T_c (and usually T_p) and the hotter the flame, the more flammable is the fiber. This generalization is typified by the natural cellulosic fibers cotton, viscose, and flax, as well as some synthetic fibers like the acrylics.

Table 4.1 Thermal Transitions of the More Commonly Used Fibers [3]

Fiber	T_g [°C] (softens)	T_m [°C] (melts)	T_p [°C] (pyrolysis)	T_c [°C] (ignition)	LOI [%]	ΔH_c [kJ/g]
Wool	-	-	245	600	25	27
Cotton	-	-	350	350	18.4	19
Viscose	-	-	350	420	18.9	19
Nylon 6	50	215	431	450	20-21.5	39
Nylon 6.6	50	265	403	530	20-21.5	32
Polyester	80-90	255	420-447	480	20-21	24
Acrylic	100	> 220	290 (with decomposi-tion)	> 250	18.2	32
Polypropylene	-20	165	470	550	18.6	44
Modacrylic	< 80	> 240	273	690	29-30	-
PVC	< 80	> 180	> 180	450	37-39	21
Oxidized acrylic	-	-	≥ 640	-	55	-
Meta-aramid (e.g., Nomex)	275	375	410	> 500	29-30	30
Para-aramid (e.g., Kevlar)	340	560	> 590	> 550	29	-
Polybenzimid-azole (PBI)	-	> 400	> 500	> 500	40-42	-

In Table 4.1 respective limiting oxygen index (LOI) values are listed, which are measures of the inherent burning characteristics of a material, and may be expressed as a percentage or decimal [4]. Fibers having LOI values of 21% (0.21) or below ignite easily and burn rapidly in air (which contains 20.8% oxygen). Those with LOI values above 21% ignite and burn more slowly, and generally when LOI values rise above about 26–28%, fibers and textiles may be considered to be flame-retardant and will pass most small-flame fabric ignition tests in the horizontal and vertical orientations. Nearly all flammability tests for textiles, whether based on simple fabric strip tests, composite tests (e.g., BS 5852:1979 and subsequent versions, and BS EN 1021-1 & -2:2014 for furnishings; and BS EN ISO 12952-1 & -2:2010 for bedding items) or more product/hazard-related tests (e.g., BS EN ISO 9239-1:2010 for carpets; BS 6341:1983 for tents; and BS EN ISO 9185:2007 for molten metal splash), are essentially ignition-resistance tests.

Within the wider community of materials fire science, it is widely recognized that under real fire conditions, it is the rate of heat release that determines burning hazard. While the heat of combustion values (ΔH_c) in Table 4.1 indicate that little difference exists between all fibers (and indeed some fibers like cotton appear to have a low heat of combustion compared to less flammable fibers like the aramid and oxidized acrylic fibers), it is the speed at which this heat is released that determines rate of fire spread and severity of burns.

Currently, only textiles used in building materials, aircraft, transport interiors, and seating are required to have minimal levels of rate of heat release. This is measured using instruments such as the cone calorimeter [5, 6] and the Ohio State University calorimeter [7] (used to assess the performance of textile-covered internal wall panels used in aircraft interiors at an incident heat flux of 35 kW/m^2). While there is currently very little published data on heat release for textiles, cone calorimetric data determined at an incident flux of 35 kW/m^2 reported by ourselves [8] is shown in Table 4.2 for a number of fabrics.

Table 4.2 Summary of Cone Calorimetric Data and Ranking Order for All the Samples [8]

Sample	Cone calorimetric values				Ranking order of FIGRA values
	TTI/ TTP [s]	PHRR [kW/m^2]	THR [MJ/m^2]	FIGRA [kW/m^{2}s]	
Lightweight cotton (87 g/m^2)	9/10	94	1.0	9.40	3
Heavyweight cotton (180 g/m^2)	14/14	128	3.2	9.10	2
(65/35 Polyester/cotton (105 g/m^2)	10/10	154	1.9	15.4	5
Acrylic (118 g/m^2)	17/18	292	4.5	16.2	6
Silk (174 g/m^2)	28/30	45	1.0	1.5	1
Wool (173 g/m^2)	16/16	171	2.9	10.6	4

Key: TTI = Time to ignition; TTP = Time to peak heat release; PHRR = Peak heat release rate; THR = Total heat release; FIGRA = Fire growth index (= PHRR/TTP); The FIGRA ranking increases from 1 (least hazard) to 6 (greatest hazard).

The time-to-ignition (TTI) values reflect the ease of ignition of each fabric, and Table 4.2 shows that lightweight cotton is the most easily ignitable fabric and that doubling its area density increases the TTI value significantly, thus demonstrating the well-known effect that high-area-density fabrics are less easily ignited than lighter ones comprising the same fiber types. While it is difficult to identify simple trends in terms of the effect of fiber type, it is noteworthy that the acrylic fabric has the highest PHRR and THR, as reflected in its low LOI and quite high ΔH_c in Table 4.1. With regard to the ranking order based on FIGRA values [9], the acrylic (rank 6) fabric in Table 4.2 again stands out as the possibly most hazardous, followed by 65/35 polyester/cotton (rank 5), wool (rank 4), lightweight cotton (rank 3), heavyweight cotton (rank 2), and heavyweight silk (rank 1).

This ranking order does, however, place wool in an unexpectedly high hazard position when intuitively it should be one of the least hazardous fabrics studied. Also, the FIGRA values do not significantly distinguish between light and heavyweight cotton specimens. The obvious reason for this is that while the PHRR value for heavier cotton is comparatively higher than that of lighter cotton, the FIGRA index is lowered because of its higher TTP value in the denominator. However, when the two parameters, PHRR and TTP, are combined to define a FIGRA index, the flammability hazards of light and heavy cotton are judged to be similar. With wool as an exception, intuitively, the FIGRA ranking places all other fabrics in a more expected hazard ranking order.

4.1.2 Effect of Fabric and Yarn Structures

The burning behavior of fabrics comprising a given fiber type or blend is influenced by a number of factors including the nature of the ignition source and time of its impingement, the fabric orientation and point of ignition (e.g., at the edge or face of the fabric, or the top or bottom), the ambient temperature and relative humidity, the velocity of the air, and (last but not least) fabric structural variables. Fabric orientation, point of ignition source and time, and the atmospheric variables are controlled in standard tests (see below). However, notwithstanding these, and as shown by Backer et al. [10], low fabric area density values and open structures aggravate burning rate and so increase the hazards of burn severity more than heavier and multi-layered constructions.

Hendrix et al. [11] have related LOI linearly with respect to area density, and logarithmically with respect to air permeability, for a series of cotton fabrics, although correlations were poor. Thus, fabric flammability is determined not only by the fiber behavior but the physical geometry of fibrous arrays in fabrics. Miller et al. [12] considered that an alternative measure of flammability was to determine the oxygen index at which the burning rate was zero. The resulting intrinsic oxygen index $(OI)_0$ for cotton is 13, which is considerably less than the quoted LOI of 18–19 (see Table 4.1).

The effect of yarn geometry and structure on burning behavior is less clear and has not been studied in depth, although the above-referenced works on fabric structure indicate that coarser yarns will have a greater resistance to ignition. This assumes that fiber type and area density remain constant (for coarser yarns, the cover factor will reduce and the air permeability will increase, which will have the converse effect). Work by Garvey et al. [13] has examined the burning behavior of blended yarns comprising modacrylic/flame-retardant viscose and wool/flame-retardant viscose, where the flame-retardant viscose is Visil® (formerly Sateri Fibers, Finland), produced either by ring-spinning and rotor-spinning, and knitted into panels (both methods result in the same nominal linear density).

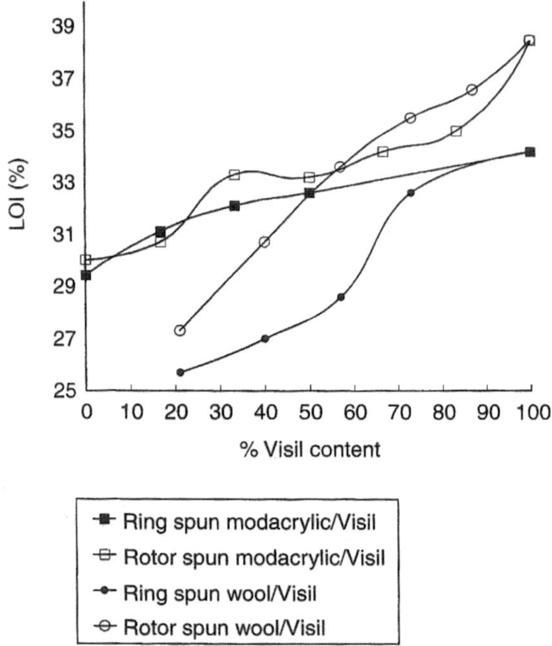

Figure 4.2 Effect of ring-spun versus rotor-spun yarn structures on fabric LOI [13]

Figure 4.2 shows the LOI results for all blends indicating that the difference in yarn structure significantly influences the fabric burning behavior. The greater flammability of the rotor-spun yarns is believed to be a consequence of the greater fiber component randomization that occurs; in ring-spun yarns, component fiber aggregation is known to be a feature.

■ 4.2 Flammability Testing of Textiles

Textile testing regulations and test methodologies have recently been comprehensively reviewed [14], and the discussion below represents a brief update.

4.2.1 Regulations

Textile-related fire regulations in different nations can seem to present an overall confusing picture in terms of the items regulated and the applications covered by them. However, in general, regulations fall into categories depending on whether they apply to: a consumer in a domestic environment; a member of the public in a public environment (such as a hotel, airport, or public building such as a hospital or prison); a worker routinely exposed to fire hazards as part of their job (for example, in the emergency and defense services); or passengers and personnel in the transport sector (where escape is restricted). With regard to child safety, there are nightwear and toy fire safety regulations at both national (e.g., UK) and economic block (e.g., EU) levels. These regulations generally cover:

- Nightwear (domestic environment)
- Upholstered furnishings (domestic and contract/public)
- Bedding (domestic and contract/public)
- Protective clothing (workplace, civil emergency, and defense)
- Transport (land, marine, and air)
- Toys

The first four of these regulation groups are (more often than not) nationally enforced, although within the EU, regulations may result from a relevant directive. For example, in 2007 a new European Standard, EN 14878:2007, was issued to cover all types of nightwear, including nightdresses, nightshirts, pyjamas, dressing gowns, and bath robes. It provides requirements for an absence of surface flash and a maximum burn rate acceptable for different categories of nightwear garments. Currently, this standard is a voluntary one, and is still under review for publication in the *Official Journal of the European Union (OJEU)* under the General Product Safety Directive (2001/95/EC).

However, within the EU, the Directive on Personal Protective Equipment (PPE) 89/686/EEC of 1989 requires that all protective standards are harmonized to ensure a high level of protection for citizens throughout Europe. Under the transport heading, marine and aviation fire standards are managed by the International Maritime Organization (IMO) and the International Civil Aviation Authority (ICAO), respectively. Land transport such as trains have fire standards determined nationally or (as has happened recently in the EU) harmonized under EN 45545-2:2013+A1:2015 *(Railway applications - Fire protection on railway vehicles. Part 2: Requirements for fire behaviour of materials and components),* in which textiles are obviously covered.

It should also be pointed out that when textiles are a part of a building such as a wall- or floor-covering, they are usually covered by the normal building fire regulations for that particular country or region.

Finally, within the EU, toys are covered by the Toy Safety Directive 2009/48/EC, and any textile (including children's party dresses and costumes) or fibrous product (e.g., wigs, raised-pile fabrics) must comply with EN 71-2:2011+A1:2014.

A comparison of regulations for each of the above areas is beyond the scope of this article, and the reader is referred to other publications [14-23].

4.2.2 Test Categorization

Table 4.3 attempts to summarize the complex range of tests currently available for textile products. Most flammability tests for common textiles are currently based on ease of ignition and/or burning rate behavior, which can be easily quantified for fabrics and composites in varying geometries. Few, however, yield quantitative and fire-science-related data, unlike the often-maligned methods using the limiting oxygen index (LOI) [4]. LOI, while it proves to be a very effective indicator of ease of ignition, has not achieved the status of an official test within the textile arena. Furthermore, sample ignition takes place at the top, with subsequent burning in a downward direction - which is not considered to be representative of most ignition geometries.

Furthermore, the exact LOI value of a fabric is also influenced by structural variables of the fabric (see above), and does not have a single value for a given fiber type or blend. However, it finds significant use in developing new flame retardants and optimizing levels of application to fibers and textiles [4].

Table 4.3 Selected Test Methods for Textiles

Nature of test	Textile type	Standard	Ignition source/ Comments
Simple fabric strip test methods and related product performance tests			
Vertical strip method based on BS 5438:1989	Curtains and drapes	BS 5867-2:1980	Small flame
	Nightwear	BS 5722:1991	Small flame
	Protective clothing (now withdrawn)	BS 6249-1:1982	Small flame
ISO vertical strip similar to Tests 1 and 3 in BS 5438	Vertical fabrics	BS EN ISO 6940:2004 and 6941:2003 BS EN ISO 15025:2016	Small flame
US strip tests	Vertical fabrics	ASTM D6413 US Federal Standard 191, Method 5136(3)	Small flame
	Horizontal 0, 45°, 60°, 90° 45°	FMVSS 302 FAR 25. 853(b) CFR-2012-16-2	Small flame

Nature of test	Textile type	Standard	Ignition source/ Comments
Textile composite and product tests			
Small-scale composite test for furnishing fabrics/fillings/ bedding materials	Furnishing fabrics	BS 5852-1 & -2:1979 (retained pending changes in legislation [2])	Cigarette and simulated match flame (20 s ignition)
		BS 5852:1990 (1998, 2006) replaces BS 5852-2	Small flames and wooden cribs applied to small and full scale tests
		ISO 8191-1 & and -2 (same as BS 5852:1990)	-
		BS EN 1021-1:2014	Cigarette
		BS EN 1021-2:2014	Simulated match flame (15 s ignition)
	Bedding items	BS 6807:2006	Ignitability of mattresses, divans, etc., by sources in BS 5852 Ignitability of bed covers and pillows by sources in BS 5852
		BS 7175:1989 (1994)	Ignitability of bedding items by cigarette and small flame sources
		BS EN ISO 12952-1 & -2:2010	Bedding items to cigarette and match
		BS EN 597-1 & -2:2015	Bedding and mattress bases to cigarette and match
Tests undertaken with the addition of radiant heat, including reaction-to-fire tests			
Use of radiant flux plus specified ignition source	Aircraft seat assemblies (so-called "Boeing" test)	ASTM E9060 1983 and FAR 25.853 Part 4, App F, uses an Ohio State University heat release calorimeter	Irradiate under 35 kW/m^2 with small flame igniter
	Carpets	BS EN ISO 9239-1:2010	Irradiate with 30° gas heated panel (~10 kW/m^2)
	All fabrics/ composites	NF P 92501-7, French "M test"	Irradiate with small burner

Table 4.3 Selected Test Methods for Textiles *(continued)*

Nature of test	Textile type	Standard	Ignition source/ Comments
Thermal protection (including protective clothing and manikin tests)			
Protective clothing	Resistance to radiant heat	BS EN ISO 6942:2002	Exposure to radiant source
	Resistance to convective heat	BS EN 367:1992	Determine heat transfer index
	Resistance to molten metal splash	BS EN ISO 9185:2007	Molten metal
	Gloves	BS EN 407:2004	Radiant, convective and molten metal
	Protection against limited heat and flame	BS EN ISO 15025:2016	Small flame
	Protection from limited flame spread	BS EN ISO 14116:2015 (uses BS EN ISO 15025 method)	Small flame
	Contact heat transmission	BS EN 702:1995	Contact temp. 100–500 °C
	Instrumented manikin testing of whole garments	BS ISO 13506:2008 and BS ISO 13506-2:2017 regarding burn prediction	Prediction of burn injury in terms of 1st, 2nd and 3rd degree burn propensity
Protective clothing composite tests	Firefighters' clothing	BS EN 469:2014	Composite standard
	Welders' clothing	BS EN ISO 11611:2015	–
	Protection against heat and flame	BS EN ISO 11612:2015	–
Durability tests			
Cleansing and wetting procedures for use in flammability tests	All fabrics	BS 5651:1989	Not applicable, but used on fabrics prior to submitting for standard ignition tests
	Commercial laundering	BS EN ISO 10528:1995	–
	Domestic laundering	BS EN ISO 12138:1997	–
	Domestic laundering & dry-cleaning	BS EN ISO 6330:2012	–

Despite the apparent complexity of the many tests available, some of which are presented in Table 4.3, it is possible to categorize them more simply as shown below:

I) Simple fabric strip tests

II) Textile composite tests

III) Protective clothing specification standards

IV) Tests undertaken with the addition of radiant heat (including reaction-to-fire tests)

V) Thermal protection (including protective clothing and manikin tests)

Many of these tests require samples to have undergone some form of durability test, especially if flame-retardant-treated textiles are present. Examples of these tests are also included in Table 4.3.

At the simplest level, most test procedures define a standard test method procedure (e. g., BS 5438 for vertical fabric strips), a specific performance level for a given product (e. g., BS 5722 uses BS 5438 to test and define performance levels for nightwear fabrics), or a combined test and performance-related set of defining criteria (e. g., BS 5852-1 & -2:1979 and EN 1021-1 & -2 for testing upholstered furnishing fabric/filling composites to simulated cigarette and match ignition sources).

Ideally, all practical tests should be based on reasonably straightforward principles, resulting in a test method that is simple and convenient to carry out. Observed parameters such as time to ignition, post-ignition after-flame times, burning rates, and nature of the damage and debris produced should be reproducibly and repeatedly measured with an acceptable and defined degree of accuracy. Figure 4.3 shows a schematic representation of the points of ignition within a typical vertical strip test such as BS 5438:1989 and BS EN ISO 6941/6942, in which a simple vertically orientated fabric is subjected to a standard igniting flame source either at the edge or on the face of the fabric for a specified time such as 10 s. For flame-retarded fabrics the properties measured after removal of the ignition source are the damaged (or char) length (*D*), size of hole if present, times of after-flame and afterglow, and nature of any debris (e. g., molten drips). For slow-burning fabrics, such as are required in nightwear, a longer fabric strip is used across which cotton trip-wires connected to timers are placed. In BS 5722, for example, these are at 300 mm and 600 mm above the point of ignition, and the time taken to cut through each thread enables an average burning rate to be determined. Tests of the type shown are simple and give reproducible and repeatable results. Furthermore, for similarly flame-retarded fabrics, the length of the damaged or char length can show semi-quantitative relationships with respect to the level of flame retardancy as determined by methods such as LOI.

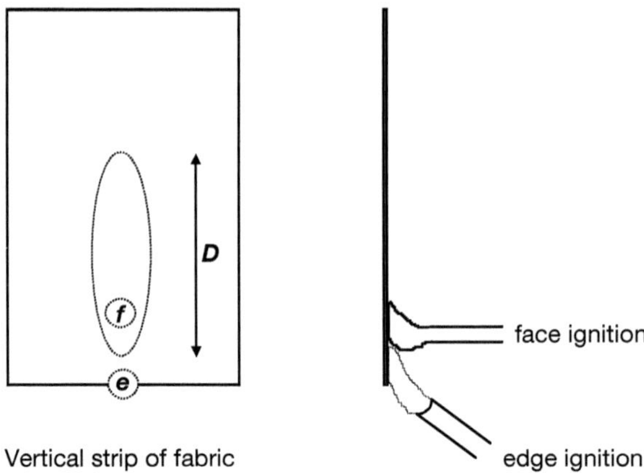

Figure 4.3 Schematic diagram of edge and face ignition geometries in vertical strip test methods such as BS 5438:1989 and BS EN ISO 6491:2003

However, in all flammability test procedures conditions should attempt to replicate real use, and so while atmospheric conditions are specified in terms of allowable relative humidity and temperature ranges, fabrics should be tested after having been exposed to defined cleansing and aftercare processes. Table 4.3 lists BS 6561 and its EN ISO derivatives as being typical here, and these standards define treatments from simple water soaking, through dry-cleaning and domestic laundering to the harsher processes used in commercial laundries and hospitals, for instance.

When textiles are used in more complex structures and more demanding environments, the associated test procedures can be more complex. This is especially the case for protective clothing, where the garment and its components have to function not only as a typical textile material but be resistant to a number of agencies including heat and flame. Table 4.3 shows tests that have recently been developed across the EU to accommodate the different demands of varying types of protective clothing and the hazards catered for (open flame, hot surface, molten metal splash, or a combination).

One relevant test only recently standardized, BS ISO 13506:2008 [24], is based on the simulation of a human torso and its reaction to a given fire environment when clothed, and is developed from the original Du Pont "Thermoman" [25] or instrumented manikin. This demonstrated a means of recording the temperature profile and simulated burn damage sustained by the torso when clothed in defined garments (usually prototype protective garments) during exposure to an intense fire source. This latter is typically a series of gas burners yielding a heat flux of 84 kW/m^2. This test is listed as a possible additional test in BS EN 469:2014, the standard for fire-fighters' clothing, and in Annex C of BS EN ISO 11612:2015, which is a composite standard for protective clothing against heat and flame. Recommended flame application times at 84 kW/m^2 heat flux are either 4 or 8 s depending on the level of protection requirement. Within the USA, however, to assess the flash-fire resistance of textile materials for both industrial and military applications, ASTM F1930 requires exposure of a fully dressed manikin to a heat flux of 84 kW/m^2 for 3 s. To pass the related performance standard US NFPA 2112, materials used in the tested flame-resistant garments should yield a body burn rating < 50%.

BS EN 469:2014 and BS EN ISO 11612:2015 are examples of composite standards that attempt to determine all the significant factors that determine the overall protective characteristics required of a firefighter's garment, for two levels of performance in the case of the former, and for various levels of protection against specific industrial hazards for the latter. Table 4.4 lists the component tests (including non-flame and heat resistance tests) within BS EN 469:2014, to exemplify the complexity of this and other composite protective clothing standards. A fuller description of these and other tests can be found elsewhere [14–23].

Table 4.4 Test Methods Within the Composite Standard for Firefighters' Clothing, BS EN 469:2014

Property tested	Standard
Flame spread (material and seams)	EN ISO 14116:2015 (uses BS EN ISO 15025)
Heat transfer (flame)	BS EN ISO 9151:2016
Heat transfer (radiant)	BS EN ISO 6942:2002 at 40 kW/m^2
Residual strength when exposed to heat	EN ISO 13934-1 or EN ISO 1421:1998, (method 1), after pretreatment to EN ISO 6942:2002 (method A) at 10 kW/m^2
Heat resistance	ISO 17493 at 180 ± 5 °C for 5 min
Tensile strength	BS EN ISO 13934-1:2013 or BS EN ISO 1421:2016 Outer seams: EN ISO 13935-2:2014
Tear strength	BS EN ISO 4674-1:2016, method B (coated); BS EN ISO 13937-2:2000 (uncoated)
Surface wetting (after washing and drying)	BS EN ISO 4920:2012
Dimensional change	BS EN ISO 5077:2008
Penetration by liquid chemicals	BS EN ISO 6530:2005 (for 40% NaOH, 36% HCl, 36% H_2SO_4 and *o*-xylene at 20 °C)
Water resistance (pressure for water entry)	BS EN 20811:1992 (ISO 811:1981)
Breathability (to water vapor)	BS EN 31092:1993+A1:2012, ISO 11092:1993
Visibility (photometric)	BS EN 20471:2013
Manikin (optional)	BS EN ISO 469:2014 Annex E (method BS ISO 13506:2008)
Marking	Clothing: EN ISO 13688:2013 Firefighters' clothing: EN 469:2011

4.3 Flame-Retardant Textiles

Table 4.5 lists all the commonly available flame-retardant and inherently flame-resistant textiles, and generally these have LOI values of 27 or greater, and will pass most vertical strip tests for all but the most lightweight fabrics. The main flame retardant treatments that may be applied to all conventional fiber-containing fabrics and blends have been comprehensively reviewed in recent years [26, 27, 28].

Table 4.5 Durably-finished and Inherently Flame Retardant Fibers in Common Use

Fiber	Flame retardant structural components	Mode of introduction
Natural		
Cotton	Organophosphorus and nitrogen-containing monomeric or reactive species, e.g., Proban® (Solvay, formerly Rhodia); Pyrovatex® CP New and CP-LF (Huntsman, formerly Ciba); Aflammit® SAP and KWB (Thor); Flacavon WP (Schill & Seilacher)	F
	Antimony-organohalogen systems, e.g., Flacavon F42 and H14 (Schill & Seilacher); Myflam™ (Lubrizol)	F
Wool	Zirconium hexafluoride (Zirpro) complexes, e.g., Aflammit® ZR and ZAL (Thor); Cetaflam® PHFZ and ZAS (Avocet Chemicals)	F
Regenerated		
Viscose	Organophosphorus and nitrogen/sulfur-containing species, e.g., Viscofil® Exolit® 5060 (Clariant) in FR Viscose (Lenzing)	A
Inherently synthetic		
Polyester	Organophosphorus species: phosphinic acidic co-monomer, e.g., Trevira CS®, (Trevira GmbH); phosphorus-containing additive, e.g., Avora CS (KoSa); phosphinate diester co-monomer, e.g., Heim/Toyobo GH (Toyobo, Japan); phosphorus-containing additive, e.g., Fidion® FR (Montefiber)	C/A
Acrylic (modacrylic)	Halogenated co-monomer (35–50% w/w) plus antimony compounds, e.g., Armora® (Aksa, Turkey); Kanecaron® (Kaneka Corp.); Sevel® FRSC (Fushun Huifu Fire Resistant Fiber Co., Japan)	C
Polypropylene	Halo-organic compounds usually as brominated derivatives, e.g., tris(tribromopentyl)phosphate (FR-370, ICL); hindered amine stabilizer, e.g., BASF Flamstab® NOR 116 (formerly Ciba), plus bromo-organic species	A
Polyhaloalkenes	Poly(vinyl chloride) or chlorofiber, e.g., Clevyl® (Rhovyl SA) Polyvinylidene chloride, e.g., Saran™ (Asahi Kasei)	H

Fiber	Flame retardant structural components	Mode of introduction
High heat- and flame-resistant (aromatic)		
Polyaramids	Poly(*m*-phenylene isophthalamide), e. g., Nomex® (DuPont), Conex® (Teijin); Poly (*p*-phenylene terephthalamide), e. g., Kevlar® (Du Pont), Twaron® (Teijin)	Ar
Poly(aramid-arimid)	e. g., Kermel® (Kermel, France)	Ar
Novoloid	e. g., Kynol® (Kynol, Japan)	Ar
Polybenzimidazole	e. g., PBI® (PBI Performance Products)	Ar
Carbonized acrylics (semicarbon)	e. g., Panox® (SGL Group)	Ar

Key: F = Chemical finish; A = Additive introduced during fiber production; C = Copolymeric modifications; H = Homopolymer; Ar = Aromatic homopolymer or copolymer.

4.3.1 Cellulosics

Flame-retardant cellulosic textiles generally fall into three groups based on fiber genus:

- Flame-retardant cotton
- Flame-retardant viscose (or regenerated cellulose)
- Blends of flame-retardant cellulosic fibers with other fibers, usually synthetic or chemical fibers

4.3.1.1 Flame-Retardant Cottons

It is most important that effective flame retardants are also effective afterglow retardants [29]. Flame-retardant cottons are usually produced by chemically after-treating fabrics during textile finishing, which (depending on chemical character and cost) results in the flame retardant properties having varying degrees of durability to various laundering processes. The flame retardants involved may be:

- Simple soluble salts, to give non-durable finishes, e. g., ammonium phosphates, polyphosphate and bromide; borate/boric acid mixtures
- Chemically reactive, usually functional finishes to give durable flame retardancy, e. g., alkylphosphonamide derivatives (Pyrovatex® (Huntsman); Aflammit® (KWB); Thor (Flacavon WP, Schill & Seilacher)); tetrakis(hydroxymethyl) phosphonium (THP) salt condensates (e. g., Proban® (Solvay); Aflammit® (SAP, Thor))
- Back-coatings, which usually comprise a resin-bonded antimony-bromine flame retardant system, and which typically contain antimony(III) oxide in

combination with decabromodiphenyl oxide or hexabromocyclododecane (both now restricted) in an Sb:Br molar ratio of 1:3 and suspended in a resin [28]. The two latter are now largely replaced either by bis(pentabromophenyl) ethane or a brominated polymeric resin component such as pentabromobenzyl acrylate.

4.3.1.2 Flame-Retardant Viscose

Flame-retardant viscose fibers usually have flame retardant additives incorporated into the spinning dopes during their manufacture, which makes the fibers durable and reduces the environmental hazard because it avoids the need for a flame retardant to be applied during the finishing process (see Table 4.5). Additives like Exolit® 5060 are phosphorus-based and so are similar to the majority of flame-retardant cotton finishes in terms of their mechanisms of activity in the condensed phase, performance and cost-effectiveness.

4.3.1.3 Flame-Retardant Cellulosic Blends

In principle, flame-retardant cellulosic fibers may be blended with any other fiber, whether synthetic or natural. In practice, technical limitations include:

- Compatibility of fibers during spinning or fabric formation
- Compatibility of fiber and textile properties during chemical finishing
- The preference for additivity/synergy in the flame retardant blend

Consequently, the current practice for achieving flame-retardancy in fiber blends is either to apply flame retardant only to the majority fiber present or to apply halogen-based back-coatings, which are effective on all fibers because of their common flame chemistries.

The prevalence of polyester–cotton blends, coupled with the apparent flammability-enhancing interaction in which both components participate (the so-called scaffolding effect, reviewed elsewhere [4, 30]), has resulted in greater attention being given to these blends than any others. However, because of the observed interaction, only halogen-containing coatings and back-coatings find commercial application to blends that span the whole composition range [28].

Durable, phosphorus-containing cellulose flame-retardants are generally only effective on cellulose-rich blends with polyester. THP-based systems like Proban® CC (Rhodia, formerly Albright and Wilson) are effective on blends containing no less than 55% cotton if a combination of flame retardancy and acceptable handling are required. Application of methylolated phosphonamide finishes (e.g., Pyrovatex CP, Huntsman) is effective on blends containing 70% or less cellulose. This is because the phosphorus present is less effective on the polyester component than in THP-based finishes [26]. The reasons for this are not clear, but are thought to be associated with vapor-phase activity of phosphorus in the latter finish on the polyester component.

4.3.2 Flame-Retardant Wool and Blends

Of the so-called conventional fibers, wool has the highest inherent flammability, and for some end-uses where high density of structure and horizontal orientation (e.g., carpets) are required in the product, wool fabrics will often pass the required flame retardancy tests untreated. Table 4.1 shows it to have a relatively high LOI of about 25 and a low flame temperature of about 680 °C. Its similarly high ignition temperature of 570–600 °C is a consequence of its higher moisture regain (8–16% depending upon relative humidity), high content of nitrogen (15–16 wt%) and sulfur (3–4 wt%), and low hydrogen content (6–7 wt%). The reviews by Horrocks [26, 27] comprehensively discuss developments in flame retardants for wool up to the late 1990s, and very little has changed since that time. Of particular note is the well-established durable "Zirpro" process developed by Benisek and coworkers (see Table 4.5), which is based upon the attachment of negatively charged zirconium or titanium complexes to positively charged wool fibers under acidic conditions at a relatively low temperature of 60 °C. Key references for this process have been reviewed [26]. Zirpro treatments can be applied to wool at any processing stage from loose fibers to fabrics, either during or after dyeing. The relatively low treatment temperature is an advantage because this limits the felting of wool. Recently, however, the process has come under the critical eye of environmentalists as a consequence of release of heavy metal ions into effluent discharges.

Wool blends pose different challenges, but given the complexity of wool and the current position of the Zirpro process as the only major durable flame retardant treatment, it is fortunate that the specificity of the process ensures that little if any transfer occurs to other fibers present. This is particularly true given that antagonisms between Zirpro-treated wool and other flame-retardant fibers were reported by Benisek in 1981 [31]. In the absence of any back-coating treatment, acceptable flame retardancy of Zirpro-treated blends is obtainable in 85:15 wool/polyester or wool/polyamide combinations. For blends containing a lower proportion of wool, and without resorting to use of alternative flame-retardant treatments, flame retardance can be maintained only if some of the Zirpro-treated wool is replaced by certain inherently flame-retardant fibers (except for Trevira CS polyester) [31]. Chlorine-containing fibers such as PVC and modacrylics are particularly effective in this respect.

4.3.3 Flame-Retardant Synthetic Fibers

These fibers may be divided into two groups - those that have been treated or coated with a flame retardant formulation, and those that are inherently flame retardant because of the introduction of an additive during their production or an inherently flame-retardant copolymeric repeat unit. Examples of the latter type predominate, and are included in Table 4.5. However, because of the general unreactivity of synthetic fiber-forming polymers such as polyacrylic and polypropylene, only polyamides and polyesters are possible candidates for durable flame-retardant treatments.

Durable flame-retardant treatments for polyester have been dominated by the Antiblaze® CU product (formerly by Rhodia and now produced in China) based on a monomeric cyclic phosphate [26, 27]. It is also claimed to be effective on polyamides and polypropylene as well as polyester, for which it was initially developed. Antiblaze CU has a high phosphorus content (21.5% w/w), and is a clear viscous liquid that is applied to polyester fabric at 3-6% (w/w), add-on buffered at pH 6.5, dried at 110-135 °C, and followed by thermofixation at 185-205 °C for 1-2 min. Thermofixation usually only results in about 80% retention of the original finish because of its volatility at high temperature. After rinsing and drying, the finish should resist 50 washes at 60 °C or 10 dry-cleaning cycles with 90% retention.

Thor's Aflammit® PE and Schill & Seilacher's Flacavon® AZ are believed to have a similar (or identical) chemical composition to Antiblaze CU.

There are other durable flame retardant finishes for polyester that may be applied directly from the dyebath, and these include Cetaflam® DB 9 (Avocet, UK).

For nylon fabrics, few flame-retardant treatments are satisfactory. Durable but fabric-stiffening flame-retardant finishes based upon methylated urea-formaldehyde with thiourea-formaldehyde have been successfully applied to nylon nets for evening wear and underskirts. For example, Thor's Aflammit NY comprises two components, Aflammit NY 1 based on an organic nitrogen/sulfur compound (probably a thiourea derivative) and Aflammit NY 2, the cross-linking methylolated urea component. In such formulations, typically about 15-20 wt% thiourea-formaldehyde precondensate is padded with ammonium chloride (1 wt% on the weight of the resin) as a latent catalyst, followed by low-temperature drying and then curing at 170 °C for 1 min.

Finally, mention must be made of back-coatings [28], which when based on brominated flame retardant/antimony synergist systems, are effective on all synthetic fiber-containing fabrics including blends. Furthermore, they may offer sufficient char-forming character and char coherence to offset fiber thermoplastic and fusion consequences, although this property is less easily achieved for polypropylene fabrics. However, the bromine-containing flame-retardant components present in back-coatings have been under environmental pressure for several years, as discussed elsewhere [32]. The most commonly used antimony-based synergist is antimony III oxide, which has known carcinogenic properties when in the bulk form. However, when dispersed and contained within a resin, its presence is considered to be safe, although it has not escaped continued attention by the environmental community. The only currently available commercial alternatives are zinc hydroxy stannate and zinc stannate, which, while having no toxicological properties of concern, are more specific in their ability to act as synergists with a range of brominated flame retardants and are more expensive [33]. However, they have smoke suppressing properties, which will often guarantee their choice for certain applications.

4.4 Inherently Flame-Retardant Synthetic Fibers

Conventional synthetic fibers may be rendered flame-retardant during production, either by incorporation of a flame retardant additive in the polymer melt/solution prior to extrusion, or by copolymeric modification. Synthetic fibers produced in these ways are often said to be inherently flame retardant. However, problems of compatibility, especially at the high temperatures used in melt-extrusion of fibers like polyamide, polyester and polypropylene, have ensured that only a few such fibers are commercially available. Table 4.5 lists examples of inherently flame-retardant synthetic fibers, in which the absence of polyamides reflects their high melt reactivities and hence poor flame retardant compatibilities.

Flame-retardant acrylics are currently not available, and modacrylics are used instead. This latter group has been commercially available for over 40 years or so, but at present few manufacturers continue to produce them. This is largely because of the success of back-coatings applied to normal acrylic fabrics, which create high levels of flame retardancy more cost-effectively. An added problem with modacrylics is the need to include antimony III oxide as a synergist at quite high levels (typically > 20%w/w) to enhance their flame retardancy.

In contrast, one group which continues to be successful are flame-retardant polyesters typified by the well-established Trevira® CS, which contains the phosphinic acid co-monomer. Other flame-retardant systems, based on phosphorus-containing polymeric melt additives such as Toyobo GH (and variants), are commercially available. The former Rhodia (now P-1045 from Unibrom Corporation, China) Antiblaze® 1045 additive is the dimer of the Antiblaze CU compound referred to above, and is melt-stable at polyester extrusion temperatures. None of these three flame-retardant polyester variants promote char, but function mainly by reducing the flaming propensity of molten drips normally associated with unmodified polyester. As yet, there are no char-promoting flame retardants for any of the conventional synthetic fibers, and this must constitute the real challenge for the next generation of acceptable inherently flame-retardant synthetic fibers.

4.5 High Heat- and Flame-Resistant Synthetic Fibers and Textiles

Inherently flame- and heat-resistant fibers and textiles, including the inherently flame-retardant viscose and synthetic fibers, comprise about 20% of total flame retardant usage.

Table 4.5 includes the main members of the group of high heat- and flame-resistant fibers, which have fundamentally combustion-resistant all-aromatic polymeric structures. Table 4.1 shows that most of these decompose above about 375 °C. Their all-aromatic structures are responsible for their low or non-thermoplasticity and high pyrolysis temperatures. In addition, their high char-forming potentials are responsible for their low flammabilities and, as established by van Krevelen [34], their high LOI values. This group is typified by the polyaramids, poly(aramid-arimids) and polybenzimidazole, which may find end-uses where high levels of heat resistance (HR) are required in addition to flame retardancy.

However, also included in Table 4.5 are the novoloid and carbonized acrylic fibers, which while having poorer fiber and textile physical properties, do have significant char-forming potentials and flame and heat resistance. These fibers tend to be used in nonwoven structures or in blends with other fibers in order to offset their less desirable textile characteristics. However, their high cost restricts their use to applications where their performance at high temperatures justifies the price. In practice, this is in high-performance protective clothing and barrier fabrics, particularly in fire-fighting, transport, and defense areas. A fuller description of this fiber group is available elsewhere [35]. However, the recently reported intumescent systems outlined below demonstrate heat barrier characteristics similar to (and in some cases superior to) those of the HR synthetic fibers.

4.6 Intumescent Applications to Textiles

Clearly, any enhancement of the char barrier in terms of thickness, strength and resistance to oxidation will enhance the flame and heat barrier performance of textiles. Generation or addition of intumescent chars as part of the overall flame-retardant property will also reduce the smoke and other toxic fire gas emissions. The application of intumescent coatings is especially beneficial to fabrics comprising fibers such as the polyolefins and polyesters, which lack any char-forming ability and where the intumescent char provides a supportive network preventing melt dripping and restricting the overall burning process. Such formulations may be intumescent in their own right, and generate carbonaceous chars independently of the surrounding polymer matrix, or they may interact with the matrix so that the flame-retardant–polymer combination together gives rise to an expanded, intumescent char when exposed to heat and flame.

The application of intumescent materials to textile materials has been reviewed [28, 36] and is exemplified particularly in the patent literature. As an example, glass-fiber-cored yarns used in a conventional, flexible woven or knitted fabric have been found to complement the flame and heat resistance of an intumescent coating. The presence of sheath fibers of a more conventional generic type ensures that the aesthetic properties of the textile can be optimized [37].

Most intumescent formulations are applied within a binder, and are based on ammonium and/or melamine phosphate-containing components in the presence of a char blowing agent such as pentaerythritol or a polyol. If the char-forming chemistry of the intumescent is similar to that of the underlying fibers such as flame-retardant viscose rayon or cotton, then a unique "char-bonded" structure results with exceptional heat barrier properties [38] that are commensurate with barrier fabrics comprising high-performance fibers such as aramids [39]. Intumescent-coated fabrics find application in barrier fabrics for mattress covers and upholstered furnishings, as well as in technical textile barrier applications.

References for Chapter 4

[1] *Fire Statistics: United Kingdom*, The Home Office, The Government Statistical Office, London, *https://www.gov.uk/government/collections/fire-statistics-great-britain.*

[2] *Consumer Protection Act (1987), The Furniture and Furnishings (Fire) (Safety) Regulations*, SI1324, HMSO, London, 1988 (currently under revision).

[3] Horrocks A.R., An introduction to the burning behaviour of cellulosic fibres, *J. Soc. Dyers Colourists* (1983), 99, pp. 191–197.

[4] Horrocks A.R., Price D., Tunc M., The burning behaviour of textiles and its assessment by oxygen index methods, *Textile Progress* (1989) 18, pp. 1–205.

[5] Babrauskas V., Grayson S.J., *Heat Release in Fires* (1992), Elsevier Applied Science, London and New York.

[6] ASTM E9060 (1983), Standard test method for heat and visible smoke release rates for materials and products; ISO 5660-1 (2015), Reaction-to-fire tests – Heat release, smoke production and mass loss rate – Part 1: Heat release rate (cone calorimeter method) and smoke production rate (dynamic measurement).

[7] US Federal Aviation Regulation FAR 23:853, Appendix F, Part IV.

[8] Nazare S., Kandola B.K., Horrocks A.R., Use of cone calorimetry to quantify the burning hazard of apparel fabrics, *Fire Mater.* (2002) 26, pp. 191–199.

[9] Sundström, B., The development of a European Fire Classification system for building products test methods and mathematical modelling, PhD Thesis (2007), Lund University, Sweden.

[10] Backer S., Tesoro G.C., Toong T.Y., Moussa N.A. (eds), *Textile Fabric Flammability* (1976) MIT Press, Cambridge, Mass., USA.

[11] Hendrix J.E., Drake G.L., Reeves W.A., Effects of fabric weight and construction on OI values for cotton cellulose, *J. Fire Flammability* (1972) 3, pp. 38–45.

[12] Miller B., Goswami B.C., Turner R., The concept and measurement of extinguishability as a flammability criterion, *Textile Res. J.* (1973) 43, pp. 61–67.

[13] Garvey S.J., Anand S.C., Rowe T., Horrocks A.R., The effect of fabric structure on the flammability of hybrid viscose blends, in *Fire Retardancy of Polymers – The Use of Intumescence*, Le Bras M.J., Camino G., Bourbigot S., Delobel R. (eds) (1998) Royal Society of Chemistry, London, p. 376.

[14] Horrocks A.R., Regulatory and testing requirements for flame retardant textile applications, in *Update on Flame Retardant Textiles: State of the Art, Environmental Issues and Innovative Solutions*, Alongi J., Horrocks A.R., Carosio F., Malucelli G. (eds) (2013) Smithers Rapra, Shawbury, UK, pp. 53–122.

[15] Horrocks A.R., Nazaré S., Kandola B., The particular flammability hazards of nightwear, *Fire Safety J.* (2004) 39, pp. 256–276.

[16] Guillaume E., Chivas C., Sainrat A., Regulatory issues and flame retardant usage in upholstered furniture in Europe, in *Proceedings of Fire and Building Safety in the Single European Market*, Edinburgh, 2008, pp. 38–48; *http://www.see.ed.ac.uk/FIRESEAT/files08/04-Guillaume.pdf.*

[17] Haase J., Flame resistant clothing standards and regulations, in *Handbook of Fire Resistant Textiles*, Selen Kilinc F. (ed) (2013) Woodhead Publishing, Cambridge, pp. 364–414.

[18] Horrocks A.R., Flame resistant textiles for transport applications, in *Handbook of Fire Resistant Textiles*. Selen Kilinc F. (ed) (2013) Woodhead Publishing, Cambridge, pp. 603–622.

[19] Sorathia U., Flame retardant materials for maritime and naval applications, in *Advances in Fire Retardant Materials*, Horrocks A.R., Price D. (eds) (2008) Woodhead Publishing, Cambridge, pp. 527–572.

[20] Nazaré S., Fire protection in military fabrics, in *Advances in Fire Retardant Materials*, Horrocks A.R., Price D. (eds) (2008) Woodhead Publishing, Cambridge, pp. 492–526.

[21] Aircraft Materials Fire Test Handbook, *http://www.fire.tc.faa.gov/handbook.stm* (see also Appendix C), Federal Aviation Administration, USA, 2006.

[22] Lyon R.E., Materials with reduced flammability in aerospace and aviation, in *Advances in Fire Retardant Materials*. Horrocks A.R., Price D. (eds) (2008) Woodhead Publishing, Cambridge, pp. 573–598.

[23] Nazaré S., Horrocks A.R., Flammability testing of fabrics, in *Textile Testing*, Hu J. (ed) (2008) Woodhead Publishing, Cambridge, pp. 339–388.

[24] Camenzind M.A., Dale D.J., Rossi R.M., Manikin test for flame engulfment evaluation of protective clothing: Historical review and development of a new ISO standard, *Fire Mater.* (2007) 31, pp. 285–295.

[25] Chouinard M.P., Knodel D.C., Arnold H.W., Heat transfer from flammable fabrics, *Textile Res. J.* (1973) 43, pp. 166–175.

[26] Horrocks A. R., Flame retardant finishing of textiles, *Rev. Progr. Coloration Related Topics* (1986) 16, pp. 62–101.

[27] Horrocks A. R., Flame retardant finishes and finishing, in *Textile Finishing* (vol. 2), Heywood D. (ed) (1999) Society of Dyers and Colourists, Bradford, UK, pp. 214–250.

[28] Horrocks A. R., Overview of traditional flame-retardant solutions, in *Update on Flame Retardant Textiles: State of the Art, Environmental Issues and Innovative Solutions*, Alongi J., Horrocks A. R., Carosio F., Malucelli G. (eds), (2013) Smithers Rapra, Shawbury, UK, pp. 123–178.

[29] Dyakonov A. J., Grider D. A., Smolder of cellulosic fabrics. I. Development of a framework, *J. Fire Sci.* (1998) 16, pp. 297–322.

[30] Miller B., Meiser C. H., Heat emission from burning fabrics; potential harm ranking, *Textile Res. J.* (1978) 48, pp. 238–243.

[31] Benisek L., Antagonisms and flame retardancy, *Textile Res. J.* (1981) 51, pp. 369–370.

[32] Horrocks A. R., Flame retardant and environmental issues, in *Update on Flame Retardant Textiles: State of the Art, Environmental Issues and Innovative Solutions*. Alongi J., Horrocks A. R., Carosio F., Malucelli G. (eds) (2013) Smithers Rapra, Shawbury, UK, pp. 207–238.

[33] Horrocks A. R., Smart G., Price D., Kandola B., Zinc stannates as alternative synergists in selected flame retardant systems, *J. Fire Sci.* (2009) 27, pp. 495–521.

[34] Van Krevelen R. W., Some basic aspects of flame resistance of polymeric materials, *Polymer* (1975) 16, pp. 615–620.

[35] Horrocks A. R., Technical fibers for heat and flame protection, in *Handbook of Technical Textiles* (2nd edition), Horrocks A. R., Anand S. C. (eds), (2016) Elsevier, Amsterdam, pp. 237–270.

[36] Horrocks A. R., Developments in flame retardants for heat and fire resistant textiles – The role of char formation and intumescence, *Polym. Degradation Stability* (1996) 54, pp. 143–154.

[37] Tolbert T. W., Dugan J. S., Jaco P., Hendrix J. E., US Patent 5,091,243 (1992).

[38] Horrocks A. R., Anand S. C., Hill B. J., UK Patent Appl. 9416189.2 (1995) and US Patent 5,645,926 (1997).

[39] Horrocks A.R., Kandola B.K., Flame retardant cellulosic textiles, in *6th European Conference on Flame Retardant Polymers,* published as: *Fire Retardancy of Polymers: The Use of Intumescence,* Le Bras M., Camino G., Bourbigot S., Delobel R. (eds) (1998) Royal Society of Chemistry, London, pp. 343–362.

5 Smoke Development and Suppression

5.1 Smoke Development and Measurement

Eric Guillaume

5.1.1 Introduction

Fire effluents consist of combustion gases and vapors, water droplets, and aerosols (solids and liquids formed during and after combustion). The aerosol component of fire effluents forms the visible portion of smoke, and is made up of particles and droplets of small size. Aerosols generated in fires are complex, non-homogeneous mixtures of liquid droplets of tar or water, solid-phase carbonaceous agglomerated soot with adsorbed organic compounds, or mineral particles. After formation through complex chemical and physical processes, fire aerosols continuously undergo changes in physical size, structure, and chemical composition, as the particles may coalesce, agglomerate, absorb gases, evaporate, or deposit on surfaces.

The nature of combustion products is less dependent on the fuel type at high fire temperatures (i.e., around 1000 °C) than at lower temperatures. The process of smoke generation appears to depend more on the local conditions during formation. It has been demonstrated that the combustion of cellulose, tobacco, and various polymers can generate approximately the same yield [1] of soot at temperatures of 1000 °C. In open, well-ventilated fires the production of soot is therefore highly dependent on the temperatures in the flame (or smoldering) zone, the oxidation conditions (e.g., the ventilation ratio of the fire), and the extent of aerosol agglomeration or re-oxidation.

The kind of descriptive data of the aerosol depends on end-use of the data. For example:

a) When studying escape from fire, as described in many fire scenarios, the impact of aerosols on visibility is one of the most important parameters in determining the ability to escape (see ISO 13571 [2]). Pre-movement actions and movement speed are strongly dependent on the visibility in smoke-filled spaces. It has been suggested that above a certain opacity of smoke, many would consider escape to be seriously impeded.

b) The direct physiological effect of aerosols on people is related to the size fractions within the aerosol and the morphology of the particles and droplets. It is therefore often more important to know the particle and aerosol distribution by size rather than the total aerosol mass for the determination of physiological effects. Unfortunately, size distribution is a parameter that is often difficult to measure, mainly due to the processes of agglomeration over short time intervals. Agglomeration can greatly affect the structure of the particles and droplets, as well as their density and optical (i.e., obscuration) properties. In addition, volatile species that may evaporate to a varying degree while being sampled and measured will influence the data recorded. Furthermore, measuring instruments operating on different principles will give different effective size fractions, and the methods cannot be assumed to be equivalent. The main parameter to consider for the physiological effect of aerosols is their deposition in the respiratory tract. Different classes of particle size, based on their relation to physiological effects, have been defined: "PM10", "PM2.5", "inhalable" ($D_p < 100$ µm), "thoracic" ($D_p < 10$ µm), "respirable" ($D_p < 4$ µm), "ultrafine" ($D_p < 0.1$ µm), and "nanoparticulates" ($D_p < 0.050$ µm).

c) In fire modeling, soot yield is an important input parameter. It has a very significant effect on the heat radiation properties of flames (often a sub-model of the main model), and therefore on the general heat transfer by radiation. This can affect escape from a fire (see ISO 13571) and significantly influence fire growth and the occurrence of "flashover".

Mass concentration and particle and droplet size fraction can also be valuable parameters for modeling of fire processes using computational fluid dynamics (CFD). This is especially relevant as models become capable of simulating the evolution of the aerosols and the effects of the evolution on flame radiation.

5.1.2 Smoke Generation

5.1.2.1 Definitions

An *aerosol* is defined as a suspension of liquid droplets or solid particles in a gas-phase matrix. Fires generate aerosols ranging in particle size from under 10 nm to over 10 µm.

A *particle* is defined as a solid-phase product present in an aerosol. Two categories of fire aerosol particles are identified: unburnt or partially burnt particles containing a high amount of carbon (i.e., "soot") and relatively completely combusted, small-size particles ("ashes"). Soot particles of small diameter (i.e., about 1 μm) typically consist of small simple spheres 10 nm to 50 nm in diameter. The formation of soot particles depends on many parameters including nucleation, agglomeration, and surface growth. Oxidation of soot particles (i.e., further combustion) may also take place.

5.1.2.2 Principles

Regarding the mechanisms of generation and evolution of aerosols, it is important to accurately measure aerosol concentrations and size distributions, to assess the scope and limitations of the apparatus and methodologies available for these measurements, and to interpret such measurements effectively, in a way consistent with the hazards and risks being evaluated. ISO 29904 [3] has been developed with the objective to comprehensively address these aspects and can be considered as the reference document in the field.

The soot contained in the smoke is mainly formed during the coking of the fuels, because the temperature in the gas phase is insufficient to oxidize all the carbon. Thus, aromatic rings join together to form compounds ranging from polycyclic aromatic hydrocarbons (PAHs) to macromolecular networks (cokes). The formation of soot therefore has a significant relationship to the presence of aromatic nuclei in the fuel and how easily they can be formed (molar ratio of carbon in the monomer, number of C–C bonds, and presence of double bonds).

The energy for these nuclei to assemble is lower than that required for their destruction, and so soot is formed preferentially. Compounds such as toluene or polystyrenes, for example, have aromatic rings in their structure, and will tend to produce large quantities of soot during combustion.

Examples of soot production for four fuels are given in Table 5.1, showing the presence of smaller or larger numbers of smoke-generating structures. For example, polyoxymethylene (POM) does not have C–C bonds and its combustion does not generate soot. Poly(methyl methacrylate) (PMMA) has C–C bonds, but some carbon atoms are already bonded to oxygen atoms and the spatial configuration of the molecule does not facilitate the folding of atoms into an aromatic ring. Polypropylene is composed of three C–C bonds arranged spatially at 120°. The monomer then already forms an aromatic half-cycle and the formation of soot is favored. Polystyrene already contains aromatic nuclei, and the energy needed to break these nuclei is not always available, so a large amount of soot is formed.

Table 5.1 Effect of Molecular Structure on Soot Generation of Polymers in Well-Ventilated Fire Conditions

Fuel	Molecule pattern	Proportion of carbon released as soot
Polyoxymethylene (POM)	$-(CH_2-O)_n-$	0.0%
Poly(methyl methacrylate) (PMMA)	$\left[-CH_2 - C(CH_3)(CO_2\text{–}H_3C) - CH_2 - C(CH_3)(CO_2\text{–}H_3C) - \right]_n$	0.3%
Polypropylene (PP)	$-\left[-CH_2 - CH(CH_3) - \right]_n-$	5.5%
Polystyrene (PS)	$\left(-CH(C_6H_5) - CH_2 - \right)_n$ (benzene ring)	18.0%

The liquid droplets present in aerosols are mainly spherical. Solid carbon particles have quasi-fractal morphology and can be present over a relatively large range of sizes. Inorganic mineral particles can have various shapes, depending on their initial shape in the fuel and on the fire “history”.

The types of soot formed often seem to depend on the local conditions (temperature, oxidant) during their formation than on the nature of the degraded products. For example, it has been shown that the combustion of cellulose, tobacco, polyethylene, and PVC produced the same quantities of soot at high temperature (1000 °C) [4]. The difference in production observed in free fires depends more on the flame temperature of the product and the oxidation or re-combustion conditions of the particles than on the chemical nature of the clean fuel. This can be explained because gaseous fuels produced by the pyrolysis of materials are similar in most cases (mostly methane, ethane and ethylene, propane).

In hydrocarbon fires or carbon-based polymers, it is accepted that soot particles consist almost exclusively of carbon (at least 90% by mass) and hydrogen. The phenomena leading to the formation of soot particles in the flame are complex. The main difficulty in understanding the phenomenon of carbon particle generation lies in the conversion of hydrocarbons or polymer pyrolysis products consisting of at most ten carbon atoms and a ratio of H/C ≈ 0.2, to particles consisting of hundreds of carbon atoms and a ratio of H/C ≈ 0.1. The phenomena involved include five main stages: nucleation, surface growth, coagulation, agglomeration, and oxidation. These stages are detailed below.

5.1.2.3 Nucleation and Surface Growth

Nucleation is the generation of primary particles. It occurs in the most reactive zone of the flame. The products resulting from the oxidation and the pyrolysis of the fuel molecules condense and form the first solid particles, called nuclei. These are formed in large numbers, but their size is less than 2 nm and they have a mass of about 1000 Daltons[1]. Figure 5.1 shows the processes involved in the formation of complex aromatic products during the combustion of hydrocarbons.

a) from C-C bond [1]

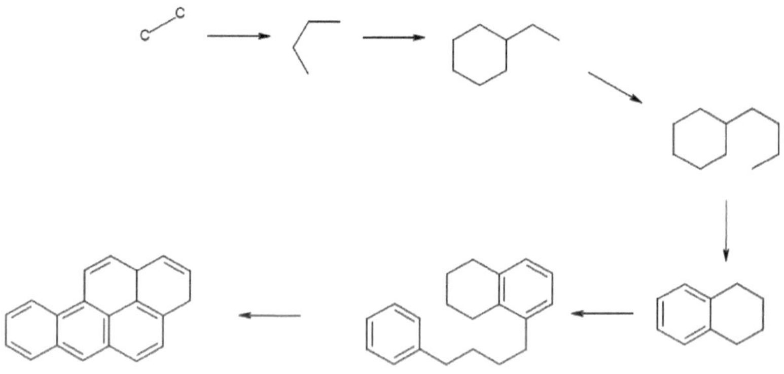

[1] The Dalton, or unified atomic mass unit, is approximately the mass of one nucleon and is numerically equivalent to g/mol.

b) from acetylene [5]

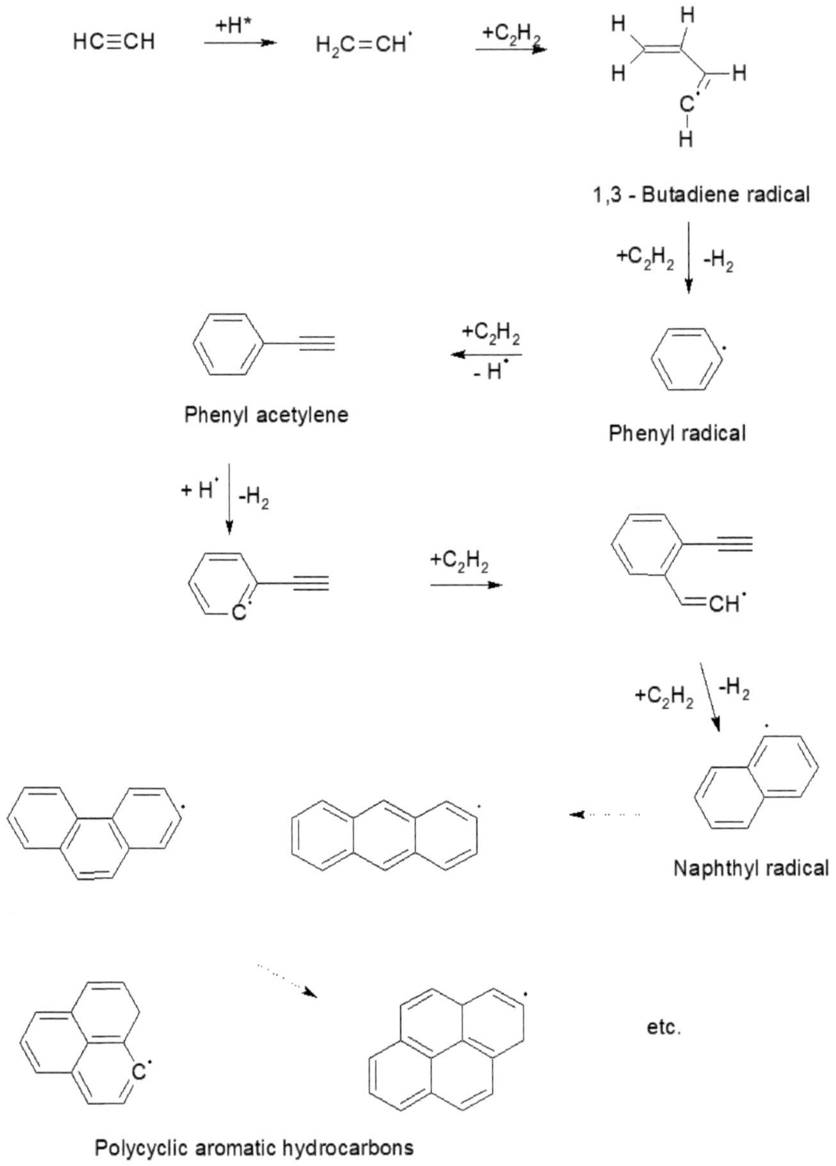

$HC{\equiv}CH \xrightarrow{+H^{*}} H_2C{=}CH^{\cdot} \xrightarrow{+C_2H_2}$ 1,3 - Butadiene radical

$+C_2H_2$ / $-H_2$

Phenyl radical

$+C_2H_2$ / $-H^{\cdot}$

Phenyl acetylene

$+H^{\cdot}$ / $-H_2$

$+C_2H_2$

$=CH^{\cdot}$

$+C_2H_2$ / $-H_2$

Naphthyl radical

etc.

Polycyclic aromatic hydrocarbons

Figure 5.1 Cyclization process and generation of hydrocarbons

During the growth stage of the surface, the gaseous species react and are integrated into the surface of the nuclei. This process continues until particles of spherical shape and diameter of between 10 nm and 50 nm are obtained. This phase represents the main stage of mass production of soot particles, and is illustrated in Figure 5.2. The size of the primary particles depends on the generation conditions (residence time, for example).

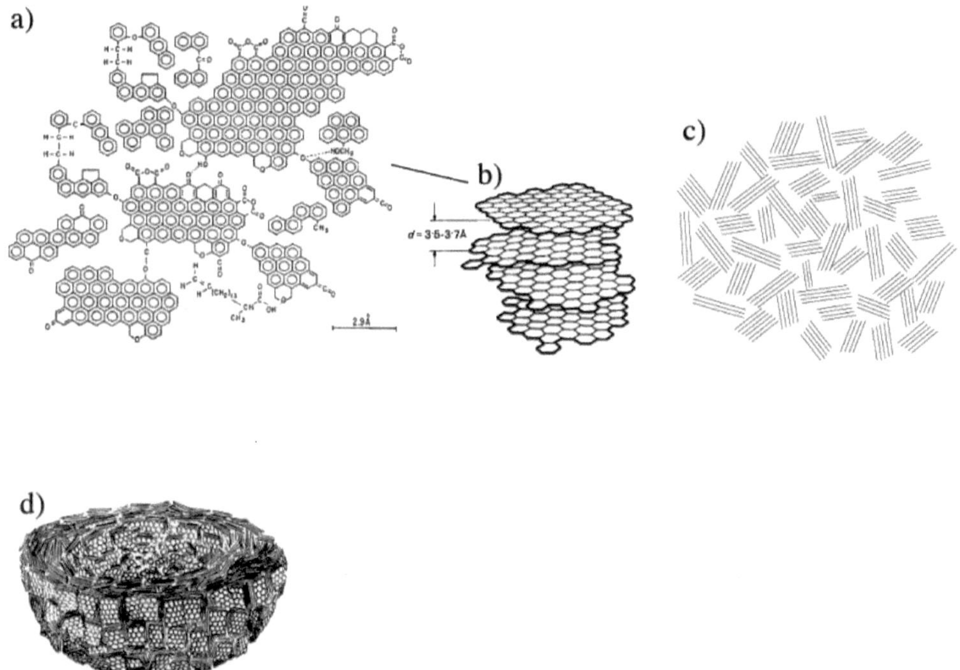

Figure 5.2 Primary particle formation process: a) Agglomeration of PAHs and cyclic compounds to form flat sheets; b) Formation of the sheets into lamellae through π linkages. These structures are called "platelets"; c) Agglomeration of groups of "platelets" into so-called "crystallite" structures, then surface growth; d) Formation of a primary particle of soot

5.1.2.4 Coagulation and Agglomeration

The essential phenomena are those related to the collision between particles. When the particles are still young in a hot and concentrated medium, and their diameter is still small (about 10 nm), they can merge by collision and reorganization of their internal structure. When the particles are older and colder, they can agglomerate without melting. The complex morphology of the particles then appears and the particle size distribution is established. This phenomenon also may take place outside of the flame zone. It comprises two major stages, which affect the morphology of the particles:

- The particle-particle effect, which consists in adding elementary particles to a set of particles. This effect tends to form relatively compact aggregates.
- The cluster-cluster effect, which involves the agglomeration of clusters of elementary particles, and leads to the formation of compact clusters.

These two phenomena generate quasi-fractal aggregates. Figure 5.3 illustrates this type of fractal object seen by transmission electron microscopy (TEM). This image shows nuclei of a few tens of nanometers, consisting of these elementary particles that agglomerate to form complex structures that can reach several micrometers. Depending on the local conditions, this process continues until quasi-fractal objects of about ten nanometers are formed, which agglomerate to create soot particles of a few micrometers. All these compounds, from nuclei to macroscopic soot particles, strongly absorb light over the whole spectrum and are the source of smoke opacity.

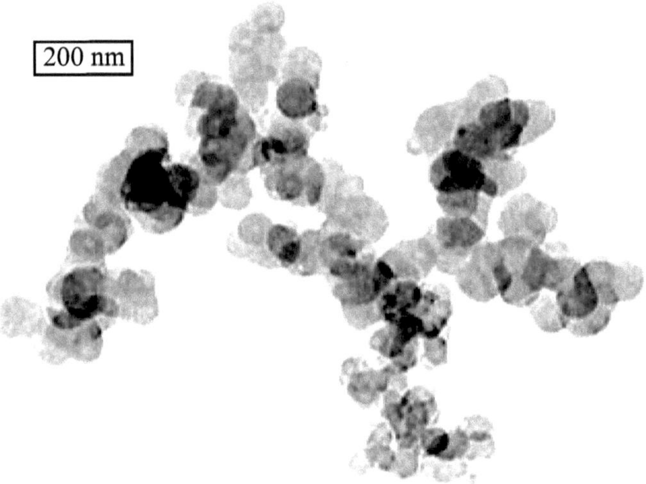

Figure 5.3 Soot particle viewed by transmission electron microscopy

5.1.2.5 Oxidation

In non-reactive areas of the flame, particle oxidation is possible by increasing the oxygen concentration. This is determined by the presence of oxygen as well as of ·OH radicals. Oxygen can diffuse into the core of the particles and thus cause internal oxidation that may lead to their disagglomeration. The ·OH radical starts surface oxidation and tends to reduce the mass of the aggregates, without leading to their disintegration.

5.1.2.6 Liquid Aerosols

The droplets present in liquid smoke aerosols are spherical, and consist of fairly heavy tars. Whether these tars are liquid or solid depends very much on temperature, but at room temperature, hydrocarbons containing more than 5 and less than 20 carbon atoms are considered to be liquid. Depending on the boiling and melting temperatures of the various hydrocarbons present in tars, the size and chemical composition of droplets is modified.

Under given conditions of pressure and temperature, water can condense on solid particles and tar aerosols, and form a mist, consisting of suspended water droplets a few micrometers in size.

5.1.3 Smoke Production from the Most Important Polymers

The smoke yield is very much dependent on combustion conditions and on the nature of fuels. For well-ventilated conditions, smoke production can be estimated indirectly by the ratio method, from the mass loss rate according to Equation (5.1). In this case, the mass loss concentration of the fuel is given by Equation (5.2). Thus, knowing the ω_{sa} factor, volume and mass loss, it is possible to recalculate the soot mass contained in the fumes. Tables are available in the literature for various products. A summary of the values is presented in Table 5.2.

$$\omega_{sa} = \frac{m_s}{m_c} \tag{5.1}$$

$$\rho_c = \frac{\rho_s}{\omega_{sa}} \tag{5.2}$$

where:

ω_{sa} Smoke yield (kg/kg)

m_s Total mass of soot generated (kg)

m_c Corresponding mass of fuel burnt (kg)

ρ_c Fuel mass loss rate, corresponding to the mass of fuel burned relative to the volume into which the smoke dilutes (kg/m^3)

ρ_s Mass of soot in the dilution volume (kg/m^3).

The values of the factor ω_{sa} are generally obtained experimentally. The fuel is weighed before and after combustion, which makes it possible to evaluate m_c. The smoke is sucked and passed through a filter, and the mass deposited on the filter represents m_s. In the presence of well-ventilated fires, common plastics have soot production rates of 1% to 10%. The average rate is of the order of 4%, with an aerosol mass concentration of 0.8 g/m^3. This corresponds to a mass loss in the order of 20 g/m^3 of air. For under-ventilated fires, the average rate of soot production doubles [16]. In this case, a soot concentration of 0.8 g/m^3 corresponds to a mass loss of 10 g/m^3. The method is very simple, but the transposition of a value of the ω_{sa} factor obtained on a test apparatus to another combustion condition is questionable - in particular, the change of scale influences the value. The data under sub-ventilated conditions are also irrelevant.

Table 5.2 Smoke Yields in Well-Ventilated Conditions, According to [16]

Product		ω_{sa} (g/g)
Natural materials	Red oak	0.015
	Brushwood	0.015
	Pure virgin wool	0.008
Rigid synthetic materials	Acrylonitrile-butadiene-styrene copolymer (ABS)	0.105
	Poly(methyl methacrylate) (PMMA)	0.022
	Polyethylene (PE)	0.060
	Polypropylene (PP)	0.059
	Polystyrene (PS)	0.164
	Silicone	0.065
	Polyester no. 1	0.091
	Polyester no. 2	0.089
	Epoxy resin	0.098
	Polyamide 6-6 (PA 6-6)	0.075
	Silicone rubber	0.005
Foams	Flexible PU	0.131-0.227
	Rigid PU	0.104-0.130
	Polystyrene	0.180-0.210
	Polyethylene	0.056-0.102
Halogenated polymers	PE with 25% Cl	0.115
	PE with 36% Cl	0.139
	PE with 48% Cl	0.134
	PVC resin	0.172
	PTFE	0.003
	ETFE	0.042

5.1.4 Smoke Measurement

5.1.4.1 Smoke Opacity

All visibility assessment models are calculated from smoke opacity, expressed as concentration, extinction coefficient, or optical density. Opacity is therefore the experimental parameter allowing the determination of visibility. The main characteristics of soot are overall mass concentration, particle size distribution, chemical composition and particle morphology. In the following discussion, the optical effect generated by the soot will be covered, but measurement of smoke particles will not be dealt with (for this, the reader is instead referred to ISO 29904).

The simplest direct measurement methods of smoke opacity are based on the principle of attenuation of a light beam. Thus, if smoke is present between a light source and a detector, part of the luminous flux is lost under the influence of several phenomena:

- The direct absorption by soot acting as a screen for the largest particles (whose sizes are in the order of a few microns to a few millimeters).
- The diffraction of the light beam on small particles and soot having a size comparable to the wavelength of the diffracted beam.
- The spectral absorption of colored gases such as nitrogen oxides, which have several absorption bands in the visible spectrum.

The measurement of the transmitted light I is compared to the transmission in the absence of smoke I_0. The value measured for I then depends on the distance between the source and the detector, the mass concentration of the smoke and the intrinsic characteristics of the smoke, namely its extinction coefficient.

Light emission from the fire, if it is too close to the detector, may contribute to illumination and therefore cause light transmission to be overestimated.

The wavelength of the light beam is therefore paramount in the measurement, which in rigorous treatments use the Beer-Lambert-Bouguer theory [7-9] as the principle of measurement. This assumes a relatively homogeneous concentration and is therefore theoretically applicable only to one wavelength. Nevertheless, it is frequently applied to polychromatic beams.

The basic experimental device (consisting of a transmitter, a receiver and a measurement zone where the beam interacts with the fumes) is called an opacimeter. There are two types of light sources for opacimeters:

- Monochromatic laser-type sources, which emit at a single wavelength (350 nm for He-Ne lasers).
- Polychromatic white light sources, which are closer to real-world situations. Rasbash [10] specifies that a stable white light source working at a color temperature of (2900 ± 100) K is ideal. To ensure the light intensity stability of the source, the power supply of the lamp must be very stable in voltage.

The measurement of smoke opacity using a monochromatic source has been carried out using a cone calorimeter to ISO 5660-1 [11], and this study illustrated the difficulty of implementing this method, as well as poor repeatability, attributable to the high dilution of the flue gases in the exhaust duct. These phenomena are due to the size of the soot particles released by the materials studied being close to the wavelength of the beam, which results in diffraction that distorts the measurement. Nevertheless, ASTM D5424 [12] specifies that the difference observed between results obtained with monochromatic and polychromatic sources may be negligible for the smoke developed by certain materials.

When using a polychromatic source in white light, the right choice of detectors is very important because they do not behave similarly at all wavelengths. Thus, detection based on a selenium oxide solar cell is best suited (by its response time and sensitivity) to slow phenomena and important optical paths, such as accumulation in a large volume (so-called "static" methods, as used in EN 61034-1 [13]. For dynamic phenomena (measurements on transmission), the detector mostly consists of a silicon photodiode, very sensitive in the red and the near-infrared up to 1100 nm. This type of highly sensitive detector has a very short response time, and is suitable for dynamic measurement over a short optical path, such as the stack-mounted measurement.

The smoke opacity can be characterized by the optical density, the extinction coefficient, or the mass concentration of soot. The simplest method for correlating the transmission measured with opacity is to convert it to optical density according to Equation (5.3). In order to normalize the length of the optical path, it can be reduced to a unit length according to Equation (5.4).

$$D = \log\left(\frac{I_0}{I}\right) \tag{5.3}$$

$$\mathrm{OD} = \frac{D}{L} \tag{5.4}$$

where:

D Optical density (dimensionless) as a function of time

I_0 Light intensity in the absence of smoke

I Light intensity measured as a function of time

OD Specific optical density for 1 meter path length

L Path length in meters.

The extinction coefficient (k, expressed in m^{-1}) is equivalent to the specific optical density, but uses a natural logarithm rather than a decimal logarithm with k = 2.303 × OD. The Beer–Lambert–Bouguer law applies according to Equation (5.5).

$$\frac{I}{I_0} = e^{-k.L} \tag{5.5}$$

It is possible to estimate the mass concentration of soot particles per unit volume from the extinction coefficient. Indeed, the extinction coefficient of light depends on the mass concentration of soot and its propensity to absorb light, characterized by its specific extinction surface σ_s (Equation (5.6)). In the literature, values of σ_s are given as σ_s = 10 m²/g [14] or more precisely as (9.6 ± 3.0) m²/g [15] for gaseous hydrocarbons. This value depends on many parameters and can only be accurately described for simple fuels under ideal combustion conditions.

$$k = \sigma_s C_s \tag{5.6}$$

where:

C_s Mass concentration of soot (kg/m^3)

σ_s Specific extinction area by mass unit of soot (m^2/kg).

The test methods used for smoke measurement are listed in Table 5.3. Some of these examples come from reference [5] and the previous edition of this chapter [6]. The application of certain methods in different contexts is detailed in subsequent chapters of this book.

Table 5.3 Main Methods Used for Smoke Opacity Measurement

Method	Type of measurement	Light source	Application	References
NBS chamber	Static	White light	Transportation	NF X 10-702 ASTM E662
ISO chamber	Static	White light		EN ISO 5659-2
XP2 chamber	Static	White light	Building (USA)	ASTM D2843
30 m^3 cabinet	Static	White light	Electric cables	EN 61034-1
Medium burning item (MBI)	Dynamic	White light	R&D	ISO 21367
Single burning item (SBI)	Dynamic	White light	Construction products (Europe)	EN 13823
Flooring radiant panel	Dynamic	White light		EN ISO 9239-1
Cone calorimeter	Dynamic	He-Ne laser	R&D Building	ISO 5660-1
Brandschacht	Dynamic	White light	Building Railways (Germany)	DIN 4102-15 DIN 5510-2
Steiner tunnel	Dynamic	White light	Building (USA)	ASTM E84
Room corner	Dynamic	White light	Real-scale tests	ISO 9705 ISO 13784-1

5.1.4.2 Other Smoke Parameters

ISO 29904 describes in-depth other smoke measurement parameters, such as granulometric distribution and aerodynamic diameter. Figure 5.4 and Table 5.4 summarize the main measurement techniques and their characteristics for aerosols in fire.

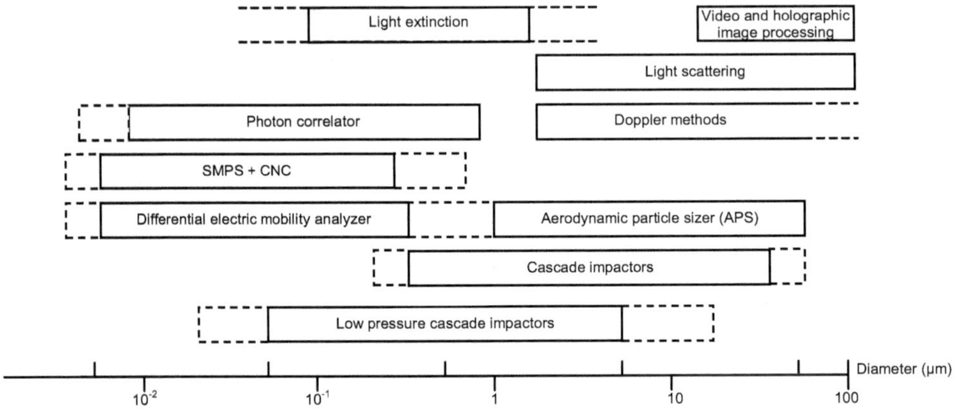

Figure 5.4 Size range of the most common aerosol measurement techniques

Table 5.4 Summary of Fire Aerosols Measurement Techniques Described in ISO 29904

Principle	Method used	Information provided	Usage	Diameter range (µm)	Resolution	Max conc.	Ease of usage [a]	Ease of interpretation
Light extinction	In situ, non-intrusive, time-dependent	Extinction coefficient, Soot concentration, soot yield	Modeling, Regulation	0.1–1	N/A	Not limited	+++	+++
Light scattering	In situ, non-intrusive, time-dependent	Extinction coefficient, particle size distribution	Modeling	> 1	N/A	Limited by signal	++	++
Microphotography/ombroscopy	In situ, non-intrusive, time-dependent	Soot concentration, size distribution	–	> 5	–	Low concentrations	++	+
Holography	In situ, non-intrusive time-dependent	Soot concentration, size distribution	–	> 10	–	Low concentrations	++	+
Photon correlation spectrometry	In situ, non-intrusive, time-dependent	Soot concentration, size distribution	Research	0.01–1	–	Low concentrations	+	+
Doppler method	In situ, non-intrusive, time-dependent	Soot concentration, size distribution	Research	> 1	–	Low concentrations	+	+
Direct gravimetric method	Extractive, integrated	Soot mean concentration, soot yield	Comparisons, modeling	Total	N/A	Limited by device	+++	+++
Cascade impactor	Extractive, integrated	Granulometric distribution (aerodynamic diameter)	Toxicity	0.3–30 [b] 0.02–30 [c]	+	0.1–1 g	++	++

Table 5.4 Summary of Fire Aerosols Measurement Techniques Described in ISO 29904 *(continued)*

Principle	Method used	Information provided	Usage	Diameter range (µm)	Resolution	Max conc.	Ease of usage [a]	Ease of interpretation
Electrical low pressure cascade impactor (ELPI)	Extractive, time-dependent	Granulometric distribution (aerodynamic diameter)	Toxicity	0.02–30	+	0.1–1 g	+	++
Tapered element oscillating microbalance (TEOM)	Extractive, time-dependent	Soot yield at a given cut-off size	Regulation	0.5–100	N/A	–	+++	+
Crystal vibration frequency	Extractive, time-dependent	Soot yield at a given cut-off size	Regulation	0.5–100	N/A	–	+++	+
Scanning mobility particle sizer (SMPS)	Extractive, time-dependent	Granulometric distribution (electric diameter)	Research, nanoaerosols	0.01–1	+++	–	+++	++
Engine exhaust particle sizer (EEPS-FMPS)	Extractive, time-dependent	Granulometric distribution (electric diameter)	Research, nanoaerosols	0.005–0.56	++	–	++	+
Aerodynamic particle sizer (APS)	Extractive, time-dependent	Granulometric distribution (aerodynamic diameter)	Research	0.5–20	++	< 1000 particles/cm^3	+++	++

[a] Flexibility to use on different fire models.
[b] Standard cascade impactors.
[c] Low-pressure impactors (DLPI).

5.1.5 Visibility Models

5.1.5.1 Definition of Visibility

Visibility, denoted as S, is essential when considering evacuation and depends on the level of contrast between an object and its environment. For an isolated object placed on a uniform background, the contrast C can be defined by Equation (5.7) [17]. In solar illumination conditions, the contrast of a black object placed on a white background is then $C = -0.02$ [18]. This value is commonly used as the visualization limit of an object. The visibility of an object is then the distance needed to reduce the contrast to the value of -0.02.

$$C = \frac{B}{B_0} - 1 \tag{5.7}$$

where:

B Luminance of the target under consideration

B_0 Background luminance.

Visibility depends on many parameters, including the illuminant considered, the extinction coefficient of smoke and the visual acuity of individuals, which is itself a function of the dilation of the pupil. Visibility is not measured directly but can be calculated by models from a measurement of density or concentration of smoke. Existing models are presented below.

5.1.5.2 Simplified Models (1950–1980)

The first models were obtained from tests carried out by Rasbash [19] and Malhotra [20] between 1950 and 1960. They show a log/log correlation between the optical density of the smoke (and thus the extinction coefficient) and the visibility. As early as 1970, Jin proposed an initial model [21], which was later refined [22], and used by Butcher and Parnell [23]. The simplified correlation they proposed is given in Equation (5.8). In addition, Butcher and Parnell indicated that a luminous object could be observed at a distance 2.5 times greater than a reflective object.

$$\mathrm{OD} \times S = 10$$

Or (5.8)

$$k \times s = 23$$

Various authors have also proposed extinction coefficient limits (Table 5.5). Jin [24] suggests that the limit of the allowable extinction coefficient is 0.15 m^{-1} (OD = 0.06 m^{-1}) for those unfamiliar with the escape route. For subjects who know the escape route, this limit is 0.5 m^{-1} (OD = 0.2 m^{-1}). He thus highlighted the importance of the occupant's knowledge of the premises in assessing visibility. Rasbash

[10] suggests a visibility limit of 10 meters, equivalent to an optical density of 0.08 m^{-1} (k = 0.19 m^{-1}). Babrauskas [25] places the escape limit at an extinction coefficient of 1.2 m^{-1} (OD = 0.5 m^{-1}).

These limits are clearly very different, and the influence of the knowledge of the escape route is obvious. Proposals from Jin (unknown escape route) and Rasbash are the safest, and set the visibility limit for uncomplicated evacuation around k = 0.15 to 0.20 m^{-1}.

Table 5.5 Limits Proposed by Jin, Rasbash, and Babrauskas

Model		Extinction coefficient (m^{-1})
Jin	Unknown escape route	0.15
	Known escape route	0.50
Rasbash		0.19
Babrauskas		1.2

5.1.5.3 Jin and Yamada Models

Later, Jin [26] studied the link between visibility and extinction coefficient for different systems. The studies performed were based on the minimum acceptable visual contrast in two configurations, resulting in two simplified relations:

- For reflective systems: $k \times S = 3$
- For light-emitting systems: $k \times S = 8$

This correlation is valid for distances between 5 and 15 meters. The data used to establish this model were obtained by observing people behind a window through smoke. The model does not take into account the irritating effect induced by fumes. This effect tends to degrade the visibility, causing blinking and production of tears due to the acid gases absorbed by the eye fluid, as well as the presence of soot deposits on the surface of the eye. Jin and Yamada [28] then studied the real effect of wood smoke on visual acuity. They put the subjects in the fire environment, with a respiratory mask but without protection of the eyes. They found that the irritation effect in the case of a wood fire was important for extinction coefficients greater than 0.25 m^{-1}.

Today, these models established by Jin and Yamada are widely used. They are implemented in fire development modeling tools, which provide output data on the "visibility" parameter. It is also necessary to define the nature of the system ("reflective" or "luminous"), as the tool used to calculate the local smoke concentration is based on the soot production rate. Then, the extinction coefficient, and based on that the visibility, can be calculated from explicit hypotheses. Numerical evacuation models also incorporate this type of model in order to consider the effect of visibility on movement [27].

5.1.5.4 ISO 13571 Model

The standard ISO 13571 proposes a model based on the minimum detectable visual contrast, given in Equation (5.9). This model considers that the occupants of a room will be unable to evacuate when they cannot see their own hands. The standard also states that the visibility limit is generally reached for a combustion of 20 grams of material per m^3 of air for a ventilated fire, and 10 grams per m^3 of air for an under-ventilated fire. Experiments have shown that the visibility threshold of reflective objects is reached at a soot mass concentration of about 0.3 g/m^3 per unit of optical path length, and for luminous objects of about 0.8 g/m^3. The optical path considered is between 5 and 15 meters, and the lower of these values is recommended for the evaluation of the visibility of stairs, doors and walls.

$$\ln(-C) = -\sigma_s \cdot M \cdot L \tag{5.9}$$

where:

C: Minimum detectable visual contrast (generally equal to -0.02)

M: Soot mass concentration, expressed in g/m^3

5.1.6 Fire Threat Related to Smoke

There are multiple mechanisms by which fire-generated aerosols affect the fire threat to people and the environment. First, small particles are respirable and can penetrate deep into the lungs. Inhaled particles themselves can be irritating, reducing the ability of people to escape from a fire [29, 30]. Next, these particles can adsorb and/or absorb toxic and irritant gases and vapors, providing a means for transport past the defences of the respiratory tract and deep into the lungs [30, 31]. Third, even less-respirable or non-respirable particles may effectively reduce the concentration of toxic gases and vapors in the fire effluent and can deposit them on surfaces. Fourth, aerosols may obscure vision, potentially reducing the ability of people to move effectively toward safety (see ISO 13571). Finally, the aerosol fraction in fire effluents also has significant potential to adversely affect the environment, particularly where the fires are large and of long duration [29, 32].

Soot, on the other hand, strongly contributes to the radiation of the fire. In short, soot behaves like a grey body of high emissivity. As a result, the soot particles radiate in all directions with an intensity proportional to the fourth power of their temperature. The phenomenon of smoke radiation is dominant in the thermal effects of fire on people. In fact, because occupants move away from the seat of the fire during evacuation, they may become more exposed to the thermal radiation of the fumes above them than to the direct radiation of the flame, which is too far away. Smoke radiation is also one of the most important criteria for the occurrence of flashover.

References for Section 5.1

[1] Harvey, R. G., *Polycyclic Aromatic Hydrocarbons: Chemistry and Carcinogenicity,* Cambridge University Press, pp. 11-15, 1991.

[2] ISO 13571. Life-threatening components of fire - Guidelines for the estimation of time to compromised tenability in fires, 2012.

[3] ISO 29904. Fire chemistry - Generation and measurement of aerosols, 2013.

[4] Hamins, A., Soot, in *Environmental Implications of Combustion Processes,* Chapter 3, pp. 71–95, CRC Press, ed. Puri, I. K., 1993.

[5] Drysdale, D., Production and measurement of smoke, in *An Introduction to Fire Dynamics* (2nd edition), Wiley, ISBN 0-471-97291-6, 2000.

[6] Koppers, M., Smoke development of fire effluents, in *Plastics Flammability Handbook* (3rd edition), pp. 614–627, ed. Troitzsch, J., Hanser, ISBN 1-56990-365-5, 2004.

[7] Bouguer, P., Essai d'optique sur la gradation de la lumière, ed. Jombert, C., Paris, 1729.

[8] Lambert, J. H., Photometria, sive de mensura et gradibus luminis, colorum et umbrae, Sumptibus Vidae Eberhardi Klett, 1760.

[9] Beer, A., Bestimmung der absorption des rothen lichts in farbigen flüssigkeiten, *Annalen der Physik und Chemie,* vol. 86, pp. 78–88, 1852.

[10] Rasbash, D. J., Sensitivity criteria for detectors used to protect life, *Fire International,* vol. 5, pp. 30–49, 1975.

[11] ISO 5660-1. Reaction-to-fire tests - Heat release, smoke production and mass loss rate - Part 1: Heat release rate (cone calorimeter method) and smoke production rate (dynamic measurement), 2015.

[12] ASTM D5424. Standard test method for smoke obscuration of insulating materials contained in electrical or optical fiber cables when burning in a vertical cable tray configuration, 2018.

[13] EN 61034-1. Measurement of smoke density of cables burning under defined conditions - Part 1: Test apparatus, 2005.

[14] Mulholland, G. W. and Choi, M. Y., Measurement of the mass specific extinction coefficient for acetylene and ethene smoke using the large agglomerate optics facility, *Proceedings of the Combustion Institute,* vol. 27, pp. 1515–1522, 1998.

[15] Putorti, A. D., Design parameters for stack-mounted light extinction measurement devices, NISTIR 6215, 1999.

[16] Tewarson, A., Generation of heat and chemical compounds in fires, in *SFPE Handbook of Fire Protection Engineering,* sect. 3, p. 92, 1995.

[17] McCartney, E. J., *Optics of the Atmosphere,* Wiley and Sons, 1976.

[18] Friedlander, S., *Smoke, Dust, and Haze,* John Wiley, p. 143, 1977.

[19] Rasbash, D. J., The efficiency of hand lamps in smoke, *Institute of Fire Engineers Quarterly,* vol. 11, pp. 46–52, 1951.

[20] Malhotra, H. L., Movement of smoke on escape routes. Instrumentation and effect of smoke on visibility, *Fire Research Notes,* nos. 651–653, 1967.

[21] Jin, T., Visibility through fire smoke, Building Research Institute, Tokyo, Report 30, 1970 and report 33, 1971.

[22] Jin, T., Visibility through fire smoke, *Journal of Fire & Flammability,* vol. 9, pp. 135–157, 1978.

[23] Butcher, E. G. and Parnell, A. C., *Smoke Control in Fire Safety Design,* Spon, 1979.

[24] Jin, T., Studies of emotional instability in smoke from fires, *Journal of Fire & Flammability,* vol. 12, pp. 130–142, 1981.

[25] Babrauskas, V., Full scale burning behaviour of upholstered chairs (Technical note 1103), National Bureau of Standards, Washington, 1979.

[26] Jin, T., Visibility Through Fire Smoke - Part 5, Allowable Smoke Density for Escape from Fire (Report No. 42), Fire Research Institute of Japan, p. 12, 1976.

[27] Jin, T., Studies on decrease of thinking power and memory in fire smoke, *Bulletin of Japanese Association of Fire Science & Engineering,* vol. 32, pp. 43–47, 1982.

[28] Jin, T. and Yamada, T., Irritating effects of fire smoke on visibility, *Fire Science & Technology,* vol. 5, pp. 79–89, 1985.

[29] Blomqvist, P., *Emissions from Fires - Consequences for Human Safety and the Environment,* Ph. D. thesis, Report 1030, Lund University, 2005.

[30] Levin, B. C. and Kuligowski, E. D., Toxicology of fire and smoke, in *Inhalation Toxicology* (2nd Edition), ch. 10, pp. 205–228. CRC Press (Taylor and Francis Group), ed. Salem, H. and Katz, S. A., 2005.

[31] Pauluhn, J., Retrospective analysis of acute inhalation toxicity studies: Comparison of actual concentrations obtained by filter and cascade impactor analyses, *Regulatory Toxicology and Pharmacology,* vol. 42, pp. 236–244, 2005.

[32] Health Hazards of Smoke, USDA Forest Service, Missoula Technology and Development Center, 9167-2809-MTDC, 1991.

5.2 Smoke Suppressants

Rudolf Pfaendner

5.2.1 Introduction

Reducing generation of smoke from plastics in case of fire through suitable additives called smoke suppressants is a potential strategy to make flame retardant polymer formulations more efficient and safer. As a matter of fact, smoke generation depends on the polymer, the flame retardants selected and any other ingredient present in the formulation to adjust polymer properties, e. g., plasticizers. Flame retardants acting in the condensed phase and forming char or intumescent layers and flame retardants acting through endothermic decomposition generating water contribute less to smoke than gas phase active flame retardants. As a consequence, the development of smoke suppressants is mainly aimed at the class of halogenated flame retardants and halogenated polymers such as PVC. Nevertheless, with increasing safety requirements and more recent fire safety standards such as EN 45545 for railways including smoke density and smoke toxicity testing, the importance of smoke suppressants in halogen-free flame-retardant formulations may be growing. Such a trend is reflected through the number of publications (patents, open literature) related to smoke suppressants: it was in the range of 10 to 30 publications annually from 1975 to 2011 and now amounts to more than 100 publications per year from 2014 onwards.

Smoke can be defined as aerosol or condensed phase components produced from combustion and fuel pyrolysis varying widely from light colored to black particles and soot [1]. The generation of smoke depends on the polymer substrate and its degradation mechanism, the flame retardants and their protection mechanism, other potential additives/ingredients as part of the polymer formulation, and the combustion conditions (temperature, oxygen availability). Combustion of a polymer substrate resulting in volatile degradation products will result in more smoke compared to a polymer which tends towards cross-linking and charring. Therefore, thermoset resins usually show less smoke than thermoplastic materials. Polymers forming aromatic units through degradation (e.g., PVC) or through depolymerization (polystyrene) show a higher smoke density than aliphatic-based structures (polyethylene, polypropylene, poly(meth)acrylates).

Flame retardants initiating or supporting charring and/or intumescence will in any case have a positive effect on reducing smoke compared to other flame retardants. Other flame retardants diluting the generated smoke through inert gases such as water ($Al(OH)_3$, $Mg(OH)_2$, AlO(OH)) or CO_2 (Hydromagnesite, Huntite) will reduce the smoke density to some extent. On the other hand, halogenated flame retardants and polymers containing halogen groups tend towards a high smoke density. Thus, it is obvious that PVC, as one of the high-volume polymer classes, is the main consumer of smoke suppressants.

Conceptually, smoke suppressants influence/change the degradation mechanism of the polymer formulation in reducing volatiles and increasing residues, preferably through catalytic processes to keep their amount and cost down. Therefore, active smoke suppressants are mainly based on catalytically active (transition) metals or metal complexes, mainly molybdates, stannates, antimonates, iron-, copper-, and bismuth oxides, and others. Most of the older literature is reviewed in overview articles and book chapters [2–4]. Details about efficiency, mechanisms, and application areas of smoke suppressants are explained in the following sections with a focus on PVC and the most important smoke suppressant classes: molybdates, stannates, and borates. In addition, smoke suppressants in combination with halogenated and halogen-free flame retardants are discussed. Due to the different evaluation criteria of smoke suppressants in publications, it is difficult to quantify the performance on a general basis, thus a qualitative description is preferred in the next section.

5.2.2 Smoke Suppressants for PVC

The main species contributing to smoke during the pyrolysis and thermooxidation of PVC are HCl and aromatics such as benzene, toluene, chlorobenzene, and divinylbenzene [5]. In plasticized PVC, the contribution to smoke by the plasticizer depends on its chemical structure and may be influenced through interactions of HCl from degradation with the plasticizer [6]. For example, phthalate plasticizers contribute more to smoke formation than phosphates [7]. Commercial smoke suppressants are traditionally based on molybdenum. Besides molybdates, only borates, stannates, and antimony oxide have some commercial importance for PVC applications.

5.2.2.1 Molybdates

The classical standard smoke suppressants for PVC and for plasticized PVC applications are based on molybdenum compounds, mainly molybdenum trioxide (MoO_3) and ammonium octamolybdate (AOM, $(NH_4)_4Mo_8O_{26}$). More recent developments aim to reduce the molybdenum concentration, to improve dispersion, and to provide more active catalytic sites through mixtures, mixed metal complexes, or by using inorganic carriers such as calcium-molybdate complexes, calcium-molybdate-zinc complexes, zinc molybdate on a magnesium silicate carrier, or zinc molybdate-zinc borate complexes. For example, a molybdate precipitate on the surface of a borate core is claimed to be efficient due to a greater surface area [8, 9]. Molybdates and modified molybdates are commercially available, e.g., from Huber (Tradename: Kemgard), Marshall (C-Tec) and others. Use concentrations showing a significant effect start at 1 phr and are used up to 15 phr in PVC formulations.

Molybdenum trioxide and its variants act in the condensed phase to promote cross-linking of PVC in the form of a redox catalyst system [10, 11]. The mechanism of activity is proven by reduction of aromatic degradation products and increase in char residue, which is in the range of 30–50% versus 10% without additive [11]. The generated smoke is reduced down to values of 20% or less compared to the formulation without smoke suppressants. Moreover, molybdenum oxide shows a pronounced concentration effect, i.e., increasing the concentration reduces the smoke density further [11].

Molybdenum variations are still of scientific interest. For example, it was recently shown that $CuMoO_4$ has the best smoke suppression performance in plasticized PVC in comparison to other molybdates [12, 13]. However, further combinations with char stabilizing agents (montmorillonite) were not successful [14]. Organic/inorganic salts such as melamine molybdate were investigated as well [15].

5.2.2.2 Stannates

Stannates are potential alternatives to molybdates as smoke suppressants in PVC. Commercial stannates are available in the form of zinc stannate ($ZnSnO_3$) and zinc hydroxystannate ($ZnSnO_3 \cdot 3H_2O$), the latter having a limited thermal stability of about 220 °C, while zinc stannate is thermally stable up to 400 °C. Suppliers of zinc hydroxystannates are, e.g., William Blythe (Flamtard), Joseph Storey (Storflam), Larderello Group (Zinflam), Marshall Additive Technologies (C-Tec), and Oceanchem. The use concentrations can be up to 15%; however, the standard range is between 1–5%.

Zinc hydroxystannates form $ZnCl_2$ and $SnCl_2$ Lewis acids promoting dehydrochlorination and volatile $SnCl_4$ under combustion conditions of PVC at a sufficiently high halogen availability [16]. Thus, a condensed phase mechanism is combined with a gas phase mechanism. For smoke reduction, at least in plasticized PVC, it is reported that zinc hydroxystannate is more efficient than zinc stannate [17]. As expected, char formation and LOI (limiting oxygen index) values are moderately increased. In earlier cone calorimeter experiments, it was found that zinc hydroxystannate is more efficient than zinc borate, at least at concentrations of 4%; an additional increase of the concentration above 2% (zinc borate) and 4% (zinc hydroxystannate) does not reduce smoke formation further [18]. The overall smoke reduction achieved is in the range of 75% at a concentration of 3–5% in comparison to PVC without stannates. Zinc hydroxystannate coated on metal hydroxides (magnesium dihydroxide (MDH) or aluminum trihydroxide (ATH)) can improve char formation and thus smoke suppression [19, 20]. Furthermore, coating of stannates on fillers such as $CaCO_3$ [16] was seen to be beneficial via a dispersion and char stabilizing effect. Copper stannate ($CuSnO_3$) was reported to be a potential alternative as a smoke suppressant to the corresponding zinc compound [21, 22].

5.2.2.3 Borates

Boric acid derivatives such as zinc borate, calcium borate, and ammonium borate act as flame retardants in polymers and other substrates such as cellulose, cotton, and lumber. Commercial products are available from Rio Tinto Minerals (Firebrake), Societa Chimica Larderello (Storflam, Zinborel), Sibelco Specialty Minerals (Portaflame), Thor (Aflammit), Oceanchem, Zibo Wuwei Industrial, and many others. Available products contain different amounts of crystal water ($2ZnO \cdot 3B_2O_3 \cdot 7H_2O$ (or $\cdot 3.5H_2O$ or $3H_2O$)), $4ZnO \cdot 3B_2O_3 \cdot H_2O$ up to the water-free product $2ZnO \cdot 3B_2O_3$. The amount of crystal water determines the thermal stability (between 170 °C and 500 °C) and, therefore, the potential use field. A zinc-free alternative (calcium borate on a silicate carrier) is offered as an antimony-trioxide- and zinc-free borate replacement (trade name: FR-1120, supplier ICL-IP [23]).

As a smoke suppressant, zinc borate acts in the condensed phase and increases the formation of char, the majority of zinc and boron remaining in the char. Thus, the char in plasticized PVC increases from 8% to 16–18% at a zinc borate concentration of 15 phr and the smoke is reduced to nearly 50% compared to a plasticized PVC without zinc borate [24, 25]. Furthermore, synergistic activity of borates in smoke suppression is reported, for example, with ATH [24, 26] and ammonium octamolybdate [27]. Zinc borate (or calcium borate) doped with ZnO and MoO_3 is claimed to be most efficient as a borate-based smoke suppressant in PVC [8].

During combustion, antimony oxide (Sb_2O_3) acts in the vapor phase through volatile halide and oxyhalide intermediates interrupting the free-radical processes. Thus, the smoke suppressant activity is generally seen to be low in comparison to molybdates and stannates. Combining the gas phase mechanism of antimony oxide with the condensed phase mechanism of zinc hydroxystannates in plasticized PVC increases synergistically the LOI values; however, there is no synergism in reducing the smoke density [28].

According to the accepted mechanism of PVC degradation, other transition-metal-containing additives are claimed to act as smoke suppressants in a similar way to molybdenum or zinc. Such additives contain iron, copper, titanium, zirconium, manganese, or bismuth, often in the form of the oxides [e.g., 21, 29, 30], sulfates [31], and in the form of mixed metals or complexes and/or combinations with other flame retardants, e.g., inorganics [32]. For example, the addition of FeOOH to PVC/ABS and PVC/SAN in a concentration of 5 phr can result in a smoke reduction of 45% [33].

5.2.3 Smoke Suppressants in Combination with Halogenated Flame Retardants

Suppression of smoke in formulations of thermoplastics or thermosets containing halogenated flame retardants by standard smoke suppressants known as being active in PVC is somewhat limited. In general, the smoke generated from combustion of the non-flame-retarded starting polymers is considerably lower compared to the flame-retarded formulations with brominated flame retardants. Thus, the addition of the smoke suppressant may lower smoke formation, however, not to the extent of the polymer without flame retardant. The polymers of interest comprise mainly unsaturated polyesters, ethylene-vinyl acetate copolymers and, more recently, polyamides. Molybdenum oxide was shown to be efficient as a smoke suppressant in polypropylene and polystyrene in combination with chlorinated and brominated flame retardants [34]. Zinc stannates and the combinations with antimony oxide have some effect as smoke suppressants in unsaturated polyesters [17, 35, 36] and are superior to antimony oxide alone. In further experiments, it was shown that zinc hydroxystannate, if coated on ATH in combination with chlorinated paraffins, reduces smoke density by 60% compared to the material without ATH [20]. Molybdenum oxide acts as a smoke suppressant in brominated polyesters [37]. Furthermore, zinc stannate is an efficient smoke suppressant [38] in polyamide 6 and polyamide 6.6, where a ratio of 2:1 Br:Sn is proposed for polyamide 6 and 4:1 Br:Sn for polyamide 6.6. Thus, smoke density is significantly lower (up to a factor of 10) in the presence of zinc stannate compared to antimony oxide as a synergist to brominated flame retardants. In a similar way, it was shown that zinc stannate reduces the smoke density of polyamide 6.6 flame retarded with brominated polyacrylate, contrary to antimony trioxide used as a synergist [39]. Very recently, it was demonstrated that graphene platelets in the presence of an $AlCl_3$ Friedel–Crafts catalyst reduce the smoke formation of brominated polystyrene [40]. Furthermore, small concentrations of fullerene (0.1–0.5%) were introduced in a decabromodiphenyl oxide/antimony oxide/high density polyethylene system with the effect of a radical scavenging mechanism in the condensed phase, somewhat reducing volatile decomposition products and smoke density [41].

5.2.4 Smoke Suppressants in Combination with Halogen-Free Flame Retardants

As expected, halogen-free flame retardants contribute less to smoke formation than halogenated ones. Nevertheless, the strategy to reduce smoke is similar to that of halogen-containing formulations, mainly to increase the char formation, e. g., by using intumescent systems based on ammonium polyphosphate (APP). Polymers mostly mentioned in combination with smoke suppressants comprise polyamides, unsaturated polyesters, polystyrene, polypropylene, epoxy resins, and polyurethanes.

In polyamide 6, zinc stannate has some positive effects in smoke reduction with selected phosphorus-containing flame retardants [42]. Here, the best performing combination with regard to smoke reduction is melamine polyphosphate and zinc stannate.

For many years, it has been known that the smoke development of burning polystyrene can be reduced by the addition of inert solids, where pyrogenic silica is claimed to be exceptionally active in the presence of air through barrier formation [43]. Compared to polystyrene without flame retardants, polystyrene with ammonium polyphosphate shows some smoke reduction, which is extremely enhanced in the presence of a specific carbonization agent (poly(1,3,5-triazin-2-aminoethanol diethylenetriamine)) [44]. Recently, several combinations based on carbon (graphene, nanotubes) were introduced as barrier materials [45, 46], as they are, or in a modified form, e. g., with molybdenum disulfide [47].

Smoke suppression of epoxy resins based on intumescent systems is improved by adding iron-containing compounds such as Ferrite Yellow (FeOOH) [48], zinc borate/antimony oxide combinations [49], or zinc hydroxystannate [50]. The approach to increase char formation or barrier formation through carbon-based additives was demonstrated through zinc hydroxystannate/graphene oxide [51], graphene/metal oxide [52] combinations, and nanotubes/polyphosphazene [53] combinations. Similarly, a flame retardant based on DOPO and *N*-(4-hydroxyphenyl) maleimide achieved an increased cross-linked char layer [54].

In polypropylene, combinations of APP and zinc hydroxystannate result in increased and more compact char formation [55]. The addition of lanthanum oxide to APP [56] or APP combined with a polymer synthesized from ethylene diamine and 4,6-dichloro-*N*-phenyl-1,3,5-triazine-2-amine showed a similar behavior [57]. Other APP combinations comprise a polymeric phosphordiamidate [58] or organically modified layered double hydroxides [59]. Even kaolin is capable of reducing the smoke formation of PP/APP systems [60]. In ATH and MDH flame-retarded PP formulations, it was found that the smoke density increases with the filler particle diameter, and that the addition of zinc borate was beneficial for further smoke reduction [61]. The barrier formation strategy was used in PP with CuO/graphene combinations, reducing smoke density by up to 50% [62].

For polyethylene formulations, the same smoke suppression strategy as that described for polypropylene is used to achieve a stable char layer and barrier. Either inorganic flame retardants such as MDH are improved through zinc borate [63], through a combination of zinc borate and talc [64], or APP is combined with a polymer synthesized from ethylene diamine and 4,6-dichloro-*N*-phenyl-1,3,5-triazine-2-amine [65]. APP in combination with zinc stannate [17] or with zinc borate [66] reduces smoke density in EVA to some extent [17].

Zinc borate and zinc stannate reduce the smoke generation of intumescent APP-based unsaturated polyesters usually containing styrene as a comonomer; the further addition of organically modified clays changes the degradation mechanism and does not contribute to smoke reduction [17, 67]. Boron phosphate was found to be an efficient smoke suppressant and is claimed to be superior to zinc borate in PET [68].

The smoke formation of polyurethane foam can be reduced through a melamine formaldehyde/red phosphorus combination as a flame retardant, contrary to red phosphorus alone [69]. However, melamine alone as one of the standard flame retardants for polyurethanes is capable of suppressing smoke through interaction with isocyanates from degradation processes [70].

5.2.5 Summary and Outlook

Smoke suppressants will continue to play an important part within the flame retardants market due to an increased awareness of smoke formation in fires and growing safety and toxicity aspects. As smoke suppressants are mainly used in PVC and halogenated flame-retarded formulations, the growth of smoke suppressants as individual additives will be influenced through replacement of PVC and the shift to halogen-free flame-retardant solutions. In recent years, the introduction of smoke suppressant products based on new chemical entities has been rather limited; however, the number of suppliers and grades still increases.

Future strategies to develop polymer formulations with less smoke are mainly aimed at modifying the polymer degradation process towards condensed phase mechanisms and/or using char- and barrier-forming flame retardants and additives.

From a scientific point of view, it appears that the interest in smoke suppression with a focus on halogen-free systems is increasing. However, this is only one important element in flame-retarded formulations and is not a key topic that is being investigated on its own.

References for Section 5.2

[1] Mulholland, G.W., Smoke production and properties in *SFPE Handbook of fire protection engineering*, 2nd edition, DiNenno, P.J. (Ed.) (1995) Society of fire protection engineering, Boston, pp. 2/217–2/227.

[2] Le Bras, M., Price, D., Bourbigot, S., Smoke development and suppression in *Plastics Flammability Handbook*, Troitzsch, J. (Ed.) (2004) Hanser, pp. 189–206.

[3] Green, J., Mechanisms for flame retardancy and smoke suppression – a review, *J. Fire Sci.* (1996) 14, pp. 426–442.

[4] Levchik, S.V., Weil, E.D., Overview of the recent literature on flame retardancy and smoke suppression of PVC. *Polym. Adv. Technol.* (2005) 16, pp. 707–716.

[5] Huggett, C., Levin, B.C., Toxicity of the pyrolysis and combustion products of poly(vinyl chlorides): a literature assessment. *Fire Mater.* (1987) 11, pp. 131–142.

[6] Michel, A., Sainrat, A., Bert, M., Role of poly(vinylchloride) and di-2-ethyl-hexyl-phthalate in the smoke formation from plasticized poly(vinylchloride), *J. Vinyl Technol.* (1981) 3, pp. 182–188.

[7] Horrocks, A.R., Smart, G., Nazaré, S., Kandola, B., Price, D., Quantification of zinc hydroxystannate and stannate synergies in halogen-containing flame-retardant polymeric formulations. *J. Fire Sci.* (2010) 28, pp. 217–248.

[8] Walker, J., Luke, E.D. (inv.), WO 2011/133621, 2011, Smoke suppressants, assigned to J.M. Huber Corporation, priority: 23.04.2010/15.04.2011.

[9] Isarov, A., Temples, D., Chen, T., Walker, J.K., Molybdate/borate complexes for enhanced cable compound fire performance. *Proc. Int. Wire and Cable Symposium* (2011) 60, pp. 250–256.

[10] Lattimer, R.P., Kroenke, W.J., The functional role of molybdenum trioxide as a smoke retarder additive in rigid poly(vinyl chloride) *J. Appl. Polym. Sci.* (1981) 26, pp. 1191–1210.

[11] Kroenke, W.J., Metal smoke retarders for poly(vinyl chloride), *J. Appl. Polym. Sci.* (1981) 26, pp. 1167–1190.

[12] Lei, L., Weihong, W., Huanhuan, X., Hongqiang, Q., A series of metal molybdates as flame-retardants and smoke suppressants for flexible PVC. *Adv. Mater. Res.* (2013) 634–638, pp. 1881–1885.

[13] Rodolfo Jr., A., Mei, L.H.I., Metallic oxides as fire retardants and smoke suppressants in flexible poly(vinyl chloride). *J. Appl. Polym. Sci.* (2010) 118, pp. 2613–2623.

[14] Rodolfo Jr., A., Mei, L.H.I., Poly(vinylchloride4/metallic oxides/organically modified montmorillonite nanocomposites: fire and smoke behavior. *J. Appl. Polym. Sci.* (2010) 116, pp. 946–958.

[15] Wu, W.H., Wu, J.H., Liu, W.H., Wang, Y.E., Liu, N., Yang, X.M., Li, Y.M., Hong, Q., Two series of inorganic melamine salts as flame retardants and smoke suppressants for flexible PVC. *Polym. Compos.* (2018) 39, pp. 529–536 doi:10.1002/pc.23965.

[16] Xu, J.Z., Jiao, Y.H., Qu, H.Q., Tian, C.M., Cai, N., Zinc hydroxystannate- or zinc stannate-coated calcium carbonate as flame retardant for semirigid poly(vinyl chloride). *J. Fire Sci.* (2006) 24, pp. 105–119.

[17] Horrocks, A.R., Smart, G., Price, D., Kandola, B., Zinc stannates as alternative synergists in selected flame retardant systems. *J. Fire Sci.* (2009) 27, pp. 495–521.

[18] Thomas, N.L., Zinc compounds as flame retardants and smoke suppressants for rigid PVC. *Plast., Rubber Compos.* (2003) 32, pp. 413–419.

[19] Qu, H., Wu, W., Xie, J., Xu, J., Zinc hydroxystannate-coated metal hydroxides as flame retardant and smoke suppression for flexible poly vinyl chloride. *Fire Mater.* (2009) 33, pp. 201–210.

[20] Cusack, P.A., Hornsby, P.R., Zinc stannate-coated fillers: novel flame retardants and smoke suppressants for polymeric materials. *J. Vinyl Addit. Technol.* (1999) 5(1), pp. 21–30.

[21] Starnes Jr, W.H., Pike, R.D., Cole, J.R., Doyal, A.S., KImlin, E.J., Lee, J.T., Murray, P.J., Quinlan, R.A., Zhang, J., Cone calorimetric study of copper-promoted smoke suppression and fire retardance of poly(vinyl chloride). *Polym. Degrad. Stab.* (2003) 82, pp. 15–24.

[22] Cassidy, D., Crossley, D.C., Bamford, A. (inv.), WO 2016/038359, 2016, assigned to William Blythe Ltd. (priority 09.09.2014/30.10.2014).

[23] Levchik, S., Moy, P., Alessio, G.R., Shawhan, G., Innes, J. (inv.), Synergized flame retarded polyolefin polymer composition, article thereof, and method of making the same, WO 2013/085788, assigned to ICL-IP America Inc., priority 09.12.2011/12.06.2012.

[24] Shen, K.K., Sprague, R.W., Recent studies on the use of zinc borate as a flame retardant and smoke suppressant in PVC. *J. Vinyl Technol.* (1982) 4(3), pp. 120–123.

[25] Shen, K.K., Kochesfahani, S., Jouffret, F., Zinc borates as multifunctional polymer additives. *Polym. Adv. Technol.* (2008) 19, pp. 469–474.

[26] Ning, Y., Guo, S., Flame-retardant and smoke suppressant properties of zinc borate and aluminum trihydrate-filled rigid PVC. *J. Appl. Polym. Sci.* (2000) 77, pp. 3119–3127.

[27] Ferm, D.J., Shen, K.K., The effect of zinc borate in combination with ammonium octamolybdate or zinc stannate on smoke suppression in flexible PVC. *J. Vinyl Addit. Technol.* (1997) 3(1), pp. 33–41.

[28] Qu, H., Wu, W., Zheng, Y., Xie, J., Xu, J., Synergistic effects of inorganic tin compounds and Sb_2O_3 on thermal properties and flame retardancy of flexible poly(vinyl chloride). *Fire Saf. J.* (2011) 46, pp. 462–467.

[29] Li, B., Wang, J., A cone calorimetric study of flame retardance and smoke emission of PVC. I. The effect of cuprous and molybdic oxides. *J. Fire Sci.* (1997) 15, pp. 341–357.

[30] Li, B., A study of the thermal decomposition and smoke suppression of poly(vinylchloride) treated with metal oxides using a cone calorimeter at a high incident heat flux. *Polym. Degrad. Stab.* (2002) 78, pp. 349–356.

[31] Tian, C., Wang, H., Ma, Z., Guo, H., Xu, J., Low melting sulfate glasses as additives to semirigid PVC and their flame retardant and smoke suppressant properties. *J. Vinyl Addit. Technol.* (2003) 9(2), pp. 69–80.

[32] Tian, C.M., Qu, H.Q., Wu, W.H., Guo, H.Z., Xu, J.Z., Metal chelates as flame retardants and smoke suppressants for flexible polyvinyl chloride. *J. of Fire Sciences* (2004) 22, pp. 41–51.

[33] Carty, P., White, S., Flammability of polymer blends. *Polym. Degrad. Stab.* (1996) 54, pp. 379–381.

[34] Cullis, C.F., Hirschler, M.M., Thevaranjan, T.R., Combinations of titanium (IV) oxide, iron (III) oxide and molybdenum (VI) oxide as flame retardants and smoke suppressants for thermoplastic polymers. *Eur. Polym. J.* (1984) 20, pp. 841–847.

[35] Cusack, P.A., Monk, A.W., Pearce, J.A., Reynolds, S.J., An investigation of inorganic tin flame retardants which suppress smoke and carbon monoxide emission from burning brominated polyester resin. *Fire Mater.* (1989) 14, pp. 23–29.

[36] Kicko-Walczak, E., New generation of fire retardant polyester resin. *Macromol. Symp.* (2003) 199, pp. 343–350.

[37] Das, M., Haines, P.J., Lever, T.J., Skinner, G.A., Thermal degradation studies of molybdenum containing polyester thermosets. *Fire Mater.* (1983) 7(1), pp. 41–48.

[38] Horrocks, A.R., Smart, G., Kandola, B., Holdsworth, A., Price, D., Zinc stannate interactions with flame retardants in polyamides; Part 1: Synergies with organobromine-containing flame retardants in polyamides 6 (PA6) and 6.6 (PA6.6). *Polym. Degrad. Stab.* (2012) 97, pp. 2503–2510.

[39] Williams, J., Rainford, C., Zinc stannate synergists in engineering thermoplastics. *Specialty Chemicals Magazine*, July 2014, pp. 14–15.

[40] Guo, Z., Ran, S., Fang, Z., Smoke suppression of graphene platelets fabricated by Friedel-Crafts reaction in brominated flame-retarded PS. *J. Therm. Anal. Calorim.* (2017) 128, pp. 1719–1730.

[41] Guo, Z., Zhao, L., Fang, Z., The flame retardant and smoke suppression effect of fullerene by trapping radicals in decabromodiphenyl oxide/Sb_2O_3 flame-retarded high density polyethylene. *Fire Mater.* (2017), pp. 1-9.

[42] Horrocks, A.R., Smart, G., Kandola, B., Price, D., Zinc stannate interactions with flame retardants in polyamides; Part 2: Potential synergies with non-halogen-containing flame retardants in polyamide-6 (PA 6). *Polym. Degrad. Stab.* (2012) 97, pp. 645–652.

[43] Chalabi, R., Cullis, C.F., Smoke formation from burning polystyrene and its suppression. *Fire Mater.* (1983) 7, pp. 25–31.

[44] Yan, Y.W., Chen, L., Jian, R.K., Kong, S., Wang, Y.Z., Intumescence: An effect way to flame retardance and smoke suppression for polystyrene. *Polym. Degrad. Stab.* (2012) 97, pp. 1423–1431.

[45] Zhou, K., Gui, Z., Hu, Y., The influence of graphene based smoke suppression agents on reduced fire hazards of polystyrene composites. *Compos., Part A* (2016) 80, pp. 217–227.

[46] Yan, L., Xu, Z., Zhang, J., Flame retardant and smoke suppression mechanism of multi-walled carbon nanotubes on high-impact polystyrene nanocomposites. *Iran Polym. J.* (2016) 25, pp. 623–633.

[47] Zhou, K., Yang, W., Tang, G., Wang, B., Jiang, S., Hu, Y., Gui, Z., Comparative study on the thermal stability, flame retardancy and smoke suppression properties of polystyrene composites containing molybdenum disulfide and graphene. *RSC Adv.* (2013) 3, pp. 25030–25040.

[48] Chen, X., Liu, L., Jiao, C., Qian, Y., Li, S., Influence of ferrite yellow on combustion and smoke suppression properties in intumescent flame-retardant epoxy composites. *High Perform. Polym.* (2015) 27, pp. 412–425.

[49] Zhang, F., Chen, P., Wang, Y., Li, S., Smoke suppression and synergistic flame retardancy properties of zinc borate and diantimony trioxide in epoxy-based intumescent fire-retardant coating. *J. Therm. Anal. Calorim.* (2016) 123, pp. 1319–1327.

[50] Formicola, C., De Fenzo, A., Zarrelli, M., Gordano, M., Antonucci, V., Zinc-based compounds as smoke suppressant agents for an aerospace epoxy matrix. *Polym. Int.* (2011) 60, pp. 304–311.

[51] Liu, X., Wu, W., Qi, Y., Qu, H., Xu, J., Synthesis of a hybrid zinc hydroxystannate/reduction graphene oxide as a flame retardant and smoke suppressant of epoxy resins. *J. Therm. Anal. Calorim.* (2016) 126, pp. 553–559.

[52] Wang, X., Xing, W., Feng, X., Yu, B., Lu, H., Song, L., Hu, Y., The effect of metal oxide decorated graphene hybrids on the improved thermal stability and the reduced smoke toxicity in epoxy resins. *Chem. Eng. J.* (2014) 250, pp. 214–221.

[53] Qiu, S., Wang, X., Yu, B., Feng, X., Mu, X., Yuen, R.K.K., Hu, Y., Flame-retardant-wrapped polyphosphazene nanotubes: A novel strategy for enhancing the flame retardancy and smoke toxicity suppression of epoxy resin. *J. Hazard. Mater.* (2017) 325, pp. 327–339.

[54] Yang, S., Wang, J., Huo, S., Cheng, L., Wang, M., Preparation and flame retardancy of an intumescent flame-retardant epoxy resin system constructed by multiple flame-retardant compositions containing phosphorus and nitrogen heterocycle. *Polym. Degrad. Stab.* (2015) 119, pp. 251–259.

[55] Su, X., Yi, Y., Tao, J., Qi, H., Synergistic effect of hydroxystannate with intumescent flame-retardants on fire retardancy and thermal behavior of polypropylene. *Polym. Degrad. Stab.* (2012) 97, pp. 2128–2135.

[56] Li, Y., Li, B., Dai, J., Jia, H., Gao, S., Synergistic effects of lanthanum oxide on a novel intumescent flame retardant polypropylene system. *Polym. Degrad. Stab.* (2008) 93, pp. 9–16.

[57] Feng, C., Zhang, Y., Liu, S., Chi, Z., Xu, J., Synergistic effect of La_2O_3 on the flame retardant properties and the degradation mechanisms of a novel PP/IFR system. *Polym. Degrad. Stab.* (2012) 97, pp. 707–714.

[58] Xu, Z. Z., Huang, J. Q., Chen, M. J., Tan, Y., Wang, Y. Z., Flame retardant mechanism of an efficient flame-retardant polymeric synergist with ammonium polyphosphate for polypropylene. *Polym. Degrad. Stab.* (2013), pp. 2011–2020.

[59] Wang, X., Spörer, Y., Leuteritz, A., Kuehnert, I., Wagenknecht, U., Heinrich, G., Wang, D. Y., Comparative study of the synergistic effect of binary and ternary LDH with intumescent flame retardant on the properties of polypropylene composites. *RSC Adv.* (2015) 5, pp. 78979–78985.

[60] Zhang, S., Jiang, P., Liu, X., Gu, X., Zhao, Q., Hu, Z., Tang, W., Effects of kaolin on the thermal stability and flame retardancy of polypropylene composite. *Polym. Adv. Technol.* (2014) 25, pp. 912–919.

[61] Liang, J., Zhang, Y., A study of the flame-retardant properties of polypropylene/$Al(OH)_3$/$Mg(OH)_2$ composites. *Polym. Int.* (2010) 59, pp. 539–542.

[62] Shi, Y., Qian, X., Zhou, K., Tang, Q., Jiang, S., Wang, B., Wang, B., Yu, B., Hu, Y., Yuen, R. K. K., CuO/graphene nanohybrids: preparation and enhancement on thermal stability and smoke suppression of polypropylene. *Ind. Eng. Chem. Res.* (2013) 52, pp. 13654–13660.

[63] Wu, Z., Hu, Y., Shu, W., Effect of ultrafine zinc borate on the smoke suppression and toxicity reduction of a low-density polyethylene/intumescent flame-retardant system. *J. Appl. Polym. Sci.* (2010) 117, pp. 443–449.

[64] Kim, S., Flame retardancy and smoke suppression of magnesium hydroxide filled polyethylene. *J. Polym. Sci., Part B: Polym. Phys.* (2003) 41, pp. 936–944.

[65] Feng, C., Liang, M., Jiang, J., Huang, J., Liu, H., Preparation and characterization of oligomeric char forming agent and its effect on the thermal degradation and flame retardant properties of LDPE with ammonium polyphosphate. *J. Anal. Appl. Pyrolysis* (2016) 119, pp. 75–86.

[66] Bourbigot, S., Le Bras, M., Leeuwendal, R., Shen, K. K., Schubert, D., Recent advances in the use of zinc borates in flame retardancy of EVA. *Polym. Degrad. Stab.* (1999) 64, pp. 419–425.

[67] Kandare, E., Kandola, B. K., Price, D., Nazaré, S., Horrocks, R. A., Study of the thermal decomposition of flame-retarded unsaturated polyester resins by thermogravimetric analysis and Py-GC/MSA. *Polym. Degrad. Stab.* (2008) 91, pp. 1996–2006.

[68] Kilinc, M., Cakai, G. O., Bayram, G., Eroglu, I., Özkar, S., Flame retardancy and mechanical properties of PET-based composites containing phosphorus and boron-based additives. *J. Appl. Polym. Sci.* (2015), p. 42016.

[69] Cao, Z. J., Dong, X., Fu, T., Deng, S. B., Liao, W., Wang, Y. Z., Coated vs. naked red phosphorus: A comparative study on their fire retardancy and smoke suppression for rigid polyurethane foams. *Polym. Degrad. Stab.* (2017) 136, pp. 103–111.

[70] Price, D., Liu, Y., Milnes, G. J., Hull, R., Kandola, B. K., Horrocks, A. R., *Fire Mater.* (2002) 26, pp. 201–206.

6 Smoke and Toxicity from Fire Effluents

Eric Guillaume

6.1 Preliminary Remarks

Smoke inhalation is the leading cause of death in fires. Many studies have shown that smoke inhalation is the primary hazard to most victims [1, 2, 3, 4, 5]. Fire effluents are hazardous due to the toxicants they contain, their high temperature, and their opacity, which impairs escape, e.g., by not finding the exit quickly enough. The following paragraphs present the major aspects associated with smoke toxicity. It should be noted that the type and concentration profile of these effluents depend on both the characteristics of the fire [2] and the nature of the combustibles, as detailed in comprehensive reviews [6, 7]. For risk assessment, particular attention should be paid to the incapacitating effects of these effluents and their contribution to lethality. The established paradigm in fire safety relies upon the generalization that the prevention of escape impairment, be it toxicant-induced incapacitation or psychological disturbance, is the basic prerequisite for precluding fatal outcomes of people subjected to enclosed fires. These outcomes depend on physical factors (heat, smoke obscuration) and irritant or asphyxiating (narcotic) toxicants. Thus, fire engineering aims at risk management strategies to limit the load of combustibles so that the normal population can successfully be evacuated from an enclosed fire before ability to escape becomes impaired.

The time available for escape of the evacuees in an enclosed fire is inversely proportional to the concentration (C) × exposure duration (t) relationship (dose) and the toxicities of the released effluents. However, this simplified paradigm requires more refinements in order to be applicable to toxicity-based risk assessments. Notably, what is referred to above as "dose" should really be the cumulative dose over time necessary to attain a specified severity of effects to which a population is exposed. Thus, the simple relationship $C \times t$ = "dose" becomes instead $C^a \times t^b$ = constant effect, in order to accommodate compound-specific changes in ventilation and inhaled dose (C^a) or time-related differences in metabolism (t^b) that may alter the cumulative dose with time, compound-dependently. Hence, individual differences of sensitivities within the human population have to be taken account of for either variable. The parameterization of the effect-based relationship is derived as illustrated in Chapter 7. As concluded there, data from rodents (rats) is given preference because they have a much higher body-weight-adjusted ventilation rate than humans. Consequently, adverse effects occur earlier and are more pronounced, with less variability when compared to humans. The time-dependent hazards affecting fire safety calculations include the accumulated fractional doses of toxicants, heat, and smoke density causing visual obscuration, and depend on the following factors:

- The variation of fire growth, which depends on the fuel mass loss rate and the volume in which the gases are dispersed.
- The quantity of toxic products released per unit mass of the burnt material.
- The toxic potential of the products, which corresponds to the exposure concentration (kg/m^3) or the exposure dose ($kg \times min/m^3$) necessary to produce toxic effects. An assessment of these effects requires knowledge of exposure concentrations or doses that reduce or prevent escape.

It is important to note that for any toxicant, the human body always has at least one detoxification process (mechanical or biochemical), and it is the prevalence of that process that determines the toxic effects of the substance. Another essential criterion is the product absorption rate by the body. Together, these factors drive the parametrization of the equations.

All these considerations seem to suggest the separation of fire risk assessment from the toxicity assessment itself. Toxicity assessment is dealt with in some more detail in Chapter 7. Fire risk assessment has to include fire-related aspects that should only be carried out at system level, i. e., related to real-scale scenarios, because the chemistry of emissions and their impact on people are highly dependent on the scenario assessed. This is stated in ISO 19706 [8] and ISO/TR 13571-2 [9]. The recent developments in fire safety engineering and related fire models increasingly allow consideration of these different aspects.

6.2 Acute Effects of Fire Effluents

The effects of gases presented in the following will be dealt with in some more detail later.

6.2.1 Asphyxiants

Carbon monoxide (CO) and hydrogen cyanide (HCN) account for most deaths from smoke poisoning. The effects of these species are summarized below:

CO is the main asphyxiant, and is responsible for 90% of deaths related to gas inhalation during a fire. Indeed, numerous post-mortem analyses have revealed carboxyhemoglobin levels sufficiently high to cause death (ISO 13571 [10], ISO 13344 [11]). In a fire, the amount of CO released is related to the amount of available oxygen and the nature of the burning material. In most situations, the rate of production of this gas is high, and the volumes produced are large. CO asphyxiation results from the complexation of this gas with blood hemoglobin to form a much more stable complex than the oxyhemoglobin normally formed with oxygen (oxyhemoglobin carries oxygen to the organs of the body by means of the blood circulation). Several studies have shown that a 50% reduction in blood oxyhemoglobin concentration results in rapid death [12]. The physical activity associated with escape is also a parameter to take into account when evaluating the quantity of inhaled CO, since the behavior of people during evacuation may double the rate of absorption of this gas. The level of carboxyhemoglobin in the blood that causes impairment of escape capacity is around 30% [4]. At an intracellular level, asphyxia also results from inactivation of certain enzymes responsible for respiration.

HCN is one of the gases that cause the most casualties in fires, always in combination with CO. This asphyxiant can even cause death several days after inhalation, or lead to chronic respiratory problems [13] and long-term neurological consequences. HCN molecules cause asphyxiation by inhibition of respiration in the mitochondria, which are the energetic organelles of the cells. The mechanism of asphyxia is in fact linked to the action of cyanide ions on cytochrome oxidase, the terminal enzyme of mitochondrial respiration.

To a lesser extent, carbon dioxide (CO_2) is also an asphyxiant to consider. This gas causes hyperventilation, which increases the amount of toxic gas inhaled. At high doses, its action produces narcotic effects similar to those of CO, the partial pressure of CO_2 being sufficient to replace the oxygen at hemoglobin level.

Asphyxiants affect the cardiovascular and nervous systems by under-oxygenation, which quickly leads to unconsciousness and death if the exposure is prolonged. At the beginning of intoxication, the effects are scarcely noticeable. However, as the duration of exposure increases, the individual is first in a state similar to drunkenness, which greatly affects their ability to escape, and this is quickly followed by loss of consciousness.

In all cases, the nervous and cardiovascular systems are the first to be affected by asphyxiants, because they are still active and consume oxygen. When cellular respiration is hampered, either by lack of oxygen supply (through the effect of CO or

CO_2), or additionally by histotoxic hypoxia of tissue cells no longer able to absorb oxygen (through the effect of HCN), the affected cells die quickly. It is important to note that the toxic effect depends on the activity of the nervous system and the cardiovascular system. Thus, the susceptibility to intoxication is very different between active and sedentary individuals. In fact, somebody who easily panics gets intoxicated even more quickly.

6.2.2 Irritants

The main irritants in smoke from fires are:

- Inorganic acids: hydrogen halides (hydrogen chloride (HCl), hydrogen bromide (HBr) and hydrogen fluoride (HF)), sulfur dioxide (SO_2) and nitrogen oxides (NO_x),
- Some organic compounds such as low molecular weight aldehydes (acrolein, formaldehyde, and acetaldehyde). Currently, more than 20 irritants have been identified in fire smoke.

When a victim is exposed to an irritant-containing atmosphere, they first feel irritation of the eyes, nose, throat, and lungs, in proportion to the concentration of irritating gas. For hydrogen chloride, it has been shown that the ability of people to escape is not compromised for concentrations between 100 and 500 µL/L. Above 1000 µL/L, escape becomes impossible [14, 15]. For higher exposures, a dose-effect appears and irritation begins in the respiratory tract.

6.3 Sub-Acute Effects of Fire Effluents

Sub-acute intoxication results from the repeated absorption of moderate doses of poison over a period of a few days to a few weeks. It is unlikely to occur in fires and therefore not dealt with here.

6.4 Chronic Effects of Fire Effluents

Chronic intoxication results from repeated and frequent exposure to low or very low doses of poison, spread over periods of a few months to a few years. This results either in the accumulation of toxicants in the body or in the addition of toxic effects (as is the case for carcinogens). These toxicants are often unknown but specific populations may frequently be exposed to them [16, 17]. This type of intoxication may lead to reversible or irreversible toxic diseases of varying degrees of severity, and even eventually to death by carcinogenesis for some particularly hazardous products [18]. Examples for exposed populations are:

- Firefighters, fire experts, and fire cleaning personnel regularly involved in fires,
- Persons regularly exposed to the consequences of a single major disaster that occurred in their environment (e.g., people living in the area contaminated by a single major fire).

Chronic exposure can be addressed with appropriate personal protection. However, chronic exposure must not be confused with delayed effects from acute exposure (which may sometimes occur several years afterwards). The major species leading to chronic toxicity in a fire are:

- Volatile hydrocarbons (VOCs), in particular monoaromatics such as benzene, toluene, styrene, and phenol,
- Polycyclic aromatic hydrocarbons (PAHs), with the reference toxicant being benzo[a]pyrene,
- Halogenated dibenzodioxins and furans, which are found in fire fumes of certain chlorinated products such as pesticides or polychlorinated biphenyls (PCBs),
- Metals, for which the toxic effect varies greatly depending on the exact compound present. For example, organometallic compounds are much better assimilated by the body than the metal alone and thus have a much greater toxic effect.

6.5 Environmental Effects of Fire Effluents

The effluents from major fires can harm the environment with substances that may have a long-term effect in the food chain. Sometimes, the deposits are found far from the original fire site. A toxic environmental effect will occur if the compound (or one of its metabolites) is a persistent pollutant, i.e., it remains stable in the environment.

In addition to the species mentioned above, two further classes of substances have to be considered:

- Endocrine disruptors, which are a family of substances mimicking hormones and having effects at very low doses. These include compounds such as nonylphenols, bisphenol-A, and phthalates, and are found in trace amounts in the fumes of many polymers. They can have unexpected consequences (e.g., causing the sex of amphibians or fish to change),
- Fire aerosols have a mechanical effect on plant leaves, reducing photosynthesis and plugging their stomata. In addition, adsorbed acid gases can chemically attack the leaf surface.

Persistent species, in particular persistent organic pollutants (POPs), are concentrated in the food chain via higher predators, but these aspects have hardly been studied so far. In addition, the dispersion of fire-extinguishing agents must be considered, since the chemical agents used can have an impact on the environment. For instance, perfluorooctane sulfonates (PFOS) are foaming additives used as extinguishing agents, and have been classified as POPs; they are currently banned from use in Europe.

6.6 Quantitative Risk Assessment of Fire Effluents

6.6.1 Introduction

The effect of a substance depends on the notion of dose, i.e., its concentration and exposure time. The effects of a toxicant can be immediate or delayed irrespective of the type of exposure considered, and correspond to tenability with respect to a given effect.

6.6.2 Direct and Indirect Methods

To correctly measure the effect of fire effluents, it is necessary to define what level to study: incapacitation or lethality. In addition, is desirable to know or search for the maximum number of potentially toxic species. For example, the most common species may represent 90% of the lethal potential of the fire effluents, but only 50% of their irritant potential.

The kinetic aspect of production is also essential. Some products emit very large quantities of toxin in the first minutes, while others release less, but over a longer time period. Unfortunately, it is difficult to determine at which time fire effluents are most harmful to people, but toxic effects can be predicted in two ways, as expressed in Figure 6.1:

- Direct methods involve exposing an animal population to combustion smoke and observing the effect. Although they provide better performance than indirect methods, they are less often used for ethical reasons,
- Indirect methods involve analyzing the fumes and then comparing the results obtained with those from models that take into account the joint effects of the different parameters.

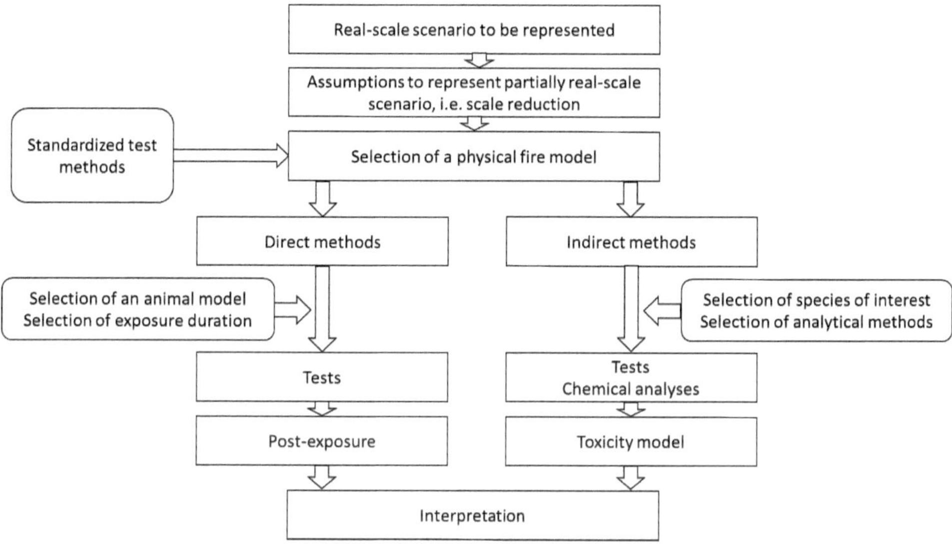

Figure 6.1 Direct and indirect methods to measure toxic effects of fire effluents

Indirect methods nevertheless have many limitations. For example, some models take into account the evolution of the doses recorded over a certain time period, while others only deal with the total quantity of each species recorded during a predefined time period. In addition, the methods used to analyze the species present in the fumes must have had their adequacy proven: the limits of detection, sensitivity, accuracy, and fidelity of the method or the existence of interferences are the key points in the choice of a method.

Finally, whether direct or indirect methods are used, it should always be kept in mind that a fundamental difference exists between small-scale fire tests in the laboratory and real fire situations, for which the variety of fuels, temperatures, chemistry, combustion and afterburning are essential factors.

6.6.2.1 Direct Methods

The toxic effect of fire effluents is difficult to evaluate directly. Basically, lethality can only be determined on animal models. However, long and costly animal experiments have become increasingly uncommon in recent years, and have even been totally banned for ethical reasons in some countries.

The German standard DIN 53436 [19] is an example of the experimental determination of the LC_{50} of a smoke gas mixture. This standard uses an apparatus to thermally decompose solid or liquid materials under certain conditions in an air flow, and rats are exposed to the toxic gases. Blood tests are then performed on the dead rats to assay for carboxyhemoglobin.

The first series of tests consists of setting the air flow in the decomposition tube at a constant value, and diluting the mixture of toxic gases from the decomposition products with fresh air. If the toxicity of this mixture is too high (100% dead animals) the amount of fresh air is increased (it should be noted that the oxygen level in the exposure chamber is always greater than 12%). The Relative Acute Inhalation Toxicity (RAIT) is determined as the ratio of the number of rats that died to the number of rats undergoing the test.

For the second series of tests, the air flow in the decomposition zone is set at a constant value. The mixture of decomposition gases is diluted by a constant amount of fresh air, different for each test. This gives different concentrations of gaseous decomposition products from which the concentration–lethality relationship is determined. This method allows the experimental determination of the LC_{50} of the material.

For tests with rats, DIN 53436 recommends to initially use the first method to set the appropriate volume of fresh air. The animals are then exposed for 30 minutes to the combustion products, and kept under surveillance for 14 days.

6.6.2.2 Indirect Methods

Indirect methods characterize smoke gases by means of appropriate analytical techniques. A model attempting to reproduce the cumulative effect of the different species analyzed is then used. It is obvious that the models are only effective:

- If the analyses allow characterization of all the species present in the smoke gases that have a toxic effect,
- If the model takes into account the relative effects of all species considered.

In addition, the model can consider the kinetic aspect of intoxication. In this case, analytical techniques used reflect the dynamics of concentration changes. At present, these indirect methods are the only ones in common use. It is important to note that there is no universal method for the analysis of toxic gases in smoke, and many different techniques are required to analyze the smoke gases. One of the key points when selecting a method is its suitability for the gaseous matrix constituting the smoke, its quantification and detection limits, and its accuracy and fidelity (repeatability and reproducibility) for the chosen area. The methods are mainly presented in ISO 19701 [20] (classical methods) and ISO 19702 [21] (Fourier transform infrared method), as well as in reference [7].

6.7 Generation of Fire Effluents: Physical Fire Models Used for Toxicity Testing

6.7.1 Introduction to Physical Fire Models

Studies of life safety in the event of fire include the evaluation of smoke toxicity and fire risk assessment. This type of determination requires data on the toxic effect of the fire effluent. Data can be provided by tests of various sizes, from laboratory scale to real scale, and these tests are performed to reflect the real fires in terms of reproducing effluent generation. Nevertheless, a real fire is characterized by important spatial and temporal variations in combustion conditions, so that small-scale tests cannot describe all fire conditions. The role of a physical fire model for generating realistic toxic effluent compositions is to reproduce on a reduced scale the essential features of the complex thermal and reactive chemical environment in real fires. The bench-scale test, representing a physical fire model, has to be defined in accordance with the knowledge of the fire stage it represents. The accuracy of the physical fire model to map a given scenario depends on how the effluent data obtained can be scaled up to reflect real fire conditions and on how well the bench-scale apparatus represents specific fire stages [8].

The physical fire model is also designed to provide data on animal exposure (direct toxicity evaluation) or to allow gas analysis (indirect toxicity evaluation). In the latter case, a small number of gases can be analyzed to represent the toxicity of the mixture, with an exposure model such as FED/FEC described in ISO 13571 [10]. Nevertheless, only animal exposure studies allow consideration of the effects of unusual or unexpected toxicants, if the latter cannot be precisely measured. In some validation cases, it is therefore essential to use animal exposure, despite all the ethical and logistical problems involved.

6.7.2 What is a Physical Fire Model?

As described in ISO 16312-1 [22], a physical fire model is characterized by requirements such as the shape of the test specimen, the operational combustion conditions, and the capability for analyzing the combustion products. For providing data for effluent toxicity assessment, the validity of a physical fire model is determined by the degree of accuracy with which it reproduces the yields of the principal toxic components in real-scale fires.

As presented in ISO/TR 16312-2 [23], there are numerous physical fire models in use today for generating some form of toxic potency information. The combustion conditions vary widely and are generally not well characterized. There has been little rigorous validation of the related equipment; in fact, there is no standard protocol for performing such a validation. One of the reasons is their design: many of them were developed to address parameters other than toxicity, like the smoke box chamber designed to measure conventionally generated opacity. Others were designed to be as simple as possible and robust enough to produce data for ranking of materials (by considering their production of gases under given conditions), and not really to produce toxicity data for fire risk assessment.

In addition, there is no standard protocol for obtaining a test specimen that will burn like the finished item it has to reproduce, nor is there quantitative information regarding the sensitivity of the toxic potency measurement to the design of the test specimen.

The physical fire model is not only characterized by the generation of gases in the location where combustion occurs, but also by the transportation and conservation of individual effluents to the place of analysis, as presented previously. It is therefore essential to determine the variations of this gas in terms of evolution with time, and to take into account phenomena like adsorption, condensation, absorption or dilution from generation to analysis. It is also essential to control the degradation conditions in terms of all parameters driving degradation. The aspects related to the operator, the robustness of the method used, and interpretation are also important factors that can help in method selection.

Another point to consider in a physical fire model is its relation to end-use. For example, a given bench-scale test can be adapted to produce input data for fire modeling, while another larger-scale test is needed to ensure validation of the numerical tool used for prediction. Forensics may require a method different from ranking materials in a robust way. All these aspects have to be considered when a method is chosen.

The diversity of methods available and presented here is to remind the user that there never will be only one bench-scale test and that a bench-scale test will never reproduce reality; it only tries to describe a given aspect of the toxicity problem.

6.7.3 Important Characteristics to be Reproduced by the Fire Model

6.7.3.1 Parameters Influencing the Nature and Proportion of Gas Release

The production of gases and derived toxic fire effluents is highly dependent on the fire conditions. The environmental conditions that characterize the stages of a fire and a physical fire model are:

- Initial temperature,
- Temperature where the combustion occurs, or heat flux to the fuel surface,
- Mass loss rate,
- Oxygen concentration at the fuel surface,
- Oxygen concentration at the flame boundaries,
- Availability of fresh oxygen to replenish that depleted by combustion (ventilation rate and mixing),
- Surface effects like charring or intumescent processes that disconnect the pyrolysis front from the surface.

Once a fire has started, the developing situation in terms of the basic fire scenario, the chemistry of combustion, and the effects on the occupants depend on the interaction of the developing fire with the fuel items first ignited and the characteristics of the fire enclosure (especially its size and ventilation). It is possible to classify fires into four main types:

- Non-flaming/smoldering fires,
- Early well-ventilated flaming fires,
- Ventilation-controlled fires (pre-flashover vitiated flaming fires),
- Ventilation-controlled fires (post-flashover vitiated flaming fires).

For the purpose of designating combustion conditions for tests, key characteristics of different fire stages are proposed in ISO 19706 [8]. All these aspects are used to define the ability of the physical fire model to reproduce useful data. Most recent physical models include the knowledge of each parameter used to represent a fire stage.

6.7.3.2 Combustion Conditions

The yields of combustion products depend on apparatus conditions such as the fuel/air equivalence ratio, whether the decomposition is flaming or non-flaming, the flaming persistence of the sample, the temperature of the specimen and the effluents produced, the stability of the decomposition conditions, and the interaction of the apparatus with the effluents and the flames in the decomposition process.

6.7.3.3 Thermal Conditions of Degradation and Equivalence Ratio

Combustion is a chemical reaction in the gas phase between a fuel gas and an oxidant, generally the oxygen of the air, in the presence of an energy source (external thermal source). Decomposition is an endothermic phenomenon, which requires energy higher than the bonding energy between atoms. The decomposition mechanism depends on weak connections in the system and the presence (or not) of oxygen. Mostly, decomposition results from the combined effects of heat and oxygen.

A distinction is made between non-oxidizing and oxidizing thermal degradation. Non-oxidizing thermal degradation occurs as depolymerization generally initiated by chain scission. In many cases, degradation is based on an oxidation mechanism implying the free radicals of the macromolecular chains, eventually releasing the very reactive radicals $H^{\bullet}$ and $OH^{\bullet}$.

The thermal degradation of the polymer matrix is governed by physical and chemical reactions in the gas phase, condensed phase or both. Considering the physical aspect, the polymer decomposes at specific temperatures and forms combustible volatile degradation products, leading to continuous chemical and physical restructuration of the material during combustion.

Flame feeding by polymer degradation products can result in their progressive transformation into a carbonaceous structure or char barrier limiting the diffusive transfer of volatiles towards the surface of the burning material. These physical actions in the burning polymer are combined with chemical interactions, which decrease the number of combustible volatiles feeding the combustion cycle.

Various thermal and oxidative degradation mechanisms take place in both phases. Factors affecting the thermal degradation are heat flux, oxygen concentration, and temperature. The presence of oxygen appears to alter to varying degrees the decomposition mechanisms in all materials studied. For most materials the yields of toxic species have been shown to depend critically on the fire conditions.

The combustion conditions (e.g., thermal boundary conditions, gasification rate, oxygen availability, and mixing) are rarely uniform along the surfaces of the test specimen and are approximated by their mean values. Two physical fire models with the same calculated global equivalence ratio are still likely to show some differences in the regions where the combustion products are actually formed. The equivalence ratio represents the availability of oxygen compared to the stochiometric need of oxygen for combustion. There are two main issues to consider when expressing the equivalence ratio in physical fire models:

- The equivalence ratio can vary with time. In the flaming regime, the availability of oxygen can be reduced and combustion conditions changed so that average values during the test may not apply within a given period of time. To limit this effect, specimen mass loss and air mass have to be measured and monitored continuously,
- Local and global equivalence ratio are often confused. The combustion conditions are variable anywhere in the flame, because a flame has no proper shape. The mixing of fuel gases and air drives combustion, so that only the global equivalence ratio is relevant. This limitation is reduced when combustion conditions are relatively constant, as with laminar flames.

6.7.3.4 Parameters Influencing the Preparation for Analysis

Parameters driving the degradation and production of effluents from a physical fire model have been presented. Fire effluents evolve from their production area to the analysis area. Proper measurements have to consider the many parameters listed below:

- The effluent is transported and can be modified by dilution. This effect is important, and the dilution factor has to be known. In addition, leak tests of the installation have to be made to limit the undesirable introduction of air or release of gases.
- The effluent can be chemically and physically modified:
  - Some products have a short lifetime so that analysis has to be made quickly after their formation. For example, formaldehyde starts to polymerize within a few minutes. If the test is made in two steps, i. e., during degradation and after analysis, or if the effluent's path to detection is too long, losses can occur.
  - If gases pass through hot areas after their production (as is often the case in tube furnaces or the cone calorimeter), their composition can be affected. Some products like aldehydes are destroyed at high temperatures, and ratios between CO and CO_2 can be modified.
  - Materials present in the gas path must not interfere with the gas. As some species present in fire effluents may react with some materials, the components should be made of inert material, like stainless steel (passivated or coated), glass or quartz (to be avoided when sampling fluorine species), or PTFE and other fluorinated polymers. Combustion reactors are often made of quartz and therefore sensitive to HF.
  - Condensation on the apparatus can limit the amount of highly hygroscopic species like HCl or NO_2. Soot deposition induces adsorption and desorption, affecting analysis dynamics and producing retention effects.

6.7.3.5 Parameters Influencing the Scale Effects

It is essential to remember that all modern bench-scale tests used for toxicity measurements are designed to represent real-scale fires. The way to correlation may be more or less complex, using similarity equations or complex fire modeling, but it is always defined regarding toxicity aspects. Several approaches to show the toxic potency of fire effluents have been proposed [24]:

- Lethality (LC_{50}) data for laboratory animals, which is the concentration of fire effluent that causes death or incapacitation of 50% of a population of a given species during a specified exposure time (typically 30 min) and post-exposure time,
- Concentration of combustion products, which is a time-dependent mass generation rate and the dispersion of that mass into a volume. The magnitude of the volume and the air flow through that volume determine the concentration as much as the source term. This is also true in a physical fire model. It is more appropriate to characterize the effluent by the set of yields (mass of a combustion product per mass of consumed fuel) of the components, and remove transport and dilution effects,
- Yield of carbon monoxide and other significant toxic combustion products. When evaluating a physical fire model, the yield of all gases that represent toxicity of the effluent must be included in the comparison with real-scale test data, and this comparison should not be limited to CO or CO_2.

The validation process based on the yields of toxicants generated during the combustion of a material is summarized below:

- The accuracy assessment should be performed for each of the fire stages according to ISO 19706 needed for fire safety engineering. An assessment for one such stage is not necessarily sufficient for appraising the value of a physical fire model for a different fire stage,
- The combustion conditions in the physical fire model should be consistent with those in the real-scale fires of reference. The use of equivalence ratios presented previously helps to quantify these terms,
- The specimens tested at both scales must be composed of exactly the same materials. This is often an issue, because bench-scale tests are more linked with material tests, while real-scale conditions are related to finished products that very often cannot have been sampled properly for bench-scale tests. For combustibles consisting of multiple materials or complex products, bench-scale test specimens should have all the materials present in a similar proportion by mass and in a similar conformation, until it is demonstrated that the yields of toxicants are not sensitive to the conformation of materials in the test

specimen. It limits very often the usage of smallest bench-scale tests to assess toxicity of real complex products,

- The list of measured toxicants should include the most important contributors. A quantitative comparison is necessary between the yield data from the physical fire model and the data from the room-scale fires. It should also include chemicals that indicate differing degrees of oxidation of the materials in the tested item. For carbon-containing materials, the list should include CO_2, CO, and aldehydes, while for nitrogen-containing materials NO_2 and HCN should be listed. Sulfur-containing compounds may produce SO_2 and a variety of other species (H_2S, COS, CS_2, etc.), while halogenated materials only generate halogenated acids (HCl, HBr, HF).

The ratio of carbon dioxide to carbon monoxide (CO_2/CO), often used as a descriptive characteristic of a fire, also depends more on the ventilation conditions of the fire than on the nature of the materials being burned [25, 26]. Studies on the dependence of CO_2/CO ratios on the equivalence ratio show that, for well-ventilated fires, i. e., $\varphi \leq 1$, essentially all the fuel carbon is oxidized to CO_2. Once the equivalence ratio has exceeded that associated with flashover ($\varphi > 1$), indicating a fuel-rich or ventilation-controlled fire, two plateaus are seen.

All these aspects are linked with the species to be analyzed. It is important to consider all the effects driving the ability of the bench-scale physical fire model to reproduce larger scales. Scale-by-scale validation is sometimes needed to allow consideration of all the parameters. These aspects are presented in Table 6.1, showing that many scales can be defined, while very few scales are representative of available robust and standardized toxicity tests.

Table 6.1 Existing Physical Fire Models for Toxicity Testing

Physical fire models	Fire stage(s), ISO 19706	Data produced	Limitations
Cup Furnace Smoke Toxicity Test Method No national or international standard	Oxidative pyrolysis Well-ventilated flaming	Measurement of: total mass loss, averaged mass consumed and mass charged concentrations, gas concentrations, and gas yields The gases to be measured are: CO_2, CO, O_2, HCN, HCl, and HBr	The toxic potency and gas yield data does not replicate real-scale post-flashover test data well. The method has not been assessed against real-scale test data for oxidative pyrolysis or well-ventilated flaming

Table 6.1 Existing Physical Fire Models for Toxicity Testing *(continued)*

Physical fire models	Fire stage(s), ISO 19706	Data produced	Limitations
Radiant Furnace Toxicity Test Method (United States) NFPA 269 and ASTM E1678	Well-ventilated flaming combustion Under-ventilated flaming (using a post-flashover correction for the yield of CO) Oxidative pyrolysis, if the sample does not autoignite	Measurement of mass loss and gas concentrations, gas yields, and atmosphere vitiation. In addition, the procedure includes measurement of the mortality of six rats and documentation of any physiological harm, performed post-mortem	The output of this method for three products has been compared to room-scale data for the same products (wall lining configuration). The post-flashover LC_{50} values were well within a factor of 2. No data were taken for pre-flashover combustion
Closed Cabinet Toxicity Test (International) ISO/TS 19021 NFPA 270 IMO FTP code part 2	Oxidative pyrolysis Well-ventilated flaming combustion Material is exposed to a radiant heater for a minimum of 10 min. A test is conducted at 25 kW/m^2 with and without pilot flame, and at 50 kW/m^2 without pilot flame	The standard procedure includes measurement of total mass loss, smoke obscuration, and specific effluent gas concentrations (CO_2, CO, HCN, HCl, HF, HBr, NO_x, SO_2) at the time when the maximum smoke concentration is reached	There are no reported comparisons of toxic gas generation with data from real-scale fire tests Equivalence ratio affected by accumulation of smoke in the test chamber Deposition of gas on soot and walls
NES 713 (United Kingdom)	Well-ventilated flaming However, this may not relate to a real fire, as the burner is actually a premixed blow-torch-type flame at about 850 °C and not a free-burning fire	The standard procedure includes measurement of CO, CO_2, formaldehyde, NO_x, HCN, acrylonitrile, phosgene, SO_2, H_2S, HCl, NH_3, HF, HBr and phenol. Corrections are applied for the concentrations of CO, CO_2 and NO_x produced by the gas flame alone burning for the same period as the test specimen	There are no reported comparisons of toxic gas generation with data from real-scale fire tests Equivalence ratio affected by accumulation of smoke in the test chamber Deposition of gas on soot and walls
Japanese Toxicity Test Method 1231 of the Japanese Ministry of Construction	Small, under-ventilated fire	Measurement of the times to incapacitation of the mice In addition, gas samples can be extracted for external analysis Following the test, blood samples can be extracted for analysis	No comparison against real-scale fire tests has been published

Physical fire models	Fire stage(s), ISO 19706	Data produced	Limitations
Cone Calorimeter (International) ISO 5660-1, ASTM E1354, NFPA 271, and NFPA 272	Oxidative pyrolysis Well-ventilated flaming Other stages possible with vitiated atmosphere module	Measurement of mass loss and effluent gas concentrations, gas yields, smoke obscuration, and exhaust gas vitiation The exhaust duct flow is pre-set	The toxic potency and gas yield data did not replicate real-scale post-flashover test data well. The method has not been assessed against real-scale test data for oxidative pyrolysis or well-ventilated flaming Losses on duct possible High dilution of effluents limits analysis possibilities
Flame Propagation Apparatus (United States/International) ISO 12136, NFPA 287, and ASTM E2058	Oxidative pyrolysis Anaerobic pyrolysis Well-ventilated flaming Under-ventilated flaming	Measurement of mass loss and effluent gas concentrations, gas yields, smoke obscuration, and exhaust gas vitiation	CO yields have been related to real-scale results
University of Pittsburgh Tube Furnace - UTPITT (USA) No national or international standard	Oxidative pyrolysis (varying fuel-to-air ratio) Well-ventilated flaming Low-ventilated flaming	Measurement of mass loss, gas concentrations, and gas yields (with additional instrumentation) The respiratory rate changes and times of death (if any occur) of four mice are determined within the exposure period The animals are observed for a 10 min post-exposure period. Any physiological effects are determined post-mortem	Efforts to validate the test results against real-scale fire data were unsuccessful due to the lack of smoke accumulation for the duration of the test and the inability to relate the time-dependent test results to the real-scale burning

Table 6.1 Existing Physical Fire Models for Toxicity Testing *(continued)*

Physical fire models	Fire stage(s), ISO 19706	Data produced	Limitations
Tube Furnace (France/NATO) NF X 70-100 AFAP 3 UITP E7	Oxidative pyrolysis (low- and well-ventilated), if sample does not autoignite Well-ventilated flaming Under-ventilated flaming Tube furnace at 350 °C, 400 °C, 600 °C, and/or 800 °C	Measurement of total mass loss, concentrations and yields of CO, CO_2, HCl, HBr, HF, HCN, SO_2, NO_x (NO and NO_2), formaldehyde and acrolein using HPIC, HPLC and classical analytical methods. FTIR can also be used	Reasonable correlations were found for toxicity of structural materials between (a) the toxicity index found with the NFX 70-100 method (at 400 °C and 600 °C) when combined with mass loss, determined using ISO 5660 and FED No knowledge of equivalence ratio within time
Tube Furnace (United Kingdom/International) BS 7990 ISO/TS 19700	Oxidative pyrolysis Well-ventilated flaming Under-ventilated flaming Steady-state degradation with a controlled equivalence ratio	Oxygen concentrations are measured to confirm the fire stage, and as input for estimation of hypoxia The concentrations of CO_2, CO, HF, HCl, HBr, NO, NO_2, HCN, SO_2, H_3PO_4, acrolein, formaldehyde, and a range of other organic species may be measured as gas concentrations in the diluted fire effluent or collected for a fixed period through bubblers This method is intended primarily for chemical analysis, animal exposure can be carried out for irritancy, acute lethality, and other toxicological investigations	Published work shows a correlation between CO yields in real-scale fires and those found in the tube furnace Possible deposition on walls of mixing chamber Complexity of usage

6.7.3.6 Parameters Influencing the Quality of the Fire Model

The assessment of any fire toxicity test method is based on four aspects:

1. Quality of the generation of gases,
2. Quality of their transport to the analytical part or to the animal exposed,
3. Quality of the analysis itself,
4. The material-related effects on the parameters measured.

This quality is driven by notions of trueness, repeatability, and reproducibility.

For the generation process itself, quality is linked with air speed, oxygen concentration, and thermal environment. The repeatability and reproducibility of the generation process can be therefore improved by suitable metrology and calibration operations. Reducing the variability in experimental parameters automatically improves reproducibility. Nevertheless, the gas generation process may be slightly different for specific species. If temperature is high enough, materials will (for example) release halogenated acids with a quite low sensitivity to the conditions, while the formation of aldehydes is very dependent on the thermal conditions.

The second aspect concerns transport. Hygroscopic species are more sensitive to moisture and can be absorbed more easily than CO or CO_2. The effect of transport is important and affects the design of test equipment: examples are devices to limit condensation or heated walls, as well as short transport time. Also important for transport are material parameters such as homogeneity, sample size, and shape.

Estimations of repeatability and reproducibility of the apparatus itself, without considering material effects, can be performed on CO/CO_2 by using products like PMMA or zinc carbonate, and checking the carbon balance. It is nevertheless difficult to perform such analyses on other species. Tests were done in the past to estimate losses of HCl when burning reference rigid PVC of known composition; they showed important deviations between laboratories.

The characteristics for a desired method can partially be found in methods available at the present time. This implies the development of well-defined normative tools to allow adequate toxicity measurement. Parameters related to scale have proven that a unique bench-scale test will never be able to reproduce real products behavior at each scale. Nevertheless, the development of an accurate test method is needed for promoting fire safety engineering.

Production and analysis of fire effluents is complex. One of the major issues identified is the limitation of repeatability and reproducibility of bench-scale testing. Improving accuracy of parameters related to the test equipment design (materials used, temperature of walls, and residence time of the effluent), ventilation (oxygen availability and air speed rates) and thermal attack (calibration in flux and temperature) will limit this uncertainty. However, it is difficult to make such calculations for each gas produced.

6.8 Chemical Analysis of Fire Effluents

6.8.1 Gases of Interest

In international and national standards dealing with the evaluation of the effects of fire effluents, the way to characterize the toxic compounds present in the effluents is to analyze for predefined gases considered as the main toxicants. For routine evaluation, a complete and exhaustive characterization of the effluents is never the objective, as it will be very difficult to achieve and time-demanding. Other gases may be considered, and especially when taking into account long-term effects, organic species such as benzene and toluene may be of interest. In addition, if the target is not acutely toxic, other gases may be taken into account; but the evaluation of such components is not yet considered in regulations related to fire. In Annex A of ISO 19701 [20] important species and minimum concentrations considered as relevant limit values for acute toxicity evaluation are listed. Examples of typical minimum concentration required for fire effluents are listed in Table 6.2.

Table 6.2 Minimum Required Concentrations for Important Species

Species	Minimum concentration required (μL/L)
Carbon monoxide	100
Carbon dioxide	500
Hydrogen fluoride	10
Hydrogen bromide	10
Hydrogen chloride	10
Hydrogen cyanide	10
Sulfur dioxide	1
Nitric oxide	10
Nitrogen dioxide	10
Acrolein	0.5
Formaldehyde	0.5
Total aldehydes	5

6.8.2 Sampling

6.8.2.1 From Fire Source to Analysis

Before considering the chemical analysis of the effluents, sampling has to be looked at with particular attention, because it is the most critical part of the procedure for analysis of fire effluents. The samples available to the analyzer should be as representative as possible of the fire atmosphere without being altered by the sampling system. The sampling procedure should be kept as simple as possible and adapted to the respective fire source. The sampling line must be capable of operating with minimal interference, and without condensation of the species to be analyzed.

When analyzing fire effluents, water evolving from combustion is an important factor to be considered. In order to avoid water condensation, heated sampling devices or traps containing drying agents should be used.

The main components for a sampling line are:

- The sampling probe usually positioned at the fire source,
- The sampling line to transport the gases to the analyzer,
- The filtering system to protect analyzers,
- A pumping device to drive the gases to the analyzer.

For integrated sampling, the type of gas collector should also be considered.

6.8.2.2 Probes

As some species present in fire effluents may react with some materials, the probes must be made from an inert material. Materials used for probes fall into three main types:

- Stainless steel (passivated or coated). Special care should be taken when sampling corrosive gases; it is recommended to check the probe for corrosion before testing,
- Glass or quartz (to be avoided when sampling fluorine species),
- PTFE and other fluorinated polymers.

When defining a sampling line, dimensions such as the length and the diameter of the probe vs. the fire source have to be taken into account. For example, if the probe is needed for large-scale fire tests, a long probe will be preferred (e.g., sampling for the ISO 9705 – Room corner test [27, 28]) while for a small-scale test a short probe (or even no probe at all) is more suitable (e.g., French tubular furnace where no sampling probe is used).

The location and size of sampling probes is influenced by the size of the test apparatus and the requirements of the analytical system. The positioning of sampling probes is specific to the apparatus. In general, the possibility of the stratification of gases in test chambers without good mixing should be considered and sampling in the proximity of a test chamber wall avoided. In order to provide the analyzer with a representative sample of the fire effluents, probes must allow sampling a homogeneous part of the fire smoke. In the vicinity of the fire source, with the exception of some small-scale test methods, stratification of smoke could affect the homogeneity of sampling; multi-hole probes positioned at different levels of the smoke layer should improve sampling homogeneity. In addition, the probe should be defined to minimize the perturbation of the effluent flow. For example, in the cone calorimeter test, a circular multi-hole probe positioned in the exhaust duct has been found suitable.

The ease of mounting and cleaning of the probe should also be considered, as particles accumulating on the walls of the probe could lead to the clogging of the probe.

Probes are usually positioned in the vicinity of the fire source. Therefore, the probes must be made of a material able to resist high temperature (or the temperature level expected at the sampling point). The probe must also be inert in regard of the fire effluents (e.g., glass cannot be used when HF is expected to be found in the fire effluents).

6.8.2.3 Filters

The filtering system protects the analyzers or part of the collecting system (e.g., frits for bubblers). The filters can range from glass wool plugs to sophisticated heated filters. The filtering media must be carefully selected. The main parameters to consider when defining a type of filtering system are:

- Amount of soot expected,
- Volume of effluent to be passed through,
- Inertia of the filtering media in regard of the fire effluents,
- Porosity of the filter,
- Whether the filter is heated or not heated,
- Size.

An important point is the compatibility of the filter with the analyzed effluents. Filters trap gaseous species, and some may be damaged by chemicals or the heat of effluents. A common example is glass fiber filters, which are sensitive to HF in effluents because HF reacts with silica. Another example is paper filters, which may release hydrochloric acid when heated. Glass fiber, stainless steel, and ceramic filters are the main types used. The two latter are reusable when heated at high temperatures, but are expensive. Another important point is the retention capacity of the filter. A high filtration surface will limit plugging but will affect the response time. Soot retained on filters may also trap species on its surface and release them slowly (adsorption/desorption phenomena).

To summarize, there is no optimum type of filter in terms of geometry or nature. Filters modify gas response curve, especially when partially or totally plugged, and this plugging also affects the sampling flow rate. Filter selection depends on effluents and soot concentration, the nature of these effluents, the sampling flow rate, and the duration of sampling.

When using a heated filter, it should be heated to a temperature of at least 100 °C to minimize the condensation of water and dissolution of gases in water evolving from combustion. In addition, when using analytical methods sensitive to water, a trap containing a drying agent (e. g., calcium sulfate or calcium chloride) or a cold trap may be placed just before the analyzer and after any heated section of the sampling device.

Filtering systems and water traps may retain some of the gases to be tested for toxicity evaluation. For example, HCl may be adsorbed on soot or dissolved in the condensed water, so that part of the HCl produced during the fire could be lost for the analysis. Depending on the type of analysis to be carried out, there are two ways to avoid the partial or total loss of gases:

- Rinsing the filters with an adequate solution to solute the components of interest,
- Analyzing separately (e. g., in different burns) the gases, with regard to their affinity to water and/or the need to protect analyzers.

6.8.2.4 Sampling Line

The sampling line is used to drive the gases from the fire source to an appropriate area for the analysis. The material constituting the line is chosen with regard to its inertia to fire effluents; however, it is nearly impossible to find a material that will have no effect on the gases collected. PTFE has been found to offer the best performance; however, stainless-steel lines may also be used. The general requirements for a line are:

- Inner diameter: 3 or 4 mm has been found suitable,
- Length: as short as possible. The length will depend on the fire model (e. g., for ISO 9705 a 4 m line is adequate, while for a small-scale test 1 m is adequate),
- Ability to be heated: for FTIR gas analysis, a sampling line temperature of at least 150 °C is recommended to avoid condensation or secondary reaction of chemicals.

When choosing a sampling line, a compromise has to be found between inner diameter and length to optimize the residence time of the gases in the system. The objective is to limit their chemical transformation.

Potential condensation of chemicals on the walls of the line is a problem. Trapping of gases, particularly of acid gases, has been demonstrated. Although saturation behavior has been described and pre-aged lines have been found to be less sensitive to trapping, it has not been characterized further and thus a recommended method of ageing the sampling line is not yet available.

6.8.2.5 Pumping System

A pump is needed to transfer the effluents to the analyzer. There is no specific pump recommended, but characteristics such as the flow rate and the pressure in the system are important. Usually a compromise between high flow rate and the clogging effect due to a high flow is needed. To avoid any cold point in the sampling device or modification of the fire effluents, it is recommended to place the pump after the analyzers. Sometimes, the pump has to be heated and designed to limit pressure variations, when used with specific techniques like FTIR. Pumps must be connected to a control system (pressure or mass flow) to bring the sampling conditions under control during the whole analysis.

6.8.2.6 Sampling Conditions

Sampling may involve either continuous, on-line analysis (e.g., FTIR gas analysis) or non-continuous batch sampling (e.g., absorption in specific solutions using bubblers). Batch-type sampling can be further subdivided into two categories: instantaneous or integrated over the duration of the test.

The distinction between the two categories is based on the time of sampling. While instantaneous sampling uses a short collection time (e.g., < 1 min), integrated sampling collection takes place over a longer period of time (i.e., a substantial portion of the total test period).

Furthermore, sampling methods can be applied to either cumulative analysis or dynamic analysis. The final use of the information is also an important factor: namely, ranking of products using conventional methods or evaluation for use in fire safety engineering.

For some analytical methods, the gases must be collected in a specific "medium", whether further preparation is then needed before analysis, or the medium is analyzed directly. For direct analysis (e.g., FTIR) the gases must be free from particles before they are introduced into the analyzer.

6.8.2.7 Gas-Solution Absorbers

Absorption of gases in solution by the use of gas-washing bottles, bubblers, impingers, etc., always relies on the same principle: the test atmosphere is drawn or pushed through the absorbing media at a measured rate for a specified period of time. At the end of the sampling period, the solution is analyzed for the species of interest (e.g., the chloride ion for absorption of hydrogen chloride gas in water). Assuming 100% efficiency (see discussion below), it is possible to calculate the concentration of the species in the gas phase, as measured in the solution.

The volume of the absorber solution and the total gas flow directly affect the ratio of the gas and solution concentrations. For a given gas concentration, the smaller the solution volume and/or the larger the gas volume sampled, the higher the solution concentration. The choice of sampling conditions is governed by the requirements of the analytical technique, including the volume and sampling rate tolerated, expected gas concentration in the test atmosphere, necessity for frequent sampling, etc.

The absorption efficiency of a gas in a liquid is affected by the following:

- Solubility of the gas in the solution,
- Physical characteristics of the absorber,
- Ratio of gas flow rate to solution volume.

Generally, absorption efficiency is estimated empirically by allowing the flow of a known concentration of the gas of interest through a series of impingers and measuring the "break-through" from the first impinger (i.e., whatever is collected in the other traps). Another check of the efficiency of a given flow/impinger system is to conduct a series of experiments with a known concentration of gas, using different impingers and various flow rates.

There are basically four types of gas-solution absorbers: simple gas-washing bottles (including midget impingers), spiral or helical absorbers, packed glass-bead columns, and fritted bubblers. The gas-washing bottles, or impingers, function by drawing the gas through a tube (usually with a constricted opening), which is immersed in the trapping liquid/solution. This type is most suitable for highly soluble gases because contact time between solution and gas is short, and bubble size is relatively large. For less soluble species, the other absorbers offer longer contact time and/or smaller bubble size (which increases relative surface contact). The spiral or helical absorbers are built in specialized shapes to allow a long contact time, but the flow rate in these bubblers is limited because of the possibility of trapping solution over-flow with high flow rates. Packed glass-bead columns allow increased gas/liquid contact by dispersing the bubbles through a bed of glass beads, and flow rates can be higher than for the spiral absorbers.

The fritted bubblers contain a sintered or fritted disc on the gas inlet tube to disperse the gas into fine bubbles (the size of the bubbles is dependent on the porosity of the frit). It is necessary to exercise caution in using such bubblers, because if frothing occurs coalescence of the fine bubbles can defeat the purpose of the frit. Also, it is necessary to filter smoky atmospheres (containing particulates or liquid aerosols) before drawing them through a fritted bubbler in order to prevent clogging of the frit (which occurs very easily). Such clogging can also occur from the build-up of wax-like deposits. Certain gas species (e.g., HCl) can be absorbed onto a filter, especially if particulates have also been trapped on the filter.

6.8.2.8 Solid Sorption Tubes

Solid sorption tubes are an alternative method to gas-solution absorbers for sampling certain gases from fire effluents. Following sampling, the species of interest is desorbed in water and analysis can be performed in a way similar to that for aqueous solution absorbers. The advantages of solid sorption tubes over solution absorbers are:

- Ease of handling,
- Compactness,
- High absorption efficiency,
- Ability to be located directly at the point of sampling.

This latter advantage can have dramatic consequences in the measurement of HF, HCl, and HBr in fire effluents because these species are easily lost to the inside surfaces of sampling lines. With solid sorption tubes (except in conditions of extreme heat), a sampling line is not necessary before the sorption tube itself. All associated hardware (e.g., valves, flowmeters, and pumps) can be located behind the tubes, even far from the sampling point. This ensures that the sample is as representative as possible of the fire atmosphere.

Much experience has been gained through using solid sorption tubes, for example in the field of atmospheric sampling and for staff exposure monitoring in the workplace. Similar tubes have been re-examined for potential use in sampling fire effluents.

Calculation of the original gas concentration (e.g., HCl) from the representative species recovered in the desorption solution (e.g., Cl^-) is the same as that described for solution absorbers. The same considerations that apply to solution absorbers (namely inefficient absorption, breakthrough, and the relationship of volume sampled to gas and solution concentration) also apply to the use of solid sorbents. Instead of bubble size, it is the particulate size of the absorbent that is important (large particles offer less surface area per unit volume and more opportunity for channeling, while smaller particles can cause the tube to plug when sampling moist gas). The tubes must be small enough (typically 100 mm long, 6 mm OD) so that two tubes can easily be placed in series to allow for the possibility of "breakthrough" from the first tube.

Solid sorption tubes are subject to clogging due to soot collection. This can be recognized during sampling by a decrease in sample flow rate. The same flow rate should be maintained over the duration of sampling using a constant flow device; otherwise, an error is introduced in the calculation of gas concentration. A glass wool plug loosely packed into the inlet of the tube reduces the tendency to blocking from soot. Thermal desorption of the adsorbed sample is also possible; the sample tube is heated in an inert gas stream thus driving off the sample without the need for a liquid solution stage.

6.8.2.9 Gas Bags

Sampling with gas bags is another possibility when dealing with fire effluents, and are used when gas-phase analytical methods are being considered (even those involving sorption tubes). However, gas bags should only be used when other sampling systems are not suitable. As the bags are generally not heated, condensation of species (e.g., water, benzene, etc.) may occur between sampling and analysis. Hygroscopic species may then dissolve in the condensed water and therefore not be available for gas analysis. In addition, secondary reactions may occur in the bag, as the temperature of the effluents will decrease before analysis (e.g., polymerization of formaldehyde).

Sampling with gas bags can be used for most analytical methods. The test atmosphere is pumped, or allowed to flow under pressure, into a gas bag at a measured constant rate for a measured time period, to give a known volume of sample in the bag. It is necessary to filter the fire effluents before passing into the bag; simple in-line glass wool filters for particulates, and calcium chloride filters for moisture, have been found to be effective. However, a calcium chloride absorbent removes water vapor and water-soluble gases. At the end of the sampling period, the bag may be stored before it is connected to the analyzer. Storage times in bags should be kept to a minimum, preferably less than 1 h. Gases such as HF and HCl can dissolve in condensed/trapped water, thus reducing the concentration sent to the analyzer.

Bags must be gas-tight and inert, and those with a lining of poly(vinyl fluoride) (PVF) are recommended.

6.8.3 Selection of Analytical Methods

A large number of analytical methods are available on the market; however, not all of them can be used for fire effluent analysis. In addition, the analytical method will vary depending on the type of sample (solutions (e.g., from bubblers as described earlier), direct measurement of the effluent gases, desorption from sorption tube, etc.). ISO 19701 and ISO 19702 standards give a description of the methods that have been proven to be efficient for fire effluent analysis. They are listed in Table 6.3 and information on their limitations is also given.

Table 6.3 Summary of Analytical Methods for Major Acute Toxicants

Analyzed gas		Method name	Method brief information	Limitations
CO_2	Carbon dioxide	NDIR	Infrared absorption of CO_2 in a wide band is compared to reference gas	Interferences by water
		FTIR	Infrared absorption of CO_2 is compared to reference spectra obtained in an interferometer	Interfering species (N_2O, H_2O)
CO	Carbon monoxide	NDIR	Infrared absorption of CO in a wide band is compared to reference gas	Interferences by water
		FTIR	Infrared absorption of CO is compared to reference spectra obtained in an interferometer	Interfering species (N_2O, H_2O, COS)
NO	Nitrogen monoxide	NDIR	Infrared absorption of NO in a wide band is compared to reference gas with beam correlation technique	Interfering species
		FTIR	Infrared absorption of NO is compared to reference spectra obtained in an interferometer	Interfering species (C_2H_4)
		Chemiluminescence	NO reacts with O_3 forming NO_2 and releasing a defined photon. Quantity of photons produced in a mixing chamber is then linked to NO concentration	Interfering species, especially nitrogenous compounds like other NO_x, HCN
NO_2	Nitrogen dioxide	HPIC	NO_2 is trapped as nitrite and nitrate ions in 0.1 M NaOH and weighed by ILC	Few coeluants
		FTIR	Infrared absorption of NO_2 is compared to reference spectra obtained in an interferometer	Interfering species (SO_2)
		Chemiluminescence	NO_2 is converted into NO by a catalytic converter, then weighed as NO	Interfering species (other NO_x, HCN ...) Conversion efficiency
HCN	Hydrogen cyanide	FTIR	Infrared absorption of HCN is compared to reference spectra obtained in an interferometer	Interfering species (C_2H_2)
		Spectrocolorimetry	HCN is trapped as CN^- in 0.1 M NaOH and weighed by various colorimetric methods: picric acid, dimedone, or chloramine-T methods	Interfering species

Analyzed gas		Method name	Method brief information	Limitations
		HPIC	HCN is trapped as CN^- in 0.1 M NaOH and weighed by ILC in amperometric detection mode	Extensive usage results in accelerated aging of columns
SO_2	Sulfur dioxide	FTIR	Infrared absorption of SO_2 is compared to reference spectra obtained in an interferometer	Interfering species (H_2O, C_2H_2, HCN, CH_4)
		HPIC	SO_2 is trapped as SO_4^{2-} in H_2O_2 and weighed by ILC	Few coeluants Partial oxidation can conduct to formation of SO_3^- ions
HF	Hydrogen fluoride	FTIR	Infrared absorption of HF is compared to reference spectra obtained in an interferometer	Interferences (H_2O) Adsorption by pipes and filters
		HPIC	HF is trapped as F^- in 1 M NaOH and weighed by ILC	Sensitive to matrix effect and difficult to separate
		ISE	HF is trapped as F^- in 1 M NaOH and weighed in a TISAB by ISE	Interfering species (OH^-)
		Spectro-colori-metry	HF is trapped as F^- in 1 M NaOH and weighed by SPADNS inverse colorimetry in presence of zirconium ions	High limit of detection
HBr	Hydrogen bromide	FTIR	Infrared absorption of HBr is compared to reference spectra obtained in an interferometer	Interferences Adsorption by pipes and filters – highly hygroscopic
		HPIC	HBr is trapped as Br^- in water and weighed by ILC	Few coeluants
		Titrimetry	HBr is trapped as Br^- in water and weighed by titration with silver nitrate solution	High limit of detection Interferences with HCl
		ISE	HBr is trapped as Br^- in water and weighed by ISE	Interferences (CN^-, Cl^-) Matrix effect

Table 6.3 Summary of Analytical Methods for Major Acute Toxicants *(continued)*

Analyzed gas		Method name	Method brief information	Limitations
HCl	Hydrogen chloride	FTIR	Infrared absorption of HCl is compared to reference spectra obtained in an interferometer	Interferences (CH_4, C–C, C–H) Adsorption by pipes and filters – highly hygroscopic
		HPIC	HCl is trapped as Cl^- in water and weighed by ILC	Separation Coeluants
		Titrimetry	HCl is trapped as Cl^- in water and weighed by titration with silver nitrate solution	High limit of detection Interferences with HBr
		ISE	HCl is trapped as Cl^- in water and weighed by ISE	Interferences Matrix effect
HCHO	Formaldehyde (methanal)	FTIR	Infrared absorption of formaldehyde is compared to reference spectra obtained in an interferometer	Interferences (HCl, CH_4, C–C, C–H) Adsorption by pipes and filters
		HPLC	Formaldehyde is trapped as hydrazone in silica grafted with DNPH. DNP-formaldehyde is extracted by acetonitrile and weighed by HPLC	Chlorides and chlorinated compounds
		HPLC	Formaldehyde is trapped as hydrazone in HCl 2 M saturated with DNPH solution and DNP-formaldehyde is weighed by HPLC	DNP-formaldehyde is in solid and liquid form and has to be extracted using chloroform liquid–liquid extraction.
CH_2CHCHO	Acrolein (2-propenal)	HPLC	Same as formaldehyde	Same as formaldehyde
		HPLC	Same as formaldehyde	Same as formaldehyde

The evaluation of the toxic potency of fire effluents is an important parameter, but obtaining the data is not always straightforward. For an analytical result to be adequate for its intended purpose, it must be sufficiently reliable that any decision based on it can be taken with confidence. In addition, the analytical methods used for fire effluents evaluation must be validated and the uncertainty on the results estimated, at a given level of confidence.

All data from a chemical analysis should be given with its uncertainty, and also information on the limits of the method used should be noted.

6.8.4 Validation of Analytical Methods

6.8.4.1 Evaluation of a Method

Depending on the toxicity aspect to be considered (incapacitation or lethality), the level of toxic gas concentrations to be measured can vary significantly. Therefore, the analytical methods to be used for the evaluation of fire effluents have to be selected in regard of their expected concentration levels and in fact be suitable for the toxicity limits specified. The ISO 12828 series of standards [29, 30, 31] have been developed to assist this evaluation.

The following parameters are valid for any analytical method and must be considered when defining an appropriate analytical method:

- Selectivity of the method (matrix effect and interference with other species),
- Sensitivity of the method,
- Range of concentration,
- Limit of detection and limit of quantification,
- Accuracy: trueness and precision,
- Availability of reproducibility and reproducibility data.

6.8.4.2 Selectivity

Selectivity refers to the extent to which a method can determine particular analytes in mixtures or matrices without interferences from other components. A fire effluent is a complex mixture of different gases, and interferences cannot be avoided whatever analytical method is used. Therefore, the knowledge of possible interferences for a specific analytical method is essential. It is often difficult to evaluate all the potential compounds that may interfere. The ISO 19701 standard lists the main known interferences for each analytical method described there.

6.8.4.3 Sensitivity

The sensitivity of an analytical method is its capability for detecting small variations in the measured concentration. The knowledge of the sensitivity of the chemical analysis is of particular importance when evaluating kinetics of fire effluents produced during combustion. When a cumulative evaluation is performed, the sensitivity of the method is not of primary importance.

6.8.4.4 Concentration Ranges

In fire effluents, depending on the stage of the fire, the product involved in the fire and ambient conditions, the quantity of a specific gas may vary significantly from low to high concentration. It is therefore necessary to choose the application range of the method regarding the expected concentration to obtain a more precise evaluation of the quantities to be measured.

As an example, NDIR equipment for CO and CO_2 analysis is multi-range and has a typical resolution range of 1%. For other types of analytical methods (e.g., FTIR, colorimetry, ion liquid chromatography (ILC), etc.), calibration curves are required and form part of the evaluation. The definition of the concentration range for the calibration curves has to be defined in regard of the expected level of concentration.

Another parameter to consider is the maximum limit of the range (and also the limit of the equipment) to avoid saturation and therefore underestimation of the concentration of the compound of interest.

For other analytical methods, the technique itself limits the concentration range to be measured (e.g., potentiometric methods), and the equipment (e.g., electrodes) has to be selected with regard to the concentration range to be measured.

Within the working range, there may be a region where the signal will have a linear relationship to the analyte concentration. To study the linearity of the range, a calibration curve with a given number of levels is carried out with a given number of repetitions for each level. The study of the linearity consists in checking whether a regression model explains the data. The study starts with a linear model, but under certain circumstances it may be better to try to adapt a non-linear curve to the data.

6.8.4.5 Detection and Quantification Limits

Especially when dealing with highly toxic gases having a significant effect in low concentrations (e.g., acrolein), measuring low concentrations needs to be considered carefully. Two parameters – limit of detection and limit of quantification – need to be considered. ISO 12828-1 [29] describes the methods available for the evaluation of limit of detection (LD) and limit of quantification (LQ) applied to fire effluent evaluation. These two limit values depend on the analyzed matrix (including all other species present), the equipment used for the analysis, and the laboratory.

6.8.4.6 Accuracy of the Method

The accuracy of a measurement is a measure of how close a result comes to the true value. Determining the accuracy of a measurement usually requires calibration of the analytical method with a known standard. It is usually defined by two parameters: trueness and precision. Trueness of an analytical method is the expression of how close the mean of a set of results is to the true value (usually represented by bias). Precision is the expression of how close the results are to each other (usually represented by the standard deviation, standard error, or confidence interval).

ISO 12828-2 [30] gives an example on how to perform an accuracy evaluation for an analytical method. The accuracy depends on the matrix, and therefore it may be necessary to perform a new evaluation each time a different matrix is suspected.

6.8.4.7 Repeatability and Reproducibility of the Method

Repeatability and reproducibility evaluation of the analytical methods used for fire effluents analysis is of primary importance, especially when the data may be used for decision-making. Such evaluations are usually performed through interlaboratory trials. The ISO 5725 [32] series of standards explains in general terms how to perform such an evaluation, while ISO/TS 12828-3 [31] describes the interpretation of round-robin experiments on fire effluents.

References for Chapter 6

[1] Sumi K., Tsuchiya Y., Evaluating the toxic hazard of fires, *Canadian Building Digest* (1978) 197.

[2] Purser D. A., Improved fire and smoke-resistant materials for commercial aircraft interiors, Publication NMAB-477-1, National Academy Press, Washington, D. C. (1995).

[3] Imbert M., Baud J. F., Richter F., Nolland X. B., Julien H., Toxicité aiguë des fumées d'incendie, in: *Encyclopédie Médico-Chirurgicale*, Toxicologie-Pathologie professionnelle, Paris, 16-359-G-10 (1997).

[4] Garnier R., Chataigner D., Efthymiou M. L., Toxicité des Produits de Dégradation thermique des principaux polymères - Données expérimentales, *Réanimation Med. Urg.* (1990), pp. 411-426.

[5] Guillaume E., Verlinden G., Gensous F., Muller A., Bonhomme M., Joyeux D., Lecoq-Jammes O., Cornillon T., Oraison J. M., Labarthe J. C., Van Niel K., Bellivier A., Chivas-Joly C., Didieux F., Rochat B., Detraz P., Chaudron B., Parisse D., Le Madec B., Lacoste J. F., Heuzé B., Rémy A., Hourqueig R., Etude de la prise en compte de la toxicité des fumées en cas d'incendie. Rapport final du Groupe de Travail « Toxicité » du Comité d'Etudes et Classification des Matériaux vis à vis du risque Incendie (CECMI - Ministère de l'Intérieur, Direction de la Sécurité Civile), *http://www.lne.fr/publications/guides-documents-techniques/rapport-toxicite-fumees-incendie-final.pdf* (2010).

[6] Advances in Combustion Toxicology, ed. G. Hartzell, Technomic Publishing Company (1989-1992). Volume 1: ISBN 87762-590-5. Volume 2: ISBN 87762-591-3. Volume 3: ISBN 87762-886-6.

[7] Fire Toxicity, ed. Stec/Hull, CRC Woodhead Publishing (2010), ISBN 1 84569 502 X, ISBN-13: 978 1 84569 502 6.

[8] ISO 19706. Guidelines for assessing the fire threat to people (2011).

[9] ISO TR 13571-2. Life-threatening components of fire - Part 2: Methodology and examples of tenability assessment (2016).

[10] ISO 13571. Life-threatening components of fire - Guidelines for the estimation of time to compromised tenability in fires (2012).

[11] ISO 13344. Estimation of the lethal toxic potency of fire effluents (2015).

[12] Troitzsch J., Fire gas toxicity and pollutants in fires - The role of flame retardants, *Proceedings of Flame Retardants Conference,* London, UK, pp. 177-184 (2000).

[13] Claire Y., Gaudin E., Rossi C., Perichaud A., Kaloustian J., El Watik L., Zineddine H., Identification of cyanide in gaseous combustion products evolving from intumescent styrene-butadiene copolymer formulations in: *Fire Retardancy of Polymers: The Use of Intumescence,* ed. Le Bras, M., Camino, G., Bourbigot, S., Delobel, R., The Royal Society of Chemistry, pp. 437-447 (1998).

[14] Purser D. A., Toxicity assessment of combustion products, pp. I-200-I-245, in: *The SFPE Handbook of Fire Protection Engineering,* ed. Beyler, C. L., National Fire Protection Association, Quincy, Massachusetts (2002).

[15] Purser D. A., Gwynne S. M. V., Identifying critical evacuation factors and the application of egress models, *Proceedings of Interflam* (2007).

[16] Austin C., Risques pour la santé des pompiers forestiers et protection respiratoire (Wildland firefighter health risks and respiratory protection), Études et Recherches/Rapport R-571, IRSST, Montréal (2008).

[17] Austin C., Goyer N., Respiratory protection for wildland firefighters – Much ado about nothing or time to revisit accepted thinking?: Wildfire – 4th International Wildland Fire Conference (2007).

[18] Straif K., Baan R., Grosse Y., Secretan B., El Ghissassi F., Bouvard V., Altieri A., Benbrahim-Tallaa L., Cogliano V., Carcinogenicity of shift-work, painting, and fire-fighting, Lancet Oncology (2007) 8, pp. 1065–1066.

[19] DIN 53436-1. Generation of thermal decomposition products from materials for their analytic-toxicological testing – Part 1: Decomposition apparatus and determination of test temperature (2015).

[20] ISO 19701. Methods for sampling and analysis of fire effluents (2013).

[21] ISO 19702. Guidance for sampling and analysis of toxic gases and vapours in fire effluents using Fourier Transform Infrared (FTIR) spectroscopy (2015).

[22] ISO 16312-1. Guidance for assessing the validity of physical fire models for obtaining effluent toxicity data for fire hazard and risk assessment. Part 1 – Criteria (2016).

[23] ISO/TR 16312-2. Guidance for assessing the validity of physical fire models for obtaining effluent toxicity data for fire hazard and risk assessment. Part 2 – Evaluation of Individual Physical Fire Models (2007).

[24] Gann R. G., Averill J. D., Marsh N. D., Nyden M. R., Assessing The accuracy of a physical fire model for obtaining smoke toxic potency data, Building and Fire Research Laboratory, National Institute of Standards and Technology, Gaithersburg, USA. (2019).

[25] Babrauskas V., Levin B. C., Gann R. C., Paabo M., Harris R. H. Jr, Peacock R. D. and Yusa S. Toxic potency measurement for fire hazard analysis (NIST Special Publication 827), National Institute of Standards and Technology, Gaithersburg, USA (1991).

[26] Babrauskas V., Harris, R. H. Jr, Braun E., Levin B. C., Paabo M., Gann R. G., The role of bench-scale test data in assessing real-scale fire toxicity (NIST Technical Note 1284), National Institute of Standards and Technology (1991).

[27] ISO 9705. Reaction to fire tests – Room corner test for wall and ceiling lining products – Part 1: Test method for a small room configuration (2016).

[28] ISO 16405. Room corner and open calorimeter – Guidance on sampling and measurement of effluent gas production using FTIR technique (2015).

[29] ISO 12828-1. Validation method for fire gas analysis – Part 1: Limits of detection and quantification (2011).

[30] ISO 12828-2. Validation method for fire gas analysis – Part 2: Intralaboratory validation of quantification methods (2016).

[31] ISO/TS 12828-3. Validation method for fire gas analysis – Part 3: Considerations related to interlaboratory trials (2020).

[32] ISO 5725-1. Accuracy (trueness and precision) of measurement methods and results – Part 1: General principles and definitions (1994).

7 Combustion Toxicology

Jürgen Pauluhn

7.1 Introduction

This chapter addresses the principles of hazard characterization and assessment of fire gases from products under specified laboratory- or bench-scale conditions. The objectives of the applied test systems focus on expedience and reproducibility rather than attempting to mirror holistically the many variables occurring in real fires. These tests provide a versatile means for the comparison of relative toxic potencies of gases evolved from decomposing materials under a given standardized condition. This is to achieve temporally and spatially stable atmospheres of fire effluents at temperatures suitable for the inhalation exposure of experimental animals and the concurrent analytical characterization of the toxic components anticipated to be present in these atmospheres. However, it must be borne in mind that their toxicity can only be identified when the concentrations of oxygen and carbon dioxide do not approach ≤ 8% and ≥ 6%, respectively. Either condition increases the toxicity of smoke atmospheres precipitously. In combustion toxicology, asphyxiation- and/or irritation-related impairment of escape (incapacitation) precede death. Although incapacitation can be examined in specialized bioassays, the outcome is highly dependent on the species, endpoint, protocol, and dose-response/effect model chosen. This includes the animals' activity when executing this assay and whether the average "time to incapacitation" (t_i) or the "median response to incapacitation" (IC_{50}) was chosen. Opposite to the lethality-based all-or-nothing binary response, continuous intensity-of-effect-based analyses are subject to high variability depending on the endpoint chosen. In common dose-response models, the independent variable is usually referred to as the dose or the concentration exposed to over a specified duration of time. However, one must recall that "dose" is defined as the product of concentration × time of exposure and not the amount of toxicant inhaled. The latter requires a clear understanding of the respiratory minute ventilation of the examined species under the given exposure situation.

The most standardized and least subjective cornerstones of acute toxicity are the median lethal concentration (LC_{50}) and the non-lethal threshold concentration

(LC_{01}). Although the LC_{01} can be derived from 1/3 × LC_{50} by default [1], experimental data should be given preference [2, 3]. As a rule, experimental results from inhalation studies with rats following globally harmonized OECD testing guidelines [4, 5] have precedence over data from less standardized or isolated and never reproduced approaches. Reliable epidemiological data and experience on the effects of substances on humans should be considered in the evaluation of human health hazards and risks. However, human data or those from more human-like species, e.g., non-human primates, will not necessarily supersede those from well-conducted rat studies, but rather should both be assessed for their quality, the robustness of their data, and the impact of potentially confounding factors.

As there are several characteristics of fires, it is unreasonable to expect any single piece of apparatus or testing approach to mimic all situations possible. The major requirement of any combustion apparatus is that the user can demonstrate a relationship to a specific state of a fire so that the performance of a material in those conditions can be predicted with reasonable accuracy. Through comprehensive work, attempts have been made to define or categorize a limited number of thermal decomposition conditions found in fires by the character of the atmospheres from their carbon monoxide (CO), hydrogen cyanide (HCN), carbon dioxide (CO_2), and oxygen (O_2) content. The effects of fire ventilation on combustion products are expressed in terms of the relationship between the concentration of products and the equivalence ratio, ϕ. The equivalence ratio, defined as the ratio of the amount of gasified material (fuel) to the amount of air, normalized by the stoichiometric fuel-to-air ratio, has been used successfully to correlate the generation of heat and products of complete and incomplete combustion and to predict the concentrations of products of fires [6]. For well-ventilated fires, $\phi < 1.0$, mostly heat and products of complete combustion (such as CO_2 and water) are generated. For ventilation-controlled fires, $\phi > 1.0$, mostly products of incomplete combustion are generated. Such products are often toxic and irritant and reduce visibility, and thus are dangerous to life and property.

Ideally, the kind of toxicants released under the conditions of any test, including their yield, should be known to predict the likely toxic outcome at the outset of testing when using animals. Predictions generally assume additivity of toxicants. However, for toxicants with independent toxicities, e.g., irritants vs. asphyxiants, or interactions (chemical or physical), simplified additivity rules may either over- or under-estimate the actual toxic hazard. A typical example is that volatile irritants that act on the "upper airways" become an alveolar irritant when partitioned/adsorbed in/by a liquid/solid aerosol. The toxicological significance of such interactions can readily be revealed by confirmatory testing in animal inhalation bioassays. The highest gain in knowledge can be attained by a comparison of hypothesis-based modeled data with a confirmatory point-estimate from acute inhalation studies on animals. Thus, animal tests may provide an invaluable piece of information for revealing cryptic hazards not yet accounted for or as yet unidentified analytical shortcomings due to interferences, unsuitable analytical methods, or simply for not having considered the most toxic constituent. Hence, animal testing should be restricted to point-estimates serving as a validating feedback loop for any predicted toxic hazards rather than as a default approach.

Any identified toxic hazards then need to be put into a risk perspective regarding the extent to which the fire scenario modeled may occur under real-life conditions and if the selected scenario limits a successful escape from fires. As conceptualized by the relationship "Risk = Toxic hazard × Exposure × Population", toxicological risk assessments have to observe at least three categories: (i) toxicity as such and whether the data generated match the fire scenario of interest with regard to the lethal/non-lethal toxic endpoints chosen; (ii) the exposure intensity and duration need to be adjusted to those of the highest concern for humans; (iii) the size and susceptibility of the human population expected to be exposed is equally important and whether it consists of healthy adults or populations more susceptible to toxicity and incapacitation. Exposures to fire effluents are short and most escape paradigms can successfully be managed by healthy adults only. Thus, infants, children, and disabled subjects require assistance. Time-extrapolations may be required to extrapolate toxicity from one exposure scenario to another to bridge data from animal models to humans subjected to differing exposure regimens. It is important to acknowledge that exposure intensity and likely exposed population are variables that cannot be judged by toxic hazard alone. This calls for an integrated risk assessment approach to be applied by the fire risk assessor.

7.2 Toxic Hazards

This overview focuses on the analysis of published approaches for assessing the occurrence of fatalities in relation to impairment of escape (incapacitation) from the most prevalent asphyxiant combustion gases - CO, HCN, and CO_2 - and whether similar approaches are applicable to respiratory tract irritants. The toxic-hazard-based calculation models of ISO 13571 [7] are currently under revision to keep abreast with progress in the science of the assessment of toxicological risks occurring in emergency situations [8, 9] and recommended for land-use planning [10–13]. When focusing on the modeling of the time available for directed escape from burning buildings by the healthy and normally performing population, one is inclined to accept that the regulatory established land-use planning approach could also be adopted in the "burning-building" planning approach as conceptualized in ISO 13571 [7]. Therefore, the principles of toxicological risk assessment already in place for land-use planning will be considered as well.

When considering a fire risk assessment, a major prerequisite is to fully understand the hazards originating from at least two components: the possibility of a fire occurring and the magnitude of the consequences of that fire under conditions of developed flaming. Regarding fire safety objectives, with the methodology of testing and analysis, multiple factors are considered. These include, but are not restricted to, ignition and flame spread in the enclosure of the origin as well as the variable stages of fire-related hazards that are associated with heat, smoke, and effluents that are toxic to the inhabitants of such enclosures [14]. The pre-flashover and post-flashover phases of the fire are distinctly different and need to be distinguished. The toxicological hazard assessment focuses on the people's successful escape from fires, which is potentially affected by the asphyxiants detailed above as well as irritants. Any "impairment of escape", either due to anxiety or other psychophysical factors, loss of motor function, memory and/or recognition is subsumed by combustion toxicologists under the term "incapacitation". The development of experimental models of anxiety that readily translate between animals and humans is required to better integrate and clarify the biological, behavioral, and cognitive mechanisms that underlie the species-specific response determining a "successful escape". In case the effects leading to incapacitation cannot reliably be derived from animal data, then 1/3 of the non-lethal threshold concentration (LC_{01}) value was shown to be a reasonable surrogate for asphyxiants [2]. Incapacitation bioassays in animals commonly focus on this endpoint only in the absence of consideration of the occurrence of post-exposure mortality, which typically occurs when exposed to irritants. Unlike incapacitation bioassays, the $1/3 \times LC_{01}$-based approach considers the more robust binary endpoint relative to the more subjective incapacitation endpoint and, therefore, decreases uncertainty.

Irritant gases are segregated into chemically reactive and/or water-soluble ones (Figure 7.1). Water-soluble irritants are retained preferentially in the mucosal layers of the eyes, nose, and upper airways (e.g., ammonia, sulfur dioxide, and halo acids), whilst the non-soluble irritants reach the lower lung with few interactions within the upper respiratory tract (e.g., phosgene, ozone, nitric oxides). No sharp line for the exact location of injury can be drawn, as very high concentrations may blur this categorization (Figure 7.1). Toxic outcomes from exposure to irritants do not occur as early and are not as clear-cut as those following exposure to asphyxiants, with deaths already occurring during or shortly after the incident. To understand the toxic modes of action of irritant gases, one should appreciate the anatomical and physiological differences of each region of the tract, which is segregated into the nasal (I), tracheobronchial (II), and pulmonary (alveolar) (III) regions (Figure 7.1).

Irritation of region I may elicit nociceptive sensations commonly perceived as pain or a pungency causing a state of annoyance and/or anxiety. Such early warning signals motivate a subject to take action to escape and to seek shelter. Tearing of eyes, nasal mucous secretion, and coughing may occur as a result of these nociceptive signals. It is important to note that such reflex-related stimulations occur immediately and are concentration-dependent, whereas any tissue injury is caused inhaled-dose ($C \times t$)-dependently. However, when such gases gain access to region II, lingering airway injuries with delayed mortalities may occur. (Hyper-)susceptible individuals inflicted with a pre-existing airway disease may experience instant severe asthma-like attacks or even death. The injury of region III is characterized by a potentially life-threatening lung edema, i. e., the lung is flooded with fluid from the vascular circulation. If severe enough, death occurs within 24 hours post-exposure; however, when survived, the lung commonly reconstitutes itself within one week or even less.

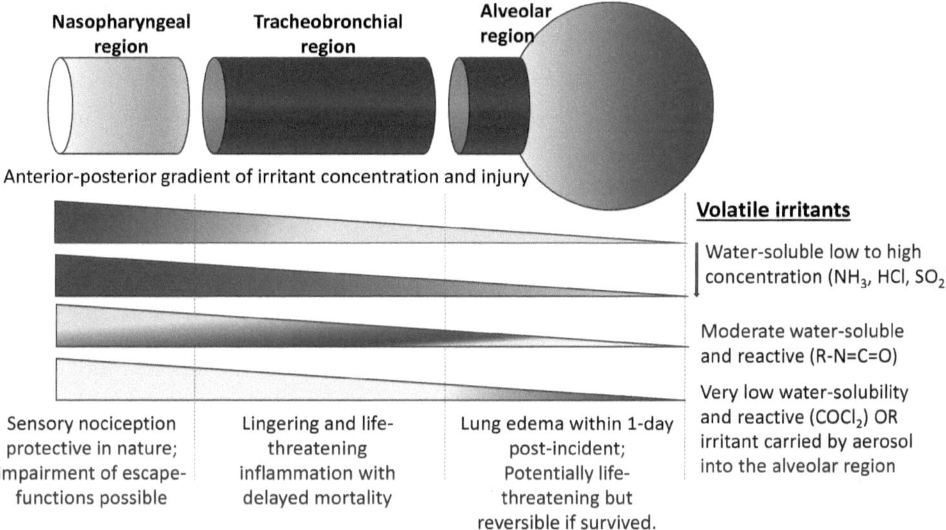

Figure 7.1 Illustration of the retention of volatile reactive and/or water-soluble/insoluble gases within the respiratory tract of mammals

The interpretation of data from complex animal bioassays requires expert knowledge and an appreciation of the established principles of "translational toxicology". This field of toxicology refers to the general approach of applying toxicological findings to human settings considering the following hallmarks: (1) assessing human exposure in the critical window of interest; (2) defining modes of action and relevance of data from animal models; (3) use of mathematical models to develop plausible predictions as the basis for: (4) derivation of exposure thresholds to protect human health. Unlike humans, rats and mice are obligate nasal breathing species with a markedly higher body-weight-adjusted respiratory ventilation than humans (see below). However, when rodents and humans are exposed to upper respiratory tract (URT) sensory irritants, such as ammonia, rats may elicit an instant reflexively-induced depression of ventilation because of nociceptive sensory irritation whereas humans respond with hyperventilation. Hence, despite the unifying mode of action of irritants, the regional dose inhaled requires thoughtful dosimetric adjustments. Likewise, it must be borne in mind that many of these reflex-related nociceptive responses serve the purpose of protection and to prompt subjects to minimize exposure to hostile environments by escape.

7.2.1 Inhalation Toxicity of HCN and CO

The key mode of action of HCN is histotoxic hypoxia of tissue cells that are unable to utilize oxygen due to a compromised mitochondrial oxidative phosphorylation [15] whereas the most prominent pathophysiological effect of CO is hypoxemia caused by avid binding of CO to hemoglobin (Hb) resulting in carboxyhemoglobin (COHb). Formation of COHb reduces the O_2-carrying capacity of blood and impairs release of O_2 from O_2Hb to tissues. In addition to tissue hypoxia, ultimate diffusion of CO to cells may adversely affect their function. Concurrent with their modes of action, the brain and heart are particularly sensitive to both HCN- and CO-induced hypoxia and cytotoxicity. The rate of COHb formation and elimination depends on many factors; however, it has limitations to reliably predict any clinical outcome [59]. The same factors that govern CO uptake determine CO elimination as well. Thus, any physiological conditions that affect these physical and physiological variables (e.g., exercise, age) will also affect the kinetics of COHb. There is a growing recognition of the role that CO may play in normal neurophysiology and in microvascular vasomotoric control. Mechanisms for these changes have been linked both to mitochondria and to a CO-mediated disturbance of intracellular "buffering" of endogenously-generated free radicals [16–21].

The predicted changes in COHb for rats exposed to the time-adjusted $1/3 \times LC_{01}$ are shown in Figure 7.2, top. These simulations match the assumption made about the inhaled dose and kinetics. As shown in Figure 7.2, bottom, a consistent relationship for the regimen-specific $1/3 \times LC_{01}$ values could not be identified for COHb [2]. This finding is consistent with published evidence [22–25]. Initial COHb does not appear to correlate well with severity of poisoning, which relates to the mechanism of toxicity of CO, that is binding of CO to the intracellular oxygen-carrying proteins (e.g., mitochondrial cytochromes) rather than solely to Hb. Metabolic acidosis was known to result from impaired systemic oxygen delivery due to COHb formation, especially with longer exposure durations. Such superimposing factors may further complicate the interpretation of COHb. The range of COHb measurements among individuals who died, however, was wide and inconsistent with the concept that there is a predictable level at which death occurs [26, 27]. While symptoms are common in CO poisoning, none could be directly correlated to COHb levels, even in a population greater than 1000 patients. The concept of a table relating specific symptoms to specific COHb levels was considered invalid [25]. Hence, the outcome of this analysis on animals is consistent with these conclusions drawn from human evidence.

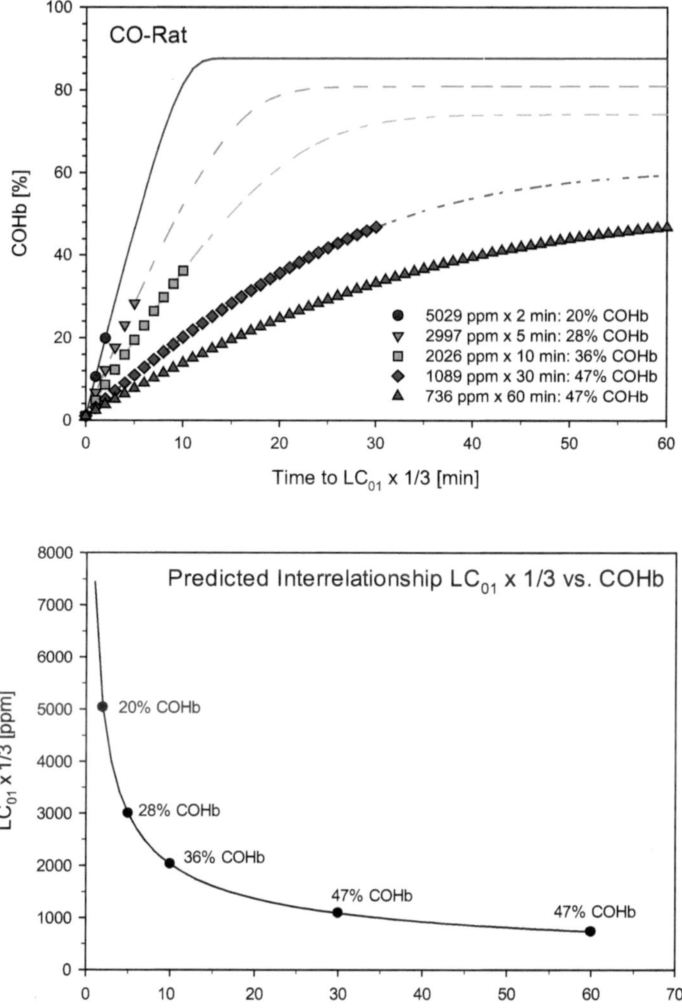

Figure 7.2 Modeled time-course changes of COHb saturation in rats exposed at 1/3 x LC_{01} equivalent concentrations and variable exposure durations. COHb levels were modeled using the Coburn-Forster-Kane equation [17, 26] (figures reproduced from Pauluhn [2])

Notably, some emergency response guidance values, including those of ISO 13571 [7] were derived on fixed COHb levels for either impairment of escape or lethality. Caution is advised when doing so because COHb is deemed to be a biomarker of exposure rather than of adversity [22–25]. Accordingly, consistent with published evidence, the toxic-load-adjusted 1/3 × LC_{01} approach [2] appears to deliver more reliable estimates for incapacitation than COHb-based alternatives.

7.2.2 Inhalation Toxicity of CO_2

Notably, atmospheres depleted of oxygen or rich in CO_2 may elicit a variety of typical rodent-specific compensatory responses that cannot directly be extrapolated to humans. Burrow dweller rodents with labile thermoregulation utilize a high rate of conductance to remove metabolic heat produced within the body in their natural microhabitat. The resultant hypothermia would be advantageous to an animal living in a poorly-ventilated burrow or any other hostile environment. Hence, amongst the most important rodent-specific responses to such stressors is hypothermia and hypoventilation and decreased metabolism resulting in less O_2 demand and CO_2 production [2, 27–29]. Due to their smaller thermal inertia, mice may attain this hibernation-like state in adaptive physiology at an even faster rate than rats [29]. This may render mice more resistant to asphyxiants. Especially for CO_2, humans might be more susceptible than rodents and human data should be given preference by default.

CO_2 in blood is dissolved in the plasma; about 20% is bound to Hb as a carbamino compound. At this binding site, CO_2 does not compete with the oxygen bound to the heme iron. Hb-bound CO_2 can act as an important homeostatic reservoir for CO_2 and contributes to its buffering capacity. Hence, it seems to be reasonable to assume that this binding site acts as an intermediate sink for dissolved CO_2, which also is catalyzed by carbonate dehydratase, an enzyme that controls the equilibrium of dissolved CO_2 and H_2CO_3 (carbonic acid). This mechanism serves as a major pathway to translocate CO_2 from tissues to blood and further into the alveolar space. Carbonic acid dissociates into bicarbonate (HCO_3^-) and carbonate. In blood, the relationship of HCO_3^- : pCO_2 is physiologically maintained at a 20 : 1 ratio, i. e., any inhalation-related increase of the partial pressure of CO_2 may affect blood gases and the pH of blood.

The systemic capillaries deliver oxygen to the tissues and remove the CO_2 produced by the oxidative phosphorylation taking place in mitochondria. This endogenous process generates a venous pCO_2 of ≈ 45 mm Hg (equivalent to ≈ 6% CO_2 in the alveolus). If the partial pressure of CO_2 in the alveolus is less than 45 mm Hg, CO_2 can readily be cleared from the systemic circulation. In contrast, when this critical cut-off is exceeded, the systemic circulation becomes overloaded with CO_2 and an ensuing decline in oxygen utilization and oxidative metabolism occurs. This, then, is the first step into a metabolic acidosis in addition to the respiratory acidosis caused rapidly by an inhaled high value of pCO_2. Panic attacks were reported to occur in humans just at the transition from 5 to 6% pCO_2 in air [30]. This physiologically estimated onset of (patho)physiological changes matches exactly that derived by Blockley and Speitel (1995) [31, 32] (Figure 7.3). Due to this threshold-like phenomenon, time-adjustments have to be applied cautiously. Hypoxemic respiratory failure is commonly characterized by an arterial oxygen tension (PaO_2) lower than 60 mm Hg (8%) at a normal arterial carbon dioxide tension. The observed precipitous increase of toxicity with lower pO_2 indicates that severe hypoxemia may often be the primary etiopathology relative to the severe hypercapnia.

Rats exposed to a range of CO concentrations (2500 to 4100 ppm) together with CO_2 (> 15000 ppm) showed evidence of an aggravation of the CO-related toxicity. The authors hypothesize that the mechanism for this synergistic interaction appears to be due to an increased ventilation rate and formation of COHb with a more severe degree of acidosis. It is conceivable that both a metabolic acidosis (resulting from exposure to CO alone) and a respiratory acidosis (resulting from inhaled CO_2) occurs [33]. Notably, animals exposed to both CO and CO_2 succumbed distinctively later than those exposed to CO alone. This finding does not necessarily support any isolated "hyperventilation" hypothesis as the sole etiopathological factor. The retrospective evaluation of acute inhalation combustion studies in rats using the DIN 53436 tube furnace method did not identify any marked modulatory effects on the acute inhalation toxicity of toxic combustion gases by CO_2 [34]. Evidence shows, however, that CO_2 creates an immediate threat to life when exceeding the critical $C \times t$-relationships of 6% (Figure 7.3).

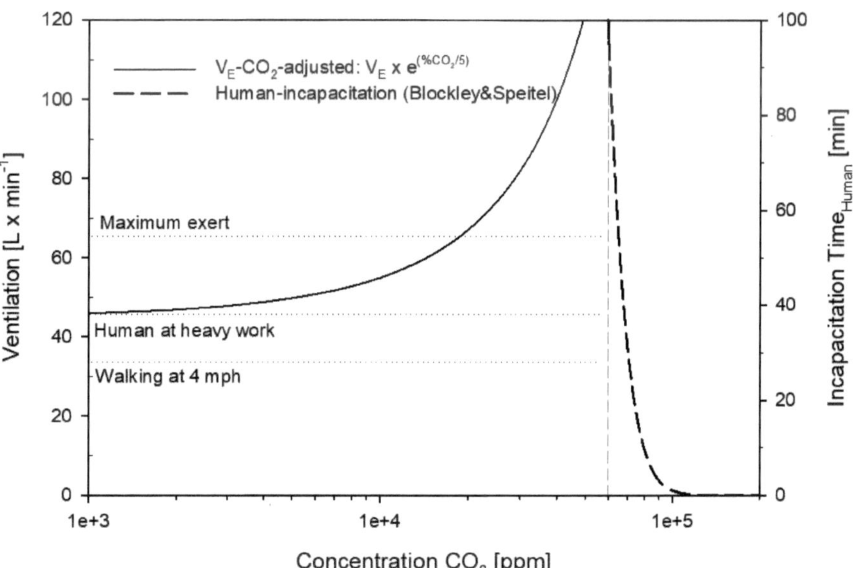

Figure 7.3 Simulation of hyperventilation as called for by ISO 13571 [7]. When this equation was applied to human ventilation at heavy work, it was closest to that of rats (see also Figure 7.4). The predicted incapacitation time of humans used the algorithm published by Blockley and Speitel [31, 32] (figure reproduced from Pauluhn [3])

It should be appreciated that with toxicants for which inhalation is the primary route of exposure, the intensity of contact is influenced by one's level of exertion (often referred to as "activity level"). Categorical exertion-level data are linked to typical respiratory ventilation patterns mirroring the activity ranges from rest, sedentary, light, moderate, and heavy activities. Each activity range is characterized by a typical range of ventilation [35–37]. This analysis attempts to interrelate the level of respiratory ventilation and exertion typically occurring in human escape situations and to explain how this ventilation compares to the animal bioassays from which points of departure (POD) are derived. Due to the higher ventilation of rats (kg body-weight-adjusted), this species is considered to receive a higher kg-body-weight-based dose compared to humans at heavy work. This exertion level is a reasonably conservative estimate for managing a successful escape (Figure 7.3 and Figure 7.4). Consequently, additional ventilation adjustments do not appear to be justified when using POD from rats. These considerations and simulations support an abandonment of generic corrections for differences or variability in respiratory ventilation rates when using state-of-the-art reference inhalation data from rats. However, such adjustments need to be considered when using larger animals with possibly more "human-like" breathing and reflecting the sedentary to moderate state of activity.

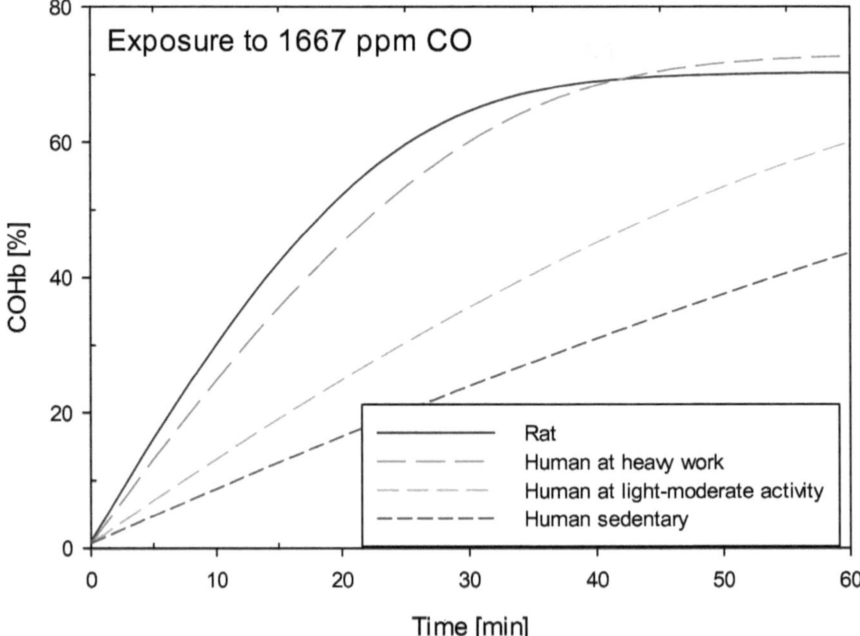

Figure 7.4 Modeling of time-course changes of COHb saturation in nose-only exposed rats relative to humans at different activity levels when exposed to carbon monoxide (CO) at 1667 ppm up to an exposure period of 60 min (for more details, see caption of Figure 7.2) (figure reproduced from Pauluhn [2])

7.2.3 Respiratory Tract Irritants

Unlike asphyxiants, where a toxicological assessment and POD derivation can rely upon a large database from experimental animals and humans, their derivation for irritant gases is more complex due to their concentration- (sensation- and localization-related) and concentration × time-dependent (injury-related) modes of action. At low concentrations, most of the water-soluble irritants cause a reflex-mediated URT sensory irritation at the lower end of the concentration–response curve and may become a cytotoxic and inflammogenic irritant at higher exposure intensities (Figure 7.1). The POD derived by the toxicologist for successful escape depends on the occurrence of effects that do not interfere with the chosen endpoint and also prevent the occurrence of irreversible outcomes or even mortality. Due to the diversity of the respiratory tract, equal concentrations of different irritants may cause different types of injuries at different sites of the tract. As delineated in Figure 7.1, each site of the tract has its site-specific vulnerability and response to injury. The commonly applied and most conservative approach is to assume that all irritants act additively and independently within the tract.

Every data set is a unique, chemical-specific source of information that reflects the investigations conducted on the chemical and the properties of the chemical. This assessment reflects a "best professional judgment" approach in the evaluation of the data adequacy. Different toxic endpoints are used for the two POD tiers characterizing the POD_1 for non-lethality (LC_{01}) first, followed by the POD_2 for a successful escape and absence of irreversible outcomes (IC_{01}). Thus, each tier requires a different data set or studies for executing a state-of-the-art weight-of-evidence approach. Due to the well-defined modes of action of asphyxiants that do not change with dose and are mostly independent, the POD_2 could expediently be derived from 1/3 × POD_1 as suggested for the derivation of Emergency Response Values (AEGL [8], ERPG [9]). However, the location-of-injury-dependent modes of action for irritants, including their dependence on the prevailing exposure regimen "$C×t$", require a more comprehensive set of data for deriving the POD_2. When extrapolating data from nasally breathing rodents to oronasally breathing humans, a reasonable fraction of the inhaled dose retained in the URT of rodents may gain access to the more susceptible structures of the tracheobronchial airways of humans (Figure 7.1). At the outset of any confirmatory analyses, this calls for an adjustment of 1/10 × POD_1 by default for irritants. Notably, while water-soluble irritants evoke marked nociceptive sensations to stimulate actions for escape, asphyxiants do not evoke such sensations and individuals are not aware of exposure to potentially noxious agents. The impact of such protective homeostatic responses on escape (perceived throat and eye irritation, lacrimation, rhinorrhea, cough, and respiratory distress to mention but a few) is controversially discussed. In case extreme URT irritation may cause psychophysical (relationship between one's internal (psychic) and external (physical) qualities) responses leading to impairment of escape, an adjustment factor of 1/30 × POD_1 appears to be toxicologically defensible. An example for tier 1 and 2 derivations of POD is given for highly water-soluble ammonia examined in rats with a focus on the LC_{01} and URT irritation [38, 39].

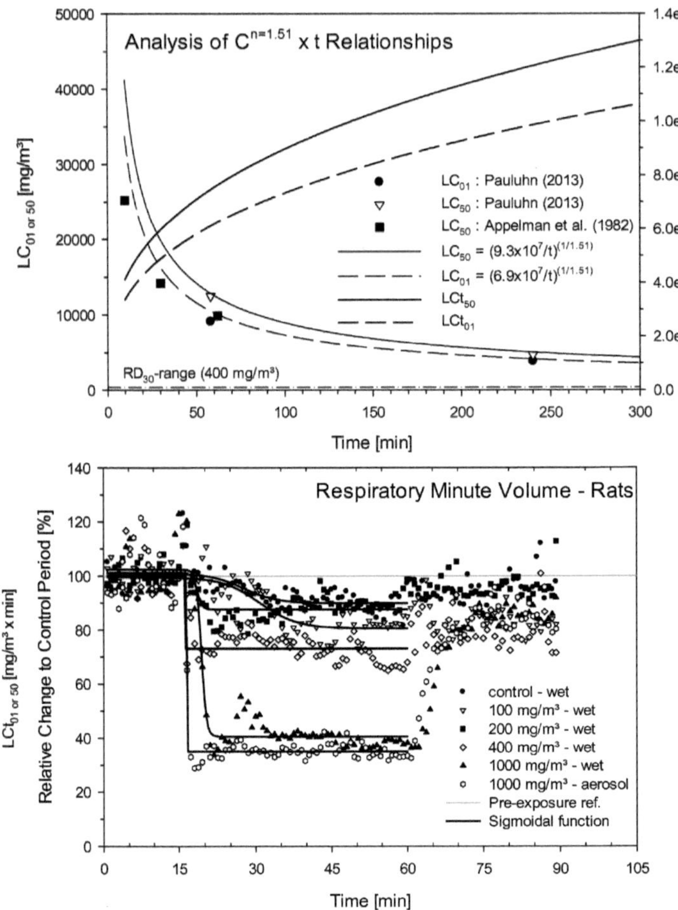

Figure 7.5 Top panel: toxic load modeled $C^n \times t$ = const. effect relationships of LC_{50} and LC_{01} as well as LCt_{50} and LCt_{01} data from rats exposed to ammonia [38–40]. The threshold for marked sensory irritation (RD_{30}) was assumed to be constant (RD_{30}: decrease in respiratory minute volume (MV) of 30%; see 400 mg/m^3, bottom panel). Bottom panel: time- and concentration-dependence of the MV of rats exposed nose-only to ammonia for 45 min to ammonia gas in dry and wet air (steam or aerosol) followed by a 30 min recovery period. All data normalized to the pre-exposure data (100%)

The evoked changes in breathing patterns resembled those known to occur following exposure to "URT sensory irritants", rapid in onset and reversibility. Reflex stimulation from the lower airways was not apparent up to the maximum concentration examined (see Figure 7.5, bottom). The authors conclude that the sensory irritation potency and location at which sensory irritation occurs were independent of the humidity of the atmospheres [38, 39]. The chemosensory effect range (lateralization thresholds) was explored in healthy volunteers exposed to ammonia vapor through one nostril and clean air through the other one at ascending concentrations (60–350 ppm) using dry or humidified air. These results show that humidity was not a critical factor in determining sensory irritation thresholds for ammonia [41]. These findings suggest that an enormous amount of ammonia becomes uploaded in URT tissues, triggering nociceptive stimulations before acute lung damage and lethality occurs. However, when concentrations are high enough to reach the lower lung, it is just one step from the LC_{01} to the LC_{50} (Figure 7.5, top).

The $LCt_{01/50}$ relationships depicted in Figure 7.5 (top) provide clear evidence that effects do not accumulate linearly with time. This is plausible for a vapor highly soluble in the fluids lining the URT airways. Therefore, the same $C \times t$ dose inhaled over a short or long period may produce higher and accumulated lower tissue doses and injuries, respectively. Any depression in ventilation at short exposures to high concentrations of URT sensory irritants may markedly reduce the actual inhaled dose of the irritant under consideration but also all other accompanying toxicants (see Figure 7.5, bottom). As toxic potencies are expressed as "exposure concentration" (external) rather than the actual "inhaled concentration over time" (internal dose), the "$C^n \times t$ = const. effect" equation may misjudge the latter because of the complete disregard of this equation for any changes in the respiratory ventilation. This, then, would result in the overestimation of the inhaled $C \times t$ (or dose) necessary to cause the lethal or incapacitating outcome and translates to an underestimation of toxicity when extrapolating from one species and scenario to another. Under such conditions the "n" is not necessarily a reflection of any substance-specific time-dependent change in toxic potency rather than a "correction factor" for the rodent-specific concentration-dependent depression in ventilation.

Unlike rats, humans were shown to respond to ammonia by hyperventilation and reflexively-induced lacrimation (eyes not directly exposed) [42]. Concentrations in the range of ≈ 1200 mg/m^3 caused an "offensive effect" characterized as an immediate cough, and throat and eye irritation. Reflex-laryngospasm cannot be excluded at the predicted concentrations occurring at exposure times < 8 min [43]. In an experimental study in volunteers consisting of an "expert" and a "non-expert" cohort, throat irritation and an urge to cough were perceived as offensive to unbearable by the non-expert group, whilst the same $C \times t$ was perceived to be just about distinctly perceptible in the expert group [44]. The expert group was familiar with the effects of ammonia and had no previous exposure to ammonia; the non-expert group of university students was unfamiliar with the effects of ammonia or with experiments in laboratory situations. As alluded to earlier, such psychophysical events cannot be claimed to be unequivocally adverse; however, they may affect escape under idealized conditions. However, such conditions do not prevail in flaming buildings.

When deriving the POD_2 equivalent from the data shown in Figure 7.5 and Figure 7.6, one has to account for the hypoventilation-related under-dosing of rats relative to the hyperventilation-related over-dosing of humans relative to their normal states of ventilation. This may call for an additional inter-species dosimetric adjustment factor of 3 at the LC_{01} level but is considered redundant at the RD_{30} level (penetration factor of 3 already considered). Interestingly, when using the $1/10 \times LC_{01}$ as the POD_2, this $C \times t$ relationship obtained from rats [38, 39] and SLOT-DTL derived for the general public [10] converge favorably. The $1/30 \times LC_{01}$ and RD_{30} for sensory irritation converge as well at exposure durations longer than $\approx$ 20 min (Figure 7.6, top). It appears to be not unreasonable to assume that the 10-times lower breathing frequency of humans relative to rats implies outcomes better modeled by the toxic load approach than the time-independent threshold approach applied by ISO 13571 [7].

Rats exposed nose-only at $\approx$ 3500 mg NH_3/m^3 for 1 hour displayed a reflexively-induced depression in ventilation without any temporal changes in nociceptor susceptibility (Figure 7.6, bottom). This nociceptor reflex is conveyed by the trigeminal nociceptive system, the same nerve that causes reflex-laryngospasms in humans [42–46]. All rats exposed at $\approx$ 8000 mg/m^3 for 4 hours displayed severe respiratory tract irritation (sensory and cytotoxic in nature); however, with the absence of mortality. Thus, despite the consistent prediction of concentrations likely eliciting tolerable sensory and reversible cytotoxic URT irritation at $1/10 \times LC_{01}$, the anecdotal evidence of reflex-revelated laryngospasms in humans at $\approx$ 1200 mg/m^3 could not be reproduced in rats. This discrepancy cannot be reconciled by rat inhalation studies but shows the difficulty of classifying isolated human evidence.

This analysis shows that the derivation of a POD_2 for incapacitation from irritant fire effluents has numerous facets that require in-depth expertise in pulmonary toxicology and physiology when translating data from animals to humans. The limited availability of data from humans mandates a transparent and comprehensible description of the course taken to derive at a scientifically defensible POD_1 and POD_2 for humans based on a set of animal bioassay data. Precise numerical approaches must be put into any modeled fire scenario perspective that is complex and largely uncontrolled. Hence, exposure situations may occur in fires where additional, even more detrimental factors must be accounted for. This complexity, including the required time for escape for a population, can only be comprehensively overseen and assessed by the fire engineer engaged with the fire safety calculations. This overall evidence supports the use of the generally applied toxic load approach as already established for an array of “once-in-a-lifetime”-occurring accidental situations [8, 10, 13].

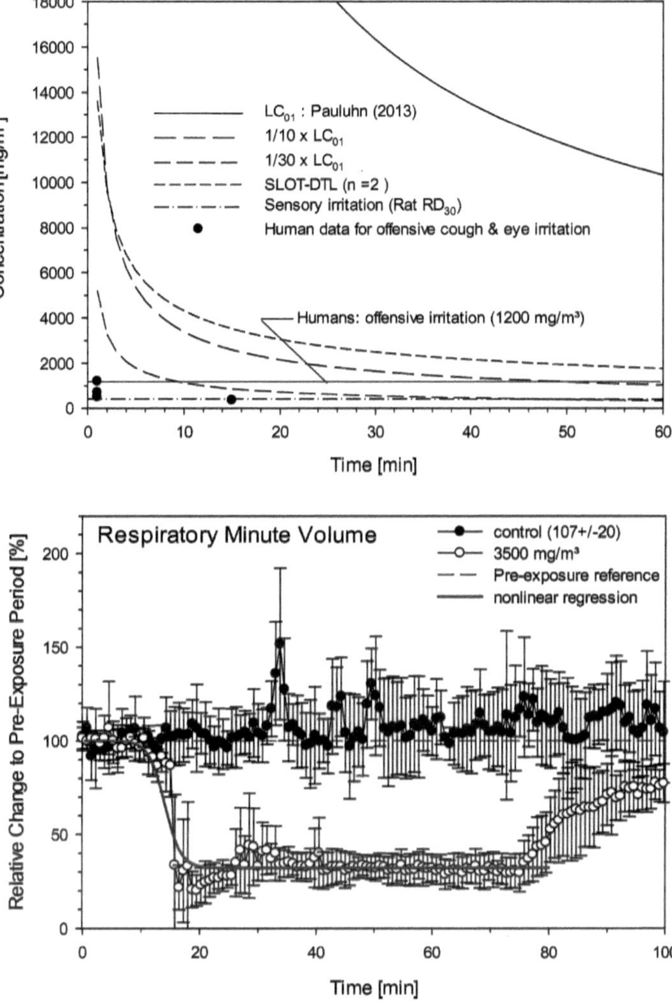

Figure 7.6 Top: LC_{01}-based $C \times t$ analyses of rats exposed to ammonia utilizing the toxic load model with $C^{1.51} \times t$ = const. effect as detailed in Figure 7.5. The coherence of the 1/10 x LC_{01} from rats with the SLOT-DTL for the general human population could be demonstrated; bottom: time-dependence of respiration of rats exposed to ≈ 3500 mg/m^3 ammonia. All data normalized to the pre-exposure data. Abbreviations: dangerous toxic load (DTL), specified level of toxicity (SLOT)

7.3 Toxicological Risk Assessment

7.3.1 Definition of Uncertainty

In combustion toxicology, an exposure dose is defined as any pre-specified amount of toxic substances exposed to over a specified duration eliciting a certain, well-defined response (binary) or effect (continuous). A response algorithm can be judged as a binary simplification of a continuous relationship. Inhalation toxicology defines the response to a given dose "$C \times t$" as the quantification of a biologically relevant effect and as such it is subject to random variation. The traditional interpretation of dose-response information is to accept the existence of a threshold level of dose that must be inhaled to produce the toxic effect. Thus, a threshold or POD exists if there is no effect below a certain exposure level, but above that level the effect is certain to occur. The POD is defined as the point on a toxicological dose-response/effect curve established from experimental data generally corresponding to an estimated low (adverse) effect level or no (adverse) effect level. As exemplified, not all effects we see are necessarily adverse. This applies especially to sensory irritants and their psychophysical sequelae. This threshold marks the beginning of extrapolation to toxicological reference concentration values that can be calculated by dividing the POD with corresponding uncertainty or adjustment factors as illustrated earlier. These factors are used to address the differences between the experimental data and the specific human situation of interest, considering the following major uncertainties in the extrapolation procedure: inter- and intraspecies differences, differences in duration of exposure of data from bioassays and targeted human exposure, issues related to dose-response, and quality of whole database.

There is a consensus amongst toxicologists and risk assessors that uncertainty factors have to be accommodated when applying PODs from animal studies to a specific population at risk. However, in the case of fire accidents, the scope is to minimize failure of an escape event experienced once in a lifetime. There is likely to be considerable variability in the escape responses of different individuals affected by such an incident. Similarly, the concentrations of toxic fire gases released over time may change dramatically with great variability from one location to another. Many of the available toxicity data are not usually adequate for predicting precise dose-time-response/effect relationships. Smoke obscuration and heat may readily become a greater threat than that originating from toxicants. Thus, the prediction of effective cumulative exposure doses for a given scenario is more complex than the concept of a fractional effective dose (FED) used as a tool to assess the toxicity-related impact on the impairment of escape as defined by the International Standard ISO 13571 [7]. This standard defines an FED (for asphyxiants) and an FEC (for irritants) value of 1.0 as the median value of a lognormal distribution of the ability to perform an escape response within a defined time window. Any failure to perform this task would inevitably result in fire-related fatalities. Post-exposure deaths or irreversible outcomes may not entirely be prevented by this standard. An FED or FEC of 0.1 translates to a ≈ 1% population response. More recently suggested modifications of this standard are detailed elsewhere [2, 3].

The wealth of physiological and toxicological information available on rats reduces the uncertainty when extrapolating findings from this species to humans. Harmonized guidance has been published for applying inter- and intraspecies uncertainty factors to adjust inhaled doses and for applying findings from this species to humans. To the contrary, data from non-human primates cannot rely upon a strong database and the species-specific inhalation methodology is less standardized and is subject to laboratory-specific outcomes. Therefore, such unique studies can only be judged as supplemental rather than core evidence. For most of the asphyxiant toxicants, similar modes of action in animals and humans can be assumed with no interspecies factor required. Dosimetrically, the rats' respiratory ventilation is high and conservative enough to omit the variability in human activity-related differences in ventilation. However, as exemplified for ammonia, for respiratory tract irritants, it appears to be indispensable to consider in detail concentration-, modes of action-, and species-specific response-based dosimetric adjustments.

7.3.2 Analysis of Time-Dose-Response Relationships

7.3.2.1 Asphyxiants

Published analyses on the time-scaling of data from acute inhalation studies on rats used the ten Berge algorithm "$C^n \times t = k$ (constant effect)" [2, 47, 58]. By introducing the toxic load exponent (n), multiple inhalation studies with variable exposure concentrations (C) and exposure times (t) can be combined to calculate the LC_{50} and LC_{01} from the entire matrix of data. The exponential weighing factor was incorporated to better describe the relative contribution of C and t mathematically; however, for practicability, either the toxic load "n" or the toxic load "k" is a combination of both. For HCN and CO, the calculated "n" was 1.64 and 1.77, respectively. The related toxic load constants for the non-lethal threshold "$k(LCt_{01})$" were 0.109×10^6 and 0.498×10^8 [$ppm^n \times min$], respectively. The median lethal concentrations "$k(LCt_{50})$" were 0.294×10^6 and 1.21×10^8 [$ppm^n \times min$], respectively (Figure 7.7). This parameterization was derived from published nose-only inhalation studies in rats and is valid for exposure durations up to 60 min [2, 47–49].

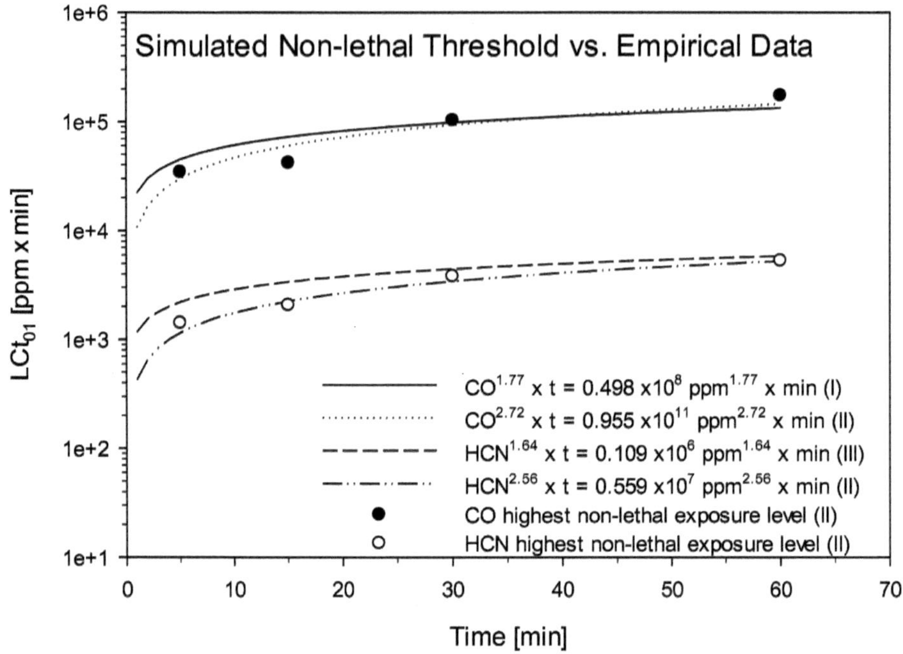

Figure 7.7 Toxic load model with C^n x t = const. effect. The non-lethal threshold (LCt_{01}) values of CO and HCN were calculated based on the multiple $C \times t$ relationships of nose-only exposed rats as detailed elsewhere [2] (figure reproduced from Pauluhn, 2016 [2])

Despite the different modes of exposure (nose-only vs. whole-body) and the applied technical standards, the calculated LC_{01} of both studies were similar at 30 and 60 min exposure durations (Figure 7.7). Differences between modes of exposure may occur due to the restraint-related higher ventilation in nose-only exposed animals. Recent OECD testing guidelines consider nose-only studies superior to whole body studies as technical mishaps are less likely to occur [4, 5]. The comparison given in Figure 7.7 illustrates that the "toxic load model" provides a versatile means to calculate the cornerstones of acute inhalation toxicity LC_{50} and LC_{01} from the entire $C \times t$ matrix examined. Therefore, the comprehensive data sets from rats were given preference to isolated data using non-standardized approaches. Concurrent with the rationale given earlier, the $1/3 \times LC_{01}$ relationship was taken as a threshold below which no impairment of escape is expected to occur [2]. As illustrated in Figure 7.8, the $C^n \times t = k$ relationship is suitable to calculate any time-adjusted LCt_{01} from any set of mortality data with multiple exposure durations. It appears to be reasonable to expand this equation to also calculate the incapacitation threshold (IC_{01}) with $IC_{01} = LC_{01} \times 1/3$ as shown below.

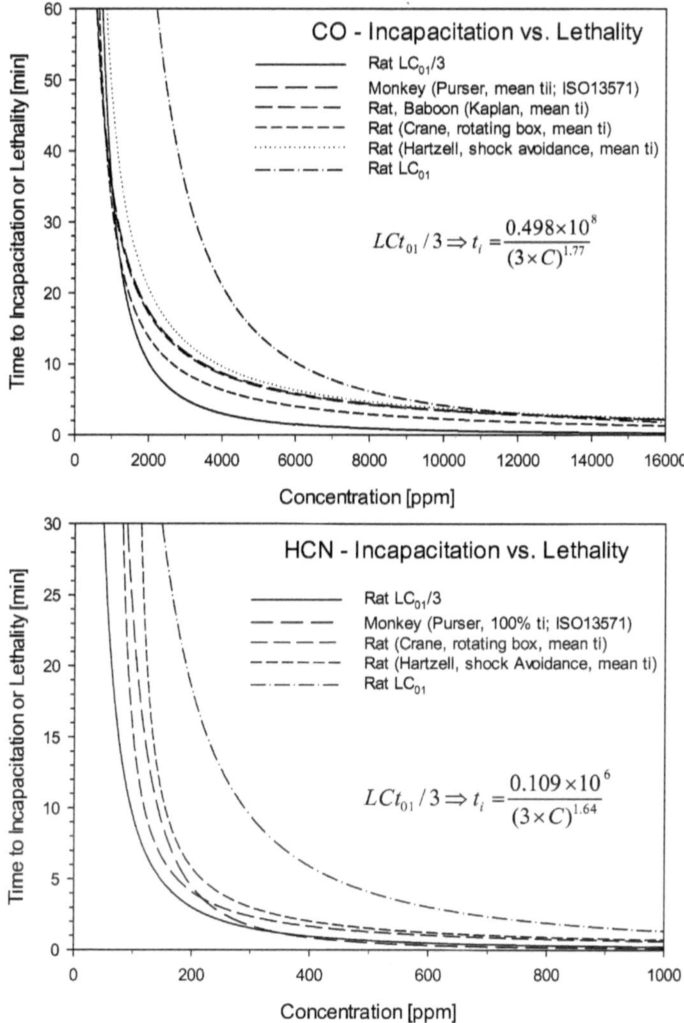

Figure 7.8 Comparison of the time-to-incapacitation in rats and non-human primates using four different endpoints to characterize incapacitation. Data were from Sweeney [48–50], Purser [51, 52], Crane [18], Kaplan [53], and Hartzell [19] as summarized by Speitel (1995) [32] (figures reproduced from Pauluhn [2])

Thus, despite their underlying different endpoints, the empirical $C \times t$ relationships support the $1/3 \times LC_{01}$ approach as sufficiently conservative. As long as this threshold is not exceeded, "impairment of escape" and post-exposure lethality will not occur. Thus, the statistically derived descriptor of toxicity POD_1 (lethality) can suitably serve as a starting point for calculating the POD_2 (incapacitation). Lethal and non-lethal POD from standardized, testing guideline-compliant acute inhalation studies on rats commonly serve as the most important cornerstones for deriving such guidance levels for chemical emergencies [8].

Mortality-based data commonly serve a broader range of exposure concentrations and durations than studies aiming solely at t_{inc}, the time to attain incapacitation. They also follow more rigid and internationally harmonized testing protocols using sufficient number of rats instead of few non-human primates with limited baseline data and benchmark validations [4, 5]. This facilitates enormously their use for hazard analysis and comparisons across different laboratories. The resultant broader and more consistent database outweighs the conceived advantage of duplicating a more "human-like" incapacitation paradigm. The surrogate endpoints used in animal bioassays for defining t_{inc} often require conditioned animals for the applied performance tests and used titration towards incapacitation. High inter-animal variability occurs at subtoxic exposure levels and either dichotomous or continuous endpoints are used. Accordingly, extrapolation errors appear to be less likely to occur when starting with a unequivocal binary endpoint to be determined simultaneously in equally exposed animals in the absence of superimposed methods of detection. These aspects give preference to the $IC_{01} = 1/3 \times LC_{01}$ approach as compared to laboratory-specific, fine-tuned, and highly specialized neurobehavioral testing batteries [2]. Titration of t_{inc} in single larger animals can hardly serve the purpose of probabilistic assessments due to animal- and method-specific variabilities.

7.3.2.2 Irritants

As referred to above, airborne chemical sensory irritants are known to evoke a burning sensation in the eyes, nose, and throat, thereby causing "impairment of escape" in an exposed individual. At high concentrations, exposure can be both incapacitating and life-threatening. Animal models were developed using the decrease in respiratory rate in mice or rats as an index of sensory irritation. Based on the concentration–response relationships, the RD_{50}, defined as the concentration causing a 50% decrease in respiratory rate, was shown to have a predictable relationship with sensory irritation in men [54–57]. The interrelationship of the RD_{50} and other descriptors of acute toxicity is compared for hydrogen hydrochloride (HCl) in Figure 7.9. Although sensory irritation occurs concentration-dependently, its severity is better described as a $C^n \times t$ dependent response [8–10]. As exemplified for HCl, the lethality-based $LC_{01}/30$ and SLOT-dangerous toxic load (DTL)/10 values correspond favorably to the AEGL-2 to AEGL-3 range (see Figure 7.9). Taken as a whole, these findings suggest that defined fractions of the LC_{01} and SLOT values are suited to serve as a basis for the estimation of threshold $C^n \times t$ values below which incapacitation can be excluded with reasonable probability.

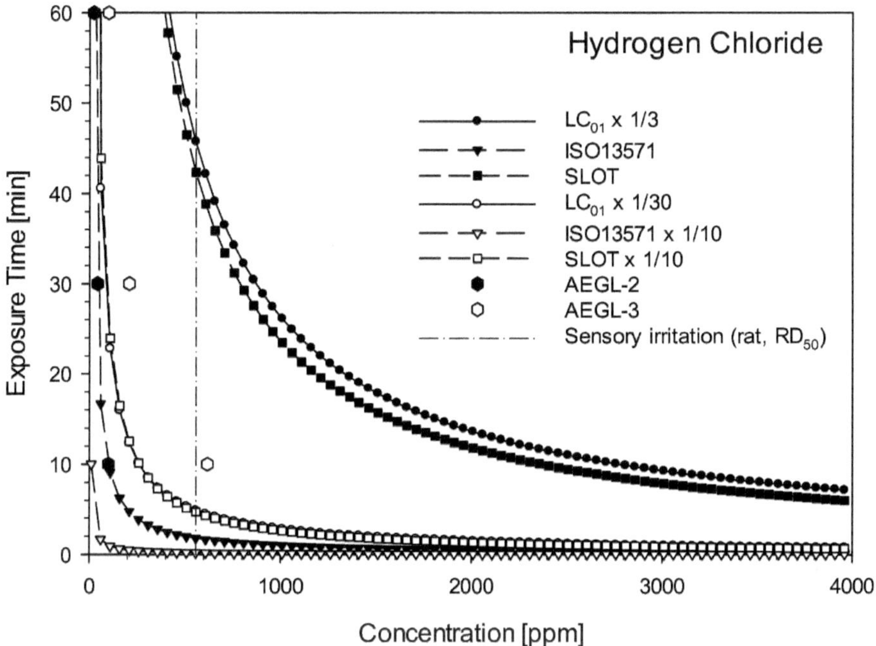

Figure 7.9 Concentration x exposure time comparisons of the irritant gas HCl. The toxic load $C^n \times t$ = const. effect model used the published parameterization of SLOT-DTLs [10] for the general population. The $1/30 \times LC_{01}$ from rats exposed to HCl was taken as a reference to demonstrate the implicit conservatism of the course taken

It is interesting to note that the SLOT (DTL) principle [10] arrives at the same conclusion for both asphyxiant and irritant chemicals by using the toxic load model but different rationales for the same uncertainty factor to account for variability of the human population (Figure 7.9) [10]. When looking at the SLOT criteria, they reflect exposure conditions just on the verge of causing a low percentage of deaths in the exposed general population. It takes around 1% mortality in animals (LC_{01}) as being representative of SLOT conditions. They are used to provide estimates of the extent (i. e., hazard ranges and widths) and severity (i. e., how many people are affected, including the numbers of fatalities) of the consequences of each identified major accident hazard [10, 11]. The comparison of the SLOT-DTL/10 with the LC_0/30 relationships given in Figure 7.9 supports the notion that SLOT-DTL/10 appears to be a defensible estimate for assessing the irritation-related threshold for incapacitation.

■ 7.4 Application of the Fractional Effective Dose (FED) Model

The focus of the modified FED concept is to predict *a priori* a ceiling (threshold) $C{\times}t$ at which 99% of the healthy and normally performing human population confined in burning buildings can execute the necessary escape paradigm within a specified duration. The interrelationship of the $C{\times}t$s predicting "incapacitation" and that characterizing the "non-lethal threshold" was dealt with elsewhere [2, 3]. From toxicant concentration as a function of time data, incremental exposure doses can be calculated and related to the specific $C{\times}t$ exposure dose required to produce the given toxicological effect at each incremental concentration. As conceptualized by Eq. (7.1) and Eq. (7.2), CO_2 can either be qualified as a modifier of inhaled dose (1) or as a separate toxicological entity (2) attributing to toxicity in its own way.

FEDs can be calculated for each asphyxiant at each discrete increment of time. The time at which their accumulated sum exceeds a specified threshold value represents the time available for escape relative to the chosen safety criteria. The parameterization of Eq. (7.1) assumes that a cumulative exposure to 35 000 ppm × min CO produces 30% COHb with incapacitation in 50% of the adult human population. The FED for HCN was based on a monkey study from Purser (for details, see ISO 13571 [7]). The course taken by this standard was criticized [2, 3, 59].

Equations suggested to calculate the FEDs for the narcotic (toxic) gases CO, HCN, and CO_2 according to ISO 13571 [7] (Eq. (7.1)) and the proposed alternative (Eq. (7.2)) [2, 3]:

$$\mathrm{FED}_{\mathrm{median}} = \left(\sum_{t_1}^{t_2} \frac{[CO]}{35000} \Delta t + \sum_{t_1}^{t_2} \frac{([HCN])^{2.36}}{1.2 \mathrm{x} 10^6} \Delta t \right) \mathrm{x}\, \mathrm{e}^{\%CO_2/5} \tag{7.1}$$

$$\mathrm{FED}_{\mathrm{threshold}} = \sum_{t_1}^{t_2} \frac{([CO \,\mathrm{x}\, 3])^{1.77}}{0.498 \mathrm{x} 10^8} \Delta t + \sum_{t_1}^{t_2} \frac{([HCN \,\mathrm{x}\, 3])^{1.64}}{0.109 \mathrm{x} 10^6} \Delta t + \sum_{t_1}^{t_2} \frac{3}{\mathrm{e}^{11.4 - 1.14 \mathrm{x} \%CO_2}} \Delta t \tag{7.2}$$

where CO (ppm) and HCN (ppm) are the averaged measured or estimated concentrations of the respective asphyxiant gas over the chosen time increment, Δt (1 min). The parameterization of the CO- and HCN-related terms was based on $C^n \times t$ = constant effect (k) analyses from multiple concentration × exposure time acute rat inhalation studies. Based on this relationship, the time-dependence of the non-lethal threshold concentration (LC_{01}) was parameterized as shown above. The 1/3 × LC_{01} value was taken as a reasonable threshold for incapacitation to occur

[2, 3]. Denominators represent the toxic load constant "*k*" and the toxic load exponent "*n*" [$ppm^n \times min$].

In ISO 13571 the FED_{median} values of $0.1 \times$ FED calculated by Eq. (7.1) translate to a 1.1% compromised tenability in the exposed population. In contrast, the $FED_{threshold}$ used the $1/3 \times LC_{01}$ threshold concentrations and precludes both incapacitation and lethality to occur at or below this threshold. When using the $1/3 \times$ SLOT-DTL approach for irritants, the same approach can be applied, i.e., the FEC approach can be abandoned.

For CO and HCN, the $1/3 \times LC_{01}$-based threshold $C \times t$s for incapacitation were calculated based on using the modified approach. When comparing the respective FED calculations according to ISO 13571 Eq. (7.1) (Figure 7.10, top panel) and the $1/3 \times LC_{01}$ Eq. (7.2) (Figure 7.10, bottom panel), it is apparent that the COHb-based ISO approach leads to a marked underestimation of the CO-related toxicity relative to HCN. In contrast, the $LC_{01} \times 1/3$-based approach demonstrates an equal contribution from both CO and HCN. The model-specific CO_2-based adjustments compared in these figures yielded total FEDs of 6 and 3–4 when using the ISO 13571- and $LC_{01} \times 1/3$-based approach, respectively. When comparing the equitoxic FEDs of $1/10 \times FED_{median}$ and $FED_{threshold}$, it appears as if the FED_{median} approach leads to overly conservative risk estimates.

Assuming CO_2 concentrations to be 1.5-fold higher than the 6% threshold (see Figure 7.3), CO_2-related fatalities cannot be excluded any longer. Under such conditions, the total FEDs yielded 12 and 22 for the ISO 13571 and the alternative LC_{01} approach, respectively (Figure 7.10). This demonstrates that the ISO 13571 approach is biased to overestimate the impact of CO_2 at non-lethal concentrations; however, at potentially lethal levels, it is biased to markedly underestimate the CO_2-related toxicity (for details, see [3]).

In summary, the risks associated with asphyxiant fire gases' tenability can be modeled by the FED approach. Based on the comparisons made, the alternative $1/3 \times LC_{01}$-approach provides the most conclusive outcome with a more distinct appreciation of the relative toxic potencies of CO, HCN, and CO_2. A ten-fold larger factor is required for irritants due to their portal of entry-related modes of action and implicit dosimetric adjustments from nasally breathing rats to oronasally breathing humans.

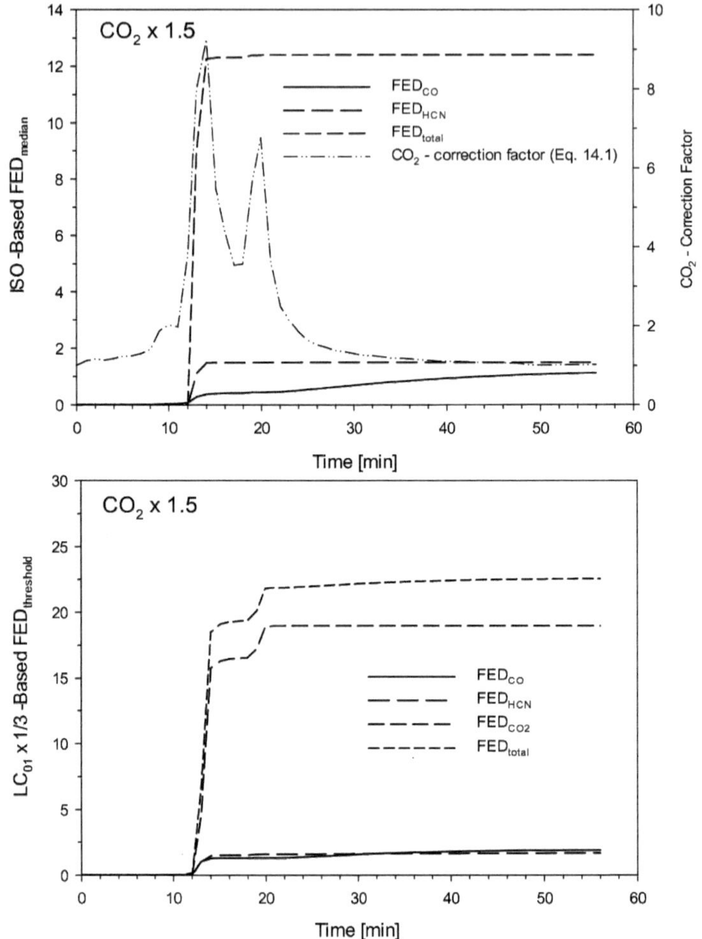

Figure 7.10 Comparison of FEDs using the ISO 13571 FED_{median} calculation method (Eq. (7.1)) and 1/3 x LC_{01} $FED_{threshold}$ calculation method (Eq. (7.2)). FED calculations assumed maximal concentrations of CO_2 in the range of 9–11% (figures reproduced from Pauluhn [3])

■ 7.5 Conclusions

This analysis compared the *C*×*t* matrix for deriving POD for the lethal and incapacitating thresholds. Data from multiple rat and non-human primate bioassays were used in combustion toxicology to define human surrogates of "impairment of escape" or "incapacitation". The findings from this analysis suggest that the rat delivers the most consistent data set. However, it must be recalled that any perception-related psychophysical variables of human behavior typical for escape cannot be modeled by any surrogate animal species. For the asphyxiant gases examined, POD characterizing "impairment of escape" were hardly qualified to distinguish unequivocally signs of "impending death" and "successful escape with post-escape survival". Thus, passing the escape paradigm as such is not equivalent to post-escape survival which is a definite weakness of the current ISO 13571 [7] approach. Similarly, consistent with published evidence from humans, COHb saturation levels appear to be unrelated to incapacitation [59]. These observations suggest that the exposure-regimen-specific profile and especially the C_{max} of CO appear to be more critical than the accumulated inhaled total $C \times t$ dose. The most consistent and conclusive endpoint to protect from the occurrence of "impairment of escape" appears to be the lethality-based biomarker of effect $1/3 \times LC_{01}$ from acute rat inhalation studies. Unlike t_{inc}, the LC_{01} endpoint can readily be derived from objectively determined binary data, ideally in context with a $C \times t$ matrix of acute inhalation studies in rats with the pure toxicant, that accommodate the time periods of interest. Thus, as exemplified by the newly generated $C \times t$ set of data for CO and HCN in rats by Sweeney et al. [48-50], a purpose-adjusted new database has opened up new avenues of modeling and animal-saving toxicological risk assessment.

References for Chapter 7

[1] Rusch, G.M., Bast, C.B., Cavender, F.L., Establishing a Point of Departure for Risk Assessment Using Acute Inhalation Toxicology Data. *Regul. Toxicol. Pharmacol.* (2009) 54, pp. 247-255.

[2] Pauluhn, J., Acute inhalation toxicity of carbon monoxide and hydrogen cyanide revisited: Comparison of models to disentangle the concentration × time conundrum of lethality and incapacitation. *Regul. Toxicol. Pharmacol.* (2016) 80, pp. 173-182.

[3] Pauluhn, J., Risk Assessment in Combustion Toxicology: Should Carbon Dioxide be Recognized as a Modifier of Toxicity or Separate Toxicological Entity? *Toxicol. Lett.* (2016) 262, pp. 142-152.

[4] *Guideline for the testing of chemicals - Guideline 403: Acute Inhalation Toxicity* (2009) OECD (Organization for Economic Cooperation and Development), Environment Directorate, Adopted 7th. September 2009. URL: *http://www.oecd-ilibrary.org/environment/test-no-403-acute-inhalation-toxicity_9789264070608-en.*

[5] *Environment, Health and Safety Publications, Series on Testing and Assessment No. 39: Guidance Document for Acute Inhalation Toxicity Testing* (2009) OECD (Organization for Economic Cooperation and Development), [ENV/JM/MONO 28; July 21, 2009]. URL: *https://ntp.niehs.nih.gov/iccvam/suppdocs/feddocs/oecd/oecd-gd39.pdf.*

[6] Tewarson, A., Ventilation effects on combustion products. *Toxicology* (1996) 115, pp. 145-156.

[7] ISO 13571: *Life threat from fires - guidance on the estimation of the time available for escape using fire data* (2012).

[8] Standing Operating Procedures for Developing Subcommittee on Acute Exposure Guideline Levels Committee on Toxicology Board on Environmental Studies and Toxicology. *Commission on Life Sciences - Acute Exposure Guideline Levels for Hazardous Chemicals* (2001) NRC (National Research Council), URL: *http://www.nap.edu.*

[9] WEELs and ERPGs, *Emergency Response Planning Guidelines (ERPG) and Workplace Environmental Exposure Levels (WEELS) Handbook* (2015) American Industrial Hygiene Association (AIHA) Guideline Foundation, Falls Church, VA.

[10] Anon., *Toxicity levels of chemicals - Assessment of the Dangerous Toxic Load (DTL) for Specified Level of Toxicity (SLOT) and Significant Likelihood of Death (SLOD)* (2015) URL: *http://www.hse.gov.uk/chemicals/haztox.htm* (accessed: August 14, 2017).

[11] Fairhurst, S., Turner, R.M., Toxicological assessments in relation to major hazards. *J. Hazard. Mater.* (1993) 33, pp. 215–227.

[12] Harper, P., *Assessment of the major hazard potential of carbon dioxide (CO2)* (2011) HSE (Health and Safety Executive), URL: *http://www.hse.gov.uk/carboncapture/assets/docs/major-hazard-potential-carbon-dioxide.pdf. First published 06/2011.*

[13] *Assessment of the Dangerous Toxic Load (DTL) for Specified Level of Toxicity (SLOT) and Significant Likelihood of Death (SLOD)* (2013) HSE (Health and Safety Executive), URL: *http://www.hse.gov.uk/chemicals/haztox.htm* and *http://www.hse.gov.uk/carboncapture/carbondioxide.htm.*

[14] Wittbecker, F.-W., Chapter 8: Methodology of Fire Testing, in *International Plastics Flammability Handbook - Principles, Regulations, Testing and Approval* (2004) 3rd ed, Troitzsch, J. (Ed.) Hanser Publishers, Munich, Vienna, New York, pp. 209–221.

[15] *Acute Exposure Guideline Levels for Selected Chemicals: Volume 2, Hydrogen Cyanide* (2002) NRC (National Research Council), pp. 211–276. URL: *https://www.epa.gov/sites/production/files/2014-09/documents/tsd6.pdf.*

[16] Anon., Are There Other Deleterious Effects of Varying Exposures To Carbon Monoxide And Hydrogen Cyanide? in *Combined Exposures to Hydrogen Cyanide and Carbon Monoxide in Army Operations* (2008) pp. 20. URL: *http://www.nap.edu/catalog/12467.html.*

[17] Coburn, R.F., Forster, R.E., Kane, B.P., Considerations of the physiological variables that determine the blood carboxyhemoglobin concentration in man. *J. Clin. Invest.* (1965) 44, pp. 1899–1910.

[18] Crane, C., Sanders, D., Endecott, B., *Inhalation Toxicology: IX. Times to Incapacitation for Rats Exposed to Carbon Monoxide Alone, to Hydrogen Cyanide Alone, and to Mixtures of Carbon Monoxide and Hydrogen Cyanide* (1989) Federal Aviation Administration, Civil Aeromedical Institute, Report Number DOT/FAA/AM-89/4, January 1989.

[19] Hartzell, G.E., Switzer, W.G., Priest, D.N., Modeling of Toxicological Effects of Fire Gases: V. Mathematical Modeling of Intoxication of Rats by Combined Carbon Monoxide and Hydrogen Cyanide Atmospheres. *J. Fire Sci.* (1985) 3, pp. 330–342.

[20] *Acute Exposure Guideline Levels for Selected Chemicals: Volume 8, Carbon monoxide* (2010) NRC (National Research Council), pp. 49–142. URL: *http://www.nap.edu/catalog/12770.html.*

[21] Pharmacokinetics and Mechanisms of Action of Carbon Monoxide, Chapter 5, in *Air Quality Criteria for Carbon Monoxide*, EPA document 600/P-99/001F, June 2000, US-EPA. URL: *https://cfpub.epa.gov/ncea/risk/recordisplay.cfm?deid=18163.*

[22] Hampson, N.B., Weaver, L.K., Carbon monoxide poisoning: a new incidence for an old disease. *Undersea Hyperbaric Med.* (2007) 34, pp. 163–167.

[23] Hampson, N.B., Dunn, S.L., Symptoms of carbon monoxide poisoning do not correlate with the initial carboxyhemoglobin level. *Undersea Hyperb. Med.* (2012) 39, pp. 657–665.

[24] Hampson, N.B., Hauff, N.M., Carboxyhemoglobin levels in carbon monoxide poisoning: do they correlate with the clinical picture? *Am. J. Emerg. Med.* (2008) 26, pp. 665–669.

[25] Hampson, N.B., Piantadosi, C.A., Thom, S.R., Weaver, L.K., Practice Recommendations in the Diagnosis, Management, and Prevention of Carbon Monoxide Poisoning. *Am. J. Respir. Crit. Care Med.* (2012) 186, pp. 1095–1101.

[26] Kimmel, E.C., Carpenter, R.L., Reboulet, J.E., Still, K.R., A physiological model for predicting carboxyhemoglobin formation from exposure to carbon monoxide in rats. *J. Appl. Physiol.* (1999) 86, pp. 1977–1983.

[27] Gordon, C.J., *Temperature Regulation in Laboratory Rodents* (1993) Cambridge University Press, New York.

[28] Gordon, C.J., *Temperature and Toxicology: An Integrative, Comparative, and Environmental Approach* (2005) CRC Press, Boca Raton, FL.

[29] Gordon, C.J., Spencer, P.J., Hotchkiss, J., Miller, D.B., Hinderliter, P.M., Pauluhn, J., Thermoregulation and its influence on toxicity assessment. *Toxicology* (2007) 244, pp. 87-97.

[30] Sanderson, W.C., Rapee, R.M., Barlow, D.H., The Influence of an Illusion of Control on Panic Attacks Induced via Inhalation of 5.5% Carbon Dioxide-Enriched Air. *Arch. Gen. Psychiatry* (1989) 46, pp. 157-162.

[31] Blockley, W., *Bioastronautics Data Book*, (1964) NASA SP-3006, pp. 5-17, pp. 103-132.

[32] Speitel, L.C., Toxicity Assessment of Combustion Gases and Development of a Survival Model (1995) Airport and Aircraft Safety Research and Development Division FAA Technical Center Atlantic City International Airport, NJ 08405, DOT/FAA/AR-95/5 Office of Aviation Research Washington, D.C. 20591. URL: *http://www.fire.tc.faa.gov/pdf/95-5.pdf.*

[33] Levin, B.C., Paabo, M., Gurman, J.L., Harris, S.E., Braun, E., Toxicological interactions between carbon monoxide and carbon dioxide. *Toxicology* (1987) 47, pp. 135-164.

[34] Pauluhn, J., A Retrospective Analysis of Predicted and Observed Smoke Lethal Toxic Potency Values. *J. Fire Sci.* (1993) 11, pp. 109-130.

[35] Astrand, I., Ovrum, P., Lindqvist, T., Hultengren, M., Exposure to butyl alcohol: Uptake and distribution in man. *Scand. J. Work Environ. Health* (1976) 3, pp. 165-175.

[36] Brake, D.J., Bates, G.P., Criteria for the design of emergency refuge stations for an underground metal mine. *The AusIMM Proceedings No 2* (1999) URL: *http://www.mvaust.com.au/assets/e02-criteria-for-the-design-of-emergency-refuge-stations-for-an-underground-metal-mine.pdf.*

[37] Chapter 6. Inhalation rates, in *Exposure Factors Handbook 2011 Edition* (Final) (2011) U.S. Environmental Protection Agency, Washington, DC, EPA/600/R-09/052F.

[38] Li, W.L., Pauluhn, J., Comparative assessment of the sensory irritation potency in mice and rats nose-only exposed to ammonia in dry and humidified atmospheres. *Toxicology* (2010) 276, pp. 135-142.

[39] Pauluhn, J., Acute inhalation toxicity of ammonia: revisiting the importance of RD_{50} and $LCT_{01/05}$ relationships for setting emergency response guideline values. *Regul. Toxicol. Pharmacol.* (2013) 66, pp. 315-325.

[40] Appelman, L.M., ten Berge, W.F., Reuzel, P.G.J., Acute inhalation toxicity study of ammonia in rats with variable exposure periods. *Am. Ind. Hyg. Assoc. J.* (1982) 43, pp. 662-665.

[41] Monsé, C., Sucker, K., Hoffmeyer, F., Jettkant, B., Berresheim, H., Bünger, J., Brüning, Th., The Influence of Humidity on Assessing Irritation Threshold of Ammonia. *BioMed Research International* (2016) Volume 2016, Article ID 6015761, 7 pages.

[42] Silverman, L., Whittenberger, J.L., Muller, J., Physiological response of man to ammonia in low concentrations. *J. Ind. Hyg. Toxicol.* (1949) 31, pp. 74-78.

[43] Anon. Chapter 4: The prediction of acute effects of ammonia upon humans from animal data, in *Major Hazard Monograph - Ammonia Toxicity* (1997) Institution of Chemical Engineers, Rugby, UK.

[44] Verberk, M.M., Effects of ammonia in volunteers. *Int. Arch. Occup. Environ. Health* (1977) 39, pp. 73-81.

[45] Shusterman, D., Matovinovic, E., Salmon, A., Does Haber's law apply to human sensory irritation? *Inhal. Toxicol.* (2006) 18, pp. 457-471.

[46] Widdicombe, J., Nasal and pharyngeal reflexes: Protective and respiratory functions, in *Respiratory function of the upper airway*, Mathew, O.P., Sant Ambragio, G. (Ed.) (1988) Marcel Dekker, New York, pp. 233-258.

[47] ten Berge, W.F., Concentration-time mortality response relationship of irritant and systemically acting vapours and gases. *J. Hazard. Mater.* (1986) 13, pp. 301-309.

[48] Sweeney, L. M., Sommerville, D. R., Channel, S. R., Impact of non-constant concentration exposure on lethality of inhaled hydrogen cyanide. *Toxicol. Sci.* (2014) 138, pp. 205–216.

[49] Sweeney, L. M., Sommerville, D. R., Channel, S. R., Sharits, B. C., Gargas, N. M., Gut, C. P., Jr., Evaluating the validity and applicable domain of the toxic load model: impact of concentration vs. time profile on inhalation lethality of hydrogen cyanide. *Regul. Toxicol. Pharmacol.* (2015) 71, pp. 571–584.

[50] Sweeney, L. M., Sommerville, D. R., Goodwin, M. R., James, A., Channel, S. R., Acute toxicity when concentration varies with time: a case study with carbon monoxide inhalation by rats. *Regul. Toxicol. Pharmacol.* (2016) 80, pp. 102–115.

[51] Purser, D. A., A bioassay model for testing the incapacitating effects of exposure to combustion product atmospheres using cynomolgus monkeys. *J. Fire Sci.* (1984) 2, pp. 20–36.

[52] Purser, D., Berrill, K., Effects of Carbon Monoxide on Behavior in Monkeys in Relation to Human Fire Hazard. *Arch. Environ. Med.* (1983) 38, pp. 308–315.

[53] Kaplan, H. L., Grand, A., Switzer, W., Mitchell, D., Rogers, W., Hartzell, G., Effects of Combustion Gases on Escape Performance of the Baboon and the Rat. *J. Fire Sci.* (1985) 3, pp. 228–244.

[54] Alarie, Y., Sensory Irritation by Airborne Chemicals. *Crit. Rev. Toxicol.* (1973) pp. 207–209.

[55] Alarie, Y., Dose-response analysis in animal studies: prediction of human responses. *Environ. Health Perspect.* (1981) 42, pp. 9–13.

[56] Kane, L. E., Barrow, C. S., Alarie, Y., A short-term test to predict acceptable levels of exposure to airborne sensory irritants. *Am. Ind. Hyg. Assoc.* (1979) 40, pp. 207–229.

[57] Barrow, C. S., Alarie, Y., Warrick, J. C., Stock, M. F., Comparison of the sensory irritation response in mice to chlorine and hydrogen chloride. *Arch. Environ. Health* (1977) 32, pp. 68–76.

[58] Pauluhn, J., Concentration × time analyses of sensory irritants revisited: Weight of evidence or the toxic load approach. That is the question. *Toxicology Letters* (2019) 316, pp. 94–108.

[59] Pauluhn, J., Estimation of time to compromised tenability in fires: is it time to change paradigms? *Regul. Toxicol. Pharmacol.* (2020), 111:104582, doi: 10.1016/j.yrtph.2020.104582.

Part II

National and International Fire Protection Regulations and Test Procedures

8 Regulations and Testing

Edith Antonatus and Jürgen Troitzsch

8.1 Introduction

The topics discussed in the following chapters mainly reflect those areas for which fire protection requirements are imposed by regulations or mandatory codes. These include building, transportation, electrical engineering, and furniture. Building is dealt with in particular detail, since this is the field where regulations, test methods, and approval procedures are most numerous, and where the greatest variety exists in different countries and regions. Furthermore, the use of plastics as combustible building materials is dealt with here from the point of view of fire protection. This field therefore forms the most important part of this book. The European Union, 14 European countries, and 8 non-European countries are covered. In addition, the activities of international standardization organizations such as ISO and CEN are dealt with in detail in Chapter 9.

In addition to legal provisions, insurance codes and standards in some countries are also of major importance for products used in building, and some of them are internationally accepted.

In transportation and the electrical industry, regulations and test methods are usually international, so that they are less numerous and less varied. They are discussed in Chapters 11 and 12.

Furniture and furnishings are covered in detail in Chapter 13. CEN and ISO standards are used, and in some countries regulatory requirements have been introduced. In a few countries, i.e., the UK and the USA (California), compulsory requirements apply for furniture and furnishings in public and residential buildings. A large number of countries only regulate special risk areas like public buildings, hospitals, prisons, etc.; they are dealt with in the respective country sections of Chapter 10.

The various fields covered in the following are dealt with according to a uniform system. This is intended to help the reader become acquainted with, and to review rapidly, the various regulations, requirements, test methods, and approval procedures for the principal areas of application of materials and products. After a short introduction to the statutory regulations or codes in force, the classifications and tests necessary for meeting the requirements are described. The tests are described according to a uniform scheme with diagrams illustrating the principles of the methods, and tables of the specifications. Further details are given where necessary for clarity. Only those methods contained in regulations or referred to in associated supporting documents are discussed here. Most of the tests are application-related and can be carried out on laboratory- or pilot-scale. Full-scale tests are increasingly required in building for specific applications such as façades. Tests which are not clearly defined such as ad-hoc investigations, voluntary tests for product development and quality control, as well as methods for determining the fire resistance of building components (which only contain plastics in exceptional cases), are not considered.

It is not the aim of this book to evaluate the fire performance of various plastics, semi-finished or finished parts on the basis of the test methods discussed in the following. After describing the test methods, a list of appointed and recognized test laboratories that carry out fire tests is given. The ways of obtaining approval and test or quality marks are described, together with the associated inspection and monitoring procedures.

Short sections on future developments follow each subject and country dealt with in the following. These briefly cover planned national and international changes or new trends in regulations, testing, and approval.

The fire protection regulations and test procedures mentioned in Chapters 9 to 13 cover the principal application areas of plastics in the main countries listed in this book. However, this vast subject cannot be dealt with exhaustively. Instead, this survey is intended to provide the reader with a comprehensive review of the most important aspects of this field, and enable her/him to become rapidly acquainted with the complex field of fire protection.

Testing and assessment of side-effects from fire, such as smoke development, toxicity, and corrosivity of fire effluents, are also described in Chapters 9 to 13. As smoke and toxicity from fire effluents lead to most fire fatalities, they are increasingly taken into account in regulations and standards, and are extensively discussed in Chapters 6 and 7.

8.2 Sets of Regulations and Codes

8.2.1 Types and Basis

The aim of fire protection regulations and codes is to minimize the fire hazard and the resulting risk for people, thus protecting life and property. Governments, as official custodians of public safety, ensure such protection via relevant legislation, which includes laws and statutory orders mainly focusing on buildings, but also covering other areas, like transportation or safety of electrical appliances. Standards and codes of practice based on recognized technical principles are the means of putting the general requirements of fire protection defined in legislation into practice. Materials, semi-finished, and finished products are tested and classified according to methods laid down in the standards. Such tests are in most cases carried out by officially recognized testing institutes, for legal purposes. In addition, general rules for building are defined to make buildings safer, i.e., requirements for escape routes, smoke detectors, compartmentation of buildings, etc.

The certificates of test results and classifications provide a basis for the use of a material or product. A test mark is frequently required as evidence of the suitability of a product and its identity with the product tested. This requires a quality check on production either by the manufacturer or an outside body.

Numerous public and private organizations contribute to the development of the rules and regulations for fire protection. In addition to governmental bodies, these include professional societies and industrial, commercial, technical, and insurance associations.

Additional rules and requirements for buildings and for electrotechnical appliances are put in place by some major property insurers. The main objective of these codes and standards is protection of property. The assessments of products by insurers are based partly on official testing standards, but also on test methods specially developed by insurers. Insurers have their own testing institutes, but in some cases also use the results from external testing houses.

In some areas, special requirements in manufacturers' purchasing specifications must, in practice, be complied with, since products that do not satisfy them have little chance of commercial success. Such requirements are, for example, imposed by the aircraft industry or by car manufacturers, with the aim of improving the fire safety of their products beyond the official regulations.

This multiplicity is necessary to create an appropriate and practical "state of the art" system of fire precautions.

Standards for testing, assessment, and classification of materials and products are a basis for all kinds of requirements in regulations or codes. They are developed by standardization organizations where representatives of government, test institutes, industry, and insurers collaborate and introduce their particular knowledge and experiences. These can work on a national, regional, and/or international level.

While the IEC is responsible for international electrical standards, almost all other technical fields are covered by the International Organization for Standardization (ISO). This organization aims to promote the worldwide development of standards in order to facilitate the exchange of goods and services and to encourage mutual cooperation in intellectual, scientific, and economic activities. ISO member committees are the national standardization organizations. Further details on ISO can be found in Section 9.2.

Additional standards are developed by regional and national standardization organizations. Often international and regional standards are adopted by national standardization institutes and used by regulatory bodies in different fields of application. Thus, a number of ISO standards are used in building and construction regulations in many European and Asian countries, as well as for regulations and codes for trains, ships, and others.

8.2.2 Application

Regulations and codes, depending on the field of application, are used nationally, in certain regions, and/or internationally.

National fire safety codes are encountered almost exclusively in building, because every country possesses its own set of rules and regulations, which have developed historically. International harmonization of regulations would in some cases have major economic implications, which would primarily be negative in the eyes of individual states. The safety concepts or fire philosophies of many countries are in some cases diametrically opposed to internationalization in certain areas.

Additional classifications for construction products and buildings formulated by property insurance associations that relate to the fire resistance of structures and the fire performance of materials, as well as specific safety requirements, are often used in different countries. They are significant because they form the basis of fire insurance policies. These codes also cover design, installation, and routine maintenance of fire protection systems and lay down testing and approval procedures. Some of them are used in many countries; others are limited to one or two countries only.

Another example of national regulations is requirements for furniture. There are European and international standards for testing of furniture, but it still is up to every country to decide which standards have to be applied and which criteria must be met. Further details on tests and regulations can be found in Chapter 13.

In other fields, regulations exist which are valid in almost all countries. This applies for transportation. For civil aviation the most important tests and requirements were developed in the USA, and thus the US Federal Aviation Regulations (FAR) of the Federal Aviation Administration (FAA) set the criteria for the corresponding internationally accepted regulations. Evidence of airworthiness must be delivered in accordance with these regulations, which have been partially or totally adopted by most countries.

Sets of regulations in international use also exist in other areas of transportation. For example, the International Convention for the Safety of Life at Sea (SOLAS) is accepted in nearly all countries. For railways in Europe, a mandatory standard has been developed by CEN which defines fire safety tests, classifications, and requirements for all new trains, replacing the former national standards and regulations. Other countries (for example, the USA) have their own national requirements for trains. For further details on transportation, see Chapter 11.

8.2.3 The Role of Standards

Standards for testing and classification of materials and products are important, because they are based on input from producers, testing houses, instrument producers, and regulators, and give principles and details which mirror the state of the art. Based on these standards, regulatory requirements can be developed and product properties can be verified.

International standards are used in many countries and many fields of application. The ISO fire test methods, for example, are used in many countries and regions for building applications as well as for transportation.

IEC has prepared specific standards for electrotechnical applications, which are used worldwide.

In addition, regional standards (for example, for construction products in the EU) and national standards are in place.

8.3 Test Methods

The fire tests discussed below are arranged according to the test criteria on which they are based and their contribution to measuring certain fundamental parameters that affect the initiation and spread of fire.

Size is one of the most important factors affecting test procedures. Tests may be carried out on a laboratory scale, pilot scale, or full scale. The first two are the most commonly used, and are discussed below. In specific applications full-scale tests are gaining more importance in the framework of regulations and insurance requirements. This is particularly true for steel-faced sandwich panels and for different types of façade insulation and cladding systems, which will be also covered here.

8.3.1 Laboratory-Scale Tests

Laboratory tests are usually performed in a defined environment. They are either executed in a closed chamber with defined dimensions and ventilation conditions, or at least in a room where the ambient temperature and the ventilation near the specimen are clearly defined.

Testing standards and in some cases additional specific product standards clearly define the type and size of the specimen to be tested, the mounting of the specimen, the type of fire exposure, and the parameters to be measured.

8.3.1.1 Test Specimens

For material tests, either flat specimens, or in some cases (i. e., some non-combustibility tests) specimens ground or cylindrical specimens are used.

In some laboratory tests composites or complete products are tested. These can be flat specimens, which are tested as an assembly and have the same thickness as the finished part. In the electrical sector, in addition to material tests, complete finished parts such as plugs and switches are frequently tested on a laboratory scale. The same applies for furniture and seats. Here a mock-up of a simulated seat or real pieces of furniture are tested.

Specimens can be positioned either horizontally (in floor or ceiling positions), vertically, or at an angle, in order to simulate the application of a product.

8.3.1.2 Ignition Sources

For laboratory fire tests a number of different ignition sources are used alone or in combination. The exposure conditions for the specimen depend on the type and size of the ignition source and on the duration of the exposure.

The most important types of ignition sources are:

- Gas burners: These exist in a number of variations. They are defined by type, size, and construction of the burner, by the type and the volume flow of gas used, as well as by the amount of air mixed into the gas. Small ignition sources such as matches are simulated by laboratory burners like Bunsen, Tyrill, or small tube burners. For larger tests (for simulating exposure in a developing fire) sand bed burners are frequently used.

- Radiative ignition sources: The most common radiative heat sources for fire tests are radiant panels (gas panels) and electrical heaters, which in many cases have a conical shape. The radiant ignition sources are often used in combination with a direct ignition source, like a gas burner or a spark ignition device.
- Wood cribs: These are used for many different types of tests in various sizes and constructions. The exposure of a specimen in this case depends on the type, size, and construction of the wood crib, as well as on the type of wood used and its humidity.
- Alternative ignition sources: Special ignition sources are used to simulate real ignition sources more precisely. These may be burning cigarettes, candles, oil burners, etc.

The list of variables above relating to specimen and ignition source illustrates how much individual tests may vary and why test results obtained by different methods cannot be compared directly. Test results can only be correlated empirically with a real fire.

8.3.1.3 Parameters Measured

The aim of the various laboratory-scale fire tests is to measure ignitability, combustibility, contribution to spread of fire, and continuous glowing combustion, as well as the various side effects such as smoke development, toxicity, and corrosivity of fire effluents.

The most important parameters for the fire performance of a product are:

- Combustibility: The goal is to determine whether a product can be ignited and will contribute to a fire under extreme exposure conditions (high temperature or radiation combined with a flaming ignition source).
- Ignitability: The goal is to determine whether a product can be easily ignited and contributes to further spread of fire. Time to ignition, duration of burning, and spread of flame on a sample exposed to a small ignition source are measured. Burning droplets are often considered.
- Contribution to fire spread: Time to ignition, fire spread, critical heat flux (CHF), heat release rate (HRR), total heat released (THR), and fire growth rate (FIGRA) are examples of the parameters used to assess the contribution to fire spread of a material or product. In these tests too, burning droplets are often considered.
- Continuous glowing and smoldering combustion (hidden fire propagation): In some tests the duration of heating up of a specimen and the propagation of heat development inside a specimen are measured either after finishing a standard test or after the end of the fire exposure in specially designed tests.

Smoke and toxic gas components are measured in static and dynamic tests. Smoke obscuration is measured as optical density (D_s) or smoke extinction area. Some test methods define smoke growth rate (SMOGRA) in addition to total smoke production (TSP) as relevant parameters.

For toxicity, in some countries the effects of combustion gases are still assessed with animal tests, but analytical methods are increasingly used. The parameters measured are concentrations of different gas components accumulated in the test chamber or continuously released in the exhaust gas flow. Based on these values, in most cases, a fractional effective dose (FED) and/or a fractional effective concentration (FEC) is determined (see Chapters 6 and 7). Some assessments are still based on maximum concentrations for defined single components.

Some test methods are designed to measure only one parameter, while others allow several to be measured simultaneously. An example of the first case is the ISO 1182 combustibility test, and of the latter the SBI test (EN 13823), in which heat release, smoke production, burning droplets, and flame spread are determined.

The parameters determined in these tests are referred to in many national tests under identical names but, due to differing test conditions, they are not comparable.

8.3.2 Large-Scale Tests

Large-scale tests are performed with the aim to simulate the performance of products as they are used in the real world, if they are exposed to realistic fire sources.

There are a number of standardized tests for rooms, sandwich panel constructions, façades, or roofs. In these tests the fire sources are mainly gas burners, wood cribs, or pool fires. The criteria for assessing the products tested are in most cases heat release, time to flashover, visual flame spread, and flame spread defined by measurement of temperatures. Burn-through, glowing, and smoldering are sometimes assessed as well. In addition, some of these tests are used to determine smoke development and toxicity.

Sometimes tests are performed on real buildings, vehicles, or furniture to provide the information necessary for research into fire behavior, and for development of meaningful laboratory tests. An example is the development of the European classification for construction products. A number of products were tested on behalf of the European Commission as wall and ceiling linings in the so-called room corner test (at that time ISO 9705 was used; now the European standard EN 14390 is used instead). The time to flashover in the room has been used as a criterion for the classification of these products. The criteria for classification of construction products with laboratory tests (except floorings and external roofing) were subsequently defined in such a way that the majority of products achieved the same classification in the room test and the related laboratory tests (for classifications see Section 10.3.1).

9 International Standardization

9.1 Introduction

Edith Antonatus

9.1.1 International Standardization

ISO (International Organization for Standardization) [1] and IEC (International Electrotechnical Commission) [2] are the most important international standardization organizations developing fire-related standards (amongst many others).

The organization known today as ISO was founded in the 1920s as the International Federation of the National Standardizing Associations (ISA). It was suspended in 1942 during World War II, but after the war ISA was approached by the newly set-up United Nations Standards Coordinating Committee (UNSCC) with a proposal to form a new global standards body. In October 1946, ISA and UNSCC delegates from 25 countries met in London and agreed to join forces to create the new International Organization for Standardization. Work started with 67 technical committees (groups of experts focusing on a specific subject). Because "International Organization for Standardization" would have different acronyms in different languages (IOS in English, OIN in French for *Organisation Internationale de Normalisation*), the founders decided to give it the short form ISO. ISO is derived from the Greek *isos*, meaning equal.

In 1883, the Société Internationale des Electriciens was formed in France, and the Elektrotechnischer Verein founded in Vienna, Austro-Hungary. The following year saw the establishment of the American Institute of Electrical Engineers. The Canadian Electrical Association appeared in 1891, followed two years later by Germany's Verband Deutscher Elektrotechniker, and in 1897 by Italy's Associazione Elettrotecnica Italiana. Electrical engineers in the early 20th century began to see the need for closer collaboration embracing terminology, testing, safety, and internationally agreed specifications. While the 19th century had been the era of electrotechnical innovation, the emphasis was now on consolidation and standardization. The series of International Electrical Congresses, particularly those between 1881 and 1900, had been solely concerned with electric units and standards. At the Congress in St. Louis, USA, held in 1904, in the interest of commercial transactions and trade, the proposal was made to set up a permanent international commission to study the unification of electrical machines and apparatus. The IEC was founded on 26–27 June 1906 in London, UK, and ever since has been developing global standards to the world's electrotechnical industries. Fire safety was always an essential part of the requirements for electrotechnical applications, and is important for IEC standardization.

Most of the fire-related standards used in the fields of construction and transportation have been developed in ISO/TC 92. They were partly based on existing national, regional, and insurance standards, but also completely new test methods were developed in ISO. Section 9.2 lists the standards developed in ISO/TC 92/SC1 dealing with reaction to fire. They are applied in many countries for construction products, as well as for furniture testing and transportation.

For plastic products, the ISO standards regarding toxicity of combustion products are also relevant. They are developed in ISO/TC 92/SC3 and include fire test models for toxicity, measurement/analytical methods, and assessment of the impact of toxic gases on humans. They are explained in detail in Chapters 6 and 7. The broadest field of application for toxicity standards are national and international regulations and codes for ships, aircraft, and railways. Some countries also apply ISO standards dealing with toxicity for selected construction products, and in Europe the need for the assessment of fire gas toxicity from building products is under review. In addition, ISO/TC 92/SC3 has started working on standards dealing with the environmental effects of fires.

Additional fire-related standards are developed in ISO/TC 92/SC2 (Fire Containment) and ISO/TC 92/SC4 (Fire Safety Engineering), which are not covered in this book.

IEC is the leading committee for fire-related standards for electrotechnical applications. When appropriate, IEC cooperates with ISO or ITU (International Telecommunication Union) to ensure that international standards fit together seamlessly and complement each other. Joint committees ensure that international standards combine all relevant knowledge of experts working in related areas. IEC and ISO/IEC standards are applied worldwide. Details about IEC standards are presented in Chapter 12 of this book.

9.1.2 Regional Standardization

In addition to the international standardization bodies, a number of regional standardization bodies play an important role.

For Europe, CEN (European Committee for Standardization) [3] and CENELEC (European Committee for Electrotechnical Standardization) [4] have been acknowledged by the European Union and EFTA.

CEN supports standardization activities in relation to a wide range of fields and sectors including: air and space, chemicals, construction, consumer products, and others.

Their fire-related standards are of special importance for construction products, as under the European Construction Products Regulation (CPR) the fire safety of such products needs to be tested and classified according to harmonized EN standards (see Chapter 10). Where suitable ISO standards are available, they are implemented as ISO/EN standards. If no adequate international standards are available, CEN develops its own test methods for construction products published as EN standards. Classification standards for construction products in Europe are also developed by CEN. They all have to be introduced in the EU member countries in the frame of the European CPR. Even now (2020) no complete harmonization of testing and classification of construction products has been achieved in Europe. For example, to assess the fire performance of façades, different full-scale test methods have been developed in European countries, while in parallel, and over many years, the European Commission has tried to develop a European harmonized assessment method.

The focus of CENELEC on electrotechnical applications is to create market access at European and international level by adopting international standards wherever possible, through its close collaboration with IEC.

9.1.3 International and Regional Fire-Related Standards for Transportation

International regulations are increasingly used for some types of transportation, while for other types of vehicles national regulations and standards are still applied.

Requirements and test methods for aircraft products are based on the methods developed by the American FAA (Federal Aviation Administration), and implemented in the CFR (Code of Federal Regulations) in the USA. Based on these, regional and national legislation is in place all around the world - for example, in Europe the EASA (European Aviation Safety Agency) is in the lead, and JAR (Joint Aviation Requirements) codes are developed in cooperation with the FAA.

In addition, the International Civil Aviation Organization (ICAO) is a UN specialized agency, established in 1944 to manage the administration and governance of the Convention on International Civil Aviation (Chicago Convention). Details are shown in Section 11.4.

For ships, since 1948 international standards have been developed. IMO (International Maritime Organization) has implemented a broad range of regulations for fire safety of ships, which are applied worldwide. Most of the fire test methods are based on ISO test methods. These codes and requirements are explained in Section 11.3.

For products used in cars, fire standards were first developed in the USA. The FMVSS 302 (Federal Motor Vehicle Safety Standard) test procedure has been copied worldwide into regional and national standards and legislation as well as car producers' specifications. Further test methods followed, for example for fuel tanks. Today, the safety requirements for individual cars are internationally regulated by the UNECE (United Nations Economic Commission for Europe). Additional tests have been added for buses in Europe. Details are presented in Section 11.1.

For trains, the UIC (Union Internationale des Chemins de Fer) [5] was founded in 1922 and sets general requirements to promote interoperability and international cooperation. Nevertheless, for many years fire safety requirements still were defined separately in all countries. In Europe, the development of common testing and assessment standards by CEN and CENELEC started in 1991. This resulted in the publication of the EN 45545 series, "Railway applications – Fire protection on railway vehicles" which is as far as possible based on ISO testing methods, but also includes some specific former national and newly developed test procedures (see Section 11.2.2). These European standards are used in some other parts of the world (for example in China), but in many other countries different national regulations and standards still apply. In the U.S. and Canada for example, the fire safety requirements for trains are based on ASTM standards only (see Section 11.2.3).

Although there is an increasing tendency to use international standards and performance requirements, many national and regional standards and codes remain. In particular, Asian countries like China, Korea, and Japan are increasingly involved in the ISO work. They propose standards that in the past have been national standards to ISO, and also try to change their national regulations towards using international standards. This process will take time – so, in spite of an increasing number of international and regional standards, national standards will still play an important role in the coming years.

References for Section 9.1

[1] *https://www.iso.org/home.html/*

[2] *https://www.iec.ch/*

[3] *https://standards.cen.eu/*

[4] *https://www.cenelec.eu*

[5] *https://uic.org/*

9.2 International Organization for Standardization (ISO)

Björn Sundström

9.2.1 Introduction

ISO's principal activity is to develop technical standards required by the market. The work is carried out by experts from the industrial, technical, and business sectors, which have asked for the standards and which will put them into use. Member bodies are standards organizations from 162 countries (2018). Of these, 120 member bodies are entitled to participate, and exercise full voting rights within ISO. The most active ISO countries in terms of holding secretariats are DIN (Germany) with 135 secretariats, and ANSI (USA), with 105 secretariats. JISC (Japan), AFNOR (France), SAC (China), and BSI (UK) are quite similar, and hold between 73 and 78 secretariats. At the end of 2018, ISO had produced more than 22 000 international standards. Information technology, mechanical engineering, and transport account for almost 50% of the standards [1].

ISO is largely financed by membership fees and royalties received from members selling ISO standards. The central secretariat in Geneva has a full-time staff of 152 (in 2018) [2].

The governance structure of ISO is shown in Figure 9.1. The general assembly and the council have ultimate control of ISO. The technical management board (TMB) manages the technical work and has responsibility for the technical committees. The central secretariat supports the work in the technical committees, and publishes standards and publications, etc. There are also advisory groups.

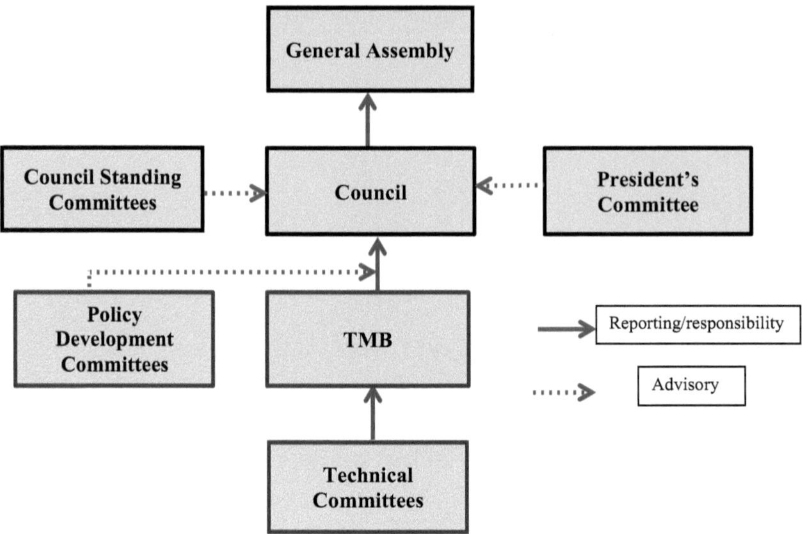

Figure 9.1 The structure of ISO

Each technical committee has a business plan, which also covers the activities of its subcommittees. The business plans should analyze the conditions and trends in the market sector served by the technical committee and will be required explicitly to link work programs and sector needs. Thus, priorities for which standards are needed can be set.

The technical work in ISO is carried out via technical committees (TCs). A TC may assign sub-committees (SCs) with working groups (WGs) to cover certain areas of work. The actual standardization work takes place in the WGs and SCs. Further general information about ISO can be found in [1] and [2].

9.2.2 Scope of ISO/TC 92 Fire-Related Activities

Fire issues appear in more than one TC, but only TC 92 is solely dedicated to the fire field. The scope of TC 92 is fire safety, and therefore its activities are intended to cover a broad spectrum of standardization issues in the fire area. As of 2017, liaison with other ISO TCs and SCs is as follows:

- ISO/TC 21: Equipment for fire protection and fire fighting
- ISO/TC 35/SC9: General test methods for paints and varnishes
- ISO/TC 59: Building and civil engineering works
- ISO/TC 61/SC4: Plastics burning behaviour
- ISO/TC 71: Concrete, reinforced concrete and pre-stressed concrete

- ISO/TC 77: Products in fibre reinforced cement
- ISO/TC 85/SC 6: Reactor technology
- ISO/TC 89: Wood-based panels
- ISO/TC 162: Doors and windows
- ISO/TC 167: Steel and aluminum structures
- ISO/TC 178: Lifts, escalators and moving walks

ISO/TC 92 also has liaison with organizations outside ISO, namely the International Maritime Organization (IMO) and the International Council for Research and Innovation in Building and Construction (CIB), and is internally built up in a network. The structure is shown in Figure 9.2.

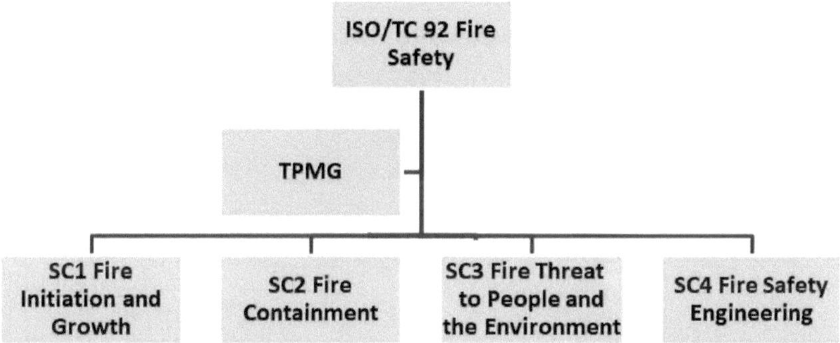

Figure 9.2 Structure of ISO/TC 92. TPMG is the technical programme management group

The major activities of TC 92 are given in Table 9.1.

Table 9.1 Major Activities of ISO/TC 92 (as of 2020)

Unit	Secretariat	Chairperson	Major activities
TC 92 Fire safety	BSI, UK/	P. Van Hees, Sweden	General management of the TC's activities
SC1 Fire initiation and growth	BSI, UK/	C. Lukas, UK	Measurement of fire initiation and growth, and standards relating to fire scenarios and characteristic fire growth of products
SC2 Fire containment	ANSI, USA/	J.D. Nicholas, USA	Measurements of fire resistance, and integration of fire resistance test standards
SC3 Fire threat to people and environment	AFNOR, France/	E. Guillaume, France	Standards relating to effects on people from toxic gases, smoke, and heat; Environmental effects of fires
SC4 Fire safety engineering	AFNOR, France/	D. Nilsson, Sweden	Developing and maintaining a set of ISO documents on the use of fire hazard and risk assessment models

The sub-committees co-operate closely in order to produce standards and supporting documents useful as a whole for the fire testing and research community. For example, the use of test data from SC1 standards is found in fire safety engineering methods developed by SC4.

Test methods for reaction to fire as well as their application to fire safety engineering are covered by SC1. There are also standards for toxicity being developed by SC3 that will be used for building products. However, they are not covered here, and further information can be found from ISO/TC 92/SC3, and in Chapters 6 and 7 of this book.

9.2.3 ISO/TC 92/SC1: Fire Initiation and Growth

The business plan for ISO/TC 92 defines the objectives for SC1 [3]:

- **Test protocols, measuring techniques, and fire scenarios in support of fire safety engineering (FSE):**
  - Test protocols, measuring techniques, and procedures for securing data of fundamental fire properties
  - Test protocols, measuring techniques, and procedures for input data to FSE models
  - Standards relating to fire scenarios and characteristic fire growth of products
- **Performance codes:**
  - Test protocols for reference scenarios
  - Test protocols, measuring techniques, and procedures for fire calorimetry
- **Maintain and improve existing SC1 standards:**
  - Updating tests already in use
- **Test validation:**
  - Protocols to determine the precision of fire test procedures
  - Test protocols for validation of fire growth predictions
- **Instrumentation:**
  - Protocols for measurement technologies used in fire test procedures

Accordingly, the activities of SC1 have led to a number of different standards for different purposes. Some are intended for prescriptive codes. The use of ISO standards in the European harmonized classification system for building products (see Section 10.3.1) and the use of ISO standards by the IMO for shipping are good examples. Other standards like the cone calorimeter (ISO 5660) find a use in fire safety engineering. Some are used as large-scale reference scenarios, for example the Room/Corner Test (ISO 9705).

The standards and technical reports issued by SC1 are in many cases accompanied by guidance documents. The intention is to provide the user of a certain standard with background information, technical details to support quality assurance in testing and scientific background to that test. In addition, bibliographies are provided. A handful of standards are of a horizontal nature, and give advice on how to use the test data from the SC1 package of standards for mathematical modelling, fire hazard analysis, and uncertainty of test data. The overall intention is not to leave the user of a SC1 standard only with a test procedure, but a deeper understanding of the technical work that underpins it, as well as how the test data can be used to mitigate hazards. The package of standards is listed in Table 9.2.

Table 9.2 Standards Under the Direct Responsibility of ISO/TC 92/SC 1 Secretariat. Withdrawn Standards, Committee Drafts, and Standards Under Development are Not Shown

Standard	Title
ISO 1182:2020	Reaction to fire tests for products - Non-combustibility test
ISO 1716:2018	Reaction to fire tests for products - Determination of the gross heat of combustion (calorific value)
ISO/TS 3814:2014	Standard tests for measuring reaction to fire of products and materials - Their development and application
ISO 5657:1997	Reaction to fire tests - Ignitability of building products using a radiant heat source
ISO/TS 5658-1:2006	Reaction to fire tests - Spread of flame - Part 1: Guidance on flame spread
ISO 5658-2:2006 ISO 5658-2:2006/Amd 1:2011	Reaction to fire tests - Spread of flame - Part 2: Lateral spread on building and transport products in vertical configuration
ISO 5658-4:2001	Reaction to fire tests - Spread of flame - Part 4: Intermediate-scale test of vertical spread of flame with vertically oriented specimen
ISO 5659-2:2017 (under the responsibility of ISO/TC 61 Plastics)	Plastics - Smoke generation - Part 2: Determination of optical density by a single-chamber test
ISO 5660-1:2015	Reaction to fire tests - Heat release, smoke production and mass loss rate - Part 1: Heat release rate (cone calorimeter method) and smoke production rate (dynamic measurement)
ISO/TS 5660-3:2012	Reaction to fire tests - Heat release, smoke production and mass loss rate - Part 3: Guidance on measurement
ISO/TS 5660-4:2016	Reaction to fire tests - Heat release, smoke production and mass loss rate - Part 4: Measurement of low levels of heat release

Table 9.2 Standards Under the Direct Responsibility of ISO/TC 92/SC 1 Secretariat. Withdrawn Standards, Committee Drafts, and Standards Under Development are Not Shown *(continued)*

Standard	Title
ISO/TS 5660-5:2020	Reaction-to-fire tests - Heat release, smoke production and mass loss rate - Part 5: Heat release rate (cone calorimeter method) and smoke production rate (dynamic measurement) under reduced oxygen atmospheres
ISO 9239-1:2010	Reaction to fire tests for floorings - Part 1: Determination of the burning behaviour using a radiant heat source
ISO 9239-2:2002	Reaction to fire tests for floorings - Part 2: Determination of flame spread at a heat flux level of 25 kW/m^2
ISO 9705-1:2016	Reaction to fire tests - Room corner test for wall and ceiling lining products - Part 1: Test method for a small room configuration
ISO/TR 9705-2:2001	Reaction to fire tests - Full-scale room tests for surface products - Part 2: Technical background and guidance
ISO/TR 11696-1:1999	Uses of reaction to fire test results - Part 1: Application of test results to predict fire performance of internal linings and other building products
ISO/TR 11696-2:1999	Uses of reaction to fire test results - Part 2: Fire hazard assessment of construction products
ISO/TR 11925-1:1999	Reaction to fire tests - Ignitability of building products subjected to direct impingement of flame - Part 1: Guidance on ignitability
ISO 11925-2:2020	Reaction to fire tests - Ignitability of products subjected to direct impingement of flame - Part 2: Single-flame source test
ISO 11925-3:1997 ISO 11925-3:1997/Cor 1:1998	Reaction to fire tests - Ignitability of building products subjected to direct impingement of flame - Part 3: Multi-source test
ISO 12136:2011	Reaction to fire tests - Measurement of material properties using a fire propagation apparatus
ISO 12863:2010 ISO 12863:2010/Cor 1:2011 ISO 12863:2010/Amd 1:2016	Standard test method for assessing the ignition propensity of cigarettes
ISO 12949:2011	Standard test method for measuring the heat release rate of low flammability mattresses and mattress sets
ISO 13784-1:2014	Reaction to fire test for sandwich panel building systems - Part 1: Small room test
ISO 13784-2:2020	Reaction to fire tests for sandwich panel building systems - Part 2: Test method for large rooms

Standard	Title
ISO 13785-1:2002	Reaction to fire tests for façades - Part 1: Intermediate-scale test
ISO 13785-2:2002	Reaction to fire tests for façades - Part 2: Large-scale test.
ISO 14696:2009	Reaction-to-fire tests - Determination of fire and thermal parameters of materials, products and assemblies using an intermediate-scale calorimeter (ICAL)
ISO 14697:2007	Reaction to fire tests - Guidance on the choice of substrates for building and transport products
ISO 14934-1:2010	Fire tests - Calibration and use of heat flux meters - Part 1: General principles
ISO 14934-2:2013	Fire tests - Calibration and use of heat flux meters - Part 2: Primary calibration methods
ISO 14934-3:2012	Fire tests - Calibration and use of heat flux meters - Part 3: Secondary calibration method
ISO 14934-4:2014	Fire tests - Calibration and use of heat flux meters - Part 4: Guidance on the use of heat flux meters in fire tests
ISO 16405:2015	Room corner and open calorimeter - Guidance on sampling and measurement of effluent gas production using FTIR technique
ISO/TR 17252:2019	Fire tests - Applicability of reaction to fire tests to fire modelling and fire safety engineering
ISO/TS 17431:2006	Fire tests- Reduced-scale model box test
ISO 17554:2014	Reaction to fire tests - Mass loss measurement
ISO/TS 19021:2018	Test method for determination of gas concentrations in ISO 5659-2 using Fourier transform infrared spectroscopy
ISO 20632:2008	Reaction to fire tests - Small room test for pipe insulation products or systems
ISO/TS 22269:2005	Reaction to fire tests - Fire growth - Full-scale test for stairs and stair coverings
ISO 24473:2008	Fire tests - Open calorimetry - Measurement of the rate of production of heat and combustion products for fires of up to 40 MW
ISO 29473:2010	Fire tests - Uncertainty of measurements in fire tests

The names of the standards often explain clearly how they are used. SC1 standards that are widely used or have specific features are further described in the subsections below, according to the following groupings:

1. General use of test data and fire safety engineering
2. Non-combustibility

3. Ignition and flame spread
4. Heat release and smoke production
5. Large-scale
6. Calibration and supporting procedures

9.2.3.1 General Use of Test Data and Fire Safety Engineering

ISO/TS 3814:2014 *Standard tests for measuring reaction-to-fire of products and materials - Their development and application*
and
ISO/TR 11696-1:1999 *Uses of reaction to fire test results - Part 1: Application of test results to predict fire performance of internal linings and other building products*
and
ISO/TR 11696-2:1999 *Uses of reaction to fire test results - Part 2: Fire hazard assessment of construction products*
and
ISO 29473:2010 *Fire tests - Uncertainty of measurements in fire tests*

This series of standards contains the principles of the SC1 tests for fire growth measurements, theoretical modelling of fire processes relating to growth and hazard analysis, as well as uncertainty of test data. Any user of the standards enumerated and described below should study this package, in order to obtain the maximum benefit from a given test standard.

9.2.3.2 Non-Combustibility

ISO 1716:2018 *Reaction to fire tests for products - Determination of the gross heat of combustion (calorific value)*

ISO 1716 determines the potential maximum total heat release of a product when it is completely burnt. The calorific potential of a material is measured in a bomb calorimeter. The powdered material is completely burned under high pressure in a pure oxygen atmosphere. This standard is also used by CEN (see Section 10.3.1).

ISO 1182:2020 *Reaction to fire tests for products - Non-combustibility test*

ISO 1182 identifies products that will not (at least to a significant degree), contribute to a fire, regardless of their end use. It was first published by ISO during the 1970s and is now well-known. It is used in various building codes, by IMO, and by CEN (see Section 10.3.1). The test apparatus is shown in Figure 9.3, and the test specifications are given in Table 9.3.

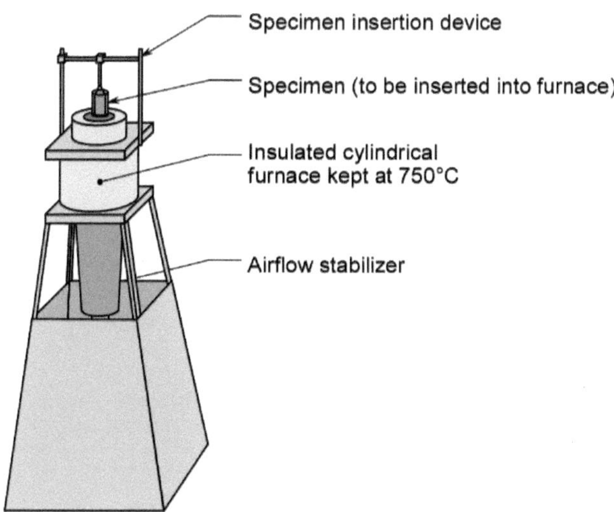

Figure 9.3 ISO 1182 Non-combustibility. Image drawn by RISE; reproduced with permission

Table 9.3 ISO 1182 Test Specifications

Specimens	Five cylindrical samples, diameter 45 mm, height 50 mm
Specimen position	Vertical specimen placed in holder in the center of the furnace
Heat source	Electrical cylindrical furnace at 750°C (measured by the furnace thermocouple)
Test duration	Depends on temperature stabilization
Conclusions	Classification is based on temperature rise due to combustion of the sample as measured by the furnace thermocouple, duration of flaming, and mass loss of the sample

9.2.3.3 Ignition and Flame Spread

ISO/TR 11925-1:1999 *Reaction to fire tests - Ignitability of building products subjected to direct impingement of flame - Part 1: Guidance on ignitability*
and
ISO 11925-2:2020 *Reaction to fire tests - Ignitability of products subjected to direct impingement of flame - Part 2: Single-flame source test*
and
ISO 11925-3:1997 *Reaction to fire tests - Ignitability of building products subjected to direct impingement of flame - Part 3: Multi-source test*

The ISO 11925 series evaluates the ignitability of a product under exposure to a flame. This series of standards covers most cases where a flame test for ignitability determination would be required. Part 1 gives guidance on the theory and use of the ignitability tests. Part 2 is a small flame test, which also appears in the European regulations. The test apparatus is shown in Figure 9.4, and the test specifications are given in Table 9.4. Part 3 is the multi-source test, in which various flame sources from small flames to very large flames are covered. Each flame source is described in terms of the real case it intends to model, for example a fire in a saucepan, a wastepaper basket or a candle-like flame.

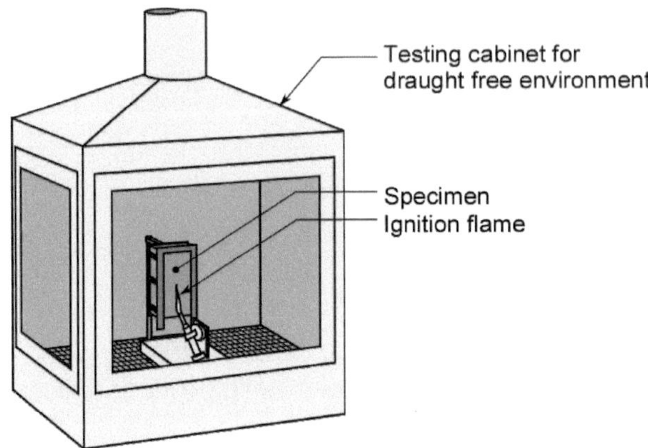

Figure 9.4 ISO 11925-2 Small flame test. Image drawn by RISE; reproduced with permission

Table 9.4 ISO 11925-2 Test Specifications

Specimens	250 mm long, 90 mm wide, maximum thickness 60 mm
Specimen position	Vertical
Ignition source	Small burner. Flame inclined 45° and impinging either on the edge or the surface of the specimen over a period of 15 s or 30 s
Conclusions	Classification is based on the extent of flame spread and occurrence of burning droplets/particles

ISO 12863:2010 *Standard test method for assessing the ignition propensity of cigarettes*

This standard assesses the capability of a cigarette to extinguish by itself when not in use, and so reduce the risk for ignition of bedding or upholstered furniture. ISO 12863 is applicable to factory-made cigarettes that burn along the length of a tobacco column. The test data has been correlated with the potential for cigarettes to ignite upholstered furniture. In USA cigarettes must meet the requirements of ASTM E 2187, which is the original standard, and which was later published as ISO 12863 with minor differences. The European Union requires that cigarettes sold in the member states fulfil criteria according to the European standard EN 16156 when tested according to ISO 12863.

ISO/TS 5658-1:2006 *Reaction to fire tests - Spread of flame - Part 1: Guidance on flame spread*

and

ISO 5658-2:2006 *Reaction to fire tests - Spread of flame - Part 2: Lateral spread on building and transport products in vertical configuration*

This standard relating to the spread of a flame has two parts. Part 1 gives advice on the use of the test and its results. Part 2 contains the description of the hardware and the actual test procedure to be followed. The lateral flame spread as a function of time is measured on a vertical 800 mm long sample that is exposed to a heat flux of 50 kW/m^2 that diminishes to 1.5 kW/m^2 along the length of the sample. This test is similar to an IMO test procedure used for approval of marine products to be used on board ships, and which also uses a thermal method for measuring the heat release rate. ISO 5658-2 is also used in the European railway standard EN 45545-2 to determine spread of flame (see Section 11.2.2). The test apparatus is shown in Figure 9.5, and the test specifications are given in Table 9.5.

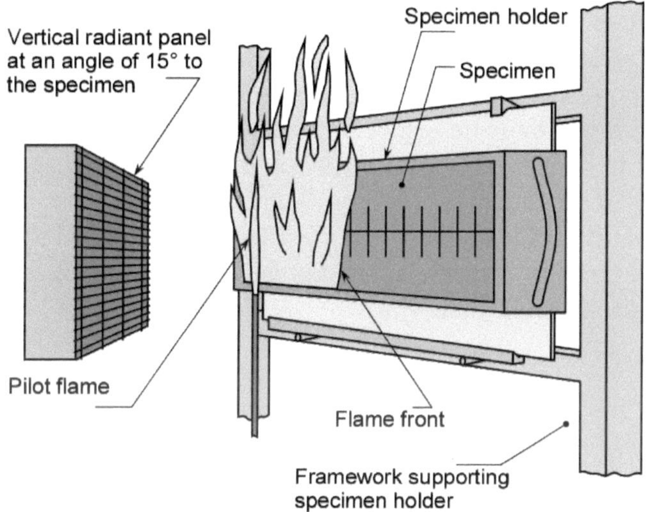

Figure 9.5 ISO 5658-2 Spread of flame. Image drawn by RISE; reproduced with permission

Table 9.5 ISO 5658-2 Test Specifications

Specimens	Three specimens 800 mm × 155 mm
Specimen position	Vertical in specimen holder
Heat source	Gas-fired radiant panel
Test duration	30 min, or earlier if flame front stops or if the flames spread to the end of the specimen
Conclusions	Average heat for sustained burning, in MJ/m^2 (= time from test start × nominal heat flux at actual flame front position) Critical heat flux at extinguishment (CFE) = nominal heat flux at the point where the flame front stops

ISO 5658-4:2001 *Reaction to fire tests - Spread of flame - Part 4: Intermediate-scale test of vertical spread of flame with vertically oriented specimen*

A large vertical sample is exposed to heat flux from a gas-fired radiant panel. Flame may spread in all vertical directions along the sample surface, and the flame spread rate and distance is measured. Due to the size of the sample, flame spread is easily observed and measured. This test may also be equipped with thermocouples for measurement of heat output from the fire.

ISO 9239-1:2010 *Reaction to fire tests for floorings - Part 1: Determination of the burning behaviour using a radiant heat source*
and
ISO 9239-2:2002 *Reaction to fire tests for floorings - Part 2: Determination of flame spread at a heat flux level of 25 kW/m²*

These standard tests evaluate the wind-opposed spread of flame of horizontally-mounted floorings exposed to a heat flux from a radiant panel in a test chamber. They evaluate the critical heat flux below which flames no longer spread. Flame spread on the floor surface of a corridor exposed to heat flux from a hot gas layer or from flames in a nearby room can be seen as the model case. In addition, smoke development can be measured. ISO 9239-1 is used for classification in European regulations. ISO 9239-2 exposes the sample to a higher heat flux, and is more versatile, as it also covers conditions in a corridor adjacent to a room containing a fully-developed fire [4]. The test apparatus is shown in Figure 9.6, and the test specifications are given in Table 9.6.

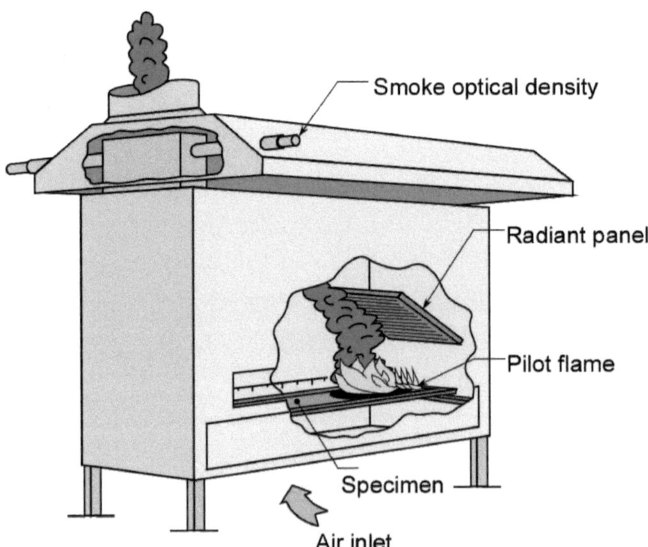

Figure 9.6 ISO 9239-1 and ISO 9239-2 floor covering test. Image drawn by RISE; reproduced with permission

Table 9.6 ISO 9239-1 and ISO 9239-2 Floor Covering Test Specifications

Specimens	1050 mm long × 230 mm wide
Specimen position	Horizontal
Ignition source	A gas-fired radiant panel that gives a heat flux to the specimen. A pilot flame impinges on the surface of the hot end of the specimen to initiate flame spread ISO 9239-1: The maximum heat flux is 11 kW/m^2, dropping to 1 kW/m^2 at the end of the specimen ISO 9239-2: The maximum heat flux is 25 kW/m^2, dropping to 2.6 kW/m^2 at the end of the specimen
Test duration	Until the flames extinguish, after 30 min or in certain cases longer
Conclusions	Classification is based on the critical heat flux below which flame spread does not occur and may include smoke density

9.2.3.4 Heat Release and Smoke Production

ISO 5660-1:2015 *Reaction-to-fire tests - Heat release, smoke production and mass loss rate - Part 1: Heat release rate (cone calorimeter method) and smoke production rate (dynamic measurement)*
and
ISO/TS 5660-3:2012 *Reaction-to-fire tests - Heat release, smoke production and mass loss rate - Part 3: Guidance on measurement*
and
ISO/TS 5660-4:2016 *Reaction-to-fire tests - Heat release, smoke production and mass loss rate - Part 4: Measurement of low levels of heat release*
and
ISO/TS 5660-5:2020 *Reaction-to-fire tests - Heat release, smoke production and mass loss rate - Part 5: Heat release rate (cone calorimeter method) and smoke production rate (dynamic measurement) under reduced oxygen atmospheres*

The cone calorimeter, originally developed by Babrauskas [5], is widely used as a tool for fire safety engineering, by industry for product development, and as a product classification tool in building (Japan) and transportation (EN 45545-2 for railways, and by the IMO for ships). It has been shown to predict large-scale test results in the Room/Corner Test [6, 7], fire growth in upholstered furniture [8], flame spread on cables [9], and in a number of other cases. ISO 5660 is one of the most versatile standards developed by SC1 over its entire period of activity.

In the cone calorimeter, ISO 5660-1, specimens measuring 0.1 m × 0.1 m are exposed to controlled levels of radiant heating. The specimen surface is therefore heated, and an external spark igniter ignites the pyrolysis gases from the specimen. The gases are collected by a hood and extracted by an exhaust fan. The heat release rate (HRR) is determined by measurements of the oxygen consumption, derived from the oxygen concentration and the flow rate in the exhaust duct.

Smoke obscuration is measured as the fraction of laser beam intensity that is transmitted through the smoke in the exhaust duct and then calculated as smoke production. The specimen is placed on a load cell during testing.

ISO/TS 5660-3 gives guidance on measurements. ISO/TS 5660-4 is a variant used for products that produce low levels of HRR when exposed to high levels of irradiance, e.g., fully developed fires. Another variant under development is ISO/TS 5660-5, which applies to measurements under controlled oxygen reduced atmospheres.

The test apparatus is shown in Figure 9.7, and the test specifications are given in Table 9.7.

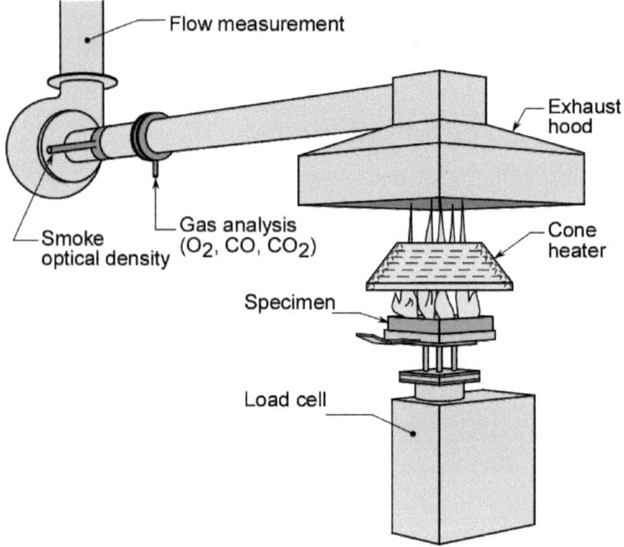

Figure 9.7 ISO 5660-1 cone calorimeter. Image drawn by RISE; reproduced with permission

Table 9.7 ISO 5660-1 Test Specifications

Specimens	Three specimens, 100 mm × 100 mm
Specimen position	Horizontal, in specimen holder
Heat source	Conical electrical heater giving a heat flux of 0–100 kW/m^2. If no specifications are given, tests at 25, 35, and 50 kW/m^2 are recommended
Test duration	32 min generally. If there is no ignition, test is stopped after 30 min
Conclusions	HRR data: Time to sustained flaming, curve of HRR versus time, 180 s and 300 s average of the HRR, peak HRR, total HRR and possibly other HRR data Sample mass loss data: Mass loss rate, total mass loss Smoke data: Graph of smoke production rate per unit area of exposed specimen as well as discrete values for non-flaming and flaming phases of total smoke production per unit area

ISO 12136:2011 *Reaction to fire tests - Measurement of material properties using a fire propagation apparatus*

This test addresses similar issues as the cone calorimeter. It is commonly used in cases of insurance of large buildings. ISO 12136 measures the flammability characteristics of materials, in relation to their propensity to support fire propagation. Output data are time to ignition, heat release rate, effective heat of combustion, heat of gasification, and smoke yield.

ISO 14696:2009 *Reaction-to-fire tests - Determination of fire and thermal parameters of materials, products and assemblies using an intermediate-scale calorimeter (ICAL)*

A vertical sample is exposed to a constant, precisely defined heat flux from a radiant panel. The heat release rate is measured by means of the oxygen-consumption calorimetry. The test is similar in technology to the cone calorimeter. However, the sample in the ICAL test is larger and vertically oriented.

ISO 5659-2:2017 *Plastics - Smoke generation - Part 2: Determination of optical density by a single-chamber test*

ISO TC 61 (plastics) has the responsibility for this test standard. IMO prescribes the standard for products to be used on-board ships. This is also the case for European trains, and the ISO standard is also published by CEN. In both cases measurements of toxic gas species are added. EN 45545-2 refers to EN ISO 5659-2, and specifies the toxicity measurements procedure in its Appendix C, and gives criteria for optical smoke density and toxic gas species for products installed in trains. The IMO regulation for fire test procedures does the same for ships. TC 92/SC1 is currently developing a procedure for measurement of gas concentrations called **ISO/DTS 19021** *Test method for determination of gas concentrations in ISO 5659-2 using Fourier transform infrared spectroscopy.* This technology is quite powerful in determining concentration of gas species, and the standard is expected to increase precision of the IMO and EN 45545-2 procedures.

A specimen is exposed to heat flux with or without an impinging pilot flame. Thus, smoke is generated both under smoldering and flaming conditions. The smoke produced from the sample is accumulated in an airtight box. The optical density through the smoke is measured by using a lamp and a photocell. Concentrations of gas species are measured by pumping gas from the chamber and analyzing them at certain intervals. The apparatus is shown in Figure 9.8, and the test specifications are summarized in Table 9.8.

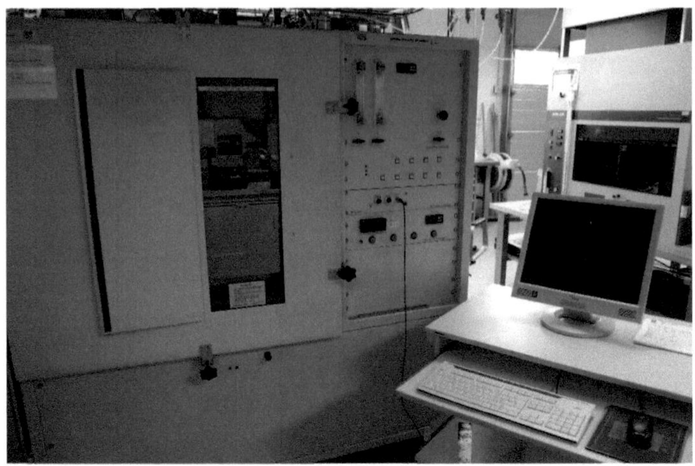

Figure 9.8 ISO 5659-2 smoke density chamber

Table 9.8 ISO 5659-2 Test Specifications

Specimens	12 specimens, 75 mm square
Specimen position	Horizontal, in a sample holder exposing 65 × 65 mm
Ignition source	Electrical conical radiator exposing the sample at 25 kW/m^2 or 50 kW/m^2, with or without a pilot flame
Test duration	Normally 10 min
Conclusions	IMO requirements are average maximum optical density ($\bar{D}_m$) ≤ 200, 400, or 500, depending on product type and use Maximum measured gas concentrations of: CO 1450 ppm, HCl 600 ppm, HF 600 ppm, NO_x 350 ppm, HBr 600 ppm, HCN 140 ppm, and SO_2 120 ppm or 200 ppm (floor coverings) Requirements according to EN 45545-2 are quite elaborate, and details can be found in the standard

9.2.3.5 Large-Scale

ISO 9705-1:2016 *Reaction to fire tests - Room corner test for wall and ceiling lining products - Part 1: Test method for a small room configuration*
and
ISO/TR 9705-2:2001 *Reaction-to-fire tests - Full-scale room tests for surface products - Part 2: Technical background and guidance*

The room/corner test (Figure 9.9) was first published by ASTM in 1982 [10], and then by NORDTEST in 1986 [11] and by CEN [12]. The international standard, ISO 9705, was first published in 1993.

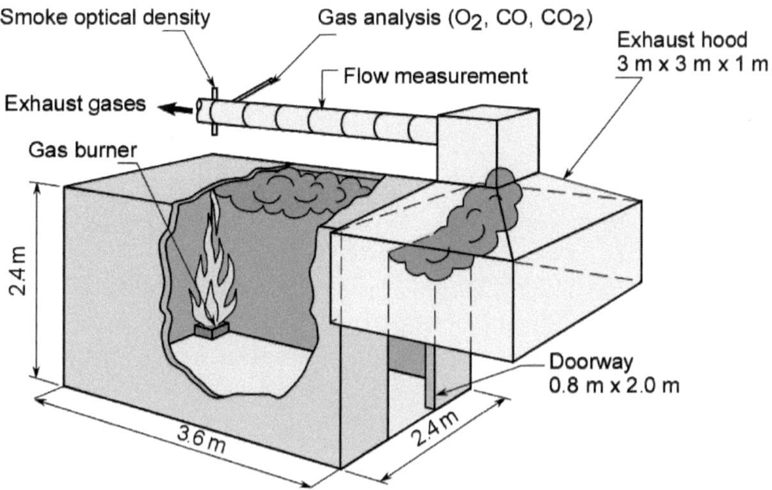

Figure 9.9 ISO 9705-1 Room/Corner Test. Image drawn by RISE; reproduced with permission

The room/corner test is a large-scale test method for measurement of the burning behavior of wall and ceiling products in a well-ventilated room. The product is mounted on three walls and on the ceiling of a small compartment. A door opening ventilates the room. The principal output is the occurrence and time to flashover. A direct measure of fire growth (Heat Release Rate, HRR), light-obscuring smoke (Smoke Production Rate, SPR) and production of toxic gas species are results from the test.

Variants of the test are used for sandwich panel building systems and pipe insulation (i.e., ISO 13784-1 and ISO 20632).

Experience on testing products has been gained during more than 30 years of work with the room/corner test. A considerable amount of information on product burning behavior, modelling results using small-scale data and general understanding of fire growth have been gathered over the years using this method [13].

ISO 9705-2 provides a valuable source of background data and theory for the test, and describes the complete heat balance of the test room. Equations for mass flow are given as well as a mapping of the heat flux from the burner. ISO 9705-2 also contains a large bibliography of technical papers related to the room/corner test.

ISO 9705-1 was the reference scenario in defining fire performance during the development of the European Union classification system for linings. The test has an important role as it validates the fire performance of the classes for reaction to fire. The tendency of a lining to cause room flashover is reflected in the criteria for the different classes when the product is tested in the so-called SBI test (see Section 10.3.1). ISO 9705-1 is also used directly as a classification tool for high-speed craft in the IMO regulations, and data obtained through use of the standard has been used extensively when developing building codes, as well as for validating mathematical fire models all over the world.

The test specifications are shown in Table 9.9.

Table 9.9 ISO 9705-1 Test Specifications

Specimens	The specimens cover three walls and the ceiling of the test room. The wall containing the doorway is not covered
Specimen position	Lined inside the room according to end use conditions
Ignition source	Burner placed on the floor in a corner directly opposite the doorway wall in contact with the specimens. Heat output from the burner: 100 kW for first 10 min of the test, increasing to 300 kW for the subsequent 10 min
Test duration	20 min or until flashover
Conclusions	A number of parameters relating to a room fire such as temperatures of the gas layers, flame spread, combustion products, and heat fluxes can be measured. However, the most important outputs are HRR, SPR, and time to (or occurrence of) flashover

ISO 20632:2008 *Reaction-to-fire tests - Small room test for pipe insulation products or systems*

The standard is a variant of ISO 9705 applied for pipe insulation or pipe installation systems when installed in a small room. Reaction to fire is determined, and data of heat release, smoke production and toxic gas species are output. The method is suitable when small-scale tests cannot be applied. An important field of application of this procedure was to serve as reference scenario for intermediate/small-scale testing and classification according to the European SBI test (see Section 10.3.1).

ISO 13784-1:2014 *Reaction-to-fire test for sandwich panel building systems - Part 1: Small room test*

This test is a variant of the room/corner test when used for the testing of sandwich panels. These panels are generally well-insulated and self-supporting. Fires in buildings made of sandwich panels may include phenomena such as burn-through of the wall and collapse of the construction. Therefore, when tested the panels themselves are used as construction materials for the test room. In other aspects, testing is performed in the same way as an ordinary ISO 9705, test and thus data can be compared with ordinary interior linings.

ISO 13784-2:2002 *Reaction-to-fire tests for sandwich panel building systems - Part 2: Test method for large rooms*

Constructions made of sandwich panels may be of considerable size. The ISO 13784-2 test set-up is a large open configuration (see Figure 9.10). The test specification is given in Table 9.10.

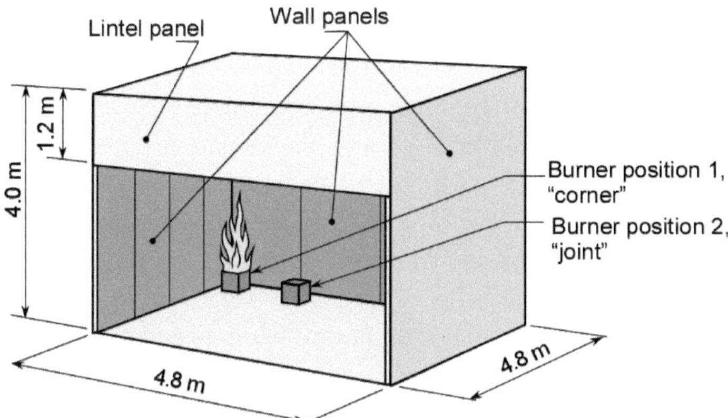

Figure 9.10 ISO 13784-2 large-scale test for sandwich panels. Image drawn by RISE; reproduced with permission

Table 9.10 ISO 13784-2 Test Specifications

Specimens for constructing the test room	Test room made up from sandwich panels. Consists of four walls at right-angles and a ceiling located on a rigid, non-combustible floor surface. The room may be located indoors or outdoors
Ignition source	Burner placed on the floor in a corner directly opposite the doorway wall in contact with the specimens. Heat output from the burner: 100 kW for first 5 min of the test, 300 kW for the subsequent 5 min, increasing to 600 kW for the remaining 5 min test duration, if ignition and sustained burning of the test specimens has not occurred
Test duration	30 min, or earlier if flashover occurs or if the structure collapses.
Conclusions	A number of parameters relating to a room fire such as temperatures, flame spread and structural failure

ISO 13785-1:2002 *Reaction-to-fire tests for façades - Part 1: Intermediate-scale test* and
ISO 13785-2:2002 *Reaction-to-fire tests for façades - Part 2: Large-scale test*

These two standards address the problem of fire spread along the surface of a façade. The reference fire for the large-scale test, Part 2, is a flashover inside a compartment in a building, followed by a large flame breaking through a window opening and subsequently impinging on the façade, or alternatively, a large fire acting on the façade from outside. The parameters measured are flame spread, and breaking and falling pieces of the façade material. The test configuration includes an internal corner of 90°, since this introduces a more intense fire exposure than on a flat façade. The test does not include façade details such as windows balconies, etc. The large-scale test is being updated, as fires on façades have become increasingly common, one reason for this being new ways of construction.

ISO 24473:2008 *Fire tests - Open calorimetry - Measurement of the rate of production of heat and combustion products for fires of up to 40 MW*

The method simulates a real-scale fire on a test object or group of objects under well-ventilated conditions. Heat release and smoke data are measured. The size of the fire varies with the object tested, and data for all stages of the fire is provided. The standard has a broad field of application, for example ISO 12949:2011 *Standard test method for measuring the heat release rate of low flammability mattresses and mattress sets*, which determines the heat release rate and total heat release from a mattress (or a mattress and foundation) when tested at full-scale.

9.2.3.6 Calibration and Supporting Procedures

ISO 14934-1:2010 *Fire tests - Calibration and use of heat flux meters - Part 1: General principles*
and
ISO 14934-2:2013 *Fire tests - Calibration and use of heat flux meters - Part 2: Primary calibration methods*
and
ISO 14934-3:2012 *Fire tests - Calibration and use of heat flux meters - Part 3: Secondary calibration method*
and
ISO 14934-4:2014 *Fire tests - Calibration and use of heat flux meters - Part 4: Guidance on the use of heat flux meters in fire tests*

An international standard for the calibration of heat flux meters is a prerequisite for the proper operation of many of the SC1 tests. Theories of ignition and subsequent flame spread rely on data for a certain heat flux. The repeatability and reproducibility of tests rely on proper calibration of radiant heat sources, etc. These packages of standards give the technical background and calibration techniques that are needed.

ISO 14697:2007 *Reaction-to-fire tests - Guidance on the choice of substrates for building and transport products*

This standard recommends substrates, representing various end use conditions, on which the product sample can be attached before testing. This allows the results of the test to be general, and so keep down the amount of testing required.

9.2.4 Future Developments

Recent decades have seen the development of many completely new test methods. One reason for this was the development of technologies to measure fire properties (e.g., oxygen consumption measurements and FTIR spectroscopy), while another driving force was the European harmonization of newly developed test methods, which used many standards from SC1. Thus, a comprehensive and competent package of tests was developed that has found use in many countries and also in other fields like transportation. However, the intensive phase of method development is over, and the existing tests are now being improved regularly according to the re-evaluation system in ISO.

New promising long-term developments would be standards for measuring parameters used in fire modelling. There have been activities in this area for some years. A problem is that fire growth models need more development in order to be shown to be accurate, but some of the parameters called for are very difficult to measure, or not even possible to quantify in a reproducible way. However, work is continuing that eventually could lead to a new generation of standards for fire parameters.

In SC3, the development of test methods for toxic gas species under controlled burning conditions is under way. This follows from a growing interest in toxicity of fire gases and their influence on humans. SC3 also has work under way on wildland fires.

In SC4 renewed packages for fire safety engineering are being developed.

References for Section 9.2

[1] ISO website, *www.iso.org.*

[2] ISO Annual Report, 2018.

[3] ISO/TC 92 Business Plan, 2014.

[4] van Hees, P., Wind-aided flame spread of floor coverings. Development and validation of large scale and small scale test methods, PhD Thesis, University of Gent, Belgium [in Dutch], 1995.

[5] Babrauskas, V., The cone calorimeter Section 3, Chapter 3, pp. 3-37 to 3-52, *The SFPE Handbook of Fire Protection Engineering* (2nd edition), National Fire Protection Association, Quincy, MA, 1995.

[6] EUREFIC - European Reaction to Fire Classification (International Seminar), InterScience Communications Limited, London, September 1991, ISBN 0 9516320 1 9.

[7] Wickström, U., Göransson, U., Prediction of heat release rates of surface materials in large-scale fire tests based in cone calorimeter results, *Journal of Testing and Evaluation,* 15: 364–370, 1987.

[8] CBUF Fire Safety of Upholstered Furniture - The final report of the CBUF research programme (EU project, Report EUR 16477 EN), edited by B. Sundström, SP Fire Technology, 1992.

[9] van Hees, P., Grayson, S., Breulet, H., Vercellotti, U., FIPEC, Fire Performance of Electric Cables (EU project, Final Report on the European Commission SMT Programme SMT4-CT96-2059), 410pp, ISBN 0 9532312 5 9, London, 2000.

[10] Proposed method for room fire test of wall and ceiling materials and assemblies, *Annual Book of ASTM Standards* (Part 18), 1982.

[11] Surface products: Room fire test in full scale (NT FIRE 025), Helsinki, 1986.

[12] Fire test - Large-scale room reference test for surface products, EN 14390:2007.

[13] Sundström, B., The development of a European fire classification system for building products test methods and mathematical modelling, PhD Thesis, Department of Fire Safety Engineering, Lund University, Sweden, 2007.

9.3 European Committee for Standardization (CEN)

Edith Antonatus

9.3.1 Introduction

The European Committee for Standardization (CEN) is one of three European Standardization Organizations (together with CENELEC and ETSI) that have been officially recognized by the European Union and by the European Free Trade Association (EFTA) as being responsible for developing and defining standards at European level. The aims set out in the European Union (EU) treaties are achieved by regulations, which are binding legislative acts and must be applied in their entirety across the EU. The EU Regulation No. 1025/2012 [1] on European standardization sets the legal framework for standardization. The members of CEN [2] are the National Standardization Bodies of 34 European countries - including all the member states of the European Union (EU) and other countries that are part of the European Single Market.

Each National Standardization Body that is part of the CEN system is obliged to adopt each European Standard as a national standard and make it available to customers in their country. The Standardization Bodies also have to withdraw any existing national standard that conflicts with the new European Standard. Therefore, one European Standard (EN) becomes the national standard in all 34 countries covered by CEN Members.

In addition, European Standards are also adopted as identical national standards by CEN Affiliates, which are the National Standards Bodies of 17 neighboring countries, and by National Standardization Bodies in other countries around the world.

9.3.2 Development of Standards

Each CEN Technical Committee has a business plan, which also covers the activities of its subcommittees. Any interested party can introduce a proposal for new work. Most standardization work is proposed through the CEN Members.

For harmonized standards under the European Standardization Regulation 1025/2012, standardization requests are issued by the European Commission to develop and adopt European standards or European standardization deliverables within a given time. They are created following a request from the European Commission. Manufacturers, other economic operators, or conformity assessment bodies can use harmonized standards to demonstrate that products, services, or processes comply with relevant EU legislation and with the legal requirements in the EU member states.

The route for developing a European standard (EN) is as follows:

- **Acceptance of the proposal**
  Once a project to develop an EN is accepted by the relevant Technical Body, or by the Technical Board, the member countries shall put all national activity within the scope of the project on hold. This means that they do not initiate new projects, or revise existing standards at national level. This obligation is called "standstill", and allows efforts to be focused on the development of the EN.
- **Drafting**
  The EN is developed by experts within a Technical Body.
- **Enquiry – Public comment at national level & weighted vote**
  Once the draft of an EN is prepared, it is released for public comment and vote. During this stage, everyone who has an interest (e.g., manufacturers, public authorities, consumers, etc.) may comment on the draft. These views are gathered by the members who then submit a national position by means of a weighted vote, which is subsequently analyzed by the CEN Technical Body. If the results of the Enquiry show approval for the EN, then the Technical Body can decide to publish it.
- **Adoption by weighted Formal Vote**
  If the results of the Enquiry show that the draft EN requires technical reworking, the Technical Body can decide to update the draft and resubmit it for another weighted vote, called the Formal Vote.
- **Publication of the EN**
  Following the approval, the EN is published. A published EN must be given the status of national standard in all member countries, who also have the obligation to withdraw any national standards that conflict with it. This guarantees that, when applying ENs, a manufacturer has easier access to the markets of all the member countries.
- **Review of the EN**
  To ensure that a EN is still current, it is reviewed within five years of its publication. This review results in the confirmation, modification, revision or withdrawal of the standard.

The technical work in CEN is carried out via technical committees (TCs). A TC may have assigned working groups (WGs) to cover certain areas of work. The actual standardization work takes place in the WGs. Further general information about CEN and the European Commission can be found in [3] and [4].

9.3.3 Fire-Related Activities in Building

Under the Construction Products Regulation (CPR) [5], the European performance tests include the examination of how construction products behave in fire, in terms of reaction to fire, resistance to fire and fire performance of roofs.

CEN/TC 127 (Fire safety in buildings) is the leading committee in CEN for fire-related standards [6]. It is a horizontal TC (Technical Committee), as the work it does is in support of TCs preparing product standards. CEN/TC 127 has to provide a means for consistently determining the fire performance characteristics of the different types and families of construction products. In addition, it supports other TCs to implement the details for correctly applying the technical specifications for fire tests and classifications in their products' standards.

The official scope of CEN/TC 127 is:

1. To develop standards that use existing work where available, e.g., in ISO, IEC, CENELEC, CEC, and EFTA for assessing the fire behavior of building products, components and elements of construction
2. To develop standards for classification of products, components and elements of construction, appropriate to the fire risks related to their application
3. To develop standards for assessing fire hazards and for providing fire safety in buildings.

If an ISO test already exists or is under development, the Vienna agreement on technical cooperation can be used to avoid duplication of effort. This is an agreement on technical cooperation between ISO and the European Committee for Standardization (CEN).

It provides three main modes of cooperation between ISO and CEN:

- Cooperation by correspondence/exchange of information
- Cooperation by mutual representation at meetings
- Cooperation by parallel approval of standards at the international and European levels.

The working groups in CEN/TC 127 and their responsibilities are shown in Table 9.11.

Table 9.11 CEN/TC 127 Working Groups

CEN/TC 127/WG 1	Structural and separating elements
CEN/TC 127/WG 2	Services
CEN/TC 127/WG 3	Fire doors
CEN/TC 127/WG 4	Reaction to fire
CEN/TC 127/WG 5	Roofs

CEN/TC 127/WG 7	Classification
CEN/TC 127/WG 8	Fire safety engineering
CEN/TC 127/WG 9	Fire protective products

The list of the most important standards regarding reaction to fire and fire performance of roofs that are either approved or under revision or development are summarized in Table 9.12. Standards which have been under the Vienna agreement are published as EN ISO standards. Section 10.3.1 describes the tests in detail.

Table 9.12 CEN Standards for Reaction to Fire and Fire Performance of Roofs

CEN/TR 16988:2016	Estimation of uncertainty in the single burning item test
CEN/TS 1187:2012	Test methods for external fire exposure to roofs
CEN/TS 15117:2005	Guidance on direct and extended application
CEN/TS 15447:2006	Mounting and fixing in reaction to fire tests under the Construction Products Directive
CEN/TS 16459:2013	External fire exposure of roofs and roof coverings - Extended application of test results from CEN/TS 1187
EN 13238:2010	Reaction to fire tests for building products - Conditioning procedures and general rules for selection of substrates
EN 13501-1:2018	Fire classification of construction products and building elements - Part 1: Classification using data from reaction to fire tests
EN 13501-5:2016	Fire classification of construction products and building elements - Part 5: Classification using data from external fire exposure to roofs tests
EN 13823:2020	Reaction to fire tests for building products - Building products excluding floorings exposed to the thermal attack by a single burning item
EN 14390:2007	Fire test - Large-scale room reference test for surface products
EN ISO 1182:2020	Non-combustibility test (ISO 1182:2010)
EN ISO 11925-2:2020	Ignitability of products subjected to direct impingement of flame - Part 2: Single-flame source test
EN ISO 1716:2018	Determination of the gross heat of combustion (calorific value)
EN ISO 9239-1:2010	Reaction to fire tests for floorings - Part 1: Determination of the burning behaviour using a radiant heat source
EN 16733:2016	Reaction to fire tests for building products - Determination of a building product's propensity to undergo continuous smouldering

References for Section 9.3

[1] Regulation (EU) No. 1025/2012 of the European Parliament and of the Council of 25 October 2012 on European standardization, *https://eur-lex.europa.eu/legal-content/EN/TXT/?uri=celex:32012R1025.*

[2] *https://standards.cen.eu/dyn/www/f?p=CENWEB:5.*

[3] *https://www.cen.eu/Pages/default.aspx.*

[4] *https://ec.europa.eu.*

[5] Construction Products Regulation – Regulation (EU) No. 305/2011 of the European Parliament and of the Council of 9 March 2011 laying down harmonized conditions for the marketing of construction products and repealing Council Directive 89/106/EEC. *Official Journal of the European Union* L 88 of 4 April 2011.

[6] CEN/TC 127 – Fire safety in buildings, https://standards.cen.eu/dyn/www/f?p=204:7:0::::FSP_ORG_ID:6109&cs=104D85C15BEEEFAC5963909130D6A7EB7.

10 Building

10.1 Introductory Remarks

Edith Antonatus

The statutory regulations and provisions relating to fire protection for buildings are furthest advanced in industrialized countries, and their comprehensive sets of regulations may differ significantly. Many industrialized countries belong to economic groupings or have other close relations, and regulations are influenced by attempts at harmonization within such groups in order to eliminate trade barriers.

In Europe, the Construction Products Regulation (CPR) [1] lays down harmonized rules for the marketing of construction products in the EU. The regulation provides a common technical language to assess the performance of construction products. It ensures that reliable information is available to professionals, public authorities, and consumers, to compare the performance of products from different manufacturers in different countries.

In this approach, harmonization is restricted to the adoption of basic requirements for safety, health, or other requirements for welfare. The basic requirements for construction works constitute the basis for the preparation of standardization mandates and harmonized technical specifications. These requirements are:

- Mechanical resistance and stability
- Safety in case of fire
- Hygiene, health and the environment
- Safety and accessibility in use
- Protection against noise
- Energy economy and heat retention
- Sustainable use of natural resources.

Harmonized European standards on construction products are developed by technical experts from the European Standardization Organizations (CEN/CENELEC). The Technical Committees of CEN and CENELEC are working on completing the necessary sets of harmonized European standards and test standards, and further improving existing ones.

In other cases, countries which are located close to each other or have a very strong relationship use similar standards and have similar but not identical regulations, for example USA and Canada, or Australia and New Zealand.

Some countries have a federal structure, which means that there are regulations from the central government as well as special requirements for buildings and for construction products defined by the federal states or regional governments.

In the USA the structure of fire regulations is partly based on codes for the United States of America (International Building Code, and the National Fire Protection Association, NFPA). In addition, regional regulations from the different states have been set.

China has regulations which in the past were mainly based on German tests and regulations. Since then, the European classification and testing system has been added, and in recent years there has been additional input from the UK and the USA. The Chinese regulations are designed for the whole country, but the provinces can still set additional requirements.

The countries dealt with in the following sections have been selected based on their industrial and commercial importance and the degree of comprehensiveness of their regulations on fire precautions. Some states have not been included either because no information was forthcoming or because fire protection regulations and test methods are still under development. In most of the countries that are not covered in this book, tests either based on international standards or on other regional/national standards are implemented.

Each country is dealt with according to a uniform scheme:

- Statutory regulations
- Classification and testing of the fire performance of building materials and components
- Official approval
- Future developments.

It is stressed again that fire resistance is not generally dealt with in this book. A bibliography for each country is given at the end of each section.

10.1.1 Classification and Testing of Fire Performance of Building Materials and Components

In many countries, regulations are mainly based on a prescriptive approach. This means that the requirements for classification of building materials and components are an important tool for defining the safety level of buildings. Material and product classifications are mainly based on reaction to fire tests. These will be explained in the following chapters regarding buildings.

In addition to requirements based on the classifications regarding reaction to fire, requirements for fire resistance of certain parts of the building and installations are defined in most regulations, as well as design requirements dealing (for example) with escape routes and access for fire brigades.

An add-on to fire safety of buildings is the increasing number of countries that require the use of smoke detectors and sprinklers.

Performance-based regulations have also been introduced in a number of countries. This means that the classification of products is not necessarily the sole basis for the assessment of safety. Performance-based fire safety design considers individual features of the building such as the height and geometry of its rooms, escape routes, as well as active and passive methods of fire prevention. As a result, this will help to accomplish the best and most cost-efficient level of fire safety available as well as significant savings in building costs. The method utilizes fire simulation software to model the behavior of fires, fire safety equipment and risks caused by fires.

For plastic products, prescriptive as well as performance-based regulations are important. In the sections below only product performance in the event of a fire is explained in detail.

10.1.1.1 Non-Combustible Materials

In most countries, an assessment of a material having zero or very low contribution to fire is based on the so-called combustibility tests. Sometimes differentiation is made between non-combustible materials and materials with a very limited contribution to fire (sometimes called quasi-non-combustible). For non-combustible materials the following test methods are used (in some countries two of these tests have to be passed and additional tests may be required):

- Test in a furnace at 750 °C, using a procedure based on ISO 1182 (or ASTM E136, which is essentially the same)
- Bomb calorimeter test for assessing the total heat of combustion (calorific value), mostly based on ISO 1716
- Heat release test, using the cone calorimeter (ISO 5660) with an exposure of the specimen of 50 kW/m^2.

10.1.1.2 Combustible Materials – Laboratory Tests

The “limited”, “normal”, and “high” classifications regarding contribution to fire of a building material are a measure of its behavior in a fire that is initiating. Materials showing a low contribution to fire usually start to participate in the fire at a later stage of initiation. Those with a normal contribution resist medium- and low-intensity ignition sources for a short time period (of the order of minutes), and those with a high contribution are set on fire by virtually any ignition source and are therefore not permitted to be used as building materials in most countries. Many national classification systems contain intermediate classes in order to be able to differentiate materials more clearly.

Test methods are available for simulating all the important phases of an initiating fire, from ignition to flashover, and can provide information on the contribution of building materials to each stage. Roughly 50 methods for simulating the initiating fire and 20 methods for describing parallel fire effects (i. e., smoke development and toxicity of fire effluents) exist in the various countries covered below. In some cases, an additional test is carried out to determine whether materials give off burning drops. In the harmonized “reaction to fire” tests used in Europe, dripping has been introduced as a test criterion.

Fire tests using low intensity ignition sources (usually small burners) measure ignitability and allow the elimination of materials with high contribution to fire.

The least stringent case is when the specimen is positioned horizontally. An example is the American ASTM D635 test. The test is more rigorous if the specimen is inclined (usually at 45°).

The toughest demands are made on a material if the specimen is positioned vertically. The igniting flame may be applied to the specimen surface and/or edge as defined in the ISO 11925-2 small flame test, which is now used as part of the European system for testing and classification of construction products. Similar tests, which in most cases use edge flame impingement only, are used for example in USA, Canada, and Switzerland.

Additional types of tests with a small ignition source for combustible materials are applied in some cases, e. g., the Limiting Oxygen Index test (LOI) according to ISO 4589 (used in China for building products). Here the ignitability of a product is tested depending on preset oxygen levels of the air in a tube around the specimen.

The above methods are used to establish the ignitability and contribution to fire spread of materials in the early phase of a fire, when only small ignition sources are likely to be involved. They are often referred to in regulations to define a minimum fire safety requirement for all building products, for example in Germany, Switzerland, and China.

Higher requirements in building regulations apply for products and materials depending on their use, as well as on the type and size of the building or special parts of it (for example escape routes). Therefore, tests suitable to assess the contribution to fire development and spread under a higher level of exposure have been defined. Such laboratory-scale tests are based on more rigorous open-flame ignition sources, and/or radiant heat sources.

Examples for open flame tests are the Single Burning Item (SBI) test to EN 13823, which is now used in the European Union and in China, and the Steiner tunnel test to ASTM E84 used in the USA, Canada, and the Emirates.

In the past, radiant panels were applied in a number of European countries as an ignition source for various tests of building products used for walls and ceilings, but these have now been replaced by the SBI. However, in Australia for example, a specific radiant panel test is still in use for all types of building products (AS 1530.3). In most countries worldwide, flooring products used in building are tested with a gas radiant panel combined with a small open flame.

The conical radiator used in the Cone Calorimeter (ISO 5660-1) is a different type of radiation source. In addition to its application for determining non-combustibility or limited combustibility, some countries use it also for assessing the fire performance of combustible materials (for example, Japan and Korea).

10.1.1.3 Roofs

For roofs, in many countries specific testing and classification procedures have been developed. The major hazard covered by building regulations is the exposure of a roof by a fire from the outside, causing penetration into a roof and/or fire spread on a roof. Various test procedures have been developed with different exposure conditions, for example the four methods covered in the European specification CEN/TS 1187. In these, roof coverings are exposed to different open fire sources, and in some cases to additional radiation and wind.

In some countries (including Germany and the USA), special requirements have been developed for the exposure of light industrial roofs (with non-fire-resistant structures) to a fire developing inside a building.

10.1.1.4 Large-Scale Tests

In addition, some countries use large-scale tests to simulate fires in a realistic environment. These are sometimes used as reference scenarios for developing and validating laboratory tests. Some countries, as well as insurers, include full-scale testing in the approval procedures for building products. One of the most frequently used tests here, the so-called room/corner test, standardized as ISO 9705, has been taken over with small variations in the USA and Europe.

Façade tests are another example for large-scale tests. Most fire scientists and regulators agree that the performance of materials is not the only relevant factor for fire spread on façades. Therefore, different façade tests have been developed worldwide (see Chapter 8, Sections 9.2, 10.2, 10.3, and 10.4).

10.1.2 Official Approval

In almost all countries requirements have been defined for laboratories and certification bodies. In some cases, the latter are government-owned, but in most cases they are private laboratories, which need to prove their independence and neutrality as well as their competence in accordance with different accreditation schemes.

Accreditation according to the ISO/IEC 17025 standard is widely accepted for testing laboratories. Further international standards have been developed for certification and inspection bodies (i.e., EN ISO/IEC 17065). In Europe, an additional notification from the local government to the European Commission and the other EU member states is needed.

In some countries, national accreditation organizations are in place (e.g., Australia) or only governmental organizations are accepted (e.g., China).

When a test has been carried out, the test laboratory issues a test certificate. This is usually accepted as sufficient evidence by the authorities, although in some cases a committee has to verify and approve it. Some countries require a test mark linked with inspection and quality control for materials subject to stringent fire protection requirements. Organizations such as the American Underwriters' Laboratories grant test marks for certain building materials and products, which require internal and independent monitoring. Further information on test marks and inspection will be found in the individual country sections.

10.1.3 Present Situation and Future Developments

The following sections summarize the current situation in building. In most countries, building regulations are revised in a more-or-less regular cycle. This is done to ensure that they reflect the state of the art and take into account new social developments and challenges.

After some major fire incidents (e.g., the Grenfell Tower Fire) in the UK and increasingly in other European countries, the focus is now on necessary changes regarding:

- Testing of façades

- Better control of building materials and products
- Continued maintenance of buildings and relevant parts and components.

Regarding full-scale tests for façades, the European Commission is currently trying to develop a harmonized full-scale test method, but an international discussion has started as well. In a number of countries, regulations for high-rise buildings are under review or have been reviewed, and the international scientific community is focusing on the topic of fire safety of façades.

Another discussion relates to toxicity of combustion gases from construction products. While in Europe, Australia, and the USA (except New York) this topic is currently not considered as relevant for fire safety, some Asian countries have introduced toxicity requirements for construction products. Most countries rely on the approach that it is key to prevent people from exposure to smoke, because smoke is always toxic. The European Commission has commissioned a study confirming this approach [2]. But the discussion is ongoing, and further requirements for combustible building products may come up in the future.

References for Section 10.1

[1] Regulation (EU) No. 305/2011 of the European Parliament and of the Council of 9 March 2011 laying down harmonised conditions for the marketing of construction products and repealing Council Directive 89/106/EEC.

[2] Study to evaluate the need to regulate within the Framework of Regulation (EU) 305/2011 on the toxicity of smoke produced by construction products in fires, European Commission, 2017.

10.2 Americas

Marc Janssens

10.2.1 United States of America

10.2.1.1 Statutory Regulations

An acceptable level of fire safety is accomplished for new or substantially renovated construction through compliance with local regulations. These regulations are based on national model building codes. Additional regulations based on a national model fire prevention code ensure that this level of fire safety is maintained during the lifespan of the building. The U.S. model codes are largely prescriptive. Acceptance criteria for construction materials, products, and assembly designs are generally based on performance in standardized fire tests. The following sub-sections provide a brief overview of the model code system in the USA.

For a more comprehensive review, the reader is referred to the book by Diamantes [1].

U.S. Model Codes

There are two model building and fire prevention code organizations in the United States; the International Code Council (ICC) [2] and the National Fire Protection Association (NFPA) [3].

ICC was formed in 1994 as an umbrella organization consisting of representatives of three older model code organizations: Building Officials and Code Administrators International (BOCA), International Conference of Building Officials (ICBO) and Southern Building Code Congress International (SBCCI). The purpose of ICC was to facilitate the development of a single set of model codes to replace the codes maintained by BOCA, ICBO and SBCCI, i. e., the National, Uniform, and Standard codes respectively. The BOCA, ICBO and SBCCI codes served for many years as the basis for local fire safety regulations in Northeastern, Southern, and Western regions of the country. Although the ICC member groups agreed to discontinue their individual codes in 2000, many local jurisdictions did not immediately transition and continued to rely on the older documents for several years. At the time of this writing, the older regional codes have been superseded by national ICC and NFPA codes throughout the U. S.

ICC promulgates the International Building Code (IBC) and the International Fire Code (IFC). The IBC and IFC are supplemented with a series of model code documents that provide more detail and additional requirements for specific types of buildings (e. g., one- and two-family dwellings) and specific building systems and components (e. g., plumbing), or to achieve specific objectives (e. g., mitigate risks to life and property from wildfires). Collectively, the IBC, IFC, and supplemental documents are referred to as the International Codes®, or I-Codes®. New editions of the I-Codes are published every three years. The I-Codes can be viewed online for free [4]. The free access is referred to as publicACCESS™. A subscription to premiumACCESS™ is required for users who want to print sections of a code, search and annotate the document, or need other advanced features. In the District of Columbia and the states of Arizona, Kansas, and Nevada the I-Codes are adopted by jurisdictions at the local level. In the remaining states, the I-Codes are adopted statewide, although local jurisdictions in some states are allowed to amend the state code and include more stringent requirements.

In 2003, NFPA published the first version of its model building code, NFPA 5000. A new edition is published in the same year as the IBC. In addition, NFPA promulgates a fire prevention code and the *NFPA 101: Life Safety Code®*. The latter is widely used throughout the country but is primarily concerned with occupant safety in specific types of buildings and does not address all building construction and fire prevention issues. NFPA also maintains codes that are included in the I-Codes by reference. The most noteworthy of these is the *NFPA 70: National Electrical Code®*. At this time NFPA 5000 has not yet been adopted by a local jurisdiction. NFPA codes and standards can also be viewed online for free [5].

Demonstrating Compliance with Building Code Fire Safety Requirements

Fire protection of buildings addresses all aspects of fire safety and consists of a combination of active and passive measures. Active fire protection devices such as sprinkler, smoke control, and detection and alarm systems require manual, mechanical, or electrical power for their operation. Testing of active fire protection devices is beyond the scope of this handbook. Passive fire protection does not require any external power. There are essentially two types of passive fire protection measures that involve fire testing of construction products, structural elements, and assemblies.

The objective of the first type of measures is to reduce the likelihood of ignition and limit the rate of fire growth to the critical stage of flashover in the compartment of fire origin. A slow-growing fire leaves more time for safe egress of building occupants, and generally results in reduced property damage at the time of manual or automatic suppression. This can be accomplished by using interior finishes and furnishings with specific ignition, flame spread, and heat release characteristics. These characteristics are determined based on performance in standardized tests that expose a specimen to the thermal environment representative of the initial stages of a fire. Standardized tests are also used to control ignition of and flame spread over exterior façade and roof surfaces. A secondary objective of the first type of measures is to control the quantities of particulate matter (which affects visibility) and of toxic products of combustion that can cause human casualties and excessive damage to equipment. The city of New York is the only jurisdiction in the U.S. that specifies a smoke toxicity requirement for construction products in its building code[1].

Despite the first type of measures, fires do grow to full involvement of the room of origin. When this happens, the focus shifts to containing the fire within a limited area, at least for a certain time. Thus, fire spread to other parts of the building or adjacent buildings is delayed or prevented. This containment process is referred to as compartmentation. It is accomplished by providing fire-resistive floor, wall, and ceiling assemblies and by protecting openings and penetrations through room boundaries. Compartmentation also involves protecting structural elements and assemblies to avoid or delay partial or total collapse in the event of fire.

[1] § 803.5 of the New York City Building Code® [6] requires that interior wall or ceiling finishes, other than textiles, upon exposure to fire, shall not produce products of decomposition or combustion that are more toxic than those given off by wood or paper when decomposing or burning under comparable conditions as tested in accordance with NFPA 269: *Standard test method for developing toxic potency data for use in fire hazard modeling.*

To demonstrate compliance with fire safety requirements in the code, the architect or builder typically needs to present a report from an accredited laboratory to the code official that confirms that the product, structural element, or assembly was tested according to the applicable standard method and meets the acceptance criteria specified in the code. Most accredited fire testing laboratories publish a directory of tested products. These directories facilitate the code official's job to verify compliance and determine which variations from the tested product, element, or assembly are acceptable (e.g., acceptable range of thicknesses, substrates, and adhesives, etc.)

Generally, there is no requirement that tested products be listed, labeled, and subject to periodic follow-up plant inspections and verification testing, although there are exceptions. For example, the IBC requires that fiber-reinforced polymers delivered to the job site shall bear the label of an approved agency showing the manufacturer's name, product listing, product identification and information sufficient to determine that the end use will comply with the code requirements.

Product manufacturers sometimes develop a new product or identify a new application for an existing product that does not meet the established building code requirements. In 2003, the ICC created a technical evaluation service (ICC-ES) [7], which determines whether such a product or application meets the intent of the building code. If that is found to be the case, ICC-ES identifies a test method or develops a calculation method to evaluate the product, and establishes acceptance criteria. The results of the technical evaluation are published in a report. Presentation of the evaluation report to the code official serves as an alternative approach to demonstrate building code compliance. Testing in support of an evaluation report shall be performed by an accredited laboratory. New acceptance criteria developed as part of a technical evaluation are published in a separate document, which facilitates future evaluations of the same type of product or application requested by other manufacturers (or later by the same manufacturer). In recent years several accredited fire testing laboratories, third-party quality assurance agencies, and fire consulting engineering firms have developed competing evaluation services.

Testing Laboratory and Third-Party Quality Assurance Agency Accreditation

Testing in support of building code compliance can only be performed by accredited laboratories. There are two independent organizations in the U.S. that provide this type of accreditation service for fire testing laboratories: the American Association for Laboratory Accreditation (A2LA) [8] and the International Accreditation Service (IAS) [9]. Both use the requirements and criteria described in ISO/IEC 17025 [10]. As a result, a laboratory has to demonstrate an internationally recognized level of competence to be accredited. A2LA and IAS accreditations only cover specific standard test procedures. The test methods for which a laboratory is accredited are listed on the certificate, which can be printed from the A2LA and IAS websites. Both A2LA and IAS also accredit agencies that offer a listing, labeling, and follow-up inspection program for products that (in accordance with the code) require third-party quality assurance (QA). In this case, the agencies are periodically assessed according to the guidelines and criteria in ISO/IEC 17020 [11]. Table 10.1 provides a list of the fire testing laboratories in the U. S. that are accredited by IAS. The table also indicates for each laboratory whether it has an accredited third-party QA program and can perform some of the most common standardized fire test methods referenced in the IBC. A complete list of tests can be found on the IAS certificate. Only the laboratories that offer QA services have a directory of tested products.

Table 10.1 IAS-Accredited Fire Testing Laboratories Located in the U. S.

Fire testing laboratory	QA	ASTM standards				NFPA standards	
		E84	E108	E119	E136	285	286
Architectural Testing [12]	✓	✓	✓	✓	✓	✓	✓
FM Approvals [13]	✓	✓	✓	✓			
Intertek Testing Services [14]	✓	✓	✓	✓	✓	✓	✓
NGC Testing Services [15]		✓	✓		✓		
QAI Laboratories [16]	✓	✓	✓		✓		
Southwest Research Institute [17]	✓	✓	✓	✓	✓	✓	✓
Underwriters' Laboratories [18]	✓	✓	✓	✓	✓	✓	✓
Western Fire Center [19]			✓	✓			

10.2.1.2 Consensus Standards

In general, the building code fire safety requirements for products, structural elements, and assemblies are based on performance in a standardized test developed according to a consensus process. The primary consensus-based organizations in the U.S. developing and maintaining fire test standards are ASTM International (previously the American Society for Testing and Materials) [20] and the National Fire Protection Association (NFPA) [3]. Several fire testing laboratories in the U.S., such as Underwriters' Laboratories (UL) [18] and FM Approvals [13], have established a consensus process that meets the requirements of the American National Standards Institute (ANSI) so that they are permitted to develop and publish American National Standards [21].

ASTM International

The American Society of Testing and Materials was formed in 1898. In 2001 the name of the organization was changed to ASTM International, consistent with the fact that the organization is actively and successfully promoting the use of its standards outside North America. The *ASTM Book of Standards* is an 80+ volume set of more than 12,000 standards that is reprinted on an annual basis. The internet has made it possible to easily obtain interim revisions of ASTM standards.

ASTM International currently consists of 143 committees, which develop and maintain test standards pertaining to a wide range of products and technologies. Committee E05 on Fire Tests is the primary committee in ASTM that develops and maintains fire test standards [22]. This activity was initiated in 1904 with the formation of committee P on Fireproofing Materials in response to the Baltimore fire that occurred earlier that year. In 1910 the designation of the committee was changed to C-5, and 35 years later it changed again to E05. A history of ASTM Committee E05 and its antecedents was compiled at the time of the committee's centennial [23]. Several material- or product-oriented ASTM committees have subcommittees that develop fire test standards as well. For example, committee D20 on Plastics [24] has a subcommittee on Thermal Properties (D20.30) that develops and maintains some fire test standards for plastics.

To ensure that ASTM standards are consensus-based, committees operate according to strict rules. For example, the membership of committees writing standards has to be balanced. For this purpose the voting membership is classified into four categories: users, producers, consumers, and general interest. The bylaws of Committee E05 require that the number of voting producers be equal to or less than the total number of voting members classified in the other three categories. For this reason there is often a waiting list for new producers who apply for voting membership in the committee or one of its subcommittees. In addition, only one voting member is allowed to represent a company or organization.

The ASTM standard development process provides three steps of peer review. A draft developed by a task group is first balloted in the subcommittee. If the subcommittee ballot is valid (requires a minimum return from the voting membership) and successful (requires a minimum number of positive votes), all negative votes (which have to be substantiated by the negative voter) need to be resolved before the document can move on to the main committee. Main committee balloting procedures are very similar to the subcommittee balloting procedures. Items submitted for main committee ballot also appear on the ASTM website for society review. Each member of the society is entitled to vote on the items. Development of an ASTM standard can be started at any time. Theoretically the development can be completed in six months, but it usually takes at least a year.

The requirements for balance of standards-writing committees and the three-step peer review process result in truly consensus-based ASTM standards. Everybody can become a member of any ASTM committee and its subcommittees. If requested, a new member is given voting privileges provided the addition of the applicant to the list of voting members does not cause the (sub)committee to become unbalanced. Annual dues for ASTM membership include a subscription to one volume in the book of standards. All standards developed and maintained by ASTM Committee E05 are included in volume 04.07, and membership in the committee therefore provides full access to nearly all ASTM fire standards.

National Fire Protection Association

NFPA was formed in 1896 to harmonize electrical codes and sprinkler standards. Today, a significant part of the mission of NFPA is to reduce the worldwide burden of fire on the quality of life by providing and advocating scientifically-based consensus codes and standards. All NFPA codes and standards are compiled annually and printed in the *National Fire Codes®*, and are also available individually and in electronic form. Any or all of the 300+ NFPA codes and standards can be purchased from the NFPA website and can be can be viewed online free of charge [5]. NFPA codes and standards are developed and maintained by approximately 279 committees. The Fire Test Committee (designation FIZ-AAA) is responsible for all NFPA fire test standards that are referenced in other codes and standards published by NFPA. Most NFPA committees have a waiting list, and those seeking members are listed on the NFPA website [25].

As in ASTM, the membership of NFPA committees has to be balanced. The voting membership is classified into nine categories. Not more than one third of the voting membership can represent a single interest. Each company or organization is entitled to a maximum of one voting member and an alternate. In November 2010, the NFPA Board of Directors approved a comprehensive set of revisions to the regulations that govern NFPA's standards development process. The new regulations are in effect since the Fall 2013 Revision Cycle and can be found on the NFPA website [26]. The NFPA code and standard development process takes 101 weeks (Annual Revision Cycle) or 141 weeks (Fall Revision Cycle), and provides many opportunities for public review.

Other Standards Development Organizations (SDOs)

There are also many fire test methods that are not developed by a consensus-based process. In the U.S. these types of test standards are typically promulgated by regional or federal government agencies. The latter are published as part of the regulations that refer to these test methods in the Code of Federal Regulations (CFR) [27], although many federal regulations rely on consensus-based test standards. The U.S. government has recognized the advantages of using test standards that have been subjected to a review and are accepted by all interest groups, and is now actively supporting the adoption of consensus-based test standards.

Examples of regional fire test standards that are not consensus-based are the California Technical Bulletins to assess the fire characteristics of upholstered furniture and mattresses. These standards can be obtained from the California Bureau of Household Goods and Services (BHGS) [28].

10.2.1.3 Test Methods Referenced in U.S. Building and Fire Safety Codes

A complete list of fire test methods referenced in the IBC, IFC, and NFPA 101 (LSC) is given in Table 10.2-Table 10.7. Despite efforts by ASTM, NFPA, and UL to avoid duplication of fire test standards, there still is a significant amount of redundancy. Consequently, several ASTM test standards have NFPA and/or UL equivalents. Each table lists a specific category of fire test standards referenced in the codes, and for each standard provides the subject, identifies similar or "equivalent" tests standards, and indicates which codes reference the test standard. A brief description of the primary test standards is provided in Sections 10.2.1.4-10.2.1.9.

Table 10.2 Small-Scale Ignition and Combustibility Test Standards

Referenced	Subject	Similar or "equivalent"	IBC	IFC	LSC
ASTM D635	Burning rate of plastics		✓		
ASTM D1929	Ignition temperature of plastics		✓		✓
ASTM D2859	Flammability of carpets and rugs	16 CFR 1630	✓	✓	✓
ASTM E136	Non-combustibility	ASTM E2652	✓		✓
ASTM E2652	Non-combustibility	ASTM E136			✓
NFPA 259	Potential heat		✓		✓
16 CFR 1500.44	Extremely flammable solids		✓	✓	
16 CFR 1630	Flammability of carpets and rugs	ASTM D2859	✓	✓	

Table 10.3 Small-Scale Reaction-to-Fire Test Standards

Referenced	Subject	Similar or "equivalent"	IBC	IFC	LSC
ASTM D2843	Smoke density from plastics		✓		
ASTM E648	Critical radiant flux of flooring	ASTM E970, NFPA 253	✓	✓	✓
ASTM E970	Critical flux of attic floor insulation	ASTM E648, NFPA 253	✓		
ASTM E1354	Cone calorimeter		✓	✓	
ASTM E2965	Low levels of heat release	ASTM E1354			✓
NFPA 253	Critical radiant flux of flooring	ASTM E648, ASTM E970	✓	✓	✓
NFPA 701	Flame propagation of textiles		✓	✓	✓

Table 10.4 Intermediate- and Large-Scale Reaction-to-Fire Test Standards

Referenced	Subject	Similar or "equivalent"	IBC	IFC	LSC
ASTM E84	Steiner tunnel test	UL 723	✓	✓	✓
NFPA 265	Room/corner test for textile wall coverings	NFPA 286, UL 1715	✓	✓	✓
NFPA 286	Room/corner test	NFPA 265, UL 1715	✓	✓	✓
NFPA 276	Heat release of roof assemblies	FM 4450	✓		
DASMA 107	Garage doors with foam plastic insulation	NFPA 286	✓		
FM 4450	Heat release of roof assemblies	NFPA 276, FM 4880	✓		
FM 4880	Industrial-scale open corner test	UL 1040	✓		✓
FM 4996	Pallets and other material handling products			✓	
UL 723	Steiner tunnel test	ASTM E84	✓	✓	✓
UL 1040	Industrial-scale open corner test	FM 4880	✓		✓
UL 1715	Room/corner test	NFPA 265, NFPA 286	✓		✓
UL 1256	Roof deck construction	FM 4880, NFPA 276	✓		✓
ULC S102.2	Steiner tunnel test for flooring	ASTM E84, UL 723	✓		

Table 10.5 Fire Test Standards for Furnishings and Contents

Referenced	Subject	Similar or "equivalent"	IBC	IFC	LSC
ASTM E1537	Upholstered furniture			✓	✓
ASTM E1590	Mattresses	16 CFR 1632		✓	✓
NFPA 260	Cigarette ignition resistance of upholstered furniture components			✓	✓
NFPA 261	Smoldering cigarette ignition of upholstered furniture mock-ups			✓	✓
NFPA 289	Individual fuel packages	UL 1975	✓	✓	✓
16 CFR 1632	Mattresses	ASTM E1590		✓	✓
FM 6921	Combustible waste containers				✓
UL 1315	Metal waste paper containers				✓
UL 1975	Foam plastics for decoration	NFPA 289	✓	✓	✓

Table 10.6 Fire Resistance Test Standards

Referenced	Subject	Similar or "equivalent"	IBC	IFC	LSC
ASTM E119	General fire resistance test	UL 263	✓		✓
ASTM E814	Penetration firestop systems	UL 1479	✓		✓
ASTM E1529	Hydrocarbon pool fire exposure			✓	
ASTM E1966	Fire-resistive joint systems	UL 2079	✓	✓	✓
ASTM E2307	Perimeter fire barrier systems		✓		✓
ASTM E2387	Joint systems between wall and floor				✓
NFPA 252	Door assemblies	UL 10 A, 10B, and 10C	✓		✓
NFPA 257	Window and glass block assemblies	UL 9	✓		✓
NFPA 275	Evaluation of thermal barriers		✓		

Referenced	Subject	Similar or "equivalent"	IBC	IFC	LSC
NFPA 288	Horizontal fire door assemblies		✓		✓
NFPA 415	Airport loading walkways				✓
UL 9	Window assemblies	NFPA 257	✓		✓
UL 10A	Tin-clad fire doors	NFPA 252	✓		
UL 10B	Door assemblies	NFPA 252	✓		✓
UL 10C	Positive pressure test of doors	NFPA 252	✓	✓	✓
UL 263	General fire resistance test	ASTM E119	✓		✓
UL 1479	Penetration firestop systems	ASTM E814	✓		✓
UL 2079	Building joint systems	ASTM E1966	✓		✓
UL 2196	Fire resistive cables		✓		

Table 10.7 Exterior Fire Exposure Test Standards

Referenced	Subject	Similar or "equivalent"	IBC	IFC	LSC
ASTM E108	Roof coverings	UL 790	✓	✓	✓
NFPA 268	Ignitability of exterior wall assemblies		✓		
NFPA 285	Fire propagation characteristics of combustible exterior curtain walls		✓		
UL 790	Roof coverings	ASTM E108	✓	✓	✓
UL 1703	Flat-plate photovoltaic panels		✓		

10.2.1.4 Summary of Small-Scale Ignition and Combustibility Tests

ASTM D635: Standard Test Method for Rate of Burning and/or Extent and Time of Burning of Plastics in a Horizontal Position

The ASTM D635 standard measures the burning rate or extent of plastics in the form of bars molded or cut from sheets, plates, or panels. The method is used in the IBC to qualify light-transmitting plastics. The apparatus is illustrated in Figure 10.1 and the test specifications are summarized in Table 10.8.

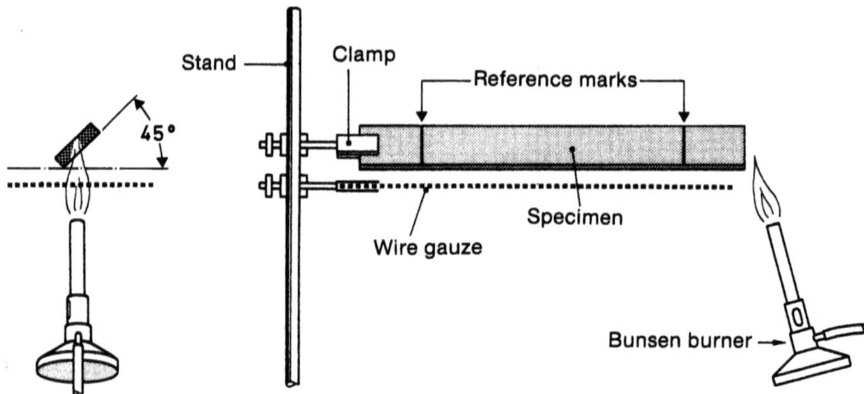

Figure 10.1 Horizontal rate of burning test fixture

Table 10.8 ASTM D635 Burning Rate of Plastics Test Specifications

Specimens	Ten specimens, 125 × 13 × 1.5 mm or thickness intended for use
Specimen position	Horizontal
Ignition source	Bunsen burner with 20 mm high blue flame applied for 30 s
Test duration	Time to extinction or time for the flame front to reach 100 mm
Conclusions	Different classification based on the extent of burning or burning rate

A conditioned bar specimen is supported horizontally at one end. The free end is exposed to a 20 mm blue flame for 30 s. The time and extent of burning are measured and reported if the specimen does not burn to the 100 mm mark. An average burning rate based on the time for the flame front to advance from the 25 mm to the 100 mm mark is reported for a material if it burns to the 100 mm mark.

In the IBC, a Class CC1 material exhibits a burning extent of 25 mm or less (with a specimen thickness of 1.5 mm, or the thickness intended for use). Class CC2 must have a burning rate of less than 1.06 mm/s at the same thickness. The practical differentiation is that more Class CC1 material may be used in a given building area than Class CC2.

ASTM D1929: Standard Test Method for Determining Ignition Temperature of Plastics

The flash and spontaneous ignition temperatures of plastics are determined according to ASTM D1929, which uses an electric hot air furnace, referred to as the Setchkin furnace (see Figure 10.2). The test specifications are summarized in Table 10.9.

Table 10.9 ASTM D1929 Flash and Spontaneous Ignition Test Specifications

Specimens	3 g of any material with density > 100 kg/m³, 20 × 20 × 50 mm for cellular materials with density ≤ 100 kg/m³
Specimen position	Pan containing specimen suspended in furnace
Ignition source	Electric furnace which can be heated to 750 °C and small pilot flame for flash ignition tests
Test duration	Each ignition test is 10 min or until ignition occurs, whichever is shorter
Conclusions	Determination of the flash and spontaneous ignition temperatures of plastics to within 10 °C

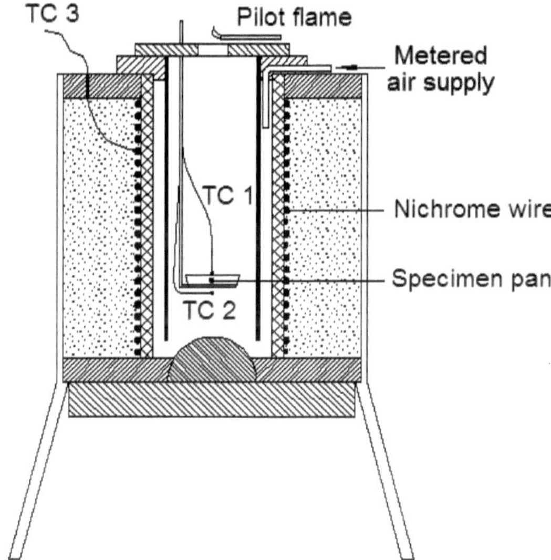

Figure 10.2
Electric hot air ignition furnace for ignition temperature of plastics

The test apparatus consists of a small tubular furnace, with an inside diameter of 76 mm and a height of 210 to 250 mm. A controlled flow of ambient air is supplied at the bottom of the apparatus. A small pilot flame is used at the top of the furnace for the flash ignition tests.

The test material can be in any form, but typically pellets or powder are normally supplied for molding, with a total mass of 3 g. For cellular materials with a bulk density of 100 kg/m³ or less, specimens shall be 20 × 20 mm with a height of 50 mm. For sheet materials with a higher density, the height of the specimen is adjusted so that its mass is equal to 3 g.

First, preliminary tests are conducted to determine the flash ignition temperature to within 50 °C. These tests are performed with the pilot flame. Generally, the first test is conducted with a furnace temperature of 400 °C. The specimen is lowered into the furnace and is observed to determine whether ignition occurs. Ignition can be monitored by watching the temperature difference between thermocouples T_1 and T_2. On ignition, T_1 will rise more rapidly than T_2. If ignition has not occurred in 10 min, the oven temperature is raised by 50 °C for the next test. This process is repeated until ignition is observed within 10 min.

Next, additional tests are conducted to determine the flash ignition temperature to within 10 °C. For the first additional test, the furnace temperature is set at 10 °C below the ignition temperature established in the preliminary tests. If ignition occurs, runs are carried out lowering the temperature each time by 10 °C until ignition no longer occurs within 10 min.

The self-ignition temperature is determined as above but without the pilot flame.

The IBC and the LSC require that light-transmitting plastics have a minimum spontaneous ignition temperature of 343 °C.

ASTM D2859: Standard Test Method for Ignition Characteristics of Finished Textile Floor Covering Materials

This standard provides a test method to determine the surface flammability of carpets and rugs when exposed to a standard small source of ignition under carefully prescribed draft-protected conditions. It is applicable to all types of carpets and rugs used as floor covering materials regardless of their method of fabrication or whether they are made of natural or synthetic fibers or combinations of these. Test specifications are listed in Table 10.10. Carpets treated with flame retardants shall be washed and dried 10 times before test specimens are cut.

Table 10.10 ASTM D2859 Methenamine Pill Test Specifications

Specimens	Eight specimens, 229 × 229 mm
Specimen position	Horizontal, covered by a 6.4 mm thick steel flattening plate with a circular opening of 203 mm in diameter in the center of the plate
Ignition source	Methenamine reagent tablet, 0.15 g
Test duration	Until flames extinguish or reach the edge of the steel frame
Conclusions	A specimen passes the test if the charred area does not extend to within 25 mm of the edge of the hole in the flattening plate, and at least seven out of the eight tested specimens are required to meet the acceptance criteria

A conditioned specimen is placed on a horizontal surface. A steel flattening plate with a hole in the center is then put on top of the specimen. A methenamine tablet is placed on the specimen in the center of the hole and lighted with a match. The test is terminated when self-extinguishment occurs, or if burning reaches the edge of the hole in the steel plate.

The method, commonly referred to as the "methenamine pill test", is also described in 16 CFR Part 1630: *Standard for the flammability of carpets and rugs*. Manufacturers and importers of carpets and rugs intended for general use in the U.S. must certify that their products comply and ensure that they are properly labeled in accordance with the provisions in 16 CFR 1630.

ASTM E136: Standard Test Method for Behavior of Materials in a Vertical Tube Furnace at 750 °C (and Related Non-Combustibility Test Standards)

Materials that are not explicitly recognized as non-combustible, such as steel and concrete, must be tested and meet specific criteria. The test procedure specified in the IBC and LSC to qualify a material as non-combustible is described in ASTM E136. The test specifications are summarized in Table 10.11.

Table 10.11 ASTM E136 Non-Combustibility Test Specifications

Specimens	Four specimens, 38 × 38 × 51 mm
Specimen position	Holder containing specimen suspended in furnace
Ignition source	Electric furnace heated to 750 °C
Test duration	Until center and surface temperatures have reached maximum values, or at least 30 min have elapsed and the center temperature has stabilized
Conclusions	A material passes the test if at least three of the four test specimens tested meet the following individual test specimen criteria: 1. Total specimen mass loss does not exceed 50%, rise in center and surface temperatures above the initial furnace temperature do not exceed 30 °C, and no flaming combustion is observed after the first 30 s; or 2. Total specimen mass exceeds 50%, center and surface temperature do not rise above the initial furnace temperature, and no flaming combustion is observed at all.

The test apparatus consists of a small tubular furnace, identical to the one described in ASTM D1929 (see Figure 10.2) but without the pilot flame. The air temperature in the furnace is 750 °C. A controlled flow of ambient air is supplied at the bottom of the apparatus.

A 38 × 38 × 51 mm specimen is inserted into the furnace from the top. Three thermocouples are used to measure the furnace air, specimen surface, and specimen center temperatures, respectively. ASTM E136 provides pass/fail criteria based on the specimen surface and center temperatures not rising more than 30 °C over the initial furnace temperature. In addition, mass loss must not exceed 50% of the initial specimen mass, and flaming combustion shall not be observed after the first 30 s of the test. If the mass loss exceeds 50%, no flaming is permitted at all and the specimen surface and center temperatures shall not exceed the initial furnace temperature at any time during the test. A material passes the test if at least three of the four test specimens tested meet the individual test specimen criteria.

To qualify a material as non-combustible, ASTM E136 now includes an alternative test procedure. This procedure is described in ASTM E2652: *Standard test method for behavior of materials in a tube furnace with a cone-shaped airflow stabilizer, at 750 °C*, and is identical to that described in EN ISO 1182 (see Section 9.2 for a summary). ASTM E2652 recommends the same pass/fail criteria when the method is used as an alternative to ASTM E136.

NFPA 259: Standard Test Method for Potential Heat of Building Materials

NFPA 5000 and the LSC make a distinction between non-combustible and limited-combustible materials. ASTM E136 is used to qualify materials as non-combustible. Limited-combustible materials must have a potential heat of 8.2 MJ/kg or less as determined by NFPA 259. According to this method, the potential heat of a material is determined as the difference between the gross heat of combustion of the material measured with an oxygen bomb calorimeter and the gross heat of combustion of its residue after heating in a muffle furnace at 750 °C for two hours.

16 CFR Part 1500.44: Method for Testing Extremely Flammable and Flammable Solids

16 CFR 1500.44 describes a test method to determine whether a solid material is extremely flammable or flammable. For testing granules, powders, and pastes, the material is packed into a flat, rectangular metal boat with inner dimensions 155 mm long × 25 mm wide × 6.4 mm deep. The dimensions of rigid and pliable solid specimens are not specified. The specimen is supported so that the major axis is oriented horizontally and the maximum surface is freely exposed to the atmosphere. The flame of a candle, at least 25 mm in diameter, is held in contact with the surface of the sample at one end of the major axis for 5 s or until the specimen ignites, whichever is less. The time of self-sustained flaming combustion is measured, or the flame is extinguished at 60 seconds. The dimensions of the burnt area are measured, and the rate of burning along the major axis calculated. 16 CFR 1500.3: *Definitions* defines extremely flammable solids as solid substances that ignite and burn at an ambient temperature of 27 °C or less when subjected to friction, percussion, or electrical spark. A flammable solid is defined as a solid substance that, when tested by the method described in 16 CFR 1500.44, ignites and burns with a self-sustained flame at a rate greater than 2.5 mm/s along its major axis.

10.2.1.5 Summary of Small-Scale Reaction-to-Fire Tests

ASTM D2843: Standard Test Method for Density of Smoke from the Burning or Decomposition of Plastics

ASTM D2843 describes a laboratory procedure for quantifying and visually observing the relative amounts of smoke produced by the burning or decomposition of plastics. The measurements are made in terms of the loss of light transmission through a collected volume of smoke produced under controlled, standardized conditions. The apparatus is constructed so that the flame and smoke can be observed during the test. The apparatus is illustrated in Figure 10.3 and the test specifications are summarized in Table 10.12.

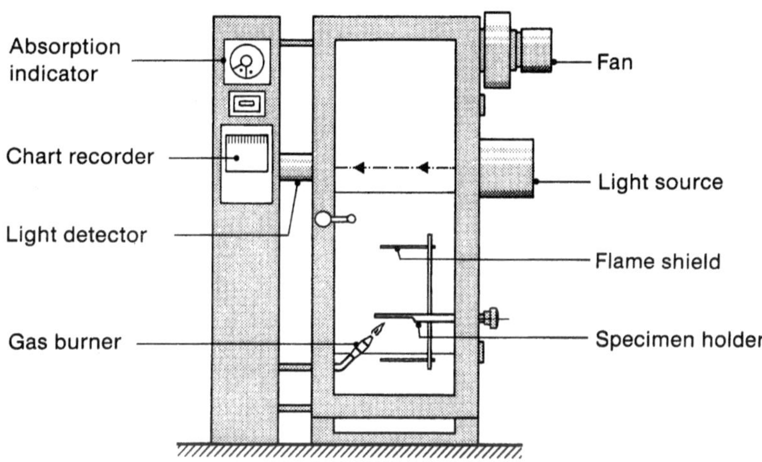

Figure 10.3 ASTM D2843 smoke density test apparatus

Table 10.12 ASTM D2843 Smoke Density of Burning or Decomposing Plastics Test Specifications

Specimens	Three specimens, 25 × 25 × 6 mm (or thickness intended for use)
Specimen position	Horizontal
Ignition source	Premixed propane-air burner flame
Test duration	The test is terminated when the flame is extinguished, the specimen is consumed, or 4 min have elapsed, whichever occurs first
Conclusions	The primary result is a smoke density rating, which ranges from 0 to 100

The 300 × 300 × 790 mm test chamber is instrumented with a white light source, photoelectric cell, and temperature-compensated meter to measure light absorption horizontally across the 300 mm light beam. An exit sign is mounted on the back wall of the chamber and can be viewed through the heat-resistant glazed front door. The chamber is closed during the 4 min test period, except for the four

25 × 230 mm ventilation openings around the bottom. The test specimen is placed on a supporting metal screen inside the chamber and exposed to a premixed propane–air burner flame for the duration of the test. The smoke generated by the burning specimen is substantially trapped in the chamber in which combustion occurs. Standard specimen size is 25 × 25 × 6 mm. Thicknesses other than 6 mm may be tested provided the specimen size is reported with the smoke density values.

The time until the specimen bursts into flame, the time for flame extinguishment or specimen consumption, and the time for obscuration of the exit sign in the test chamber are recorded. A curve is plotted of light absorption (in %) averaged over 15 s versus time (in min), to calculate a smoke density rating (SDR). The latter is the total smoke produced (area under the curve) in %·min divided by four. The IBC uses ASTM D2843 to control the use of plastic materials in light-transmitting applications, and specifies a maximum SDR of 75.

ASTM E648: Standard Test Method for Critical Radiant Flux of Floor-Covering Systems Using a Radiant Heat Energy Source (and Related Test Standards)

The National Bureau of Standards (NBS, currently the National Institute of Standards and Technology, or NIST) conducted a series of full-scale fire tests in the 1970s to investigate the fire hazard of floor coverings. The main concern was flame spread over the floor from a fire room to a connected corridor. This work resulted in the development of the radiant flooring panel test, which was subsequently published as ASTM E648. The standard describes a procedure for measuring the critical radiant flux (CRF) of horizontally mounted floor-covering systems exposed to a flaming ignition source in a graded radiant heat energy environment. The test specifications are summarized in Table 10.13.

Table 10.13 ASTM E648 Radiant Flooring Panel Test Specifications

Specimens	Three specimens, approximately 250 × 1050 mm × thickness of use
Specimen position	Horizontal
Ignition source	Premixed gas-fired radiant heater, 305 × 457 mm, producing a heat flux to the specimen varying from 1.0 to 0.1 W/cm^2, inclined at 30° to the specimen and lower edge 140 mm above the specimen Pilot flame on pivot, stainless steel tube with perforations
Test duration	5 min preheat, 5 min pilot burner contact, up to 5 min longer if specimen does not propagate flame; otherwise until flames go out
Conclusions	Critical radiant flux measurement

The test apparatus is shown in Figure 10.4. The test chamber is about 1400 × 500 × 710 mm. Inside the chamber, the horizontally-mounted floor covering system specimen is exposed to the heat flux from a premixed gas-fueled radiant heat energy panel inclined at 30°. The area of the specimen exposed to the heat flux is 200 mm wide and 1000 mm long. The radiant panel generates a radiant heat flux distribution ranging along the 1 m exposed length of the test specimen from a nominal maximum of 1.0 W/cm^2 (or 10 kW/m^2) at the panel end to a minimum of 0.1 W/cm^2 (or 1.0 kW/m^2) at the opposite end (see Figure 10.5). The distribution is established from calibrations involving heat flux measurements at 100 mm increments along the length of a dummy specimen.

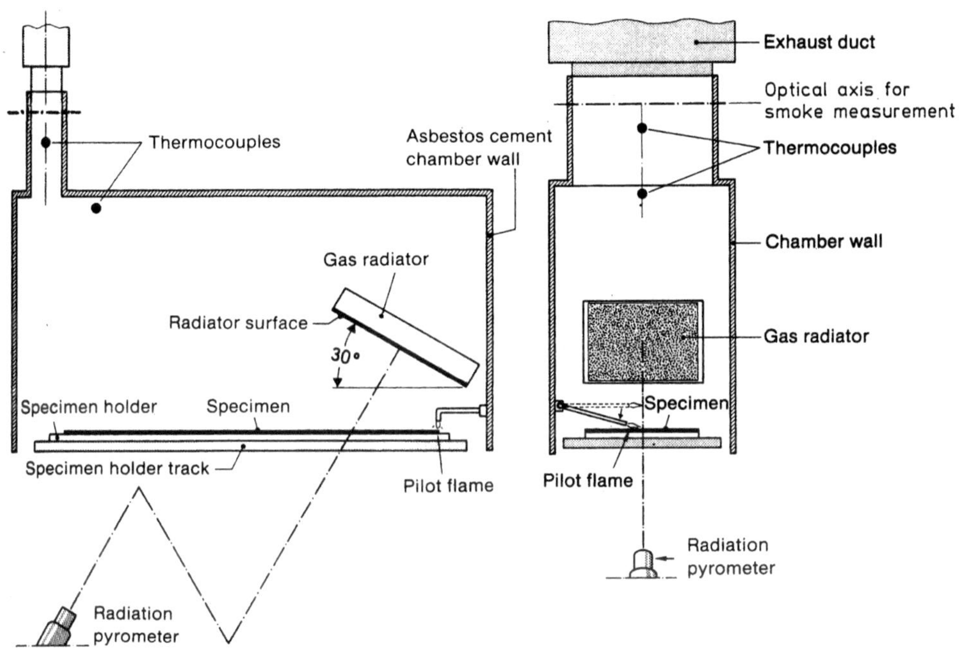

Figure 10.4 Radiant flooring panel test rig

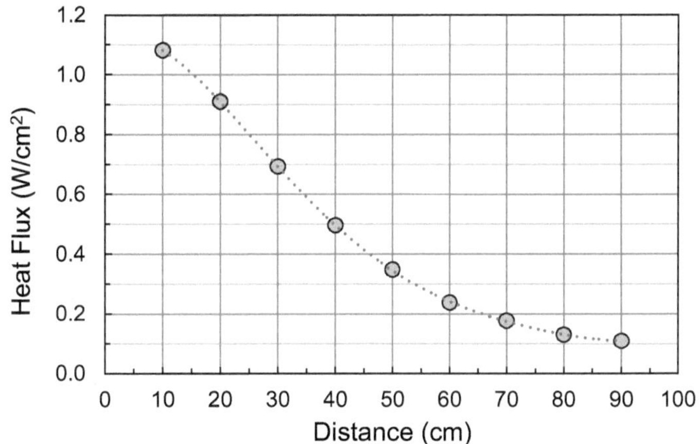

Figure 10.5 ASTM E648 radiant heat flux profile

A specimen is tested over a simulated concrete structural floor (fiber-cement board) or otherwise mounted in a representative way. The test is carried out by exposing the test specimen to the radiation from the panel for 5 min and then applying the pilot flame for 5 min to the narrow edge nearest the radiant panel. If the specimen has ignited, the test is continued until the flames extinguish, otherwise it is terminated after the specified time. The heat flux at the farthest distance from the panel end where the flame ceases to further advance and eventually extinguishes is determined from the distance and the heat flux profile in Figure 10.5, and reported as the CRF in W/cm².

ASTM E970: *Standard test method for critical radiant flux of exposed attic floor insulation using a radiant heat energy source* and NFPA 253: *Standard method of test for critical radiant flux of floor covering systems using a radiant heat energy source* are virtually identical to ASTM E648. However, ASTM E970 is used for measuring the critical radiant flux of exposed attic floor insulation and includes detailed instructions for evaluating loose-fill insulation that are not in ASTM E648.

The IBC requires that "non-traditional" interior floor finish materials be tested according to ASTM E648 or NFPA 253. The code defines two classes. Class I materials have a CRF $\geq$ 0.45 W/cm² and Class II materials have a CRF $\geq$ 0.22 W/cm². Exposed insulation materials installed on attic floors shall have a CRF $\geq$ 0.12 W/cm² when tested in accordance with ASTM E970.

ASTM E1354: Standard Test Method for Heat and Visible Smoke Release Rates for Materials and Products Using an Oxygen Consumption Calorimeter (and Related Test Standards)

The cone calorimeter is the most commonly used bench-scale fire test apparatus to obtain ignition and heat release, mass loss, and smoke production rate data of planar materials. The apparatus and test procedure are standardized in the U.S. as ASTM E1354. A schematic of the apparatus is shown in Section 9.2, Figure 9.7. The test specifications are summarized in Table 10.14.

Table 10.14 ASTM E1354 Cone Calorimeter Test Specifications

Specimens	Three specimens for each heat flux, 100 × 100 mm × thickness of use (≤ 50 mm)
Specimen position	Horizontal
Ignition source	Truncated conical electrical heater producing a pre-set heat flux to the specimen between 0 and 100 kW/m², electric spark ignition pilot
Test duration	10 min if ignition does not occur; otherwise until 2 min after flame-out
Conclusions	Time to ignition, heat release rate, mass loss rate, and smoke production rate as a function of time at the specified heat flux

A square specimen of 100 × 100 mm is exposed to the radiant flux of an electric heater. The heater has the shape of a truncated cone (hence the name of the instrument) and is capable of providing heat fluxes up to 100 kW/m^2. Prior to testing, the heater temperature is set at the appropriate value resulting in the desired heat flux.

At the start of a test, the specimen in the appropriate holder is placed on the load cell, which is located below the heater. An electric spark is used to ignite the pyrolysis products released by the specimen. As soon as sustained flaming is observed, the electric spark igniter is removed. All combustion products and entrained air are collected in a hood and extracted through an exhaust duct. At a sufficient distance downstream from a mixing orifice, a gas sample is taken and analyzed for oxygen concentration. Measurements of the gas temperature and differential pressure across the orifice plate are used for calculating the mass flow rate of the exhaust gases. The heat released rate can be determined based on the oxygen depletion and the mass flow rate in the exhaust duct. Most cone calorimeters include instrumentation for measuring light extinction in the exhaust duct, using a laser light source. Instruments to measure concentrations of soot, carbon dioxide, carbon monoxide, and other gases are often added.

ASTM E2965: *Standard test method for determination of low levels of heat release rate for materials and products using an oxygen consumption calorimeter* is a modified version of ASTM E1354. It is intended for use on materials and products that contain only small amounts of combustible ingredients or components, such as test specimens that yield a peak heat release < 200 kW/m^2 and total heat release < 15 MJ/m^2. The modifications are as follows: (1) shorter data collection (1 s instead of 3 s in ASTM E1354), (2) lower noise and drift required for the oxygen analyzer (30 ppm instead of 50 ppm), (3) heat release calibration at 1 kW (versus 5 kW in ASTM E1354), (4) larger specimen size (150 × 150 mm versus 100 × 100 mm), (5) lower duct flow rate (0.012 m^3/s versus 0.024 m^3/s), and (6) different end-of-test criteria.

The IFC and LSC impose restrictions on the use of combustible materials for waste containers by requiring that the peak heat release rate of the material be 300 kW/m^2 or less when tested according to ASTM E1354 at 50 kW/m^2. The LSC qualifies a material as "limited combustible" if the peak heat release rate does not exceed 150 kW/m^2 for longer than 10 s, and the total heat released does not exceed 8 MJ/m^2 when tested for 20 min according to ASTM E2965 at 75 kW/m^2.

NFPA 701: Standard Methods of Fire Tests for Flame Propagation of Textiles and Films

NFPA 701 covers textile applications, as well as decorations and trim. In some occupancies, this standard is required for curtains, draperies, hangings, and other decorative materials suspended from walls or ceilings. These can include textiles and other plastic materials. There is an additional limitation for the amount of flame-resistant materials covering a certain percentage of wall space depending on occupancy type. There are two test methods in this standard. Test Method 1 is called for with single-layer fabrics or multi-layer assemblies of curtains, draperies, or window treatments of less than 700 g/m², while Test Method 2 is used for materials of greater than 700 g/m² or for vinyl-coated fabric blackout linings. Table 10.15 gives the test specifications for Test Method 1 of this standard.

Table 10.15 NFPA 701 Test Method 1 Specifications for Flame-Resistant Textiles and Films

Specimens	Ten specimens, 150 × 400 mm
Specimen position	Vertical
Ignition source	Meeker burner with grid top channels, methane-air
Test duration	Until flaming ceases
Conclusions	A material passes the test if the average mass loss is < 40% for ten specimens and the average burn time of fragments on floor is < 2 s

One of ten specimens is mounted with a pin bar in the test chamber (see Figure 10.6) and binder clips on the bottom two corners. The burner is adjusted with a methane flow of 1205 ml/min and air flow of 895 ml/min, and positioned 25 mm from the bottom edge of the specimen. The burner is ignited and allowed to burn for 45 s before it is moved away from the specimen and turned off. The after-flame time of the specimen is measured as well as the burning time of material that may have fallen to the floor during the test.

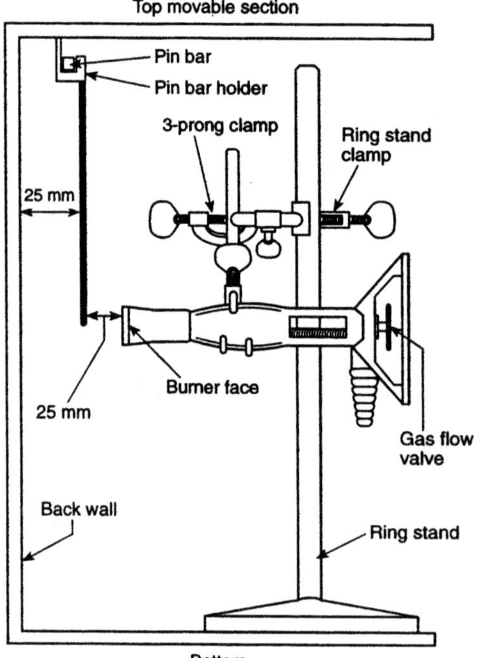

Figure 10.6
Test apparatus and specimen placement for NFPA 701 Test Method 1

Performance criteria for NFPA 701 Test Method 1:

- Average mass loss of 10 specimens must be less than 40%.
- Fragments that fall to the chamber floor must not continue to burn for more than 2 s on average.
- The mass loss of no individual specimen (in percent) can deviate more than three standard deviations from the mean.
- When a retest is needed, no individual specimen in the second set can have a mass loss percent that deviates from the mean value by more than three standard deviations calculated for the second set.

10.2.1.6 Summary of Intermediate and Large-Scale Reaction-to-Fire Tests

ASTM E84: Standard Test Method for Surface Burning Characteristics of Building Materials (and Related Standards)

ASTM E84 is probably the most used or referenced fire test standard, and the generally most important in the building codes to characterize flammability of plastics. The test fixture is known as the Steiner tunnel and was originally developed by UL. The present UL tunnel test standard, UL 723: *Test for surface burning characteristics of building materials,* is essentially identical to ASTM E84.

ASTM E84 is used to assess the comparative surface burning behavior of building materials and is applicable to exposed surfaces such as walls and ceilings, although building codes rely on tunnel test data for most cellular product insulation materials even if they are used behind a barrier. The test is conducted with the specimen in the ceiling position, with the surface to be evaluated exposed face-down to the ignition source. The material, product, or assembly must be capable of being mounted in the proper test position during the test. Thus, the specimen must either be self-supporting by its own structural quality, held in place by added supports along the test surface (steel rods or chicken wire), or secured from the back side. The purpose of this test method is to determine the relative burning behavior of the material by observing the flame spread along the specimen. Flame spread and smoke developed indices are reported. The test specifications are summarized in Table 10.16.

Table 10.16 ASTM E 84 Flame Spread and Smoke Development Test Specifications

Specimens	At least one specimen, 0.51 × 7.32 m × thickness of use up to 155 mm
Specimen position	Horizontal, facing down
Ignition source	79 kW natural gas or methane diffusion flame burner located 190 mm below the specimen at 305 mm from the fire end of the test chamber
Test duration	10 min
Conclusions	Flame spread index (FSI) and smoke developed index (SDI)

The test apparatus consists of a long tunnel-like enclosure measuring 8.7 × 0.45 × 0.31 m (see Figure 10.7). The test specimen is 7.3 m long and 0.51 m wide and is mounted in the ceiling position. It is exposed at one end, designated as the burner end, to a 79 kW gas burner. There is a forced draft through the tunnel from the burner end with an average initial air velocity of 1.2 m/s. The measurements consist of flame spread over the surface, i. e., visual observation of flame tip location as a function of time, and light obscuration in the exhaust duct of the tunnel. The latter is measured with a smoke photometer, which consists of a white light source on one side of the duct and a photocell with the sensitivity of the human eye on the opposite side of the duct. Test duration is 10 min.

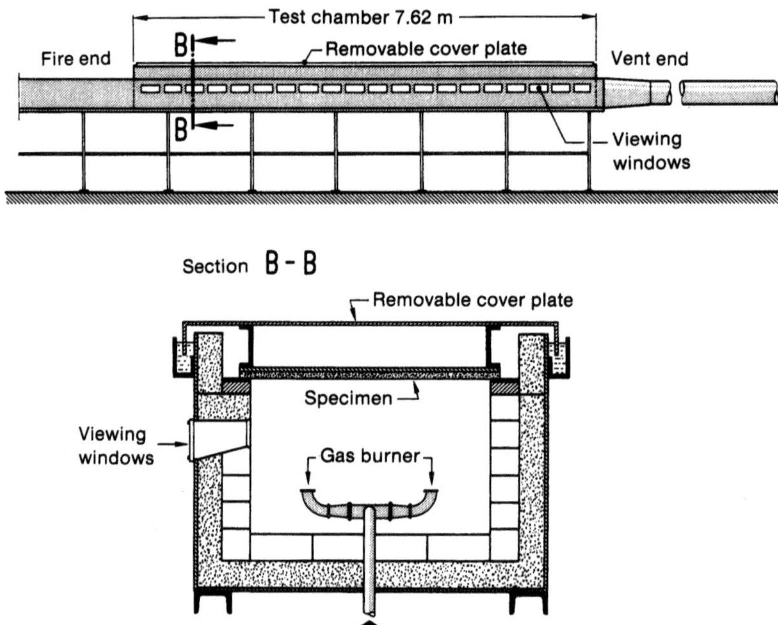

Figure 10.7 The 25-foot Steiner tunnel

The flame-spread index (FSI) is calculated based in the area under the curve of the visually observed flame tip location versus time, A_T. If A_T is less than or equal to 1783 m·s, then:

$$\mathrm{FSI} = 0.0282\, A_T$$

If A_T is greater than 1783 m·s, then

$$\mathrm{FSI} = 89611/(3566 - A_T)$$

The FSI is rounded to the nearest multiple of 5. A_T is calculated based on the flame extension, i. e., the distance of the flame tip from the burner minus the length of the burner flame. It is important to note that flame recession is ignored in the calculation of A_T. Periodic calibrations are performed with fiber cement board and red oak flooring. The FSI for fiber cement board should be equal to 0. Red oak flooring is used to verify the burner gas flow settings. The settings are within tolerance if the flame tip in a test on red oak flooring reaches the end of the tunnel in 5 min 30 s ± 15 s. The corresponding FSI of red oak flooring is between 90 and 95.

The smoke developed index (SDI) is equal to 100 times the ratio of (a) the area under the curve of light obscuration versus time for the 10 min test duration to (b) the area under the curve for a small heptane pool fire on the tunnel floor. Thus, the SDI of the heptane pool fire is 100, by definition. The SDI for fiber cement board is between 0 and 5. For an SDI less than 200, the calculated value is rounded to nearest multiple of 5. For an SDI of 200 or more, the calculated value is rounded to nearest multiple of 50.

The classification in the codes is based on the FSI and the SDI. There are three classifications: Class A for products with FSI ≤ 25, Class B for products with 25 < FSI ≤ 75, and Class C for products with 75 < FSI ≤ 200. For all classes the SDI shall not exceed 450. Class A products are generally permitted in enclosed vertical exits. Class B products can be used in exit access corridors, and Class C products are allowed in other rooms and areas. The general requirement in the IBC for plastic insulation and foam plastic cores of manufactured assemblies, unless otherwise indicated, is an FSI of 75 or less and an SDI of 450 or less.

In recent years, the following standards have been developed to provide guidance for ASTM E84 specimen preparation and mounting of specific types of products, and are referred to in the codes:

- ASTM E2404: *Standard practice for specimen preparation and mounting of textile, paper or vinyl wall or ceiling coverings to assess surface burning characteristics*
- ASTM E2573: *Standard practice for specimen preparation and mounting of site-fabricated stretch systems to assess surface burning characteristics*
- ASTM E2579: *Standard practice for specimen preparation and mounting of wood products to assess surface burning characteristics*
- ASTM E2599: *Standard practice for specimen preparation and mounting of reflective insulation, radiant barrier and vinyl stretch ceiling materials for building applications to assess surface burning characteristics.*

ASTM E2768: *Standard test method for extended duration surface burning characteristics of building materials (30 min tunnel test)* is an extended duration version of ASTM E84 used to qualify fire-retardant-treated wood for use in buildings of non-combustible construction.

CAN/ULC S 102.2: *Standard method of test for surface burning characteristics of flooring, floor coverings and miscellaneous materials and assemblies* is a variation of the Canadian version of the Steiner tunnel test. In this variation, the burner is turned upside down so that the burner flame hits the floor of the tunnel test apparatus. This variation is used to test flooring, floor coverings, loose fill insulation, etc., and is described in Section 10.2.2.2.

NFPA 286: Standard Methods of Fire Test for Evaluating Contribution of Wall and Ceiling Interior Finish to Room Fire Growth (and Related Test Standards)

Significant inconsistencies have been found between the ASTM E84 FSI and SDI classification and real fire performance of certain products such as foam plastics and textile wall coverings. To address these inconsistencies, the U.S. model building codes require that these materials pass a room/corner test such as NFPA 286: *Standard methods of fire test for evaluation contribution of wall and ceiling interior finish to room fire growth.* The test specifications are summarized in Table 10.17.

Table 10.17 NFPA 286 Evaluation of Fire Growth Contribution of Interior Finish Test Specifications

Specimens	Typical configuration involves test material on both 2.44 × 3.66 m side walls and 2.44 × 2.44 m back wall
Specimen position	Vertical
Ignition source	0.3 × 0.3 m propane diffusion "sand box" burner, 40 kW heat release rate for 5 min followed by 160 kW for 10 min
Test duration	15 min
Conclusions	The IBC acceptance criteria for NFPA 286 are as follows: 1. Flames shall not spread to the ceiling during the 40 kW exposure 2. Flames shall not spread to the outer extremities of the walls 3. Flashover shall not occur 4. Peak heat release rate shall not exceed 800 kW 5. Total smoke released shall not exceed 1000 m^2

The NFPA 286 test apparatus consists of a room measuring 3.66 × 2.44 × 2.44 m high, with a single ventilation opening (doorway) measuring approximately 0.76 m × 2.03 m high in the front wall. Typically, only the interior surfaces of the sidewalls and the back wall are covered with the test material. Alternatively, only the ceiling is covered for materials that are used on the ceiling alone, and the sidewalls, back wall, and ceiling are lined for materials that are used on walls and ceiling. A schematic of the NFPA 286 test apparatus is shown in Figure 10.8.

The test material is exposed to the flame of a 0.3 × 0.3 × 0.15 m propane diffusion "sand box" burner, located with the top surface 0.3 m above the floor in one of the rear corners of the room opposite the doorway. Propane is supplied at a specified rate so that a net heat release rate of 40 kW is achieved for the first 5 min of the test, followed by 160 kW for the remaining 10 min (15 min test duration unless terminated when flashover occurs). Thermocouples are installed at five locations under the ceiling (center and four quadrants) and below the top of the doorway.

A heat flux gauge is located at the center of the floor, and paper target flashover indicators (crumpled newsprint) are placed along the floor centerline at 1.22 m from the front and back walls. The products of combustion generated in the fire are collected in a hood and extracted through a duct, which is instrumented to measure heat release rate based on oxygen consumption calorimetry, and smoke production rate using a white light or laser photometer.

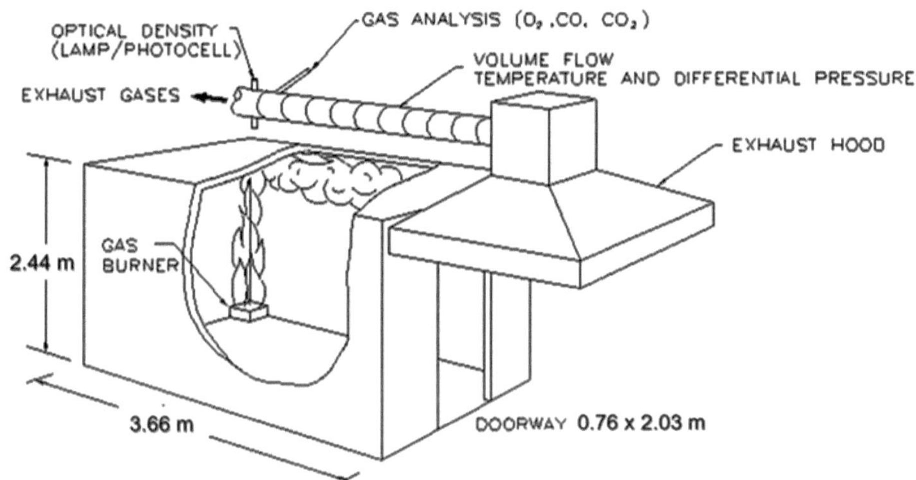

Figure 10.8 The NFPA 286 room/corner test apparatus

A material fails the test if (1) flames spread to the ceiling during the 40 kW exposure, (2) flames spread to the extremities of the walls, (3) flashover is observed at any time during the test, (4) peak heat release rate exceeds 800 kW, or (5) the total amount of smoke produced at the end of the 15 min test exceeds 1000 m^2. Flashover is considered to have occurred if any two of the following conditions have been attained: (1) heat release rate exceeds 1 MW, (2) heat flux to the floor exceeds 20 kW/m^2, (3) upper layer temperature exceeds 600 °C, (4) flames exit the doorway, or (5) ignition of a paper target on the floor.

NFPA 265: *Standard methods of fire tests for evaluating room fire growth contribution of textile wall coverings on full height panels and walls* is similar to but predates NFPA 286 and was developed to address the recognized hazard of using carpet-like textile materials for wall coverings. Initially the standard provided two options; a screening protocol with partially lined walls (Method A) and a room/corner test with three fully-lined walls as in NFPA 286 (Method B). In 2007 Method A was moved to a non-mandatory Annex. In 2011 the scope of the standard was revised to include expanded vinyl wallcoverings. The test method described in the current version of the standard (2015) is nearly identical to that in NFPA 286. The principal differences are as follows: (1) the heat release rate of the burner is 40 kW for 5 min followed by 150 kW (compared to 160 kW in NFPA 286) for 10 min, (2) the burner is at 51 mm from the back wall and one of the sidewalls, (3) it includes more detailed guidance for specimen mounting and installation, and (4) it specifies slightly different acceptance criteria.

ANSI/DASMA 107: *Room fire test standard for garage doors using foam plastic insulation* is designed to evaluate the fire growth characteristics of foam plastic-insulated garage doors. An insulated garage door 2.4 m wide by 2.13 m high is mounted to the back wall of the NFPA 265 test room. The NFPA 265 burner is located in the left rear corner 51 mm from the left sidewall and the test specimen in the back of the room. Acceptance criteria are as follows: (1) maximum net peak heat release rate < 250 kW, (2) flames shall not propagate over the full width of the specimen, and (3) total smoke production < 60 m^2 at 5 min and < 150 m^2 at 7.5 min.

UL 1715: *Fire test of interior finish material* is a room/corner test conducted in a room identical to that used in NFPA 286, NFPA 265, and ANSI/DASMA 107. A 13.6 kg wood crib is used as the ignition source. Specimens of the test material are attached to cover the entire back wall and a 2.44 m section of a sidewall. The following is reported: (1) time to ignition of the test material, (2) the extension of the flame front versus time, (3) flashover (described as flameover) time, and (4) smoke development as a function of time. Heat release rate measurements are optional.

NFPA 276: Standard Method of Fire Tests for Determining the Heat Release Rate of Roofing Assemblies with Combustible Above-Deck Roofing Components

NFPA 276 describes a method for determining the heat release rate from below the deck of a roofing assembly that has combustible above-deck roofing components, when the assembly is exposed to a fire from below the roof deck. It is essentially identical to the test method to determine whether a roof system with foam plastic insulation meets the combustibility requirements in FM 4450: *Approval standard for class 1 insulated steel deck roofs*. The roof assembly is usually made up of a steel deck, with the insulation material fastened to the deck followed by weatherproof covering installed on top of the insulation. The final covering can be fastened or ballasted in position. The NFPA 276 test specifications are given in Table 10.18.

Table 10.18 NFPA 276 Test Specifications for Heat Release Rate of Insulated Steel Roof Decks

Specimens	One roof deck specimen, 1.41 × 1.55 m (1.22 × 1.22 m exposed), one blank test specimen, and one calibration specimen (same size)
Specimen position	Horizontal, facing down
Ignition source	Three main heptane burners and three auxiliary propane burners
Test duration	30 min
Conclusions	Maximum average specimen heat release rate in kW/m^2 for any 3, 5, and 10 min period, and average heat release rate for the entire test

The test furnace is shown in Figure 10.9. The furnace is equipped with two sets of three burners. The main burners are spray nozzles supplied with heptane, and the auxiliary burners are premixed propane burners. The heat release rate of the specimen is measured based on the substitution method, which requires two runs. In the first run the specimen is exposed for 30 min to a pre-calibrated heptane fire. A blank refractory concrete specimen is used in the second run. The same heptane flow rate is used as in the first run, and the propane flow rate to the auxiliary burners is continuously adjusted during the 30 min test duration so that the temperature in the furnace exhaust duct matches that measured in the first run. The heat release rate of the specimen in the first run is then determined (as a function of time) from the product of the propane flow rate in the second run and the net heat of combustion of the propane. Periodic calibrations are performed with a calibration specimen that consists of wood fiber insulation placed on a steel deck with a steel cover. For foam plastic-insulated roof assemblies that pass this test, the building codes do not require a thermal barrier underneath the insulation.

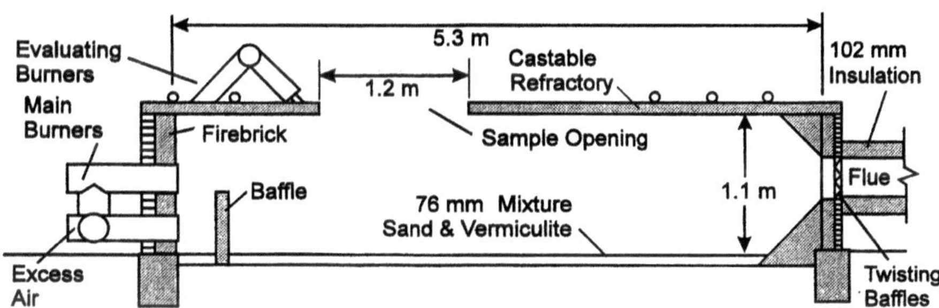

Figure 10.9 NFPA 276 furnace (also referred to as FM Construction Materials Calorimeter)

FM 4880: Approval Standard for Class 1 Fire Rating of Building Panels or Interior Finish Materials (and Related Test Standards)

FM Approvals uses industrial-scale open corner tests to evaluate insulated wall and roof/ceiling assemblies, plastic interior finish materials, plastic exterior building panels, wall/ceiling coating systems, and interior or exterior finish systems. The test requirements are described in FM 4880, and involve an open corner test that is 7.6 or 15.2 m (25 or 50 ft) tall, depending on the height of the installation. The ignition source in both tests consists of 340 kg of wood pallets. A product fails the test if flames propagate to the outer edges of the test structure. A schematic of the 7.6 m corner test setup is shown in Figure 10.10. The test specifications are summarized in Table 10.19.

Table 10.19 FM 4880 Test Specifications for 7.6 m Building Corner Test Class 1 Approval

Specimens	One specimen
Specimen position	Vertical for wall materials, horizontal for ceiling/roof assemblies
Ignition source	340 kg wood crib
Test duration	End of burn, or when fire reaches structure limits
Conclusions	Must not support self-propagating fire reaching any structure limits

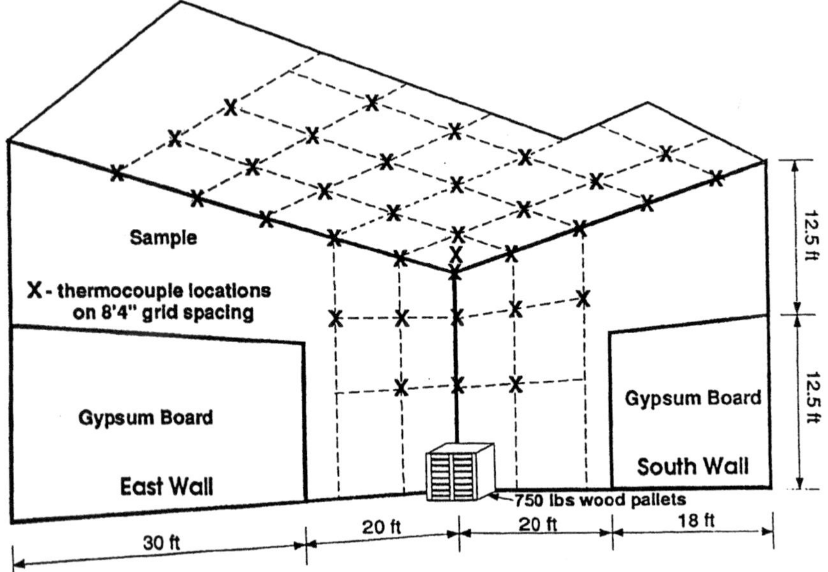

Figure 10.10 7.6 m (25 ft) FM 4880 corner test structure

FM Global recently developed an intermediate-scale fire test as a screening tool for the 7.6 and 15.2 m corner tests. The scale of the test is large enough for the tested materials to reveal their behavior in a full-scale fire, but still small enough to provide substantial cost savings compared to corner tests. The test consists of parallel panels of the material assemblies being evaluated, which are each 1.07 m wide and 4.9 m high, and are separated by 0.53 m. A 360 kW propane sand burner is used as the ignition source. The parallel panel test is conducted under a 5 MW calorimeter hood to measure fire heat release rate. Maximum heat release rate in the parallel panel test correlates with fire propagation behavior in the corner tests as follows:

- Fire will not propagate to the end of the test array in the 7.6 m corner test with combustible wall panels and a non-combustible ceiling if the heat release rate in the parallel panel test is less than 1100 kW;
- Fire will not reach the top of the test array in the 15.2 m corner test if the heat release rate in the parallel panel test is less than 830 kW; and

- Fire propagation will not reach the ends of the horizontal ceiling in the 7.6 m corner test with both combustible wall and ceiling panels if the heat release rate in the parallel panel test is less than 830 kW.

FM Approval Standard 4880 has been revised to allow for the use of the parallel panel test to qualify materials, systems, and assemblies with the appropriate pass/fail criteria as a cost-effective alternative to the 7.6 and 15.2 m corner tests. In addition to the parallel panel or corner tests, for some types of insulation (e.g., thermoset foam cores) a small-scale flammability characterization is required. All assemblies also need to pass the ISO 9705 or similar room/corner test (see Section 9.2 for a description).

UL 1040: *Fire test of insulated wall construction* describes a test procedure and requirements applicable to insulated wall constructions by use of a standardized large-scale corner test. The setup consists of two 6.10 × 9.14 m tall walls at a 90° angle forming an open corner, covered with a 6.10 × 6.10 m uninsulated flat metal roof. The walls are exposed to a 347 kg wood crib fire for 30 min. Acceptance criteria are as follows: (1) burning shall not extend beyond 5.49 m from the corner, and (2) damage shall diminish from the area directly exposed to the fire.

FM 4996: American National Standard for Classification of Pallets and Other Material Handling Products as Equivalent to Wood Pallets

Approval standard FM 4996 describes the test requirements for the fire hazard classification of plastic pallets as equivalent to wood pallets. To meet the requirements of the standard, eight 1.2 m stacks of the plastic pallets in a 2 × 2 × 2 array shall be tested in a 20 MW calorimeter provided with a water application apparatus. To demonstrate equivalency to wood pallets, (1) the fire must be controlled when a water application density of 5.9 mm/min is applied, and (2) the controlled fire will not continue to grow after 10 min of testing. The standard also requires that flammability characterization tests (ASTM D1929 and ASTM E2058) be conducted as an aid in quality control of the production process, and to determine the anticipated performance of future changes in the pallet resin formulation.

UL 1256: Fire Test of Roof Deck Construction

UL 1256 is referenced by the codes as an alternate means to FM 4450 for approving the use of roof assemblies without a thermal barrier. However, the two standards rely on different tests. There are currently two methods listed in the UL standard. The first method is now listed as Part II and is based on the Steiner tunnel test described in UL 723, which is essentially identical to ASTM E 84 (see Figure 10.7). Part II is not discussed in this section. The other method, now listed as Part I, utilizes a large-scale tunnel-type structure. This larger structure was originally used by UL as a tool to investigate safer roofing assemblies after a very large automotive plant fire in Livonia, Michigan in 1953, which destroyed the 14 hectare plant. A third part of UL 1256 based on an intermediate-scale room fire test for roof decks was withdrawn in March 2000. The test specifications for Part I of the standard are given in Table 10.20.

Table 10.20 UL 1256 Test Specifications for Fire Tests of Roof Deck Constructions

Specimens	One specimen, 6.1 × 30.5 m roof deck assembly
Specimen position	Horizontal
Ignition sources	Two heptane spray burners, maximum flow rate 0.17 l/s
Test duration	30 min
Conclusions	Limited flame propagation, limited distance of flaming droplets, limited thermal degradation of roof components

The roof assembly is fastened to the large test structure measuring 30.5 m long, 6.1 m wide and 3.1 m high. The two heptane burners are at the closed end, and the opposite end is open. Defined steel roof supports are used throughout the length of the structure. A series of observation windows are located down the side of the structure and thermocouples are used in various locations. Air and heptane sprays are introduced. The heptane flow is increased from 0.06 l/s at time zero to 0.17 l/s at 17 min into the test, and that flow is then maintained until the end of the 30 min test. The heptane flow was chosen so that flames from it alone do not go past about 6 m into the tunnel from where the burners are located.

Conditions of acceptance for UL 1256, Part I are as follows:

- The maximum sustained flame front within the structure due to underdeck propagation must not exceed 18.3 m from the fire end of the structure during the 30 min test.
- The flaming of any molten material falling from the roof deck must not exceed 18.3 m from the fire end during the 30 min test.
- Intermittent underdeck flaming must not exceed 21.9 m from the fire end during the 30 min test.
- Examination of the roof deck after the test must show that combustion damage has decreased at increasing distances from the fire end.
- Thermal degradation of the roof components must not have extended to the extremity of the structure.

Part I of this standard was added to UL 1256 in the late 1990s as a result of additional testing of some foam plastic roof assemblies. Historically, for foamed plastics, it was thought that only some thermoset plastic insulation materials, e.g., isocyanurate board, could pass this type of test. However, it was found that under the test conditions outlined above, some thermoplastic insulation materials could pass if the roof deck assembly was able to vent combustion products through the steel decking towards the fire end of the structure because of heat warping of the steel deck sections. There are no acceptance criteria for above-deck flaming or smoke.

10.2.1.7 Summary of Fire Tests for Furnishings and Contents

A summary of the fire tests for furnishings and contents referenced in the IBC, IFC, and/or LSC is given in Section 13.3.

10.2.1.8 Summary of Fire Resistance Tests

ASTM E119: Standard Test Methods for Fire Tests of Building Construction and Materials

ASTM E119 provides procedures for measuring the fire resistance of wall and floor/ceiling assemblies, roof structures, beams, and columns. Wall assemblies, floor/ceiling assemblies, and roof structures are mounted in a vertical or horizontal frame. The frame is placed against an open wall furnace or on top of an open ceiling furnace and is exposed to the standard fire. Columns and beams are (partially) immersed in the furnace. The ASTM E119 test specifications are given in Table 10.21. The UL equivalent to ASTM E119 is UL 263: *Fire tests of building construction and materials*. The most recent version of the NFPA equivalent of ASTM E119 (NFPA 251: *Standard methods of tests of fire resistance of building construction and materials*) was published in 2006. To minimize duplication between ASTM, NFPA, and UL standards, NFPA 251 has since been withdrawn.

Table 10.21 ASTM E119 Test Specifications for Fire Resistance of Building Elements and Assemblies

Specimens	One specimen: Walls: width and height ≥ 2.7 m, area ≥ 9 m^2 Columns: height ≥ 2.7 m Floors and roofs: width and length ≥ 3.7 m and area ≥ 16 m^2 Beams: clear span ≥ 3.7 m
Specimen position	Vertical for walls and columns; horizontal for floors, roofs, and beams
Ignition sources	Premixed gas burners (typically), heat output adjusted to follow the specified furnace temperature-time curve
Test duration	Until failure or the desired fire resistance rating has been achieved
Conclusions	Fire resistance rating, i. e., time rounded to the nearest min when any of the applicable end-point criteria (heat transmission, integrity, or load-bearing capacity) is exceeded

The standard fire is quantified by a specified furnace temperature-time curve (see Figure 10.11). The furnace temperature is measured with no fewer than nine thermocouples (not fewer than eight for columns), distributed to show the temperature near the specimen. The thermocouples are located in an Inconel pipe. This is an important difference between ASTM E119 and the ISO 834-based fire resistance test standards, which are used throughout most of the rest of the world and rely on plate thermometers to measure the furnace temperature. As a result, because the ASTM E119 and ISO 834 temperature–time curves are nearly identical during the first 10 min of the test, the exposure conditions in an ASTM E119 are slightly more severe than in an ISO 834 test due to the slower response of the ASTM E119 furnace probes. Later in the test the ISO 834 curve is slightly higher than the ASTM E119 curve, which compensates for the initial discrepancy. For test durations of 30 min or greater, the differences between the thermal exposure conditions in ASTM E119 versus ISO 834 have a very small to negligible effect on the performance of the test specimen.

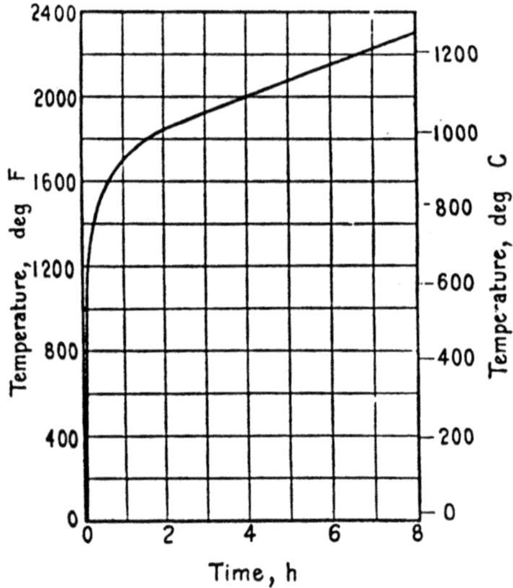

Figure 10.11 ASTM E119 furnace temperature–time curve

Three end-point criteria are used to determine the fire resistance of a test specimen:

- *Heat transmission:* Transmission of heat through a wall, floor, or roof specimen is quantified based on the unexposed-side temperature measurements. ASTM E119 specifies that at least nine thermocouples must be attached to and distributed over the unexposed surface of the specimen. The end point for heat transmission is reached when the average of all thermocouple readings reaches 140 °C over the initial average temperature or when any of the individual thermocouple readings exceeds 180 °C over the initial temperature.
- *Integrity:* The integrity end point is reached when gases or flames emerging through cracks in the specimen are hot enough to ignite a cotton pad on the unexposed side.

- *Load-bearing capacity:* Fire resistance furnace test standards require that load-bearing assemblies and structural members be loaded so that the maximum stresses in the specimen are equal to the allowable design stresses. Structural collapse is an obvious indication that the specimen is no longer able to support the load. Fire resistance tests are usually terminated long before collapse occurs, as ASTM E119 specifies a maximum deflection to define structural failure. It should be noted that steel-frame assemblies and steel elements can be tested under restrained or unrestrained conditions. For the former, the test specimen is mounted in a stiff frame that resists the thermal expansion of the specimen or its steel members, which may be more representative of real conditions but makes the test significantly more stringent.

The end-point criteria for heat transmission and integrity do not apply to columns and beams. The criterion for load-bearing capacity does not apply to non-load-bearing walls (referred to as partitions). The primary result of an ASTM E119 test is the fire resistance rating, i. e., time rounded to the nearest min when any of the applicable end-point criteria (heat transmission, integrity, or load-bearing capacity) is exceeded. To obtain the corresponding fire resistance classification, the rating is rounded down to 20 min, ½ hour, 1 hour, 1½ hours, 2 hours, 3 hours, 4 hours, etc.

ASTM E119 prescribes a supplemental hose stream test procedure to evaluate the ability of the construction to resist disintegration under adverse conditions. The hose stream test is either performed after termination of the fire resistance test, or on a duplicate specimen that has been exposed to the standard fire for half the duration of the desired fire resistance classification. A hose stream test is not required for columns, floor/ceiling assemblies, roof structures, and wall assemblies with a fire resistance rating of less than 1 hour. The hose stream requirement is unique to North America and is described in a separate standard, ASTM E2226.

Several furnace standards have been developed that provide specific details for fire resistance testing of doors, windows, and other types of building elements that are not covered by ASTM E119. A list of standards referenced in the IBC, IFC, and/or LSC is provided below.

- ASTM E814: *Standard test method for fire tests of penetration firestop systems*
  - UL equivalent: UL 1479: *Fire tests of penetration firestops*
- ASTM E1529: *Standard test methods for determining effects of large hydrocarbon pool fires on structural members and assemblies*
- ASTM E1966: *Standard test method for fire-resistive joint systems*
  - UL equivalent: UL 2079: *Tests for fire resistance of building joint systems*
- ASTM E2837: *Standard test method for determining the fire resistance of continuity head-of-wall joint systems installed between rated wall assemblies and nonrated horizontal assemblies.*

- NFPA 252: *Standard methods of fire tests of door assemblies*
  - Similar: UL 10A: *Tin-clad fire doors*
  - Similar: UL 10B: *Fire tests of door assemblies*
  - Equivalent: UL 10C: *Positive pressure fire tests of door assemblies*
- NFPA 257: *Standard for fire test for window and glass block assemblies*
  - UL 9: *Fire tests of window assemblies*
- NFPA 275: *Standard method of fire tests for the evaluation of thermal barriers*
- NFPA 288: *Standard methods of fire tests of horizontal fire door assemblies installed in horizontal fire-resistance-rated floor systems*
- NFPA 415: *Standard on airport terminal buildings, fueling ramp drainage, and loading walkways*
- UL 2196: *Tests for fire resistive cables*

ASTM E1529 is similar to ASTM E119, but prescribes a more severe thermal exposure characterized by a rapid-rise furnace temperature-time curve representative of a large hydrocarbon pool fire. ASTM E814 and ASTM E1966 provide the option to use the ASTM E119 or the ASTM E1529 temperature–time curve.

ASTM E119 does not specify furnace pressure, but the commentary section in the standard states that it is generally negative (relative to the pressure in the laboratory). This implies that for a wall assembly that is not entirely sealed, air will be drawn into the furnace through leakage paths in the specimen. Wall assemblies are usually sufficiently tight so that the cooling effect due to air leaking into the furnace is negligible. However, this may not be the case for door and window assemblies. Consequently, both NFPA 252 and NFPA 257 include an option to perform the test under positive pressure. NFPA 252 specifies that the neutral plane shall be located approximately 1 m above the bottom of the door. For a normal size door (~2 m in height), the top half will be exposed to positive furnace pressure. NFPA 257 also has a positive pressure option, and requires that at least two-thirds of the window assembly be exposed to positive furnace pressure. Furthermore, NFPA 257 includes a procedure to measure the transmission of radiant heat through and re-radiation by a fire window or glass block assembly. The heat flux measurement is optional and is performed when the code requires it.

The IBC requires that foam plastic insulation and metal composite materials (MCMs) be separated from the building interior by an approved thermal barrier. The latter is a panel material that is tested in accordance with and meets the acceptance criteria of the Temperature Transmission Fire Test (Part I) and the Integrity Fire Test (Part II) of NFPA 275. Part I is a reduced-scale ASTM E119 test of a specimen consisting of a wood frame (minimum dimensions 0.91 × 0.91 m) protected with the barrier material, i.e., the barrier material is tested without foam plastic insulation or MCM backing. The material qualifies as a thermal barrier if the heat transmission criterion based on thermocouples attached to the unexposed surface of the barrier material is not exceeded within 15 min of furnace exposure. Part II evaluates the ability of the thermal barrier to provide protection from ignition of the foam plastic insulation or MCM based on a standard corner test, i. e., NFPA 286, FM 4880, UL 1040, or UL 1715 (see Section 10.2.1.6 for a summary of these test methods).

ASTM 2387: Standard Test Method for Determining Fire Resistance of Perimeter Fire Barriers Using Intermediate-Scale, Multi-Story Test Apparatus

ASTM E2387 was developed to evaluate the fire resistance of perimeter fire barriers, i. e., the joint between an exterior wall assembly and the floor system. This test standard does not involve thermal exposure in a furnace but uses a slightly modified version of the two-story test structure described in NFPA 285: *Standard method of test for the evaluation of flammability characteristics of exterior non-load-bearing wall assemblies containing combustible components using the intermediate-scale, multistory test apparatus* (see Section 10.2.1.9 for a summary description). ASTM E2387 evaluates the ability of the fire barrier at the perimeter of the second-story floor slab to maintain a seal and prevent fire spread to the second story while being exposed for 30 min to the standard ASTM E119 fire in the first-story compartment as well as the flame and plume emitted from a window burner below.

10.2.1.9 Summary of Exterior Fire Exposure Tests

ASTM E108: Standard Test Methods for Fire Tests of Roof Coverings (and Related Test Standards)

ASTM E108 covers the measurement of the relative fire characteristics of roof coverings under simulated fire originating outside the building. The UL equivalent is UL 790: *Standard test methods for fire tests of roof coverings*. The standard is applicable to roof coverings intended for installation on either combustible or non-combustible decks, when applied as intended for use. The following fire test methods are included:

- Spread of flame test
- Intermittent flame exposure test
- Burning brand test
- Flying brand test.

When a roof covering is restricted for use on non-combustible decks, only the spread of flame test is required. When a roof covering is not restricted for use on non-combustible decks, the spread of flame, intermittent flame, and burning brand tests are required. The flying brand tests are required when there is a possibility that the roof covering will break into pieces of flying, flaming brands or particles that continue to glow after reaching the floor of the test facility. Three classes of fire test exposure are described:

- Class A tests are applicable to roof coverings that are effective against severe test exposure, afford a high degree of fire protection to the roof deck, do not slip from position, and do not present a flying brand hazard.
- Class B tests are applicable to roof coverings that are effective against moderate test exposure, afford a moderate degree of fire protection to the roof deck, do not slip from position, and do not present a flying brand hazard.
- Class C tests are applicable to roof coverings that are effective against light test exposure, afford a light degree of fire protection to the roof deck, do not slip from position, and do not present a flying brand hazard.

The procedures measure the surface spread of flame and the ability of the roof covering material or system to resist fire penetration from the exterior to the underside of a roof deck under the conditions of exposure. The tests are conducted with a gas burner flame or with burning brands (cribs) of Douglas fir. Figure 10.12 shows a cross-section of the test structure set up with a gas burner. See Table 10.22 for burning brand test specifications of a Class A roof.

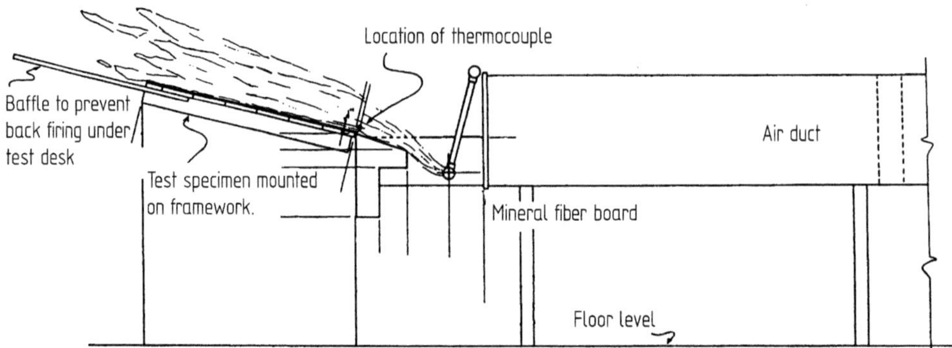

Figure 10.12 Cross-section of ASTM E108 roof deck test structure

Table 10.22 ASTM E108 Burning Brand Test Specifications for Class A Roof Coverings

Specimens	Minimum of four specimens, each consisting of the roof covering mounted on a 1.0 × 1.3 m deck of combustible or non-combustible construction (depends on end-use application of the roof covering)
Specimen position	Inclined at an angle depending on end-use application of the roof covering
Ignition sources	Burning brand, which is a wood crib consisting of 36 Douglas fir strips, 305 × 19 × 19 mm in size, placed in three layers

Test duration	Until the brand is consumed and all evidence of flame, glow and smoke has disappeared from the exposed surface and the underside of the deck
Conclusions	At no time during or after the burning brand tests shall: 1. Flaming or glowing brands or particles of the roof covering or deck that fall off the test deck continue to glow after reaching the floor 2. The roof deck be exposed outside the area directly underneath the burning brand 3. Sustained flaming occur on the underside of the deck

ASTM E108 specimens first need to be subjected to rain tests whenever the fire-retardant characteristics of the roof covering materials or construction has the potential to be adversely affected by water. The rain tests are described in ASTM D2898: *Standard test methods for accelerated weathering of fire-retardant-treated wood for fire testing*. In addition, weathering tests are required for wood shakes and shingles, or when the fire-retardant characteristics of the roof covering materials or construction has the potential to be adversely affected by weathering outdoors.

NFPA 268: Standard Test Method for Determining Ignitability of Exterior Wall Assemblies Using a Radiant Heat Energy Source

Building codes address the problem of fire spread from one building to an adjacent building due to radiant ignition of combustible exterior façades by specifying a minimum distance to the property line. This distance is based on the assumption that the exterior cladding is wood. The commonly accepted threshold for piloted ignition of wood is 12.5 kW/m^2. However, many different types of exterior wall claddings are now available in the marketplace. To ensure that the building code provisions are adequate for these materials, it is necessary to verify that their ignition threshold is equal to or higher than that for wood. NFPA 268 describes a test method that can be used to perform this verification. The NFPA 268 test specifications are listed in Table 10.23.

Table 10.23 NFPA 268 Test Specifications for Ignitability of Exterior Wall Assemblies

Specimens	One specimen, 1.22 × 2.44 m × use thickness of the exterior wall assembly
Specimen position	Vertical
Ignition source	Propane-fired radiant panel, 0.91 × 0.91 m, imposing a heat flux of 12.5 kW/m^2 over the central 0.3 × 0.3 m region of the specimen
Test duration	20 min
Conclusions	A wall assembly passes if ignition does not occur during the 20 min test

The apparatus consists of a vertical 0.91 × 0.91 m propane-fired radiant panel that exposes the 1.22 × 2.44 m specimen to a radiant heat flux that is equal to approximately 12.5 kW/m^2 over a central 0.3 × 0.3 m region. A spark igniter is mounted on the vertical centerline of the test specimen at a point 0.46 m above its horizontal centerline, and at 16 mm from its surface.

A conditioned specimen is mounted securely in a trolley assembly and positioned directly in front of the radiant panel in a vertical position and at the predetermined distance to expose the central area of the specimen to the specified heat flux. The specimen and panel are oriented in a parallel plane configuration. The radiation emitted by the panel is measured with a heat flux meter before the test begins. Also, before the beginning of the test a radiation shield is placed between the specimen and radiant panel. At the start of the test, the spark igniter is energized, and the radiation shield is removed. The time to sustained ignition is recorded when it occurs. If no ignition occurs, the test is terminated at 20 min by placing the radiation shield in back in position. An exterior wall system passes the test if no sustained ignition is observed during the 20 min exposure period.

NFPA 285: Standard Fire Test Method for the Evaluation of Fire Propagation Characteristics of Exterior Nonload-Bearing Wall Assemblies Containing Combustible Components (and Related Test Standards)

Exterior Insulation Finish Systems (EIFS) are very common in the construction of exterior walls for high-rise buildings. Because the systems typically consist of plastic foam insulation and other combustible components, the potential exists for rapid upward flame spread over façades to stories above the fire room. NFPA 285 describes a test method intended to evaluate the capability of such wall assemblies to resist vertical and, to some extent, lateral flame propagation over the exterior and interior faces and within the core of the assembly. The NFPA 285 test specifications are listed in Table 10.24.

Table 10.24 NFPA 285 Test Specifications for Evaluation of Combustible Exterior Wall Assemblies

Specimens	One wall assembly specimen, minimum 4.1 m wide × 5.3 m high
Specimen position	Vertical
Ignition sources	Main gas burner used to create the exposing fire in the bottom room, additional burner near the top of the first floor window opening
Test duration	30 min
Conclusions	Extent of vertical and lateral flame propagation over the exterior and interior faces and within the core of the wall assembly

The NFPA 285 test structure consists of two stories. The interior dimensions of the first and second floor rooms are identical and equal to 3.05 × 3.05 × 2.13 m high.

Both rooms are constructed of concrete block walls and concrete slabs, and are fully enclosed except for the front side. The interior surfaces of the bottom room are protected with gypsum board and ceramic fiber insulation. The test assembly is mounted against the front of the test structure and covers both stories. There is a window opening of 1.96 × 0.76 m high at 0.76 m from the floor of the first floor room. The test assembly is instrumented with an extensive number of thermocouples to monitor vertical and lateral flame propagation over the exterior and interior faces, and within the core of the wall assembly.

Two gas burners are used to create the exposing fire. The main burner is located inside the first-floor burn room and is supplied with gas so that its heat output increases according to a prescribed regime from approximately 700 kW at the start of the test to approximately 900 kW at the end of the 30 min test. The intent is to develop a temperature-time curve in the bottom compartment that is comparable to the standard ASTM E119 curve. A second burner is located inside the window opening so that flames hit the window head, which is the most vulnerable part of the exterior wall assembly for flame penetration into the core. The heat output of the window burner increases from 0 kW at the start of the test to approximately 400 kW at the end of the 30 min test. Acceptance criteria for NFPA 285 include:

- Flame propagation must not occur either vertically or laterally beyond the area of flame impingement on the exterior wall.
- Flame propagation must not occur either vertically or laterally through the core components.
- Flame propagation must not occur laterally beyond the limits of the burn room.
- Temperatures 25 mm from the interior surface of the test wall assembly within the second floor area must not exceed 278 °C above the initial ambient temperature.
- Flames shall not occur in the second floor room.

Visual observations and specific thermocouple temperatures are used to determine when any of the acceptance criteria have been exceeded.

UL 1703: Flat-Plate Photovoltaic Modules and Panels

UL 1703 describes a series of performance tests for flat-plate photovoltaic modules and panels such as a set of electrical performance tests, an impact test, a water spray test, an accelerated aging test, a corrosive atmosphere test, etc. The standard includes two sets of fire test requirements:

- To label a module or panel as a specific type, independent of the mounting system and roof covering, it must be subjected to the flame spread test and the burning brand test in UL 790, which is equivalent to ASTM E108 described earlier in this subsection. The standard defines 15 types depending on frame material, and substrate and encapsulant materials and thickness. For each type the standard specifies the UL 790 pass/fail criteria and the wood crib to be used in the burning brand test.

- The standard also describes in detail how the UL 790 flame spread and burning brand tests shall be performed on a module or panel with its mounting system in combination with a specific roof covering, and specifies the corresponding acceptance criteria to obtain a "System Fire Class Rating" for the system.

10.2.1.10 Future Developments

The most significant change in the 2021 IBC is the introduction of three new types of mass timber construction, HT-A, HT-B, and HT-C. Buildings of these types are constructed with large load-bearing floor, wall and ceiling panels.

All mass timber buildings require non-combustible protection for the exterior walls. Type HT-A buildings can be up to 18 stories for low hazard occupancies. For Type FT-B and Type HT-C, the limits are 12 and 9 stories, respectively. Type IV-A is fully protected by non-combustible material inside (typically fire-rated gypsum board). Type IV-B allows limited exposed mass timber interior members. Type IV-C does not require protection of the interior members.

Cross-laminated timber (CLT) is the most common type of wood panels in mass timber buildings. These panels consist of several layers of wood boards stacked in alternating directions, glued together with structural adhesives. Of particular concern is delamination of the surface layer of a CLT panel during the cooling phase of a fire, which exposes fresh wood and may result in fire re-growth and a second flashover. A full-scale room fire test was developed to identify adhesives that are likely to fail during the cooling phase [29]. The adhesive qualification test is described in the ANSI/APA PRG 320 CLT performance standard [30], which is referred to in the IBC.

References for Section 10.2.1

[1] Diamantes, D., *Fire Prevention: Inspection and Code Enforcement* (fourth edition), CreateSpace Independent Publishing Platform, 2015.

[2] *https://www.iccsafe.org/*

[3] *https://www.nfpa.org/*

[4] *https://codes.iccsafe.org/codes/i-codes*

[5] *https://www.nfpa.org/Codes-and-Standards/All-Codes-and-Standards/Free-access*

[6] *https://www1.nyc.gov/site/buildings/codes/2014-construction-codes.page#bldgs*

[7] *http://www.icc-es.org/*

[8] *https://www.a2la.org/*

[9] *https://www.iasonline.org/*

[10] ISO/IEC 17025:2017, General requirements for the competence of testing and calibration laboratories, International Organization for Standardization, Geneva, Switzerland.

[11] ISO/IEC 17020:2012, Conformity assessment - Requirements for the operation of various types of bodies performing inspection, International Organization for Standardization, Geneva, Switzerland.

[12] *http://www.archtest.com/*

[13] *https://www.fmapprovals.com/*

[14] *http://www.intertek.com/building/fire-testing/*

[15] *http://www.ngctestingservices.com/fire.html*

[16] *https://qai.org/*

[17] *https://www.swri.org/industries/fire*

[18] *https://www.ul.com/*

[19] *http://www.westernfire.com/*

[20] *https://www.astm.org/*

[21] *https://www.ansi.org/american-national-standards/info-for-standards-developers/accreditation*

[22] *https://www.astm.org/COMMITTEE/E05.htm*

[23] Hall, J., *A Century of Fire Standards, Committee E05 1904-2004*, ASTM International, West Conshohocken, PA, 2004.

[24] *https://www.astm.org/COMMITTEE/D20.htm*

[25] *https://www.nfpa.org/Codes-and-Standards/Standards-development-process/Technical-Committees/Committees-seeking-members*

[26] *http://www.nfpa.org/regs*

[27] *https://www.govinfo.gov/app/collection/cfr*

[28] *https://bhgs.dca.ca.gov/*

[29] Janssens, M. and Joyce, A., "Development of a Fire Performance Assessment Method for Qualifying Cross-Laminated Timber Adhesives", Interflam 2019, 15th International Fire Science & Engineering Conference, Royal Holloway College, July 1-3, 2019, pp. 61-72.

[30] "ANSI/PRG 320: Standard for Performance-Rated Cross-Laminated Timber", APA - The Engineered Wood Association, Tacoma, WA, 2018.

10.2.2 Canada

10.2.2.1 Statutory Regulations

The regulation of buildings in Canada is the responsibility of provincial and territorial governments. Each of the 10 Provinces and 3 Territories enacts its own building code. All of the provincial and territorial building codes, however, are based on a single model code, the National Building Code of Canada (NBCC) [1]. Although the NBCC is intended to establish a minimum standard of fire safety for the construction of new buildings or the renovation of existing buildings, several provinces make amendments to the NBCC that render their codes somewhat more demanding. The National Research Council of Canada (NRC) offers electronic access to the NBCC and all provincial codes at *https://cnrc.canada.ca/en/research-development/products-services/shop-nrc.*

The NBCC is normally published on a five-year cycle. The last edition was published in 2015 and the next will be published in 2021. The 2015 NBCC is an objective-based code. It establishes requirements to address the following four objectives that are described in Division A of the code: (1) safety, (2) health, (3) accessibility for persons with disabilities, and (4) fire and structural protection of buildings. Division A also defines the scope of the NBCC and presents the functions the building must perform to help satisfy the objectives. Each code requirement is linked to at least one of the four objectives. Division B of the code identifies a set of "Acceptable Solutions", which are based on the requirements in the 2005 IBCC, i.e., the last prescriptive version of the NBCC. Division C of the code contains administrative provisions related to the application of the code.

For housing and small buildings, the fire safety requirements for building materials, products, and assemblies are contained in several parts of the NBCC. Part 9 of the NBCC applies to houses and small buildings of three stories or less in building height, having a building area not exceeding 600 m^2, and used for major occupancies classified as: (1) residential occupancies, (2) business and personal services occupancies, (3) mercantile occupancies, or (4) medium- and low-hazard industrial occupancies. For all other buildings, the fire safety requirements are contained in Part 3 "Fire Protection, Occupant Safety and Accessibility". The explicit NBCC fire safety requirements depend upon the intended use and the size of a building. Separate sections address fire resistance ratings required for building assemblies, and the fire performance requirements of combustible building products such as interior finishes.

10.2.2.2 Testing of the Fire Performance of Building Materials and Components

The fire test methods cited in the NBCC were all developed by Underwriters Laboratories of Canada (ULC), and are listed in Volume 2 of the NBCC. They can be obtained from: Underwriters Laboratories of Canada, 7 Crouse Road, Scarborough, Ontario, Canada M1R 3A9.

The following standards relate to the fire resistance of building assemblies:

- CAN/ULC-S101: Fire test methods to evaluate the fire resistance of building construction and materials [2].
- CAN/ULC-S104: Fire test methods for door assemblies [3].
- CAN/ULC-S105: Specification for fire door frames meeting the performance required by CAN/ULC-S104 [4].
- CAN/ULC-S106: Fire test methods for window and glass block assemblies [5].
- CAN/ULC-S110: Fire test methods for air ducts [6].
- CAN/ULC-S111: Fire test methods for air filter units [7].

- CAN/ULC-S112: Fire test method for fire-damper assemblies [8].
- CAN/ULC-S112.1: Fire test method for leakage-rated dampers for use in smoke control systems [9].
- CAN/ULC-S112.2: Fire test method for ceiling firestop flap assemblies [10].
- CAN/ULC-S113: Specification for wood core doors meeting the performance required by CAN/ULC-S104 for twenty minute fire rated closure assemblies [11].
- CAN/ULC-S115: Fire test methods of firestop systems [12].
- CAN/ULC-S124: Fire test method for the evaluation of protective coverings for foamed plastic [13].
- CAN/ULC-S144: Fire resistance test method for grease duct assemblies [14].

Fire resistance test methods are not dealt with further in this chapter.

Whether a building is classified as combustible or non-combustible is determined by subjecting it to CAN/ULC-S114 [15], which differs somewhat from ASTM E136 and ISO 1182 (see Sections 10.2.1.4 and 9.2, respectively, for a description of these methods). Several methods for testing the fire performance of combustible materials are prescribed in the NBCC. The most important method used for testing interior finishes is CAN/ULC-S102 [16], which is very similar to ASTM E84 (see Section 10.2.1.6 for a description of the test method in ASTM E84). Floor coverings and materials that melt add drip in the CAN/ULC-S102 method, such as thermoplastic materials, are tested using CAN/ULC-S102.2 [17], and light diffusers are tested according to ULC-S102.3 [18]. Both methods are derived from CAN/ULC-S102. They have different mounting systems (floor or ceiling, respectively). An additional test for non-melting building materials which exhibit a “flash” behavior is ULC-S127 [19], which is referenced within CAN/ULC-S102 and CAN/ULC-S102.2 for materials that exhibit a rapid advance in the flame front followed by a rapid reduction in the rate of flame advancement (i. e., flash behavior). This is often observed with thermosetting foamed plastic materials. In certain cases, the NBCC requires fabrics and films to be flame-tested to ULC-S109 [20].

The test methods listed in the previous paragraph are discussed in some detail in Section 10.2.2.3. The remaining fire test standards referenced in the NBCC are listed below.

- CAN/ULC-S107: Fire test method for roof coverings [21]. This method is essentially identical to that described in ASTM E108 (see Section 10.2.1.9).
- CAN/ULC-S126: Test method to evaluate fire spread under roof-deck assemblies [22]. This method uses the tunnel test apparatus described in CAN/ULC-S102 [16] to assess the fire spread of roof-deck assemblies using metallic or non-metallic components exposed to fire in a building.

- CAN/ULC-S134: Fire test method for exterior wall assemblies [23].
- CAN/ULC-S135: Test method for the determination of combustibility parameters of building materials using an oxygen consumption calorimeter (cone calorimeter) [24].
- CAN/ULC-S138: Test method for fire growth of insulated building panels in a full-scale room configuration [25].
- CAN/ULC-S139: Fire test method for the evaluation of integrity of electrical power, data and optical fibre cables [26]. This method is identical to ANSI/UL 2196.
- CAN/ULC-S143: Fire test methods for non-metallic electrical and optical fibre cable raceway systems [27].

Conceptually, CAN/ULC-S134 [23] is similar to the method described in NFPA 285 (see Section 10.2.1.9 for a summary of NFPA 285). However, the NFPA 285 test wall is shorter (4 × 5.33 m versus 5 × 7 m), and the heat flux to the exterior wall from the flame and plume above the window is significantly higher but of shorter duration in the Canadian test (average 45 kW/m^2 over 15 min at 0.5 m above the window versus average 25 kW/m^2 over 30 min at 0.6 m).

CAN/ULC-S135 [24] was first approved in 1992 and is used to determine the degrees of combustibility of building materials. A new test method was deemed necessary to replace the existing test ULC/CAN-S114 [15], which determines whether a material is combustible or non-combustible. In CAN/ULC-S135, the heat release rate of materials exposed to specified conditions in a slightly modified version of the cone calorimeter is used to define degrees of combustibility of building materials. The cone calorimeter is described in ASTM E1354 (see Section 10.2.1.5). A material is permitted to be used in a non-combustible construction provided that, when tested in accordance with CAN/ULC-S135 at a heat flux of 50 kW/m^2, (1) its average total heat release is not more than 3 MJ/m^2, (2) its average total smoke extinction area is not more than 1.0 m^2, and (3) the test duration is extended beyond the time stipulated in the referenced standard until it is clear that there is no further release of heat or smoke. Further, if a material tested to CAN/ULC-S135 consists of a number of discrete layers and testing reveals that the surface layer (or layers) protect the underlying layers such that complete combustion of the underlying layers does not occur, the test is repeated by removing the outer layers sequentially until all layers have been exposed during testing, or until complete combustion has occurred. The acceptance criteria for a material tested in this way is based on the cumulative emissions from all layers.

CAN/ULC-S138 [25] describes a test method used to determine the contribution to fire growth provided by a pre-fabricated non-loadbearing insulated building panel installed in a sprinklered or unsprinklered room configuration. The test method determines the time to flashover under specified test conditions, and measures ignitability, smoke obscuration, rate of fire growth, and rate of heat release of the test panels.

10.2.2.3 Summary of Selected Material Flammability Tests Referenced in the NBCC

CAN/ULC-S114: Standard Method of Test for Determination of Non-Combustibility in Building Materials

The non-combustibility of building materials is determined employing CAN/ULC-S114 [15]. The apparatus is shown in Figure 10.13, and the test specifications are summarized in Table 10.25.

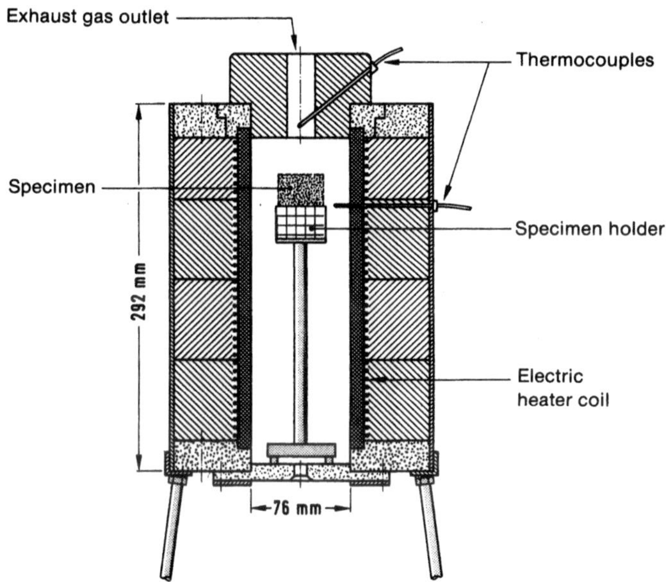

Figure 10.13 ULC non-combustibility furnace

Table 10.25 Test Specifications for Determining Non-Combustibility of Building Materials to CAN/ULC-S114

Specimens	Three specimens, 38 mm × 38 mm × 51 mm
Specimen position	Vertical
Ignition source	Electric furnace heated to 750 °C
Test duration	15 min (maximum)
Conclusions	The material passes if the following criteria are met: 1. The three specimens do not contribute on average more than 36 °C to the temperature rise in the exhaust gas outlet 2. No specimen burns 30 s after the start of the test 3. No specimen exceeds a maximum weight loss of 20%

CAN/ULC-S102: Standard Method of Test for Surface Burning Characteristics of Building Materials and Assemblies

The surface burning characteristics of building materials are tested by CAN/ULC-S102 [16], which is similar to ASTM E84. The apparatus and test specifications correspond to those given in Figure 10.7 and Table 10.16, respectively, in Section 10.2.1.6, except that the windows are flush-mounted to the outside face of the tunnel furnace, creating cavities that provide turbulence and mixing of the air and combustion gases. As a consequence, the ULC versions of the tunnel test do not require the turbulence bricks. The experimental results, however, are evaluated differently. The results from at least three runs are used to determine the Flame Spread Classification as FSC_1 or FSC_2. The distance traveled by the flame front is plotted graphically as a function of time and the total area A_T under the flame spread time-distance curve is calculated. The Flame Spread Classification FSC_1 is calculated in terms of A_T as follows:

- if $A_T \leq 29.7$ (m·min), then $FSC_1 = 1.85 \cdot A_T$
- if $A_T > 29.7$ (m·min), then $FSC_1 = 1640/(59.4 - A_T)$

During the testing of some materials, particularly those of low thermal inertia, the flame front may advance rapidly during the initial stages of the test, and subsequently slow down or even fail to reach the end of the specimen. In such cases the Flame Spread Classification FSC_2 is calculated according to the following equation:

- $FSC_2 = 92.5 \cdot d/t$

where t is the time in minutes required by the flame front to travel d meters, where this is the distance at which a marked slow-down of the flame front occurs. If the flame spread behavior is such that it is difficult to ascertain the point at which the speed of the flame front starts to decrease, then FSC_2 is determined by consideration of test results obtained according to ULC-S127 (see the relevant section below).

The Flame Spread Classification is the greater of FSC_1 or FSC_2. Smoke Developed Classifications are also determined as in ASTM E84.

CAN/ULC-S102.2: Standard Method of Test for Surface Burning Characteristics of Flooring, Floor Covering, and Miscellaneous Materials and Assemblies

The surface burning characteristics of floorings and materials that melt and drip onto the floor of the tunnel apparatus, such as thermoplastic materials, are tested according to CAN/ULC-S102.2 [17]. The method is also used for materials that cannot be tested when affixed to the ceiling, such as thermoplastic and loose-fill materials.

The test rig is identical to that for CAN/ULC-S102, but the test specimen lies on the floor and the burners are directed downwards at an angle of 45° to the specimen. Diagrams are shown in Figure 10.14.

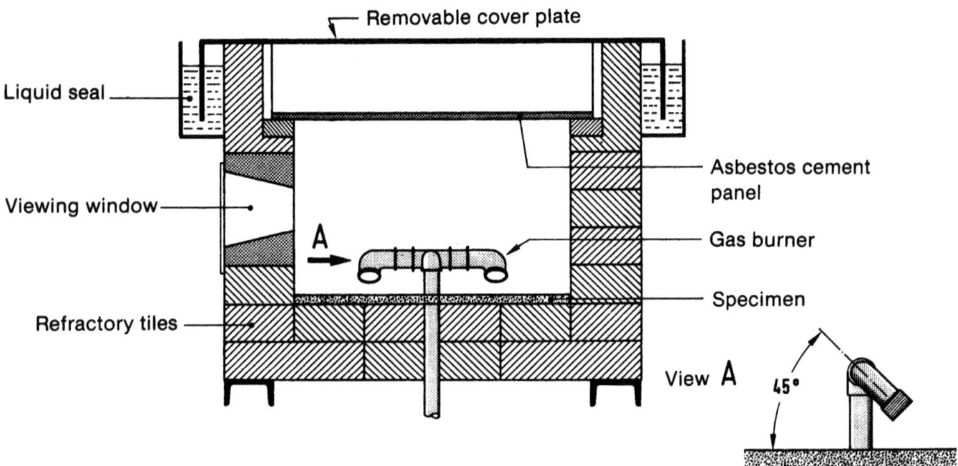

Figure 10.14 Modified tunnel for testing floor coverings and miscellaneous materials

The test specifications are similar to those described in CAN/ULC-S102. The specimen, however, is placed on the floor of the tunnel and is therefore narrower, with dimensions of 0.44 m × 7.32 m × usual thickness.

The values obtained from tests are used to compute the Flame Spread and Smoke Developed Classifications.

CAN/ULC-S102.3: Standard Method of Fire Test of Light Diffusers and Lenses

The tendency of light diffusers and lenses to become loose and fall out of their holders when exposed to fire is assessed by ULC-S102.3 [18] using the tunnel described in CAN/ULC-S102 and ASTM E84.

The test procedure is similar to CAN/ULC-S102 except that two tests are carried out on specimens half the length (3.66 m). In one case, the specimen is simply placed on the ledges of the furnace and in the second it is clamped in position. Note is made of whether and when the specimens fall from position and whether or not ignition occurred before the material fell.

CAN/ULC-S127: Standard Corner Wall Method of Test for Flammability Characteristics of Non-Melting Foam Plastic Building Materials

The ULC-S127 corner wall test method [19] is used in addition to the CAN/ULC-S102 and CAN/ULC-S102.2 to calculate FSC_2 when these two methods do not give clear data on the advance of the flame front. The test rig is illustrated in Figure 10.15 and Figure 10.16, and the test specifications are summarized in Table 10.26.

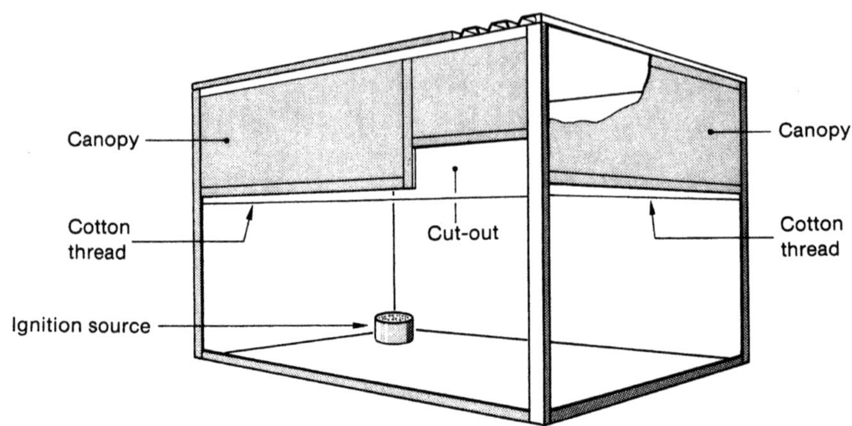

Figure 10.15 ULC-S127 corner wall test. Fire chamber

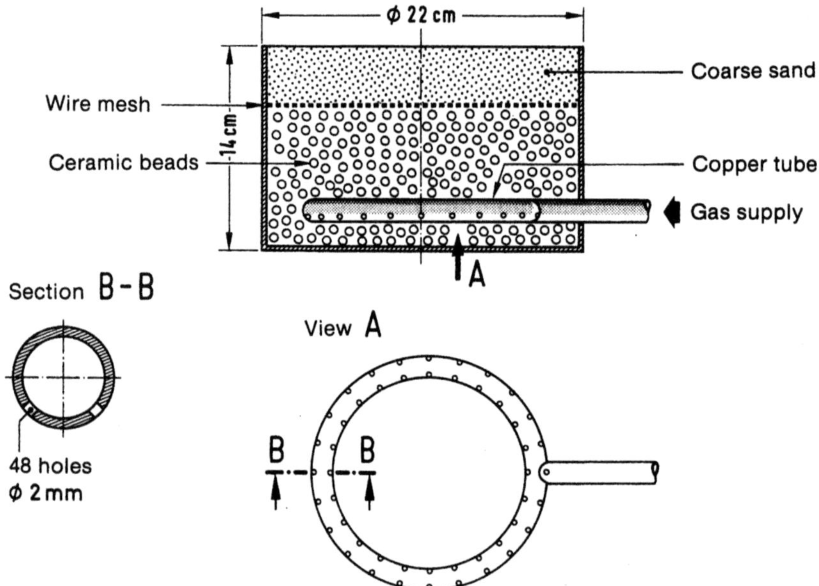

Figure 10.16 ULC-S127 corner wall test. Ignition source

Table 10.26 ULC-S127 Corner Wall Test Specifications

Specimens	The test material lines the ceiling, two walls and the canopy of the fire test chamber.
Fire chamber	Cube-shaped, with internal dimension of 1.3 m, a canopy measuring 0.52 m from top to bottom, and with a 0.475 × 0.15 m cut-out in front of the canopy.
Ignition source	Cylindrical tray (diameter 0.22 m, depth 0.14 m), filled with sand and ceramic beads. In the lower part of the bead layer, natural gas with a heat output of 2 MJ/min flows through a ring-shaped copper tube with 48 holes; gas is ignited by a small burner located above the sand bed. The tray is located on the floor in the corner formed by the two walls.
Test duration	Test is terminated when flames issue beneath the canopy or the cotton thread stretched 25.4 mm beneath it breaks; maximum 5 min after lighting ignition source.
Conclusions	Measures time until flames issue beneath the canopy or the cotton thread breaks; determines Flame Spread Classification FSC_2.

The time t at which flames appear from beneath the canopy or the cotton thread breaks is recorded and is converted to the Flame Spread Classification FSC_2 with the aid of the following equation:

- $FSC_2 = 51.47 \cdot t^{-1.215}$

CAN/ULC-S109: Standard Method for Flame Tests of Flame-Resistant Fabrics and Films

The CAN/ULC-S109 [20] method for testing of flame resistant fabrics and films involves two tests with ignition sources of differing intensity (small and large flame). In both cases the flame is applied vertically to the specimen with a Bunsen burner. The apparatus for the small flame test is shown in Figure 10.17. The test specifications are summarized in Table 10.27.

Table 10.27 Small Flame Test Specifications

Specimens	Ten specimens: 70 × 254 mm × usual thickness; five specimens in direction of warp and five in direction of weft. Specimens are clamped in a metal frame so that an area of 50 × 254 mm is tested.
Specimen position	Vertical in combustion chamber 305 × 305 × 750 mm; lower end of specimen 19 mm above tip of burner.
Ignition source	Bunsen burner with luminous 38 mm long flame with air supply shut off; burner located vertically beneath specimen or at an angle of 25° from vertical if dripping expected.
Test duration	Flame applied for 12 s; after-flame time and charring noted.
Conclusions	The material passes the test if all the following criteria are met: 1. After-flame time ≤ 2 s; 2. No burning drops or material falling; 3. Length of char due to vertical flame spread and afterglow does not exceed specified limits (see below).

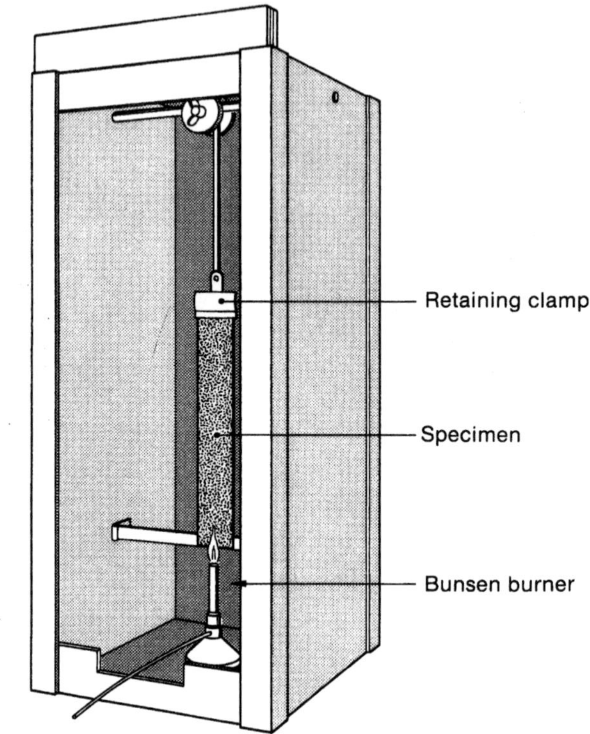

Figure 10.17 Flammability test for fabrics and films

Charring of the specimen caused by vertical flame spread and afterglow must not exceed the values given in Table 10.28. The length of char is determined by tearing.

Table 10.28 Classification of Fabrics and Films by Their Extent of Destruction

Weight of material [g/m^2]	Mean tear length (10 specimens) [mm]	Maximum tear length (individual specimens) [mm]
> 340	90	115
> 200 to ≤ 340	115	140
≤ 200	140	165

The large flame test is similar to the small flame test but employs flat and folded specimens, up to 2 m long. Normally, however, flat specimens need not be longer than 760 mm and folded specimens not longer than 1065 mm in order to establish compliance with the test requirements. The maximum lengths are necessary when it is desirable to ascertain the extent of destruction following a failed test. The specifications for the large flame test are given in Table 10.29.

Table 10.29 Large Flame Test Specifications

Specimens	Flat: Ten specimens, 127 × 762 mm; up to 2134 mm if necessary. Folded: Four specimens, 635 × 762 mm; up to 2134 mm if necessary. Each specimen is folded lengthwise four times so that folds are approximately 127 mm apart.
Test stack	305 × 305 × 2134 mm on 305 mm high supports, open at top and bottom.
Specimen position	Vertical, bottom of specimen 100 mm above tip of burner; folded specimens set up with folds approximately 127 mm apart.
Ignition source	Bunsen burner with 280 mm long oxidizing flame; burner vertical under specimen or at an angle of 25° to vertical if dripping expected.
Test duration	Flame applied for 2 min; after-flame time and charring noted.
Conclusions	The material passes the test if all the following criteria are met: 1. After-flame time ≤ 2 s; 2. No burning drops or material falling; 3. Length of char due to vertical flame spread and afterglow does not exceed 254 mm.

10.2.2.4 NBCC Fire Performance Requirements for Interior Finish Materials

Subsection 3.1.12 of the NBCC requires that the Flame Spread Classification (referred to as the *flame-spread rating* in the code) and Smoke Developed Classification (referred to as the *smoke developed classification* in the code) of wall, ceiling, and floor finish materials be assessed by CAN/ULC-S102 or CAN/ULC-S102.2. Flame-spread ratings and smoke developed classifications of some common interior finish materials are presented in Section D-3 of Appendix D of the NBCC.

NBCC requirements for the flame-spread rating and the smoke developed classification of the interior finish of walls and ceilings in housing and small buildings are specified in Part 9 of the NBCC. The general principle is that the flame spread requirements are more stringent for exits and public corridors than for other spaces in the building. The exposed surface of every interior wall and ceiling, including skylights and glazing, shall have a surface flame-spread rating of not more than 150. The flame-spread rating and smoke developed classification requirements for larger buildings are specified in Part 3 of the NBCC. In larger buildings the flame spread rating for wall and ceiling finishes shall not exceed 25 in exits, lobbies, covered vehicular passageways, and vertical service spaces. In all other locations the flame spread rating of wall and ceiling finishes shall not exceed 150 if the building is sprinklered, and 75 if the building is not sprinklered. In general, up to 10% of the wall and ceiling surfaces are permitted to be covered with a material that exceeds the maximum allowed flame-spread rating for the wall or ceiling finish. The NBCC does not regulate the flame-spread rating of flooring, with the exception of certain essential areas in high buildings such as exits, corridors not within suites, elevator cars, and service spaces. Light diffusers and lenses are assessed by CAN/ULC-S102.2 and CAN/ULC-S102.3 and can have a flame-spread rating of 250 or less and a smoke developed classification not more than 600.

10.2.2.5 Accredited Testing Laboratories

Laboratories that perform fire tests and publish test reports that are used to demonstrate compliance with the provincial building code requirements must be accredited by the Standards Council of Canada (SCC). A directory of accredited laboratories can be found on the SCC website (see *https://www.scc.ca/en/search/palcan*). Canada is a signatory to the International Laboratory Accreditation Cooperation (ILAC). As a result, the SCC also recognizes laboratories that are accredited by agencies in other countries that are signatory to ILAC, e.g., the International Accreditation Service in the U.S. (see *https://www.iasonline.org/*).

Underwriters Laboratories of Canada (ULC) and Intertek Testing Services (ITS) are accredited by the SCC to provide testing and certification services related to the fire performance of building products. Products that have been certified by ULC are identified by the ULC Mark (which consists of ULC in a circle) or by a ULC Certificate. Products certified by ITS are identified by the Warnock Hersey Listed Mark. Both ULC and ITS publish directories of listed products.

10.2.2.6 Future Developments

The government of Canada has now provided funding to enable free access to the downloadable and online formats of the National Building Code, the National Fire Code, and other codes published by the National Research Council of Canada (NRCC) [28]. This initiative will reduce barriers to trade and encourage provinces and territories to align building codes across the country.

As in the IBC, the 2020 NBCC allows for the construction of taller mass timber buildings. However, with a maximum of 12 stories, the allowable height increase of wood structures is more modest than for mass timber buildings in the USA.

References for Section 10.2.2

[1] National Building Code of Canada 2015 (Fifteenth Edition), Canadian Commission on Building and Fire Codes, National Research Council Canada, Ottawa.

[2] CAN/ULC-S101-14. Standard methods of fire endurance tests of building construction and materials.

[3] CAN/ULC-S104-15. Standard method for fire tests of door assemblies.

[4] CAN/ULC-S105-16. Standard specification for fire door frames meeting the performance required by CAN/ULC-S104 fourth edition.

[5] CAN/ULC-S106-15. Standard method for fire tests of window and glass block assemblies.

[6] CAN/ULC-S110-13. Standard methods of fire test for air ducts.

[7] CAN/ULC-S111-13. Standard method of fire tests for air filter units.

[8] CAN/ULC-S112-10. Standard method of fire test of fire-damper assemblies.

[9] CAN/ULC-S112.1-10. Standard Method of Fire Test of Leakage Rated Dampers for Use in Smoke Control Systems.

[10] CAN/ULC-S112.2-07. Standard method of fire test of ceiling firestop flap assemblies.

[11] CAN/ULC-S113-16. Standard specification for wood core doors meeting the performance required by CAN/ULC-S104 for twenty minute fire rated closure assemblies.

[12] CAN/ULC-S115-11. Fire tests of firestop systems.

[13] CAN/ULC-S124-06. Standard method of test for the evaluation of protective coverings for foamed plastic.

[14] CAN/ULC-S144-12. Standard method of fire resistance test - Grease duct assemblies.

[15] CAN/ULC-S114-05. Standard method of test for determination of non-combustibility in building materials.

[16] CAN/ULC-S102-10. Standard method of test for surface burning characteristics of building materials and assemblies.

[17] CAN/ULC-S102.2-10. Standard method of test for surface burning characteristics of flooring, floor covering, and miscellaneous materials and assemblies.

[18] CAN/ULC-S102.3-07. Standard method of fire test of light diffusers and lenses.

[19] CAN/ULC-S127-14. Standard corner wall method of test for flammability characteristics of non-melting foam plastic building materials.

[20] CAN/ULC-S109-14. Standard method for flame tests of flame-resistant fabrics and films.

[21] CAN/ULC-S107-10. Standard methods of fire tests of roof coverings.

[22] CAN/ULC-S126-14. Standard method of test for fire spread under roof-deck assemblies.

[23] CAN/ULC-S134-13. Standard method of fire test of exterior wall assemblies.

[24] CAN/ULC-S135-04. Test method for the determination of combustibility parameters of building materials using an oxygen consumption calorimeter (cone calorimeter).

[25] CAN/ULC-S138-06. Standard method of test for fire growth of insulated building panels in a full-scale room configuration.

[26] CAN/ULC-S139-12. Standard method of fire test for evaluation of integrity of electrical power, data and optical fibre cables.

[27] CAN/ULC-S143-14. Standard method of fire tests for non-metallic electrical and optical fibre cable raceway systems.

[28] *https://nrc.canada.ca/en/certifications-evaluations-standards/codes-canada/codes-guides-online-library.*

10.3 Europe

10.3.1 European Union

Björn Sundström

10.3.1.1 Introduction

Currently, the EU has 27 members (with five candidates for membership), and about 500 million inhabitants. The GDP (in 2017) was €15.3 trillion, which is bigger than, for example, the USA. In the EU's single market, people, goods, services, and money can move around as freely as within a single country. EU citizens can study, live, shop, work, and retire in any EU country - and purchase products from all over Europe [1]. Four countries in Europe (Iceland, Norway, Lichtenstein, and Switzerland) are not members of the EU but form the European Free Trade Association (EFTA). Through the European Economic Area (EEA), three EFTA states (Iceland, Liechtenstein, and Norway) are included in the EU single market, and are governed by the same basic rules as the EU member states, whereas Switzerland has a bilateral agreement with the EU.

EU legislation is drafted and proposed by the European Commission. The elected European Parliament together with the Council (governments of the member states) then approves the legislation. When drafting laws, the Commission consults widely with national regulators, technical experts, industry groups, etc. before making a proposal. That process was followed when creating the Construction Products Regulation (see below), which opens up a free market in the EU for construction products by a common declaration of properties, as shown through the CE mark.

The Construction Products Regulation

The European Commission published the Construction Products Directive (89/106/EEG) in 1989. The directive was replaced by the Construction Products Regulation (CPR) in 2011 [2]. The regulation contains seven basic requirements that apply to construction work. One of these requirements is safety in case of fire.

The function of the CPR relies on a number of specifications, with harmonized test standards and the manufacturer's declaration of performance according to these standards being of primary importance when considering fire performance. Therefore, construction products must have a fire classification based on the same standards throughout Europe. A member state that regulates for a certain safety level will be able to identify the fire properties of a construction product corresponding to that level. By drawing up the declaration of performance, the manufacturer assumes responsibility for the conformity of the construction product with the declared performance. Such products can then be CE-marked.

The Role of Regulators, Notified Bodies, and Standardization Bodies

The European standards (EN) for reaction-to-fire are published by the European Committee for Standardization (CEN). Draft standards are indicated by the acronym prEN. CEN works out the test standards based on specifications given in a mandate from the European Commission. An expert group set up by the European Commission, the Fire Regulators Group (FRG), worked out the basis for the European system of classification and the request for test methods. In particular, the work of defining the Euroclasses was performed by FRG. These specifications were then taken over by CEN, and the classification system is now a harmonized EN-standard.

There is a continuous need for quality assurance work in operating the system. This can include interpretation of test procedures, extended application of test data, technical co-operation between test laboratories, agreements of practice between certification bodies, etc. CE-marked products are also subject to various levels of surveillance - for example, by third parties - and there is also market control, with responsibility being held by the authorities in the member states. A major role for keeping the system in work is carried out by the Fire Sector Group of notified bodies. This group consists of the notified bodies for testing and certification throughout Europe. A notified body is a body (for example, a test laboratory or a certification organization) that a member state has notified to the European Commission as competent to perform testing/certification under the European system.

Technical work, such as development of good technical practice in testing, relies heavily on EGOLF (European Group of Organisations for Fire Testing, Inspection and Certification) and European industrial organizations.

The Packages of Standards Used for Product Classification and Testing

There are separate standards for classification and testing. In some cases the number of tests of individual products in a family of products can be very large and create an undue burden for manufacturers and test laboratories. In those cases, rules for extended application of test data (EXAP) have been created. An EXAP procedure reduces considerably the amount of testing needed to classify, for example, a family of cables. In other cases, there are products having generic fire behavior. They can then be classified without further testing (CWFT). For example, ordinary untreated wood is classified without testing. However, the products must fulfill criteria given in a document for CWFT of wood products.

10.3.1.2 The Euroclasses: Classification Systems and Fire Tests

The European Commission first published classification criteria for harmonized European reaction-to-fire classes in 2000, the so-called Euroclasses. The Euroclasses and the associated test procedures introduced a completely new concept compared to previous procedures in Europe. Now they are EN standards, and are handled by CEN.

10.3.1.2.1 Interior Products

Fire Classification

EN 13501-1:2018: Fire classification of construction products and building elements – Part 1: Classification using data from reaction-to-fire tests

The reaction-to-fire classification procedure for all construction products, including products incorporated within building elements, is covered. There are, however, specific standards for roof coverings and cables (see Sections 10.3.1.2.2 and 10.3.1.2.3, respectively). Products are considered in relation to their end application, and there are seven main classes: A1, A2, B, C, D, E, and F. A1 and A2 represent different degrees of limited combustibility. For linings, B–E represent products that can support various degrees of fire growth. F means that the product does not fulfill any requirements. Additional classes for smoke and the occurrence of any burning droplets are also included. EN 13501-1 applies to three categories, which are treated separately:

- Construction products, excluding floorings and linear pipe thermal insulation products (Table 10.30);
- Floorings (Table 10.31);
- Linear pipe thermal insulation products (Table 10.32).

Table 10.30 Classes of Reaction-to-Fire Performance for Construction Products Excluding Floorings and Linear Pipe Thermal Insulation Products

Class	Test method(s)	Classification criteria	Additional classification
A1	EN ISO 1182 [a] *And*	$\Delta T \leq 30$ °C *and* $\Delta m \leq 50\%$ *and* $t_f = 0$ (i. e., no sustained flaming)	-
	EN ISO 1716	PCS ≤ 2.0 MJ/kg [a] *and* PCS ≤ 2.0 MJ/kg [b, c] *and* PCS ≤ 1.4 MJ/m^2 [d] *and* PCS ≤ 2.0 MJ/kg [e]	-
A2	EN ISO 1182 [a] *Or*	$\Delta T \leq 50$ °C *and* $\Delta m \leq 50\%$ *and* $t_f \leq 20$ s	-
	EN ISO 1716 *And*	PCS ≤ 3.0 MJ/kg [a] *and* PCS ≤ 4.0 MJ/m^2 [b] *and* PCS ≤ 4.0 MJ/m^2 [d] *and* PCS ≤ 3.0 MJ/kg [e]	-
	EN 13823 (all products)	FIGRA ≤ 120 W/s *and* LFS < edge of specimen *and* $THR_{600\,s} \leq 7.5$ MJ	Smoke production [f] *and* Flaming droplets/ particles [g]
B	EN 13823 *And*	FIGRA ≤ 120 W/s *and* LFS < edge of specimen *and* $THR_{600\,s} \leq 7.5$ MJ	Smoke production [f] *and* Flaming droplets/ particles [g]
	EN ISO 11925-2 [i] (Exposure = 30 s)	Fs ≤ 150 mm within 60 s	
C	EN 13823 *And*	FIGRA ≤ 250 W/s *and* LFS < edge of specimen *and* $THR_{600\,s} \leq 15$ MJ	Smoke production [f] *and* Flaming droplets/ particles [g]
	EN ISO 11925-2 [i] (Exposure = 30 s)	Fs ≤ 150 mm within 60 s	

Table 10.30 Classes of Reaction-to-Fire Performance for Construction Products Excluding Floorings and Linear Pipe Thermal Insulation Products *(continued)*

Class	Test method(s)	Classification criteria	Additional classification
D	EN 13823 *And*	FIGRA ≤ 750 W/s	Smoke production [f] *and* Flaming droplets/ particles [g]
	EN ISO 11925-2 [i] (Exposure = 30 s)	Fs ≤ 150 mm within 60 s	-
E	EN ISO 11925-2 [i] (Exposure = 15 s)	Fs ≤ 150 mm within 20 s	Flaming droplets/ particles [h]
F	EN ISO 11925-2 [i] (Exposure = 15 s)	Fs > 150 mm within 20 s	-

[a] For homogeneous products and substantial components of non-homogeneous products.

[b] For any external non-substantial component of non-homogeneous products.

[c] Alternatively, any external non-substantial component having a PCS ≤ 2.0 MJ/m², provided that the product satisfies the following criteria of EN 13823: FIGRA ≤ 20 W/s; and LFS < edge of specimen; and $THR_{600\,s}$ ≤ 4.0 MJ; and s1; and d0.

[d] For any internal non-substantial component of non-homogeneous products.

[e] For the product as a whole.

[f] s1 = SMOGRA ≤ 30 m²/s² and $TSP_{600\,s}$ ≤ 50 m²; s2 = SMOGRA ≤ 180 m²/s² and $TSP_{600\,s}$ ≤ 200 m²; s3 = not s1 or s2.

[g] d0 = No flaming droplets/particles in EN 13823 within 600 s; d1 = No flaming droplets/particles persisting longer than 10 s in EN 13823 within 600 s; d2 = not d0 or d1; Ignition of the paper in EN ISO 11925-2 results in a d2 classification.

[h] Pass = no ignition of the paper (no classification); Fail = ignition of the paper (d2 classification).

[i] Under conditions of surface flame attack and, if appropriate to end application of product, edge flame attack.

Table 10.31 Classes of Reaction-to-Fire Performance for Floorings

Class	Test method(s)	Classification criteria	Additional classification
$A1_{fl}$	EN ISO 1182 [a] *And*	ΔT ≤ 30 °C *and* Δm ≤ 50% *and* t_f = 0 (i. e., no sustained flaming)	-
	EN ISO 1716	PCS ≤ 2.0 MJ/kg [a] *and* PCS ≤ 2.0 MJ/kg [b] *and* PCS ≤ 1.4 MJ/m² [c] *and* PCS ≤ 2.0 MJ/kg [d]	-

Class	Test method(s)	Classification criteria	Additional classification
$A2_{fl}$	EN ISO 1182 [a] *Or*	$\Delta T \leq 50\,°C$ *and* $\Delta m \leq 50\%$ *and* $t_f \leq 20$ s	-
	EN ISO 1716 *And*	PCS ≤ 3.0 MJ/kg [a] *and* PCS ≤ 4.0 MJ/m² [b] *and* PCS ≤ 4.0 MJ/m² [c] *and* PCS ≤ 3.0 MJ/kg [d]	-
	EN ISO 9239-1 [e] (all products)	Critical flux [f] 8.0 kW/m²	Smoke production [g]
B_{fl}	EN ISO 9239-1 [e] *And*	Critical flux [f] 8.0 kW/m²	Smoke production [g]
	EN ISO 11925-2 [h] (Exposure = 15 s)	Fs ≤ 150 mm within 20 s	
C_{fl}	EN ISO 9239-1 [e] *And*	Critical flux [f] 4.5 kW/m²	Smoke production [g]
	EN ISO 11925-2 [h] (Exposure = 15 s)	Fs ≤ 150 mm within 20 s	
D_{fl}	EN ISO 9239-1 [e] *And*	Critical flux [f] 3.0 kW/m²	Smoke production [g]
	EN ISO 11925-2 [h] (Exposure = 15 s)	Fs ≤ 150 mm within 20 s	
E_{fl}	EN ISO 11925-2 [h] (Exposure = 15 s)	Fs ≤ 150 mm within 20 s	–
F_{fl}	EN ISO 11925-2 [h] (Exposure = 15 s)	Fs > 150 mm within 20 s	–

[a] For homogeneous products and substantial components of non-homogeneous products.
[b] For any external non-substantial component of non-homogeneous products.
[c] For any internal non-substantial component of non-homogeneous products.
[d] For the product as a whole.
[e] Test duration = 30 min.
[f] Critical flux is defined as the radiant flux at which the flame extinguishes or the radiant flux after a test period of 30 min, whichever is the lower (i.e., the flux corresponding with the furthest extent of spread of flame).
[g] s1 = Smoke ≤ 750% · min; s2 = not s1.
[h] Under conditions of surface flame attack and, if appropriate to the end application of the product, edge flame attack.

Table 10.32 Classes of Reaction-to-Fire Performance for Linear Pipe Thermal Insulation Products

Class	Test method(s)	Classification criteria	Additional classification
$A1_L$	EN ISO 1182 [a] *And*	$\Delta T \leq 30\,°C$ *and* $\Delta m \leq 50\%$ *and* $t_f = 0$ (i. e., no sustained flaming)	–
	EN ISO 1716	PCS ≤ 2.0 MJ/kg [a] *and* PCS ≤ 2.0 MJ/kg [b] *and* PCS ≤ 1.4 MJ/m^2 [c] *and* PCS ≤ 2.0 MJ/kg [d]	–
$A2_L$	EN ISO 1182 [a] *Or*	$\Delta T \leq 50\,°C$ *and* $\Delta m \leq 50\%$ *and* $t_f \leq 20$ s	–
	EN ISO 1716 *And*	PCS ≤ 3.0 MJ/kg [a] *and* PCS ≤ 4.0 MJ/m^2 [b] *and* PCS ≤ 4.0 MJ/m^2 [c] *and* PCS ≤ 3.0 MJ/kg [d]	–
	EN 13823 (all products)	FIGRA ≤ 270 W/s *and* LFS < edge of specimen *and* $THR_{600\,s} \leq 7.5$ MJ	Smoke production [e] *and* Flaming droplets/particles [f]
B_L	EN 13823 *And*	FIGRA ≤ 270 W/s *and* LFS < edge of specimen *and* $THR_{600\,s} \leq 7.5$ MJ	Smoke production [e] *and* Flaming droplets/particles [f]
	EN ISO 11925-2 [h] (Exposure = 30 s)	Fs ≤ 150 mm within 60 s	–
C_L	EN 13823 *And*	FIGRA ≤ 460 W/s *and* LFS < edge of specimen *and* $THR_{600\,s} \leq 15$ MJ	Smoke production [e] *and* Flaming droplets/particles [f]
	EN ISO 11925-2 [h] (Exposure = 30 s)	Fs ≤ 150 mm within 60 s	–

Class	Test method(s)	Classification criteria	Additional classification
D_L	EN 13823 *And*	FIGRA ≤ 2100 W/s $THR_{600\,s}$ ≤ 100 MJ	Smoke production [e] *and* Flaming droplets/particles [f]
	EN ISO 11925-2 [h] (Exposure = 30 s)	Fs ≤ 150 mm within 60 s	–
E_L	EN ISO 11925-2 [h] (Exposure = 15 s)	Fs ≤ 150 mm within 20 s	Flaming droplets/particles [g]
F_L	EN ISO 11925-2 [h] (Exposure = 15 s)	Fs > 150 mm within 20 s	–

[a] For homogeneous products and substantial components of non-homogeneous products.
[b] For any external non-substantial component of non-homogeneous products.
[c] For any internal non-substantial component of non-homogeneous products.
[d] For the product as a whole.
[e] s1 = SMOGRA ≤ 105 m^2/s^2 and $TSP_{600\,s}$ ≤ 250 m^2; s2 = SMOGRA ≤ 580 m^2/s^2 and $TSP_{600\,s}$ ≤ 1600 m^2; s3 = not s1 or s2.
[f] d0 = No flaming droplets/particles in EN 13823 within 600 s; d1 = No flaming droplets/particles persisting longer than 10 s in EN 13823 within 600 s; d2 = not d0 or d1; Ignition of the paper in EN ISO 11925-2 results in a d2 classification.
[g] Pass = no ignition of the paper (no classification); Fail = ignition of the paper (d2 classification).
[h] Under conditions of surface flame attack and, if appropriate to end application of product, edge flame attack.

Symbols

The characteristics are defined with respect to the appropriate test method.

ΔT	Temperature rise [K]
Δm	Mass loss [%]
t_f	Duration of flaming [s]
PCS	Gross calorific potential [MJ/kg] or [MJ/m^2]
FIGRA	Fire growth rate index
$THR_{600\,s}$	Total heat release within 600 s [MJ]
LFS	Lateral flame spread [m]
SMOGRA	Smoke growth rate index
$TSP_{600\,s}$	Total smoke production within 600 s [m^2]
Fs	Flame spread [mm]

Definitions

Material: A single basic substance or uniformly dispersed mixture of substances, e. g., metal, stone, timber, concrete, mineral wool with uniformly dispersed binder, or polymers.

Homogeneous product: A product consisting of a single material, of uniform density and composition throughout the product.

Non-homogeneous product: A product that does not satisfy the requirements of a homogeneous product. It is a product composed of one or more components, substantial and/or non-substantial.

Substantial component: A material that constitutes a significant part of a non-homogeneous product. A layer with a mass per unit area $\geq$ 1.0 kg/m^2 or a thickness $\geq$ 1.0 mm is considered to be a substantial component.

Non-substantial component: A material that does not constitute a significant part of a non-homogeneous product. A layer with a mass per unit area < 1.0 kg/m^2 and a thickness < 1.0 mm is considered to be a non-substantial component.

Two or more non-substantial layers that are adjacent to each other (i.e., with no substantial component(s) in-between the layers) are regarded as one non-substantial component when they collectively comply with the requirements for a layer being a non-substantial component.

Internal non-substantial component: A non-substantial component that is covered on both sides by at least one substantial component.

External non-substantial component: A non-substantial component that is not covered on one side by a substantial component.

A Euroclass is intended to be declared in the format (for example) B-s1,d0. B stands for the main class, s1 stands for smoke Class 1 and d0 stands for droplet/particle Class 0. This gives theoretically a total of 40 classes of linings and linear pipe thermal insulation products, and 11 classes of floor coverings. However, in reality only a few classes are used by the member states. For example, class B-s1, d0 is quite common as a requirement for linings in escape routes out of the six possible class B combinations.

EN 13501-1 gives further details on classification. One important principle is that the products end-use conditions are included in the Euroclass. For example, wallpaper has very different burning behavior depending on what substrate it is applied to. Therefore, not only the properties of the wallpaper itself, but also how it is attached to the substrate and what the substrate is, are relevant. This means that the same product can have different Euroclasses depending on how it is used. However, non-combustibility and gross calorific potential are material properties and so independent of the end use.

The test methods are called for in the classification standard, but are described in separate test standards. In many cases, ISO test methods are used. These are well known, and some of them have been in use in various countries for many years. In liaison with CEN, ISO/TC92/SC1 has actively been involved in the development of European standards. These standards are called "EN ISO" to indicate that they apply world-wide as well as in Europe.

Reaction-to-Fire Test Standards

EN ISO 1716:2018: Reaction to fire tests for products - Determination of the gross heat of combustion (calorific value)

EN ISO 1716 determines the potential maximum total heat release of a material when completely burning, regardless of its end use. The test is relevant for classes A1, A2, $A1_{fl}$, $A2_{fl}$, $A1_L$, and $A2_L$.

The calorific potential of a material is measured in a bomb calorimeter (Figure 10.18). The powdered material is completely burned under high pressure in a pure oxygen atmosphere.

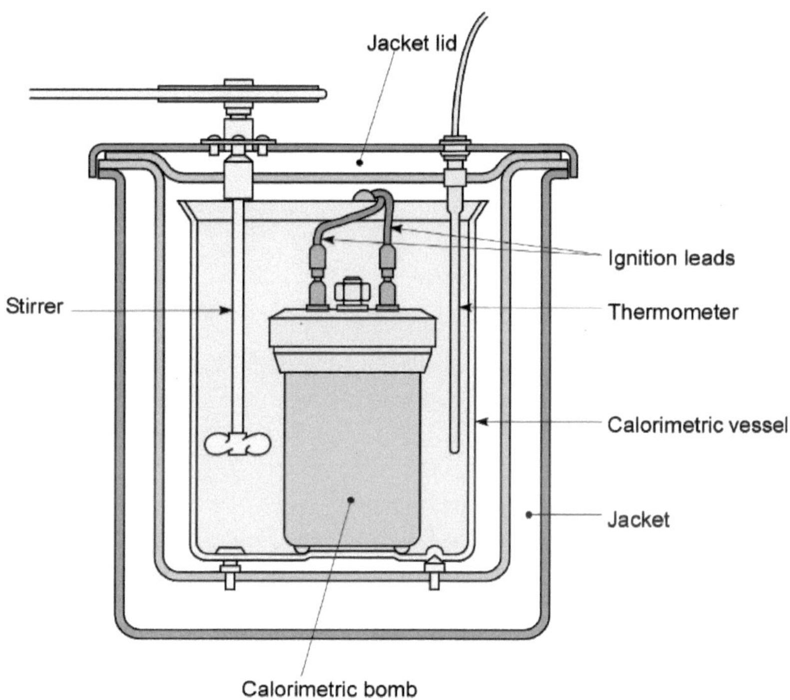

Figure 10.18 EN ISO 1716 test for calorific value. Image drawn by RISE; reproduced with permission

EN ISO 1182:2020: Reaction to fire tests for products - Non-combustibility test

EN ISO 1182 identifies products that will not, or significantly not, contribute to a fire, regardless of their end use. The test is relevant for classes A1, A2, $A1_{fl}$, $A2_{fl}$, $A1_L$, and $A2_L$.

EN ISO 1182 is a pure material test, and a product cannot be tested in end-use conditions. Therefore only homogenous building products or homogenous components of a product are tested. EN ISO 1182 was first published by ISO during the 1970s, and is well-known. The EN ISO version is shown in Figure 10.19, and the test specifications are given in Table 10.33.

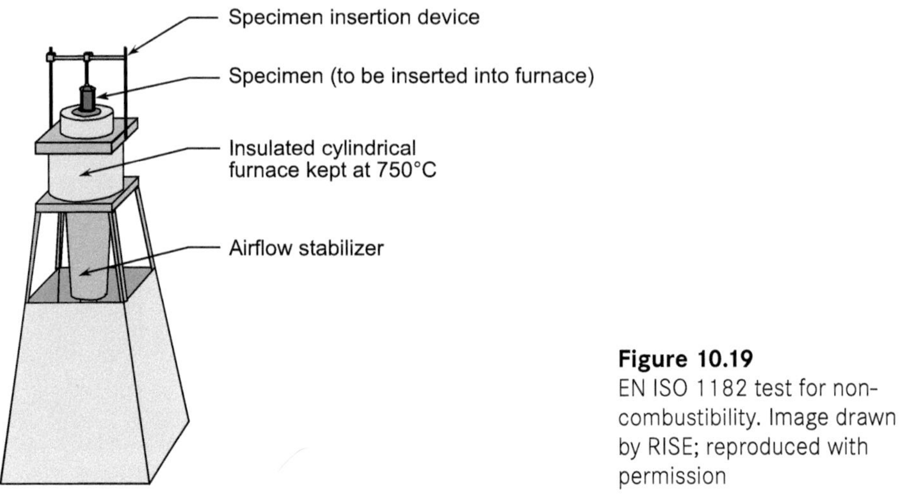

Figure 10.19
EN ISO 1182 test for non-combustibility. Image drawn by RISE; reproduced with permission

Table 10.33 EN ISO 1182 Test Specifications

Specimens	Five cylindrical samples, diameter 45 mm, height 50 mm
Specimen position	Vertical in specimen holder in the center of the furnace
Heat source	Electrical cylindrical furnace at 750 °C (measured by the furnace thermocouples)
Test duration	Depends on temperature stabilization
Conclusions	Classification is based on temperature rise as measured by the furnace thermocouple, duration of flaming and mass loss of the sample

prEN 13238:2018: Reaction to fire tests for building products - Conditioning procedures and general rules for selection of substrates

EN 13238 is applicable to tests where the product's end-use conditions are relevant when testing. Conditioning of samples and standard substrates on which the product sample can be attached before testing are described. The standard substrates represent various end-use conditions, so the test results are general in nature, and the amount of testing can be kept down.

EN 14390:2007: Fire test - Large-scale room reference test for surface products (ISO 9705 Room/corner test)

This room fire test is similar to the room/corner test ISO 9705 with only minor differences. Test specifications are given in Table 10.35. EN 14390 constitutes the reference scenario for room surface products for the Euroclasses originally developed using ISO 9705 [3, 4, 5].

Reference scenarios (Figure 10.20) are basically used to validate the Euroclasses by relating them to fire situations.

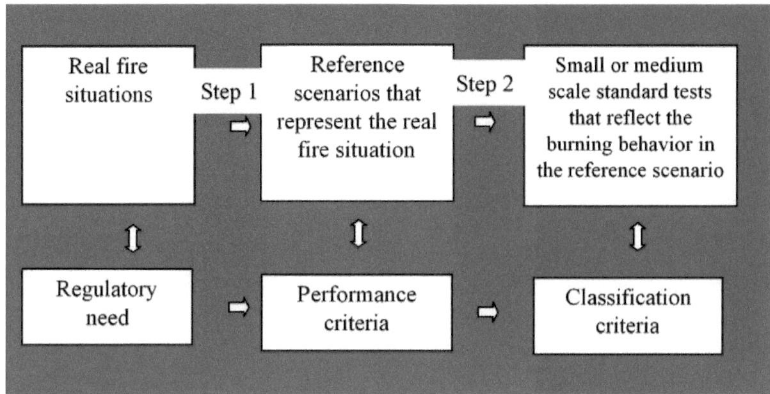

Figure 10.20 The use of reference scenarios for determining classification criteria for the Euroclasses. Image drawn by RISE; reproduced with permission

There is a regulatory need to limit the occurrence and consequences of real fires. A first step was to identify reference scenarios that could represent real fire situations of widespread relevance. Such a situation is a room fire standardized as the room/corner test, and the selected performance criteria were related to flashover. EN 13823, the SBI test procedure, was then used with classification criteria defined so as to correspond to the burning behavior of the product in the reference scenario. Reference scenarios are provided for linings, linear pipe insulations, and cables. A detailed description of how this system works, correlation data, theory, and literature references is available [6].

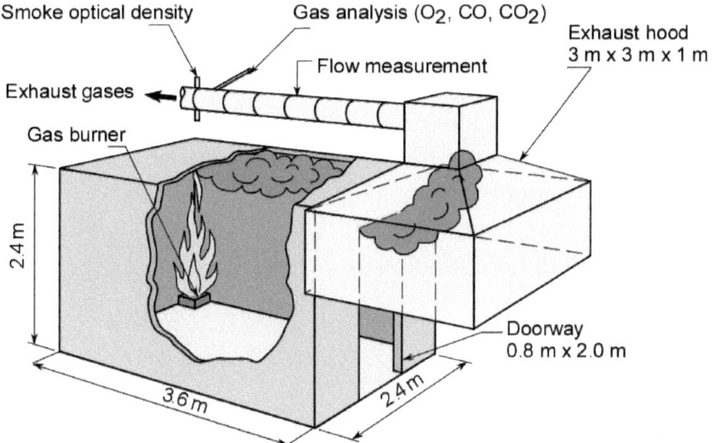

Figure 10.21 EN 14390 large-scale room reference test for surface products (the room/corner test). Image drawn by RISE; reproduced with permission

The room/corner test was first published by ASTM in 1982 [7] and then by NORDTEST in 1986. The international standard, ISO 9705, was published in 1993, and was followed by EN 14390 in 2007.

The room/corner test is a large-scale test method for measurement of the burning behavior of building products (linings) in a room scenario. The principal output is the occurrence and time to flashover. A direct measure of fire growth (heat release rate, HRR) and light-obscuring smoke (smoke production rate, SPR) are also results from the test. The product is mounted on three walls and on the ceiling of a small compartment. A door opening ventilates the room.

Experience on testing products has been gained during more than 30 years of work with the room/corner test. A considerable amount of information on product burning behavior in this method is available, and the thermal conditions during a test fire have been carefully mapped [8].

The development of EN 13823, the SBI test, included testing of 30 building products across Europe. The same products were tested according to the reference scenario in the room/corner test. The subsequent analysis then resulted in a correlation between FIGRA (ISO 9705) and FIGRA (SBI) [3]. The correlation between the tests also relates flashover in ISO 9705 to certain Euroclasses (see Table 10.34).

Table 10.34 Description of the Euroclasses as the Tendency of a Lining to Reach Flashover in the Large-Scale Reference Scenario, the Room/Corner Test

Euroclass	Limit value for FIGRA (SBI test) (W/s)	Expected burning behavior in the room/corner test
A2	120	No flashover
B	120	No flashover
C	250	No flashover at 100 kW ignition source
D	750	No flashover before 2 min at 100 kW ignition source
E	> 750	Flashover before 2 min

It can be seen how safety was one of the reasons behind selecting the specific class limits. The classes for linear pipe insulation have limit values for FIGRA that were developed using a reference scenario which is a variant of EN 14390 whereby the pipe insulation was mounted so as to simulate a real installation. A large test program carried out in co-operation with industry, which compared large-scale results with SBI data, resulted in the Euroclasses [9, 10].

Table 10.35 EN 14390 Room/Corner Test, Specifications for Linings

Specimens	Sample material enough to cover three walls and the ceiling of the test room. The wall containing the doorway is not covered
Specimen position	Forms a room lining
Ignition source	Gas burner placed in one of the room corners. The burner heat output is 100 kW for the first 10 min and then 300 kW for another 10 min
Test duration	20 min or until flashover
Conclusions	The most important outputs are HRR, SPR and time to (or occurrence of) flashover

prEN 13823:2018: Reaction to fire tests for building products - Building products excluding floorings exposed to the thermal attack by a single burning item

EN 13823 (the Single burning item test, SBI) is relevant for classes A1, A2, $A2_L$, B, B_L, C, C_L, D, and D_L.

The SBI is the major test procedure for classification of linings and linear pipe insulation. The test specifications are summarized in Table 10.36.

The SBI is of intermediate size (Figure 10.22 and Figure 10.23). When testing linings, two samples, 0.5 m × 1.5 m and 1.0 m × 1.5 m, are mounted in a corner configuration where they are exposed to a gas flame ignition source. The test specifications are shown in Table 10.36. As in EN 14390, a direct measure of fire growth (heat release rate, HRR) and light-obscuring smoke (smoke production rate, SPR) are principal results from a test. Other properties such as the occurrence of burning droplets/particles and maximum flame spread are observed.

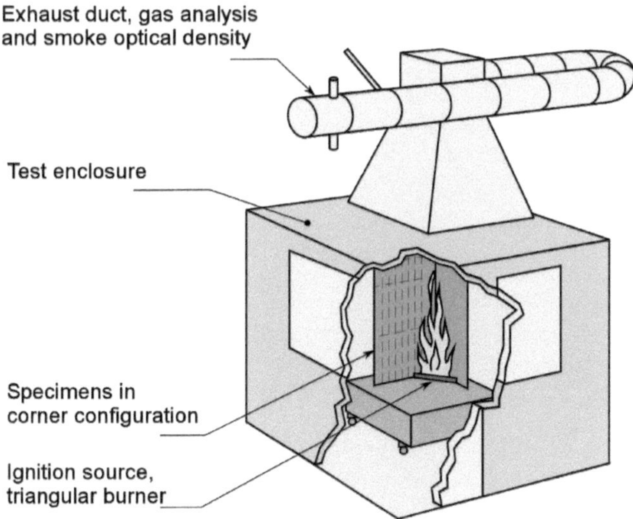

Figure 10.22 EN 13823 (the SBI test) - specimen configuration for testing of linings. Image drawn by RISE; reproduced with permission

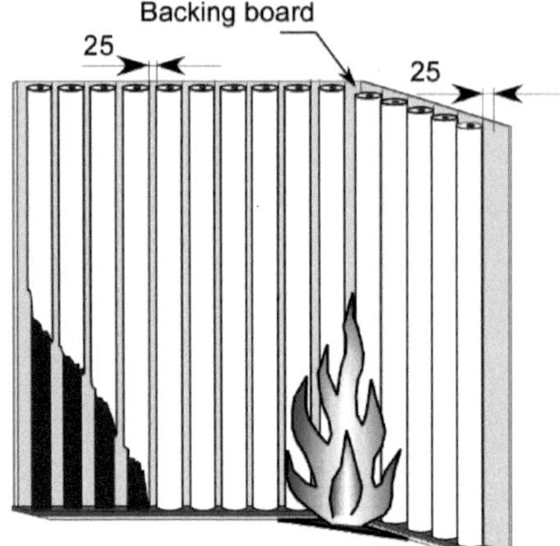

Figure 10.23
EN 13823 specimen configuration for testing of linear pipe insulation. The spacing of the pipe insulation units is 25 mm. Image drawn by RISE; reproduced with permission

The parameter FIGRA (FIre Growth Rate) was first introduced in 1998 as a way of classifying the product based on its tendency to support fire growth. Thus, FIGRA is a measure of the greatest growth rate of the fire during a SBI test, as seen from the start of the test (Figure 10.24). FIGRA is calculated as the maximum value of the function (heat release rate/elapsed test time), and units are W/s. Filtering of the data avoids extreme values from small fires and rapidly fluctuating HRR.

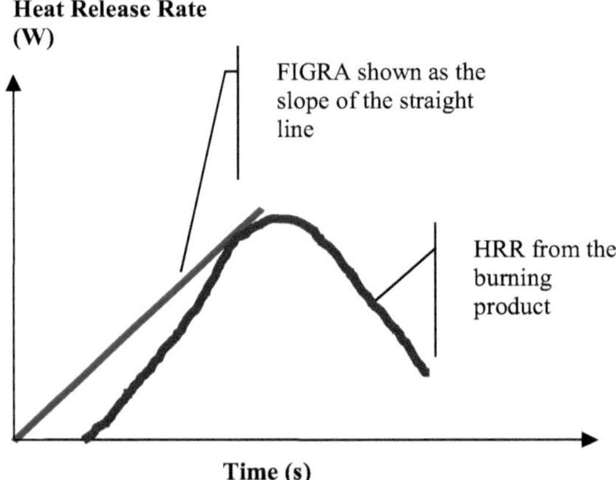

Figure 10.24 The parameter FIGRA is the maximum value of the function (heat release rate/elapsed test time); shown as the slope of the straight line in the figure. Image drawn by RISE; reproduced with permission

The additional classification for smoke is based on the index SMOGRA (SMOke Growth Rate). This index is based on similar principles as FIGRA.

The detailed definitions of FIGRA and SMOGRA can be found in EN 13823 (SBI).

Table 10.36 EN 13823 SBI Test Specifications

Specimens	Samples for at least three tests. Each test requires samples covering 0.5 m × 1.5 m and 1.0 m × 1.5 m
Specimen position	Forms a vertical corner
Ignition source	Gas burner of 30.7 kW heat output placed in corner
Test duration	20 min
Conclusions	Classification is based on FIGRA, $THR_{600\,s}$ and maximum flame spread Additional classification is based on SMOGRA, $TSP_{600\,s}$ and droplets/particles. Details are given in Table 10.30 and Table 10.32

EN ISO 11925-2:2020: Reaction to fire tests - Ignitability of products subjected to direct impingement of flame - Part 2: Single-flame source test

EN ISO 11925-2 evaluates the ignitability of a product under exposure to a small flame. The test is relevant for classes B, C, D, E, F, B_{fl}, C_{fl}, D_{fl}, E_{fl}, F_{fl}, B_L, C_L, D_L, E_L, and F_L.

Test rig and specifications are shown in Figure 10.25 and Table 10.37.

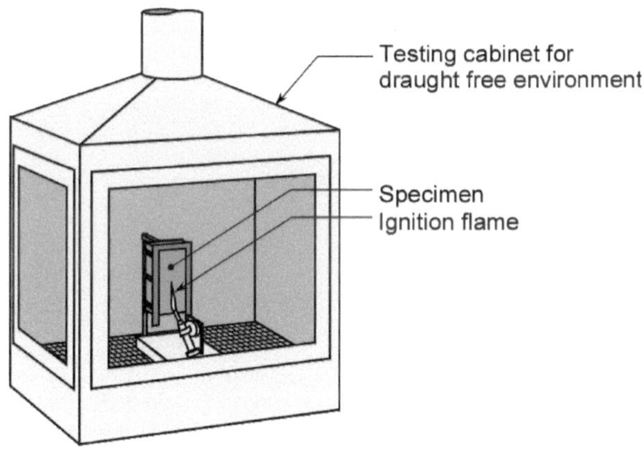

Figure 10.25 EN ISO 11925-2 Small flame test. Image drawn by RISE; reproduced with permission

Table 10.37 EN ISO 11925-2 Small Flame Test. Test Specifications

Specimens	250 mm long, 90 mm wide, thickness maximum 60 mm
Specimen position	Vertical
Ignition source	Small burner. Flame inclined at 45° and impinging either on the edge or the surface of the specimen for 15 s or 30 s
Conclusions	Classification is based on the time for flames to spread and occurrence of droplets/particles

EN 16733:2016: Reaction to fire tests for building products - Determination of a building product's propensity to undergo continuous smouldering

EN 16733 evaluates the propensity of a building product to smolder continuously when exposed to an open flame under natural convective airflow.

It is intended for all building products classified according to EN 13501-1. However, smoldering propensity does not appear as a special class in EN 13501-1. Instead, the field of application, and mounting and fixing of a product for test, are given in relevant product standards.

The test is shown in Figure 10.26 and the test specifications are given in Table 10.38.

Figure 10.26
EN 16733 Smoldering test. Photo: Technical University of Munich, Holzforschung München (HFM)

Table 10.38 Smoldering Test. Test Specifications

Specimens	800 mm long × 300 mm wide × 100 mm thick
Specimen position	Vertical
Ignition source	Gas burner, exposure time 15 min
Test duration	Maximum 6 h or based on temperature rise
Conclusions	Propensity for continuous smoldering (or not) based on visual observations and/or measured temperature rise

EN ISO 9239-1:2010: Reaction to fire tests for floorings – Part 1: Determination of the burning behaviour using a radiant heat source

EN ISO 9239-1 evaluates the critical radiant flux below which flames no longer spread over a horizontal flooring surface. The test is relevant for classes $A2_{fl}$, B_{fl}, C_{fl}, and D_{fl}. The test apparatus is shown in Figure 10.27 and the test specifications are given in Table 10.39.

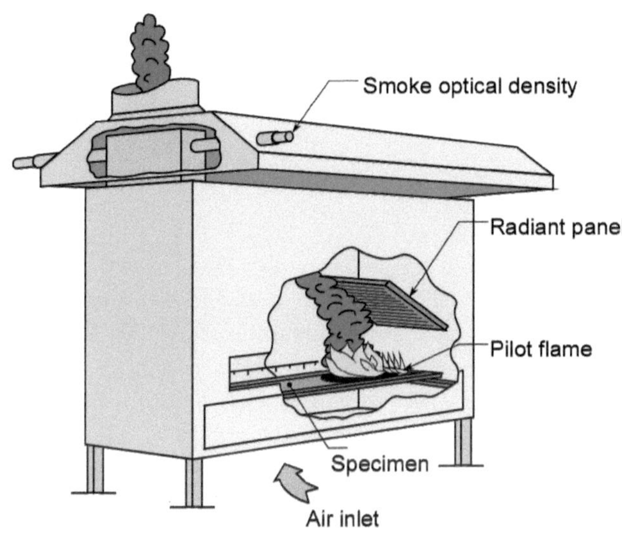

Figure 10.27 EN ISO 9239-1 floor covering test. Image drawn by RISE; reproduced with permission

Table 10.39 EN ISO 9239-1 Floor Covering Test. Test Specifications

Specimens	1050 mm long × 230 mm wide
Specimen position	Horizontal
Ignition source	A gas fired radiant panel that gives a heat flux to the specimen. The maximum heat flux is 11 kW/m², dropping to 1 kW/m² at the end of the specimen. A pilot flame impinges on the surface of the hot end of the specimen to initiate any flame spread
Test duration	30 min
Conclusions	Classification is based on the critical heat flux below which flame spread does not occur. Additional classification is based on smoke density

10.3.1.2.2 Roofs

EN 13501-5:2016: Fire classification of construction products and building elements – Part 5: Classification using data from external fire exposure to roofs tests

There are four methods to choose from when testing the fire performance of roofs or roof coverings - CEN/TS 1187 test 1, test 2, test 3, and test 4. There is no direct correlation between the tests and no hierarchy of classification. Therefore, which method to choose depends on the market where the roof/roof covering will be used. However, all four tests address related fire situations (see Table 10.40).

Table 10.40 Fire Scenarios Specified in CEN/TS 1187

Standard CEN/TS 1187	Roof/Roof covering	Thermal attack
Test 1	Roof	Burning brands
Test 2	Roof covering	Burning brands and wind
Test 3	Roof	Burning brands, wind, and radiant heat
Test 4	Roof	Burning brands, wind, and radiant heat

The classes for the four tests are defined in Table 10.41. As can be seen, there are 13 possible classes for roofs/roof coverings under the CPR. The classes essentially address similar fire phenomena. Tests 1, 3, and 4 measure the fire spread across the external surface of a roof construction, the fire spread within the roof construction, the fire penetration, and the production of flaming droplets or debris falling from the underside of the roof or from the exposed surface. Test 2 measures the fire spread on the surface and on the substrate of a roof covering.

Table 10.41 Classes of External Fire Performance for Roofs/Roof Coverings

Test method	Class	Classification criteria
CEN/TS 1187 Test 1	B_{ROOF} (t1)	All of the following conditions shall be satisfied for any one test: ▪ external and internal fire spread upwards < 0.700 m ▪ external and internal fire spread downwards < 0.600 m ▪ maximum burned length external and internal < 0.800 m ▪ no burning material (droplets or debris) falling from exposed side ▪ no burning/glowing particles penetrating the roof construction ▪ no single through opening > 25 mm^2 ▪ sum of all through openings < 4500 mm^2 ▪ lateral fire spread does not reach the edges of the measuring zone ▪ no internal glowing combustion ▪ maximum radius of fire spread on "horizontal" roofs, external and internal < 0.200 m
	F_{ROOF} (t1)	No performance determined

Test method	Class	Classification criteria
CEN/TS 1187 Test 2	B_{ROOF} (t2)	For both test series at 2 m/s and 4 m/s wind speed: ▪ mean damaged length of the roof covering and substrate < 0.550 m ▪ max. damaged length of the roof covering and substrate < 0.800 m
	F_{ROOF} (t2)	No performance determined
CEN/TS 1187 Test 3	B_{ROOF} (t3)	External fire spread time, $T_E \geq 30$ min and time to fire penetration, $T_P \geq 30$ min
	C_{ROOF} (t3)	$T_E \geq 10$ min and $T_P \geq 15$ min
	D_{ROOF} (t3)	$T_P > 5$ min
	F_{ROOF} (t3)	No performance determined
CEN/TS 1187 Test 4	B_{ROOF} (t4)	No penetration of roof system within 1 h In preliminary test, after withdrawal of the test flame, specimens burn for < 5 min In preliminary test, flame spread < 0.38 m across region of burning
	C_{ROOF} (t4)	No penetration of roof system within 30 min In preliminary test, after withdrawal of the test flame, specimens burn for < 5 min In preliminary test, flame spread <0.38 m across region of burning
	D_{ROOF} (t4)	Roof system is penetrated within 30 min but is not penetrated in the preliminary test In preliminary test, after withdrawal of the test flame, specimens burn for < 5 min In preliminary test, flame spread < 0.38 m across region of burning
	E_{ROOF} (t4)	Roof system is penetrated within 30 min but is not penetrated in the preliminary test Flame spread is not controlled
	F_{ROOF} (t4)	No performance determined

CEN/TS 1187:2012: Test methods for external fire exposure to roofs

The procedures for the test methods are summarized in Table 10.42. These test methods are quite elaborate, and there may be special requirements that are not mentioned in the table, for example when testing roof lights.

Table 10.42 Summary of Test Procedures for CEN/TS 1187 tests 1–4

Test	Principle	Specimen and special conditions of test	Ignition source	Test time
Test 1 Burning brand	The roof deck construction is tested including insulation, layers, joints etc. Pitch 15° or 45° depending on use	Specimen size 0.8 m × 1.8 m Normally four tests	600 g of wood wool in a wire basket 300 mm × 300 mm and 200 mm deep placed on the deck almost centrally	Normally 60 min
Test 2 Burning brand and wind	The roof covering is tested fixed to a substrate representing end use conditions Pitch 30°	Specimen size 0.4 m × 1.0 m Three tests for each wind speed Wind speed 2 m/s and 4 m/s	Wooden crib 100 mm × 100 mm Weight 40 g placed close to the lower edge of the sample	15 min or earlier if total flame spread occurs
Test 3 Burning brand, wind and radiant heat	The roof deck construction is tested including insulation, layers, joints etc. Pitch 5° or 30° depending on use	Specimen size 1.2 m × 3.0 m Two tests Wind speed 3 m/s at the lower area of the specimen Heat flux on the lower area of the specimen 10–12.5 kW/m^2 as measured with a water-cooled heat flux meter	Wooden crib 55 mm × 55 mm and 32 mm high made from four pieces of fiber board Two cribs soaked in heptane are used for each test	30 min

Test	Principle	Specimen and special conditions of test	Ignition source	Test time
Test 4 Method with two stages: (1) Burning brand; (2) Wind and radiant heat	The roof deck construction is tested including insulation, layers, joints etc. Pitch 0° or 45° depending on use	Specimen size 0.84 m × 0.84 m (1) One test. (2) Three tests with wind and radiant heat; simulated wind speed of 6.7 m/s by applying under-pressure to the underside of the test specimen; heat flux to the specimen 12 kW/m^2 as measured with a water-cooled heat flux meter	Simulated town-gas flame (55.2% hydrogen, 27.4% natural gas, 17.4% nitrogen) Flame length 230 mm from a 9.5 mm nozzle	(1) 1 min test time if no ignition otherwise until flaming ceases (2) Test time 1 h of radiant heat exposure. After 5 min of testing the specimen is exposed to the gas flame during 1 min at various locations

10.3.1.2.3 Cables

EN 13501-6:2018: Fire classification of construction products and building elements – Part 6: Classification using data from reaction to fire tests on electric cables

In a similar way as for construction products, cables are classified in seven main classes with sub-classes for smoke production and formation of burning droplets. However, there is also a sub-class for acidity, which addresses whether a cable can produce acid gases like (for example) HCl when burning. The classification table is shown in Table 10.43.

Table 10.43 Classes of Reaction-to-Fire Performance for Electric Cables

Class	Test method(s)	Classification criteria	Additional classification
A_{ca}	EN ISO 1716	PCS ≤ 2.0 MJ/kg [a]	-
$B1_{ca}$	EN 50399 (30 kW flame source) *And*	Fs ≤ 1.75 m *and* $THR_{1200\,s}$ ≤ 10 MJ *and* Peak HRR ≤ 20 kW *and* FIGRA ≤ 120 W/s	Smoke production [b, e] *and* Flaming droplets/particles [c] *and* Acidity [d]
	EN 60332-1 & -2	H ≤ 425 mm	

Table 10.43 Classes of Reaction-to-Fire Performance for Electric Cables *(continued)*

Class	Test method(s)	Classification criteria	Additional classification
$B2_{ca}$	EN 50399 (20.5 kW flame source) *And*	Fs ≤ 1.5 m *and* $THR_{1200\,s}$ ≤ 15 MJ *and* Peak HRR ≤ 30 kW *and* FIGRA ≤ 150 W/s	Smoke production [b, f] *and* Flaming droplets/ particles [c]. *and* Acidity [d]
	EN 60332-1 & -2	H ≤ 425 mm	
C_{ca}	EN 50399 (20.5 kW flame source) *And*	Fs ≤ 2.0 m *and* $THR_{1200\,s}$ ≤ 30 MJ *and* Peak HRR ≤ 60 kW *and* FIGRA ≤ 300 W/s	Smoke production [b, f] *and* Flaming droplets/ particles [c] *and* Acidity [d]
	EN 60332-1 & -2	H ≤ 425 mm	
D_{ca}	EN 50399 (20.5 kW flame source) *And*	$THR_{1200\,s}$ ≤ 70 MJ *and* Peak HRR ≤ 400 kW *and* FIGRA ≤ 1300 W/s	Smoke production[b, f] *and* Flaming droplets/ particles [c] *and* Acidity [d]
	EN 60332-1 & -2	H ≤ 425 mm	-
E_{ca}	EN 60332-1 & -2	H ≤ 425 mm	-
F_{ca}	EN 60332-1 & -2	H > 425 mm	-

[a] For the product as a whole, excluding metallic materials, and for any external component (i. e., sheath) of the product.

[b] s1 = $TSP_{1200\,s}$ ≤ 50 m^2 and Peak SPR ≤ 0.25 m^2/s. s1a = s1 and transmittance in accordance with EN 61034-2 (≥ 80%). s1b = s1 and transmittance in accordance with EN 61034-2 (≥ 60% and < 80%). s2 = $TSP_{1200\,s}$ ≤ 400 m^2 and Peak SPR ≤ 1.5 m^2/s. s3 = not s1 or s2.

[c] d0 = No flaming droplets/particles within 1200 s; d1 = No flaming droplets/particles persisting longer than 10 s within 1200 s; d2 = not d0 or d1.

[d] EN 60754-1, EN 60754-2: a1 = conductivity < 2.5 μS/mm and pH > 4.3; a2 = conductivity < 10 μS/mm and pH > 4.3; a3 = not a1 or a2.

[e] The smoke class declared for class $B1_{ca}$ cables shall originate from the test according to EN 50399 (30 kW flame source).

[f] The smoke class declared for class $B2_{ca}$, C_{ca}, D_{ca} cables shall originate from the test according to EN 50399 (20.5 kW flame source).

The extra sub-classes imply that there are a vast number of possible combinations to create classes. However, in reality only a few are used since a number of the combinations reflect similar fire properties. There is no obligation to declare a3 if acidity is not determined.

Reference Scenario and Fire Situations for Cables

Cables have reference scenarios typical of cable installations. The testing technology in EN 50399 and its links to reference scenarios was first developed in the large project funded by the EU called FIPEC (FIre Performance of Electric Cables) [11], and developed by CENELEC/TC 20/WG 10. Techniques for extended application of test data was then developed in the CEMAC project funded by Europacable [12].

There are two full-scale reference scenarios, a horizontal and a vertical one. The horizontal reference scenario (see Figure 10.28) comprises three cable ladders, each about 3 m, mounted horizontally on top of each other. They are ignited in one end, with a burner giving 40 kW for 5 min, then 100 kW for another 10 min and finally 300 kW for 10 min. The supply of air is by natural ventilation.

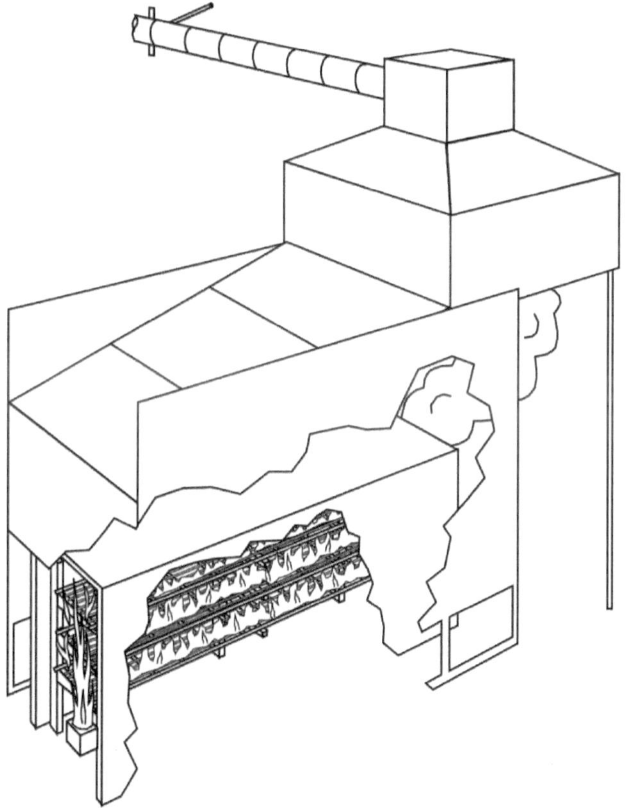

Figure 10.28 Horizontal reference scenario for cables [11]. Image drawn by RISE; reproduced with permission

A cable ladder mounted in a corner comprises the vertical reference scenario shown in Figure 10.29. The heat output from the burner is the same as for the horizontal scenario.

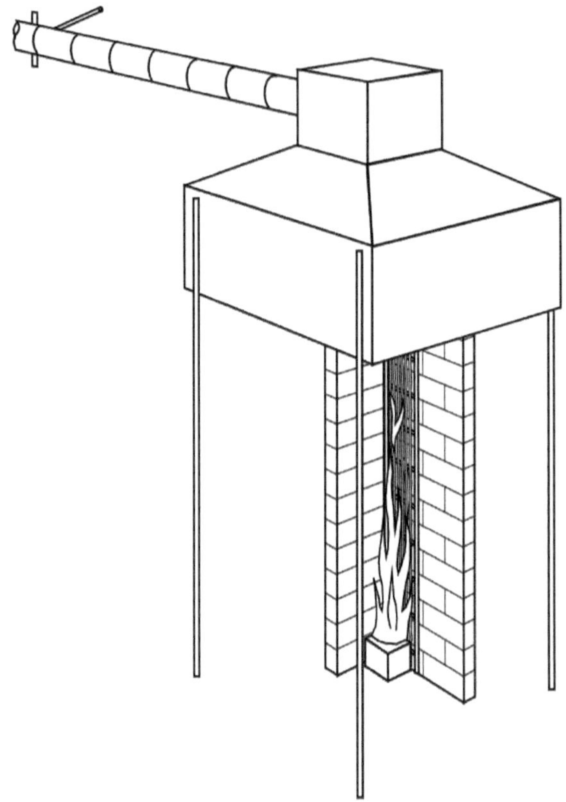

Figure 10.29 Vertical reference scenario for cables [11]. Image drawn by RISE; reproduced with permission

The burning behavior of the cables in different classes can approximately be described as follows.

- Class A_{ca}: Level of highest performance corresponding to products that in practice cannot burn, i. e., ceramic products.
- Class $B1_{ca}$: Products that are combustible but show no or very little burning when exposed to both the reference scenario tests and the classification test procedure EN 50399 (30 kW flame source).
- Class $B2_{ca}$ and Class C_{ca}: Products that do not give a continuous flame spread when exposed to the 40–100 kW ignition source in the horizontal reference scenario, and that do not give a continuous flame spread, show a limited fire growth rate and show a limited heat release rate when tested according to EN 50399 (20.5 kW flame source).

- Class D_{ca}: Products that show a fire performance approximately like wood when tested in the reference scenarios. When tested according to EN 50399 (20.5 kW flame source) the products show a continuous flame spread, a moderate fire growth rate, and a moderate heat release rate.
- Class E_{ca}: Products where a small flame attack does not cause large flame spread.
- Class F_{ca}: Products where a small flame attack can cause large flame spread.

EN 50399:2011 and A1:2016: Common test methods for cables under fire conditions – Heat release and smoke production measurement on cables during flame spread test – Test apparatus, procedures, results

The major standard for testing cables is EN 50399 (Figure 10.30). It is originally from IEC 60332-3, but has improved, in particular by adding measurements of HRR and smoke production.

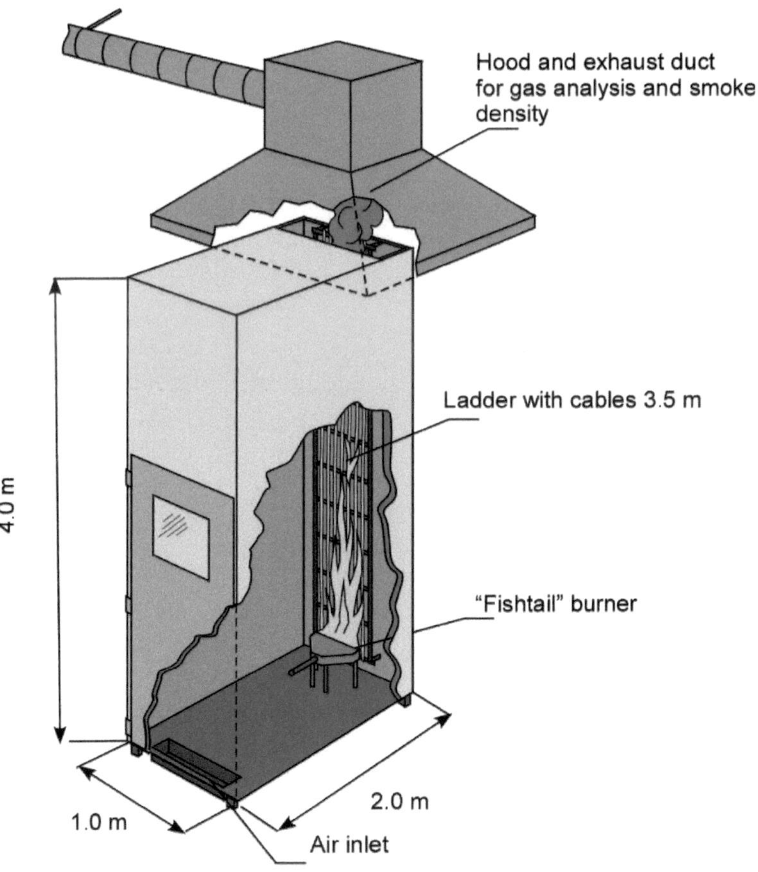

Figure 10.30 EN 50399 for testing of cables. Image drawn by RISE; reproduced with permission

The cables are 3.5 m long and mounted on a cable ladder 0.5 m wide. The number of cables is varied to cover similar widths (approximately). Small cables are bundled. The ignition source is a gas burner. Test specifications are shown in Table 10.44.

Table 10.44 Test Conditions for Classification of Cables

Class $B1_{ca}$	▪ 30.0 kW ignition source ▪ 20 min testing time ▪ A backing board is mounted behind the ladder
Class $B2_{ca}$ Class D_{ca}	▪ 20.5 kW ignition source ▪ 20 min testing time ▪ There is no backing board behind the ladder

HRR and smoke density are measured using the same principles as in ISO 9705, the room fire test. In addition, flame spread along the cables and the occurrence of flaming droplets is determined by visual observation. The test data is then used to derive the classification parameters (see Table 10.43).

The reason for the differences in test procedure between class $B1_{ca}$ and the other classes mainly has to do with the need for a class of very high performance and therefore a more severe test. The top class A_{ca} is for products that practically cannot burn, i. e., made of ceramic materials.

Cables appear in families containing individual cables that are of similar construction, although they are not identical. In CEMAC [12] and later projects, a large number of cables of different families was tested in order to find a relation between results for a specific family that would allow results to be predicted based only on a few tests. Such rules for extended application (EXAP) are necessary as testing of each individual cable in a cable family is unpractical. It was found that a parameter χ could be used to determine what cables to test. χ is defined as:

$$\chi = \frac{c}{d^2} V_{combust}$$

where
d [m] = Outer diameter
$V_{combust}$ [m^2] = Non-metallic volume per meter ladder
c [dimensionless] = Number of conductors in one cable.

In most cases, only two tests are needed as basis for an EXAP. All cables within the same family that have a value of the cable parameter between the lowest and highest value of the cable parameters of the tested cables are included in the EXAP. Classification is based on the maximum measured value plus a safety margin. There are two documents describing the EXAP procedures:

CENELEC Technical Specification 50576:2016: Electric cables – Extended application of test results for reaction to fire and

NB-CPR/SH02-16/681rev3:2017: Extended field of application (EXAP) for reaction-to-fire Euro-classification of copper communication cables (CCC)

The latter document is an approved guidance from SH02 - the Fire Sector Group of Notified Bodies for the Construction Products Regulation. This document is to be used until there is a full standard available from CENELEC. The CENELEC technical specification also covers EXAP for the small flame and smoke test described below.

EN 60332-1-2:2004+A1:2015+A11:2016: Tests on electric and optical fibre cables under fire conditions - Parts 1-2: Test for vertical flame propagation for a single insulated wire or cable - Procedure for 1 kW pre-mixed flame (IEC 60332-1-2:2004/A1:2015)

This is a small flame test of similar type as for other construction products but adapted for cables. A single cable 600 mm long is exposed to a premixed flame from a Bunsen-type burner that gives 1 kW of heat release rate. The criteria for all classes except class F_{ca} is flame spread less than 425 mm.

EN 61034-2:2005/A1:2013: Measurement of smoke density of electric cables burning under defined conditions - Part 2: Test procedure and requirements

A cable sample, 1 m long, is made from a certain number of cable pieces depending on cable diameter. The sample is mounted horizontally in a box having the dimensions 3 m × 3 m × 3 m (see Figure 10.31).

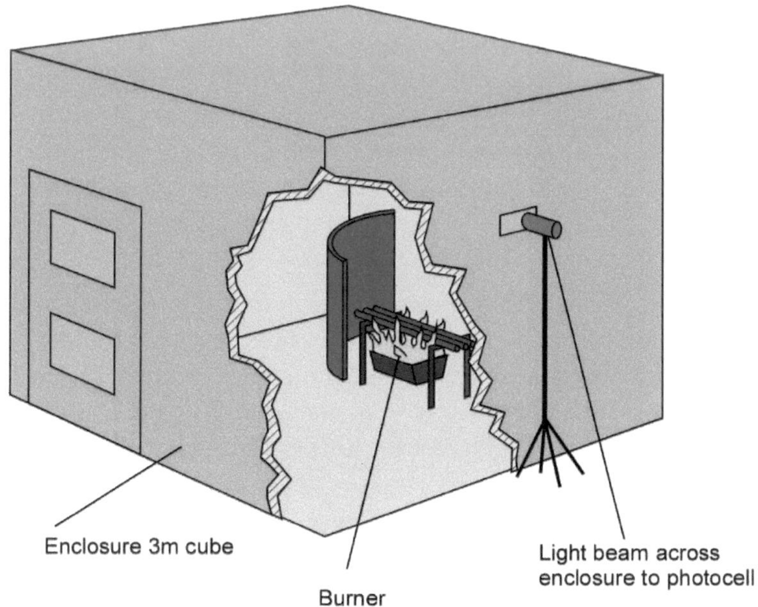

Figure 10.31 The smoke box, so called 3 m cube, according to EN 61034-2. Image drawn by RISE; reproduced with permission

The sample is exposed to flames from 1 L of alcohol burning in a tray. A white light photocell/lamp system is used to measure light transmittance through the upper part of the box. Two classes of smoke production are defined based on light transmittance (I): I ≥ 80% or 60% ≤ I < 80%. This is then combined with the smoke production classes as measured in EN 50399, giving a total of five possible classes for smoke production of classes, s1, s1a, s1b, s2, and s3.

EN 60754-1:2014: Tests on gases evolved during combustion of materials from cables - Part 1: Determination of the halogen acid gas content (IEC 60754-1:2011 + corrigendum Nov. 2013)
and
EN 60754-2:2014: Tests on gases evolved during combustion of materials from cables - Part 2: Determination of acidity (by pH measurement) and conductivity (IEC 60754-2:2011)

For determining the sub-classes a1, a2, and a3 only EN 60754-2 is used. A small sample of the cable material is burned in a tubular furnace. The gas is then passed through a sampling bottle with water at a certain rate. Electrical conductivity and pH-value are then determined for the liquid, and used for classification. Typically, products containing halogens will be detected.

10.3.1.3 Future Developments

The reaction-to-fire tests used for the CPR have been established over a number of years, and are constantly being improved by CEN. The set of tests and especially the EXAP standards for cables are relatively recent. Therefore, there is some focus on improving them and establishing good practice in their use, through the use of interlaboratory calibrations as well as interpretations of the standard in new situations.

The CPR does not include any requirements for the toxic gas species in the smoke from fires. However, such requirements are found in shipping and in Europe for products used in trains. Some new studies on test methods are made and the Commission has sponsored a study on toxicity.

As a result of many casualties in some very large fires in high-rise buildings with façades, a major effort is currently being put into developing a large-scale test standard for façades, harmonized in all EU countries. A common project to develop a test is continuing, with participants from almost all EU countries. A large-scale test needs not only to include the façade surface but also penetrations, window details, etc. As a result of these efforts, it is expected that a large-scale test will become a European standard in the near future.

References for Section 10.3.1

[1] *https://www.europa.eu/.*

[2] Regulation (EU) No. 305/2011 of The European Parliament and of The Council, of 9 March 2011, laying down harmonised conditions for the marketing of construction products and repealing Council Directive 89/106/EECA.

[3] Sundström, B., "Full-scale fire testing of surface materials. measurements of heat release and production of smoke and gas species", Technical Report SP-RAPP1986:45, Borås, 1986.

[4] Sundström, B., van Hees, P., Thureson, P., "Results and analysis from fire tests of building products in ISO 9705, the room/corner test: The SBI research programme" SP Report 1998:11, ISBN 91-7848-716-1.

[5] EUREFIC Seminar Proceedings, Interscience Communications Ltd, London, September 1991, ISBN 0 9516320 19.

[6] Sundström, B., "The development of a European fire classification system for building products test methods and mathematical modelling" (PhD thesis), Department of Fire Safety Engineering, Lund University, Sweden, 2007, ISBN 978-91-628-7243-4.

[7] Proposed method for room fire test of wall and ceiling materials and assemblies, *Annual Book of ASTM Standards* (part 18), 1982.

[8] ISO/TR 9705-2:2001: Reaction-to-fire tests - Full-scale room tests for surface products - Part 2: Technical background and guidance.

[9] Axelsson, J., Sundström, B., Rohr, U., "Development of a common European system for fire testing and classification of pipe insulation", Ninth International Interflam Conference, Edinburgh, September 2001, Volume 1, pp. 485–494, Interscience communications Ltd, ISBN 0 9532312 8 3.

[10] Sundström, B., Axelsson, J., "Development of a common European system for fire testing of pipe insulation based on EN 13823 (SBI) and ISO 9705 (room/corner test)" SP Report 2002:21, ISBN 91-7848-871-0.

[11] Grayson, S., Van Hees, P., Vercellotti, U., Breulet, H., Green, A., "Fire Performance of Electric Cables - New test methods and measurement techniques: Final report of EU SMT project SMT4-CT96-2059", Interscience Communications UK, London, 2000, ISBN 0 9532312 5 9.

[12] CEMAC - CE-marking of cables, SP Report 2012:27, ISBN 978-91-86319-65-6.

10.3.2 National Fire Regulations and Tests

Edith Antonatus

The European classification and test system for the reaction to fire of construction products as described in Section 10.3.1 has been introduced in all EU member countries, as required by the Construction Products Regulation (CPR). All products for which harmonized European product standards have been published must be tested and classified according to the European Standards. For specific construction products, no harmonized product standards have been developed. These continue to be subject to the national fire performance tests and classifications in some European countries.

There is a division of powers between the EU and EU countries: the EU deals only with the single market access rules and not with requirements for the performance of products themselves. While the CPR sets out methods and criteria for assessing and expressing the performance of construction products, as well as the conditions for the use of CE marking, the EU countries, on the other hand, are responsible for fire safety, mechanical resistance and stability, environmental, energy, and other requirements applicable to construction works [1].

Consequently, all EU member countries still have their own national building regulations defining the fire performance levels required for the different products used in building.

Where no European test and classification methods have been developed, it is still possible to define national methods. An example are the different full-scale façade tests developed in most European countries.

The following sections give an overview of the building regulations, fire performance classification and testing of construction products in some of the most important European countries. Further information about the old national fire performance requirements and test methods can be found in the third edition of this book [2].

References for Section 10.3.2

[1] Summary of regulation (EU) No. 305/2011 – Harmonised conditions for the marketing of construction products, *https://eur-lex.europa.eu/legal-content/EN/TXT/HTML/?uri=LEGISSUM:mi0078&from=EN.*

[2] *Plastics Flammability Handbook: Principles, Regulations, Testing, and Approval* (3rd edition), Troitzsch, J. (editor), Carl Hanser Verlag, Munich, Germany, Chapter 10, 2004.

10.3.2.1 Austria

Dieter Werner

10.3.2.1.1 Statutory Regulations

Austria is a federal republic, consisting of nine provinces ("Laender"). The building regulations are issued by each province independently, which means that there has never been one single Austrian building law, but at least nine systems each consisting of a building law and related orders [1]. Given the size of Austria, one may question the need for nine separate building regulations, so since 1948 efforts have been made to unify them.

In 2000 the Austrian Laender established an expert group to draft a new proposal for the harmonization of building regulations in cooperation with the Austrian Institute of Construction Engineering (OIB). According to the European New Approach, legislative provisions should be restricted to essential requirements, whereas technical details should be transferred to referenced non-legal documents [2].

This resulted in a two-tier approach for the new Austrian building regulations, namely that the regulations contained in laws and/or orders shall be restricted to purely functional requirements, and any performance-based or prescriptive requirements shall be laid down in the OIB-Guidelines.

When a building is designed in accordance with the OIB-Guidelines, it is deemed to satisfy the functional requirements of the legislative provisions. It is also possible to deviate from the OIB-Guidelines, as long as an equivalent level of safety can be demonstrated [3].

In 2007, the first generation of the OIB-Guidelines was issued, with revisions being made in 2011, 2015, and 2019. The current OIB-Guidelines are organized according to the basic requirements for construction works of the European Construction Products Regulation (see Table 10.45) [4]. (No OIB-Guideline exists for the basic requirement "Sustainable use of natural resources".)

Table 10.45 OIB-Guidelines

OIB-Guideline No.	OIB-Guideline title
1	Mechanical resistance and stability
2	Safety in case of fire [5]
2.1	Safety in case of fire in operational structures [6]
2.2	Safety in case of fire in garages, roofed parking spaces, and multi-storey car parks [7]
2.3	Safety in case of fire in buildings with a fire escape level in excess of 22 m [8]
3	Hygiene, health and preservation of the environment
4	Safety in use and accessibility
5	Protection against noise
6	Energy saving and heat insulation

All nine Austrian Laender declared these OIB-Guidelines as binding in their building codes, which means that the technical requirements concerning building in Austria are now harmonized.

The requirements in the guidelines are mostly descriptive, but sometimes also performance-based. Wherever possible, the requirements concerning construction products and their behavior in case of fire are based on European harmonized standards, mainly the classifications defined in the EN 13501 series.

10.3.2.1.2 Classification and Testing of the Fire Performance of Building Materials and Components

Construction Products in General

In Austria, since 2010, all construction products have to be tested and classified in accordance with the EN 13501 series. With the exception of the non-harmonized construction products described below, all national standards that dealt with the testing of fire behavior of the same construction products have been withdrawn (for details see the third edition of this book [9]).

There is no direct correlation between the "old" Austrian combustibility classes and the European reaction-to-fire-classes defined in EN 13501-1, because the test procedures changed. Nevertheless Austrian experts have developed a comparison table to show possible connections between the classes (see Table 10.46 and Table 10.47).

Table 10.46 Interpretation of Reaction-to-fire Performance According to the European Classification System (Construction Products without Floorings)

Class acc. to EN 13501-1	Provisions in former Austrian regulations	
A1 A2-s1,d0	Non-combustible Low smoke production Non-dripping	Class A
B-s1,d0 C-s1,d0	Low combustibility Low smoke production Non-dripping	Class B1, Q1, Tr1
A2-s2,d0 A2-s3,d0 B-s2,d0 B-s3,d0 C-s2,d0 C-s3,d0	Low combustibility Non-dripping	Class B1, Q2 or Q3, Tr1
A2-s1,d1 A2-s1,d2 B-s1,d1 B-s1,d2 C-s1,d2 C-s1,d2	Low combustibility Low smoke production	Class B1, Q1, Tr2 or Tr3
A2-s3,d2 B-s3,d2 C-s3,d2	Low combustibility	Class B1, Q2 or Q3, Tr2 or Tr3

Class acc. to EN 13501-1	Provisions in former Austrian regulations	
D-s1,d0	Moderately combustible Low smoke production Non-dripping	Class B2, Q1, Tr1
D-s2,d0	Moderately combustible Non-dripping	Class B2, Q2 or Q3, Tr1
D-s3,d0		
E		
D-s1,d2	Moderately combustible Low smoke production	Class B2, Q1, Tr2 or Tr3
D-s2,d2	Moderately combustible	Class B2, Q2 or Q3, Tr2 or Tr3
D-s3,d2		
E-d2		
F	Highly combustible	Class B3

Table 10.47 Interpretation of Reaction-to-fire Performance According to the European Classification System (Floorings)

Class acc. to EN 13501-1	Provisions in former Austrian regulations	
$A1_{fl}$	Non-combustible Low smoke production	Class A
$A2_{fl}$-s1		
B_{fl}-s1	Low combustibility Low smoke production	Class B1
C_{fl}-s1		
D_{fl}-s1	Moderately combustible Low smoke production	Class B2
$A2_{fl}$-s2	Moderately combustible	
B_{fl}-s2		
C_{fl}-s2		
D_{fl}-s2		
E_{fl}		
F_{fl}	Highly combustible	Class B3

These tables offer the possibility to transfer European classes into the old Austrian system, and were officially released by OIB as attachments to the Austrian building material lists [10, 11].

Curtains and Drapes

Testing and classification of curtains and drapes may be required according to the standards EN 13772, EN 1101, and EN 1102 (testing) and EN 13773 (classification). Table 10.48 shows the connection between the European classification system and the Austrian classification described in the withdrawn ÖNORM B 3820.

Table 10.48 Interpretation of Reaction-to-fire Performance According to the European Classification System (Curtains)

Class according to EN 13773	Provisions in former Austrian regulations
Class 1 or 2	Low combustibility B1
Class 3 or 4	Moderately combustible B2
Class 5	Highly combustible B3

Non-Harmonized Construction Products

For testing the reaction-to-fire performance of non-harmonized construction products that must fulfill requirements regarding their fire behavior, such as tarpaulins or mobile materials, the class B1 test (low combustibility) of ÖNORM B 3800-1 was kept as national standard ÖNORM A 3800-1 [12].

The test apparatus, named the Schlyter test rig, is generally used to provide evidence that materials are of low combustibility. The test equipment is illustrated in Figure 10.32, and the test specifications are summarized in Table 10.49.

Figure 10.32 Schlyter test to ÖNORM A 3800-1

Table 10.49 Schlyter Test Specifications

Specimens	Per test two specimens, each 800 mm × 300 mm; the test has to be performed three times
Specimen position	Two vertical specimens parallel to and at a distance of 50 mm from each other (surface to surface), one specimen displaced 50 mm downwards
Ignition source	Row of six gas jets at 30 mm intervals; flames impinge on lower specimen at an angle of 30° to the horizontal; distance of jets from flamed specimen 40 mm; flame length in vertical position 120 mm; gas supply 3.5 parts N_2 to 1 part propane (21 MJ/m^3)
Test duration	15 min
Results	Combustibility class B1 is achieved if: ▪ The specimen that is not directly exposed to the flame does not ignite ▪ After removal of ignition source, the specimen exposed to the flame continues to burn ≤ 1 min and to glow ≤ 5 min Unburnt length of flamed specimen ≥ 400 mm

Façades

In Austria, fire protection requirements concerning façades include demands for the reaction-to-fire performance, as well as for the restriction of fire spread along a façade.

On the one hand, requirements for the reaction-to-fire class of the façade system apply for all building classes (classes to EN 13501-1: the higher the building, the more stringent the required reaction to fire class; for high-rise-buildings with an escape level > 22 m, only materials with classes A1 or A2 are allowed).

On the other hand, façades on 4-storey buildings and higher have to meet additional requirements concerning the fire spread along the façade. This can either be confirmed by a positive test according to ÖNORM B 3800-5 (large-scale fire test) [13] or in a different way, if the same level of protection is demonstrated.

For the test to ÖNORM B 3800-5, the façade construction has to be built up under original installation conditions on a 6 m high test rig (Figure 10.33). The façade system is thermally stressed by a defined fire load under natural ventilation conditions. The thermal attack is simulated by a fire from a 25 kg spruce crib (Figure 10.34). The wood crib is located in an inner corner of the façade behind a virtual window. As a result, the greatest possible thermal attack is simulated by the flow conditions that cause an extensive flame in the inner corner. The wood crib is ignited using isopropanol and then impinges on the façade system under specified ventilation conditions over a period of 20 to 25 minutes (until the crib collapses).

The test result is considered positive if all three of the following conditions are met:

- No fire spread has occurred, which means
  - No fire damage of the test specimen
  - No visible sustained flames outside the area of flames caused by the crib fire.
- The temperature criterion is fulfilled (temperatures behind the façade surface during the test must not exceed those in front of the façade surface; the measurements are compared for defined heights)
- No large parts (> 5 kg or > 0.4 m^2) or burning parts fall off during the test.

Note: The test to ÖNORM B 3800-5 is very similar to the one defined in German DIN 4102-20.

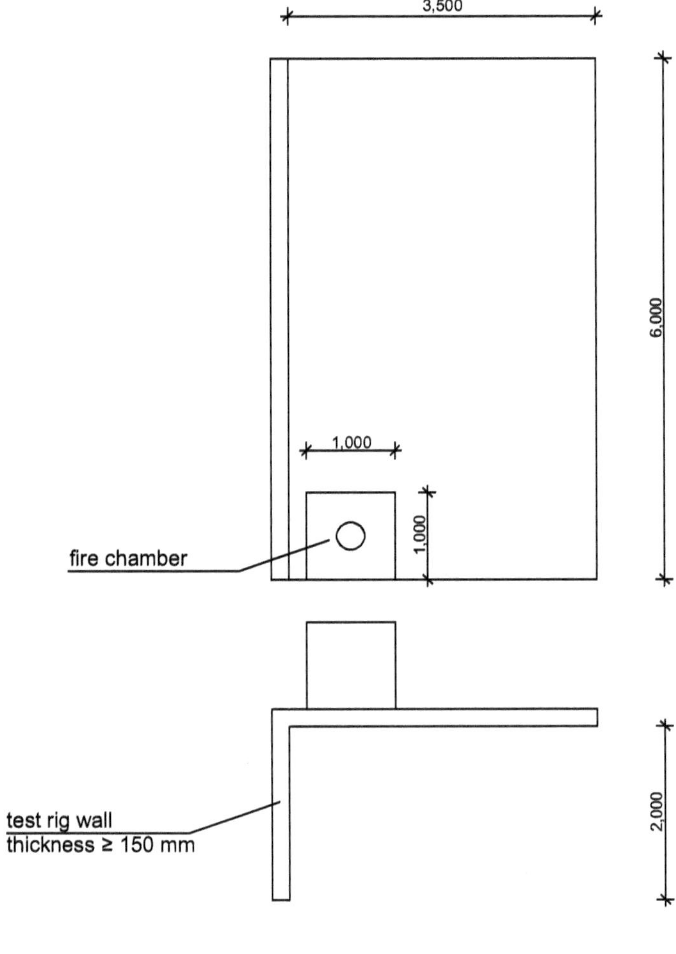

Figure 10.33 Rig for testing in accordance with ÖNORM B 3800-5

Figure 10.34 Crib fire during an ÖNORM B 3800-5 test

Summary

A concise overview of the classification and testing of the fire performance of building materials and components in Austria is shown in Table 10.50.

Table 10.50 Relevant Standards for Testing Fire Performance of Building Materials in Austria

Building materials	Fire behavior	Smoke production	Flaming droplets
Construction products	EN 13501-1	EN 13501-1	EN 13501-1 (except floorings)
Curtains and drapes	EN 13773	ÖNORM A 3800-1	EN 13772
Upholstered furniture	ÖNORM B 3825	ÖNORM A 3800-1	–
Decorative articles	ÖNORM B 3822	ÖNORM A 3800-1	ÖNORM B 3822
Non-harmonized construction products	ÖNORM A 3800-1	ÖNORM A 3800-1	ÖNORM A 3800-1
Façades (ETICS, rear-ventilated systems)	EN 13501-1 ÖNORM B 3800-5	EN 13501-1	EN 13501-1 ÖNORM B 3800-5

10.3.2.1.3 Official Approval

Two institutes in Austria are accredited for the whole range of test methods (fire resistance as well as reaction-to-fire tests), based mainly on European standards:

- Prüf-, Inspektions- und Zertifizierungsstelle der Stadt Wien – MA 39
  Research Centre, Laboratory and Certification Services
  A-1110 Vienna, Rinnböckstraße 15
  phone: +43 1 4000 8039
  fax: +43 1 4000 99 8039
  e-mail: *post@ma39.magwien.gv.at*
  web: *https://www.wien.gv.at/forschung/laboratorien/index.html*
- IBS-Institut für Brandschutztechnik und Sicherheitsforschung GesmbH
  A-4020 Linz, Petzoldstraße 45
  phone: +43 732 7617-250
  fax: +43 732 7617-119
  e-mail: *office@ibs-austria.at*
  web: *https://www.ibs-austria.at/*

Both are accredited for testing, inspection, and certification services according to EN ISO IEC 17020, 17025, and 17065, and notified product certification bodies and test labs according to CPR. Both are also members of EGOLF, the European Group of Official Laboratories for Fire Testing.

Some institutes are accredited for special methods, especially for testing the reaction-to-fire of specific products:

- OFI Technologie & Innovation GmbH
  A-1030 Vienna, Franz-Grill-Strasse 5, Objekt 213
- Staatliche Versuchsanstalt am Technologischen Gewerbemuseum
  A-1200 Vienna, Wexstraße 19-23
- OETI – Institute for Ecology, Technology and Innovation
  A-1050 Vienna, Spengergasse 20

After the first complete test with positive results, the test laboratory provides a test report giving the test results, and a classification report for the building material according to the applied test and classification standards. Note: If tests were carried out according to national standards, the classification of the product tested is also given in the test report, not in a separate classification report.

The classification and test reports form the basis for the use of the product in Austria and/or for obtaining marking (CE and national marking ÜA) and approvals. For marking and approvals, OIB is in a key position, as the institute is a European technical assessment body and national approval body for construction products. As the market surveillance authority, the OIB ensures that the construction products on the Austrian market fulfil all legal requirements and do not endanger health and safety. As a product contact point, the OIB provides information about the currently valid technical requirements for construction products in Austria.

10.3.2.1.4 Future Developments

Wherever possible, Austria has already progressed from national fire performance standards to harmonized European standards in testing and classification, as well as in building code requirements.

National standards still only exist for a few products, the most prominent amongst these being for façade systems such as ETICS, or rear-ventilated systems. Here, future developments may lead to a harmonized European façade standard including the Austrian/German fire testing scenario. In supporting this scenario, Austria wants to play an important role in this European standard development process, which has already been started with the organization of a theoretical round-robin on the proposed assessment method throughout the member states.

References for Section 10.3.2.1

[1] Gesetz vom 1. Oktober 1920 (BGBl. Nr. 1), womit die Republik Österreich als Bundesstaat eingerichtet wird (Bundes-Verfassungsgesetz) in der Fassung der Wiederverlautbarung des Bundes-Verfassungsgesetz vom 7. Dezember 1929, Wien, 1929.

[2] Council Resolution of 7 May 1985 on a new approach to technical harmonization and standards (85/C 136/01).

[3] Mikulits, R.: *The legal two-tier approach in the new Austrian Building Codes*, Austrian Institute of Construction Engineering (OIB), 2008.

[4] Regulation (EU) No 305/2011 of the European Parliament and of the Council of 9 March 2011 laying down harmonised conditions for the marketing of construction products and repealing Council Directive 89/106/EEC.

[5] OIB-330.2-012/19: OIB-Richtlinie 2: Brandschutz, Austrian Institute of Construction Engineering (OIB), 2019.

[6] OIB-330.2-013/19: OIB-Richtlinie 2.1: Brandschutz bei Betriebsbauten, Austrian Institute of Construction Engineering (OIB), 2019.

[7] OIB-330.2-014/19: OIB-Richtlinie 2.2: Brandschutz bei Garagen, überdachten Stellplätzen und Parkdecks, Austrian Institute of Construction Engineering (OIB), 2019.

[8] OIB-330.2-015/19: OIB-Richtlinie 2.3: Brandschutz bei Gebäuden mit einem Fluchtniveau von mehr als 22 m, Austrian Institute of Construction Engineering (OIB), 2019.

[9] Pöhn, C.: "Austria" (Section 10.5), in: *Plastics Flammability Handbook*, 3rd Edition, Troitzsch, J. (ed.), Hanser, Munich, 2004.

[10] OIB-095.2-026/13: Konsolidierte Fassung der Liste der Bauprodukte und der Anlagen A und B der Baustoffliste ÖE (4. Ausgabe der Baustoffliste ÖE), Austrian Institute of Construction Engineering (OIB), 2013.

[11] OIB-095.1-015/15: Verordnung des Österreichischen Instituts für Bautechnik (OIB) über die Baustoffliste ÖA (Neufassung 2015), Austrian Institute of Construction Engineering (OIB), 2015.

[12] ÖNORM A 3800-1:2005-11-01: Burning behaviour of materials excluding building products - Part 1: Requirements, tests and evaluations, Austrian Standards International, 2005.

[13] ÖNORM B 3800-5:2013-04-15: Fire behaviour of building materials and components – Part 5: Fire behaviour of façades - Requirements, tests and evaluations, Austrian Standards International, 2013.

10.3.2.2 Belgium

Bart Sette

10.3.2.2.1 Statutory Regulations

Overview of Authorities

Belgium has evolved into a federal structure [1-3]. This occurred through six state reforms (in 1970, 1980, 1988-89, 1993, 2001, and 2012-14). As a result, the first Article of the Belgian constitution reads today: "Belgium is a federal state, composed of communities and regions".

The power to make decisions is no longer the exclusive privilege of the federal government and the federal parliament. The leadership of the country is now in the hands of various partners, namely: the Federal government; the Flemish, French, and German communities; and the Flemish, Walloon, and Brussels-Capital region. The Federal government, the Communities, and the Regions independently exercise their authority within their domains.

In addition, there are the provinces and municipalities that are decentralized territorially and that are under administrative supervision. So in total, there are five authority instances/levels in Belgium with respect to the fire safety regulations in buildings, as described in Table 10.51.

With respect to fire safety, the federal government establishes the minimum requirements which have to be applied generally for new buildings - the "Basic Standards". The other authority instances/levels can add specific, additional requirements.

Federal Government

The minimal requirements which a building has to satisfy have been published by means of a Royal Decree (RD) dated July 7, 1994, resulting in the "Basic Standards for the prevention of fire and explosion". New buildings have to comply with these Basic Standards. This RD has been changed by RDs dated December 19, 1997; April 4, 2003; June 13, 2007; March 1, 2009; July 12, 2012; and December 7, 2016.

The federal government can also decide on specific regulations for those buildings for which it has responsibility. Football stadiums are also the responsibility of the federal government.

Table 10.51 Overview of Authority Instances/Levels

Governmental instance/level	Authority concerning fire safety requirements
Federal government	General legislation applicable to all buildings (basic regulations); specific legislation for buildings for which the federal government has responsibility (football stadiums, etc.); occupational safety and health
Three communities: ▪ The Flemish community (Fl) ▪ The French community (Fr) ▪ The German-speaking community (G)	Legislation concerning buildings for public and cultural activities, including care homes, kindergartens, and schools
Three regions: ▪ The Flemish Region ▪ The Walloon Region ▪ Brussels-Capital Region	Legislation related to regional activities, including tourism (hotels, B&Bs, and youth hostels), environment, town planning and urban development, and housing
Ten provinces	Additional requirements concerning fire safety based on province laws and the urban law
581 municipalities	Additional requirements concerning fire safety based on municipal law and the urban law

Another responsibility of the federal government is the fire safety regulation for safety at work: the law on Occupational Safety and Health: OSH. This legislation addresses the safety, health, and welfare of people engaged in work or employment. Accident prevention at the workplace includes prevention of fire and explosion, and as a consequence, fire safety requirements in addition to the "Basic Standards". Both regulations are complementary, but by their nature are very different: the building regulations are predominantly prescriptive, whilst the OSH is risk management/performance-based. The OSH legislation replaces progressively over time the former ARAB (General Regulations for Labor Protection).

Regions and Communities

The communities and regions have been established to take into account respectively the cultural and economic differences in Belgium. They have each their own responsibilities - for example, for matters relating to persons and culture. For fire safety, this means that the communities are responsible for specific requirements for buildings that have a people-focused function, examples being care homes, hospitals, kindergartens, and schools.

Some of the legislation is as follows:

- Care homes for the elderly:
  - Fl: BVR 09/09/2011 (BS 09/03/2012)
  - Fr: Arrêté du 03/12/1998
  - G: Erlass vom 03/12/1997

- Sheltered housing:
  - Fl: BVR 09/12/2011
  - Fr: Arrêté du 03/12/1990
  - G: Erlass vom 03/12/1997
- Hotels, guest houses, and holiday homes:
  - Fl: Decision of the Flemish Government of 17/03/2017
  - Fr: Le Code wallon du Tourisme (CWT), originally published on 17/05/2010
  - Brussels: Decision of the Brussels-Capital Region Government of 4 March 2016 implementing the Ordinance of 8 May 2014 on tourist accommodation
  - G: Décret vom 09/05/1994 + Erlass 13/04/2000
- Camping sites:
  - Fl: Besluit van 11/09/2009
  - Fr: Arrêté du 09/12/2004
- Non-commercial sleeping accommodation:
  - Fl: Decreet van 18/07/2003 en 03/03/1993
  - Fr: Décret du 18/12/2003
- Schools:
  - The "Basic Standards" are applied, and the standard NBN S21-204 (now being revised) is a specific extension to it
  - At the regional level (regions) additional requirements can be established with respect to the environment and the urban laws, and an example of such legislation can be found in the Flemish environmental laws (VLAREM).

Provinces and Communes

The provinces and communes can establish additional fire safety requirements based on the province law and the municipal law.

Apart from the legislation, a number of Belgian standards include rules for fire safety, but these have no regulatory status, and include so-called "rules for good practice". They concern specific buildings, and include specific requirements - for this reason, it is recommended that they are followed (they are only obligatory if cited in a law or RD). They are issued by NBN, Avenue de la Brabançonne 29, B-1040 Brussels. Examples are provided in Table 10.52.

Table 10.52 Overview of Belgian Standards

Reference	Application area
NBN S21-203 (2018)	Fire protection in buildings - Ventilation openings in internal staircases
NBN S21-204 (1982)	Fire safety of buildings - School buildings - General requirements for reaction-to-fire
NBN S21-208-1 (1995)	Fire safety of buildings - Design and calculation of installation for smoke and heat extractions - Part 1: Large non-divided internal spaces with one level of construction
NBN S21-208-2 (2014)	Fire protection in buildings - Design of smoke and heat exhaust ventilation systems for enclosed car parks

10.3.2.2.2 Classification and Testing of Fire Performance

Construction Products

- Building products are defined to conform to item 1 of article 2 of the EU Regulation No. 305/2011. Classification and testing requirements of the fire performance of building products are specified in the national Basic Standards [2].
- The previous national classification and test standards for construction products have been replaced by technical specifications in conformity with the EU Directive 89/106/EEG of 21/12/1988 and the EU Regulation 305/2011 of the European Parliament and the Council of 09/03/2011. See Section 10.3.1.
- Some regulations and standards have not yet been updated. Furthermore, for building permits issued before 01/12/2012 the former national classification and test system remains valid. See the previous edition of this handbook (Section 10.6) [4].
- The reaction-to-fire classification is defined in EN 13501-1:2018 (Fire classification of construction products and building elements - Part 1 - Classification using data from reaction to fire tests).
    - This classification standard refers to the relevant test standards, listed in Section 10.3.1. For construction products, excluding floorings, the classes A1, A2, B, C, D, E and F are defined, including an additional s-classification for smoke production and an additional d-classification for flaming droplets and particles.
    - For floorings, the relevant classes are:
      $A1_{fl}$, $A2_{fl}$, B_{fl}, C_{fl}, D_{fl}, E_{fl}, and F_{fl} including an additional s-classification for smoke.
    - For linear pipe thermal insulation products, the classes $A1_L$, $A2_L$, B_L, C_L, D_L, E_L, and F_L supplemented by an s- and a d-class are specified.
- Exposed surfaces below elevated floors are also subject to class B or C requirements.

- An area up to 10% of the exposed area of any wall, ceiling, or floor is not subjected to these class requirements. For façade claddings, a maximum of 5% is not subjected to the requirements for façades. A proposal for amendment also excludes glass, but at the time of writing (April 2020) this has not yet been published.
- For roofs, reference is made to EN 13501-5 (Fire classification of construction products and building elements - Part 5 - Classification using data from external fire exposure to roof tests); see Section 10.3.1. The class B_{ROOF} (t1) is relevant. Particular prescriptive rules apply for green roofs.
- The façade cladding of low-rise buildings needs to have class D-s3,d1, while the façade cladding of medium- and high-rise buildings needs to have class B-s3, d1. A maximum of 5% of the visible surface of the façades is not subject to this requirement. Adjustments are under development and are expected to be published in 2021.
- In the “Basic Standards” no requirements impose fire classes for the fire behavior of electric or data cables. It is the AREI (the General Regulations for electrical installations) that does so, and this refers to the cable classification according to EN 13501-6. The AREI that was in force from 1981 has been reworked completely, and will be applicable from 1 June 2020. The new AREI consists of three thematic books:
  - Book 1: Low voltage and very low voltage electrical installations
  - Book 2: High voltage electrical installations
  - Book 3: Installations for transmission and distribution of electrical energy.

Upholstered Furniture and Bedding

The following EN standards are applied.

- EN 1021:2014 - Assessment of the ignitability of upholstered furniture
  - Part 1: Ignition source smouldering cigarette
  - Part 2: Ignition source match flame equivalent
- NBN EN 597-1:2016 - Furniture - Assessment of the ignitability of mattresses and upholstered bed bases
  - Part 1: Ignition source: smouldering cigarette
- NBN EN 597-2:2015 - Furniture - Assessment of the ignitability of mattresses and upholstered bed bases
  - Part 2: Ignition source: match flame equivalent.

Curtains and Drapes

The following standards are applied.

- NBN EN 1101:1995 completed by NBN EN 1101/A1:2005 - Textile and textile products - Burning behaviour - Curtains and drapes - Detailed procedures to determine the ignitability of vertically oriented specimens (small flame)
- NBN EN 1102:2016 - Textiles and textile products - Burning behaviour - Curtains and drapes - Detailed procedure to determine the flame spread of vertically oriented specimens.

10.3.2.2.3 Official Approval

Accredited and notified reaction-to-fire laboratories in Belgium are:

WFRGENT NV
Ottergemseteenweg-Zuid 711
B-9000 Gent
Belgium
Tel. +32 9 243 77 50
Fax. +32 9 243 77 51
E-mail: *info@wfrgent.com*
Web: *www.wfrgent.com, www.element.com*
Accreditation: BELAC 196 - TEST
Notification: BPV/RPC 1173

CENTEXBEL
Montoyerstrat 242
B-1000 Brussel
Belgium
Tel. +32 9 220 41 51
Fax. +32 9 220 49 55
E-mail: *testing@centexbel.be*
Web: *www.centexbel.be*
Accreditation: BELAC 056 - TEST
Notification: BPV/RPC 0493

ISSEP
Rue du Chéra, 200
B-4000 Liège
Belgium
Tel. +32 4 229 83 11
Fax. +32 4 252 46 65
E-mail: *direction@issep.be*
Accreditation: BELAC 060 - TEST
Notification: BPV/RPC 2659

10.3.2.2.4 Future Developments

Adjustments for façade cladding are under development and expected to be published in 2021.

References for Section 10.3.2.2

[1] *Belgium, a federal State, https://www.belgium.be/en/about_belgium/government/federale_staat,* April 2020.

[2] Basic standards for fire safety, Royal Decree (RD) 7/7/94, 19/12/1997, RD 04/04/2003, RD 13/06/2007, RD 01/03/2009, RD 12/07/2012 and RD 07/12/2016, *https://www.besafe.be/nl/wetgeving/koninklijk-besluit-7-december-2016-basisnormen-voor-de-preventie-van-brand-en-ontploffing,* April 2020.

[3] De Saedeleer, J., *Fire Safety & Legislation - Legal framework,* Ghent University, 2019.

[4] van Hees, P., "Belgium" (Section 10.6), in: *Plastics Flammability Handbook,* 3rd Edition, Troitzsch, J. (ed.), Carl Hanser Verlag, Munich, 2004.

10.3.2.3 France

Eric Guillaume

10.3.2.3.1 Statutory Regulations

French law is complicated because, depending on the type of building, several fire regulations apply under different ministries, and because building regulations for the public sector differ from those for the private sector. French fire regulations relate mainly to high-rise buildings, buildings open to the public, residential buildings, working places, and classified installations.

Regulations for buildings open to the public ("établissements recevant du public", E.R.P.) are laid down in sections R.123-1 to R.123-55 of the building and housing code [1] ("code de la construction et de l'habitation"). The implementation of these safety regulations is covered in the decree of 25 June 1980, and in several modifying and supplementing decrees [2]. The modified decree of 31 January 1986 [3] concerns fire protection of residential buildings. Regulations for high-rise buildings ("immeubles de grande hauteur", I.G.H.) are contained in sections R.122-1 to 122-29 and R.152-1 to 152-3 of the building and housing code, and in the modified decree of 30 December 2011 concerning the construction of high-rise buildings and anti-fire and anti-panic measures [4]. Table 10.53 lists the main regulations and authorities according to the purpose of the building.

Table 10.53 Regulations and Authorities According to the Purpose of the Building

Kind of building	French acronym	Limits	Decree	Authority in charge
Buildings open to the public	E.R.P.	Height < 28 m	25 June 1980 [2]	Ministry of the Interior, DGSCGC
Residential buildings	-	Height < 28 m for first three categories	31 January 1986 [3]	Ministry of the Environment, DHUP
	I.M.H.	Height 28–50 m for the fourth category	[20]	Ministry of the Environment, DHUP Ministry of the Interior, DGSCGC
High-rise buildings	I.G.H.	Height > 28 m except residential buildings (28–50 m)	30 December 2011 [4]	Ministry of the Interior, DGSCGC
Working places	E.R.T.	All working areas without public	[5]	Ministry of the environment, DGT
Classified installations	I.C.P.E.	Classified installations, depending on the application, e. g. substances stored or processes used	Various	Ministry of the Environment, DGPR
	I.N.B. I.N.B.S.	Nuclear installations (civil and defence)	7 February 2012 [6]	Depends on the case involved, but mainly "Autorité de sûreté nucléaire" (ASN) as an independent authority

10.3.2.3.2 Classification and Testing of the Fire Performance of Building Materials and Components

10.3.2.3.2.1 Implementation of Euroclasses

For reaction-to-fire of products a distinction is made between construction products that are covered by CE marking, other construction products, and furniture. All these distinctions and classifications are stated in the decree of 21 November 2002 [7]. This decree implements the Euroclass system, but also covers national classification (M-classification), conventional classifications and a table of correspondence between a requirement expressed as M-classification and the corresponding acceptable Euroclass level of performance. It does not show equivalence between both systems, but is just a way to accept Euroclasses with regulation provisions that have not been updated. Table 10.54 indicates various application cases of the decree.

Table 10.54 Specific Applications Mentioned in the 21 November 2002 Decree

Category	Products relevant to CPR with harmonized standard	Products not covered by CE marking	Other products
Example of products	Construction products	Light compartmentation Furniture and furnishings	Curtains Sunscreens Hanging textiles
Tests applicable	Euroclasses	Euroclasses or M-classification (rigid products)	M-classification (flexible products)
Validity of test and classification	None (certification applies)	5 years	5 years

10.3.2.3.2.2 M-Classification

The fire performance of materials covered by the M-classification is divided into the five classes M0 (no contribution to fire), M1, M2, M3, and M4. Materials are divided into two groups for test purposes:

- Flexible materials up to 5 mm thick,
- Flexible materials more than 5 mm thick, and rigid materials of any thickness.

The previous edition of this chapter [8] fully describes tests methods and classification for French methods (M-classification). A primary test is carried out in both cases to ascertain whether the material can be classified in categories M1 to M3. Complementary tests are carried out for classification in class M4 or, in certain cases, to confirm classification in classes M1 to M3. If there are doubts as to which group the material belongs, both types of test are carried out and the least favorable result is evaluated. Where possible, materials are tested as in end-use with an appropriate substrate.

10.3.2.3.3 Other Products and Special Provisions for Testing

10.3.2.3.3.1 Façades

In French regulations a façade is not considered as a construction product, but as part of the construction. For public buildings, façades are covered by the "Instruction Technique 249" [10] detailing what systems need approvals and which shall be approved without further testing. This supplements the Euroclasses of components at system level.

The reference test is called "LEPIR2" and described in [11]. This is a dummy standard façade of two and half storeys (Figure 10.35). The fire is started in the lower compartment using two wood cribs of 300 kg each. Criteria are based on flame propagation and fire penetration.

Figure 10.35 French façade test LEPIR2

10.3.2.3.3.2 Roofs

For roofs, several systems are approved without further testing. If that is not the case, test 3 of CEN/TS 1187 [12] as well as the classification to EN 13501-5 [13] apply in a large majority of cases, in addition to the Euroclasses. The test was formerly known in France as T-classification (often T30-1 or T15-1). It combines wind, radiation, and fire brands.

10.3.2.3.3.3 Upholstered Furniture in Public Buildings

Theater seats and seats in rows that are attached to the ground or that may hinder evacuation in public buildings are tested for their fire hazard. The requirements are covered by article AM18 of the amended decree of 25 June 1980 [2], as well as by the corresponding technical instruction [14].

The seat structure is subjected to the M-classification for rigid materials. The seat or a mockup of equivalent constitution (Figure 10.36) are tested using a gas burner simulating a 20 g paper cushion to NF D 60-013 [15]. The criteria are based on the mass loss limited to 300 g and the limitation of the lateral flame propagation.

Figure 10.36 French NF D 60-013 test on a theater seat

10.3.2.3.3.4 The Use of Synthetic Materials

The problem of secondary fire effects and, in particular, of fire gases has achieved prominence due to various spectacular fire incidents. This has led to official controls on the use of synthetic materials in buildings open to the public. The regulations governing the use of such materials in buildings open to the public laid down in the amended decree of 4 November 1975 [16] specify that the total amounts of nitrogen (N) and chlorine (Cl) contained in synthetic materials which can be released as HCN or HCl must not exceed 5 g and 25 g respectively per m^3 of enclosed space. The test method used is the French Tube furnace [17]. Materials that achieve classes M0 and M1 are not taken into account. This decree nowadays applies to a reduced number of cases, including interior insulation in public buildings. Since the latest, recently published version of the AM18 article, this provision is no longer applied to upholstered furniture.

10.3.2.3.3.5 Openings for Fire Safety Engineering at Product Level

In France, regulations are open to fire safety engineering-based approaches. Such an approach is applicable to insulation materials since the 2004 revision of article AM8 of the 25 June 1980 decree. A construction system including combustible insulation may be accepted in a public building after a risk analysis assesses requirements such as additional smoke control or a minimum ceiling height.

There have been discussions about extending this innovative approach to combustible materials more generally, including wood linings for use in several parts of a building.

10.3.2.3.4 Official Approval

10.3.2.3.4.1 General Cases

For construction products listed in the CPR and subject to CE marking, there is no specific approval and the list kept by the European Commission on the NANDO platform is applicable. Nevertheless, additional documents may be requested for implementation of the product on construction sites, as described below.

For products that are not covered by CE marking, tests performed may be based on Euroclasses or M-classification, mainly depending on the application. If a Euroclass report comes from foreign countries or non-recognized laboratories, it has to be validated first by a recognized French laboratory. The test and the classification reports are valid for 5 years.

M-classification still may apply to several products such as large furniture, curtains, or products that cannot be classified as construction products according to the CPR. The test report and classification ("Procès Verbal") can only be delivered by a recognized laboratory. The test and the classification reports are valid for 5 years; there is no factory production monitoring requested.

10.3.2.3.4.2 "Avis Technique" (AT) and "Appréciation Technique d'Expérimentation" (ATEx)

For many construction systems, the CE certificate is completed by a technical opinion ("Avis Technique") delivered by the "Commission Chargée de Délivrer les Avis Techniques" (CCFAT). This document is not mandatory and explains how to implement the construction product at construction sites in a general way. It is used by insurance companies to classify risks as "standard techniques" for non-traditional construction systems. Traditional ones are covered by DTU ("Document Technique Unifié"), which is largely similar to national standards.

ATEx is a rapid technical evaluation procedure formulated by a group of experts on any innovative product, process, or equipment. This evaluation is often used either prior to an "Avis Technique" (because it allows initial feedback on the implementation of processes), or for a single project. Examples of ATEx include: light façades, glass roofs, reversible floors, waterproofing of roofs, and reinforcement of structures.

10.3.2.3.4.3 "Avis de Chantier" (Site Notice on Fire Resistance)

The request for an "Avis de chantier" procedure takes place before the construction phase of a structure. It is usually given to the holder following the request of a control body assigned to a site. This is an expert opinion on a process whose fire behavior has generally already been evaluated elsewhere, but under conditions that are not those of the target site. It relies on the analysis of a technical file supported by experimental data, and possibly calculations and numerical simulation results. It avoids carrying out a new fire resistance test. The request for an "Avis de chantier" is made in accordance with the provisions of Article 14 of the amended Decree of 22 March 2004. It can only be performed by a laboratory approved for fire resistance.

10.3.2.3.4.4 "Appréciation de Laboratoire" (Laboratory Assessment of Fire Resistance)

An "Appréciation de laboratoire" certifies the fire performance of a product, building component, or structure. It can be formulated for different cases:

- The classification extension of a construction element that is the topic of a report. The impact of the proposed modifications is evaluated on the basis of a rationale presented in the laboratory assessment without further testing.
- The justification of the fire performance of an innovative process, during the examination of a Technical Assessment (ATec) or a Technical Application Document (DTA), when this justification is not provided by the designer, or where the available test results do not cover the claimed area of use. Many processes are involved: new concretes, thermal breakers, insulation techniques, metal structures, wood assemblies, etc.

To certify the fire performance of the process, the laboratory assessment may be based on experimental results, results from calculations, and numerical simulations or knowledge acquired during actual fires. The request for an "Appréciation de laboratoire" is made according to the cases that are in accordance with the provisions of Article 13 or Article 18 and Annex 4 of the amended Order of 22 March 2004 [9]. It can only be performed by a laboratory approved for fire resistance.

10.3.2.3.4.5 "Avis sur Étude" – Expert Opinion Study on Fire Resistance or Reaction-to-Fire

When used for assessing engineering methods for the fire behavior of products, building components, and works, the study must result in a favorable assessment by an approved laboratory before it can be applied in building construction. This assessment then takes the form of a study opinion. Fire behavior engineering consists in determining the realistic thermal effects resulting from the examination of the fire scenarios exploiting information on the nature (quantity and heat release) of the combustible materials likely to be involved in the fire and of the ventilation conditions. The response of the structure to these thermal effects is calculated, and if necessary reinforcement measures are proposed. In this context, the study opinion allows new insights by analyzing the targeted safety objectives, the selected performance criteria, the fire scenarios taken into account, and evaluating the relevance of the assumptions used and the models carried out. The study opinion conclusions are then communicated by the applicant to the appropriate safety committees.

The request for a study opinion is carried out in accordance with the provisions of article 15 of the amended decree of 22 March 2004 [9] or article 105 of the amended decree of 31 January 1986. It can only be performed by a laboratory approved for fire resistance and/or reaction-to-fire.

10.3.2.3.4.6 "Visas de Façades"

Façades of high-rise buildings must meet requirements covered by "Visa de façade" in accordance with the provisions of Article GH12 of the Decree of 30 December 2011 [4]. This "Visa de façade" is mandatory and can only be delivered by CSTB or EFECTIS France.

10.3.2.3.4.7 Pilot Projects and Project-by-Project Assessment

Provisions already exist in the code for pilot projects, with the support of the safety commission, e.g., application of R123-13 law for public buildings.

Article 26 of ESSOC law [21] prepares a case-by-case assessment methodology for all constructions. It is nowadays open for specific openings of fire safety engineering mentioned in previous sections.

10.3.2.3.4.8 List of Recognized Laboratories

The list of recognized laboratories is kept by the Ministry of Interior as revision of Decree of 5 February 1959 (modified) [18].

The laboratories that are allowed to carry out the reaction to fire tests defined by article R. 121-5 of the code of construction and by the modified decree of 21 November 2002 [7] relative to the classification against fire are the following:

- Centre Scientifique et Technique du Bâtiment (CSTB)
- Laboratoire National de Métrologie et d'Essais (LNE)
- Laboratoire Central de la Préfecture de Police (LCPP)

- Institut Français du Textile et de l'Habillement (IFTH)
- Institut Technologique FCBA (Forêt, Cellulose, Bois-construction, Ameublement)
- Centre de Recherche et d'Études sur les Procédés d'Ignifugation des Matériaux (CREPIM)
- EFECTIS France.

The laboratories that are allowed to carry out the fire resistance tests defined by article R. 121-5 of the code of the construction and by the modified decree of 22 March 2004 [9] relative to the classification against fire are the following:

- Centre Scientifique et Technique du Bâtiment (CSTB)
- EFECTIS France
- Centre d'Études et de Recherches de l'Industrie du Béton (CERIB).

10.3.2.3.5 Future Developments

The highest level of regulations related to construction is the "Code de la Construction et de l'Habitation" (CCH). This has been recently modified to express requirements in terms of performance, including fire-related requirements. Adaptations in terms of requirements are planned for the upcoming years [19].

Project-by-project assessment will increase in the framework of ESSOC article 26 law [21] and future application decrees. One of the objectives is to open the use of fire safety engineering wider than the domains already regulated.

References for Section 10.3.2.3

[1] Code de la Construction et de l'Habitation. Livre I: Dispositions générales. Titre II: Sécurité et Protection Contre l'incendie. *https://www.legifrance.gouv.fr/affichCode.do;jsessionid=C32E6767EC159B0915304750BFCBEB99.tpdila17v_3?idSectionTA=LEGISCTA000006160485&cidTexte=LEGITEXT000006074096* (accessed May 2019).

[2] Arrêté du 25 juin 1980 modifié – Dispositions générales du règlement de sécurité contre les risques d'incendie et de panique dans les établissements recevant du public (ERP). *https://www.legifrance.gouv.fr/affichTexte.do;?cidTexte=LEGITEXT000020303557* (accessed May 2019).

[3] Arrêté du 31 janvier 1986 modifié – Arrêté relatif à la protection contre l'incendie des bâtiments d'habitation.

[4] Arrêté du 30 décembre 2011 portant règlement de sécurité pour la construction des immeubles de grande hauteur et leur protection contre les risques d'incendie et de panique.

[5] Articles R4216-1 à R4216-34 du Code du Travail. *https://www.legifrance.gouv.fr/affichCodeArticle.do?cidTexte=LEGITEXT000006072050&idArticle=LEGIARTI000018488756&dateTexte=&categorieLien=cid* (accessed May 2019).

[6] Arrêté du 7 février 2012 fixant les règles générales relatives aux installations nucléaires de base.

[7] Arrêté du 21 novembre 2002 relatif à la réaction au feu des produits de construction et d'aménagement.

[8] Bonnaire, T. and Touchais, G., France (Section 10.8), in *Plastics Flammability Handbook* (3rd edition) pp. 297–308, ed. Troitzsch, J., Hanser, 2004, ISBN 1-56990-365-5.

[9] Arrêté du 22 mars 2004 relatif à la résistance au feu des produits, éléments de construction et d'ouvrages.

[10] Instruction technique no. 249 Relative aux façades (Arrêté du 24 mai 2010).

[11] Arrêté du 10 septembre 1970 relatif à la classification des façades vitrées par rapport au danger d'incendie, pour l'évaluation du C + D et du comportement au feu de l'accrochage des façades aux planchers.

[12] CEN/TS 1187:2012. Test methods for external fire exposure to roofs.

[13] EN 13501-5:2016. Fire classification of construction products and building elements - Part 5: Classification using data from external fire exposure to roofs tests.

[14] Instruction technique relative au comportement au feu des sièges rembourrés (Prise pour l'application de l'article AM 18 du règlement de sécurité contre les risques d'incendie et de panique dans les établissements recevant du public) (Arrêté du 6 mars 2006).

[15] NF D 60-013. Protocol for assessment of the ignitability of upholstered furniture - Ignition source equivalent to a burning 20 g paper cushion - Coverings and upholstery materials (June 2006).

[16] Arrêté du 4 novembre 1975 portant réglementation de l'utilisation de certains matériaux et produits dans les établissements recevant du public, modified by Instruction du 1 décembre 1976 and arrêté du 26 juin 2008.

[17] NF X 70-100-2. Essais de comportement au feu - Analyse des effluents gazeux - Partie 2: Méthode de dégradation thermique au four tubulaire (2006).

[18] Arrêté du 5 février 1959 portant agrément des laboratoires d'essais sur le comportement au feu des matériaux.

[19] Ordonnance n° 2020-71 du 29 janvier 2020 relative à la réécriture des règles de construction et recodifiant le livre Ier du code de la construction et de l'habitation. JORF n°0026 du 31 janvier 2020. *https://www.legifrance.gouv.fr/affichTexte.do?cidTexte=JORFTEXT000041506557&dateTexte=&categorieLien=id* (accessed September 2020).

[20] Arrêté du 7 août 2019 relatif aux travaux de modification des immeubles de moyenne hauteur et précisant les solutions constructives acceptables pour les rénovations de façade. *https://www.legifrance.gouv.fr/affichTexte.do?cidTexte=JORFTEXT000038906964&dateTexte=20200907* (accessed September 2020).

[21] Loi n°2018-727 du 10 août 2018 pour un Etat au service d'une société de confiance. *https://www.legifrance.gouv.fr/jorf/id/JORFTEXT000037307624/* (accessed September 2020).

10.3.2.4 Germany

Edith Antonatus

10.3.2.4.1 Statutory Regulations

In Germany, building codes are based on the following principles regarding fire safety:

- Structural installations are to be laid out, erected, and maintained in such a way that the initiation and spread of fire and smoke are prevented and that in case of fire, rescue of humans and animals as well as effective extinguishing is possible.
- Building materials that are highly flammable must not be used in the erection and installation of structures; this does not apply to building materials if they are not highly flammable when used in combination with other building materials.

The German building regulations are based on model building codes (MBOs, Musterbauordnungen), which are issued by the conference of ministers for construction from the federal states. Based on these model codes, every federal state issues its own building ordinance (LBO, Landesbauordnung). In these codes, modifications of the MBO and additional requirements have been introduced.

The model building codes are the general building code and additional codes and guidelines such as:

- Garages (Garagenverordnung; MGarVO)
- Shops and stores (Verkaufsstättenverordnung; MVkVO)
- Assembly places (Versammlungsstättenverordnung; MVStättV)
- Industrial buildings (Industriebaurichtlinie; M-IndBauRL)
- Schools (Muster-Schulbau-Richtlinie; MSchulbauR)
- High-rise buildings (Musterhochhausrichtlinie)

These have not been introduced in all federal states. In some of the states, additional codes for special types of buildings have been introduced (for example, for hotels).

In addition to the building codes, administrative regulations are in place, which clarify and detail requirements of the building codes and define additional technical rules.

The German model building code (MBO) was completely revised in 2017 [1] to achieve better implementation of the requirements of the Construction Products Regulation. In addition, the new Model Administrative Provisions - Technical Building Rules "Muster-Verwaltungsvorschrift Technische Baubestimmungen" (MVV TB) were published in 2019 [2]. This document gives details regarding interpretation and execution of the building codes and replaces the former list B rules for construction products (Bauregelliste B). General fire safety requirements are defined in chapter A 3. Annex C defines requirements for construction products, which are not yet harmonized and part D lists construction products for which no special proof of applicability is needed.

Requirements for construction products are defined based on the type and height of buildings.

Buildings are classified as classes 1 to 5 depending on their size and height. Examples of additional special buildings (Sonderbauten) are: high-rise buildings (highest occupied floor > 22 m above ground), assembly places, restaurants, hospitals, prisons, etc.

Depending on these building classifications, requirements for reaction to fire of construction products are defined.

10.3.2.4.2 Classification and Testing of the Reaction to Fire of Construction Products

All products for which harmonized European standards or European assessment documents have been published have to comply with the European testing and classification standards. For products that are not yet covered by a European specification, the national DIN standards can still be applied.

Table 10.55 DIN 4102 (Parts 1–23)

DIN 4102-	Fire behaviour of building materials and building components
	1: Reaction to fire of building materials, concepts, requirements, and tests (1998-05)
	2: Building components; definitions, requirements, and tests (1977-09)
	3: Fire walls and non-load bearing external walls (1977-09)
	4: Synopsis and application of classified building materials, components, and special components (2016-05)
	5: Fire barriers, barriers in lift wells, and glazing resistant against fire; definitions, requirements, and tests (1977-09)
	7: Roofing; definitions, requirements, and testing (1998-07; new draft: 2018-03)
	8: Small-scale test furnace (2003-10)
	9: Seals for cable penetrations; concepts, requirements, and testing (1990-05)
	11: Pipe encasements, pipe bushings, service shafts and ducts, and barriers across inspection openings; terminology, requirements, and testing (1985-12)
	12: Circuit integrity maintenance of electrical cable systems; requirements and testing (1998-11)
	13: Fire-resistant glazing; concepts, requirements, and testing (1990-05)
	14: Determination of the burning behaviour of floor covering systems using a radiant heat source (1990-05)
	15: "Brandschacht" (1990-05)
	16: "Brandschacht" tests (2015-09)
	17: Melting point of mineral wool insulating materials - terms and definitions, requirements, and test (2017-12)
	18: Fire barriers, verification of automatic closure (continuous performance test) (1991-3)
	20: Complementary verification for the assessment of the fire behaviour of external wall claddings (2017-10)
	DIN V 4102-21: Assessment of the performance of fire-resistant air ducts (2002-08)
	DIN SPEC 4102-23: Application rules for test results of roofing to DIN CEN/TS 1187, test method 1, and DIN 4102-7 (only in German) (2018-07)

For testing and classification according to the hitherto existing national standards, test methods and classifications are defined in the different parts of DIN 4102. Table 10.55 shows the most important standards for testing and classification according to the national rules in Germany. This series includes testing and classification for reaction to fire, resistance to fire, and roofing (fire from outside). In part 4, a compilation of classified building materials, components, and special components is given.

The classification of reaction to fire according to DIN 4102 consists of five classes (class A1 and A2 for "non-combustible" products and classes B1, B2, and B3 for combustible products). Due to their organic structures, plastics usually only achieve class B. In certain cases, for example as composites with inorganic materials, they can be classified as class A2.

Table 10.56 indicates which tests have to be performed for the different reaction to fire classifications according to the national testing methods.

Table 10.56 DIN 4102-1 Classification and Test Methods for Reaction to Fire of Building Materials

Building material class	Building inspection designation	Test method
A	**Non-combustible**	
A1[(1)]		▪ Furnace test 750 °C and all tests needed for A1
A2		▪ Furnace test (750 °C) or calorific potential (DIN 51900-1) and heat release test (DIN 4102-8) ▪ Brandschacht test ▪ Smoke density Toxicity testing is only done on a voluntary basis (no longer part of legal requirements)
B	**Combustible**	
B1[(2)]	Low flammability	▪ Brandschacht (including smoke density and burning drops) ▪ Small burner test ▪ For floor coverings: radiant panel test - DIN 4102-14
B2	Moderately flammable	▪ Small burner test ▪ Special case textile floor coverings: small burner test - DIN 54332
B3	Highly flammable	▪ No tests, no compliance with B2

[(1)] Class A2 requirements must also be satisfied.
[(2)] Class B2 requirements must also be satisfied.

The test methods for non-combustible and combustible building materials are described in DIN 4102-1, sections 5 and 6, DIN 4102-14, and DIN 4102-16.

As these DIN test methods are increasingly replaced by European methods (see Section 10.3.1), details are not described here, but can be found in the third edition of this book [3].

10.3.2.4.3 Legal Provisions for the Application of Reaction to Fire Classifications of Construction Products in Germany

In Germany, building regulations and codes do not specifically mention reaction to fire classifications. They use a description of the performance of these products, which is equivalent to the classifications as shown in Table 10.57.

Table 10.57 Assignment of Classifications to Legal Requirements According to DIN 4102 and DIN EN 13501-1

Legal requirement in building codes and regulations	Minimum requirement according to DIN EN 13501-1			Requirement according to DIN 4102
Non-combustible[(1)(2)]	A2 - s1,d0[(1)]	$A2_L$ - s1,d0[(1)]	$A2_{fl}$ - s1	A2[(2)]
Difficult to ignite and no burning droplets or falling parts[(1)]	C - s2,d0[(1)]	C_L - s2,d0 [(1)]	-	B1 and no burning droplets or falling parts and smoke obscuration (I < 400% min)
Difficult to ignite and limited smoke generation[(1)]	C - s1,d2 [(1)]	C_L - s1,d2 [(1)]	C_{fl} - s1	B1 and smoke obscuration (I < 100% min)
Difficult to ignite and no burning droplets or falling parts and limited smoke generation[(1)]	C - s1,d0 [(1)]	C_L - s1,d0 [(1)]	-	B1 and no burning droplets or falling parts and smoke obscuration (I < 100% min)
Difficult to ignite	C - s2,d2[(1)]	C_L - s2,d2[(1)]	C_{fl} - s1	B1 and smoke obscuration (I < 400% min)
Moderately flammable, no burning droplets	E	E_L	E_{fl}	B2 and no burning droplets
Moderately flammable	E - d2	E_L - d2	E_{fl}	B2
Highly flammable	F (or NPD [(3)])	F (or NPD)	F (or NPD)	B3

[(1)] In some cases, additional requirements regarding continuous smoldering and glowing.
[(2)] In some cases, additional requirements for melting point.
[(3)] NPD = no performance determined.

10.3.2.4.4 Requirements Regarding the Reaction to Fire of Materials and Products

The building codes and administrative codes mentioned earlier define the requirements regarding the reaction to fire for different types of buildings and applications. The most important requirements are listed below.

(a) Non-combustible materials and products (nichtbrennbar).

The general building codes for a number of applications stipulate the use of non-combustible materials. Non-combustible products are required for applications that are deemed especially critical for the safety of occupants in case of fire.

This applies to the structural parts of outer walls, ceilings between floors, and a number of further applications, where fire compartmentation/fire resistance is required. For stairs and staircases, as well as for escape routes in buildings with a height of > 7 m, all products for walls and ceilings have to be non-combustible. For ventilated façades, insulation products have to be non-combustible above a building height of 7 m.

For high-rise buildings, in addition, all products used for insulation or cladding of outer walls have to be non-combustible.

Additional exigencies for the use of non-combustible materials have been specified in a number of special building codes of the federal states; for example, for theatres and other assembly places, hospitals, stores, etc.

(b) Materials and products that are difficult to ignite (schwerentflammbar).

For building façades above a building height of 7 m up to 22 m, all cladding and insulation products have to be difficult to ignite. This requirement also applies to photovoltaic modules.

Here too, specific building codes define additional requirements.

Floor coverings in assembly rooms and escape routes also need to be difficult to ignite.

(c) Moderately flammable materials and products (normalentflammbar).

All materials and products used in buildings, for which no special rules apply, must at least be moderately flammable. Prefabricated parts are the only exemption: in no case may they be cut or separated as they may contain highly flammable components or materials.

10.3.2.4.5 External Fire Exposure – Roofs

Almost all building roof coverings have to resist burning brands and other external fire sources. This requirement applies to the complete roof build-up, which may be a combination of an outer water-proofing layer, insulation products, vapor barriers, and a substrate similar to the end-use roof construction.

The national test procedure is defined in DIN 4102-7 and the application of test results described in DIN 4102-2. CEN/TS 1187-1 defines a test method that is almost identical to this national German standard. Therefore, test results according to the European classification B_{ROOF} (t1) to EN 13501-5 are fully accepted in Germany and used by most producers. For more test details, see Section 10.3.1: European Union under EN 13501-5.

The extended application of test results (exchange of individual components of the tested build-up, e.g., thickness and density of insulation products, use of products with the same characteristic, but from a different producer, etc.) for DIN classifications is defined in DIN SPEC 4102-23: application rules for test results of roofing. For products tested and classified to the European standards, a German testing house needs to define the field of application in a separate "AbP" (allgemeines bauaufsichtliches Prüfzeugnis = general building inspectorate test certificate).

10.3.2.4.6 Additional Requirements for Façades

In Germany, for buildings requiring class B1 (corresponds to Euroclasses C or D), additional full-scale tests are necessary to assess the fire performance of the complete façade system (e.g., an ETICS system or a ventilated cladding system).

The test procedure has been defined in DIN 4102-20. The test build-up consists of a wall in a corner configuration with a height of ≥ 5.5 m and a width of ≥ 2.5 m (long wing) and ≥ 1.5 m (short wing).

At the bottom of the long wing, a combustion chamber (simulating a window opening in the wall) with a gas burner or a wood crib as an ignition source is located. The test rig is shown in Figure 10.37. The testing time is 20 minutes, followed by an observation time of at least 40 minutes. Temperatures in front of the test rig and inside the cladding (for example, in the middle of the insulation layer) have to be recorded and flame lengths as well as dripping and melting leading to a significant secondary fire have to be observed and noted. Continuous smoldering and glowing after the end of the test is also taken into account.

The standard does not include classification criteria – the final evaluation of the test results has to be done by the DIBt (German Institute of Building Technology) and their experts.

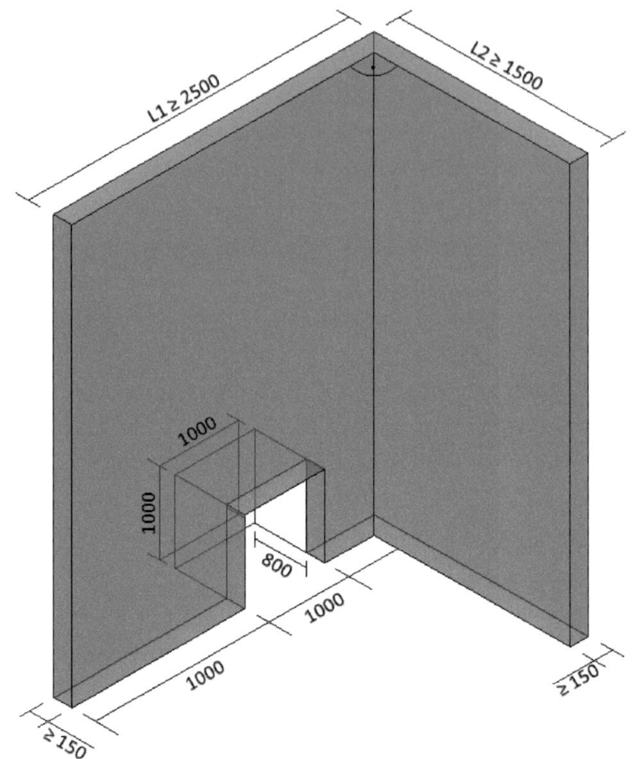

Figure 10.37 Test rig for façades according to DIN 4102-20. Dimensions in mm

10.3.2.4.7 Testing and Official Approval

Testing houses in Germany have to be acknowledged by the German DIBt (Deutsches Institut für Bautechnik) for testing and certification according to the German standards and building codes.

On its website *(www.dibt.de)* [4] the DIBt keeps a list called "Verzeichnis der Prüf-, Überwachungs- und Zertifizierungsstellen nach den Landesbauordnungen" (list of the testing, monitoring, and certification bodies according to the federal state building regulations), which is continuously updated.

For testing and classification according to the European standards, the laboratories have to be accredited according to DIN EN ISO/IEC 17025 (testing) and EN ISO/IEC 17065 (certification). Based on this accreditation, they can apply for European notification (through DIBt).

The list of notified bodies can be found on the official NANDO website of the European Commission. Fire testing laboratories can also be found in a list on the website of EGOLF (European Group of Organizations for Fire Testing, Inspection and Certification).

In Germany, fire performance and secondary fire effects of building materials are tested by the following recognized institutions [5] (status January 2019):

BAM
Federal Institute for Materials Research and Testing
Unter den Eichen 87
12205 Berlin
Germany

DMT GmbH & Co. KG
DMT-Anlagen- und Produktsicherheit (APS4)
Tremoniastrasse 13
44137 Dortmund

Ift Rosenheim GmbH
Theodor-Gietl-Str. 7–9
83026 Rosenheim

MFPA Leipzig
Hans-Weigel-Str. 2 b
04319 Leipzig

MPA Bau Hannover
Nienburger Strasse 3
30167 Hannover

MPA Braunschweig
Beethovenstr. 52
38106 Braunschweig
Germany

MPA Dresden GmbH
Fuchsmühlenweg 6F
09599 Freiberg
Germany

MPA NRW
Auf den Thränen 2
59597 Erwitte

Materialprüfungsanstalt Universität Stuttgart
Pfaffenwaldring 4g
70569 Stuttgart

Prüfinstitut Hoch
Lerchenweg 1
97650 Fladungen

TU München
HFM
Winzererstrasse 45
80797 München

10.3.2.4.8 Future Developments

In its judgement C-100/13 of 16 October 2014, the European Court of Justice declared that Germany's action to make additional national requirements for construction products with CE marking was inadmissible. This caused a review of the complete legal system for construction products, which resulted in the new model building code MBO and the MVV TB, replacing the old "Bauregellisten". Now, the federal states have to implement these changes in their building codes (LBOs), and authorities and industries have to apply these new rules. It is expected that this change will not be easy and will cause difficulties for producers as well as authorities. Therefore, the MVV TB will continuously be reviewed based on first experiences with the new system.

For products used in façades, there is an ongoing discussion about future requirements in Germany. To date, for façade cladding and insulation systems in medium height buildings (up to 22 m), the test to DIN 4102-20 is used, simulating a fire started in a room. Now, a standard for an additional test, simulating a fire from outside a building, is under preparation. It is not yet clear whether this test will be used in the future for regulatory purposes or research only. In parallel, the European Commission is currently trying to harmonize full-scale test methods for façades in Europe, based on the DIN 4102-20 and the British BS 8414 tests. If this approach is successful, further changes in regulations for façades have to be expected.

References for Section 10.3.2.4

[1] Musterbauordnung – MBO – Fassung November 2002 – Zuletzt geändert durch Beschluss der Bauministerkonferenz vom 13.05.2016, published under *https://www.is-argebau.de/verzeichnis.aspx?id=991&o=991.*

[2] Muster-Verwaltungsvorschrift Technische Baubestimmungen (MVV TB) – this document is updated continuously. The latest version can be found under: *https://www.dibt.de/de/DIBt/DIBt-EuGH-Urteil.html.*

[3] *Plastics Flammability Handbook. Principles, Regulations, Testing, and Approval*, 3rd Edition, Troitzsch, J. (Ed.) (2004) Carl Hanser Verlag, Munich, Germany, Section 10.9, pp. 309–324.

[4] *https://www.dibt.de/fileadmin/dibt-website/Dokumente/Referat/P4/LBO/PUEZ-Stellen_Verzeichnis_2017.pdf.*

[5] EGOLF website: *https://www.egolf.global/member-profiles.html.*

10.3.2.5 Hungary

István Móder, Imre Juhász

10.3.2.5.1 Statutory Regulations

The building regulations in Hungary are based on the Decree 30/2019 of the Ministry of Interior regarding the 2019 modification of the Hungarian Fire Safety Code (called "OTSZ"). It was officially published on 26 July 2019 as an amendment of the Decree No. 54/2014 of 5 December 2014 of the Ministry of Interior on National Fire Prevention Regulations.

10.3.2.5.2 Classification and Testing of the Fire Performance of Building Materials and Components

Since the 3rd edition of this Handbook, the MSZ 14800 standard series - the national fire safety standards - has been withdrawn and replaced by the appropriate test methods referred in the EN 13501 series (this took place in 2011). The nomenclature of the EN 13501 standards was fully accepted and adopted into the Hungarian fire code in 2008.

Two test methods from the MSZ 14800 standard series were not replaced, and therefore remained valid - MSZ 14800-16:1992 (Fire resistance tests. Determination of ignition temperature of solid materials) and MSZ 14800-6:2009 (Fire resistance tests. Part 6: Fire propagation for building façades) [1, 2]. One new Hungarian test method, MSZ 14890:2014 (Test for ignitability danger of building materials) [3], was adopted recently. The importance and details of the latter two standards are discussed in the following.

Ignitability requirements are excluded from the current fire code; however, fire propagation limits, depending on building height, are cited [4].

Fire Propagation on Façades

There are three main causes for fire propagation on façades. The first and most important one described in the Hungarian test method models is when a fully developed fire spreads from one compartment through a window to the façade. The second one, which seldom occurs in Hungary, is when goods and materials catch fire in front of a building. The third one is fire spread from one building to another; this is covered in the Fire Code by determining safety distances [5].

To date, several national test methods have been developed within Europe to assess fire propagation on façades, however no harmonized "EN-test method" is currently available. The national test methods therefore have to be compared using criteria such as their incorporation into a national regulatory framework, and the duration and size of the test ("large-scale" or "small-scale") [6]. The usage of window openings should also be included as an additional parameter.

MSZ 14800-6:2009 Fire Resistance Test

MSZ 14800-6:2009 is a large-scale test for façades using two window openings, and includes details of testing. This test and assessment method is unique in Europe, and is therefore presented here in some more detail.

The scope is to determine the vertical and horizontal fire propagation of combustible finishes (like ETICS) on ventilated or unventilated cladding faces of façades with openings. This test method emphasizes the importance of the details of the (window) openings and screens and the effect of the window details to the system.

The test concept is to model a fully developed fire in one compartment of a building and analyze the effect of fire spread on the outer skin of the building. The contribution of the skin layers to fire spread (by recording and analyzing temperature measurements) or of skin material peeling off (weight measurements) is considered and determined.

The three-storey test rig is located outdoors. The first floor is the fire chamber with a volume of 48.73 m^3. The second and third floors are the observation floors. The surface of the test rig is nearly 32 m^2. The test method uses two window openings, which are on top of each other, but on different storeys. The wall structure of the test rig can be "combustible" or "non-combustible" depending on the end-use application of the materials being tested.

A 650 kg timber crib consisting of first-class pinewood roof battens with dimensions of 25 × 50 × 1500–2000 mm is used as the ignition source. Equipped with a manual ventilation system, the timber crib allows the requirements of the fire curve determined in EN 1363-1 to be met for over 45 min within the test chamber [7]. From there the fire spreads onto the face of the test rig through the window opening. One 1.20 × 1.20 m rotated side-hung timber window is installed in the wall of the test chamber.

On the second floor, the temperature measuring points are located. These points are located 10 cm away from the second window opening – this one does not contain window frames or any installations. The purpose of the third floor is to screen the parapet wall details.

The evaluation method and the criteria for all systems to be tested are described below:

- Vertical flame spread criterion: flame spread reaches the upper edge (parapet wall) of the test specimen;
- Horizontal flame spread criterion: flame spread reaches 1.50 m width from both sides of the fire chamber's window opening at least in one direction;
- Falling parts criterion: weight of falling pieces of the tested product exceeds 5 kg;

- Temperature criterion: the following calculated temperature difference is recorded over a period of 2 min [2].

$$T_{lz} - T_{any} \leq 300 \qquad (10.1)$$

where:

T_{lz} is the temperature measured in front of the upstand (usually the temperature of the plume) [K];
T_{any} is the temperature measured in the observation floor's window opening [K].

As the height of the upstand has an effect on the observation floor temperature, this effect can be used as a screening and indirect evaluation criterion.

The fire spread limit (T_h) on façades is classified as 15 min, 30 min, or 45 min. These classification periods are stated as requirements in the fire code.

MSZ 14890:2014 Ignitability Test

Even though the EN 13823 (SBI) test method allows assessment of burning droplets from construction products or kits, the end-use application is not always vertical, as in the SBI, but horizontal in an overhead position [8]. MSZ 14890:2014 differs from the SBI in allowing testing to be carried out in this configuration, and in screening the ignitability effect of melting test specimens.

The scope of MSZ 14890:2014 is to test a horizontally positioned test specimen and to determine the presence of burning parts and their effect, grouped into three categories:

- "no burning parts or elements fall down",
- "burning, but not causing ignition, parts or elements fall down",
- "burning, causing ignition, parts or elements fall down".

The test concept is to determine and evaluate the ignitability property of horizontal ceiling-like construction products, structural parts, or structures classified as B–E according to EN 13501-1, namely thermal and sound insulation materials, ceiling cladding, canvas structures, etc. This test is not identical or interchangeable with the SBI test, where the test specimen is in a vertical position and the burning droplet criterion is classified according to EN 13501-1 [9]. The test specifications are shown in Table 10.58.

Table 10.58 Specifications for the Ignitability Test

Specimen	1300 × 1800 mm
Specimen position	Horizontal
Ignition source	8 L ethanol
Test duration	20 ± 5 min
Conclusions	Determine and evaluate the ignitability property of horizontal ceilings, and classify them as g2, g1, or g0

The test rig details are: The ground floor area consists of a 1160 × 330 mm burning chamber and a 1810 × 1330 mm test chamber. The fuel tank is placed in the burning chamber at a height of 1500 mm, with the heating source being 8 L of ethanol. The burning chamber is closed from that point upwards with a quarter-section of steel of 450 mm radius and a flame spreader insulated with 40 mm thick A1 insulation material. The test specimen is at the level of the upright end of this steel section. The size of the test specimen is 1300 × 1800 mm. Immediately before the test, a piece of paper with density $\rho = 80 \pm 10$ g/m^2 is placed onto the floor of the test rig. The paper sheets shall be conditioned for constant mass (or for at least 24 h before the test) at 23 ± 2 °C and 50 ± 5% relative humidity.

The 8 L ethanol provides the following heating-curve for a 20 ± 5 min test [3]:

$$T_K = 5(2 - t) + 300 \log(50t + 1) \tag{10.2}$$

where t is the time in minutes.

Three test specimens are prepared for testing, and the test is carried out on two samples. If the two test results do not provide the same classification category (see evaluation below), the third sample is tested to determine the ignition class of the test element. The following three categories can be determined for one product/kit.

- During the test, if flaming and igniting parts or elements fall down onto the floor of the test rig, and the paper on it ignites and sustained flaming occurs, the test specimen is classified as highly ignitable (sign: g2).
- During the test, if flaming but not igniting parts or elements fall down onto the floor of the test rig and neither ignition nor sustained flaming of the paper occurs, the test specimen is classified as moderately ignitable (sign: g1).
- During the test, if neither flaming nor igniting parts or elements fall down onto the floor of the test rig, the test specimen shall be classified as not ignitable (sign: g0).

10.3.2.5.3 Official Approval

At the time of publication of this book there are two laboratories accredited to MSZ 14800-6:2009, namely EMI Nonprofit Kft. and TÜV Rheinland InterCert Kft. Only EMI Nonprofit Kft. is accredited to MSZ 14890:2014.

- ÉMI Építésügyi Minoségellenorzo Inn
  Tuzvedelmi Vizsgalo Laboratorium / Fire Testing Laboratory
  Dozsa Gyorgy ut 26
  H-2000 Szentendre
  Hungary
  Phone: +36 30 220 2955
  Web: *www.emi.hu*

- TÜV Rheinland Intercert Kft.
  Vàci út 48/a-b
  H-1132 Budapest
  Hungary
  Phone: +36 1 461-1100

The all-time valid accreditation list of testing laboratories can be checked on the homepage of the National Accreditation Authority *(www.nah.gov.hu)*.

10.3.2.5.4 Future Developments

The requirements and test methods of EN 13501 series have been adopted and the formerly existing national test methods have been withdrawn. Two national test methods were introduced in this section, MSZ 14800-6:2009 (Fire resistance tests. Part 6: Fire propagation for building façades) and MSZ 14890:2014 (Test for ignitability danger of building materials). Both test methods model the real end-use applications of products and kits, but also provide sophisticated evaluation methods. MSZ 14800-6 is under revision.

References for Section 10.3.2.5

[1] MSZ 14800-16:1992: *Fire resistance tests. Determination of ignition temperature of solid materials.*

[2] MSZ 14800-6:2009: *Fire resistance tests. Part 6: Fire propagation for building façades.*

[3] MSZ 14890:2014: *Test for ignitability danger of building materials.*

[4] *National Fire Code,* issued by Ministry of Internal Affairs 54/2014 (XII. 5.).

[5] Móder et al., *Brief summary of the Hungarian test method (MSZ 14800-6:2009) of fire propagation on building façades* (2016) MATEC Web of Conferences 46, 01002.

[6] Smolka et al., *Semi-natural test methods to evaluate fire safety of wall claddings: Update* (2016) MATEC Web of Conferences 46, 01003.

[7] MSZ EN 1363-1:2013: *Fire resistance tests. Part 1: General Requirements.*

[8] MSZ EN 13823:2010 + A1:2015: *Reaction to fire tests for building materials. Building products excluding floorings exposed to the thermal attack by a single burning item.*

[9] MSZ EN 13501-1:2007+A1:2010: *Fire classification of construction products and building elements. Part 1: Classification using data from reaction to fire tests.*

10.3.2.6 Italy

Eleonora Anselmi

10.3.2.6.1 Statutory Regulations

In the past, the classification of fire behavior and the certification of materials for the purpose of fire prevention was regulated by the Decree of the Ministry of the Interior issued on the 26 June 1984 [1], containing test procedures enabling the fire performance of materials to be determined and classified.

In the 2000s, the rules were separated: the definition of technical rules (standards) became a task of the standardization body (in Italy UNI), while the safety levels and the technical rules for the use of products were to be fixed by the regulators.

The concept of risk has always been highlighted in the Italian Technical Standards of Fire Prevention; the Ministerial Decree D. M. 03/08/2015 "Approval of fire prevention technical standards, pursuant to article 15 of the legislative Decree 8th March 2006, no. 139" ("Approvazione di norme tecniche di prevenzione incendi, ai sensi dell'articolo 15 del decreto legislative 8 marzo 2006, n.139") is based on this principle.

The Unique Fire Prevention Code (Testo Unico di Prevenzione Incendi) is based on five articles and a substantial technical annex; its biggest innovation is the implementation of FSE (Fire Safety Engineering) as a projecting instrument, and the introduction of the possibility for the professional to select a "prescriptive approach" or a "performance approach".

One of the essential requirements deals with reaction to fire but, differently from other prescriptions, it applies to the building ("opera"), not directly to the product. This characteristic of reaction to fire is intended to be the degree of contribution of a combustible material to the fire to which it is subjected.

10.3.2.6.2 Classification and Testing of the Fire Performance of Building Materials and Components

In 2005, the need to adopt the European Construction Products Directive 89/106/CEE (CPD) and Decisions 2000/147/CE and 2003/632/CE resulted in the publication of two decrees at national level:

- D. M. 10/03/2005, which introduces the new test system and the new classification by incorporating the European test methods;
- D. M. 15/03/2005, which gives the acceptance criteria for the new classes.

Table 10.59–Table 10.61 compare the Italian classification with the European classification.

Table 10.59 Flooring End Use

	Italian Class (Classe Italiana)	European Class (Classe Europea)
I	Classe 1	($A2_{FL}$-s1), ($A2_{FL}$-s2), (B_{FL}-s1), (B_{FL}-s2), (C_{FL}-s1)
II	Classe 2	(C_{FL}-s2) (D_{FL}-s1),
III	Classe 3	(D_{FL}-s2)

Table 10.60 Wall End Use

	Italian Class (Classe Italiana)	European Class (Classe Europea)
I	Classe 1	(A2-s1-d0), (A2-s2-d0), (A2-s3-d0), (A2-s1-d1), (A2-s2-d1), (A2-s3-d1), (B-s1-d0), (B-s2-d0), (B-s1-d1), (B-s2-d1)
II	Classe 2	(A2-s1-d2), (A2-s2-d2), (A2-s3-d2), (B-s3-d0), (B-s3-d1), (B-s1-d2), (B-s2-d2), (B-s3-d2), (C-s1-d0), (C-s2-d0), (C-s1-d1), (C-s2-d1)
III	Classe 3	(C-s3-d0), (C-s3-d1), (C-s1-d2), (C-s2-d2), (C-s3-d2), (D-s1-d0), (D-s2-d0), (D-s1-d1), (D-s2-d1)

Table 10.61 Ceiling End Use

	Italian Class (Classe Italiana)	European Class (Classe Europea)
I	Classe 1	(A2-s1-d0), (A2-s2-d0), (A2-s3-d0), (A2-s1-d1), (A2-s2-d1), (A2-s3-d1), (B-s1-d0), (B-s2-d0), (B-s3-d0)
II	Classe 2	(B-s1-d1), (B-s2-d1), (B-s3-d1), (C-s1-d0), (C-s2-d0), (C-s3-d0)
III	Classe 3	(C-s1-d1), (C-s2-d1), (C-s3-d1), (D-s1-d0), (D-s2-d0),

These tables were implemented and slightly modified with the subsequent release of Ministerial Decree D. M. 16/02/2009.

The adoption of the CPD and the transition to the CPR resulted in the marketplace requesting the European classification, whereby each type of building has its own specific potential fire hazards and fire risks linked to the permanent elements of the building (i. e., construction products and overall design) but also to its content including furniture, paper, clothes, domestic and leisure articles.

A special case-by-case approach should be used in Italy for the so-called "historical buildings".

Historical Buildings

According to the Ministerial Decree D. M. 03/08/2015, the reaction to fire measures the extent that passive protection plays in the first phase of a fire, with the aim of avoiding ignition of the materials and the propagation of the fire. It refers to the fire behavior of the materials in the actual final conditions of application, with particular regard to their degree of fire contribution under standardized test conditions.

For the classification of the materials already in use before 2005, the withdrawal of these classifications must take place under the control of C.S.E. (Centro Studi e

Esperienze Anticendi) or, upon request, of the provincial fire department responsible for the territory, if the classification is required by it.

UNI EN ISO 1182: Non-Combustibility Test

The test method UNI EN ISO 1182 is equivalent to the non-combustibility test of building materials described in Section 9.2.2.2.

UNI 8456: Small Burner Test (Edge Application of Flame)

Method UNI 8456 [2] with its updated version of July 2010, is used to determine afterflame time, afterglow time, and the extent of damage of materials, and to test for the presence of flaming droplets from items such as curtains and awnings, which may be subjected to the effects of a flame from both sides. Droplets are defined as all burning debris. The method is essentially similar to the EN ISO 11925-2 small flame test (see Section 10.3.1), but the test specifications are somewhat different, and are summarized in Table 10.62.

Table 10.62 UNI 8456 Small Burner Test Specifications

Specimens	340 mm × 104 mm
	Two series of 10 specimens (5 in the direction of warp and 5 in the direction of weft)
Specimen position	Vertical
Ignition source	Small propane gas burner inclined at 45°, flame length 40 mm, edge application of flame
Flame application	12 s
Test duration	10 min
Conclusions	Classification into various categories taking into consideration afterflame time, afterglow, extent of damage and flaming droplets

The flame is applied to the specimen for 12 s. The afterflame time, afterglow time, extent of damage and flaming droplets are observed and recorded. The four parameters above are divided into three grades as shown in Table 10.63.

Table 10.63 Grading of Materials to UNI 8456

Grade	Afterflame time [s]	Afterglow time [s]	Extent of damage [mm]	Time for drippings to extinguish [s]
1	≤ 5	≤ 10	≤ 150	Non-burning
2	> 5 to ≤ 60	> 10 to ≤ 60	> 150 to ≤ 200	≤ 3
3	> 60	> 60	> 200	> 3

In order to establish the category which serves as the basis for classifying the building material, the grades of the four parameters are multiplied with weighting factors (Table 10.64). Materials which are consumed in less than 17 s (12 s for flame application and 5 s for afterflame) are placed in the worst category (i. e., IV); if they burn for more than 10 min, they are classified in grade 3 for extent of damage.

Table 10.64 Determination of Material Categories

Parameter	Weighting factor
Afterflame time	2
Afterglow time	1
Extent of damage	2
Dripping	1
Category	**Weighted sum of grades (grade × weighting)**
I	6–8
II	9–12
III	13–15
IV	16–18

UNI 8457: Small Burner Test (Surface Application of Flame)

Method UNI 8457 [3], with the update UNI 8457 (July 2010), is used to determine afterflame time, afterglow time, the extent of damage to materials, and the presence of flaming droplets, and hence the fire performance of linings and materials for floors, walls, and ceilings by surface application of a flame from one side.

This test method closely resembles the EN ISO 11925-2 small flame test (see Section 10.3.1). The Italian version, however, does not include a basket with a filter paper to test for burning droplets. The test specifications are summarized in Table 10.65.

The flame is applied to the specimen for 30 s. The four parameters (afterflame time, afterglow time, extent of damage and flaming droplets) are observed and recorded. These parameters are divided into three grades. The evaluation criteria are identical to those in Table 10.63 except for extent of damage, which is placed in grade 3 if ≥ 200 mm.

The procedure for establishing the material category is the same as that given for UNI 8456 in Table 10.64. Similarly, materials which are consumed within 35 s are placed in category IV and those which burn for longer than 10 min are placed in grade 3 as regards the extent of damage.

Table 10.65 UNI 8457 Small Burner Test Specifications

Specimens	340 mm × 104 mm Two series of 10 specimens (5 in the direction of warp and 5 in the direction of weft)
Specimen position	Vertical
Ignition source	Small propane burner, inclined at 45°, flame height 20 mm (flame applied to surface of specimen 40 mm above lower edge)
Flame application	30 s
Test duration	10 min
Conclusion	Classification into various categories taking into account afterflame and afterglow times, extent of damage and flaming droplets

UNI 9174: Spread of Flame Test

The UNI 9174 spread of flame test for building materials [4], with the update UNI 9174 (July 2010), is carried out with a small flame and a radiant panel. The intensity of this ignition source system corresponds to the advanced phase in the initiating fire. By varying the position of the specimen, materials and finished components can be tested under conditions simulating use in floors, walls, and ceilings as shown in Figure 10.38.

This test method is similar to ISO 5658-2 (spread of flame). A diagram of the spread of flame test is shown in Section 9.2.3.3. The test specifications are given in Table 10.66.

Table 10.66 UNI 9174 Spread of Flame Specifications

Specimens	Three specimens 800 mm × 155 mm for each orientation (wall, floor, ceiling)
Specimen position	End of exposed area nearest to radiator at a distance of 100 mm ▪ Floor: 450 mm side of radiant panel vertical, specimen horizontal and located centrally in the plane of the bottom of the radiant panel, long (800 mm) side at 90° to radiant panel ▪ Wall: 450 mm side of radiant panel horizontal, specimen vertical, long (800 mm) horizontal side at 45° to radiant panel ▪ Ceiling: 450 mm side of radiant panel vertical, specimen horizontal and located centrally in the plane of the top of the radiant panel, long (800 mm) side at 90° to radiant panel
Ignition sources	▪ Vertical variable propane radiant panel 300 mm × 450 mm, radiation intensity 62 kW/m^2, surface temperature 750 °C ▪ Variable propane pilot flame, length 80 mm, impinges on specimen 20 mm from edge nearest to radiator
Test duration	Test terminated when flame front stops or spreads to the end of the specimen
Conclusions	The maximum distance (in mm) of flame travel and the time to extinction (if the end of the specimen is not reached) are recorded

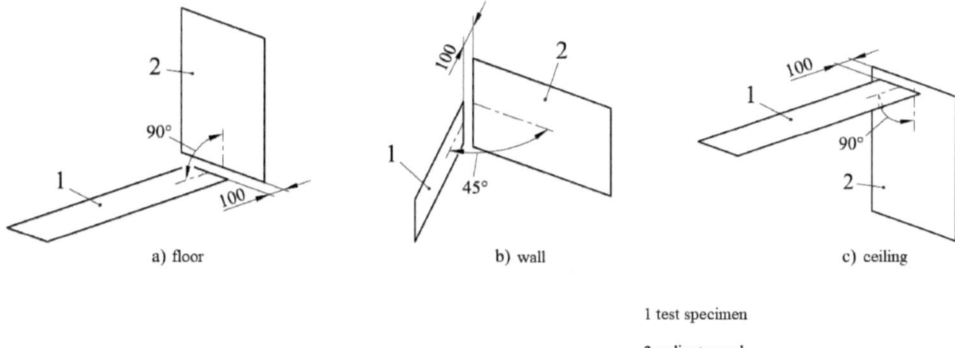

Figure 10.38 UNI 9174 specimen position

The specimen is exposed to the radiant panel and pilot flame. Note is made of whether the specimen selfignites and of the rate of spread of flame, extent of damage, afterglow, and the presence of flaming droplets. The four parameters are each divided into three grades (Table 10.67). The rate of spread of flame is not measurable if the flame front does not reach the 150 mm mark.

The building material category on which classification is based is determined by multiplying the above levels with weighting factors which, in the case of flaming droplets, distinguish between floor, wall, and ceiling use (Table 10.68).

If a material burns more than 60 min, it is placed in extent of damage grade 3. If the rate of flame spread is ≥ 200 mm/min and the extent of damage ≥ 650 mm, the material is classified in the worst category, i.e., IV. If the rate of flame spread is ≥ 30 mm/min and the extent of damage is ≤ 200 mm, the material is usually placed in grade 2 for the rate of flame spread.

Table 10.67 Grading of Materials According to UNI 9174

Grade	Spread of flame rate [mm/min]	Extent of damage, maximum length [mm]	Afterglow time [s]	Time for drippings to extinguish [s]
1	Not measurable	≤ 300	≤ 180	Non-burning
2	≤ 30	≥ 350 to ≤ 600	> 180 to ≤ 360	≤ 3
3	> 30	≥ 650	> 360	> 3

Table 10.68 Determination of Material Categories

Parameter	Weighting
Rate of spread of flame	2
Extent of damage	2
Afterglow	1

Table 10.68 Determination of Material Categories *(continued)*

Parameter		Weighting		
Dripping:	Floor	0		
	Wall	1		
	Ceiling	2		
Category Floor		Weighted sum of grades (grade × weighting)		
		Wall	Ceiling	
I		5–7	6–8	7–9
II		8–10	9–12	10–13
III		11–13	13–15	14–17
IV		14–15	16–18	18–21

10.3.2.6.3 Italian Classification for Materials

The categories obtained from the tests laid down in the old and new published decrees form the basis for classifying building materials as shown in Table 10.69, as provided in UNI 9177.

Building materials are classified by testing according to UNI EN ISO 1182, UNI 8456, UNI 8457, and updated versions. Two, five, and seven equivalent combinations of the categories obtained are available for classification in classes 2, 3, and 4, respectively. The fire hazard increases from class 0 (lowest hazard) to class 5 (highest hazard).

Table 10.69 Classification of Building Materials According to UNI EN ISO 1182, UNI 8456, UNI 8457, and UNI 9174

Class	Test method	Category						
0	UNI EN ISO 1182	–						
1	UNI 9174	I						
	UNI 8456 or UNI 8457	I						
2	UNI 9174	II	I					
	UNI 8456 or UNI 8457	I	II					
3	UNI 9174	III	II	I	II	II		
	UNI 8456 or UNI 8457	II	III	III	I	II		
4	UNI 9174	IV	III	III	IV	II	IV	I
	UNI 8456 or UNI 8457	III	IV	III	II	IV	I	IV
5	UNI 9174	IV						
	UNI 8456 or UNI 8457	IV						

UNI 9175: Fire Reaction of Upholstered Furniture under Small Flame Action

The UNI 9175 [5] spread of flame test for upholstered furniture is under revision within UNI CT 011 GL01; the current UNI 9175 (July 2010), is carried out with a small flame. The test device is shown in Figure 10.39, and the specifications are summarized in Table 10.70.

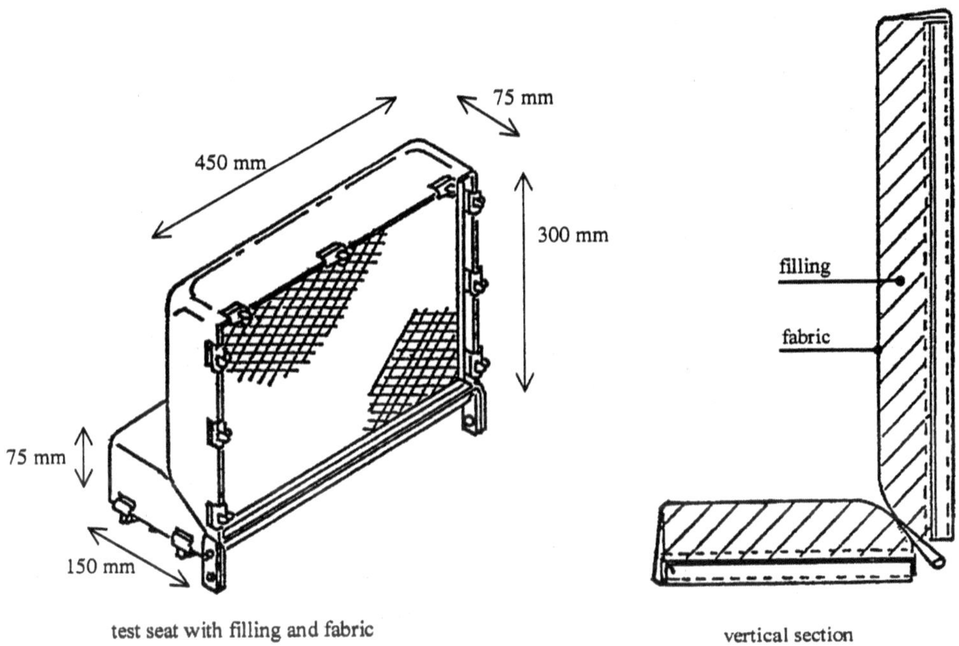

Figure 10.39 UNI 9175 test specimen

Table 10.70 UNI 9175 Upholstered Furniture Reaction-to-Fire Specifications

Specimens	Seat: 450 mm × 150 mm × 75 mm (5 series)
	Back: 450 mm × 300 mm × 75 mm (5 series)
	Fabric: 800 mm × 650 mm (3 series)
Specimen position	Seat: horizontal
	Back: vertical
Ignition source	Small propane gas burner, flame length 40 mm
Flame application	20 s/80 s/140 s
Test duration	280 s for first part, 600 s for second part
Conclusions	Classification into various categories based on flame duration

First, the test is carried out only on the filling (two tests). The flame application time is 20 s. If one of the two tests is negative, no more tests are made and no classification given.

Second, if positive results are obtained for the filling, the test is performed on a complete specimen with filling, interliner (if present), and fabric with three flame applications (20 s, 80 s, and 140 s) in three tests. If the first application is negative, no classification is given.

The test result is negative if the specimen does not stop burning within 120 s after flame removal, or if it is completely burnt within 120 s. The obtainable classifications are listed in Table 10.71.

10.3.2.6.4 Italian Classification for Upholstered Furniture

As shown in Table 10.71, upholstered furniture is classified according to three categories; the fire performance increases from category 3 IM (worst performance) to category 1 IM (best performance).

Table 10.71 Grading of Materials According to UNI 9175

Category	Positive results for
1.IM	First, second, and third flame application (20 s/80 s/140 s)
2.IM	First and second flame application (20 s/80 s)
3.IM	First flame application only (20 s)

10.3.2.6.5 Specimen Preparation Before Testing

The specimen preparation before testing is done according to UNI 9176 (second edition – January 1998).

Method A: For Textile Materials Susceptible to the Effects of Flames from Both Sides, To Be Tested to UNI 8456 and UNI 9174

Wash five times with a standard washing machine. Each time the procedure is: prewash (35–40 °C, detergent 0.5 g/l), wash (40 °C, detergent 1.5 g/l), dry in stove (60 °C, 2 hours) and iron with a steam press or standard iron.

Method C: For All Materials To Be Tested to UNI 8457 and for Non-Textile Materials To Be Tested to UNI 8456

Procedure for specimens: Brush 200 times for materials for walls and ceilings, and for materials susceptible to be attacked by both sides; brush 5000 times for flooring. The material is put into a detergent solution.

Method D: For Non-Textile Materials To Be Tested to UNI 9174, for Textile Materials Subjected to the Effects of Flames from One Side To Be Tested to UNI 9174, and for All Covering Materials To Be Tested to UNI 9175

Procedure for specimens: Brush 200 times for materials for walls and ceilings, and for materials susceptible to be attacked by both sides; brush 5000 times for flooring. The material is put into a detergent solution.

10.3.2.6.6 Official Approval

Classification and certification of the fire behavior and the official attestation ("omologazione") of materials are governed by the old and the new published decrees. Whereas classification and certification is performed by officially recognized or authorized laboratories and test institutes, the "omologazione" for the production of the material tested is granted by the Ministry of the Interior. The material must then be marked or delivered with an attestation of conformity. The tests for classification of fire behavior can be carried out and certified by:

State test institute:

- C.S.E. (Centro Studi ad Esperienze dei Vigili del Fuoco), I-00178 Capanelle (Roma)

Notified authorized laboratories:

- C.S.I. (Servizi dei laboratori di analisi del comportamento al fuoco), Viale Lombardia, 20, I-20021 Bollate (MI)
- Istituto Giordano S.p.A. (Centro Politecnico di ricerche e certificazioni), Via Rossini, 2, I-47041 Bellaria (RN)
- CNR Invalsa (Istituto per la valorizzazione del legno e delle specie arboree), Via Biasi, 75, I-38010 S. Michele all'Adige (TN)
- LAPI (Laboratorio di Prevenzione Incendi), Via Della Quercia, 11, I-59100 Prato (PO)
- L. S. Fire Testing Institute s.r.l. (Laboratorio di Studi e ricerche sul Fuoco), Via della Bonifica, 4, I-64010 Controguerra (TE)
- RINA Services S.p.A., Via Corsica, 12, I-16128 - Genova (GE)
- t²i - Trasferimento Tecnologico e Innovazione s. c. a r. l., Via Roma, 4, I-31050 - Lancenigo di Villorba (TV)

Other authorized laboratories:

- Istituto Sperimentale delle Ferrovie dello Stato,
- MIT International Testing s.r.l.
- Labortec
- CATAS
- Centro Tessile Serico - Società consortile per azioni

- Istituto per le Tecnologie della Costruzione del Consiglio Nazionale delle Ricerche - ITC-CNR

10.3.2.6.7 Future Developments

Like the other member countries of the European Union, Italy will adopt the Harmonized European Standards, and the national fire test methods described here will be withdrawn after period to be defined.

References for Section 10.3.2.6

[1] Ministero dell'Interno: Decreto Ministeriale 26 giugno 1984: Classificazione di reazione al fuoco ed omologazione dei materiali ai fini della prevenzione incendi.

[2] UNI 8456. Reazione al fuoco dei materiali sospesi e suscettibili di essere investiti da una piccola fiamma su entrambe le facce. Oct. 1987.

[3] UNI 8457. Reazione al fuoco dei materiali che possono essere investiti da una piccola fiamma su una sola faccia. Oct. 1987.

[4] UNI 9174. Reazione al fuoco dei materiali sottoposti all'azione di una fiamma d'innesco in presenza di calore radiante. Oct. 1987.

[5] UNI 9175. Reazione al fuoco di manufatti imbottiti sottoposti all'azione di una piccola fiamma - Metodo di prova e classificazione. Oct. 1987.

[6] UNI 9177. Classificazione di reazione al fuoco dei prodotti combustibili 2008.

10.3.2.7 The Netherlands

R. J. M. (Rudolf) van Mierlo

10.3.2.7.1 Statutory Regulations

The Dutch building regulations are based on the "Woningwet" (housing law), which refers to the "Bouwbesluit" (building decree) [1] for all technical building requirements and the safe use of buildings. The building decree is updated on an approximately yearly basis, with interim additions and modifications published in the "Regeling Bouwbesluit" (ministerial addition to the building decree). The requirements are differentiated by the type of building (residential, offices, hotels, schools, industry buildings, etc.) and whether the building is newly built or renovated, or whether it is an "existing building".

The requirements formulated for these existing buildings are minimum requirements. The actual requirements for an existing building depend on the history of permits obtained, but shall be between the requirement levels defined in the building decree for newly built and existing buildings.

The building decree contains functional (non-quantified) requirements for all building types. These requirements are further specified quantitatively in (mainly) performance-based requirements. Alternative requirements can be used as long as an equivalent level of safety is reached. As a consequence, all Dutch requirements described here shall be interpreted in the sense of meeting these requirements, or providing proof that the same level of safety is reached by other means.

The requirements in the building decree refer to assessment methods given in technical standards from ISO, CEN or NEN, and occasionally other sources. NEN, the "Nederlands Normalisatie-Instituut", is the Dutch standardization body.

10.3.2.7.2 Classification and Testing of the Fire Performance of Building Products and Components

Classification and Testing for Reaction to Fire

Since 2003, the building decree has referred to the European classification standard EN 13501-1 for reaction-to-fire performance of building products in new buildings. For renovation of buildings and existing buildings, the "old" Dutch reaction-to-fire classes that were in use before the European classification system are still an alternative to the European classes.

The Dutch test and classification methods are described in four standards: NEN 6064 (non-combustibility), NEN 1775 and NEN 6065 (contribution to fire propagation of floorings and of other building products, respectively), and NEN 6066 (smoke production). A description of the four Dutch standards was given in the 3rd edition of the Plastics Flammability Handbook [2]. In summary, the standards imply the following:

- NEN 1775 (1997) [3] corresponds to the DIN 4102-14, but without a smoke production measurement.
- NEN 6064 (2001) [4] corresponds in principle to EN ISO 1182.
- NEN 6065 (1997) [5] is a combination of a flame spread test (which corresponds to the BS 476 Part 7, but with a continued pilot flame) and a flashover test. The second test measures the level of electrical power needed to ignite a specimen in a box in a certain time when an identical specimen is ignited first.
- NEN 6066 (1997) [6] corresponds in principle to the ISO/TR 5924, in which the smoke production is measured at radiant levels of 20, 30, 40, and 50 kW/m^2.

Other Important Fire-Related Classification and Test Methods

Roofs are tested on their external fire performance in NEN 6063 "Bepaling van het brandgevaarlijk zijn van daken" (Fire behavior of roofs on external exposure to flying brands) (2008) [7], which is nearly identical to the German DIN 4102-7 (1998) and test 1 of the CEN/TS 1187 (2012). This test involves the assessment of damaged surface area, the damaged area of inner layers and fire penetration aspects when the roof is exposed to burning brands. The result is reported as pass or fail.

Aspects of resistance to fire are covered by test and calculation methods, the main test standard being NEN 6069 "Beproeving en klassering van de brandwerendheid van bouwdelen en bouwproducten" (Testing and classification of resistance to fire of building products and building elements) (2019) [8]. This method uses a dual system in which both European and Dutch test and calculation methods are referred to. For building products in new buildings for which product CE marking is mandatory, the European methods have to be used. In other cases the Dutch methods are still an alternative.

Requirements

The reaction-to-fire requirements for building products may be summarized (and simplified) as follows. Table 10.72 gives an overview of the minimum performance expressed in the "old" Dutch classes, which are still an alternative for existing buildings and those under renovation.

Table 10.72 Alternative National Dutch Fire Performance Classes for Existing Buildings and Those under Renovation

Application of building product	Classification according to EN 13501-1	Old Dutch standard	Classification to Dutch standard
Near fireplaces	A1	NEN 6064	Non-combustible
Façades (general requirement)	B	NEN 6065	Class 2
Parts of façades between 2.5 and 13 m (under certain conditions, e.g., when the floors are part of the same fire compartment)	D	NEN 6065	Class 4
Escape route floorings (depending on the escape route status)	C_{fl}- or D_{fl}-s1_{fl}	NEN 1775 NEN 6066	T1 (high status) or T3 (other escape routes) Smoke production ≤ 10 m^{-1}
Other internal building products (depending on the escape route status)	B-s2 or D-s2	NEN 6065 NEN 6066	Class 2 or 4 Smoke production: ≤ 2.2 m^{-1} if class 2, or ≤ 5.4 m^{-1} if class 1
Other external building products (depending on the escape route status)	C or D	NEN 6065	Class 2 or 4
Basic requirement for floorings	D_{fl}-s1_{fl}	NEN 1775 NEN 6066	T3 Smoke production ≤ 10 m^{-1}
Basic requirement for other building products	D-s2	NEN 6065 NEN 6066	Class 4 Smoke production ≤ 10 m^{-1}

For completeness, a special aspect of the Dutch fire resistance requirements is mentioned here. The requirements are given as a level of resistance (Dutch “WBDBO”) to a fire moving from one room to another by any route of fire propagation, either inside or outside the building. The contribution of constructions is their standard fire resistance. The spatial separation of façade openings contributes to the resistance to fire spread between the rooms by reducing the radiation from façade openings and external flames. The fire resistance contributions of constructions and spatial separation of façade openings along the route between the rooms may be summed to satisfy the requirement.

10.3.2.7.3 Official Approval

The tests described above and the resultant classifications provide the basis for official approval by the local authority. In the Netherlands, fire performance testing of building products is carried out by two Notified Bodies (notified within the context of the CPR, the Construction Products Regulation):

Efectis Nederland BV
Brandpuntlaan Zuid 16, NL-2665 NZ, Bleiswijk
[or PO Box 554, NL-2665 ZN, Bleiswijk]
The Netherlands
Tel. + 31 88 3473 723
E-mail: *nederland@efectis.com*

Peutz BV
Lindenlaan 41, Mook
[or PO Box 66, NL-6585 ZH, Mook]
The Netherlands
Tel. + 31 24 357 0707
E-mail: *info@peutz.nl*

10.3.2.7.4 Future Developments

In the Netherlands, a new building decree (“Besluit bouwwerken leefomgeving”) is currently under development, which will not differ substantially from the current building decree as far as reaction to fire is concerned. However, in this field the specific European classification and test methods for cables and linear pipe insulation will be added. Requirements for façades of some high-rise buildings might change due to discussions that at the time of writing (2020) are still ongoing.

References for Section 10.3.2.7

[1] Dutch Ministry of the Interior and Kingdom Relations, *Bouwbesluit 2012*, version 2020.

[2] Van Hees, P., Section 10.12.1 The Netherlands, in *Plastics Flammability Handbook* (3rd Edition), Troitzsch, J. (Ed.), Carl Hanser Verlag, Munich, 2004.

[3] NEN 1775:1991+A1:1997, *Bepaling van de bijdrage tot brandvoortplanting van vloeren* (Contribution to fire propagation of floorings), NEN, 1997.

[4] NEN 6064:1991+A2:2001, *Bepaling van de onbrandbaarheid van bouwmaterialen* (non-combustibility), NEN, 2001.

[5] NEN 6065: 1991+A1:1997, *Bepaling van de bijdrage tot brandvoortplanting van bouwmateriaal (combinaties)* (Contribution to fire propagation of building material (combinations)), NEN, 1997.

[6] NEN 6066:1991+A1:1997, *Bepaling van de rookproductie bij brand van bouwmateriaal (combinaties)* (Smoke production of building material (combinations)), NEN, 1997.

[7] NEN 6063:2008, *Bepaling van het brandgevaarlijk zijn van daken* (Test method for external exposure to roofs), NEN, 2008.

[8] NEN 6069+A1+C1:2019, *Beproeving en klassering van de brandwerendheid van bouwdelen en bouwproducten* (Testing and classification of resistance to fire of building products and building elements), NEN, 2019.

10.3.2.8 Nordic Countries

Björn Sundström

10.3.2.8.1 Statutory Regulations

The Nordic countries, Denmark, Finland, Iceland, Norway, and Sweden, have a long tradition of co-operation around building regulations. The results are procedures and regulations that have much in common in all the Nordic countries. The Nordic building regulations are performance-based and allow for fire safety engineering. The relevant national building regulations are:

- Denmark: *Bygningsreglement 2018, BR18.* Section 5, § 82–§ 158, covers fire safety [1]
- Finland: *Suomen Rakentamismääräyskokoelma* (revised 2018) [2]
- Iceland: *Byggingarreglugerð nr. 112/2012* and *Lög um mannvirki* (law on structures) *2010 nr. 160 28. desember* [3]
- Norway: *Plan og bygningsloven* (building law) *LOV-2008-06-27-71* and associated regulation; *Byggteknisk forskrift (TEK17, 2017)* [4]
- Sweden: *Boverkets byggregler, BBR 2017* [5]

10.3.2.8.2 Classification and Testing of the Fire Performance of Building Materials and Components

The CPR applies in the Nordic countries and, therefore, test methods and product classification follow the harmonized CEN standards. There are requirements for linings, floor coverings, pipe insulation, and cables and therefore the classification standards EN 13501-1 and EN 13501-6 and the associated standards apply. For details, see Section 10.3.1 European Union. For roof coverings, the Nordic building regulations use CEN/TS 1187 test method 2 and the products are classified according to 13501-5. Details are given below.

CEN/TS 1187: 2012

Test methods for external fire exposure to roofs

Test method 2

The test apparatus is shown in Figure 10.40 and the test applications in Table 10.73.

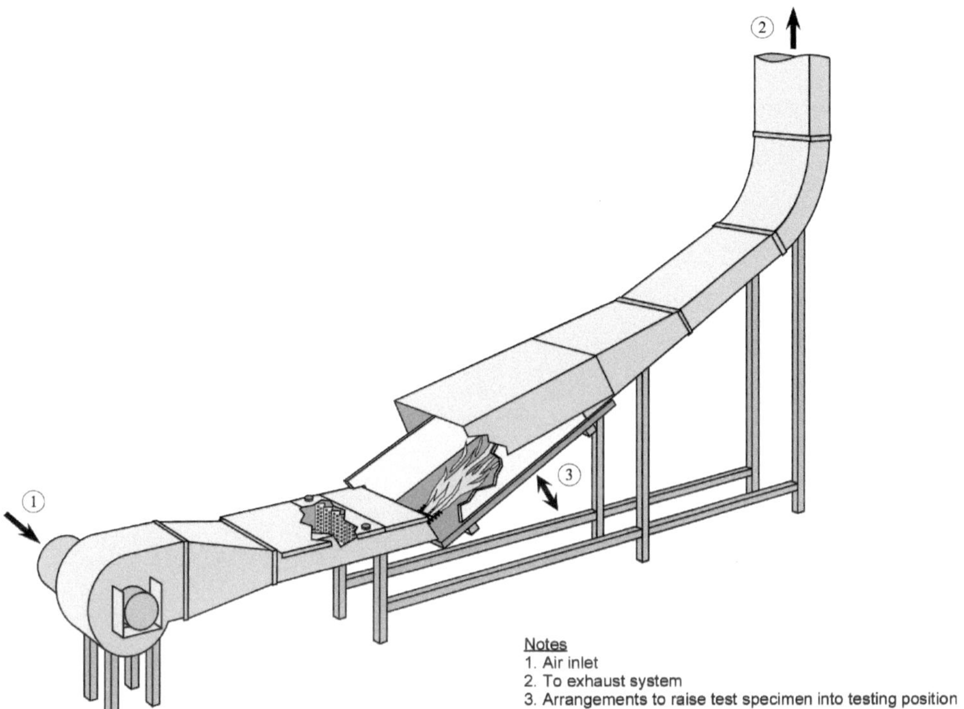

Figure 10.40 CEN/TS 1187 test method 2 for roof coverings

Table 10.73 Specifications for Testing the Fire Spread on Roof Coverings

Specimens	6 specimens 1000 mm × 400 mm (3 specimens for each wind speed)
Specimen position	Inclined at 30° to horizontal between two air ducts, upper duct with suction fan, lower duct with blower
Ignition source	Wood crib 100 mm × 100 mm consisting of 8 pieces of pine wood 10 mm × 10 mm × 100 mm, ignited by a special gas burner, and placed on the test specimen on the center line 100 mm from the lower edge
Test duration	15 min or when the flame front has reached the upper end of the specimen
Conclusion	Classification is based on the fire damage on the surface and on the substrate. There are two classes: B_{ROOF} (t2), with limited fire damage, and F_{ROOF} (t2), if the requirements for B_{ROOF} (t2) are not met.

10.3.2.8.3 Official Approval

The tests described above for determining the fire performance of building products can be carried out by officially accredited test institutes in each country except for Iceland, due to the lack of a suitable establishment. The major institutes in each Nordic country are:

DBI – The Danish Institute of Fire and Security Technology
Jernholmen 12
2650 Hvidovre
Denmark
https://brandogsikring.dk/en/

Eurofins Expert Services Ltd. (previously known as VTT Expert Services Ltd)
P. O. Box 47
FI-02151 Espoo
Finland
https://www.eurofins.fi

RISE Fire Research AS
Visiting address:
Tillerbruvegen 202
7092 Tiller (Trondheim)
Norway
https://risefr.no/

RISE Research Institutes of Sweden AB
Box 857
SE-501 15 Borås
Sweden
https://www.ri.se/en

These organizations issue test reports, certificates, and handle production control according to the CPR regulation. They are under surveillance by the Nordic accreditation authorities, SWEDAC-Sweden, DANAK-Denmark, NA-Norway, and FINAS-Finland.

10.3.2.8.4 Future Developments

The building regulations in all Nordic countries are adapted to the European harmonization. The available product standards and the associated CE-marking are implemented. In cases not covered by harmonized standards regulations, the European set of test methods generally apply. The old national systems in the Nordic countries used sets of test methods published by NORDTEST and referenced NT FIRE XX. This system is not used for construction products anymore. However, for example, for large scale testing of furniture the NORDTEST method NT FIRE 032 still applies. The old Nordic systems are described in the third edition of this handbook [6].

The institutes mentioned above work actively to support high quality of harmonized testing and certification throughout Europe. There are various groups to work with; for example, the EU-commission, in EU research projects for pre-normative research, with EGOLF (European Group of Organizations for Fire Testing, Inspection, and Certification) for interlaboratory calibrations, with the group of notified bodies for interpretation of test standards and certification procedures. These activities are ongoing and improving the tests and regulations used for reaction to fire in Europe.

Nordic building regulations also specify performance criteria. This opens possibilities for fire safety design based on calculations. Some of the international reactions to fire tests are also useful for fire modelling, for example, the cone calorimeter. The concept of large or full-scale fire scenarios is also seen to be useful in performance-based design.

References for Section 10.3.2.8

[1] Bygningsreglement. Issued by Trafik-, Bygge- og Boligstyrelsen (The Danish Transport, Construction, and Housing Authority), Edvard Thomsens Vej 14, 2300 København S. *http://bygningsreglementet.dk/*

[2] Suomen rakentamismääräyskokoelma issued by the Ministry of the Environment, Aleksanterinkatu 7, P.O. Box 35, 0023 Helsinki, Finnish Government. *https://www.ym.fi/en-US/Land_use_and_building/Legislation_and_instructions/The_National_Building_Code_of_Finland*

[3] Byggingarreglugerð no. 112/2012 and Lög um mannvirki (law on structures) 2010 issued by Mannvirkjastofnun (the Iceland Construction Authority), Skúlagata 21 - 101 Reykjavík. *http://www.mannvirkjastofnun.is/english*

[4] Plan og bygningsloven (building law) LOV-2008-06-27-71 and Byggteknisk forskrift (TEK17, 2017) issued by Direktoratet for byggkvalitet, Postboks 8742, Youngstorget, 0028 OSLO. *https://dibk.no/*

[5] Boverkets byggregler, BBR, BFS 2011:6 changed BFS 2019:2. Issued by Boverket, Box 534, 371 23 Karlskrona. *www.boverket.se*

[6] Troitzsch, J., *Plastics Flammability Handbook: Principles, Regulations, Testing, and Approval*, Hanser, Munich, 2004, ISBN 3-446-21308-2.

10.3.2.9 Poland

Jadwiga Fangrat

10.3.2.9.1 Statutory Regulations

The basic documents regarding the fire safety of buildings in Poland are as follows:

- Act of 31 July 2019 amending an Act on fire protection [1]
- Act of 13 June 2019 amending an Act on nuclear law and an Act on fire protection [2]
- Act of 22 February 2019 amending an Act on construction law [3]
- Act of 25 June 2015 amending an Act on building products, an Act on construction law, and an Act on conformity systems [4]

- Notice of the Ministry of Investment and Development of 8 April 2019, on a unified text for a Decree of the Ministry of Infrastructure, regarding technical requirements for buildings and their location [5]
- Decree of the Ministry of Internal Affairs and Administration of 11 January 2019 amending a Decree on fire protection of buildings, other construction objects, and building grounds [6]
- Regulation (EU) No. 305/2011 of the European Parliament and of the Council [7].

More documents are listed at the end of this section [8–10].

According to [5], the parts (and rooms) of buildings are divided on the basis of their function into the following five categories of life hazard:

- ZL I: Public buildings or their parts, where over 50 persons can be present
- ZL II: Buildings or their parts intended for use by people with low mobility
- ZL III: Public buildings not classified as ZL I or ZL II
- ZL IV: Residential buildings
- ZL V: Collective residences not classified as ZL I or ZL II.

Additionally, the following groups of buildings are distinguished according to their height:

- Low (N): ≤ 12 m, or residential with < 4 storeys over ground level
- Medium height (SW): ≥ 12 m to < 25 m, or residential with > 4 to < 9 storeys over ground level
- High (W): ≥ 25 m to < 50 m, or residential with > 9 to < 18 storeys over ground level
- High-rise (WW): ≥ 55 m over ground level.

The basic technical requirements on fire safety of buildings are given in [5], Part VI: Fire Safety.

Table 10.74 Required Fire Classes of ZL Category Buildings

Building	ZL I	ZL II	ZL III	ZL IV	ZL V
Low (N)	B	B	C	D	C
Medium height (SW)	B	B	B	C	B
High (W)	B	B	B	B	B
High-rise (WW)	A	A	A	B	A

According to [6], there are five fire classes of buildings from A to E (Table 10.75). In addition, the following general requirements are given in [7]:

- Building construction must fulfill the load-bearing criteria for a certain period of time
- Fire initiation, as well as fire and smoke spread in the building, is limited within the room of origin and beyond it
- Fire spread to the neighboring buildings is limited
- Occupants must be able to leave the building or to be otherwise rescued
- The safety of rescue teams is ensured.

Table 10.75 Fire Classes of Buildings and Requirements for Fire Resistance of Their Elements

Fire Class	Fire resistance of building elements					
	Main construction	Roof construction	Ceiling	Exterior wall	Interior wall	Roof covering
A	R 240	R 30	REI 120	EI 120	EI 60	RE 30
B	R 120	R 30	REI 60	EI 60	EI 30	RE 30
C	R 60	R 15	REI 60	EI 30	EI 15	RE 15
D	R 30	-	REI 30	EI 30	-	-
E	-	-	-	-	-	-

Some important detailed requirements regarding fire properties of building products are given below.

- Readily ignitable materials are not allowed along escape routes.
- In rooms with capacity for over 50 persons (and production rooms), the use of partitions, fixed finished coatings and equipment, and floor coverings made of readily ignitable materials is forbidden.
- The use of readily ignitable floor coverings is not allowed:
  - Inside the fire zone of category ZL II
  - In production and storage rooms
  - In rooms with raised floors.
- Ceiling coverings and suspended ceilings must be made of non-combustible or non-ignitable materials, which should not drip or fall in the event of fire (this requirement is not valid for dwellings).
- The free space between a suspended ceiling and floor shall be subdivided into areas not exceeding 1000 m^2. Corridors must be divided by non-combustible barriers placed every 50 m.
- In bathrooms and saunas with gas or electric heaters, the use of combustible wall coverings is permissible if the distance from apparatus to the covering is at least 0.3 m.

- The use of wall coverings made of readily ignitable materials is forbidden in bathrooms and saunas with solid fuel furnaces.

10.3.2.9.2 Classification and Testing of the Fire Performance of Building Materials and Components

The methods for testing the fire properties of materials that are subject to the Polish Standards are basically the same as in the CEN standard system [11] (see Section 10.3.1). The numbering of each standard is also the same, but with additional letters "PN" (for instance, PN EN ISO 1182). Standards have titles in Polish, and are translated into Polish, with only a few exceptions.

The terms describing material properties (e.g., "readily ignitable" or "non-spreading fire") that are used in the regulations are converted into the Euroclasses as explained in ITB Guide [12]. The former national classification and testing system is described in the 3rd (2004) edition of this handbook, under Section 10.14.2.

Regulation of Polish building law regarding fire spread over exterior walls is not covered by the CEN system. Therefore, such material properties are examined according to the Polish national standard, as described below.

Testing and Classification of the Rate of Fire Spread over Exterior Walls of Buildings on the External Side

The procedure for examining and classifying the exterior walls of buildings on the external side in terms of fire spread is provided in the Polish Standard PN-B-02867:2013-06 [13], while the evaluation of the exterior walls of buildings on the internal side is covered by the European classification system according to EN 13501-1 [11]. The Standard [13] does not apply to walls in which each separate component has a fire reaction class of at least A2-s3,d0; such walls are considered to be "non-spreading fire" without the need for testing.

Specimens 1.8 m wide and 2.3 m high that are representative of classified exterior walls are exposed to a standard fire (20 kg of pine wood) under conditions corresponding to the initial period of fire development. Flame propagation range, temperature rise, and visual observations are registered during the test.

The testing conditions are a temperature of 20 ± 10 °C, no rain, freezing rain, or frost, and a wind speed up to 2 m/s. An electric ventilator is used to obtain the required wind speed. The testing period is 30 min (15 min fire exposure and 15 min observation). Specimens are conditioned to 15 ± 10 °C prior to the test. Visual observation should take place throughout, and photographs shall be taken during and after the test. The test arrangement is shown in Figure 10.41. The classification criteria based on the test results are summarized in Table 10.76.

Figure 10.41 PN-B-02867:2013-06 Test arrangement

Table 10.76 Classification Criteria of Exterior Walls Tested for Fire Spread According to [13]

Type	Temperatures measured at specified points		Combustion during test at certain zones	Combus-tion after test period	Burn-ing drips
Non-spreading fire (NRO)	450 °C	350 °C	No	No	No
Hardly spreading fire (SRO)	No limits	350 °C	No limits	No	No
Readily spreading fire (SIRO)	No limits				

10.3.2.9.3 Official Approval

The Building Research Institute (Instytut Techniki Budowlanej, ITB) at Warsaw is an accredited governmental institution for the issue of certificates and technical assessments for building products. All necessary tests can be performed in its accredited laboratories. The fire laboratory of the Building Research Institute is able to carry out all standard tests currently required by law in Poland, and in many other countries including EU countries and beyond Europe (like US or United Arab Emirates).

Enquiries should be made to the following departments:

- **Fire testing, classification, and technical judgments:**
  Instytut Techniki Budowlanej (Building Research Institute)
  Fire Research Department
  Ksawerów 21, 00-611
  Warsaw
  Poland
  Tel. +48 22 853 34 27, +48 22 566 42 84
  E-mail: *fire@itb.pl*
- **Certificates:**
  Instytut Techniki Budowlanej (Building Research Institute)
  Certification Department
  Filtrowa 1, 00-611
  Warsaw
  Poland
  Tel. +48 22 579 61 67
  Fax. +48 22 579 62 95
  E-mail: *certyfikacja@itb.pl*
- **European and national technical assessments:**
  Instytut Techniki Budowlanej (Building Research Institute)
  Technical Assessment Department
  Filtrowa 1, 00-611
  Warsaw
  Poland
  Tel. +48 22 579 62 94
  E-mail: *kot@itb.pl, eta@itb.pl*

More information is available on the ITB website: *www.itb.pl.*

10.3.2.9.4 Future Developments

The system of Euroclasses is fully implemented in Poland. Currently, only requirements on fire spread over exterior walls, which are not covered by the CEN system, are subjected to a Polish national standard.

References for Section 10.3.2.9

[1] Act of 31 July 2019 amending Act on fire protection. *Ustawa z dnia 31 lipca 2019 r. o zmianie ustawy o ochronie przeciwpożarowej (Dz.U. 2019 poz. 1518).*

[2] Act of 13 June 2019 amending Act - Nuclear Law, and Act on fire protection. *Ustawa z dnia 13 czerwca 2019 r. o zmianie ustawy - Prawo atomowe oraz ustawy o ochronie przeciwpożarowej (Dz.U. 2019 poz. 1593).*

[3] Act of 22 February 2019 amending Act - Construction Law. *Ustawa z dnia 22 lutego 2019 r. o zmianie ustawy - Prawo budowlane (Dz.U. 2019 poz. 695).*

[4] Act of 25 June 2015 amending the Act on building products, Act - Construction Law, and Act on conformity systems. *Ustawa z dnia 25 czerwca 2015 r. o zmianie ustawy o wyrobach budowlanych, ustawy - Prawo budowlane oraz ustawy o zmianie ustawy o wyrobach budowlanych oraz ustawy o systemie oceny zgodności (Dz.U. 2015 poz. 1165).*

[5] Notice of Ministry of Investment and Development of 8 April 2019 on unified text of Decree of Ministry of Infrastructure on technical requirements for buildings and their location. *Obwieszczenie Ministra Inwestycji i Rozwoju z dnia 8 kwietnia 2019 r. w sprawie ogłoszenia jednolitego tekstu rozporządzenia Ministra Infrastruktury w sprawie warunków technicznych, jakim powinny odpowiadać budynki i ich usytuowanie (Dz.U. 2019 poz. 1065).*

[6] Decree of Ministry of Internal Affairs and Administration of 11 January 2019 amending decree on fire protection of buildings, other construction objects and grounds. *Rozporządzenie Ministra Spraw Wewnętrznych i Administracji z dnia 11 stycznia 2019 r. zmieniające rozporządzenie w sprawie ochrony przeciwpożarowej budynków, innych obiektów budowlanych i terenów (Dz.U. 2019 poz. 67).*

[7] Regulation (EU) No. 305/2011 of the European Parliament and of the Council of 9 March 2011 laying down harmonised conditions for the marketing of construction products and repealing Council Directive 89/106/EEC. *Rozporządzenie Parlamentu Europejskiego i Rady (UE) NR 305/2011 z dnia 9 marca 2011 r. ustanawiające zharmonizowane warunki wprowadzania do obrotu wyrobów budowlanych i uchylające dyrektywę Rady 89/106/EWG.*

[8] Decree of Ministry of Infrastructure and Development of 22 September 2015 amending decree on detailed scope and form of building project. *Rozporządzenie Ministra Infrastruktury i Rozwoju z dnia 22 września 2015 r. zmieniające rozporządzenie w sprawie szczegółowego zakresu i formy projektu budowlanego (Dz.U. 2015 poz. 1554).*

[9] Decree of Ministry of Infrastructure of 12 April 2002 on technical requirements for buildings and their location. *Rozporządzenie Ministra Infrastruktury z dnia 12 kwietnia 2002 r. w sprawie warunków technicznych, jakim powinny odpowiadać budynki i ich usytuowanie (Dz.U. 2002 nr 75 poz. 690).*

[10] Decree of Ministry of Internal Affairs and Administration of 27 April 2010 amending decree on the specification of products for public safety or protection of health, life and property as well as the rules of authorisation of these products for use. *Rozporządzenie Ministra Spraw Wewnętrznych i Administracji z dnia 27 kwietnia 2010 r. zmieniające rozporządzenie w sprawie wykazu wyrobów służących zapewnieniu bezpieczeństwa publicznego lub ochronie zdrowia i życia oraz mienia, a także zasad wydawania dopuszczenia tych wyrobów do użytkowania (Dz.U. 2010 nr 85 poz. 553).*

[11] Introducing EN 13501-1: 2018 Fire classification of construction products and building elements - Part 1: Classification using data from reaction to fire tests. *PN-EN 13501-1:2019-02 Klasyfikacja ogniowa wyrobów budowlanych i elementów budynków - Część 1: Klasyfikacja na podstawie badań reakcji na ogień.*

[12] ITB Guide No. 401/2004 "Assignment of the terms in the technical and construction regulations of reaction to fire classes according to PN-EN", Warsaw, 2004. *Instrukcja ITB nr 401/2004 "Przyporządkowanie określeniom występującym w przepisach techniczno-budowlanych klas reakcji na ogień wg PN-EN", Warszawa, 2004.*

[13] Fire protection of buildings - Method of testing rate of fire spread over the exterior walls of buildings on the external side and rules of classification. *PN-B-02867:2013-06 Ochrona przeciwpożarowa budynków - Metoda badania stopnia rozprzestrzeniania ognia przez ściany zewnętrzne od strony zewnętrznej oraz zasady klasyfikacji.*

10.3.2.10 Spain

Antonio Galán Penalva

10.3.2.10.1 Statutory Regulations

In Spain, the requirements for building products regarding the reaction to fire are covered by two regulations:

- The Technical Building Code - Basic document for safety in case of fire (CTE DB SI) [1] applies to buildings with public or residential character such as office buildings, homes, hospitals, etc.
- The Fire Safety Regulation in Industrial Facilities (RSCIEI) [2, 3] applies to industrial facilities. However, it is common to find other types of areas within industrial facilities such as offices, commercial premises, etc. These types of areas are regulated by the RSCIEI as long as they meet the minimum requirements described in this regulation. If this is not the case, they are regulated by CTE DB SI.

The characteristics of electric cables are included in the Low Voltage Electrotechnical Regulation (Royal Decree 842/2002) [4]. The reaction-to-fire requirements for these types of products are covered in an explanatory note from the Ministry of Economy, Industry, and Competitiveness, which states that cables with a reaction-to-fire class of at least C_{ca}-s1b,d1,a1 meet the requirements of the regulation for low-voltage electrotechnical cables [5].

Technical Building Code

The Technical Building Code (CTE) is the regulatory framework that establishes the requirements that buildings must meet in relation to the basic requirements of safety and habitability established in the Law of Construction Planning (LOE 38/1999 of November 5). It applies to new buildings, as well as refurbishments of existing buildings, such as expansions, modifications, or change of use. The CTE can be completed with regional and local regulations issued by the competent Authorities.

An interesting aspect introduced by the CTE with respect to the previous legislation for buildings in Spain is performance-based design. The CTE determines the criteria that buildings must meet but also allows different solutions for proving the performance. This allows the configuration of a regulated environment in a more flexible way. The CTE is divided into two parts.

The first section details all the safety requirements (structural safety, fire safety, safety of use) and habitability (health, hygiene and protection of the environment, protection against noise and energy saving).

The second part contains the *basic documents* (DB), which are technical texts that transfer the requirements detailed in the first part of the CTE into practical application. The basic documents include the limits and quantification of the basic requirements, and a list of procedures that allow compliance with the requirements. The basic documents are:

- DB SE: Structural safety.
- DB SI: Safety in case of fire
- DB SUA: Safety of use and accessibility
- DB HE: Energy saving
- DB HR: Protection against noise
- DB HS: Health, hygiene, and protection of the environment.

In the basic documents, a series of official *complementary documents* (which are not statutory documents) are included to help to understand and put into practice the DB with comments, supporting documents, etc.

The regulation is completed by *recognized documents,* which are texts of a technical nature without regulatory power. Recognized documents can be technical guides or codes of good practice for design procedures, calculation, execution, maintenance, and conservation of products, elements, and building systems. Evaluation methods and constructive solutions, including building software as well as other types of documents, may also belong to this group.

The objective of the basic requirement “safety in case of fire” is to reduce the risk that building users suffer harm from an accidental fire, within acceptable limits. These harms may be the consequence of incorrect building project planning, construction, use, and maintenance characteristics. The basic document for “safety in case of fire” specifies the objective parameters and procedures, compliance with which ensures that the basic requirements (and the minimum quality levels of the basic requirement “safety in case of fire”) are satisfied [5].

The basic document was approved in 2006, and the last modification was published in 2019 [6, 7]. It is updated every 6 months with new comments to explain aspects needing clarification.

Fire Safety Regulation for Industrial Facilities

The aim of this regulation [8] is to establish and define fire safety requirements, which prevent the start of fires and improve fire response by limiting their spread and providing a way of extinguishing them. The goal is to minimize the damage or loss that a fire may cause to people or property.

This regulation is applicable to newly-built industrial facilities, and to existing facilities that move, change, or modify their activity, including refurbishments that involve an increase in their occupied area or an increase in the level of intrinsic risk. Nevertheless, the region’s competent authorities may require, if deemed appropriate, the application of the regulation to fire compartments, or even to the industrial facilities as a whole.

In certain circumstances and depending on use, the application of the CTE DB SI is possible in industrial facilities. The application conditions are defined in the RSCIEI.

10.3.2.10.2 Classification and Testing of the Fire Performance of Building Materials and Components

The reaction-to-fire requirements vary depending on the legislation involved.

Basic Document for Safety in Case of Fire: Internal Fire Spread

In order to comply with the basic requirement SI 1 regarding internal fire spread, the building products have to meet the following reaction-to-fire requirements.

For decorative and furniture products, the minimum classifications are shown in Table 10.77.

Table 10.77 Reaction-to-Fire Classes of Building Products [1]

Location	Coverings [a]	
	Ceilings and walls [b, c]	Floors [b]
Occupiable areas	C-s2,d0	E_{FL}
Corridors and stairways protected	B-s1,d0	C_{FL}-s1
Parking lots and special risk enclosures	B-s1,d0	B_{FL}-s1
Plenum air spaces (suspended ceilings, raised floors, material susceptible to begin or spread a fire)	B-s3,d0	B_{FL}-s2 [d]

[a] This table applies when a product is used on more than 5% of the total surface area of all the walls, the set of ceilings, and the set of floors of the enclosure considered.

[b] Includes the pipes and ducts that pass through the indicated areas without fire-resistant coating. In the case of pipes with linear thermal insulation, the reaction to fire will be the one indicated, but incorporating the subscript L.

[c] Includes those materials that constitute a layer contained inside the ceiling or wall and that is not protected by a layer of at least EI 30.

[d] Refers to the lower part of the air cavities. For example, in the air cavities of the suspended ceilings it refers to the material located on the upper face of the membrane. In spaces with clear vertical configuration (as well as when the suspended ceiling comprises a reticule or open framework with an acoustic, decorative function, etc.), this condition is not applicable.

Enclosures made of textile elements, such as tents, have to pass class T2 according to UNE-EN 15619:2014 “Rubber or plastic coated fabrics – Safety of temporary structures (tents) – Specification for coated fabrics intended for tents and related structures” or C-s2,d0 according to UNE-EN 13501-1:2007 “Fire classification of construction products and building elements – Part 1: Classification using data from reaction to fire tests”.

In buildings and establishments for public use, the decorative and furniture elements have to meet the following conditions:

a) Upholstered fixed seats and seats in cinemas, theatres, auditoriums, etc. have to pass the tests according to UNE-EN 1021-1: 2015 "Furniture - Assessment of the ignitability of upholstered furniture - Part 1: Ignition source smouldering cigarette", and UNE-EN 1021-2: 2006 "Furniture - Assessment of the ignitability of upholstered furniture - Part 2: Ignition source match flame equivalent".

b) Suspended textile elements, such as curtains, etc. have to fall within Class 1 of UNE-EN 13773: 2003 "Textiles and textile products - Burning behaviour - Curtains and drapes - Classification scheme".

Basic Document for Safety in Case of Fire: External Fire Spread

The reaction-to-fire class of materials that occupy more than 10% of the surface area are as follows, based on the total height of the façade:

- D-s3,d0 on façades up to 10 m high
- C-s3,d0 on façades up to 18 m high
- B-s3,d0 on façades higher than 18 m.

The classification must consider the end use application of the construction system, including the materials contained in the interior layers if they are not protected by a layer with a fire resistance of EI 30 as minimum.

Insulation products located inside ventilated cavities must have at least the following reaction-to-fire class, based on the total height of the façade:

- D-s3,d0 on façades up to 10 m high
- B-s3,d0 on façades up to 28 m high
- A2-s3,d0 on façades higher than 28 m.

The fire spread in the ventilation gaps of façades must be limited. An option could be the use of E30 fire barriers.

In façades of buildings with a height equal to or lower than 18 m that are accessible to people, the construction systems as well as those located inside ventilated façades must be at least B-s3,d0 up to 3.5 m height as minimum.

Roof coverings or coatings, as well as skylights and other elements for lighting or ventilation, and the upper side of cantilevers exceeding 1 m, must meet class B_{ROOF} (t1), if the roof is located less than 5 m away from the vertical projection of any façade area of the same or another building that does not have a fire resistance of at least EI 60.

Fire Safety Regulation in Industrial Facilities

The fire behavior requirements of building products are defined according to the classes shown in UNE-EN 13501-1 for those materials for which a harmonized standard exists and the "CE" marking is already in force. The conditions of reaction to fire applicable to the building elements are justified as follows:

a) Through the class that appears first in each case, according to the new European classification.

b) Through the class that appears second in brackets, according to the national classification established by the UNE-23727 standard. The building products whose classification to UNE 23727:1990 is valid for these applications may continue to be used after the end of their coexistence period, until a new reaction-to-fire regulation is established for these applications based on their specific risk scenarios.

The products used as coverings should be:

- In floors: C_{FL}-s1 (M2) or better
- In walls and ceilings: C-s3,d0 (M2) or better.

Skylights that are not continuous, or installations for smoke control systems that are installed on the roofs, shall be at least class D-s2,d0 (M3) or better. The materials of continuous roof skylights have to be B-s1,d0 (M1) or better.

The exterior cladding products of façades shall be C-s3,d0 (M2) or better. When a product constituting a layer contained in a floor, wall, or ceiling has a lower class than that required for the exterior cladding product, the product as a whole shall be at least EI 30 (RF-30). This is not required for products used in industrial sectors classified as low intrinsic risk, located in type B or type C buildings. Here, the classification D-s3,d0 (M3) or better will be sufficient for materials that are part of wall products.

Products located inside suspended ceilings or raised floors used for thermal and/or acoustic insulation, as well as those that constitute or have air conditioning or ventilation ducts, etc., must be of class B-s3,d0 (M1) or better.

Cables shall at least meet class C_{ca}-s1b,d1,a1 [4]. Stone-built, ceramic, and metal building products, as well as glass panes, mortar, concrete, or plaster, shall be classified as class A1 (M0).

As can be seen, the requirements are mainly based on in the European system of reaction to fire described in the standards EN 13501-1 and EN 13501-6. But for specific situations the classifications based on the national standard UNE 23727:1990 "Reaction to fire test of building materials. Classification of buildings materials" are required.

In UNE 23727, the reaction to fire of building materials is divided into five classes: M0, M1, M2, M3, and M4. Materials showing a high flame spread and not meeting class M4 are "Not Classifiable" (N.C.). A short summary of this classification system and related tests is shown below. A detailed description of the fire test specifications and test devices can be found in the 3rd Edition of this book [9].

The classification depends on the results of the tests laid down in the following standards:

- UNE 23102 "Determination of non-combustibility of building materials"
- UNE 23721 "Radiation test for building materials (rigid materials of any thickness, materials to be used in rigid substrates and flexible materials more than 5 mm thick) (main test)"
- UNE 23723 "Electrical burner test for building materials (flexible materials up to 5 mm thick) (main test)"
- UNE 23724 "Flame propagation and persistency test for building materials (complementary test)"
- UNE 23725 "Dripping test for building materials (complementary test)"
- UNE 23726 "Radiant panel test for floor coverings (complementary test)"
- UNE 23735 part 1 "Quick wear and ageing procedures: General requirements"
- UNE 23735 part 2 "Quick wear and ageing procedures: Textile materials not exposed to weathering".

Depending on the product or material, the Spanish standards specify different test methods:

- Flexible materials up to 5 mm thick are tested to UNE 23723 or, if necessary, to UNE 23724 and UNE 23725
- Flexible materials more than 5 mm thick and rigid materials of any thickness are tested to UNE 23721 and, if necessary, to UNE 23724 and UNE 23725
- Floor coverings are tested to UNE 23721 and, if necessary, to UNE 23726.

The requirements which must be met in these tests to achieve the various classifications are summarized in Table 10.78 and Table 10.79.

Table 10.78 Classification of Non-Combustible Building Materials

Tests	Requirements
Radiation test UNE 23721	Meeting class M. 1 prior to non-combustibility test (see Table 10.79)
Non-combustibility test UNE 23102	Testing of five specimens. Mean values are used to classify: ▪ The increase in temperature compared to the initial furnace temperature for the thermocouple associated to the tests. This must be ≤ 50 °C ▪ The duration of flames. This must be in the range >5 to ≤20 s ▪ The mean mass loss. This must be ≤ 50%.
	M0

The main method for classifying rigid materials, materials on rigid substrates or flexible materials more than 5 mm thick is the test to UNE 23721. Four indices are calculated from the test data:

- Flammability index, i
- Flame spread index, s
- Index of maximum flame length, h
- Heat release evolution index, c.

The classification criteria are summarized below:

Table 10.79 Classification of Combustible Rigid Building Materials and Flexible Materials above 5 mm Thick

Tests	Requirements/Classification				
Radiation test [a] UNE 23721	s = 0	s < 0.2	s < 1	s < 1	s < 5
	h = 0	h < 1	h < 1	h < 1.5	h < 2.5
	c < 1	c < 1	c < 1	c < 1	c < 2.5
	i = 0	Any i	i < 1	Any i	i < 2
	M1	M2		M3	
Flame spread test UNE 23724	Materials other than those in the classes mentioned *and* flame spread speed < 2 mm/s		Materials other than those in the classes mentioned *and* flame spread speed > 2 mm/s		
	M4		N.C.		

[a] If the material presents special behavior (melting, shrinkage, disappearing), see Table 10.81.

The other main method to classify flexible materials less than 5 mm thick is the UNE 23723 test. The following criteria are considered from the test data:

- Ignition duration
- Flammable dripping
- Destroyed material surface (length and width).

Table 10.80 Classification of Combustible Flexible Building Materials up to 5 mm Thick

Tests	Requirements/Classification				
Dripping test UNE 23725	–	No ignition of cotton pad	No ignition of cotton pad	Ignition of cotton pad	Ignition of cotton pad
Electrical burner test UNE 23723	No dripping	Non-flaming dripping	Non-flaming dripping	Non-flaming dripping	Flaming dripping
Flaming ≤ 5 s [a]	M1	M1	M1	M4	M4

Tests	Requirements/Classification				
Destroyed length < 350 mm	M2	M2	M3	M4	M4
If destroyed length < 600 mm, destroyed width < 90 mm	M3	M3	M4	M4	M4
Fire spread test UNE 23724	Materials other than those in the classes mentioned *and* flame spread speed < 2 mm/s		Materials other than those in the classes mentioned *and* flame spread speed > 2 mm/s		
	M4		N.C.		

[a] If the material presents special behavior (melting, shrinkage, disappearing), see Table 10.81.

Materials that shrink or disappear rapidly on exposure to thermal radiation must also be tested to UNE 23724 (which becomes the main test) and UNE 23725. Classification is made on the basis of the requirements shown in Table 10.81.

Table 10.81 Classification of Materials that Melt, Shrink, Disappear, etc.

Tests	Requirements/Classification				
Dripping test UNE 23725	–	No ignition of cotton pad	No ignition of cotton pad	Ignition of cotton pad	Ignition of cotton pad
Persistent flame test UNE 23724	No dripping	Non-flaming dripping	Non-flaming dripping	Non-flaming dripping	Flaming dripping
No persistency	M1	M1	–	M4	M4
Persistency ≤ 5 s	M2	M2	M3	M4	M4
Persistency > 5 s	M3	M3	M4	M4	M4

Note: Persistency is the average value of the maximum persistency value obtained from each specimen in the same serial test.

For flooring materials, there are special requirements, listed in Table 10.82. The radiant panel test only gives a distinction between levels M3 and M4.

Table 10.82 Classification of Materials for Floor Coverings

Test radiant panel test to UNE 23726		Final classification according to UNE 23721 and UNE 23726 results
Destroyed length	Destroyed length at 1 min test	
–	–	M1 or M2
≤ 300 mm	–	M3
> 300 mm	≤ 100 mm	M4
> 300 mm	> 100 mm	N.C.

10.3.2.10.3 Official Approval

The Technical Building Code specifies that when a construction product does not have CE marking, the tests have to be conducted by accredited laboratories with an officially recognized entity in accordance with the Royal Decree 2200/1995 as modified by the Royal Decree 411/1997, Royal Decree 338/2010, Royal Decree 1715/2010, Royal Decree 239/2013, and Royal Decree 1072/2015.

The website of ENAC (the National Accreditation Body in Spain) contains a list of the fire laboratories accredited to conduct reaction-to-fire tests [10]. This information is shown below:

- **Acondicionamiento Tarrasense (LEITAT Technological Center)**
  Innovació, 2 (Parc Audiovisual) 08225 – Terrassa (Barcelona)
  Phone: 937 882 300
  Email: *calidad@leitat.org*
  Web: *www.leitat.org*
- **Asociación de Investigación de la Industria Textil (AITEX)**
  Plaza Emilio Sala, 1 03801 Alcoy (Alicante)
  Phone: 965 542 200
  Email: *sfigueres@aitex.es*
  Web: *www.aitex.es*
- **Asociación para el Fomento de la Investigación y la Tecnología de la Seguridad contra Incendios (AFITI)**
  Camino del Estrechillo 8 28500 Arganda del Rey (Madrid)
  Phone: 902 112 942
  Email: *afiti@afiti.com*
  Web: *www.afiti.com*
- **Ensatec, S.L.**
  Polígono Lentiscares, Avda. Lentiscares, 4 y 6 26370 – Navarrete (La Rioja)
  Phone: 941 250 466
  Email: *info@ensatec.com*
- **Fundación Gaiker**
  Parque Tecnológico. Edificio 202 48170 – Zamudio (Bizkaia)
  Phone: 946 002 323
  Email: *mark@gaiker.es*
  Web: *www.gaiker.es*
- **Fundación Tecnalia Research & Innovation**
  Parque Tecnológico de Bizkaia. C/ Geldo, Edificio 700 48160 – Derio (Bizkaia)
  Phone: 902 760 000
  Email: *pablo.etxeberria@tecnalia.com*

- **LGAI Technological Center, S. A. (Applus)**
  Campus de la U. A.B. Ronda de la Font del Carme, s/n 08193 – Bellaterra (Barcelona)
  Phone: 935 672 000
  Email: *info@appluslaboratories.com*
  Web: *www.applus.com*

10.3.2.10.4 Future Developments

Currently (January 2020), the regulation for industrial facilities is under revision, but there is no public document available so far.

References for Section 10.3.2.10

[1] Código Técnico de la Edificación. Documento Básico Seguridad en caso de incendio. (CTE DB SI). December 2019.

[2] Real Decreto 2267/2004, de 3 de diciembre, por el que se aprueba el Reglamento de seguridad contra incendios en los establecimientos industriales.

[3] Corrección de errores y erratas del Real Decreto 2267/2004, 3 de diciembre, por el que se aprueba el Reglamento de seguridad contra incendios en los establecimientos industriales.

[4] Real Decreto 842/2002, de 2 de agosto, por el que se aprueba el Reglamento electrotécnico para baja tensión, Published on 18/09/2002.

[5] Ley 38/1999, de 5 de noviembre, de Ordenación de la Edificación, 06/11/1999.

[6] Real Decreto 314/2006 de 17 de marzo, por el que se aprueba el Código Técnico de la Edificación.

[7] Real Decreto 732/2019, de 20 de diciembre, por el que se modifica el Código Técnico de la Edificación, aprobado por el Real Decreto 314, de 17 de marzo.

[8] Nota aclaratoria sobre la aplicación al Reglamento de Seguridad Contra Incendios en los Establecimientos Industriales (Real Decreto 2267/2004) del Reglamento Delegado 2016/364, que establece las clases posibles de reacción al fuego de los cables eléctricos (3 abril 2017).

[9] Garcia Alba, S. Section 10.15 Spain in Plastics Flammability Handbook: Principles, Regulations, Testing, and Approval (3rd Edition), Troitzsch, J. (Ed.), Hanser, 2004.

[10] ENAC website *(www.enac.es)* (accessed January 2020).

10.3.2.11 Switzerland

Marcel Donzé

10.3.2.11.1 Statutory Regulations

According to the federal constitution of Switzerland, the 26 cantons are responsible for matters regarding the public fire authorities, and they enact their own laws. Regarding the "Agreement between all cantons to remove technical barriers to trade" [1], in their laws the cantons have to adopt the fire protection regulation issued by the VKF (Vereinigung Kantonaler Feuerversicherungen, Association of Cantonal Fire Insurances). The intended purpose of this regulation is the safety of people and property in the event of fires and explosions. The regulation consists of the VKF fire protection standard 1-15 [2] and the VKF fire protection guidelines. The fire protection standard specifies the basic principles. Requirements and measures are listed in the guidelines, thematically divided into 19 documents.

10.3.2.11.2 Classification of Reaction to Fire of Building Materials

Within the VKF fire protection regulations, all materials and construction components in buildings with requirements regarding reaction to fire are referred to as building materials. The important criteria for reaction to fire are the burning behavior, smoke production, flaming droplets, and (for cables) corrosivity. Classification is possible according to EN 13501 or a test according to the national VKF guidance “Building material and construction components, part B: Testing regulations” [3].

Classification According to EN 13501

Construction products are classified according to EN 13501. In addition to the reaction-to-fire classification according to EN 13501-1, classifications for roof coverings (according to EN 13501-5) and cables (according to EN 13501-6) apply. In addition, the rules for a “classification without further testing” (CWFT) can also be applied. Details of the tests, the classification procedures and the “CWFT” are described in Section 10.3.1.

Classification According to the National VKF Guidance [3]

The reaction to fire is assessed according to the burning behavior, the degree of smoke development, and the presence of flaming droplets, and is classified by a fire coding. The fire coding is established by standardized tests.

Burning behavior

In the sense used in this assessment, the burning behavior of a material is defined by its flammability and the burning rate, and is substantiated by testing. The classification of building materials is based on combustibility grades 3 to 6. Materials of combustibility grades 1 and 2 are not approved as building materials. Details are shown in Table 10.83.

Table 10.83 Combustibility Grades 1 to 6

Combustibility grade	Burning behavior	Example
1	*Extremely easy to ignite and extremely fast burning*	Nitrocellulose
2	*Easy to ignite and fast burning*	Celluloid
3	*High combustibility* Building materials with high combustibility: burn spontaneously and fast without additional heat supply.	Plastic foams without flame retardants
4	*Average combustibility* Building materials with average combustibility: continue to burn spontaneously for a longer period of time without additional heat supply.	Conditioned white wood
5	*Low combustibility* Building materials with low combustibility: continue to burn slowly or carbonize only with additional heat supply. After removal of the heat source, the flames must go out within a short time interval and afterglowing must cease.	Plastics containing flame retardants
5 (200 °C)	*Low combustibility at 200 °C* Building materials fulfilling the requirements of combustibility grade 5 even under the effect of an elevated ambient temperature of 200 °C.	Rigid PVC
6q	*Quasi non-combustible* Building materials containing a low content of combustible components, which are classified as non-flammable for non-combustible application purposes.	Several mineral wool products
6	*Non-combustible* Building materials without combustible components, which do not ignite, carbonize or incinerate.	Concrete

Determination of combustibility grade

The application of the combustibility grade test described below is normal practice. Special regulations exist for certain materials such as flooring and textile materials.

The test is conducted using a standardized test apparatus shown in Figure 10.42. The test criteria are summarized in Table 10.84.

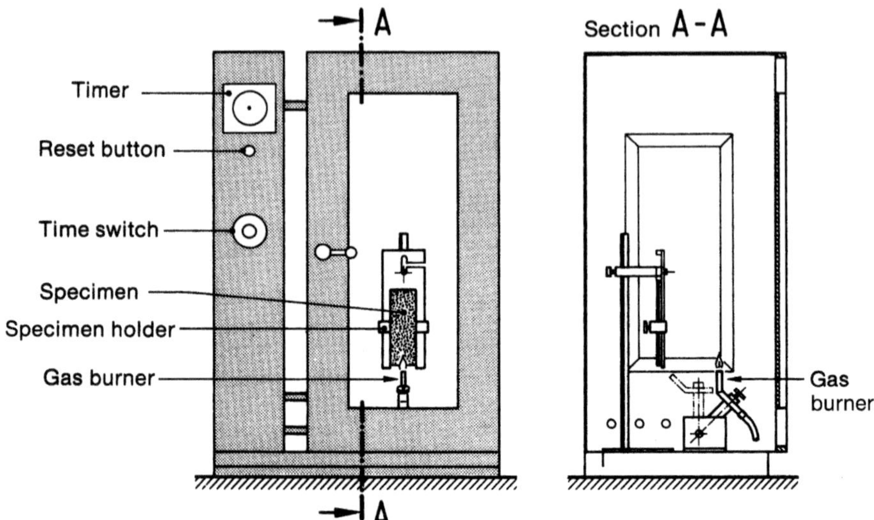

Figure 10.42 Combustibility test apparatus

Table 10.84 Combustibility Test Criteria

Specimens	Six specimens: ▪ Compact materials: 160 mm × 60 mm × 4 mm ▪ Foams: 160 mm × 60 mm × 6 mm
Specimen position	Vertical
Ignition source	Propane-operated displaceable burner, flame tip temperature approx. 900 °C, flame length 20 mm, inclined at 45° to horizontal
Test duration	Flame application 15 s on lower front edge until the flame has reached the upper part of the specimen holder or until extinction
Conclusion	According to test result, classification into combustibility classes 3–5 as specified below

Test procedure at room temperature

A minimum of three tests is conducted. If these three tests do not result in the same classification, the number of tests is increased to six, always deleting the highest and the lowest results. The remaining worst result is the one that determines the classification.

A conditioned specimen of the building material is mounted in a vertical position on the test rig, and a standardized ignition source is applied in the center of the lower front edge.

Test procedure at an ambient temperature of 200 °C

In the test apparatus, which can be heated, the temperature is increased to 200 °C until constant conditions have been reached. The specimen is clamped into the specimen holder. After the specimen has been heated up for 5 min, the tests are conducted as described above.

Classification: The decisive criterion for the classification is the time elapsed from the start of the flame application until the tip of the flame reaches the upper part of the specimen holder (150 mm from the lower edge of the specimen) (referred to as "time") or until the flame extinguishes (referred to as "burning time").

If rising of the tip of the flame to the upper part of the specimen holder is not unambiguously observed, a cotton thread according to the SN 198 898 standard must be tensioned at this level, and the time taken to burn it measured. For classification purposes, the test using the cotton thread has priority over a visual observation. Details of classification and requirements are summarized in Table 10.85.

Table 10.85 Classification of Building Materials Based on Combustibility Test Requirements

Classification	Requirements
Combustibility grade 3	▪ Time to reach upper part of the specimen holder: 5–20 s
Combustibility grade 4	▪ Time to reach upper part of the specimen holder: > 20 s ▪ Burning time: > 20 s
Combustibility grade 5	▪ The flame does not reach the upper part of the specimen holder (150 mm) ▪ Burning time: ≤ 20 s
Combustibility grade 5 (200 °C)	▪ The flame does not reach the upper part of the specimen holder (150 mm) ▪ Burning time at temperature of 200 °C: ≤ 20 s
Combustibility grades 6q and 6	▪ No ignition, incineration, or carbonizing and non-combustibility test

Non-combustibility test

Non-combustibility is tested according to DIN 4102-1 (version 1998), Chapter 5. For details, see Section 10.9.2.1 in the 3rd edition of this Handbook [4].

Classification: The decisive criteria for classification are flame duration, temperature increase in the non-combustibility furnace, and/or level of lower calorific value of the tested building material. Details are shown in Table 10.86.

Table 10.86 Classification of Non-Combustible Building Materials

Classification	Requirements
Combustibility grade 6q	Flaming: ≤ 20 s and Temperature increase (ΔT): ≤ 50 K *or* Calorific value, lower (LHV): ≤ 4200 kJ/kg
Combustibility grade 6	No flaming and Temperature increase (ΔT): ≤ 50 K

Radiant panel test for floor coverings

The combustibility of floor coverings is tested with the radiant panel test. Test apparatus and specifications are based on DIN 4102-14 (version 1990), and are shown in Section 10.9.2.1 in the 3rd edition of this Handbook [4].

The test chamber has a temperature of 18 ± 5 °C. The air throughput rate of the chamber is approx. 170 m^3/h. The incident heat flow radiated from the following distances by the radiant panel onto the plane of the specimen must range between:

- at 200 mm: 0.87 to 0.95 W/cm^2
- at 400 mm: 0.48 to 0.52 W/cm^2
- at 600 mm: 0.22 to 0.26 W/cm^2.

All values from 100 to 900 mm measured for the heat flow - plotted as a function of the distance - result in the heat flow profile required for the assignment of heat flow densities (W/cm^2).

Classification: To specimens which do not ignite (or burn to a width of < 10 cm), a heat flow density of > 1.1 W/cm^2 is assigned.

Specimens burning to a width of more than 90 cm have a lower heat flow density compared to the calibration value at 90 cm. In all other cases, a heat flow density corresponding to the burning distance is assigned to the specimens on the basis of the heat flow density profile.

The value critical for classification is found by averaging the heat flow densities of three specimens (1050 × 250 mm). The classification criteria are summarized in Table 10.87.

Table 10.87 Classification Criteria for Floor Coverings

Classification	Requirements
Combustibility grade 3	Heat flow density: < 0.25 W/cm^2
Combustibility grade 4	Heat flow density: 0.25–0.49 W/cm^2
Combustibility grade 5	Heat flow density: ≥ 0.5 W/cm^2

Smoke-developing behavior

The test is conducted in a standardized test box based on the American XP2 chamber to ASTM D2843. Test apparatus and specifications are described in Section 10.2.1.5. Six specimens are tested, with the required dimensions listed in Table 10.88.

Table 10.88 Dimensions of Test Specimens for Smoke Test (in mm, ±10% tolerance)

	Compact materials	Foams	Composite flooring materials
Length	30	60	30
Width	30	60	30
Thickness	4	25	Original thickness

Three tests are conducted. The specimen is placed on a defined wire netting and is burnt by means of a flame of 150 mm length. Any melting material is exposed to a flame in a metal sheet cup according to DIN 4102-1, item 5.1.2.2 (version 1977). Flame exposure is continued until complete combustion of the specimen.

Classification: The decisive criterion for classification is the maximum light absorption. The requirements are shown in Table 10.89.

Table 10.89 Classification Criteria for Smoke-Developing Behavior

Classification	Requirements
Smoke grade 1	Maximum light absorption: > 90%
Smoke grade 2	Maximum light absorption: > 50 to 90%
Smoke grade 3	Maximum light absorption: 0 to 50%

Fire coding

The combustibility and smoke grades established on the basis of the test results are expressed as "fire coding", which is a combination of the combustibility and the smoke-developing classification.

For example, a fire coding of 4.1 means that the building material has an average combustibility (grade 4) and develops heavy smoke when burnt (grade 1).

Flaming droplets

A filter paper is placed at the bottom of the test apparatus for the combustibility test. If the filter paper is ignited during the test by a flaming droplet, the product receives the additional assessment "Building material with flaming droplets".

Reaction to Fire Groups

Based on the reaction-to-fire classification according to EN 13501 or the national VKF guidance, building materials are divided into four reaction to fire groups [acronym = RF (from the French "réaction au feu")]:

- RF1 (no contribution to fire)
- RF2 (low contribution to fire)
- RF3 (acceptable contribution to fire)
- RF4 (unacceptable contribution to fire).

Building materials with a critical behavior are additionally denoted by (cr) [acronym = cr (from French "comportement critique")]. In the event of a fire, these materials can quickly lead to an unacceptable hazard to persons due to a strong smoke production, flaming droplets, or their corrosivity.

Assignment of Materials to the Reaction-to-Fire Groups

The following tables show how the different reaction-to-fire classifications are assigned to the reaction-to-fire groups. Assignment tables from the VKF fire protection guideline 13-15 "Building materials and construction components" [5] are shown in Table 10.90–Table 10.93.

Table 10.90 Assignment of Building Materials with a Classification According to EN 13501-1 [a]

Reaction-to-fire group	Critical behavior	Classification according to EN 13501-1		
RF1	–	A1 A2-s1,d0	$A1_L$ $A2_L$-s1,d0	$A1_{fl}$ $A2_{fl}$-s1
RF2	–	A2-s1,d1 A2-s2,d0 A2-s2,d1	$A2_L$-s1,d1 $A2_L$-s2,d0 $A2_L$-s2,d1	–
		B-s1,d0 B-s1,d1 B-s2,d0 B-s2,d1	B_L-s1,d0 B_L-s1,d1 B_L-s2,d0 B_L-s2,d1	B_{fl}-s1
		C-s1,d0 C-s1,d1 C-s2,d0 C-s2,d1	C_L-s1,d0 C_L-s1,d1 C_L-s2,d0 C_L-s2,d1	C_{fl}-s1
	cr	*A2-s1,d2* *A2-s2,d2* *A2-s3,d0* *A2-s3,d1* *A2-s3,d2*	*$A2_L$-s1,d2* *$A2_L$-s2,d2* *$A2_L$-s3,d0* *$A2_L$-s3,d1* *$A2_L$-s3,d2*	–
		B-s1,d2 *B-s2,d2* *B-s3,d0* *B-s3,d1* *B-s3,d2*	*B_L-s1,d2* *B_L-s2,d2* *B_L-s3,d0* *B_L-s3,d1* *B_L-s3,d2*	*B_{fl}-s2*
		C-s1,d2 *C-s2,d2* *C-s3,d0*	*C_L-s1,d2* *C_L-s2,d2* *C_L-s3,d0*	*C_{fl}-s2*

Reaction-to-fire group	Critical behavior	Classification according to EN 13501-1		
		C-s3,d1 *C-s3,d2*	*C_L-s3,d1* *C_L-s3,d2*	
RF3	–	D-s1,d0 D-s1,d1 D-s2,d0 D-s2,d1	D_L-s1,d0 D_L-s1,d1 D_L-s2,d0 D_L-s2,d1	D_{fl}-s1
	cr	*D-s1,d2* *D-s2,d2* *D-s3,d0* *D-s3,d1* *D-s3,d2*	*D_L-s1,d2* *D_L-s2,d2* *D_L-s3,d1* *D_L-s3,d2* *D_L-s3,d0*	*D_{fl}-s2*
		E *E-d2*	*E_L* *E_L-d2*	*E_{fl}*
RF4	-	–		
Not a building material	-	F	F_L	F_{fl}

[a] Entries in italics: Restriction of use due to critical behavior or due to unacceptable contribution to fire.

Table 10.91 Assignment of Building Materials with a Classification According to the National VKF Guidance [6] [a]

Reaction-to-fire group	Critical behavior	Classification according to the national VKF guidance
RF1	–	6.3 6q.3
RF2	–	5(200 °C).3 5.3 5(200 °C).2 5.2
	cr	*5(200 °C).1* *5.1*
RF3	–	4.3 4.2
	cr	*4.1*
RF4	cr	*3.3* *3.2* *3.1*

Table 10.91 Assignment of Building Materials with a Classification According to the National VKF Guidance [6] [a] *(continued)*

Reaction-to-fire group	Critical behavior	Classification according to the national VKF guidance
Not a building material	–	2.3 2.2 2.1 1.3 1.2 1.1

[a] Entries in italics: Restriction of use due to critical behavior or due to unacceptable contribution to fire.

Table 10.92 Assignment of Roof Coverings [a]

Reaction-to-fire group	Critical behavior	Classification according to EN 13501-5 (using data from external fire exposure to roofs tests)
RF1	–	–
RF2	–	–
	cr	*B_{ROOF} (t1)* *B_{ROOF} (t2)* *B_{ROOF} (t3)* *B_{ROOF} (t4)*
RF3	–	–
	cr	*C_{ROOF} (t3)* *C_{ROOF} (t4)* *D_{ROOF} (t3)* *D_{ROOF} (t4)*
RF4	*cr*	*E_{ROOF} (t4)*
Not a building material	–	F_{ROOF} (t1) F_{ROOF} (t2) F_{ROOF} (t3) F_{ROOF} (t4)

[a] Entries in italics: Restriction of use due to critical behavior or due to unacceptable contribution to fire.

Table 10.93 Assignment of Cables [a]

Reaction-to-fire group	Critical behavior	Classification according to EN 13501-6 (using data from reaction to fire tests on power, control, and communication cables)		
RF1	-	A_{ca}		
RF2	-	$B1_{ca}$-s1,a1,d0	$B2_{ca}$-s1,a1,d0	C_{ca}-s1,a1,d0
		$B1_{ca}$-s1,a2,d0	$B2_{ca}$-s1,a2,d0	C_{ca}-s1,a2,d0
		$B1_{ca}$-s1a,a1,d0	$B2_{ca}$-s1,a1,d1	C_{ca}-s1,a1,d1
		$B1_{ca}$-s1a,a2,d0	$B2_{ca}$-s1,a2,d1	C_{ca}-s1,a2,d1
		$B1_{ca}$-s1b,a1,d0	$B2_{ca}$-s1a,a1,d0	C_{ca}-s1a,a1,d0
		$B1_{ca}$-s1b,a2,d0	$B2_{ca}$-s1a,a2,d0	C_{ca}-s1a,a2,d0
		$B1_{ca}$-s1,a1,d1	$B2_{ca}$-s1a,a1,d1	C_{ca}-s1a,a1,d1
		$B1_{ca}$-s1,a2,d1	$B2_{ca}$-s1a,a2,d1	C_{ca}-s1a,a2,d1
		$B1_{ca}$-s1a,a1,d1	$B2_{ca}$-s1b,a1,d0	C_{ca}-s1b,a1,d0
		$B1_{ca}$-s1a,a2,d1	$B2_{ca}$-s1b,a2,d0	C_{ca}-s1b,a2,d0
		$B1_{ca}$-s1b,a1,d1	$B2_{ca}$-s1b,a 1,d1	C_{ca}-s1b,a1,d1
		$B1_{ca}$-s1b,a2,d1	$B2_{ca}$-s1b,a2,d1	C_{ca}-s1b,a2,d1
		$B1_{ca}$-s2,a1,d0	$B2_{ca}$-s2,a1,d0	C_{ca}-s2,a1,d0
		$B1_{ca}$-s2,a2,d0	$B2_{ca}$-s2,a2,d0	C_{ca}-s2,a2,d0
		$B1_{ca}$-s2,a1,d1	$B2_{ca}$-s2,a1,d1	C_{ca}-s2,a1,d1
		$B1_{ca}$-s2,a2,d1	$B2_{ca}$-s2,a2,d1	C_{ca}-s2,a2,d1
	cr	*$B1_{ca}$-s1,a3,d0*	*$B1_{ca}$-s3,a3,d2*	*C_{ca}-s1,a3,d0*
		$B1_{ca}$-s1a,a3,d0	*$B2_{ca}$-s1,a3,d0*	*C_{ca}-s1,a3,d1*
		$B1_{ca}$-s1b,a3,d0	*$B2_{ca}$-s1,a3,d1*	*C_{ca}-s1,a1,d2*
		$B1_{ca}$-s1,a3,d1	*$B2_{ca}$-s1a,a3,d0*	*C_{ca}-s1,a2,d2*
		$B1_{ca}$-s1,a1,d2	*$B2_{ca}$-s1a,a3,d1*	*C_{ca}-s1,a3,d2*
		$B1_{ca}$-s1,a2,d2	*$B2_{ca}$-s1,a1,d2*	*C_{ca}-s1a,a3,d0*
		$B1_{ca}$-s1,a3,d2	*$B2_{ca}$-s1,a2,d2*	*C_{ca}-s1a,a3,d1*
		$B1_{ca}$-s1a,a3,d1	*$B2_{ca}$-s1,a3,d2*	*C_{ca}-s1a,a1,d2*
		$B1_{ca}$-s1a,a1,d2	*$B2_{ca}$-s1a,a1,d2*	*C_{ca}-s1a,a2,d2*
		$B1_{ca}$-s1a,a2,d2	*$B2_{ca}$-s1a,a2,d2*	*C_{ca}-s1a,a3,d2*
		$B1_{ca}$-s1a,a3,d2	*$B2_{ca}$-s1a,a3,d2*	*C_{ca}-s1b,a3,d0*
		$B1_{ca}$-s1b,a3,d1	*$B2_{ca}$-s1b,a3,d0*	*C_{ca}-s1b,a3,d1*
		$B1_{ca}$-s1b,a1,d2	*$B2_{ca}$-s1b,a3,d1*	*C_{ca}-s1b,a1,d2*
		$B1_{ca}$-s1b,a2,d2	*$B2_{ca}$-s1b,a1,d2*	*C_{ca}-s1b,a2,d2*
		$B1_{ca}$-s1b,a3,d2	*$B2_{ca}$-s1b,a2,d2*	*C_{ca}-s1b,a3,d2*
		$B1_{ca}$-s2,a3,d0	*$B2_{ca}$-s1b,a3,d2*	*C_{ca}-s2,a3,d0*
		$B1_{ca}$-s2,a3,d1	*$B2_{ca}$-s2,a3,d0*	*C_{ca}-s2,a3,d1*
		$B1_{ca}$-s2,a1,d2	*$B2_{ca}$-s2,a3,d1*	*C_{ca}-s2,a1,d2*
		$B1_{ca}$-s2,a2,d2	*$B2_{ca}$-s2,a1,d2*	*C_{ca}-s2,a2,d2*

Table 10.93 Assignment of Cables [a] *(continued)*

Reaction-to-fire group	Critical behavior	Classification according to EN 13501-6 (using data from reaction to fire tests on power, control, and communication cables)		
	cr	*$B1_{ca}$-s2,a3,d2*	*$B2_{ca}$-s2,a2,d2*	*C_{ca}-s2,a3,d2*
		$B1_{ca}$-s3,a1,d0	*$B2_{ca}$-s2,a3,d2*	*C_{ca}-s3,a1,d0*
		$B1_{ca}$-s3,a1,d1	*$B2_{ca}$-s3,a1,d0*	*C_{ca}-s3,a1,d1*
		$B1_{ca}$-s3,a2,d0	*$B2_{ca}$-s3,a1,d1*	*C_{ca}-s3,a1,d2*
		$B1_{ca}$-s3,a2,d1	*$B2_{ca}$-s3,a2,d0*	*C_{ca}-s3,a2,d0*
		$B1_{ca}$-s3,a3,d0	*$B2_{ca}$-s3,a2,d1*	*C_{ca}-s3,a2,d1*
		$B1_{ca}$-s3,a3,d1	*$B2_{ca}$-s3,a3,d1*	*C_{ca}-s3,a2,d2*
		$B2_{ca}$-s3,a3,d0	*$B2_{ca}$-s3,a1,d2*	*C_{ca}-s3,a3,d0*
		$B1_{ca}$-s3,a1,d2	*$B2_{ca}$-s3,a2,d2*	*C_{ca}-s3,a3,d1*
		$B1_{ca}$-s3,a2,d2	*$B2_{ca}$-s3,a3,d2*	*C_{ca}-s3,a3,d2*
RF3	-	D_{ca}-s1,a1,d0	D_{ca}-s1a,a1,d1	D_{ca}-s2,a1,d0
		D_{ca}-s1,a2,d0	D_{ca}-s1a,a2,d1	D_{ca}-s2,a2,d0
		D_{ca}-s1,a1,d1	D_{ca}-s1b,a1,d0	D_{ca}-s2,a1,d1
		D_{ca}-s1,a2,d1	D_{ca}-s1b,a2,d0	D_{ca}-s2,a2,d1
		D_{ca}-s1a,a1,d0	D_{ca}-s1b,a1,d1	
		D_{ca}-s1a,a2,d0	D_{ca}-s1b,a2,d1	
	cr	*D_{ca}-s1,a3,d0*	*D_{ca}-s1b,a3,d0*	*D_{ca}-s3,a1,d0*
		D_{ca}-s1,a3,d1	*D_{ca}-s1b,a3,d1*	*D_{ca}-s3,a2,d0*
		D_{ca}-s1,a1,d2	*D_{ca}-s1b,a1,d2*	*D_{ca}-s3,a3,d0*
		D_{ca}-s1,a2,d2	*D_{ca}-s1b,a2,d2*	*D_{ca}-s3,a1,d1*
		D_{ca}-s1,a3,d2	*D_{ca}-s1b,a3,d2*	*D_{ca}-s3,a2,d1*
		D_{ca}-s1a,a3,d0	*D_{ca}-s2,a1,d2*	*D_{ca}-s3,a3,d1*
		D_{ca}-s1a,a3,d1	*D_{ca}-s2,a2,d2*	*D_{ca}-s3,a1,d2*
		D_{ca}-s1a,a1,d2	*D_{ca}-s2,a3,d0*	*D_{ca}-s3,a2,d2*
		D_{ca}-s1a,a2,d2	*D_{ca}-s2,a3,d1*	*D_{ca}-s3,a3,d2*
		D_{ca}-s1a,a3,d2	*D_{ca}-s2,a3,d2*	*E_{ca}*
RF4	*cr*	-		
Not a building material	-	F_{ca}		

[a] Entries in italics: Restriction of use due to critical behavior or due to unacceptable contribution to fire.

10.3.2.11.3 Requirements for Building Materials

General Requirements

The VKF fire protection fire safety guideline 14-15 “Application of building materials” [6] determines the minimum requirements for building materials. The principles are explained first in the guideline. The most important principles regarding the reaction-to-fire groups are listed below:

- Building materials with a critical behavior (cr) need a full-faced coverage if the material is used inside a building. The minimum material thickness of the coverage is:
  a) For group RF1: 0.5 mm
  b) For group RF2: 3 mm
  c) For group RF3: 5 mm.
- For some applications such as paints, penetration seals, linear joint seals and some others, the use of building materials with a critical behavior (cr) inside a building without coverage is possible.
- An application for a RF4 (cr) building material is only possible if the material is covered on all sides (and without a cavity) by a K 30 fire protection panel.

Below, detailed requirements are defined for exterior walls and roof constructions, for interior fittings of buildings such as internal walls, ceilings, and flooring materials, and for building technologies like cables, pipes, and their accompanying insulation.

External Wall Covering Systems (Façades)

The requirements for external wall covering systems are defined in the VKF fire protection guidelines 14–15, chapter 3.2 [6]. These depend on the use and the height of the building, the distance between buildings, and the presence of an extinguishing system. An external wall covering system essentially consists of external wall cladding and thermal insulation (and/or an intermediate layer). The substructure of the cladding in ventilated façades is also considered as part of the external wall covering system.

General requirements for medium-height buildings (11–30 m total height)

If combustible building materials are used for external wall cladding and/or thermal insulation, access for the fire brigade to the respective façade surfaces for fire-fighting purposes (e. g., pressure pipes, mobile water cannon) must be ensured. Combustible external wall cladding and/or thermal insulation must be structurally subdivided in such a way that before the fire brigade's fire-fighting operations begin, a fire on the wall cannot spread by more than two storeys above the floor where it started.

General requirements for high-rise buildings (> 30 m total height)

The external wall and external wall covering systems of high-rise buildings must be made of RF1 building materials.

Ventilated façades

On buildings of low height (≤ 11 m total height), building materials of RF3 (cr) can be used for external wall claddings. The requirements for the thermal insulation and intermediate layers depend on the use of the building. In an accommodation that houses people who depend on outside help, building materials of RF1 must be used. For all other buildings, the use of materials of RF3 (cr) is possible.

Ventilated façades on buildings of medium height (11–30 m total height) that use combustible building materials in the external wall cladding, and/or insulating materials or flat layers in the ventilation area, must be constructed using a construction recognized by VKF or equivalent. Otherwise, the external wall covering systems (without the linear substructures) have to consist of RF1 building materials.

The requirements for ventilated façades of high-rise buildings (> 30 m total height) are equal to the general requirements for external wall covering systems: RF1 building materials must be used, and this also applies to the linear substructures.

Linear substructures made of RF3 (cr) building materials are permitted for fixing of external wall claddings on low- and medium-height buildings.

For all buildings (including high-rise buildings), back-anchoring of ventilated façades located within the thermal insulation must consist of building materials of grade RF3 (cr) or better.

For breather façade membranes and thermal insulation that are either against the ground or up to 1.0 m above may consist of building materials of RF3 (cr) on all buildings (including high-rise buildings).

10.3.2.11.4 VKF Fire Protection Register

The VKF maintains the fire protection register at *www.bsronline.ch* [7] for the simple application of the Swiss fire protection regulations. It provides a comprehensive and current overview of suitable building products from industry and professional associations.

VKF issues nationally recognized documents for building products and specialized companies that meet the fire protection requirements. In the case of non-harmonized products and specialized companies, VKF issues a “VKF-recognition” (VKF-Anerkennung). For harmonized products, these documents are called “VKF technical information” (VKF Technische Auskunft). Due to the voluntary entry in the fire protection register, all users can use the “VKF-recognition” or the “VKF technical information” as a basis for the fire-protection-compliant application of products throughout Switzerland. The VKF fire protection register is therefore an effective means of sales promotion for manufacturers and suppliers, as well as being an important way of supporting the correct use of the fire protection products for engineers, quality assurance officers for fire protection, and fire protection authorities.

10.3.2.11.5 Future Developments

Since the 1980s, the VKF fire protection regulations have been revised approximately every 10 years. The current regulations are from 2015.

For the forthcoming new version, it is desired to check the regulations not just in regard to their topicality. Rather, a comparison with changes in politics and society regarding the accepted risk is sought. Therefore, the new fire protection regulations are to be revised based on a risk-oriented approach with the aim of deregulation, simplification of the regulations, and a more uniform enforcement. In an initial step, the fire protection objectives will be developed using a stakeholder process, to ensure that the accepted risk is neither significantly lower nor significantly higher than the accepted risk in other areas of life in Switzerland. This basis will then be used to completely restructure the regulations.

References for Section 10.3.2.11

[1] *Agreement between all cantons to remove technical barriers to trade* [Interkantonale Vereinbarung zum Abbau technischer Handelshemmnisse (IVTH)], Conference of cantonal governments [Konferenz der Kantonsregierungen] (1998).

[2] *Fire Protection Standard 1-15* [Brandschutznorm 1-15], VKF (2015).

[3] *Building material and construction components, part B: Testing regulations* [Wegleitung für Feuerpolizeivorschriften, Baustoffe und Bauteile, Teil B: Prüfbestimmungen], VKF (2005).

[4] Troitzsch, J. (ed.), *Plastics Flammability Handbook: Principles, Regulations, Testing, and Approval*, 3rd edition, Hanser, Munich (2004).

[5] *Fire Protection Guideline 13-15 "Building materials and construction components"* [Brandschutzrichtlinie 13–15, Baustoffe und Bauteile], VKF (2015).

[6] *Fire Protection Guideline 14-15 "Application of building materials"* [Brandschutzrichtlinie 14–15, Verwendung von Baustoffen], VKF (2015).

[7] *Fire Protection Register*, VKF, *www.bsronline.ch*.

10.3.2.12 United Kingdom

Edith Antonatus

10.3.2.12.1 Statutory Regulations

The following building regulations apply in the various parts of the United Kingdom.

- England: The Building Regulations 2010 [1] with Amendments 2011, 2012, 2013, 2014, 2015, 2016, 2017, 2018, 2019, and 2020 [2].
- Scotland: The Building Standards (Scotland) Regulations 2019 [3].
- Northern Ireland: The Building Regulations (Northern Ireland), 2012 [4] with Amendments, 2013, 2014, and 2016 [5].
- Wales: The building regulations Wales 2006 with Amendments 2010 and 2016 [6].

Technical provisions for use and fire performance of building materials and components are given in the supporting documents to the Building Regulations, e.g., under Approved Document B for England and Wales. The technical provisions for England and Wales, and Northern Ireland, are virtually identical. Those for Scotland differ in minor aspects only. All the technical provisions are based on the same test methods specified in British Standards or the harmonized European standards. These documents are published on the regulatory websites [7].

The provisions of Approved Document B are given as guidance and provide one method of showing compliance with the Regulations. There is no obligation to follow that method, provided compliance with the Regulations can be satisfactorily demonstrated.

10.3.2.12.2 Classification and Testing of the Reaction to Fire of Construction Products

Compliance with the fire behavior of building materials and components identified in the Building Regulations and supporting documents is related to performance in certain standard tests. In most cases, classifications based on the European harmonized test and classification standards have to be used. National methods can still be used to prove the performance of construction products for products which are not subject to CE marking. These methods are described in British Standards BS 476: Parts 3 to 33 [8]. These Standards are issued by and are available from the British Standards Institution, 389 Chiswick High Road, London, W4 4AL. This section gives details about the British national test methods. European tests and classifications are covered in detail in Section 10.3.1.

The 2019 Approved Document B [7] edition details the transposition of the European classifications into national classes, as shown in Table 10.94.

Table 10.94 Reaction to Fire Classifications: Transposition to National Class

BS EN 13501-1 classification	Transposition
A1	Material that, when tested to BS 476-11, does not either: a) flame b) cause a rise in temperature on either the thermocouple at the center of the specimen or in the furnaces.
A2-s1, d0	None
A2-s3, d2	Material that meets either of the following. a) Any material of density 300 kg/m³ or more, which, when tested to BS 476-11, complies with both of the following: I) does not flame II) causes a rise in temperature on the furnace thermocouple not exceeding 20 °C.

BS EN 13501-1 classification	Transposition
	b) Any material of density less than 300 kg/m^3, which, when tested to BS 476-11, complies with both of the following: I) does not flame for more than 10 s II) causes a rise in temperature on the thermocouple at the center of the specimen or in the furnace that is a maximum of 35 °C and on the furnace thermocouple that is a maximum of 25 °C.
B-s3, d2	Any material that meets both of the following criteria. a) Class 1 in accordance with BS 476-7. b) Has a fire propagation index (I) of a maximum of 12 and a sub-index (i_1) of a maximum of 6, determined by using the method given in BS 476-6. The index of performance I relates to the overall test performance, whereas sub-index i_1 is derived from the first 3 min of the test.
C-s3, d2	Class 1 in accordance with BS 476-7
D-s3, d2	Class 3 in accordance with BS 476-7

NOTE: The national classifications do not automatically equate with the transposed classifications in the "BS EN 13501-1 classification" column, therefore products cannot typically assume a European class unless they have been tested accordingly.

NOTE: A classification of s3, d2 indicates that no limit is set for production of smoke and/or flaming droplets/particles. If a performance for production of smoke and/or flaming droplets/particles is specified, then only the European classes can be used. For example, a national class may not be used as an alternative to a classification which includes s1, d0.

Approved Document B also contains a transposition table for roof coverings, as summarized in Table 10.95.

Table 10.95 Roof Covering Classifications: Transposition to National Class

BS EN 13501-5 classification	Transposition to BS 476-3 classification
B_{roof} (t4)	AA, AB, or AC
C_{roof} (t4)	BA, BB, or BC
D_{roof} (t4)	CA, CB, or CC
E_{roof} (t4)	AD, BD, or CD
F_{roof} (t4)	DA, DB, DC, or DD

Standard BS 476 "Fire Tests on Building Materials and Structures" consists of many parts covering fire behavior and fire resistance [8]. BS 476: Part 3 describes the classification and method of test for external fire exposure to roofs, which is similar to the test method 4 (T4) in CEN/TS 1187 and the classification based on this test in EN 13501-5 (see Section 10.3.1).

BS 476: Parts 20 to 24 describe the testing of fire resistance of elements of construction, and BS 476: Parts 10, 12, 13, 32, and 33 are not referred to in the Building Regulations or their supporting documents; neither are relevant in the context of this book and therefore not further dealt with.

Since 2019, separation into combustible materials and materials classified as A2 is based on the harmonized European standards, and no longer on BS tests. Details on the non-combustibility test are described in Section 10.3.1. For non-combustible products (European class A1) it is still possible to apply alternatively the national classification according to BS 476-11.

Thermoplastic materials, which cannot be assessed to BS 476: Part 7, can be classified as "TP(a) rigid", "TP(a) flexible", or "TP(b)" according to a series of methods, or based on generic acceptance.

Limitations on the use of combustible building materials apply for external walls, compartment walls or floors, separating walls, fire-protecting suspended ceilings, internal wall and ceiling linings, and window openings.

No requirements exist at present for testing combustible floor coverings and for determining smoke evolution or toxicity of fire gases.

In the following, the national BS 476 test methods still in use are presented.

BS 476: Part 6 Fire Propagation Test

This method of test, the result being expressed as a fire propagation index, provides a comparative measure of the contribution to the growth of fire made by an essentially flat material, composite, or assembly. It is used primarily for the assessment of the performance of wall and ceiling linings. A diagram of the equipment and test specifications are reproduced in Figure 10.43 and summarized in Table 10.96, respectively.

Fire propagation is determined by exposing the specimen to the 14 jets of a gas pipe burner at a distance of 3 mm. The heat evolved is 530 J/s. The two electric elements with a total output of 1800 W are switched on after 2 min 45 s. Their output is reduced to 1500 W after 5 min and maintained constant until the end of the test (20 min).

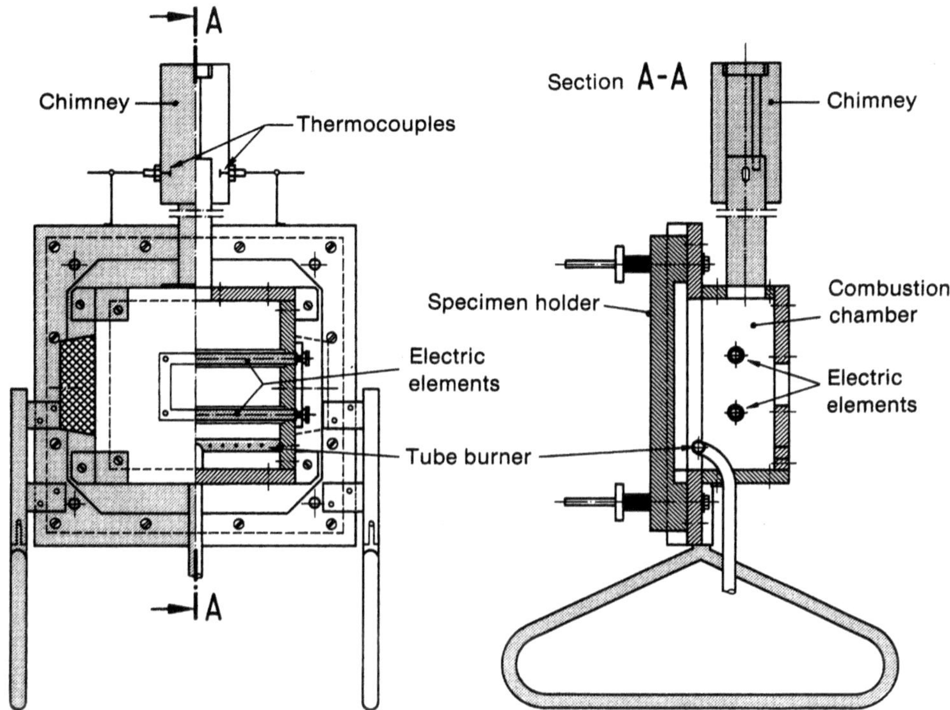

Figure 10.43 Fire propagation test apparatus

Table 10.96 Fire Propagation Test Specifications

Test unit	0.4 m × 0.4 m × 0.75 m high
Specimens	Minimum of three and maximum of five specimens 225 mm × 225 mm × max. 50 mm
Specimen position	Vertical
Ignition source	▪ Two 100 W electric elements with variable output (1800 W after 2 min 45 s and 1500 W after 5 min); distance from specimen: 45 mm ▪ Gas pipe burners (internal diameter 9 mm) with 14 holes (internal diameter 1.5 mm) at 12.5 mm centers, distance from specimen 3 mm, flame applied at 25 mm above the bottom of the exposed face of the specimen for 20 min
Test duration	20 min
Conclusions	Results are reported as sub-indices derived using the equations given below and a total fire propagation index (I)

To evaluate a material, the difference between ambient temperature and that in the chimney is recorded continuously using thermocouples and compared with a calibration curve derived in a similar manner using a non-combustible board of a specific density. The two curves are evaluated using the temperatures from the calibration curve and test curve at 30 s intervals from the start of the test to 3 min, at 1 min intervals from 4 to 10 min and at 2 min intervals from 12 to 20 min.

The individual indices are calculated as follows from these values:

$$i_1 = \sum_{1/2}^{3} \frac{\theta_m - \theta_c}{10t}, \quad i_2 = \sum_{4}^{10} \frac{\theta_m - \theta_c}{10t}, \quad i_3 = \sum_{12}^{20} \frac{\theta_m - \theta_c}{10t}$$

The total index $I = i_1 + i_2 + i_3$

where:

i_1, i_2, and i_3 are the sub-indices at 3 min, 10 min, and 20 min, respectively

t is the time in min from the origin at which the reading is taken

θ_m is the temperature in °C of the test curve at time t

θ_c is the temperature in °C of the calibration curve at time t.

If four or five specimens are needed to obtain three valid results, a suffix "R" is added.

The fire propagation test serves mainly to investigate whether materials that have achieved Class 1 according to BS 476: Part 7 also conform to Class 0. To achieve Class 0, a material must be of limited combustibility to BS 476: Part 11 or must meet the requirements of Class 1 to BS 476: Part 7 and have a sub-index $i_1 \leq 6$ and a total index $I \leq 12$ using BS 476: Part 6.

The requirements of Class 0 make increased demands on the fire performance of Class 1 building products and differentiate between them with regard to fire hazard. This test method is also used to examine the suitability of building materials for certain applications. For example, the external cladding of walls for assembly or recreation buildings which are higher than 18 m, must have a total index I of not more than 20.

BS 476 Part 7 Surface Spread of Flame Test

The lateral spread of flame along the surface of a specimen of a product orientated in a vertical plane is determined by this test method using the apparatus shown in Figure 10.44. The test specifications are summarized in Table 10.97. The test provides data suitable for comparing the performances of essentially flat materials, used primarily as the exposed surface of walls or ceilings.

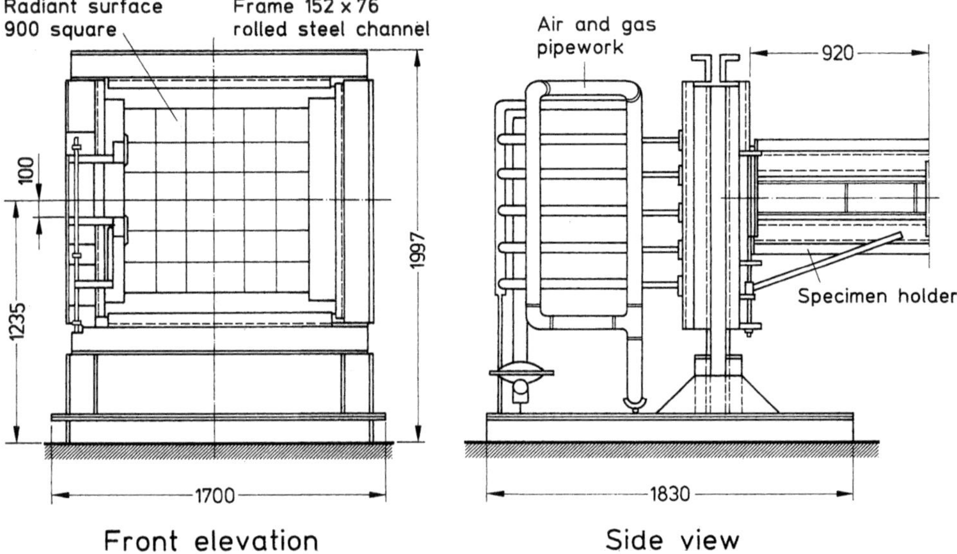

Figure 10.44 Surface spread of flame test apparatus. All dimensions in mm

The specimen is mounted in a water-cooled holder and is exposed to a radiant panel for 10 min. In addition, a pilot flame is applied to the bottom corner of the specimen during the first minute of the test. The time required for the flame front to reach reference marks on the specimen is noted, together with the extent of flame spread at 1½ min and at the end of the test. Observations are also made of the burning behavior, and a suffix "Y" is added to the result if softening or slumping occurs. If more than six specimens are required to obtain a result, a Suffix "R" is added, and if the product is tested in a modified form, a prefix "D" is added.

Table 10.97 Surface Spread of Flame Test Specifications

Test unit	1.7 m × 1.83 m × 2 m high
Specimens	Minimum of six and maximum of nine specimens 885 mm × 263 mm × maximum 50 mm
Specimen position	Vertical, longitudinal axis (885 mm) perpendicular to the radiant panel
Ignition source	▪ Gas-fired radiant panel, radiation intensity 75 mm from radiant panel surface: 32.5 kW/m^2. ▪ Gas pilot flame (height: 75 to 100 mm) impinging on the specimen on the same side as the radiant panel
Test duration	10 min
Conclusions	Classification in classes 1 to 4 depending on test performance

Materials are classified according to test performance as shown in Table 10.98. Class 4 materials are considered high risk, and their use is not permitted.

Table 10.98 Spread of Flame Classification

Classification	Spread of flame at 1.5 min		Final spread of flame	
	Limit [mm]	Tolerance for one specimen [mm]	Limit [mm]	Tolerance for one specimen [mm]
Class 1	165	25	165	25
Class 2	215	25	455	45
Class 3	265	25	710	75
Class 4	Exceeding the limits for Class 3			

BS 476: Part 11 Heat Emission Test

The assessment of heat emission from building materials is determined by ISO EN 1182 and BS 476: Part 4. A diagram of the furnace and the test specifications are shown in Section 10.3.1.

The method is applicable to simple materials that are reasonably homogeneous. Non-homogeneous materials can be tested provided irregularities within the material are small. It is not normally suitable for assessing composites or assemblies.

Typical requirements are:

- *For structural frames of the above walls:* The material must not flame and there must be no rise in temperature on either thermocouple.
- *For linings to such frames:* The material, having a minimum density of 300 kg/m^3, must not flame, and the rise in temperature on the furnace thermocouple must not be more than 20 °C.
- *For cavity insulation within such walls:* The material, of density less than 300 kg/m^3, must not flame for more than 10 s, the rise in temperature on the specimen thermocouple must not be more than 35 °C, and the rise in temperature on the furnace thermocouple must not be more than 25 °C.

Methods of Test for Thermoplastics

The methods of test for fire performance referred to in the technical provisions of the Building Regulations, in general, are for all types of building materials and not specific groups. However, where the usual test methods have been found inappropriate (for example, with some thermoplastics), use is made of quality control test methods to provide information on the performance of the plastics to allow them to be used. The use of such materials is, however, subject to restrictions on size, thickness, and separation from each other depending on their application.

The procedures identified in connection with the Building Regulations for testing thermoplastics are BS 2782: Part 0 Method 508A [9]. BS EN 6941 test 2 [10] is also used, to show compliance with the requirements given for Type C in BS 5867: Part 2 [11].

Thermoplastics can be classified as "TP(a) rigid", "TP(a) flexible", or "TP(b)", these classifications being specified for certain use areas such as windows, rooflights, and lighting diffusers.

- "TP(a) rigid" applies to any solid PVC, and solid polycarbonate of 3 mm thickness and above, and multi-skinned PVC or polycarbonate that is Class 1, or any other thermoplastic that meets the requirements of BS 2782: Part 0 Method 508A.
- "TP(a) flexible" applies to products not more than 1 mm thick that comply with Type C requirements in BS 5867: Part 2 as tested to BS 5438 Test 2.
- "TP(b)" applies to all other PVC or polycarbonate products or other thermoplastics that have a rate of burning of less than 50 mm/min when tested to BS 2782: Part 0 Method 508A.

BS 2782 Method 508A, Rate of Burning

The rate of burning (for production control purposes only) is determined by BS 2782: Part 0 Method 508A using the apparatus illustrated in Figure 10.45. The test specifications are summarized in Table 10.99.

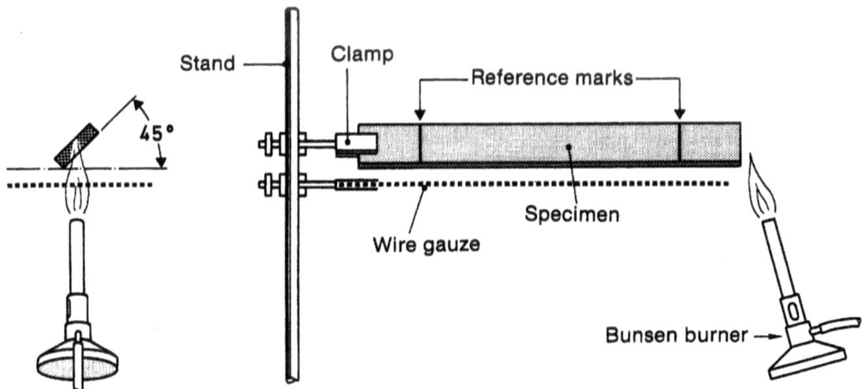

Figure 10.45 Rate of burning apparatus. Left: Front view. Right: Side view

Table 10.99 Rate of Burning Test Specifications

Specimens	Three specimens 150 mm × 13 mm × 1.5 mm Reference lines at 25 mm and 125 mm
Specimen position	Parallel to the free end of the specimen longitudinal axis (150 mm), horizontal, short edge (13 mm) inclined at 45° to horizontal Alcohol or Bunsen burner (flame length 13 mm to 19 mm)
Ignition source	Flame applied to free end of specimen 10 s, specimen monitored
Test duration	Until flame extinguishes
Conclusions	Performance criteria in connection with Building Regulations are: (a) burnt length ≤ 25 mm and residual flaming ≤ 5 s (TP(a) rigid) (b) rate of burning ≤ 50 mm/min (TP(b))

A non-luminous alcohol or Bunsen flame is applied for 10 s to the free end of the specimen. The times at which the first and second reference lines are reached by the flame front are noted and the rate of burning computed in mm/min. If the flame does not reach the first line, the duration of residual flaming or afterglow is measured.

10.3.2.12.3 Legal Provisions for Application of Classifications of Reaction to Fire of Construction Products in the UK

The requirements in UK, Wales, Ireland, and Scotland can be met with products classified according to BS standards or harmonized European standards.

The comparison of these classifications is shown in Table 10.100. These are not necessarily directly translatable and the BS classifications are now more and more removed from these regulations.

Table 10.100 Comparison of European and BS Classifications in UK Building Regulations

European classification	British Standard requirement
A1	Non-combustible
A2	Limited combustibility
B	0
C	1
D	3
E	4
F	Unclassifiable or no performance determined

10.3.2.12.4 Additional Legal Requirements for Façades

The basic requirement for external walls of a building in the UK regulations is to adequately resist the spread of fire over the walls and from one building to another, having regard to the height, use and position of the building.

According to the Building Regulations Approved document B 2010 [12], external walls for "relevant" buildings could either be made of materials of limited combustibility or meet the performance criteria given in the BRE Report "Fire performance of external thermal insulations for walls of multistorey buildings" (BR135) [13] for cladding systems using full-scale test data from BS 8414-1:2020 [14] or BS 8414-2:2020 [15]. At the time of writing (March 2019) this is still possible in Scotland and Ireland.

In 2018, the regulations in England and Wales for façades of buildings [12] were amended as follows for buildings with a storey (not including roof-top plant areas or any storey consisting exclusively of plant rooms) of at least 18 m above ground level, and containing one or more dwellings, an institution or a room for residential purposes (excluding any room in a hostel, hotel or boarding house):

- Materials which become part of an external wall, or specified attachment of a relevant building must be of European Classification A2-s1, d0 or Class A1.
- In other buildings (buildings that are higher than 18 m but that are not residential, institutional, or dwelling), and in any buildings that do not have a habitable floor more than 18 m above ground) the guidance in the Approved Document is less clear. There is currently still the option to assess the whole façade system in a BS 8414 test and meet the performance criteria set out in BR 135.

The full-scale tests described in BS 8414-1 and BS 8414-2 are the most demanding fire tests for façade claddings currently used in Europe.

The test rig is built up as a corner configuration.

- Minimum height of sample:
  - 7.7 m above combustion chamber opening on the main face
  - 9.7 m ground to full height on both wings.
- Width:
  - ≥ 2.6 m main face
  - 1.5 m short wing.
- Combustion chamber:
  - 2 m high
  - 2 m wide.

The fire is simulated by a burning wood crib in a combustion chamber, producing a nominal total heat output of 4500 MJ over the course of a 30-minute test and a peak output of 3 MW. The wood crib is extinguished 30 min after ignition, and the test runs for 60 min. Observations have to be noted regarding surface spread, release of incandescent or burning material, dripping, etc., and thermocouples are used to take temperature readings in predetermined areas (level 1 at 2.5 m above the upper edge of the combustion chamber, level 2 at 2.5 m above level 1).

The temperatures at level 1 are used to define the start of the test. Data measured at level 2 are used for classification according to fire-spread criteria.

According to BR 135, the primary pass/fail criteria are:

- External fire spread: Failure due to external fire spread is deemed to have occurred if any of the external thermocouples at level 2 exceeds 600 °C, for a period of at least 30 s, within 15 min of the start time t_s.
- Internal fire spread: Failure due to internal fire spread is deemed to have occurred if the temperature of any of the thermocouples at level 2 exceeds 600 °C, for a period of at least 30 s, within 15 min of the start time t_s for more than 30 s.
- Burn-through: Failure is deemed to have occurred if flaming is seen on the rear of the façade system (must be for more than 60 s).

10.3.2.12.5 Official Approval

To be eligible for designation as a UK Notified Body for the purposes of the Regulation, official testing laboratories must be a legal entity in the UK. They have to be accredited by UKAS (United Kingdom Accreditation Service). For the European harmonized test methods, they need to apply to MHCLG (Ministry of Housing, Communities and Local Government) for notification.

Note: Because of Brexit, from 2021 onwards, the manufacturers, importers, distributors or authorized representatives for construction products that are currently reliant on a UK "notified body" may need to either:

- arrange for the transfer of their files and the corresponding certificates from the UK "notified body" (a "notified body" registered in the UK) to an EU-27 "notified body", or
- apply for a new certificate with an EU-27 "notified body".

Some laboratories in the UK have started to move the activities relevant for CE marking to other countries, assuming that from 2021 on it will not be possible to perform the tasks of a notified body for construction products.

Accredited laboratories which are notified bodies for reaction to fire are:

- BRE Global Ltd (CE marking activities will be transferred to Ireland)
Bucknalls Lane
Garston
Watford
Herts.
WD25 9XX
UK

- Warringtonfire (CE marking activities will be transferred to the Netherlands)
  Holmesfield Road
  Warrington
  Cheshire
  WA1 2DS
  UK

Further fire testing laboratories with UKAS accreditation are:

- FM Approvals UK
  Maidenhead
  Berkshire
  SL6 2PJ
  UK
- Efectis UK/Ireland Limited
  Jordanstown Campus
  Block 27
  Shore Road, Newtownabbey
  BT 37 0QB
  UK
- Fire Protection Association
  London Road
  Moreton in Marsh
  Gloucestershire
  GL56 0RH
  UK

10.3.2.12.6 Future Developments

Currently in the UK the focus is on further development of the regulations for high buildings. Discussions are ongoing whether the limit for buildings with increased requirements for fire safety of external cladding needs to be lowered from 18 m to 11 m and whether additional requirements shall be included. In Scotland and Ireland, changes for façade requirements are under discussion as well.

References for Section 10.3.2.12

[1] The Building Regulations 2010, England.

[2] The Building Regulations 2010, England, Amendments 2011, 2012, 2013, 2014, 2015, 2016, 2017, 2018, 2019, and 2020.

[3] The Building Standards (Scotland) technical handbook, 2019.

[4] The Building Regulations (Northern Ireland), 2012.

[5] The Building Regulations (Northern Ireland), Amendments, 2013, 2014, and 2016.

[6] The Building Regulations Wales 2010 (last updated 2016) and Amendment 2018.

[7] Approved Document B: *https://www.gov.uk/government/collections/approved-documents, https://gov.wales/topics/planning/buildingregs/approved-documents/part-b-fire/?lang=en, https://www.housing.gov.ie/housing/building-standards/tgd-part-b-fire-safety/technical-guidance-document-b-fire-safety.*

[8] BS 476. Fire tests on building materials and structures:
Part 3: 2012. Classification and method of test for external fire exposure to roofs.
Part 4: 1970. Non-combustibility test for materials (withdrawn).
Part 6: 1989 + A1 2009. Method of test for fire propagation for products.
Part 7: 1997. Method for the classification of the surface spread of flame of products.
Part 8: 1972. Test methods and criteria for the fire resistance of elements of building construction.
Part 10: 2009. Guide to the principles, selection, role and application of fire testing and their outputs.
Part 11: 1982. Method for assessing the heat emission from building materials.
Part 12: 1991. Method of test for ignitability of products by direct flame impingement.
Part 13: 1987. Method for measuring the ignitability of products subjected to thermal irradiance.
Part 20: 1987. Method for determination of the fire resistance of elements of construction (general principles).
Part 21: 1987. Methods for determination of the fire resistance of loadbearing elements of construction.
Part 22: 1987. Methods for determination of the fire resistance of non-loadbearing elements of construction.
Part 23: 1987. Methods for determination of the contribution of components to the fire resistance of a structure.
Part 24: 1987. Method for the determination of the fire resistance of ventilation ducts.
Part 31: 1983. Method for measuring smoke penetration through doorsets and shutter assemblies.
Part 31.1: 1983. Method of measurement under ambient temperature conditions.
Part 32: 1989. Guide to full scale tests within buildings.
Part 33: 1993. Full-scale room test for surface products.

[9] BS 2782-0:2011. Methods of testing plastics, introduction

[10] BS EN ISO 6941:2003. Textile fabrics. Burning behaviour. Measurement of flame spread properties of vertically oriented specimens.

[11] BS 5867-2:2008. Fabrics for curtains, drapes and window blinds. Flammability requirements. Specification.

[12] The Building Regulations 2010 Approved Document B: Fire safety. Volume 2 – Buildings other than dwelling houses, 2019 edition.

[13] BR 135. Fire performance of external thermal insulation for walls of multistorey buildings. 3rd edition. *https://www.brebookshop.com/details.jsp?id=327137.*

[14] BS 8414-1:2020. Fire performance of external cladding systems. Test method for non-loadbearing external cladding systems applied to the masonry face of a building.

[15] BS 8414-2:2020. Fire performance of external cladding systems. Test method for non-loadbearing external cladding systems fixed to and supported by a structural steel frame.

10.4 Asia/Pacific

10.4.1 China

Jun-Sheng Wang and Xing Jin

10.4.1.1 Statutory Regulations

The fire safety of buildings in the People's Republic of China (PRC) is generally addressed in the Construction Law of 1997 [1] (revised 2019) and the Fire Protection Law (enforcement 1 July 2009, revised 2019) [2] of the PRC. Mandatory technical regulations and standards have the force of law when enforced by laws and administrative regulations, and concern the protection of human health, personal property, and safety. All standards that fall outside these characteristics are considered to be voluntary standards. National Standards are often referred to as "GB standards". They are consistent across all of China and are developed for technical requirements. A prefix is used to identify Chinese national standards as mandatory (GB) or recommended (GB/T).

The Code of Design on Building Fire Protection and Prevention to GB 50016 [3] (the revised version was implemented in 2018) gives details of the reaction-to-fire requirements for building materials and components, as shown for insulation materials in Table 10.101.

Table 10.101 Requirements for the Fire Performance Properties of Thermal Insulation Materials on Façades in GB 50016-2014

Exterior wall type	Occupant type	Height limits (H)	Fire performance of insulation materials
Exterior wall with no curtain wall	Residential	$H > 100$ m	A
		$27\text{ m} < H \leq 100\text{ m}$	A or B1 [a]
		$H \leq 27$ m	A, or B1 [a], or B2 [a]
	Other than residential	$H > 50$ m	A
		$24\text{ m} < H \leq 50\text{ m}$	A or B1
		$H \leq 24$ m	A, or B1 [a], or B2
Exterior wall with curtain wall	All	$H > 24$ m	A
		$H \leq 24$ m	A or B1

[a] If B1 or B2 is used, surface protection shall be provided, and horizontal fire barriers shall be installed at each floor. The combustibility of horizontal fire barriers shall be class A, with a height of at least 300 mm. The thickness of the non-combustible protection layer on the insulation surface shall be at least 15 mm on the first floor, while the thickness shall be at least 5 mm on the second floor and above.

The Code for Fire Prevention in Design of Interior Decoration of Buildings to GB 50222-2017 [4] (implemented in 2018) contains fire safety requirements for classification (Table 10.102) and combustion performance classes of decoration materials for different kinds of buildings (Table 10.103, Table 10.104, and Table 10.105).

Table 10.102 Classification of the Fire Performance of Decoration of Buildings in GB 50222-2017

Class	Fire performance of decoration of buildings
A	Non-combustible
B1	Flame-retardant ("Difficult-flammability")
B2	Flammable
B3	Inflammable

Table 10.103 Fire Performance Classes of Interior Decoration Materials in Various Parts of Single-Storey and Multi-Storey Civil Buildings in GB 50222-2017

No.	Buildings and places	Building scale and nature	Combustion performance classes of decoration materials							
			Ceiling	Wall	Flooring	Partition	Fixed furniture	Decorative fabric		Other decoration materials
								Curtain	Drapery	
1	Waiting halls, VIP lounges, ticket offices, shops, dining places, etc. in airport buildings	–	A	A	B1	B1	B1	B1	–	B1
2	Waiting rooms, shops, dining places, etc. at bus stations, railway stations and ferry terminals	Floor area > 10,000 m²	A	A	B1	B1	B1	B1	–	B2
		Floor area ≤ 10,000 m²	A	B1	B1	B1	B1	B1	–	B2
3	Auditoriums, conference halls, function rooms, waiting rooms, etc.	Floor area of each hall > 400 m²	A	A	B1	B1	B1	B1	B1	B1
		Floor area of each hall ≤ 400 m²	A	B1	B1	B1	B2	B1	B1	B2
4	Gymnasiums	> 3000 seats	A	B1	B1	B1	B1	B1	B1	B2
		≤ 3000 seats	A	B1	B1	B1	B2	B2	B1	B2

No.	Buildings and places	Building scale and nature	Combustion performance classes of decoration materials							
			Ceiling	Wall	Flooring	Partition	Fixed furniture	Decorative fabric		Other decoration materials
								Curtain	Drapery	
5	Business offices at stores	Area per floor > 1500 m² or total floor area > 3000 m²	A	B1	B1	B1	B1	B1	–	B2
		Floor area per floor ≤ 1500 m² or total building area ≤ 3000 m²	A	B1	B1	B1	B2	B1	–	–
6	Guest rooms, public activity rooms, etc. at hotels and restaurants	With central air-conditioning systems	A	B1	B1	B1	B2	B2	–	B2
		Others	B1	B1	B2	B2	B2	B2	–	–
7	Residences and activity places at nursing homes, nurseries, and kindergartens	–	A	A	B1	B1	B2	B1	–	B2
8	Wards, treatment areas, and operation zones at hospitals	–	A	A	B1	B1	B2	B1	–	B2
9	Teaching places and teaching laboratories	–	A	B1	B2	B2	B2	B2	B2	B2
10	Public activity places at memorials, exhibition halls, museums, libraries, archives, etc.	–	A	B1	B1	B1	B2	B1	–	B2

Table 10.103 Fire Performance Classes of Interior Decoration Materials in Various Parts of Single-Storey and Multi-Storey Civil Buildings in GB 50222-2017 *(continued)*

No.	Buildings and places	Building scale and nature	Combustion performance classes of decoration materials							
			Ceiling	Wall	Flooring	Partition	Fixed furniture	Decorative fabric		Other decoration materials
								Curtain	Drapery	
11	Storage facilities for artifacts, commemorative exhibits, important books, archives, and materials	-	A	A	B1	B1	B2	B1	-	B2
12	Recreation and entertainment venues	-	A	B1	B1	B1	B1	B1	B1	B1
13	A-class and B-class electronic information system rooms, and rooms with important machines and instruments	-	A	A	B1	B1	B1	B1	B1	B1
14	Dining places	Business floor area > 100 m^2	A	B1	B1	B1	B2	B1	-	B2
		Business floor area ≤ 100 m^2	B1	B1	B1	B2	B2	B2	-	B2
15	Workplaces	With central air-conditioning systems	A	B1	B1	B1	B2	B2	-	B2
		Others	B1	B1	B2	B2	B2	-	-	-
16	Other public places	-	B1	B1	B2	B2	B2	-	-	-
17	Residences	-	B1	B1	B1	B1	B2	B2	-	B2

Table 10.104 Fire Performance Classes of Interior Decoration Materials in Various Parts of High-Rise Civil Buildings in GB 50222-2017

No.	Buildings and places	Building scale and nature	Combustion performance classes of decoration materials									
			Ceiling	Wall	Flooring	Partition	Fixed furniture	Decorative fabric				Other decoration materials
								Curtain	Drapery	Bed spread	Furniture wrapper	
1	Waiting halls, VIP lounges, ticket offices, shops, dining places, etc. at airport terminal buildings	–	A	A	B1	B1	B1	B1	–	–	–	B1
2	Waiting rooms, shops, dining places, etc. at bus stations, railway stations, and ferry terminals	Floor area > 10000 m²	A	A	B1	B1	B1	B1	–	–	–	B2
		Floor area ≤ 10000 m²	A	B1	B1	B1	B1	B1	–	–	–	B2
3	Auditoriums, conference halls, function rooms, waiting rooms, etc.	Floor area of each hall > 400 m²	A	A	B1	B1	B1	B1	B1	–	B1	B1
		Floor area of each hall ≤ 400 m²	A	B1	B1	B1	B2	B1	B1	–	B1	B1
4	Business offices at stores	Area per floor > 1500 m² or total building area > 3000 m²	A	B1	B1	B1	B1	B1	B1	–	B2	B1
		Area per floor ≤ 1500 m² or total floor area ≤ 3000 m²	A	B1	B1	B1	B1	B1	B2	–	B2	B2

Table 10.104 Fire Performance Classes of Interior Decoration Materials in Various Parts of High-Rise Civil Buildings in GB 50222-2017 *(continued)*

No.	Buildings and places	Building scale and nature	Combustion performance classes of decoration materials									
			Ceiling	Wall	Flooring	Partition	Fixed furniture	Decorative fabric				Other decoration materials
								Curtain	Drapery	Bed spread	Furniture wrapper	
5	Guest rooms, public activity rooms, etc. at hotels and restaurants	Category I construction	A	B1	B1	B1	B2	B1	–	B1	B2	B1
		Category II construction	A	B1	B1	B1	B2	B1	–	B2	B2	B2
6	Residences and activity places at nursing homes, nurseries, and kindergartens	–	A	A	B1	B1	B2	B1	–	B2	B2	B1
7	Wards, treatment areas and operation zones at hospitals	–	A	A	B1	B1	B2	B1	B1	–	B2	B1
8	Teaching places and teaching laboratories	–	A	B1	B2	B2	B2	B1	B1	–	B1	B2
9	Public activity places at memorials, exhibition halls, museums, libraries, archives, etc.	Category I construction	A	B1	B1	B1	B2	B1	B1	–	B1	B1
		Category II construction	A	B1	B1	B1	B2	B1	B2	–	B2	B2

No.	Buildings and places	Building scale and nature	Combustion performance classes of decoration materials									
			Ceiling	Wall	Flooring	Partition	Fixed furniture	Decorative fabric				Other decoration materials
								Curtain	Drapery	Bed spread	Furniture wrapper	
10	Storage facilities for artifacts, commemorative exhibits, important books, archives, and materials	–	A	A	B1	B1	B2	B1	–	–	B1	B2
11	Recreation and entertainment venues	–	A	B1	B1	B1	B1	B1	B1	B1	B1	B1
12	A-class and B-class electronic information system rooms, and rooms with important machines and instruments	–	A	A	B1	B1	B1	B1	B1	–	B1	B1
13	Dining places	–	A	B1	B1	B1	B2	B1	–	–	B1	B2

Table 10.105 Fire Performance Classes of Interior Decoration Materials in Various Parts of Underground Civil Buildings in GB 50222-2017

No.	Buildings and places	Combustion performance classes of decoration materials						
		Ceiling	Wall	Flooring	Partition	Fixed furniture	Decorative fabric	Other decoration materials
1	Auditoriums, conference halls, function rooms, waiting rooms, business offices at stores, etc.	A	A	A	B1	B1	B1	B2
2	Guest rooms, public activity rooms, etc. at hotels and restaurants	A	B1	B1	B1	B1	B1	B2
3	Treatment areas and operation zones at hospitals	A	A	B1	B1	B1	B1	B2
4	Teaching places and teaching laboratories	A	A	B1	B2	B2	B1	B2
5	Public activity places at memorials, exhibition halls, museums, libraries, archives, etc.	A	A	B1	B1	B1	B1	B1
6	Storage facilities for artifacts, commemorative exhibits, important books, archives, and materials	A	A	A	A	A	B1	B1
7	Recreation and entertainment venues	A	A	B1	B1	B1	B1	B1
8	A-class and B-class electronic information system rooms, and rooms with important machines and instruments	A	A	B1	B1	B1	B1	B1
9	Dining places	A	A	A	B1	B1	B1	B2

No.	Buildings and places	Combustion performance classes of decoration materials						
		Ceiling	Wall	Flooring	Partition	Fixed furniture	Decorative fabric	Other decoration materials
10	Workplaces	A	B1	B1	B1	B1	B2	B2
11	Other public places	A	B1	B1	B2	B2	B2	B2
12	Garages and motor repair shops	A	A	B1	A	A	–	–

For practical applications, the performance requirements and the application fields for materials are specified more clearly in the relevant laws and regulations, such as the classification for the burning behavior of building materials and products to GB 8624-2012 [5]. GB 8624-2012 was issued by the General Administration of Quality Supervision, Inspection and Quarantine of the PRC, and is a classification standard, in which a classification of the fire properties of building materials is specified, and in which the technical parameters (as well as the test methods that determine in detail whether the materials reach the relevant class) are described.

10.4.1.2 Classification and Testing of the Fire Performance of Building Materials and Components

Classification of the Fire Performance of Building Materials

GB 8624-2012 is based on EN 13501-l:2007, and provides methods to classify and evaluate the fire performance of building materials and products for structural and decorative materials used in all kinds of industrial and public buildings.

The classes, designations, and test methods of the fire performance of building materials and products as specified in GB 8624-2012 are summarized in Tables 10.106 to 10.112.

Table 10.106 Classification of Composite Materials (Sandwich Panels)

Classification		Test methods	Pass criteria
A	A1	GB/T 5464	▪ Average furnace temperature rise does not exceed 30 °C ▪ Average mass loss rate is not more than 50% ▪ Mean duration of sustained flaming does not exceed 0 s
		GB/T 14402	▪ Gross calorific potential of substantial components ≤ 2.0 MJ/kg ▪ Gross calorific potential of non-substantial components ≤ 1.4 MJ/m^2
	A2	GB/T 5464 or GB/T 14402	▪ Average furnace temperature rise does not exceed 50 °C ▪ Average mass loss rate is not more than 50% ▪ Mean duration of sustained flaming does not exceed 20 s Or else, ▪ Gross calorific potential of substantial components ≤ 3.0 MJ/kg ▪ Gross calorific potential of non-substantial components ≤ 4.0 MJ/m^2
		GB/T 20284	▪ Fire growth rate index $FIGRA_{0.2\,MJ}$ ≤ 120 W/s ▪ Total heat release $THR_{600\,s}$ ≤ 7.5 MJ ▪ No sustained flames reaching the far edge of the long wing
B1[a]	B	GB/T 20284	▪ Fire growth rate index $FIGRA_{0.2\,MJ}$ ≤ 120 W/s ▪ Total heat release $THR_{600\,s}$ ≤ 7.5 MJ ▪ No sustained flames reaching the far edge of the long wing
		GB/T 8626	None of the six specimens reach the reference mark of 150 mm within 60 s after ignition for flame application 30 s, without filter paper ignited by flaming droplets
	C	GB/T 20284	▪ Fire growth rate index $FIGRA_{0.4\,MJ}$ ≤ 250 W/s ▪ Total heat release $THR_{600\,s}$ ≤ 15 MJ ▪ No sustained flames reaching the far edge of the long wing
		GB/T 8626	None of the six specimens reach the reference mark of 150 mm within 60 s after ignition for flame application 30 s, without filter paper ignited by flaming droplets
B2[b]	D	GB/T 20284	Fire growth rate index $FIGRA_{0.4\,MJ}$ ≤ 750 W/s
		GB/T 8626	None of the six specimens reach the reference mark of 150 mm within 60 s after ignition for flame application 30 s, without filter paper ignited by flaming droplets
	E	GB/T 8626	None of the six specimens reach the reference mark of 150 mm within 20 s after ignition for flame application 15 s, without filter paper ignited by flaming droplets
B3	F	-	No performance requirements

[a] For thermal insulation materials, OI ≥ 30 is required in addition to the requirements for B1.
[b] For thermal insulation materials, OI ≥ 26 is required in addition to the requirements for B2.

Table 10.107 Classification of Flooring Materials

Classification		Test methods	Pass criteria
A	A1	GB/T 5464	▪ Average furnace temperature rise does not exceed 30 °C ▪ Mean duration of sustained flaming does not exceed 0 s ▪ Average mass loss rate is not more than 50%
		GB/T 14402	▪ Gross calorific potential of substantial components ≤ 2.0 MJ/kg ▪ Gross calorific potential of non-substantial components ≤ 1.4 MJ/m^2
	A2	GB/T 5464 or GB/T 14402	▪ Average furnace temperature rise does not exceed 50 °C ▪ Average mass loss rate is not more than 50% ▪ Mean duration of sustained flaming does not exceed 20 s Or else, ▪ Gross calorific potential of substantial components ≤ 3.0 MJ/kg ▪ Gross calorific potential of non-substantial components ≤ 4.0 MJ/m^2
		GB/T 11785	Critical radiant flux CHF ≥ 8.0 kW/m^2
B1	B	GB/T 11785	Critical radiant flux CHF ≥ 8.0 kW/m^2
		GB/T 8626	None of the six specimens reach the reference mark of 150 mm within 20 s after ignition for flame application 15 s
	C	GB/T 11785	Critical radiant flux CHF ≥ 4.5 kW/m^2
		GB/T 8626	None of the six specimens reach the reference mark of 150 mm within 20 s after ignition for flame application 15 s
B2	D	GB/T 11785	Critical radiant flux CHF ≥ 3.0 kW/m^2
		GB/T 8626	None of the six specimens reach the reference mark of 150 mm within 20 s after ignition for flame application 15 s
	E	GB/T 11785	Critical radiant flux CHF ≥ 2.2 kW/m^2
		GB/T 8626	None of the six specimens reach the reference mark of 150 mm within 20 s after ignition for flame application 15 s
B3	F	-	No performance requirements

Table 10.108 Classification of Linear Pipe Thermal Insulation Materials

Classification		Test methods	Pass criteria
A	A1	GB/T 5464	▪ Average furnace temperature rise does not exceed 30 °C ▪ Average mass loss rate is not more than 50% ▪ Mean duration of sustained flaming does not exceed 0 s
		GB/T 14402	▪ Gross calorific potential of substantial components ≤ 2.0 MJ/kg ▪ Gross calorific potential of non-substantial components ≤ 1.4 MJ/m^2
	A2	GB/T 5464 or GB/T 14402	▪ Average furnace temperature rise does not exceed 50 °C ▪ Average mass loss rate is not more than 50% ▪ Mean duration of sustained flaming does not exceed 20 s Or else, ▪ Gross calorific potential of substantial components ≤ 3.0 MJ/kg ▪ Gross calorific potential of non-substantial components ≤ 4.0 MJ/m^2
		GB/T 20284	▪ Fire growth rate index $FIGRA_{0.2\ MJ} \leq 270$ W/s ▪ Total heat release $THR_{600\ s} \leq 7.5$ MJ ▪ No sustained flames reaching the far edge of the long wing
B1	B	GB/T 20284	▪ Fire growth rate index $FIGRA_{0.2\ MJ} \leq 270$ W/s ▪ Total heat release $THR_{600\ s} \leq 7.5$ MJ ▪ No sustained flames reaching the far edge of the long wing
		GB/T 8626	None of the six specimens reach the reference mark of 150 mm within 60 s after ignition for flame application 30 s, without filter paper ignited by flaming droplets
	C	GB/T 20284	▪ Fire growth rate index $FIGRA_{0.4\ MJ} \leq 460$ W/s ▪ Total heat release $THR_{600\ s} \leq 15$ MJ ▪ No sustained flames reaching the far edge of the long wing
		GB/T 8626	None of the six specimens reach the reference mark of 150 mm within 60 s after ignition for flame application 30 s, without filter paper ignited by flaming droplets
B2	D	GB/T 20284	▪ Fire growth rate index $FIGRA_{0.4\ MJ} \leq 2100$ W/s ▪ Total heat release $THR_{600\ s} < 100$ MJ
		GB/T 8626	None of the six specimens reach the reference mark of 150 mm within 60 s after ignition for flame application 30 s, without filter paper ignited by flaming droplets
	E	GB/T 8626	None of the six specimens reach the reference mark of 150 mm within 20 s after ignition for flame application 15 s, without filter paper ignited by flaming droplets
B3	F	-	No performance requirements

Table 10.109 Classification of Textiles for Curtains and Furniture

Classification	Test methods	Pass criteria
B1	GB/T 5454	Oxygen index ≥ 32
	GB/T 5455	▪ Destroyed length ≤ 150 mm ▪ Afterflame time ≤ 5 s ▪ Afterglow time ≤ 15 s Without filter paper ignited or afterglow by flaming droplets
B2	GB/T 5454	Oxygen index ≥ 26
	GB/T 5455	▪ Destroyed length ≤ 200 mm ▪ Afterflame time ≤ 15 s ▪ Afterglow time ≤ 30 s Without filter paper ignited or afterglow by flaming droplets
B3	–	No performance requirements

Table 10.110 Classification of Plastic Conduit Pipes for Wires and Cables, and Outer Shell and Accessories for Electric Appliances

Classification	Products	Test methods	Pass criteria
B1	Plastic conduit pipes for wires and cables	GB/T 2406.2	Oxygen index ≥ 32
		GB/T 2408	Vertical burning V-0
		GB/T 8627	Smoke density rating ≤ 75
	Outer shell and accessories for electric appliances	GB/T 5169.16	Vertical burning V-0
B2	Plastic conduit pipes for wires and cables	GB/T 2406.2	Oxygen index ≥ 26
		GB/T 2408	Vertical burning V-1
	Outer shell and accessories for electric appliance	GB/T 5169.16	Vertical burning V-1
B3	-	-	No performance requirements

Table 10.111 Classification of Cellular Plastics for Electrical and Furniture Products

Classification	Test Methods	Pass criteria
B1	GB/T 16172	Peak heat release rate per unit area ≤ 400 kW/m^2 when exposed to an irradiance of 30 kW/m^2
	GB/T 8333	▪ Mean burning time ≤ 30 s ▪ Mean burning height ≤ 250 mm
B2	GB/T 8333	▪ Mean burning time ≤ 30 s ▪ Mean burning height ≤ 250 mm
B3	-	No performance requirements

Table 10.112 Classification of Upholstered Furniture and Hard Furniture

Classification	Products	Test methods	Pass criteria
B1	Upholstered furniture	GB/T 27904	▪ Peak heat release rate ≤ 200 kW/m² ▪ Total heat release within 5 min ≤ 30 MJ ▪ Maximum smoke density ≤ 75%
		GB 17927.1	No progressive smoldering or flaming is detected within 1 h after applying the cigarette
	Upholstered mattresses	GB 8624 Appendix A	▪ Peak heat release rate ≤ 200 kW/m² ▪ Total heat release within 10 min ≤ 15 MJ
	Hard furniture	GB/T 27904	▪ Peak heat release rate ≤ 200 kW/m² ▪ Total heat release within 5 min ≤ 30 MJ ▪ Maximum smoke density ≤ 75%
B2	Upholstered furniture	GB/T 27904	▪ Peak heat release rate ≤ 300 kW/m² ▪ Total heat release within 5 min ≤ 40 MJ ▪ No complete combustion occurs
		GB 17927.1	No progressive smoldering or flaming is detected within 1 h after applying the cigarette
	Upholstered mattresses	GB 8624 Appendix A	▪ Peak heat release rate ≤ 300 kW/m² ▪ Total heat release within 10 min ≤ 25 MJ
	Hard furniture	GB/T 27904	▪ Peak heat release rate ≤ 300 kW/m² ▪ Total heat release within 5 min ≤ 40 MJ ▪ No complete combustion occurs
B3	-	-	No performance requirements

Generally Used Fire Performance Tests and Classifications for Building Materials

The standards for methods and criteria mentioned in GB 8624-2012 are compiled below. Most of the respective test equipment sketches for standards identical to ISO or EN can be found under Sections 8.2.2.2 and 9.3.1.

GB/T 5464-2010: Non-combustibility test method for building materials

This standard is identical to EN ISO 1182:2002. The test specifications are summarized in Table 10.113.

Table 10.113 Test Specifications of Non-Combustibility Test for Building Materials

Specimens	Five cylindrical samples, volume (76 ± 8) cm^3, external diameter (45^{+0}_{-2}) mm, height (50 ± 3) mm
Furnace temperature	(750 ± 5) °C
Test duration	30 min

GB/T 14402-2007: Reaction-to-fire tests for building materials and products - Determination of the heat of combustion

This test is identical to ISO 1716:2002. The test specifications are shown in Table 10.114.

Table 10.114 Test Specifications of Calorific Potential of Building Materials

Specimens	Three samples, homogeneous powder 0.5 g
Test apparatus	As for ISO 1716
Test duration	10 min after the water temperature of inner water jacket reaches the highest point t_m

GB/T 20284-2006: Single burning item test for building materials and products

This test is equivalent to EN 13823:2002 (single burning item test). The test specifications are shown in Table 10.115.

Table 10.115 Test Specifications of Fire Growth Rate and Total Heat Release of Building Materials

Specimens	Three samples, short wing: (495 ± 5) mm × (1500 ± 5) mm × max. 200 mm, long wing: (1000 ± 5) mm × (1500 ± 5) mm × max. 200 mm
Test apparatus	As for EN 13823
Test duration	20 min

GB/T 11785-2005: Reaction-to-fire tests for floorings - Determination of the burning behavior using a radiant heat source

This test is identical to ISO/DIS 9239-1:2002. The test specifications are shown in Table 10.116.

Table 10.116 Test Specifications of Critical Radiant Flux of Floor Coverings Using a Radiant Heat Source

Specimens	(1050 ± 5) mm × (230 ± 5) mm (three samples in production direction; three samples in vertical direction)
Test apparatus	As for ISO/DIS 9239
Flame application time	10 min

GB/T 8626-2007: Test method of flammability for building materials

This standard is identical to EN ISO 11925-2:2002. The test specifications are summarized in Table 10.117.

Table 10.117 Flammability Test Specifications

Specimens	Six specimens, 250 mm × 90 mm × max. thickness 60 mm
Ignition source	Inclined at 45°, flame height (20 ± 1) mm
Test duration	20 s or 60 s (including 15 s or 30 s for flame application, respectively)

Additional Fire and Smoke Tests for Materials and Products Used in Building

In the following, additional test methods used in China for determining the burning behavior of wall insulation foam, and the flammability of plastic conduits for wires and cables, the outer shell and accessories of electric appliances, textiles, and plastic foams, are dealt with and their pass criteria described.

GB/T 2406.2-2009: Plastics - Determination of burning behavior by oxygen index - Part 2: Ambient-temperature test

This standard is identical to ISO 4589-2:1996, the limiting oxygen index. The test specifications are shown in Table 10.118.

Table 10.118 Test Specifications of the Burning Behavior of Plastics by the Oxygen Index Test

Specimens	15 specimens, (70–150) mm × (6.5–52) mm × (3–10.5) mm thickness
Test apparatus	As for ISO 4589
Test method	Determines oxygen index as minimum O_2 percentage in an O_2/N_2 gas mixture to support flaming combustion

GB/T 2408-2008: Plastics - Determination of burning characteristics - Horizontal and vertical test

This standard is identical to IEC 60695-11-10:1999, which is a Bunsen burner test derived from the UL 94 horizontal burning (HB) and vertical (V) tests. The test criteria are shown in Table 10.119.

Table 10.119 Test Specifications of Flammability Characteristics of Plastics - Vertical Test

Specimens	(125 ± 5) mm × (13.0 ± 0.5) mm × thickness (3.0 ± 0.2) to < 13 mm, five samples
Test apparatus	As for ISO 1210
Flame application time	(10 ± 0.5) s

GB/T 8627-2007: Test method for density of smoke from the burning or decomposition of building materials

This standard is based on ASTM D2843-1999. The test apparatus is shown in Section 10.2.1.5, and the test specifications are summarized in Table 10.120.

Table 10.120 Test Specifications of Smoke Density Test

Specimens	Three specimens, (25.4 ± 0.3) mm × (25.4 ± 0.3) mm × (6.2 ± 0.3) mm Thickness: min. 6.2 mm, max. 25 mm, or in accordance with the original thickness in practical use
Test apparatus	As for ASTM D2843
Test duration	4 min

GB/T 5169.16-2017: Fire hazard testing for electric and electronic products - Part 16: Test flames - 50 W horizontal and vertical flame test methods

This standard is identical to IEC 60695-11-10:2003. The test is a Bunsen burner test with a vertical sample. The test specifications are summarized in Table 10.121.

Table 10.121 Test Specifications of Flammability Test for Outer Shell and Accessories of Electric Appliance - Vertical Burning Test

Specimens	Five samples, (125 ± 5) mm × (13.0 ± 0.5) mm × thickness (3.0 ± 0.2) to < 13 mm
Test apparatus	As for ISO 1210
Flame application time	(10 ± 0.5) s

GB/T 5454-1997: Textiles - Burning behavior - Oxygen index method

The burning behavior of textiles is measured with the limiting oxygen index test. The standard is not equivalent to ISO 4589:1984. The specifications are summarized in Table 10.122.

Table 10.122 Test Specifications of Burning Behavior of Textiles by the Oxygen Index Test

Specimens	150 mm × 58 mm (15 samples in warp direction; 15 samples in weft direction)
Test apparatus	As for ISO 4589
Test method	Determines oxygen index as minimum O_2 percentage in an O_2/N_2 gas mixture to support flaming combustion

GB/T 5455-2014: Textiles - Burning behavior - Determination of damaged length, afterglow time and afterflame time of vertically oriented specimens

The standard is based on the Japanese JIS 1091 test. This test is a Bunsen burner test with a vertical sample. The specifications are shown in Table 10.123.

Table 10.123 Test Specifications of Burning Behavior of Textiles – Vertical Test

Specimens	300 mm × 89 mm (five samples in warp direction; five samples in weft direction)
Test apparatus	As for JIS 1091
Flame application time	12 s

GB/T 8333-2008: Test method for flammability of rigid cellular plastic – Vertical burning method

The standard is based on the ASTM D3014-04a test. This test is a Bunsen burner test with a vertical sample. The test specifications are shown in Table 10.124.

Table 10.124 Test Specifications of Flammability Test for Cellular Plastics – Vertical Burning Test

Specimens	Six samples, (254 ± 1) mm × (19 ± 1) mm × (19 ± 1) mm
Flame application time	10 s

GB/T 16172-2007: Test method for heat release rate of building materials

This standard is identical to ISO 5660-1: 2002 (Reaction-to-fire tests – Heat release, smoke production and mass loss rate – Part 1: Heat release rate (cone calorimeter method)). The test specifications are summarized in Table 10.125.

Table 10.125 Test Specifications of Heat Release by Combustion of Cellular Plastics for Electrical and Furniture Products

Specimens	Three samples, 10 mm × 10 mm × max. 50 mm
Test apparatus	As for ISO 5660
Flame application time	Until sustained flaming occurs, max. 30 min

Further Tests Still in Use That Do Not Cover the Pass Criteria of Classification for Building Materials and Products

GB/T 14403-2014: Test method for combustion heat release of building materials

GB/T 8332-2008: Test method for flammability of cellular plastics – horizontal burning method

GB/T 8625-2005: Test method of difficult-flammability for building materials

GB/T 20285-2006: Toxic classification of fire effluents hazard for materials.

This test method is derived from the standards ISO TR 9122-2 and 9122-4, DIN 53436, and JIS A 1321. The test specifications are shown in Table 10.126.

Table 10.126 Test Specifications of the Smoke Toxicity of Fire Effluents

Specimens	Three to five samples, max. length 400 mm, determination of the mass of the specimens based on the designed smoke concentration
Test apparatus	▪ Smoke-generating apparatus identical to DIN 53436, mouse rotary cage identical to JIS A 1321 ▪ Methods of test and evaluation identical to ISO/TR 9122-2 ▪ Terms identical to ISO/TR 9122-4
Test duration	30 min exposure of mice to toxicity, 14 d observation after toxicity exposure

GB/T 29416-2012: Test method for fire-resistant performance of external wall insulation systems applied to building façades

The test method GB/T 29416-2012, equivalent to the British standard BS 8414 described under Section 10.3.2.12, is valid. But it cannot replace the pass criteria of classification. Special requirements for composite materials and surface coatings apply. No matter what thickness the composite building materials have, or by what kind of procedure they are produced, their burning behavior should be tested as a whole and an overall evaluation be given.

Fire Performance Tests and Classification for Furniture

GB 50222-2017: Code for Fire Prevention in Design of Interior Decoration of Buildings

This standard covers furniture such as fixed furniture, partition decorations, curtains, drapery, bedspreads, furniture wrappers and fixed ornaments for civil buildings, and covers partition decorations in industrial factory buildings. The required classification of furniture used in various buildings is specified in detail. The classification and test method for furniture is covered in GB 8624-2012.

GB 17927.1-2011: Upholstered furniture - Assessment of the resistance to ignition of mattress and sofa - Part 1: Smoldering cigarette

This standard is based on ISO 8191-1:1988 (Furniture - Assessment of the ignitability of upholstered furniture - Part 1: Ignition source: smoldering cigarette). The test specifications are summarized in Table 10.127.

Table 10.127 Smoldering Cigarette Ignition Test Specifications

Specimens	Mattress or sofa
Smoldering cigarette application time	Max. 1 h

GB/T 27904-2011: Testing method for fire characteristics of furniture and subassemblies exposed to a flaming ignition source

This standard is based on NFPA 266:1998 (Standard method of test for fire characteristics of upholstered furniture exposed to flaming ignition source). The test specifications are shown in Table 10.128.

Table 10.128 Test Specifications of Fire Characteristics Test for Furniture and Subassemblies

Specimens	Furniture and subassemblies
Test apparatus	Same as NFPA 266 (19 kW gas burner)
Flame application time	5 min for upholstered furniture; 10 min for other furniture and subassemblies

10.4.1.3 Official Approval

The two major official institutions for product quality supervision and testing are:

- **China National Center for Quality Supervision and Testing of Fixed Fire-fighting Systems and Fire-resisting Building Components (CNCF)**
  Managed by Tianjin Fire Science and Technology Research Institute of Ministry of Emergency Management of the People's Republic of China (MEM)
  2 Fuxing Road
  Jinlai Road
  Xiqing District
  Tianjin, 300382
  Tel: +86 22 58387888
  Fax: +86 22 58387858
  E-mail: *cncf@tfri.com.cn*
- **National Center for Quality Supervision and Testing of Fire Building Materials (NFTC)**
  Managed by Sichuan Fire Science and Technology Research Institute of MEM
  69 Jinke South Road
  Jinniu District
  Chengdu
  Sichuan, 611830
  Tel: +86 28 87516225
  Fax: +86 28 87516330
  E-mail: *nftc@fire-testing.net*

Other institutions are:

- **China National Fire-Fighting Equipment Quality Supervision Testing Center (CFE)**
  Managed by Fire Science and Technology Research Institute of MEM
  391 Xinzhuang West Ring Road
  Minhang District
  Shanghai
  Tel: +86 21 54959866
  E-mail: *fireshnc@sh163.net*

- **China National Supervision and Test Center for Fire Electronic Product Quality (NFFE)**
  Managed by Shenyang Fire Science and Technology Research Institute of MEM
  218-20 Wenda Road
  Huanggu District
  Shenyang
  Liaoning
  Tel: +86 024 31535801
  E-mail: *jyglb@efire.cn*

Other institutions may also offer these services.

10.4.1.4 Future Developments

In China, for building and construction, smoke and toxicity tests are mandatory for internal insulation and decorations, in addition to the reaction-to-fire tests derived from those used in the EU. However, the smoke and toxicity tests are outdated and may be replaced by international ISO-based tests in the future. For façade exterior insulation systems, the full-scale fire test to GB/T 29416 might be required for high-rise buildings in the future.

References for Section 10.4.1

[1] Construction Law of the People's Republic of China, promulgated in 1 November 1997, revised on 23 April 2019.

[2] Fire Protection Law of the People's Republic of China (2008 revision, effective 1 May 2009, revised on 23 April 2019).

[3] GB 50016-2014 (2018 edition): Code of design on building fire protection and prevention.

[4] GB 50222-2017: Code for fire prevention in design of interior decoration of buildings.

[5] GB 8624-2012: Classification for burning behavior of building materials and products.

10.4.2 Japan

Hideki Yoshioka

10.4.2.1 Statutory Regulations

The Building Standard Law of Japan came into effect on 16 November 1950. The aim of this legislation is "to protect the lives, health, and wealth of citizens, and thus contribute to the prosperity of the community, by laying down guidelines and standards for plots of land, building design, furnishing and use."

In 1998, a revision of the main parts of the BSL (Building Standard Law), including the fire safety design systems, was published, and this came into effect on 1 June

2000 [1]. This revision has started a process of changing from a specification-based to a performance-oriented design.

The revision defines:

Basic Requirements – Definition of Categories of Building Parts and Materials

- Fire-resistive construction, quasi-fire-resistive construction, fire preventive construction, non-combustible materials, etc. (BSL Article 2).
- Quasi-non-combustible materials, fire retardant materials, and so forth (Enforcement Order Article 1).

Performance Criteria Required for Defined Building Parts and Materials

- Fire-resistive performance, quasi-fire-resistive performance, and fire preventive performance (Enforcement Order Article 107, 107-2, 108).
- Non-combustible performance, quasi-non-combustible materials, and fire retardant materials (Enforcement Order Article 1, 108-2).

Approval of Building Parts and Materials with the Required Performance

- Approvals for "performance evaluation report" of tested materials (based on specific test methods and technical criteria) are submitted by designated examination bodies: (BSL Article 68-24).
- Among the specification-based materials listed in the previous notifications, those that proved to satisfy the new fire performance requirements are presented in the new regulation system.

Details are contained in Notifications No. 1399 for fire-resistive construction, No. 1358 for quasi-fire-resistive construction, No. 1359 for fire preventive construction, No. 1400 for non-combustible materials, No. 1401 for quasi-non-combustible materials, and No. 1402 for fire retardant materials.

10.4.2.2 Classification and Testing of the Fire Performance of Building Materials and Components

The reaction-to-fire test methods used in Japan are partially summarized in JIS A 1321:1994 [2]. However, even in the latest version of JIS A 1321, the ISO 5660-1 cone calorimeter test [3], which is most frequently used in Japan for the classification of reaction-to-fire performance of building materials and components, is not included. Therefore, instead of just referring to JIS A 1321, the current situation in Japan will be covered in the rest of this chapter. In Table 10.129, the test methods and criteria for reaction-to-fire performance of building products are summarized.

Table 10.129 Reaction-to-fire Performance of Building Materials, Test Methods, and Criteria in Japan [4]

Classification	Required performance criteria		Fire performance		(BSL 68-24)
	Performance	Duration of test	Test method: (a) or (b)	Performance criteria [a]	Fire gas toxicity [b]
Non-combustible materials (BSL)	1. No burning 2. No deformation, melting, cracking, or other damage detrimental to fire prevention 3. No generation of smoke or gas detrimental to evacuation (E. O. 108-2)	20 min (E. O. 108-2)	(a) Non-combustibility test (based on ISO 1182) (b) Heat release test (based on ISO 5660-1, radiant heater 50 kW/m²)	(a) Temperature increase ≤ 20 K; mass loss ≤ 30% (b) THR ≤ 8 MJ/m²; MHR ≤ 200 kW/m² (≥ 10 s)	Toxicity test in Japan (using eight mice), introduced in ISO/TR 16312-2
Quasi-non-combustible materials (E.O. 1)		10 min (E. O. 1)	(a) Heat release test (based on ISO 5660-1, radiant heater 50 kW/m²) (b) Model box test (based on ISO/TS 17431, propane burner 40 kW)	(a) THR ≤ 8 MJ/m²; MHR ≤ 200 kW/m² (≥ 10 s) (b) THR ≤ 30 MJ (excluding burner); THR ≤ 50 MJ (including burner); MHR ≤ 140 kW (≥ 10 s); no through cracks or holes	
Fire retardant materials (E. O. 1)		5 min (E. O. 1)	(a) Heat release test (based on ISO 5660-1, radiant heater 50 kW/m²) (b) Model box test (based on ISO/TS 17431, propane burner 40 kW)	(a) THR ≤ 8 MJ/m²; MHR ≤ 200 kW/m² (≥ 10 s) (b) THR ≤ 30 MJ (excluding burner); THR ≤ 40 MJ (including burner); MHR ≤ 140 kW (≥ 10 s); no through cracks or holes	

BSL = Building Standard Law; E. O. = Enforcement Order; THR = Total heat release; MHR = Max. rate of heat release

[a] Applied by performance evaluation bodies designated by the Ministry of Land, Infrastructure, and Transport of Japan.

[b] Materials with a higher content of organic components than the specified amount (200 g/m2) shall pass the fire gas toxicity test.

Building materials are tested for *non-combustibility* with an electric furnace operating at 750 °C. The apparatus and test method are based on ISO 1182 [5] and described there (see also Section 9.2.3.2). The maximum temperature rise allowed is 20 K, and the mass loss of the specimen may not be more than 30%.

Alternatively, these materials can be tested in the cone calorimeter, based on ISO 5660-1 [6]. Details of the apparatus and specifications are given in ISO 5660-1 (see also Section 9.2.3.4). The test is carried out for 20 min at a radiant heat level of 50 kW/m². Within this time, the total heat release (THR) is not allowed to exceed 8 MJ, and the maximum rate of heat release (RHR) may not exceed 200 kW/m².

Quasi-non-combustible materials are tested in the cone calorimeter, or in the ISO/TS 17341 "model box test" [7], which was originally developed in Japan as a potential screening test method for the ISO 9705-1 room/corner test [8]. Details of the "model box test" are given below.

The pass criteria in the cone calorimeter test are the same as for non-combustible materials, but for a shorter test period of 10 min. If the model box test is used, a THR of 30 MJ and a RHR of 140 kW may not be exceeded within the 10 min of the test, and no burn-through is accepted.

Fire retardant materials are also tested in the cone calorimeter or in the model box test. The criteria are the same as above but for a period of 5 min.

For the fire protective materials (all three levels in the classification), an additional toxicity test is required, whereby eight mice in rotating cages are exposed to the decomposition products from building products. A test method based on determining the incapacitation point of mice (average time for eight mice to stop moving) is used, the details of which were introduced in ISO/TR 16312-2 [9]. The test apparatus is shown in Figure 10.46. If the content of organic components in a tested specimen is 200 g/m^2 or less, toxicity tests are not necessary.

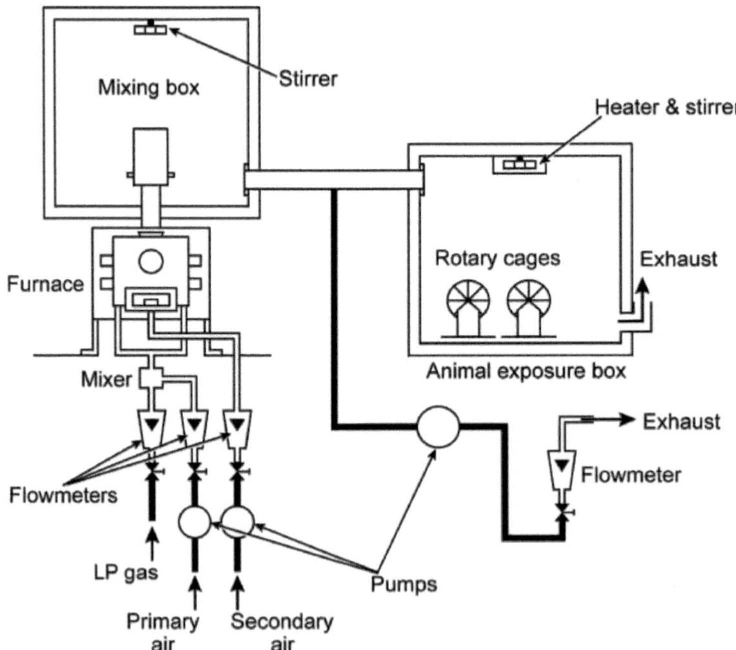

Figure 10.46 Toxicological inhalation apparatus used in Japan

The Model Box Test (ISO/TS 17431) for determining the maximum value of heat release rate and the total value of heat released uses a box-shaped specimen with a gas burner as fire source. The product to be tested shall be cut to make four panels to cover the ceiling, two side walls, and the end wall of the inside surfaces of the combustion chamber. In the standard specimen configuration, three walls and the ceiling of the combustion chamber are covered with the panels.

The dimensions of the panels shall be such that, when they are assembled and inserted into the combustion chamber, the combustion chamber has the internal dimensions listed in Table 10.130. The test rig is shown in Figure 10.47.

The ignition source is a propane gas burner having a 0.17 m × 0.17 m square-top surface layer of a porous, inert material, for example, sand or ceramic beads. The heat output of the burner is 40 kW during the test period.

Specimens have passed the model box test if in each of two tests the following criteria are met throughout the defined test duration:

- The maximum value of heat release rate during the test shall not exceed 140 kW.
- The total value of heat released during the test shall not exceed 30 MJ, excluding the propane gas burner.
- The box specimen shall not lead to flashover which would present a significant problem in terms of fire safety. No holes are allowed to burn through the walls and no significant cracks are allowed to occur.

Figure 10.47 ISO/TS 17431 Model Box Test

Table 10.130 ISO/TS 17431 (Model Box Test) Test Specifications

Specimens	Width: 0.84 m, Length: 1.72 m, Height: 0.92 m (internal dimensions)
Specimen position	The product to be tested shall be cut to make four panels supported by a sheet and frames of steel to cover the ceiling, two side walls, and the back wall of the inside surfaces of the combustion chamber.
Ignition source	Propane gas burner with 0.17 m × 0.17 m square top surface layer of a porous, inert material, e.g., sand. Heat output 40 kW during the test period. Burner placed on the floor in a corner directly opposite the doorway wall in contact with the specimens.
Test duration	15 min
Conclusions	This test method is especially suitable for products for which the full-scale room test (ISO 9705-1) has to be terminated before the room is fully affected by fire because of the occurrence of flashover or other safety reasons.

10.4.2.3 Recent Developments of New Standard Test Methods as JIS (Japanese Industrial Standard)

Most of the reaction-to-fire tests for building materials and products in Japan have been performed using ISO 5660-1 cone calorimeter tests, for which it is almost impossible to reflect the effects of construction details such as edges and joints in the actual fire scenarios at the full-scale situation. To address this, a collaboration of fire experts and researchers from industry, academia, and government, with the help of international fire experts, have developed two newly standardized reaction-to-fire test methods, namely JIS A 1310:2015 "Test method for fire propagation over building façades" and JIS A 1320:2017 "Reaction to fire test for sandwich panel building interior systems - Box test".

Even though neither of these methods have yet been cited in the Building Standard Law of Japan as a "mandatory test", many private-sector organizations have intensively used both JIS test methods for technically understanding the fire behavior of construction products that could not be fully assessed using small-scale fire tests such as ISO 5660-1 cone-calorimeter tests. More information about these two JIS test methods is given below.

10.4.2.3.1 JIS A 1310:2015 Test Method for Fire Propagation Over Building Façades

Up to the end of 2017, there was not any specific regulation or criteria for combustible building façade elements installed on the exterior side of fire-resistant walls in Japan according to its Building Standard Law. This is because only the fire-resistance performance (insulation, integrity, and stability) is taken into account with respect to the fire safety performance of building exterior walls in Japan, and fire propagation over façades was not considered. However, a large number of massive devastating façade fires have occurred globally, and there is also potential for façade fires in Japan. Based on the ISO 13785-1 and -2 reaction-to-fire tests for façades (intermediate and large-scale) performed at the Building Research Institute (BRI) and Tokyo University of Science (TUS), and detailed consideration of their advantages and disadvantages [10, 11], a completely new test method (JIS A 1310:2015) was developed, and has been published both in Japanese and English [12–14].

A schematic drawing of the test setup is shown in Figure 10.48, and the test specifications are given in Table 10.131. The entire configuration combines the combustion chamber (containing the gas burner) and the façade, with both elements sharing a 910 mm square unglazed opening. The combustion chamber is 1350 mm × 1350 mm × 1350 mm (height × width × depth), and its interior is finished with ceramic fiber blankets (25 mm thick).

In JIS A 1310, temperature and incident heat flux are measured, and visual observation noted. Figure 10.49 shows the installation positions for both heat flux meters and thermocouples. In case of "real-scale" tests on combustible façades, heat flux meters are installed at the top of the main façade surface (to measure the heat flux into the main façade surface at the top height level), as well as 2 m from the main façade surface, at heights of 500, 900, 1500, 2000, and 2500 mm above the opening top level (for measuring the heat flux at the assumed position of an adjacent building). For calibrating the heat output of the gas burner in the combustion chamber, heat flux meters are installed at the main façade surface at the heights mentioned above. In the current revision process for JIS A 1310, for calibration tests, averaged incident heat fluxes on the main façade surface should be (30 ± 5) kW/m^2 at a height of 900 mm, and (15 ± 5) kW/m^2 at a height of 1500 mm above the opening top level. A heat release of around 900 kW from the gas burner would be suitable for satisfying these heat fluxes. The thermocouples should be installed on the façade surface and at 50 mm from it, at the heights mentioned above. Since 2017, collaborative research between Hungary and Japan has been taking place to find the correlation between JIS A 1310 and MSZ 14800-6. The latter is a full-scale façade test pertaining to exactly the same combustible façade specimens tested in JIS A 1310 [15].

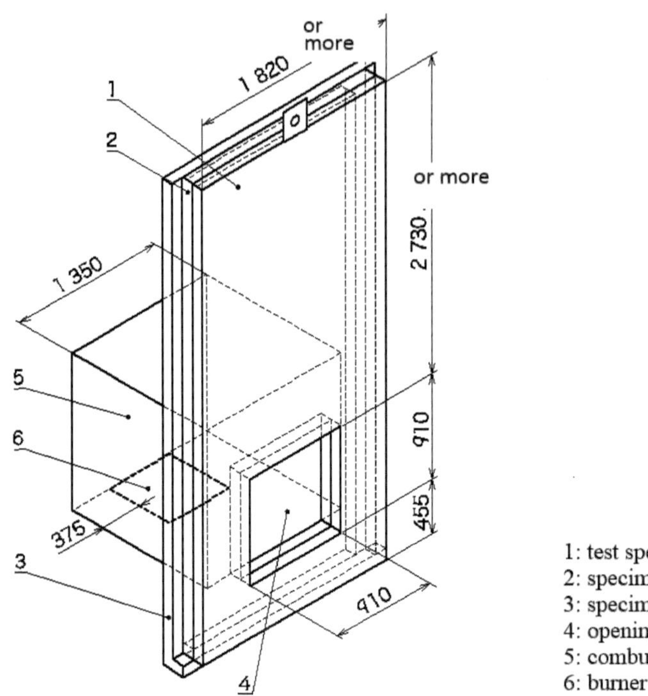

Figure 10.48 Test apparatus and test specimen in JIS A 1310

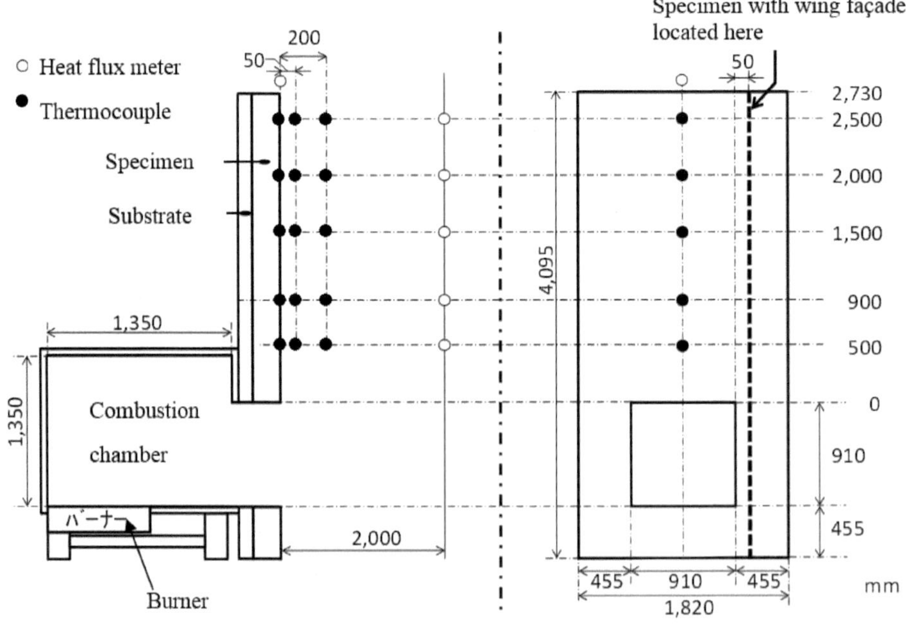

Figure 10.49 Position of measuring apparatus (thermocouples and heat flux meter) (Left: side view, right: front view of main façade)

Table 10.131 JIS A 1310 (Façade Test) Test Specifications

Specimens	Façade (Width: 1.82 m, Height: 4.095 m)
Specimen position	Vertical
Ignition source	Propane gas burner located inside the combustion chamber.
	Approximately 900 kW is found to be suitable for satisfying the requirements in the calibration test.
Test duration	20 min
Conclusions	This is a façade test, which evaluates whether a façade specimen causes the fire to spread widely over the façade, when a flame emerges from an opening of the combustion chamber.

10.4.2.3.2 JIS A 1320:2017 Reaction-to-Fire Test for Sandwich Panel Building Interior Systems – Box Test

As already mentioned, in Japan almost all of the classifications for building interior materials and components are based on the ISO 5660-1 cone calorimeter test. However, in case of sandwich panel specimens containing combustible core material sandwiched by non-combustible surface sheets (such as steel sheets), it is highly likely that most of them will be classified as "non-combustible material", even though they could easily lead to flashover in real-scale room tests such as ISO 13784-1 [16].

The ISO/TS 17431 model box test can still be used in Japan for the classification of "quasi-non-combustible material" and "fire retardant material"; it was originally developed as a screening test method, not for ISO 13784-1 (room test for sandwich panel products), but for ISO 9705-1 (room/corner test). Furthermore, the construction technique at the joints between the adjacent panels cannot be reproduced or evaluated by ISO/TS 17431, which means that it is technically almost impossible to adequately evaluate the reaction-to-fire performance of sandwich panel products by ISO/TS 17431. This is why JIS A 1320:2017 was developed [17, 18] and published [19] by a collaboration of industry, academia, and government fire researchers and experts, with help from international fire professionals. It is based on ISO/TS 17431, but with some technical changes and improvements for the sandwich panel products, with the aim to use it as the screening method for ISO 13784-1. The test specifications are shown in Table 10.132, and a schematic of the specimen box is shown in Figure 10.50.

The internal dimensions for the specimen box in JIS A 1320 (see Table 10.132) are exactly the same as in ISO/TS 17431. One of the major differences is that in JIS A 1320 all the panel joints are finished in the same manner as in the actual construction. Furthermore, in ISO/TS 17431, the external enclosure size is fixed, which means that the thickness of the sandwich panels is limited to approximately 70 mm. On the other hand, in JIS A 1320 (see Figure 10.50), the size of external enclosure is flexible and dependent on the external size of specimen box, where the gap distance between external enclosure and specimen box must be between 40 and 100 mm. This is why there is no physical limitation for the thickness of sandwich panels to be tested by JIS A 1320.

Other major differences between JIS A 1320 and ISO/TS 17431 relate to the propane gas burner, even though the same burner is used. In ISO/TS 17431, the *bottom* of the gas burner is flush with the floor on which the specimen box is located, while in JIS A 1320, the *top* surface of the gas burner is flush with the floor on which the specimen box is located, and the entire gas burner is located beneath the floor. Furthermore, in ISO/TS 17431, the heat input from the gas burner is continuously 40 kW throughout the test, while in JIS A 1320, it is 15 kW for the first 10 min and 30 kW for the last 10 min.

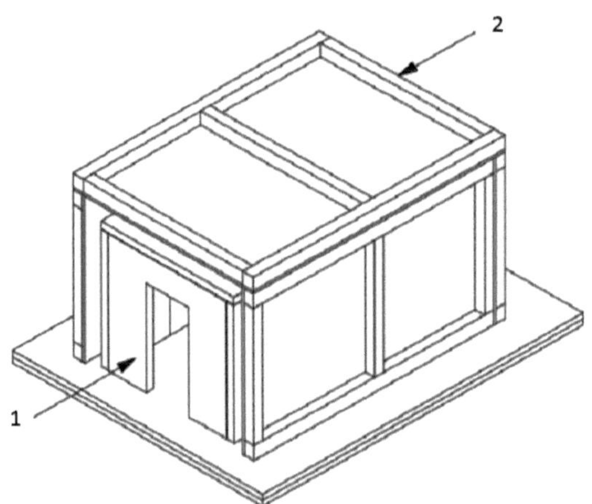

1: Specimen "box" (free-standing or frame-supported) made by sandwich panels with actual construction techniques used at the panel joints, etc.
Internal size: 0.84 (W) × 1.68 (D) × 0.84 m (H), with opening of 0.3 (W) × 0.67 m (H).

2: External "enclosure" made by calcium silicate boards supported by a steel frame.
Its size is flexible and dependent on the external size of the specimen box.

Figure 10.50 Specimen box (made by sandwich panels) and external enclosure in JIS A 1320

Table 10.132 JIS A 1320 (Box Test for Sandwich Panels) Test Specifications

Specimens	Width: 0.84 m, Length: 1.68 m, Height: 0.84 m (internal dimensions)
Specimen position	Box-shaped, with opening of 0.3 (W) × 0.67 m (H)
	The external enclosure covers the entire specimen box
	The gap between the external enclosure and the specimen box is 40–100 mm
Ignition source	The propane gas burner is located in a corner directly opposite the doorway wall.
	Heat output 15 kW for the first 10 min and 30 kW for the last 10 min
Test duration	20 min
Conclusions	Screening test method for reaction-to-fire properties of a box-shaped structure made up of sandwich panels intended for use by producers to reduce the burden of testing in ISO 13784-1

10.4.2.4 Official Approval

10.4.2.4.1 Approval for the Reaction-to-Fire Performance of Building Materials

The following research institutes do not grant approvals, but have academic research activities and give professional advice on reaction-to-fire performance of building materials and components to private-sector organizations who intend to perform indicative fire experiments prior to visiting the designated testing laboratories, where the tests will be performed and final classifications will be given.

- Building Research Institute (BRI)
  Tachihara 1, Tsukuba-shi, Ibaraki-ken, 305-0802 Japan
- National Institute for Land and Infrastructure Management (NILIM),
  Ministry of Land, Infrastructure, Transport and Tourism (MLIT)
  Tachihara 1, Tsukuba-shi, Ibaraki-ken, 305-0802 Japan
- Forest Research and Management Organization
  1 Matsunosato, Tsukuba, Ibaraki, 305-8687 Japan

Official testing and classification of building materials is performed by "performance evaluation bodies" designated by the Ministry of Land, Infrastructure, Transport and Tourism (MLIT) of Japan. Some of them are listed below.

- Japan Testing Center for Construction Materials (JTCCM)
  5-21-20, Inari, Soka-Shi, Saitama-Ken, 304-0015 Japan
- Center for Better Living (BL)
  Tachihara 2, Tsukuba-shi, Ibaraki-ken, 305-0802 Japan
- General Building Research Corporation of Japan (GBRC)
  5-8-1 Fujishirodai, Suita, Osaka, 565-0873 Japan
- Northern Regional Building Research Institute,
  Building Research Department,
  Hokkaido Research Organization (HRO)
  3-1-20, Higashi-ichi-jo, Midori-ga-oka, Asahikawa-shi, Hokkaido,
  078-8801 Japan
- Japan Housing and Wood Technology Center
  3-4-2, Shin-suna, Koutou-ku, Tokyo, 136-0075 Japan

A test report and an evaluation of the test results are issued by the test laboratories described above. The client then brings the results and the evaluation to the Ministry of Land, Infrastructure, Transport and Tourism (MLIT), in order to obtain the approval for the product.

10.4.2.4.2 Approval for Fire Resistance

The procedures and criteria for fire resistance including fire doors have been changed according to the ISO 834 series. There are only small differences left between ISO and Japanese standards. The laboratories listed above also cover fire resistance.

10.4.2.5 Future Developments

Japan is participating actively in the development of ISO standards, such as ISO/TC 92 (SCs 1–4), and is trying as much as possible to cover its regulatory needs based on these standards. The development of a performance-based system will be continued and additional methods introduced. In addition, two major academic fire research projects with international collaboration are currently underway in Japan, with potential scope for future standardization. One project supported by international experts on fire smoke toxicity addresses the development of the evaluation of the toxicity of gases generated from fires occurring inside buildings. Currently, animal testing with mice is still used in Japan, as described above under Section 10.4.2.2. In the research project, alternative analytical methods are investigated. This includes FTIR (Fourier Transform Infrared Spectroscopy) combined with the ISO 5659-2 smoke chamber [20] and the ISO/TS 19700 tube furnace [21].

Another project deals with the weathering issue of FRT (fire-retardant-treated) wood installed at the building façade [22, 23]. As the reaction-to-fire performance of FRT wood deteriorates greatly over time, there is a need for a new test method to evaluate the reaction-to-fire performance after an accelerated weathering process. After studying and performing the already standardized test methods on accelerated weathering such as NT FIRE 053 [24] and CEN/TS 15912 [25], additional improvements have been made to the existing method JSTM J 7001 [26], and it is combined with the JIS A 1310 façade test method. In 2018, standardization work for a new JIS test method was initiated regarding the fire performance of FRT wooden façades after an accelerated weathering process, and the new standard was issued as JIS A 1326:2019 [27]. Hopefully, this new standard test will be broadly used both in academia and industry.

References for Section 10.4.2

[1] Law for Amendment of the Building Standard Law 98 (Law No. 100, 12 June 12 1988), Building Center of Japan (BCJ).

[2] JIS A 1321:1994. Testing method for incombustibility of internal finish material and procedure of buildings.

[3] ISO 5660-1:2015. Reaction-to-fire tests – Heat release, smoke production and mass loss rate – Part 1: Heat release rate (cone calorimeter method) and smoke production rate (dynamic measurement).

[4] "Fireproof performance test and evaluation procedure" is used by "performance evaluation bodies" designated by the Ministry of Land, Infrastructure, Transport and Tourism (MLIT) of Japan, such as the Japan Testing Center for Construction Materials (JTCCM) [in Japanese].

[5] ISO 1182:2010. Reaction to fire tests for products – Non-combustibility test.

[6] Sugahara, S., Yoshida, M., and Ueda, K., "A study on the structure of cone calorimeter as the authorized apparatus", *Interflam conference,* UK (2001).

[7] ISO/TS 17431:2006. Fire tests – Reduced-scale model box test.

[8] ISO 9705-1:2016. Reaction to fire tests – Room corner test for wall and ceiling lining products – Part 1: Test method for a small room configuration.

[9] ISO/TR 16312-2:2007. Guidance for assessing the validity of physical fire models for obtaining fire effluent toxicity data for fire hazard and risk assessment – Part 2: Evaluation of individual physical fire models.

[10] ISO 13785-2:2002. Reaction-to-fire tests for façades – Part 2: Large-scale test.

[11] Yoshioka, H., Ohmiya, Y., Noaki, H., and Yoshida, M., "Large-scale façade fire tests conducted based on ISO 13785-2 with noncombustible façade specimens", *Fire Science and Technology,* 31, pp. 1–22 (2012).

[12] JIS A 1310:2015. Test method for fire propagation over building façades.

[13] Nishio, Y., Yoshioka, H., Noguchi, T., Kanematsu, M., Ando, T., Hase, Y., and Hayakawa, T., "Fire Spread Caused by Combustible Façades in Japan", *Fire Technology,* 52, pp. 1081–1106 (2016).

[14] Zhou, B., Yoshioka, H., Noguchi, T., and Ando, T., "Effects of opening edge treatment and EPS thickness on EPS external thermal insulation composite systems (ETICS) façade reaction-to-fire performance based on JIS A1310 standard façade fire test method", *Fire and Materials*, 42, pp. 537–548 (2018).

[15] Bánky, T., Yoshioka, H., Tóth, P., Nishio, Y., Noguchi, T., Kobayashi, K., Kanematsu, M., Ando, T., Hase, Y., and Hayakawa, T., "Different standard tests of façade fire spread applied for identical aluminum composite panel specimens – MSZ 14800-6 (full-scale) and JIS A 1310 (intermediate-scale)", *Proceedings of 3rd International Seminar for Fire Safety of Façades,* pp. 95–103, Paris, France (2019).

[16] ISO 13784-1:2014. Reaction to fire test for sandwich panel building systems – Part 1: Small room test.

[17] Yoshioka, H., Tanaka, Y., Nishio, Y., Noguchi, T., Kobayashi, K., Ohmiya, Y., Kanematsu, M., Ando, T., Naruse, T., Kagiya, K., and Hayakawa, T., "Self-standing compartment fire tests on sandwich panels", *Fire Science and Technology,* 35, pp. 19–38 (2016).

[18] Yoshioka, H., Noguchi, T., Kobayashi, K., Kanematsu, M., Ando, T., Pareek, S., and Hayakawa, T., "Development of new intermediate-scale box test on sandwich panel products pertaining to the correlation with ISO 13784-1 room test", *Proceedings of Interflam 2019,* pp. 13–24, Interscience Communications (2019).

[19] JIS A 1320:2017. Reaction to fire test for sandwich panel building interior systems – Box test.

[20] Yoshioka, H., Hayakawa, T., Fujimoto, S., Naruse, T., Zhao, X., Tanaike, Y., Noguchi, T., Yoshida, K., and Hase, Y., "Use of FTIR combined with small-scale fire tests as screening test to toxicity test", *Proceedings of the Fire and Materials Conference,* pp. 378–390, San Francisco, USA (2017).

[21] Zhao, X., Yoshioka, H., Noguchi, T., Fujimoto, S., Tanaike, Y., Hayakawa, T., Hase, Y., and Naruse, T., "Fundamental study of gas toxicity with respect to fire stages", *Fire Science and Technology*, 36, pp. 11–24 (2017).

[22] Nakamura, M., Yoshioka, H., Kanematsu, M., Noguchi, T., Hagihara, S., Yamaguchi, A., Shimizu, K., Sugita, T., Matsumoto, Y., Nishio, Y., and Hayakawa, T., "Reaction-to-fire performance of fire-retardant treated wooden façades in Japan with respect to accelerated weathering", *Proceedings of 2nd International Seminar for Fire Safety of Façades,* Lund, Sweden (2016).

[23] Yoshioka, H., Nakamura, M., Kanematsu, M., Nishio, Y., Noguchi, T., and Ando, T., "Experimental study combining accelerated weathering test with fire test regarding fire-retardant-treated wooden façades in Japan", *Proceedings of 3rd International Seminar for Fire Safety of Façades*, pp. 105–113, Paris, France (2019).

[24] NT FIRE 053:2003. Accelerated weathering of fire-retardant treated wood for fire testing.

[25] CEN/TS 15912:2012. Durability of reaction to fire performance – Classes of fire-retardant treated wood-based product in interior and exterior end use applications.

[26] JSTM J 7001:1996. Test method for thermal deformation and durability of real-scale exterior wall, Japan Testing Center for Construction Materials (JTCCM) [in Japanese].

[27] JIS A 1326:2019. Test method for accelerated weathering of fire-retardant treated wood products for façades [in Japanese].

10.4.3 Republic of Korea

Jungmin Choi

10.4.3.1 Statutory Regulations

In Korea, the fire safety regulations for buildings fall under the remit of two different authorities, which are the Ministry of Land, Infrastructure and Transport (MOLIT), and the National Fire Agency (NFA).

BUILDING ACT

The Building Act [1] specifies regulations for the safety, function, environment, and appearance of buildings. Article 52 of the Act stipulates finishing materials for buildings, and that to meet the standards prescribed by MOLIT [2], the following finishing materials shall be fireproof:

Part (a): Interior finishing materials, such as walls and ceilings (and roofs in cases where a building does not have a ceiling). These apply to any building, limited in use and scale as prescribed by Presidential Decree.

Part (b): Finishing materials used for the outer walls of buildings. These shall be prescribed by Presidential Decree.

Part (c): Floor finishing materials for bathrooms, toilets, public baths, etc. (these additionally shall meet the criteria for anti-slip surfaces prescribed by MOLIT).

The Presidential Decree [2] defines for which type of buildings Article 52 (a) of the Building Act is valid. Excluded are buildings whose main structural part is fireproof or made of non-combustible materials, and which have fireproof partitions separating every 200 square meters the floor area or living space, if no sprinklers – or similar automatic fire extinguishing equipment – are installed.

ACT ON FIRE PREVENTION AND INSTALLATION, MAINTENANCE, AND SAFETY CONTROL OF FIRE-FIGHTING SYSTEMS [3]

This Act describes the responsibilities and obligations of the State and local governments for fire prevention and safety control, the installation and maintenance of fire-fighting systems, and the safety control of objects by fire services in order to protect lives and property against fires, disasters, calamities, and other emergencies.

Article 12 of the Act regulates the flame retardancy of objects used for interior decoration, which must meet or exceed certain performance standards as prescribed by the Presidential Decree of fire service.

A Notice of the National Fire Agency specifies the standard of flame retardant performance relevant to flame retardant products regulated by the Act. Products which shall have fire retardant performance according to the Act are:

> Carpets, curtains, blinds, wall coverings and films, wood products, synthetic resin boards and sheets, sofas and chairs, and interior decoration regulated by other relevant acts.

10.4.3.2 Fire Performance Classification and Testing of Building Materials

10.4.3.2.1 Finishing Materials for Buildings

Test methods and performance criteria for building finishing materials are regulated by a Notice of MOLIT [4].

Table 10.133 Fire Classification and Test Methods Applied to Finishing Materials for Buildings

Class	Test methods and test conditions		
	KS F ISO 1182 (Non-combustibility test)	KS F ISO 5660-1 (Heat release test)	KS F 2271 (Fire gas toxicity test)
Non-combustible	Test duration: 20 min	-	Test duration: 15 min
Semi-non-combustible	-	Test duration: 10 min Incident heat flux: 50 kW/m²	Procedures are identical to Japanese mouse test (described in ISO/TR 16312-2)
Fire retardant	-	Test duration: 5 min Incident heat flux: 50 kW/m²	

Fire performance criteria are as follows:

- Non-combustible materials
  a) In-furnace temperatures shall not exceed the final equilibrium temperature by more than 20 °C within the first 20 min of heating

b) The mass loss rate of each specimen after the end of heating shall be 30% or less.

- Semi-non-combustible materials

  a) The total heat released within 10 min of the start of heating shall be 8 MJ/m^2 or less

  b) Within 10 min of the start of heating, no cracks or holes that are harmful in terms of fire preventative performance shall penetrate to the rear side

  c) Within 10 min of the start of heating, the maximum heat release rate shall not exceed 200 kW/m^2 for a period of more than 10 s.

- Fire retardant materials

  a) The total heat released within 5 min of the start of heating shall be 8 MJ/m^2 or less

  b) Within 5 min of the start of heating, no cracks or holes that are harmful in terms of fire preventative performance shall penetrate to the rear side

  c) Within 5 min of the start of heating, the maximum heat release rate shall not exceed 200 kW/m^2 for a period of more than 10 s.

The criteria for the toxicity test are as follows:

A material is not classified as toxic if the average time to the termination of mouse activity (X_s) is 9.0 min or more. This is calculated as $X_s = X - \sigma$, where:

X: Average time (minutes) to termination of activity by 8 mice (if mouse activity does not terminate, the time is set at 15 min)

σ: Standard deviation (minutes) of time to termination of activity by 8 mice (if mouse activity does not terminate, the time is set at 15 min)

10.4.3.2.2 Flame Retardant Materials

The test methods and the criteria for flame-retardant performance are specified by a Notice of the NFA [4]. The combustion apparatus is the same for all materials, but different types of burners and flame application times are applied. The test apparatus is shown in Figure 10.51, and performance specifications are given in Table 10.134.

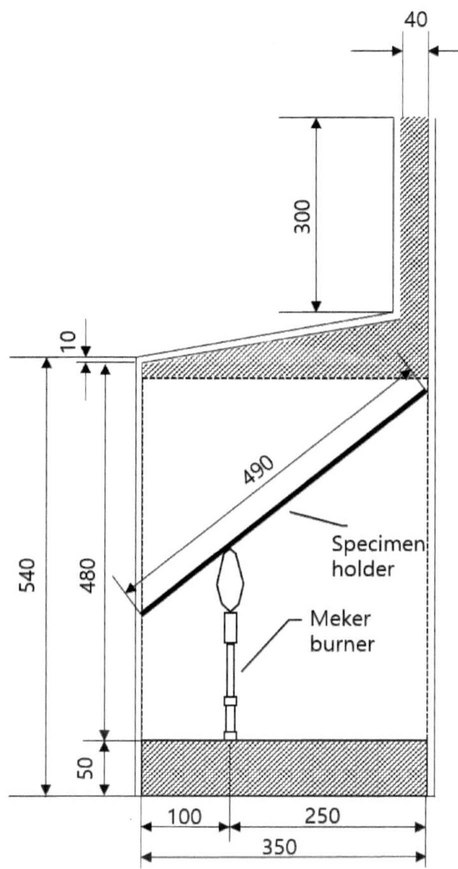

Figure 10.51
Sketch of flame retardant test apparatus using a Meker burner. Measurements are in mm

Table 10.134 Flame Retardant Performance Specifications

Items	Carpet	Thin fabric	Thick fabric	Synthetic resin board, plywood, fiberboard, wood
Specimen dimensions (mm)	400 × 220	350 × 250	350 × 250	190 × 290
Ignition	Air-mix burner LP gas Flame height 24 mm	Micro burner LP gas Flame height 45 mm	Meker burner LP gas Flame height 65 mm	Meker burner LP gas Flame height 65 mm
Burner position	Horizontal	Vertical	Vertical	Vertical
Test duration (s)	30	60	120	120

For carpets, an air-mix burner fed by liquified petroleum gas is used, in which a flame with a length of 24 mm is applied for 30 s. Six specimens of size 400 mm × 220 mm are tested.

Thin fabric is exposed to the flame of a micro-burner for 1 min, while thick fabric is exposed to the flame of a Meker burner for 2 min. If the fabric burns, the flame is removed after 3 s for thin fabric, and 6 s for thick fabric. Three specimens of size 350 mm × 250 mm are tested. Samples that melt are tested in an additional flame contact test after rolling the specimen into a cylinder with a diameter of 100 mm.

The Meker burner is used for testing synthetic resin board, plywood, fiberboard, and wood for a duration of 2 min. Three specimens of size 290 mm × 190 mm are tested.

Sofas and chairs are tested by the glowing cigarette method for 1 h. They must not burn or generate any smoke. The test is identical to the Japanese test method [5] of the Japanese Fire Retardant Association, JFRA. The specimen is 300 mm × 300 mm × 75 mm and the cigarette is located 5 cm from the edge of the specimen. The specimen shall not burn and generate any smoke. In addition, a flame test is made using an air-mix burner for 30 s. Details are included in Notice 2019-2 of the NFA [6].

All objects in Table 10.135 must be tested to their smoke density using ASTM E662 (see Section 11.4.3.3), with a radiant heat source of 2.5 W/cm^2. The maximum specific optical density of the smoke (D_s) should be calculated as the mean of at least four replicate samples.

Table 10.135 Flame-Retardant Performance of Interior Materials

Object	After-flame time (s)	Afterglow time (s)	Char area (cm^2)	Char length (cm)	Smoke density D_s
Carpet	≤ 20	–	–	≤ 10	≤ 400
Thin fabric	≤ 3	≤ 5	≤ 30	≤ 20	≤ 200
Thick fabric	≤ 5	≤ 20	≤ 40	≤ 20	≤ 200
Synthetic resin board	≤ 5	≤ 20	≤ 40	≤ 20	≤ 400
Plywood, fiber-board and wood	≤ 10	≤ 30	≤ 50	≤ 20	≤ 400
Sofas and chairs	≤ 120	≤ 120	–	–	≤ 400

10.4.3.3 Official Approval

The laboratories that can carry out the fire classification tests for building finishing materials as defined by Article 52 of the Building Act are as follows:

- Quality test laboratories, according to Article 25 of the Act on Managing Construction Techniques
- Testing laboratories accredited by the Korea Laboratory Accreditation Scheme (KOLAS).

These laboratories belong to the following organizations:

- FITI Testing & Research Institute (FITI)
- Korea Conformity Laboratories (KCL)
- Korea Testing & Research Institute (KTR)
- Fire Insurers Laboratories of Korea (FILK)
- Korea Marine Equipment Research Institute (KOMERI)
- Korea Quality Institute of Construction Industry (KQICI)
- Korea Construction Material Inspection Company (KCMIC).

Test reports are valid for 1 year after being issued.

10.4.3.4 Future Developments

The limitations of small-scale fire tests are currently being addressed, in particular for building claddings.

Firstly, a new test method and criteria regarding flame spread for building exterior finishing, technically based on the existing British BS 8414 series, was recently published as KS F 8414. New fire performance criteria using this procedure are under development.

Secondly, a large-scale test method for sandwich panels based on ISO 13784-1 is being considered. It is not yet determined whether this test method will replace the conventional small-scale test method and criteria using KS F ISO 5660-1 cone calorimeter, or whether there will be compensation tests in parallel according to manufacturers' requests. In the near future, the Notice of MOLIT is to be revised to improve the fire safety level of buildings, including the large-scale fire test methods.

References for Section 10.4.3

[1] *Korea Building Act*, 2017, *https://elaw.klri.re.kr/kor_service/lawView.do?hseq=45865&lang=ENG.*

[2] *Enforcement Decree of the Building Act*, 2018, *https://elaw.klri.re.kr/kor_service/lawView.do?hseq=46640&lang=ENG.*

[3] Installation, Maintenance, and Safety Control of Firefighting Systems Act, amended 2011.

[4] Notice 2018-771 of Ministry of Land, Infrastructure, and Transport, *Criteria of Fire Performance of Building Finishing Material and Fire Prevention Structure*, 2018.

[5] Clause 5 (Upholstered furniture) of Fire Retardant Product Test Method, Japanese Fire Retardant Association (JFRA), 2018 *https://www.jfra.or.jp/member/pdf/nintei_shokitei/07.pdf.*

[6] Notice 2019-2 of National Fire Agency, *Standards of Flame Retardant Performance*, 2019.

10.4.4 Australia

Alex Webb

10.4.4.1 Statutory Regulations

The Commonwealth of Australia consists of six States - New South Wales, Queensland, South Australia, Tasmania, Victoria, and Western Australia - and two Territories - the Australian Capital Territory where the capital, Canberra, is situated, and the Northern Territory.

Responsibility for the regulation of Australian building resides in the State and Territory Governments. A 1992 agreement between these Governments and the Commonwealth facilitated creation of the Australian Building Codes Board (ABCB) with responsibility to develop a National Construction Code (NCC) [1, 2]. Under this agreement each Government enacted regulations adopting this national code:

- Building Regulations (Australian Capital Territory) [3]
- Building Regulations (Northern Territory) [4]
- Building Regulations (Victoria) [5, 6]
- Local Government Approvals Regulation (New South Wales) [7]
- Building Regulations (Queensland) [8]
- Building Regulations (Western Australia) [9]
- Development Regulations (South Australia) [10]
- Building Regulations (Tasmania) [11].

The NCC is a performance-based code, setting out objectives to be achieved rather than prescribing construction methods. The emphasis is on how a building and its components must perform as opposed to how the building must be designed and constructed. It is structured into Objectives and Performance Requirements (which are mandatory). These are supported by Deemed-to-Satisfy provisions and Verification Methods, which although non-mandatory, if followed result in compliance with the Building Code.

Volume 1 of the NCC applies to larger buildings (building code of australia (BCA) Class 2 to 9 buildings) and Volume 2 applies to detached domestic housing. Volume 3 covers plumbing and drainage for all classes of building. The NCC was amended annually until 2016, when it changed to a three year revision cycle and interim amendments if required. The latest revision published is 2019 Amendment 1.

10.4.4.2 Classification and Testing of the Fire Performance of Building Materials and Components

There are two paths to meeting the Performance Requirements of the BCA:

1. Deemed-to-Satisfy (DtS) Solution. The BCA contains DtS Solutions provisions that are deemed to comply with Performance Requirements.
2. Performance Solution - To use the performance path for materials, it is necessary to demonstrate that a particular material or system achieves the required level of safety. One option is to prove equivalence to the same "level of fire safety in use" as a material or system that meets the DtS provisions. This can be achieved by carrying out comparative tests in any agreed large-scale or small-scale test. Reliance may be on a test method that is not referred to in the BCA.

The test methods used to ascertain whether building materials and components meet the DtS provisions for fire safety are described in the following BCA primary reference standards ("Australian Standards"):

- Combustibility - AS 1530.1 [12]
- Flexible membranes - AS 1530.2 [13]
- Wall and ceiling linings - AS 5637.1 [14] (a classification standard)
- Flooring - AS ISO 9239.1 [15]
- Structural fire resistance - AS 1530.4 [16]
- Insulation and other materials - AS/NZS 1530.3 [17]
- Bushfire - AS 3959 [18, 19].

The standards are available from Standards Australia, GPO Box 476, Sydney, NSW 2001, Australia. Full contact details are available on their website, *www.standards.org.au*.

Other test methods may be applicable for products via secondary or tertiary standards such as the requirements for ductwork testing to UL181 [20] via AS 4254.1 [21].

In AS 1530.2 and AS/NZS 1530.3, various indices enable fire performance to be gradually differentiated. AS 1530.2 applies to thin flexible materials such as roof sheeting, curtains, or wall coverings, while AS/NZS 1530.3 covers insulation and all other materials. AS/NZS 1530.3 is a joint Australian/New Zealand standard.

AS 1530.4 is a test for determining the fire resistance of building components and Fire Resistance Levels (FRLs) provisions in the BCA, and is not dealt with in this book. The method is very similar to the equivalent standards ISO 834, ASTM E119, and BS 467-20. In three specific cases, however, it is of importance for combustible building materials:

1. The BCA Specification C.10 Clause 7 Note 3 permits the use of materials that do not meet the DtS provisions when tested according to Clause C10, provided they are protected on all sides from exposure to air by material that is not combustible and the composite passes a 10 min furnace test. In this test, 1 m^2 specimens of such composite systems are heated for 10 min according to the AS 1530.4 time–temperature curve. This is not a determination of fire resistance, but a demonstration that a combustible core of a composite system will not be involved in a fire for 10 min.
2. Electrical cables that are required to maintain circuit integrity in the event of a fire must be assessed in a test, AS/NZS 3013 [22, 23],which is based on AS 1530.4. The cables, with all connections, junction boxes, etc., are fixed to a concrete slab and exposed to the furnace conditions, where their ability to maintain circuit integrity is assessed.
3. Fire-separating building elements that have been penetrated by plastic pipes or conduits are required to be protected with a penetration system that would reinstate the fire resistance level of the penetrated building elements. The performance of a prototype of the building element with the penetration system is assessed in a fire resistance test, AS 4072.1 [24], which is based on AS 1530.4. In AS 4072.1 it is noted that the material properties of the various plastics impact on the performance of penetration protection devices. The standard therefore requires all pipe sizes and pipe types to be individually tested.

AS 1530.1 Combustibility Test for Materials

The AS 1530.1 test apparatus and procedure for combustibility is essentially the same as ISO 1182 (discussed in Section 10.3.1). However, there are clauses regarding the equipment, method, terminology, and the criteria used for “combustibility”, which may diverge with each revision of the ISO method.

The standard specifies criteria of failure for assessment of combustibility not specified in ISO 1182 [25]. These are based upon three out of the five criteria recommended in ISO 1182.

However, the other two criteria, namely the temperature rise of the thermocouple in the specimen center and the mass loss, are required to be reported in the standard, in order to maintain compatibility with ISO 1182.

The term “non-combustible” is not mentioned in the title or text of the standard. This is to avoid any possible interpretation of the test results that may imply that a material which passes the test is in all circumstances “non-combustible” or is “inert” in every sense of the term.

AS 1530.2 Test for Flammability of Materials

The flammability of thin flexible building materials - in this case fabrics and films - is tested by AS 1530.2. A diagram of the apparatus and the test specifications are given in Figure 10.52 and Table 10.136, respectively.

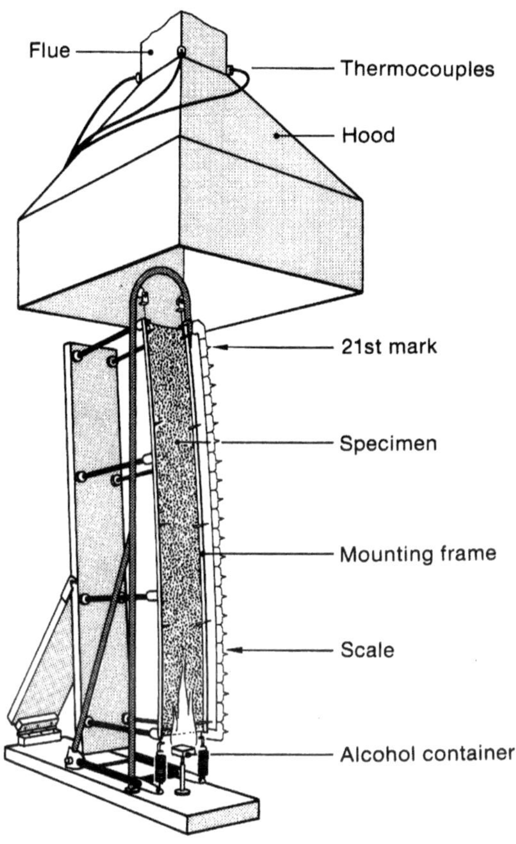

Figure 10.52 Flammability apparatus for fabrics and films

Table 10.136 Flammability Test Specifications

Specimens	Nine specimens, 535 mm × 75 mm
Specimen position	Stretched over a slightly convex frame at 3° to 4° to the vertical
Ignition source	Alcohol flame (0.1 L pure alcohol in copper container) 13 mm below lower edge of specimen
Test duration	Max. 160 s
Conclusion	Flammability index computed from a "speed" factor, heat factor and spread factor

If the material has an asymmetric weave, nine specimens are tested in the warp and nine in the weft direction. If both sides have differing surface structure, then two sets of six specimens are tested with the different surfaces facing outwards.

The material stretched over the frame is ignited by burning 0.1 L of pure ethanol in a copper container 13 mm below the lower edge of the specimen. The temperature of the fire gases is measured by thermocouples positioned 570 mm above the top edge of the specimen in a flue. Observations are made for a maximum of 160 s after ignition of the specimen. If the flame does not reach the highest (21st) mark, the maximum flame height is noted. The temperature of the fire gases is measured at least every 5 s over a period 180 s from ignition of the specimen. The following are also recorded:

- The time taken for the flame to reach the 21st mark, if this occurs before 160 s have elapsed
- The area between the recorded temperature curve for the combustion gases and the ambient temperature curve over the 180 s period.

Factors calculated from the test data are used to compute the flammability index and thus classify the material. These are the Speed factor, Heat factor, and Spread factor, which are calculated as follows:

- The Speed factor S is expressed as:

  $$S = 60 - \frac{3t}{8}$$

  where t is the time in seconds taken by the flame to reach the 21st mark. The mean value from six tests is used in the equation. If the flame does not reach the 21st mark on any specimen, the speed factor is recorded as 0.
- The Heat factor H is obtained from:

  $$H = 0.24A$$

  where A is the mean value determined from six tests of the area between the temperature curve of the combustion gas and the ambient temperature curve over the 180 s test period.
- The Spread factor E is only calculated if the speed factor S equals 0, and is defined as:

  $$E = \frac{20}{9}(D - 3)$$

  where D is the mean scale mark (0–21) reached by the flame determined for six specimens.
- The Flammability index I is obtained via one of the following equations:

  $$I = H + E \text{ (used if the flame does not reach the } 21^{st} \text{ mark)}$$

  $$I = H + S \text{ (used if the flame does reach the } 21^{st} \text{ mark)}$$

AS/NZS 1530.3 Simultaneous Determination of Ignitability, Flame Propagation, Heat Release and Smoke Release (Known as the Early Fire Hazard Test)

The fire behavior of building materials is assessed by AS/NZS 1530.3. The apparatus is illustrated in Figure 10.53 and the test specifications are given in Table 10.137. Normally, nine specimens are tested.

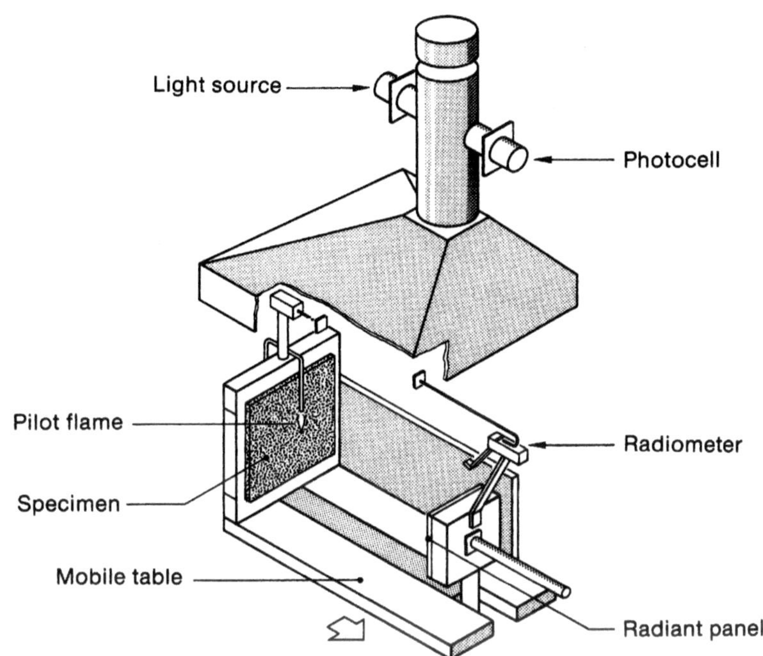

Figure 10.53 Early Fire Hazard Test apparatus

Table 10.137 AS/NZS 1530.3 Specifications

Specimens	Nine specimens, 600 mm × 450 mm × usual thickness
Specimen position	Vertical mobile specimen holder, positioned parallel to radiant panel, moving during test from 850 mm to 175 mm from radiant panel
Ignition source	Vertical gas radiator 300 mm × 300 mm, radiation intensity 850 mm from radiator: 2.4 kW/m^2 Two gas pilot flames at a distance of 15 mm from specimen, length of luminous flame portion 12 mm; first pilot flame 50 mm above center of test specimen, second variable for igniting decomposition gases
Test duration	Variable, depending on ignition time and emitted radiation
Conclusion	Increase rate Four indices - Ignitability, Spread of flame, Heat evolved, Smoke developed

In the test, a specimen is vertically positioned on a mobile holder opposite a vertical gas-fired radiant panel. The specimen is moved in steps every 30 s towards the radiant panel for 12 min or until ignition occurs. Two pilot flames are placed near the surface of the specimen to ignite pyrolysis gases produced by the heated specimen. The specimen holder has side shields placed on either side of the specimen to restrict the effects of draught on the burning behavior.

A radiometer viewing the heated face of the specimen records the radiation produced by ignition of the specimen. The radiometer moves with the specimen holder so that a set distance is maintained between the two. The smoke is collected by a hood, located above the radiant panel and specimen, and rises through a vertical duct where its optical density is recorded.

A light source and photoelectric cell are mounted on the flue above the test apparatus, such that a 305 mm long light beam is directed through the flue. The sensitivity of the photocell corresponds to that of the human eye (greatest sensitivity in the range 500–600 nm; only 50% of maximum sensitivity in the range < 400 nm and > 700 nm). The attenuation of the light beam due to smoke gases is recorded every 3 s during the test.

The test results are used to calculate the maximum smoke density for any 1 min period using the following formula:

$$D = \frac{1}{L} \log_{10} \frac{100}{100 - R}$$

where L is the length of the light path (305 mm) and R is the maximum value of the average reduction in percentage transmission of light for any 1 min period of the test. The mean of the optical densities so calculated is used to determine the Smoke developed index (see below).

The time elapsed from the start of the test until ignition is measured. Ignition is considered to have occurred if the flame continues for ≥ 10 s. Four indices are calculated from the test results. These are: Ignitability, Spread of flame, Heat evolved, and Smoke developed.

- Ignitability: The index for the ignitability is 20 minus the time in minutes until ignition of the material. If fewer than three specimens ignite, the Ignitability index is 0.
- Spread of flame: If the radiation intensity increases by more than 1.4 kW/m^2 in less than 203 s from the time of ignition, this time is used to determine the Spread of flame index. (If the ignitability index is zero or the radiation intensity increases by less than 1.4 kW/m^2 within 203 s, the Spread of flame index is 0.)
- Heat evolved: The radiant heat evolved for each sample is calculated as the integral of the difference between the instantaneous radiation intensity and that

just before ignition, for a period of 2 min after ignition. The index is determined from the mean radiation intensity integral. If five or more specimens fail to ignite, the Heat evolved index is 0.

- Smoke developed: The mean of the calculated optical densities is used to determine the Smoke developed index.

If some of the specimens do not ignite, the index is calculated for each of the cases of ignition and non-ignition.

A report listing the results and the indices obtained based on test performance is issued. The ignitability indices run from 0–20 while the other indices run from 0–10. In practice only the Smoke developed and Spread of flame indices are considered in the BCA for the approval of building materials.

AS 5637.1 Determination of Fire Hazard Properties. Wall and Ceiling Linings

AS 5637.1 is applied for the determination of group number, smoke growth rate index ($SMOGRA_{RC}$) and average specific extinction area (ASEA). These parameters are required by the NCC for control of wall and ceiling linings.

The standard is not a test standard and instead references AS/NZS 3837 (Method of test for heat and smoke release rates for materials and products using an oxygen consumption calorimeter) and AS ISO 9705 (Fire tests - full-scale room fire test for surface products). The standard provides calculation methods, guidance on selection of test method, and further guidance on testing specifically for the purposes of determining compliance with the NCC.

The group number of a material is assigned as follows, based on the time to flashover in the room fire test, or a prediction of the time to flashover using the small-scale test method and a numerical model:

- Group 1 - Material that does not reach flashover when exposed to 100 kW for 600 s followed by exposure to 300 kW for 600 s.
- Group 2 - Material that reaches flashover following exposure to 300 kW within 600 s after not reaching flashover when exposed to 100 kW for 600 s.
- Group 3 - Material that reaches flashover in more than 120 s but within 600 s when exposed to 100 kW.
- Group 4 - Material that reaches flashover within 120 s when exposed to 100 kW.

ISO 9705 (1993 or 2003 version) is explicitly referenced. The standard specifies that the Annex A.1 has to be used and that the material must be installed on three walls and the ceiling.

One acceptable prediction method for AS 3837 is based on the work of Kokkala [26]. AS 3837 is very similar to the ISO 5660 cone calorimeter method, with some differences such as determination of end of test time, etc.

AS ISO 9239.1 Reaction to Fire Tests for Floorings – Part 1: Determination of the Burning Behavior Using a Radiant Heat Source

The AS version of this standard is identical to ISO 9239.1. The BCA regulates both the Critical radiant flux and the Smoke development rate. The test is described in Section 10.3.1.

AS 5113 Fire Propagation Testing and Classification of External Walls of Buildings

The standard [27] was written by Standards Australia in response to the issue of combustible façade fires. The standard was first referenced in the BCA Volume 1 2016 Amendment 1 (an out-of-cycle amendment) under verification method CV3 (not a DtS provision). It was revised in 2018, and this included a change to the title [28].

The standard classifies external building wall systems for fire spread by testing in accordance with ISO 13784-2 or BS 8414. The standard details some clarifications to the test method, specifies additional instrumentation, and provides the classification criteria. The acceptance criteria for the fire spread classification EW include:

- Limit of air temperatures to 600 °C 50 mm out from the face at a distance above the test fire
- Limit of internal layer/cavity temperatures to 250 °C at a distance above the test fire
- Flaming or temperature rise on the non-fire side of the specimen
- Flames beyond the extent of the specimen
- Flaming debris burning for more than 20 s
- Total mass of debris fallen from the specimen not more than 2 kg.

The standard also classifies the building-to-building fire spread into four categories (BB80, BB40, BB20, and BB10) based on the ability for the wall system to resist ignition from radiant heat levels. The test method specified is AS1530.4 Appendix B, and the radiant heat level exposures are drawn from BCA Verification methods CV1 and CV2.

AS 3959 Construction of Buildings in Bushfire-Prone Areas

This standard [18, 19] specifies methods of construction and test methods for testing materials for use in bushfire zones. The standard provides a method for a site assessment for a building to calculate the Bushfire Attack Level (BAL), of which there are 6 (BAL Flame Zone, BAL 40, BAL 29, BAL 19, BAL 12.5, and BAL Low). The numerical values relate to the calculated potential radiant heat flux (kW/m^2) from the fire front based on vegetation type, ground slope, etc. The standard specifies the method of construction for each BAL, and products and systems must be tested to conform to the BAL. The fire test methods referenced are:

- AS 1530.8.1 [29] for BAL 12.5 to BAL 40
- AS 1530.8.2 [30, 31] for flame zone.

The standards provide methods for "determining the fire performance of external construction elements when exposed to radiant heat, burning embers, and debris". In Part 1, a representative element is exposed to radiant heat which increases to the BAL level in 20 s and remains at that level for 2 min before decreasing over 10 min. This is achieved by moving the test specimen toward and away from a radiant heat source in a controlled manner. Debris and embers are simulated by application of a timber crib and a hand-held pilot flame.

10.4.4.3 Official Approval

According to the Deemed-to-Satisfy (DtS) provisions of the BCA, one or more of the tests described above must be carried out in order to ascertain the fire performance of building materials. The tests can be performed by an "Accredited Testing Laboratory" which by definition is:

- an organization accredited by the National Association of Testing Authorities (NATA) to undertake the relevant tests; or
- an organization outside Australia accredited to undertake the relevant tests by an authority recognized by NATA through a mutual recognition agreement; or
- an organization recognized as being an Accredited Testing Laboratory under legislation at the time the test was undertaken.

NATA is the National Association of Testing Authorities, Australia. Approval involves regular inspection of the test laboratories by NATA experts. A NATA directory listing recognized test laboratories is available [32]. Laboratories that test fire performance of plastics are not listed separately.

The test laboratory issues a NATA-endorsed report containing the test results. This document is sufficient proof of performance of the building materials. Traditionally only a few specialist products (such as fire doors) required labelling, and most building materials are not marked in Australia. Due to recent events, elements such as glazing, and most recently aluminum cladding panels, now require labelling.

The CodeMark Certification Scheme is a voluntary third-party building product certification scheme that authorizes the use of new and innovative products in specified circumstances in order to facilitate compliance with the Building Code of Australia (BCA). Alternatively, a certification body or a State or Territory accreditation authority can issue a certificate stating that the properties and performance of a building material fulfill specific requirements of the BCA. Plastics for electrical use must also conform to the "Wiring Rules" [33, 34], and are generally required to pass appropriate IEC tests or equivalent.

Two other standards are used in the fire testing and specification of plastics materials, but are not called-up in the BCA and are therefore not mandatory. Both methods are based on research carried out at the CSIRO Division of Building, Construction and Engineering. The first, AS 2122.1 [35], assesses the flame propagation properties of rigid and flexible cellular plastics, in the form of small bars. The method is similar to ASTM D 3014, in that the specimen is examined in a vertical orientation in a metal "chimney" with a Bunsen burner type ignition source. The second, AS 2122.2 [36], assesses the minimum oxygen concentration (LOI) required for flame propagation on small specimens of plastics bars and sheets and textile fabrics. The method is similar to the ISO version of the oxygen index test, except that an incandescent electrically heated igniter is used, instead of a flame, for ignition of the specimen.

Building fire tests for plastics carried out by laboratories in Australia are listed below (the list is a brief summary only, for the full scope visit *nata.com.au*). All are NATA-approved for the tests required by the BCA as noted. The list of tests performed includes some required by building authorities outside Australia.

- Australian Wool Testing Authority (AWTA)
  Textiles Testing Division
  191 Racecourse Road, Flemington, Victoria
  Phone: +613 9371 2400
  E-mail: *awtainfo@awta.com.au*
  Tests: AS 1530.2; AS/NZS 1530.3; AS ISO 9239.1; ISO 5660-1;
  ISO 5660-2 (2003); AS/NZS 3837
- CSIRO
  Infrastructure Technologies
  14 Julius Avenue, North Ryde
  PO Box 310, North Ryde, NSW 1670
  Phone: +612 9490 5666
  Fax: +612 9490 5777
  E-mail: *firesafety@csiro.au*
  Tests: AS 1530.1; AS 1530.2; AS/NZS 1530.3; AS 1530.4; AS/NZS 3013;
  AS/NZS 3837, ISO 5660-1; ISO 5660-2, AS 4072.1; AS ISO 9705;
  UL 181 part 9, AS ISO 9239.1, ASTM E648, AS1530.8 parts 1 and 2, AS 5113,
  AS1905.1
- Warringtonfire
  Unit 2, 409-411 Hammond Road, Dandenong
  Melbourne, Victoria 3175
  Phone: +613 9767-1000
  Fax: +613 9767-1001

E-mail: *globalfire@exova.com*
Tests: AS 1530.4; AS 4072.1; AS/NZS 3013, AS1530.8 parts 1 and 2, AS 5113, AS1905.1

10.4.4.4 Future Developments

AS 3959 is currently under review and may be republished with amendments in 2020.

References for Section 10.4.4

[1] *National Construction Code Series Volume 1, Building Code of Australia 2019 Amendment 1, Class 2 to 9 Buildings,* Australian Building Codes Board, 2019.

[2] *National Construction Code Series Volume 1, Building Code of Australia 2016, Class 2 to 9 Buildings,* Australian Building Codes Board, 2016.

[3] *Building (General) Regulations 2008,* Australian Capital Territory Government, 2008.

[4] *Building Regulations,* Northern Territory Government, 2016.

[5] *Building Regulations 2018,* Victorian State Government, 2018.

[6] *Building Regulations 2006 - S. R. No. 68/2006,* Victorian State Government, 2006.

[7] *Local Government (Approvals) Regulations,* New South Wales Government, 1999.

[8] *Building Regulation 2006,* Queensland Government, 2006.

[9] *Building Regulations 2012,* Government of Western Australia, 2012.

[10] *Development Regulations 2008,* Government of South Australia (Attorney General's Department), 2008.

[11] *Building Regulations 2016,* Tasmanian Government, 2016.

[12] *AS1530.1:1994 (R2016) Methods for fire tests on building materials, components and structures Combustibility tests for materials,* Standards Australia, 2016.

[13] *AS 1530.2 Methods for fire tests on building materials, components and structures - Part 2: Test for flammability of materials,* Standards Australia, 1993.

[14] *AS 5637.1 - Determination of fire hazards properties,* Standards Australia, 2015.

[15] *Australian Standard 9239.1:2003 Reaction to fire test for flooring - determination of the burning behaviour using a radiant heat source,* Standards Australia, 2003.

[16] *AS 1530.4:2014 Methods for fire tests on building materials, components and structures - Part 4: Fire Resistance tests for elements of construction,* Standards Australia, 2014.

[17] *AS/NZS 1530.3:1993 (R2016) Methods for fire test on building materials, components and structures - Simultaneous determination of ignitability, flame propagation, heat release and smoke release,* Standards Australia, 2016.

[18] *Australian Standard 3959:2018 Amd 1 Construction of buildings in bushfire-prone areas,* Standards Australia, 2019.

[19] *Australian Standard 3959 Construction of buildings in bushfire-prone areas,* Standards Australia, 2009.

[20] *UL 181 Factory-Made Air Ducts and Air Connectors,* Underwriters Laboratories, 1996.

[21] *Australian Standard 4254.1 Ductwork for air-handling systems in buildings, Part 1: Flexible ductructures,* Standards Australia, 2012.

[22] *AS/NZS 3013:2005 Electrical installation - Classification of the fire and mechanical performance of wiring system elements,* Standards Australia, 2005.

[23] *AS/NZS 3013:1995 Electrical installation - Classification of the fire and mechanical performance of wiring systems,* Standards Australia, 1995.

[24] *AS4072.1:2005 (R2016) Components for the protection of openings in fire-resistant separating elements - Service penetrations and control joints,* Standards Australia, 2005.

[25] *ISO 1182:2020 Reaction to Fire Tests for Products - Non-combustibility test,* International Organization for Standardization, 2020.

[26] Kokkala, M., Thomas, P., and Karlsson, B., Rate of heat release and ignitability indices for surface linings. *Fire and Materials,* 1993, vol. 17: p. 209–216.

[27] *AS 5113:2016 Fire propagation testing and classification of external walls of buildings,* Standards Australia, 2016.

[28] *AS 5113:2016 (Incorporating Amendment 1) Classification of external walls of buildings based on reaction-to-fire performance,* Standards Australia, 2018.

[29] *AS 1530.8.1 Methods for fire tests on building materials, components and structures - Tests on elements of construction for buildings exposed to simulated bushfire attack - Radiant heat and small flaming sources,* Standards Australia, 2018.

[30] *AS 1530.8.2 Methods for fire tests on building materials, components and structures - Tests on elements of construction for buildings exposed to simulated bushfire attack - Large flaming sources,* Standards Australia, 2018.

[31] *AS 1530.8.2:2007 Methods for fire tests on building materials, components and structures - Tests on elements of construction for buildings exposed to simulated bushfire attack - Large flaming sources,* Standards Australia, 2007.

[32] *Proficiency Testing Directory,* National Association of Testing Authorities, 2018, p. 47.

[33] *AS/NZS 3000:2018 Amd1:2020 Electrical Installations (aka AS/NZS Wiring Rules),* Standards Australia, 2020.

[34] *AS/NZS 3000:2007 Wiring Rules,* Standards Australia, 2007.

[35] *AS/NZS 2122.1:1993 Combustion characteristics of plastics - Determination of flame propagation - Surface ignition of vertically oriented specimens of cellular plastics,* Standards Australia, 1993.

[36] *AS 2122.2:1999 Methods of test for determining combustion propagation characteristics of plastics - Method 2: Determination of minimum oxygen concentration for flame propagation following top surface ignition of vertically oriented specimens,* Standards Australia, 1999.

10.4.5 New Zealand

Peter Whiting

10.4.5.1 Statutory Regulations

The building control system in New Zealand comes under the Building Act 2004 ("the Building Act") [1], which provides the mandatory framework to be followed when undertaking building work in New Zealand.

The Building Act established a three-part framework for building controls:

1. The Building Act, which contains the provisions for regulating building work in New Zealand.
2. The various Building Regulations that contain the mandatory New Zealand Building Code (NZBC) define key terms such as "change the use" and "moder-

ate earthquake", and give details about the processing of building approvals and determinations.

3. The Building Code [2] is contained in Schedule 1 of the Building Regulations 1992. It sets performance criteria that all building work must meet. It covers aspects such as structural stability, fire safety, access, moisture control, safety of users, durability, services and facilities, and energy efficiency.

The technical clauses in the Building Code each have three levels that describe the requirements for the clause:

- Objective: Social objectives the building must achieve.
- Functional requirement: Functions the building must perform to meet the Objective.
- Performance: The performance criteria the building must achieve. By meeting the performance criteria, the Objective and Functional requirement can be achieved.

The New Zealand Building Code is therefore a performance-based code. It sets out objectives to be achieved rather than prescribing construction methods. The emphasis is on how a building and its components must perform, as opposed to how the building must be designed and constructed.

As a consequence, Acceptable Solutions and Verification Methods are not mandatory, so long as compliance with the Building Code is achieved. There is at least one Acceptable Solution or Verification Method for each Building Code clause. In the fire area, Acceptable Solutions (AS) and Verification Methods (VM) set out detailed requirements for Code Clause C1 to C6 (Protection From Fire). The documents are:

- C/VM1 Verification Method for Solid Fuel Appliances and C/AS1 Acceptable Solution for Buildings with Sleeping (residential) and Outbuildings (Risk Group SH)
- C/AS2 Acceptable Solution for Buildings other than Risk Group SH
- C/VM2 Verification Method: Framework for Fire Safety Design.

All NZBC documents including technical guidance notes are free to download from the NZ Government, Ministry of Business Innovation and Employment (MBIE) website, *www.building.govt.nz/building-code-compliance.*

The specific fire related Acceptable Solutions and Verification Methods documents may be downloaded from *www.building.govt.nz/building-code-compliance/c-protection-from-fire/c-clauses-c1-c6/acceptable-solutions-and-verification-methods.*

10.4.5.2 Classification and Testing of the Fire Performance of Building Materials and Components

Appended to the Acceptable Solutions and Verification Methods are listed suitable test and classification methods to confirm construction materials, components, and systems to satisfy the provisions of the Acceptable Solution or Verification Method. Relevant standards referenced by the NZBC are listed below.

Flammability of floor coverings is tested according to ISO 9239-1:2010 (described in Section 9.2). The critical radiant flux is defined in clause C3 of the building code depending on the use and size of the building area.

Flammability of suspended flexible fabrics and membrane structures is tested according to the Australian standard AS 1530.2 (described in Section 10.4.4).

10.4.5.2.1 Properties of Lining Materials

Combustibility of materials is tested according to AS 1530.1 or ISO 1182 (commonly referred to as the non-combustibility test).

Tests for internal surface linings (walls and ceilings) are ISO 5660 Parts 1 and 2, or ISO 9705 (or AS ISO 9705), ISO 13784 Part 1, and EN 13501 Part 1.

Ductwork for air-handling systems in buildings (alternative) is tested to AS 4254 Part 1 (flexible) or Part 2 (rigid). This references AS/NZS 1530.3 and UL 181 for the fire testing requirements.

Classification of Material Group Numbers for Internal Surface Linings

Testing in accordance with ISO 9705 is considered the reference test. The resulting Group Number is determined based on the time to flashover. For the purposes of this test method, flashover is defined to be a total heat release rate (including that of the burner) exceeding 1000 kW). The Group Number for a material or assembly is determined based on the "time to flashover" as follows:

- Group 1 - Does not reach flashover during the test
- Group 2 - Reaches flashover after 10 min, and before 20 min
- Group 3 - Reaches flashover after 2 min, and before 10 min
- Group 4 - Reaches flashover within 2 min.

The average smoke production rate (SPR) is determined over the period from 0 to 20 min for a Group 1 material, and over the period from 0 to 10 min for a Group 2 material. The SPR is not used in relation to a Group 3 or 4 material. Tests in which the calculated smoke production rate is not greater than 5 m^2/s are denoted with an "s" suffix, for example 1-s or 2-s. In buildings where the lining materials require Group Numbers, products achieving the "s" suffix can typically be used in buildings without sprinkler protection.

Alternatively, the determination of a Group Number may be achieved from testing in accordance with ISO 5660 Parts 1 and 2. This test method is only suitable for largely homogenous lining materials. Examples of products that are not suitable for ISO 5660 testing are those that melt or shrink back on exposure to heat, are reliant on jointing details, or are foil-faced. The Group Number is derived from the heat release rate data and the smoke production “s” suffix is determined from the average specific extinction area (ASEA).

Specific details for the calculation method are given in NZBC C/VM2 Appendix A.

10.4.5.2.2 Exterior Cladding Materials

Exterior cladding materials are tested to ISO 5660 Part 1 or AS/NZS 3837, or NFPA 285. Where the combustibility of a timber product is modified through the application of a fire retardant treatment to meet the fire requirements, it is to be subjected to pre-test accelerated weathering in accordance with ASTM D 2898 Method B.

Application of Testing for Exterior Surface Cladding Materials

Testing, when carried out in accordance with ISO 5660 (or AS/NZS 3837), determines the peak heat release rate and total heat release. The test duration is set by the NZBC Acceptable Solutions to be 15 min. The performance of cladding materials is regulated based on building heights and or proximity to relevant boundaries. The permitted application is then restricted in accordance with limits set in NZBC Acceptable Solutions Section 5.8.1 for the relevant Risk Group.

10.4.5.2.3 Foamed Plastics

AS1366 is a classification standard for foamed plastics that references reaction-to-fire testing in accordance with AS 2122. The method is divided into Part 1 (polyurethane), Part 2 (polyisocyanurate), Part 3 (polystyrene - moulded), and Part 4 (polystyrene - extruded).

10.4.5.3 Official Approval

10.4.5.3.1 Fire Testing and Classifications

One or more of the tests described above must be carried out on building materials if conformity to the Acceptable Solutions is to be demonstrated. The tests can be performed by any laboratory. In New Zealand, most laboratories are approved by International Accreditation New Zealand (IANZ). Approval involves regular inspection of the test laboratories by IANZ experts. An IANZ directory that lists recognized test laboratories is published regularly. This can be searched online at *www.ianz.govt.nz*.

Laboratories that test fire performance of plastics are not listed separately. A mutual recognition agreement exists between IANZ and NATA such that tests performed by NATA-accredited laboratories in Australia are also acceptable. Internationally, IANZ is a member of the International Laboratory Accreditation Cooperation (ILAC), such that tests performed by laboratories accredited by other ILAC-recognized accreditation bodies are also acceptable.

The test laboratory issues an IANZ-endorsed report containing the test results. This document is sufficient proof of performance of the building materials. Building materials are not marked or inspected in New Zealand.

10.4.5.3.2 IANZ-Accredited Fire Testing Laboratories

Fire tests carried out by the following laboratories in New Zealand are IANZ-approved for the tests accepted in the Acceptable Solutions unless otherwise indicated.

- BRANZ Ltd.
  1222 Moonshine Road, RD 1, Porirua 5381
  Private Bag 50 908, Porirua 5240
  Phone: +64 4 237 1170
  Fax: +64 4 237 1171
  E-mail: *branz@branz.co.nz*
  Web: *www.branz.co.nz*
  Tests: AS 1530 Part 4, BS 476 Parts 20 to 24, ISO 5660 Parts 1 and 2, AS/NZS 3837, ISO 9705, AS ISO 9705.
- NZ Wool Testing Authority Ltd (Textile testing)
  Corner Bridge and Lever Streets, Ahuriri, Napier 4110
  PO Box 12065, Ahuriri, Napier 4144
  Phone: +64 6 835 1086
  Web: *www.nzwta.com*
  Tests: AS 1530.2.

10.4.5.4 Codemark

Products which have been evaluated and awarded a New Zealand Codemark are deemed to comply with the performance requirements in the NZBC as stated in the Codemark. A Codemark assures the end user/installer of a product that it is fit for purpose, that all necessary testing has been completed satisfactorily, and that the manufacturer/supply chain has been adequately audited to ensure quality is consistent with the samples that were tested.

10.4.5.5 Future Developments

The New Zealand Building Code Acceptable Solutions are continuously updated to reflect advances in technology and changes in standards.

References for Section 10.4.5

[1] Building Act 2004, Public Act 2004 No. 72, 24 August 2004, *http://www.legislation.govt.nz/act/public/2004/0072/latest/DLM306337.html.*

[2] C/AS2, Acceptable Solution for Buildings other than Risk Group SH For New Zealand Building Code Clauses C1–C6 Protection from Fire. Ministry of Business, Innovation and Employment, 27 June 2019, *https://www.building.govt.nz/assets/Uploads/building-code-compliance/c-protection-from-fire/asvm/cas2-protection-from-fire-1st-edition-2019.pdf.*

11 Transportation

11.1 Motor Vehicles

Anja Hofmann and Jonas Brandt

11.1.1 Introduction

The amount of plastic and synthetic materials is still growing in all sectors of everyday life. In the automotive industry, plastic materials are now widely used in vehicles because of their excellent mechanical properties and their light weight combined with low production costs. However, plastic materials are flammable and can generate toxic smoke gases which can endanger passengers' lives in the event of fire.

In road traffic, vehicle fires occur relatively frequently, but fortunately without serious injuries in most cases. Statistics show that fires after a collision tend to be especially hazardous for the passengers. This also was true for two bus fires in France (2015) and Germany (2017) with 43 and 19 fatalities, respectively. In recent years, additional fire safety measures for buses have come into force (e.g., smoke detectors and engine fire suppression systems), which is a big step forward regarding bus fire safety. However, regulations for automotive interior materials have not been adapted to cover the changes in materials in this fast-developing sector. In a severe bus fire 2008 in Germany, the fire possibly started in a cable supplying the on-board kitchen, and developed in the luggage and toilet area. After the toilet door was opened, the fire spread incredibly fast through the cabin, resulting in the death of 20 of the 32 passengers.

Comparing the fire safety regulations for automotive interior materials with regulations for other transport sectors (such as trains, ships, and airplanes) shows that regulations for road vehicle materials are on a much lower safety level. For example, European train regulations are based on a holistic fire safety concept which is lacking for buses, although city buses and trams, coaches, and long-distance trains operate under similar conditions. For cars, fire safety requirements are even more lenient than for buses, perhaps because incidents with cars usually involve far fewer people than for buses. However, findings in the U.S. show that car fires after collisions are much more hazardous and deadly than other car fires. Additionally, a recent fire in Liverpool in a car park destroyed about 1300 vehicles in a short time, and large parts of the building as well. Based on these findings, it appears that the fire safety requirements of cars need re-evaluation.

11.1.2 Background – Vehicle Fires

In 2017, 168,000 highway vehicles (e.g., automobiles, buses, trucks, etc.) caught fire in the U.S., causing 400 fatalities, 1,370 injuries, and leading to $1.4 billion in direct property damage [1]. Since there are so many passenger cars in modern society, it is not surprising that they are the type of vehicle involved in most fires. However, fires in buses or coaches are about 10 times more likely than in passenger cars or trucks. Studies show that fire incidents occur annually in 0.5–1% of all registered buses/coaches [2, 3, 4].

Vehicle fires are typically caused by technical failures of the vehicle, by arson, or as a result of a crash. For vehicles, like for everything else, the general fire conditions presented in the fire triangle apply (Figure 11.1). A fire starts and grows when enough oxygen, fuel, and heat are present.

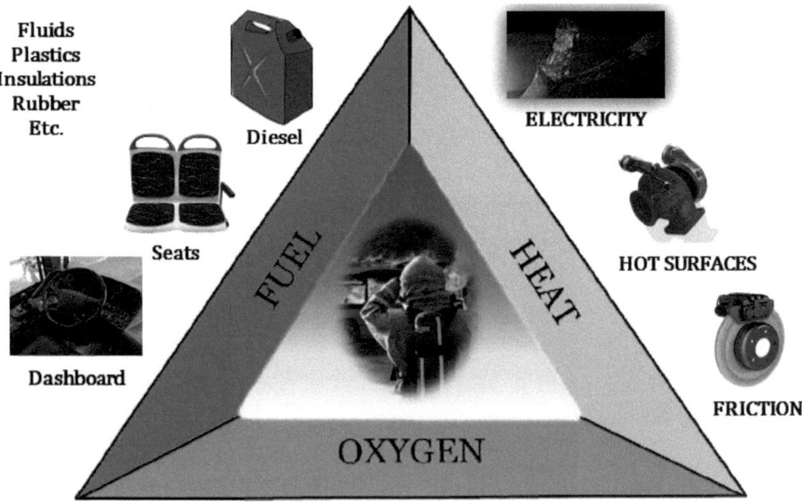

Figure 11.1 The fire triangle with oxygen, fuel, and heat, each representing one side of the triangle

In a vehicle, the oxygen supply will normally be sufficient to maintain a fire. Only fires starting in well-enclosed spaces (e.g., inside a cabin) may self-extinguish due to lack of oxygen. Vehicles also carry a lot of "fuel" in the form of combustible items such as seats, interiors, tyres, foam insulation, cargo, plastic tubes and containers, as well as oil, lubricants and the actual fuel used to propel the vehicle. Ignition sources are parts or components producing heat either at normal operation or when malfunctioning. Anything in the vehicle that can produce enough heat may become an ignition source and start a fire.

In a properly functioning vehicle, heat and fuel are separated. If this separation fails, causing heat and fuel to come into contact with each other, a fire may start. Such failures can (for example) be cable insulations worn so thin that the cable may short-circuit to a grounded part, leaking fuel coming into contact with a hot turbocharger, or frictional heating from malfunctioning brakes. Excluding cases of arson, most vehicle fires start due to electrical failures, hot surfaces in the exhaust system, or through frictional heating from moving parts. Post-collision fires may start following crash-induced failures such as ruptured fuel hoses or electrical short circuits, in addition to the frictional heating from the crash itself.

Many vehicle fires start in the engine compartment. For buses in public transport, up to 60–70% of all fires are likely to be associated with this part of the vehicle. The remaining ones arise in, for example, the tyres, cabin, driver compartment, and luggage compartment [3]. While these statistics are mainly based on vehicles with combustion engines, fires in electrical vehicles are to a greater extent related to batteries.

11.1.3 Statutory Regulations

Generally, the safety requirements for motor vehicles are internationally regulated by the UNECE (the United Nations Economic Commission for Europe) which harmonizes international economic standards under the administrative direction of the United Nations headquarters. In the Sustainable Transport Division of the UNECE, there is a working party called the World Forum for Harmonization of Vehicle Regulations (WP.29) responsible for vehicle regulations. The core of their work is based around the "1958 Agreement", which is a legal framework wherein participating countries (contracting parties) agree on a common set of technical prescriptions for type approval of vehicles and components, officially entitled "UN Regulations", and each contracting party's type approvals are recognized by all other contracting parties.

Most European countries (including the European Union) have signed the agreement, as well as eastern countries like Russia, Ukraine, Croatia, and Turkey but also other countries such as South Africa, Australia, New Zealand, Japan, South Korea, Thailand, and Malaysia. Many countries, even if not formally participating in the 1958 agreement, recognize the UN Regulations and either mirror their content in their own national requirements, or permit the import, registration, and use of UN type-approved vehicles. The United States and Canada are the two significant exceptions, which generally do not recognize UN regulations.

Vehicle fire safety is dealt with in several UN Regulations: Regulation 34 covers the protection of liquid fuel tanks; Regulations 67 and 110 cover gas fuel cylinders; Regulation 107 contains a wide range of fire safety requirements for buses and coaches, including materials and technical equipment for the fire protection; and Regulation 118 addresses the requirements and test methods for bus interior materials. The current fire safety regulations for buses are summarized in Table 11.1, and further details can be found in the next section.

Table 11.1 Fire Safety Requirements for Buses on an International Level

International regulation	Description/title
UN-ECE-R 36	Uniform provisions concerning the approval of large passenger vehicles with regard to their general construction (including fire extinguishers, fuel containers, and fuel feed pipes)
UN-ECE-R 67	Uniform provisions concerning the approval of: I. Specific equipment of vehicles of category M and N [a] using liquefied petroleum gases in their propulsion system II. Vehicles of category M and N fitted with specific equipment for the use of liquefied petroleum gases in their propulsion system with regard to the installation of such equipment
UN-ECE-R 107	Uniform provisions concerning the approval of category M2 or M3* vehicles with regard to their general construction (incl. fire extinguisher, engine compartment and allowed materials in the engine compartment, heat sources, electricity)
UN-ECE-R 110	Uniform provisions concerning the approval of: I. Specific components of motor vehicles using compressed natural gas (CNG) in their propulsion system II. Vehicles with regard to the installation of specific components of an approved type for the use of compressed natural gas (CNG) in their propulsion system
UN-ECE-R 118	Uniform technical prescriptions concerning the burning behaviour of materials used in the interior construction of certain categories of motor vehicles

[a] According to the UNECE Regulations, buses are defined as being vehicles belonging to one of the following categories: **Category M2:** Vehicles used for the carriage of passengers, comprising more than eight seats in addition to the driver's seat, and having a maximum mass not exceeding 5 tonnes; **Category M3:** Vehicles used for the carriage of passengers, comprising more than eight seats in addition to the driver's seat, and having a maximum mass exceeding 5 tonnes. For vehicles of category M2 and M3 having a capacity exceeding 22 passengers in addition to the driver, there are three classes of vehicles to which they belong:
Class I: Vehicles constructed with areas for standing passengers, to allow frequent passenger movement;
Class II: Vehicles constructed principally for the carriage of seated passengers, and designed to allow the carriage of standing passengers in the gangway and/or in an area which does not exceed the space provided for two double seats. For vehicles of category M2 and M3 having a capacity not exceeding 22 passengers in addition to the driver, there are two further classes of vehicles: **Class A:** Vehicles designed to carry standing passengers; a vehicle of this class has seats and shall have provisions for standing passengers. **Class B:** Vehicles not designed to carry standing passengers; a vehicle of this class has no provision for standing passengers.

Within the United States, the National Highway Traffic Safety Administration (NHTSA) issues and enforces Federal Motor Vehicle Safety Standards (FMVSS) that establish performance criteria for new motor vehicles and vehicle equipment. The FMVSS most related to fire safety is FMVSS 302, which includes a fire test for interior materials, developed as long ago as the 1960s. The purpose of introducing the test was to reduce vehicle fires, especially those originating in the interior of the vehicle from sources such as matches or cigarettes. Since then, plastic materials have become basic materials in the automotive industry, but the flammability and the burning behavior of plastic materials has not yet been taken into account in the fire safety requirements. In a modern bus, the fire load of plastic parts installed in the passenger compartment exceeds (for example) the fire load of the filled diesel tanks, but these tanks are well shielded against fire compared to the bus interior materials. The prescribed fire tests for interior materials only consider small ignition sources like cigarettes or lighters, although vehicle fires are mostly the result of defects in the engine compartment, collisions, or defects in electrical and electronical equipment.

Some other FMVSS deal with fire safety aspects. These include FMVSS 217, which sets requirements for bus emergency exits, and FMVSS 304 for gas fuel tank integrity, which includes a bonfire test of CNG fuel tanks. However, in addition to legal requirements, some additional fire safety requirements are specified, for example by transit authorities in the procurement of buses for public transport, or by insurance companies as a requirement for a fire insurance.

11.1.4 Requirements and Tests of Interior Materials for Buses and Cars

The fire tests for bus interior materials in the EU are similar to fire tests for interior materials of road vehicles around the world, since they are all based on the same set of test methods. In Table 11.2 the fire tests for vehicle interior materials according to UN-ECE-R 118 (as well as national and international standards, and manufacturers' specifications) are summarized based on the respective test procedures. For cars, only the test to determine the horizontal burning rate of interior materials is used.

Table 11.2 Fire Tests for Vehicle Interior Materials according to UN-ECE-R 118

Test procedure		Specified in	
		National and international standards	Manufacturers' specifications
Appendix VI	Test to determine the horizontal burning rate of materials	ISO 3795 (Int.) 95/28/EC DIN 75200 (Germany) FMVSS 302 (USA) U.T.A.C. 18-502/1 (France) BS AU 169 (Great Britain) JIS D 1201 (Japan)	GS 97038 (BMW) DBL 5307 (Daimler) FLTM-BN 24-2 (Ford) GM 6090 M (GM) MES DF 050D (Mazda) ES-X60410 (Mitsubishi) PTL 8501 (Porsche) D45 1333 (Renault) STD 5031.1 (Volvo) TL 1010 (VW)
Appendix VII	Test to determine the melting behavior of materials	NF P92-505 (France) U.T.A.C. 18-502/2 (France)	-
Appendix VIII	Test to determine the vertical burning rate of materials	EN-ISO 6941 (Int.)	-

Fire safety requirements for interior materials of motor vehicles are much less stringent than in other transport areas. Although city buses, trams, long-distance buses (coaches), and trains operate similarly, the materials used are quite different. Table 11.3 shows an overview of reaction-to-fire tests used in the different transport sectors, and the parameters addressed by the tests. The table gives an overview of the situation before the recent amendments discussed in 11.1.6.

This comparison shows that most of the required reaction-to-fire tests for interior materials of other transport sectors are not mandatory for buses and cars. The horizontal fire test, which is the basis of the fire safety requirements for buses and cars, is not requested in the other transport sectors. In the other transport sectors, a vertical test is the minimum requirement, and this ensures a basic level of fire safety for all materials that have to meet higher fire performance levels. The heat release rate, smoke production and smoke toxicity are not considered in the regulations for bus and car interiors. Although heat release is a key factor for fire propagation, smoke production and toxicity can result in incapacitation and prevent escape. Extra fire safety requirements for bus seats (e.g., against arson) do not exist.

Table 11.3 Overview of the Fire Tests for Interior Materials in Different Transport Means

	Cars	Buses [UN-ECE-R 118]	Rail vehicles [EN 45545-2]	Ships [SOLAS Chapter II-2]	Aircraft [FAR/JAR/ CS 25.853]
Horizontal burning rate	ISO 3795	ISO 3795 (horizontally mounted components)	No test	No test	FAR/JAR/CS 25.853 b(5) (cabin and cargo compartment)
Vertical burning rate	No test	ISO 6941 (vertically mounted components)	EN ISO 11925-2 (filter materials)	ISO 6940/41 (drapes and hangings)	FAR/ JAR/ CS 25.853 b(4) (cabin and cargo compartment)
Heat release rate	No test	No test	ISO 5660-1 (most materials)	ISO 5660-1 (fire-restricting materials in high-speed craft)	FAR/JAR/CS 25.853(d) (cabin compartment)
Flame spread/ Burning through	No test	No test	ISO 5658-2 (most materials)	ISO 5658-2	FAR/JAR/CS 25.855
Smoke density	No test	No test	ISO 5659-2 (most materials)	ISO 5659-2 (most materials)	FAR/JAR/CS 25.853(d) (cabin compartment)
Smoke gas toxicity	No test	No test	ISO 5659-2 (most materials)	ISO 5659-2 (most materials)	BSS 7239/ ABD 0031 (cabin compartment)
Seats	No test	No test	ISO 9705-2 (passenger seats)	ISO 8191-1/-2 (upholstered furniture)	FAR/JAR/CS 25.853(c) (upholstered furniture)
Flooring	No test	No test	ISO 9239-1 (flooring materials)	IMO Resolution 653 (flooring materials)	FAR Part 25 Appendix F Part VI (flooring materials)

Test to Determine the Horizontal Burning Rate of Materials

The basic reaction-to-fire test for interior materials of buses is a horizontal test procedure to limit the horizontal burning rate of a small flame. The test apparatus is shown in Figure 11.2, and the test specifications are summarized in Table 11.4. The principle of the test is: A horizontal sample is exposed to a Bunsen burner flame in a combustion chamber. The test determines if and when the flame extinguishes, or the time in which the flame passes a measured distance. The specimens are tested in their end-use orientation and conditioned (23 °C/50% relative humidity) for at least 24 hours prior to the test.

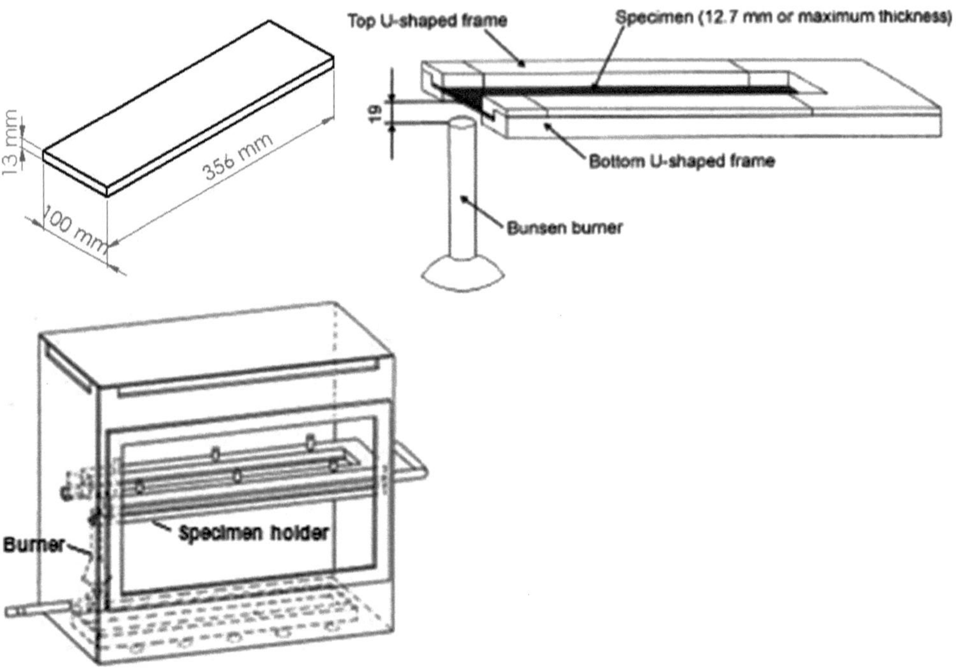

Figure 11.2 Test sample and test rig for Appendix VI of UN-ECE-R 118

Table 11.4 Test Specifications for Appendix VI of UN-ECE-R 118

Specimens	Five specimens, 356 mm × 100 mm × thickness in use Reference marks 38 mm, 292 mm, measured length 254 mm
Specimen position	Horizontal
Ignition source	Bunsen burner, 9 mm diameter, flame height 38 mm
Duration of flame application	15 s
Conclusions	Rate of flame spread over measured length; max. permitted rate of flame spread 100 mm/min

The specimens are fixed in U-shaped specimen holders (see Figure 11.2, right) and are marked at a distance of 38 mm and 292 mm from the leading edge. During the test procedure the horizontal burning rate is determined. The holder is horizontally inserted in a combustion box (see Figure 11.2, bottom) and is exposed to the flame on the leading edge for 15 s by a 38 mm flame from a Bunsen burner. The burning rate between the two marks is determined. If the flame does not reach the second mark the burning distance has to be measured. A specimen passes the burning test if the worst result does not exceed a horizontal burning rate of 100 mm/min.

Test to Determine the Melting Behavior of Materials

Ceiling materials and bordered parts must pass an additional drip test, which is focused on the melting and production of burning drips of the material. The principle of the test is: A sample (conditioned under reference conditions (23 °C/50% relative humidity for at least 24 hours beforehand) is placed in a horizontal position and exposed to an electrical radiator. A receptacle is positioned under the specimen to collect resultant drops. Some cotton wool is put into this receptacle in order to verify if any drop is flaming. In the drip test, the specimen is located on a horizontal grid, which is fixed 300 mm over the bottom and 30 mm below the radiator (an electrical heater that radiates a thermal intensity of 30 kW/m^2 to the specimen; see Figure 11.3 right).

During the test period the ignition and the drip behavior of the specimen and the cotton wool below are monitored for 10 min. If the specimen ignites in the first 5 min of the test, the radiator has to be immediately removed within 3 s until the flame vanishes. After these 5 min of the test or after the flame vanishes, the specimen is exposed to the radiation for another 5 min without stop when the sample ignites again. The requirements are fulfilled if the cotton wool is not ignited by burning drips during the test.

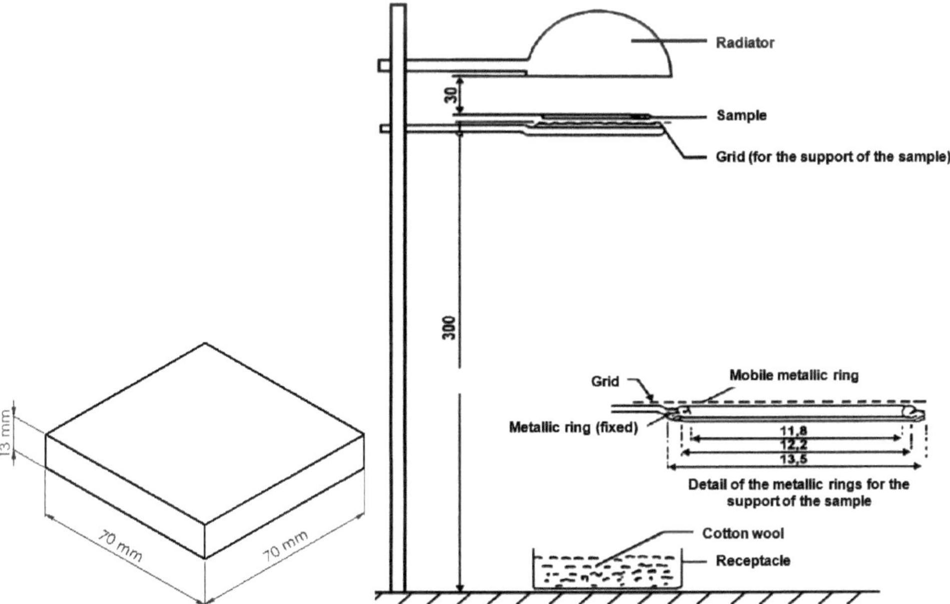

Figure 11.3 Test sample and test rig for Appendix VII of UN-ECE-R 118

Table 11.5 Test Specifications for Appendix VII of UN-ECE-R 118

Specimens	Four specimens, 70 mm × 70 mm × 13 mm, minimum weight 2 g
Specimen position	Horizontal on a grid
Ignition source	Horizontal electric radiator 500 W, radiation intensity on specimen (30 mm from radiator) 30 kW/m²
Receptacle for catching droplets	Contains cotton wool, located 300 mm below the grid
Test duration	10 min
Conclusions	Test passed if cotton wool is not ignited by burning drips during the test

Test to Determine the Vertical Burning Rate of Materials

Drapes, jalousies, hangings, and, after recent amendments, all vertically mounted materials must pass an additional test which limits the vertical burning rate. The dimensions are shown in Figure 11.4, and the test specifications are summarized in Table 11.6. The specimens shall be conditioned in reference conditions (23 °C/50% relative humidity for at least 24 hours) in preparation for the test.

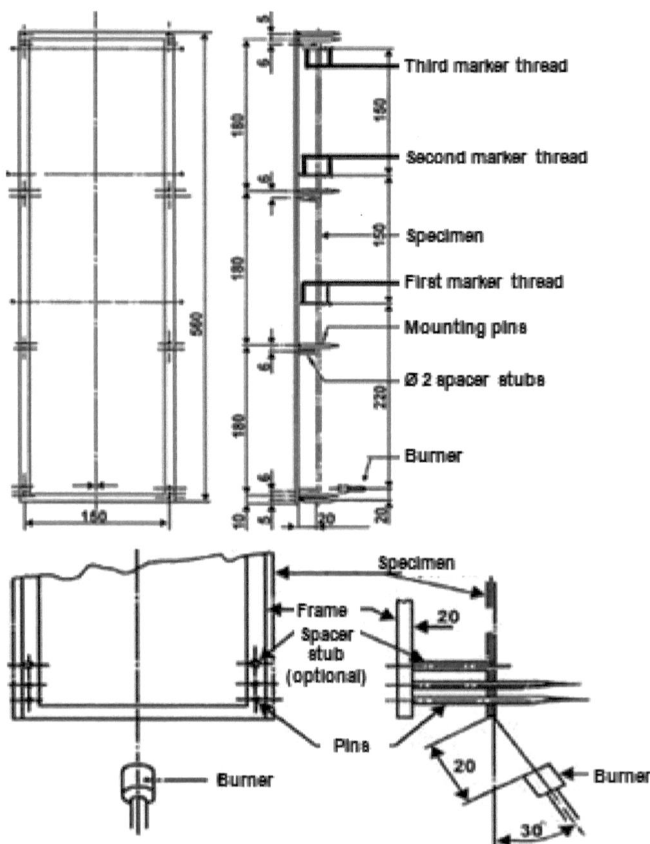

Figure 11.4 Test sample and test rig for Appendix VIII of ECE-R 118

Table 11.6 Test Specifications to Appendix VIII of ECE-R 118

Specimens	Four specimens, 560 mm × 170 mm × max. thickness 13 mm Reference marks 220 mm, 370 mm, 520 mm
Specimen position	Vertical
Ignition source	Small flame burner, inclined at 30°, flame height 40 mm
Duration of flame application	5 or 15 s
Conclusions	Rate of flame spread over measured length; max. permitted rate of flame spread 100 mm/min between both lower marks, or flame extinguishes before reaching last mark

During the test procedure, a 40 mm burner flame is directed towards the bottom edge (see Figure 11.4) for 5 s. If the specimen does not ignite, a second specimen has to be exposed to the flame for 15 s. The sample is marked at heights of 220 mm, 370 mm and 520 mm (see Figure 11.4). The requirements of the test are passed if the fastest vertical burning rate between the both lower marks is less than 100 mm/min, or if the flame extinguishes before reaching the last measuring point.

Alternative Test Procedure Based on ISO 5658-2

To approximate the testing requirements for buses to the more elaborate prescriptions for trains and ships, a recent revision of UN-ECE-R 118 allows the use of the test method ISO 5658-2 relating to the lateral flame spread on building and transportation products in a vertical configuration (see Section 9.2).

The test specimen is mounted vertically on a metal frame, and exposed to a hot radiator panel at an angle of 15° to the specimen. A pilot flame is used to ignite the volatile gases released from the material when heated by the panel.

Materials achieving an average CFE (critical heat flux at extinguishment) value greater or equal to 20 kW/m^2 fulfill the requirements, provided no burning drops are observed when taking the worst test results into account.

Capability of Materials to Repel Fuel or Lubricant

This test is described in Annex 9 of UN-ECE-R 118, and is employed for evaluating the capability of materials to repel liquid. Although this test is not a fire test, it contributes to fire safety by ensuring that insulation materials installed in the engine compartments of buses will not absorb flammable fuel or lubricant.

The material is fixed under a metal cylinder with an inner diameter of 120 mm filled with a standardized liquid (diesel fuel). After 24 hours the liquid is removed and the amount of absorbed liquid is measured.

Examples of products that can be tested with this method are insulation materials installed in the engine compartment and any separate heating compartment.

The requirement for approving the fire performance of a material according to this method is that the increase of weight of the test samples shall not exceed 1 g.

Resistance to Flame Propagation for Electrical Cables

This test, which is part of UN-ECE-R 118, is specified in ISO 6722, and evaluates the fire performance of electrical cables.

The cable of a fixed length is installed on a fixture inclined at 45°. A small flame is applied to the cable, and the burning time and the damaged length of the cable are evaluated.

The combustion flame of the insulating material shall extinguish within 70 s, and a minimum of 50 mm insulation at the top of the test sample shall remain unburned.

11.1.5 Other Fire Safety Requirements for Vehicles

This section describes fire safety requirements that are not directly related to fire testing of interior materials.

UN Regulation No. 34 includes requirements for fire protection of liquid fuel tanks. The tanks must be protected from the consequences of a collision to the front or the rear of the vehicle, and there shall be no protruding parts, sharp edges, etc. near the tank. If the fuel tank is made of plastic, it must withstand a fire test.

The fire performance test is illustrated schematically in Figure 11.5. The fuel tank is filled with fuel to 50% of its nominal capacity, and installed in a test rig or in the vehicle assembly (which should include any parts which may affect the course of the fire in any way). The filler pipe is closed and the venting system should be operative. A flame is applied to the fuel tank in four phases from an open fire trough filled with petrol:

- Phase A (Preheating): The fuel in the fire trough is ignited, and burns for 60 s at a distance of at least 3 m from the tank.
- Phase B (Direct flame exposure): The fuel tank is exposed to the flames of the fully developed fuel fire for 60 s.
- Phase C (Indirect flame exposure): Immediately after completing Phase B, a defined screen is placed between the fire trough and the tank. The latter is subjected to this reduced fire exposure for a further 60 s.
- Phase D (End of test): The fire trough covered by the screen is brought to the initial position. The tank and fire trough are extinguished.

The requirements of the test are met if, at the end of the test, the tank is still in its mount and does not leak.

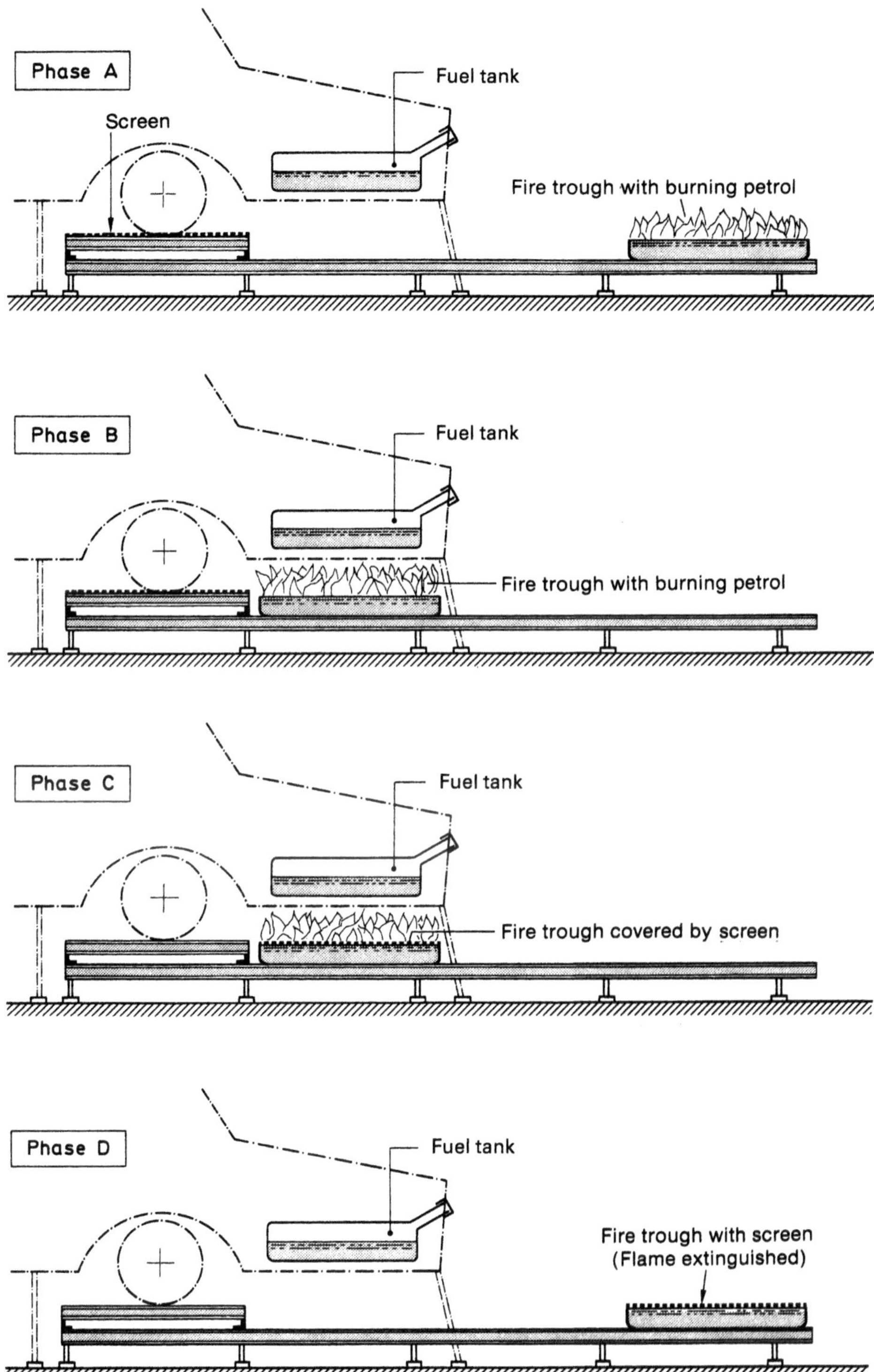

Figure 11.5 Testing the fire performance of plastic fuel tanks

UN Regulations No. 67 and No. 110 include requirements on gas fuel cylinders, such as LPG (Liquified Petroleum Gas), CNG (Compressed Natural Gas), and LNG (Liquified Natural Gas) cylinders. The cylinders must be equipped with pressure release devices, which will release the gas in event of excessive temperature or (in some cases) excessive pressure. The regulations contain fire tests that aim to ensure that the gas tanks do not burst during a fire.

UN Regulation No. 107 covers a wide range of topics for buses, many of them related to fire safety. The main fire-related requirements are:

- The use of sound-proofing materials in bus and coach engine compartments: these shall be non-flammable and not absorb liquid.
- The presence of heat-resistant partitioning between engine compartment (or any heat source) and rest of the bus.
- The safe construction and installation of cables.
- Requirements for fuses.
- The presence of an isolating switch for circuits with a voltage exceeding 100 V.
- The safe and accessible installation of the battery.
- The provision of a space for fire extinguishers and first-aid kits.
- The absence of flammable materials within 10 cm of any potential heat source, such as (for example) exhaust systems or high voltage equipment.
- The use of alarm systems detecting excess temperature in the engine and combustion heater compartment (if the engine compartment is located to the rear of the driver), and alarm systems detecting either smoke or excess temperature in toilet compartments, driver's sleeping compartments, and other separate compartments.
- The use of an automatic fire suppression system in the engine compartment and combustion heater compartment of buses with more than 22 passengers and with the combustion engine to the rear of the driver.

11.1.6 Changes in Regulations

In recent years, the fire safety regulations of buses have been amended several times, especially UN-ECE-R 107 and UN-ECE-R 118. In Table 11.7, recent amendments of these regulations are summarized.

Table 11.7 Amendments of UNECE Bus Fire Safety Regulations

Topic [Provision]	Status [Scheduling]	Implementation dates for new approvals/ registrations
Engine compartment with fire detector [UN-ECE-R 107]	Adopted [WP. 29 March 2010]	31 December 2012/ 31 December 2013
Fire behavior of bus interior (electrical cable, capability of insulation materials to repel fuel) [UN-ECE-R 118]	Adopted [WP. 29 March 2010]	09 December 2012/ 09 December 2015
Fire and/or smoke detector in enclosed compartments (except luggage compartment) [UN-ECE-R 107]	Adopted [WP. 29 November 2011]	26 July 2014/26 July 2015
Fire behavior of bus interior (fire tests in installation position, material tested according to ISO 5658-2 allowed without further testing) [UN-ECE-R 118]	Adopted [WP. 29 November 2011]	26 July 2016 (components) and 26 July 2016 (complete vehicle)/26 July 2017
Fire suppression system in engine compartment [UN-ECE-R 107]	Adopted [WP. 11 July 2016]	18 June 2018/ 18 June 2019

The severe bus fires mentioned previously, combined with several research projects, have led to several changes in the fire safety regulations for buses. As around 70% of fires start in the engine area, fire detectors within the engine compartment are now mandatory. These detectors alert the bus driver in the event of fire, but cannot stop fires in the engine compartment. As a consequence, fire suppression systems in engine compartments became mandatory in 2018/2019.

Fires in hidden areas can be very hazardous. Insulation materials in the engine compartment can absorb combustible liquids, which make them very flammable. This was addressed by regulations regarding the capability to repel fuel. Since it was acknowledged that regulations for train interior materials are stricter than for bus materials, it also became possible to use (without further testing) materials approved to ISO 5658-2 in buses.

11.1.7 Potential Future Changes in Regulations

The introduction of the engine fire suppression systems in 2016 as part of the UN ECE regulations has led to a significant improvement of fire safety in buses. However, fires that do not start in the engine compartment but develop in the cabin – although rare – are extremely dangerous, as fire and smoke spread very rapidly because of the interior materials used. Over recent decades, the amount of plastic in the cabin has grown significantly. In a modern coach, the biggest fire load is often no longer the fuel, but the interior materials [5, 6, 7].

Fire safety requirements for bus interior materials are on a significantly lower level than for materials in other transport vehicles such as trains, ships or airplanes. Although a city bus, a tram, a coach, and a long-distance train operate similarly, the requirements are completely different – for example, interior materials for trains and airplanes have to meet strict requirements for flammability as well as for smoke production and toxicity. For buses the requirements for interior materials are based on regulations developed in the 1960s, when the main concern was ignition due to smoking, leading to tests with relatively small ignition sources.

Ignition from smoking was most relevant some 40 years ago when the test method ISO 3795 was defined, but today other fire sources are more important. Examples of typical fire sources in bus fires today are electrical failures, tyre fires, and fires in the engine compartment. The fact that the ignition source in ISO 3795 is small means that many materials will self-extinguish or not ignite at all. This is indeed good news for materials exposed to small ignition sources, but does not consider fires where the heat attack is larger or large amounts of hot gases enter the passenger compartment. In such cases, most materials will ignite, and the most important point to consider will be how far and how fast the fire will spread. To date, no limits for heat or smoke production or smoke toxicity have been implemented into the regulations, although higher heat release values promote more rapid fire spread. Smoke production and toxicity are also key factors, because smoke reduces the visibility, and together with toxicity may make escape from a vehicle impossible [7, 8].

Recent fires in semi-open car parks have shown that current assumptions for fire safety design in such car parks should be investigated and possibly re-assessed. These assumptions are that fire severity is mitigated by the high ventilation of the premises, and that fires spread to adjacent vehicles every 12 minutes. As a result of these assumptions, there is currently no fire protection for structural steel elements in open car parks, even though severe fires may result in compromising the structural integrity of the building.

In October 2002, a fire occurred near Schiphol airport, with around 30 cars on fire at the same time. In October 2013, a fire raged in the open car park near the

Aquatic Centre of the Sydney Olympic Park, with a hundred cars damaged and 47 completely burnt. Other faster-spreading fires were reported in semi-open car parks and in open air parking lots, the most prominent one probably being the car park fire in Liverpool on New Year's Eve in 2017 [9].

Leakage of the fuel tank after a collision can also lead to a very rapid fire growth. This might have been a factor in the severe bus fire in France in 2015 (with 43 fatalities), and the fire in 2017 on motorway A 9 in Germany. In several bus fires, the reported time from fire detection to full development of the fire has been less than 5 minutes. This time span is too short to allow escape (especially for people with reduced mobility, or when a fire follows a collision), and in most cases insufficient to allow firefighters to intervene.

Given the vast experience with real cases and results from research projects it seems necessary to develop a holistic fire safety concept for buses, as it currently exists for other modes of transport like railways. Such a fire safety concept should include vehicle configuration and design, as well as areas of use, for example, in cities, long-distance traffic on motorways, and tunnels.

References for Section 11.1

[1] Evarts, B., Fire loss in the United States during 2017, NFPA, 2018.

[2] Sundström, B., et al., Bus fire safety (SP Report 2008:4), 2008.

[3] Rakovic, A., Försth, M., and Brandt, J., Bus fires in Sweden 2005–2013 (SP Report 2015:43) , 2015.

[4] Dülsen, S., and Hofmann, A., Study on smoke production, development and toxicity in bus fires (BAM final report, research ref. no. FE 82.0377/2009), 2013.

[5] Johansson, P., and Axelsson, J., Fire safety in Buses – WP2 report: Fire safety review of interior materials in buses (SP report), 2006.

[6] Tewarson, A., Quintiere, J., and Purser, D., FM Global technical report (project ID 0003018009): Post collision motor vehicle fires (Volume I), Theory and testing for the fire behavior of materials for the transportation industry (Volume II), Thermophysical and fire properties of automobile plastic parts and engine compartment fluids (Volume III), 2005.

[7] Försth, M., Modin, H., and Sundström, B., A comparative study of test methods for assessment of fire safety performance of bus interior materials. *Fire Mater.* 2013, 37: 350–357.

[8] Digges, K.H., Gann, R.G., Grayson, S.J., Hirschler, M.M., Lyon, R.E., Purser, D.A., Quintiere, J.G., Stephenson, R.R., and Tewarson, A., Human survivability in motor vehicle fires. *Fire Mater.* 2008, 32: 249–258.

[9] Márton, T., Dederichs, A., and Giuliani, L., Modelling of fire in an open car park. *Applications of Structural Fire Engineering*, 15–16 October 2015, Dubrovnik, Croatia.

11.2 Rail Vehicles

11.2.1 Introduction

Torben Kempers

Over the years, weight reduction has become increasingly important for railway rolling stock. This might sound surprising, because a sturdy and heavy undercarriage is essential for stability. However, in many countries and on many specific tracks, maximum weights of the full trains (and the individual railway carriages) are applied. The most important reasons for these limitations are to reduce wear of the tracks and to use energy more efficiently.

Nowadays, an ever-increasing amount of equipment is built into the railway carriages, such as heating, ventilation, and air conditioning systems, travelers' infotainment systems, etc. This adds additional weight to the carriage, so in order to stay below the maximum allowed weight, there is a need to reduce weight on other parts inside (and outside) the railway carriages.

One of the obvious ways to reduce weight is to produce as many parts as possible out of plastic materials, since plastics have a much lower density than most metals and glass. Examples include seat parts, tables, partition walls, side claddings, etc. However, in turn this will result in increased fire loads inside railway rolling stock.

Fires caused by technical defects generally occur in the externally located traction unit, and normally break out when the train is not in service. As a result, the financial loss can be large, but people are not affected.

Fires inside the carriages are typically started intentionally by passengers as acts of arson or vandalism, using ignition sources such as newspapers and rubbish. Such fires can occur on or below passenger seating, or in luggage situated in luggage racks. Sometimes the textile of the seat is cut open to reveal the foam inside, in order to increase the chance of starting the fire.

In order to reduce the risk of fires inside the railway carriages, many different standards and regulations have been developed that set limits on the flammability of the combustible materials. Typical properties that are covered are surface spread of flame, heat release, smoke opacity (smoke density), smoke toxicity, and secondary effects such as the occurrence of burning drips.

In **Europe,** for many years national fire safety standards have been applied, differing from country to country. Common requirements in these standards include flammability or flame spread, smoke density, and also smoke toxicity. With an ever-increasing number of trains crossing national borders during operation, the need for a Europe-wide harmonized standard became obvious. Under the mandate of the European Commission, the national standards have been harmonized into one set of standards: the EN 45545 series of standards. This can be regarded as a significant step forward towards a more unified fire safety approach in Europe. This series of standards will be further developed and fine-tuned in the years to come.

In the **U.S.** and **Canada,** not much change has been seen since the implementation of the requirements laid out in 49 CFR Part 238 and NFPA 130. These requirements focus on flame spread and smoke density, and in some specific cases on heat release. No requirements are listed relating to smoke toxicity, although large railway manufacturers have developed their own set of requirements, such as Bombardier's SMP 800-C (Toxic Gas Generation from Material Combustion). It is expected that not much will change in the coming years.

In the early days, **China** heavily relied on European train manufacturers, and thereby national fire safety standards have been applied – typically BS 6853 (UK), DIN 5510 (Germany), and NF F 16-101 (France). More recently, China has started to develop its own set of railway standards, building upon the known European national standards. Again, requirements for materials in railway carriages focus on flame spread, smoke density, smoke toxicity, etc. With the implementation of the harmonized railway standard EN 45545 in Europe, China is actively looking into incorporating (parts of) this standard series into their own national TB/T standards. This will be an ongoing process.

The same story also applies to **India** (not covered in this edition), where originally heavy focus was put on fire safety standards such as UIC 564, BS 6853, and DIN 5510. Nowadays, aspects of the EN 45545 series of standards are increasingly being incorporated into national Indian standards.

The testing of materials and components in **Japan** focuses on two types of test methods, involving determination of flammability and smoke density (using the 45° burn test) as well as heat release (using a cone calorimeter). New test methods are being developed, where inputs from existing European test methods are being taken into account. Also, the use of low-smoke seating materials is being investigated, again looking at current European standards.

Fire safety standards for railroads in **Korea** have also been modified, based on the concept of the European EN 45545 series of standards. The major focus is on flame spread, heat release, smoke density and toxicity, and Limited Oxygen Index (a measure of the self-extinguishing capability of a material). As an ongoing process, every year more specific items are being added to the list of items that need to be subjected to fire testing.

Historically, fire safety standards in **Australia** were based on existing standards BS 6853 (UK) and NFPA 130 (U.S.). But with the implementation of the new Australian standard AS 7529, the testing of the fire performance of materials references EN 45545-2 as main source - again underlining the importance of this new fire safety standard. It is expected that national developments will be in line with further developments of the EN 45545 series of standards.

11.2.2 Europe

Torben Kempers

For many years, most countries in Europe had their own national railway standards, describing the requirements with respect to fire safety for each country. In these standards, reference was made to the actual application in the railway rolling stock, as well as the operation mode of the train itself. Such standards differentiated between (for instance) commuter trains, high-speed intercity trains, and trains running underground or in tunnels.

With the formation of the European Community and the fact that an increasing number of trains actually cross several borders within Europe, the need for a harmonized railway standard became clear. In 1991, the European Committee for Standardization (Comité Européen de Normalisation, CEN) and the European Committee for Electrotechnical Standardization (CENELEC) joined forces to develop a single series of harmonized European railway standards. This resulted in the publication of the EN 45545 series, "Railway applications - Fire protection on railway vehicles" [1].

The former national standards dealing with the fire protection of railway vehicles have now mostly been withdrawn or are only used for very limited applications within the specific countries (for example refurbishment of old trains or local transportation networks). Therefore, they are no longer dealt with in this section. For the most important countries in Europe (United Kingdom, Germany, France, and others) the national standards can be viewed in the 3rd edition of this book under Section 11.2 [2].

In the following sections, the harmonized EN 45545 series and related standards will be dealt with in more detail.

11.2.2.1 Harmonized Railway Standard EN 45545

Since 1991, the European Committee for Standardization (CEN) has worked on the development of a single harmonized railway standard for Europe. The aim was to replace all existing national standards by one single standard, covering all aspects related to the fire protection on railways. The underlying thought was that such a harmonized standard would support the free trade within the European Union, and become a means to improve the interlinking and interoperability of the national rail networks.

In the following sections, we will explain the development process, the general structure of the standard, the details of the two most important parts, the listed products and requirements, and finally the next steps that are being undertaken.

11.2.2.1.1 Development Process

The actual work to develop the harmonized railway standard has been undertaken within technical committees CEN/TC 256 "Railway applications" and CLC/TC 9X "Electrical and electronic applications for railways", through the creation of a Joint Working Group (JWG). In this JWG various experts from industry, railway operators, and testing institutes joined forces to define the proper test methods and corresponding specification limits.

In parallel, supporting research work was done in projects funded by the European Commission, such as the FIRESTARR ("Fire Standardisation Research in Railways") research project on the reaction-to-fire performance of products in European trains (funded in 1997), and the TRANSFEU project, focusing on the development of a fire safety-performance based design methodology (funded in 2009).

Because of the complexity of the project, it was not until 2009 that CEN/CENELEC decided to publish the harmonized standard, initially as a technical specification, to gather feedback from industry and rail operators. This harmonized set of specifications was the CEN/TS 45545 series, consisting of seven parts.

In April 2010 the draft standard prEN 45545 series was published, for review and comments from the CEN/CENELEC member countries. In total, more than 500 comments were received, which were all addressed by the JWG in a period of 9 months. This resulted in the publishing of the Final Draft FprEN 45545 series in 2012.

Finally, in March 2013 the final version of the EN 45545 series of standards [1] was published. The so-called "date of withdrawal" was set at 3 years, meaning that EN 45545 would become official in March 2016, at the same time replacing all existing national standards.

Amendments to parts 2 and 5 of the series were published in 2015, comprising some minor editorial and technical modifications or clarifications, without changing the overall requirements as listed in the original 2013 editions.

Technical Specification for Interoperability

The EN 45545 is a *voluntary* series of standards. Only after it is referenced in one of the Technical Specifications for Interoperability (TSIs) it becomes mandatory. These TSIs are law in Europe.

The TSI active from 2014, "rolling stock – locomotives and passenger rolling stock" ("TSI-LOC&PAS") [3], referred to EN 45545-2 (from 2013) in the clause regarding "Measures to prevent fire" and the corresponding Appendix J-1 index 58 (material requirements) and 59 (flammable liquids).

This TSI-LOC&PAS also had a clause (7.1.1.5) on "Transitional measure for fire safety requirement". This clause stipulated that during a transitional period ending three years after the date of publication of this TSI (i.e., ending 31 December 2017), it was permitted to still refer to the former regional railway standards – even though they became obsolete with the publication of EN 45545-2. As of 1 January 2018, this transitional measure for fire safety requirement was no longer valid. As of that date, materials for all new projects plus refurbishments (e.g., new parts, designs, or systems) needed to comply with the EN 45545-2 requirements.

The new version of the TSI-LOC&PAS, published in 2019 [4], only allows the application of current EN 45545-2:2013+A1:2015.

11.2.2.1.2 Structure of EN 45545-1 and -2

As mentioned before, the EN 45545 series consists of seven parts:

- Part 1: General
- Part 2: Requirements for fire behaviour of materials and components
- Part 3: Fire resistance requirements for fire barriers
- Part 4: Fire safety requirements for railway rolling stock design
- Part 5: Fire safety requirements for electrical equipment including that of trolley buses, track guided buses and magnetic levitation vehicles
- Part 6: Fire control and management systems
- Part 7: Fire safety requirements for flammable liquid and flammable gas installations.

Of this series, Parts 1 and 2 are the most relevant to plastics in general and their fire behavior in particular, since they describe (amongst others) measures to minimize the possibility of ignition of materials installed on railway vehicles due to accidents or vandalism.

EN 45545-1

Part 1 covers the principal definitions used throughout the whole series, the Operation Categories and Design Categories, the fire safety objectives, and the general requirements for fire protection measures.

Railway vehicles are classified according to four Operation Categories:

- Operation Category 1: Vehicles for operation on infrastructure where railway vehicles may be stopped with minimum delay, and where a safe area can always be reached immediately (example: urban rail)
- Operation Category 2: Vehicles for operation on underground sections, tunnels and/or elevated structures, with side evacuation available, and where there are

stations or rescue stations that offer a place of safety to passengers, reachable within a short running time

- Operation Category 3: Vehicles for operation on underground sections, tunnels and/or elevated structures, with side evacuation available, and where there are stations or rescue stations that offer a place of safety to passengers, reachable within a long running time
- Operation Category 4: Vehicles for operation on underground sections, tunnels and/or elevated structures, without side evacuation available, and where there are stations (example: London Underground).

Note: the boundary between short and long running times is 4 min.

Additionally, railway vehicles are classified under the following Design Categories:

- A: Vehicles forming part of an automatic train having no emergency-trained staff on board
- D: Double-decked vehicles
- S: Sleeping and couchette vehicles
- N: All other vehicles (standard vehicles).

Clause 4 states that the objectives of EN 45545 are to minimize the probability of a fire starting, to control the rate and extent of fire development, and through this, to minimize the impact of the combustion products on passengers and staff. Here the standard distinguishes between fires resulting from accidental ignition or arson, fires resulting from technical defects, and fires resulting from larger ignition models.

Further, EN 45545-1 defines that when vehicles are being maintained and/or repaired, all items replaced shall either comply with the requirements of the EN 45545 series or shall, as a minimum, be of equivalent performance to the item replaced; all parts and components replaced during refurbishment shall comply with the requirements of the EN 45545 series.

EN 45545-2

In EN 45545-2, Hazard Levels (HL1 to HL3) are determined according to Table 11.8, using the definitions on Operation Categories and Design Categories, as given in Part 1.

Table 11.8 Hazard Level Classification

Operation Category	Design Category			
	N	A	D	S
1	HL1	HL1	HL1	HL2
2	HL2	HL2	HL2	HL2
3	HL2	HL2	HL2	HL3
4	HL3	HL3	HL3	HL3

In practice, around 70–90% of the commercial market is covered by HL2.

In EN 45545-2, the essential fire safety requirements are described as follows: "the design of rolling stock and the products used shall incorporate the aim of limiting fire development should an ignition event occur so that an acceptable level of safety is achieved". The reaction-to-fire performance of materials and components depends on the nature of the base material, but also on the location of the products, their shape and layout, the surface exposed, and the mass plus thickness of the materials. For that reason, all known applications in railway rolling stock have been listed in the table called "Requirements of listed products", which also includes the corresponding set of requirements that these products need to fulfill. In this table the listed products have been classified and differentiated into subgroups, depending on their general location (interior or exterior) and specific use (e.g. furniture, electrotechnical equipment, mechanical equipment).

For those products that have not been listed in the table "Requirements of listed products" in the standard, one either has to follow the so-called "grouping rules", or refer to Table 11.9.

Table 11.9 Requirements for Non-Listed Products According to the Exposed Area and Location in The Vehicle

Exposed area	Location	Requirement set
> 0.20 m^2	Interior	**R1**
	Exterior	**R7**
≤ 0.20 m^2	Interior	**R22**
	Exterior	**R23**

In EN 45545-2, some general principles are also given. Specifications apply for (as examples) cables, multilayer laminates, coatings, etc. Noteworthy in this respect is the following principle: *A test which qualifies any product or surface shall also qualify any product or surface which differs in color and/or pattern.*

This principle is different from the one given in the previous national railway standards. In the past the qualification had to be done on each material + color + thickness combination, making the amount of test work very large. This is now no longer required.

In EN 45545-2, the normative Annexes A through D play an important role. For seats, Annex A describes the standard vandalism test for seat coverings, and Annex B is devoted to the fire test method for complete seats.

Annex C describes in detail the test methods for determination of toxic gases from railway products, and Annex D gives the protocol how to prepare the test specimen for the various tests. These Annexes are an integral part of the 2013 version of EN 45545-2, including its Amendment A1 of 2015.

11.2.2.1.3 Requirements

As mentioned before, EN 45545-2 lists a large number of known applications (products) plus their corresponding sets of requirements. In the following section a selection of products and their requirement sets is described.

Wall Claddings, Ceilings, Partition Walls

Products with a relatively large surface area, such as wall claddings and partition walls (IN1A - Interior vertical surfaces) and ceilings (IN1B - Interior horizontal downward-facing surfaces) are linked to the **R1** set of requirements - which is the most stringent requirement set of this standard. Table 11.10 lists the tests that need to be performed and the classification limits for **R1.**

Table 11.10 **R1** Set of Requirements

Reference	Test	Standard	Parameter	Test criteria	HL1	HL2	HL3
T02	Spread of Flame	ISO 5658-2	*CFE* [kW/m²]	Minimum	20	20	20
T03.01	Heat Release	ISO 5660-1 at 50 kW/m²	*MARHE* [kW/m²]	Maximum	-	90	60
T10.01	Smoke Density	ISO 5659-2 at 50 kW/m²	$D_{s,4}$ [-]	Maximum	600	300	150
T10.02			VOF_4 [min]	Maximum	1200	600	300
T11.01	Toxicity		CIT_G [-]	Maximum	1.2	0.9	0.75

For these large surfaces both the spread of flame (according to ISO 5658-2 [5]) and the heat release (according to ISO 5660-1 [6]) performance are critical, as are the smoke density and toxicity (both according to ISO 5659-2 [7]).

Strips

When smaller strips of material (IN3A - Strips) are mounted on the walls, the spread of flame performance of the material can still be considered as critical, but the heat release performance no longer is. This is reflected in the **R3** set of requirements, listed in Table 11.11, which does not mention any heat-release requirements.

Table 11.11 **R3** Set of Requirements

Reference	Test	Standard	Parameter	Test criteria	HL1	HL2	HL3
T02	Spread of Flame	ISO 5658-2	*CFE* [kW/m²]	Minimum	13	13	13
T03.01	Heat Release	ISO 5660-1 at 50 kW/m²	*MARHE* [kW/m²]	Maximum	-	-	-
T10.01	Smoke Density	ISO 5659-2 at 50 kW/m²	$D_{s,4}$ [-]	Maximum	-	480	240
T10.02			VOF_4 [min]	Maximum	-	960	480
T11.01	Toxicity		CIT_G [-]	Maximum	1.2	0.9	0.75

Seat Shells

When one considers seat shells (F1C – Passenger seat shell – Base, and F1D – Passenger seat shell – Back), it becomes clear that heat release is a critical parameter, but the spread of flame is not. This is because the individual seats are not likely to be able to spread the fire easily. Table 11.12 covers the corresponding **R6** set of requirements, which does not list any spread-of-flame requirements.

Table 11.12 **R6** Set of Requirements

Reference	Test	Standard	Parameter	Test criteria	HL1	HL2	HL3
T02	Spread of Flame	ISO 5658-2	*CFE* [kW/m^2]	Minimum	–	–	–
T03.01	Heat Release	ISO 5660-1 at 50 kW/m^2	*MARHE* [kW/m^2]	Maximum	90	90	60
T10.01	Smoke Density	ISO 5659-2 at 50 kW/m^2	$D_{s,4}$ [-]	Maximum	600	300	150
T10.02			VOF_4 [min]	Maximum	1200	600	300
T11.01	Toxicity		CIT_G [-]	Maximum	1.2	0.9	0.75

Electrotechnical Equipment

For most electrotechnical equipment, including connectors, the dimensions and weights are relatively small. Therefore, it does not make sense to focus on spread of flame and/or heat release performance. Instead, for this class of products a different set of requirements has been defined – **R22.** Table 11.13 lists the requirements for **R22.**

Table 11.13 **R22** Set of Requirements

Reference	Test	Standard	Parameter	Test criteria	HL1	HL2	HL3
T01	Oxygen Index	ISO 4589-2	*OI* [%]	Minimum	28	28	32
T10.03	Smoke Density	ISO 5659-2 at 25 kW/m^2	$D_{s,max}$ [-]	Maximum	600	300	150
T12	Toxicity	NF X 70-100 at 600 °C	CIT_{NLP} [-]	Maximum	1.2	0.9	0.75

In this set of requirements, the Oxygen Index – measured according to ISO 4589-2 [8] – has been listed to characterize the flammability performance of the used material. Spread of flame and heat release are not taken into account.

In this case, the toxicity performance of the material is determined using the French tube furnace test methods NF X 70-100-1 [9] and NF X 70-100-2 [10].

11.2.2.1.4 Test Methods

In this section some of the most important test methods are described, as used in the various sets of requirements according to EN 45545-2.

Spread of Flame

The spread-of-flame test method as described in ISO 5658-2 [5] is based on the test method of the International Maritime Organization (IMO) published as IMO Resolution A.653 (16) [11], and has been developed as an International Standard in order to allow its wider use. The major differences between ISO 5658-2 and the IMO test are that ISO 5658-2 is limited in scope to testing the spread of flame over vertical specimens and does not include the stack for estimating heat release rate. It provides a simple method by which lateral surface spread of flame on a vertical specimen can be determined.

A test specimen is 800 mm long, 155 mm wide, and has a thickness of maximum 50 mm. It is placed in a vertical position adjacent to a gas-fired radiant panel where it is exposed to a defined field of radiant heat flux, with a maximum of 50 kW/m². A pilot flame is situated close to the hotter end of the specimen to ignite volatile gases issuing from the surface (Figure 11.6).

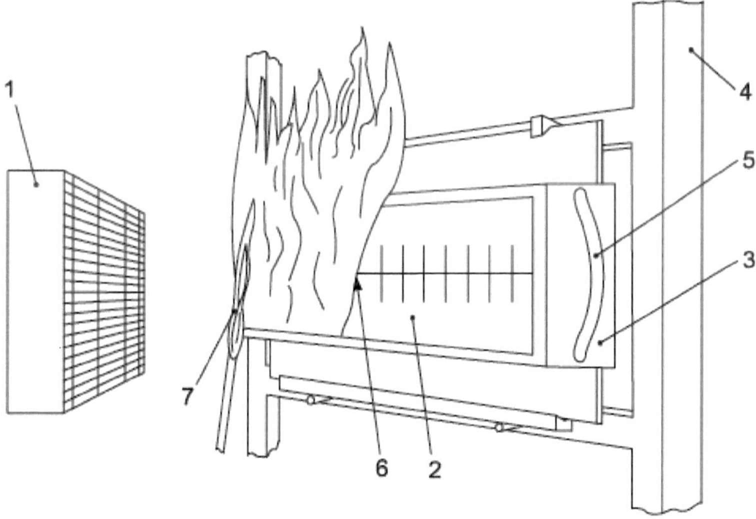

Key
1 vertical radiant panel at an angle of 15° to the specimen
2 specimen
3 specimen holder
4 framework supporting specimen holder
5 handle
6 flame front
7 pilot flame

Figure 11.6 Schematic of test

Following ignition, a record is made of the progression of the flame front horizontally along the length of the specimen in terms of the time it takes to travel various distances. The results are expressed in terms of flame spread distance versus time, flame front velocity versus heat flux, the Critical Heat Flux at Extinguishment *(CFE)* and the average heat for sustained burning.

In the case of EN 45545-2, only the *CFE* value is relevant for obtaining the specific **"R"** classification. It is determined during the test using the graph of heat flux versus distance (Figure 11.7).

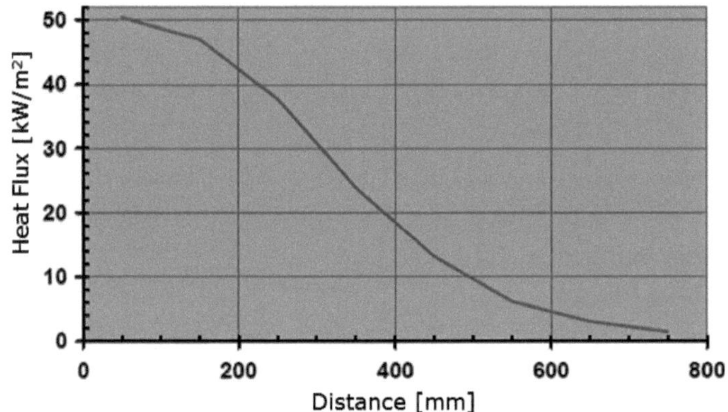

Figure 11.7 Heat flux versus distance in ISO 5658-2

Please note that in the case of spread of flame testing according to ISO 5658-2, the higher value for *CFE* corresponds to the better result. Therefore, in the various sets of requirements *minimum* values for *CFE* are listed.

Heat Release

ISO 5660-1 [6] specifies a method for assessing the heat release rate and dynamic smoke production rate of specimens exposed in the horizontal orientation to controlled levels of irradiance with an external igniter. The heat release rate is determined by measurement of the oxygen consumption, derived from the oxygen concentration and the flow rate in the combustion product stream. The time to ignition (sustained flaming) is also measured in this test.

The dynamic smoke production rate is calculated from measurement of the attenuation of a laser light beam by the combustion product stream. Smoke obscuration is recorded for the entire test, regardless of whether the specimen is flaming or not. For classification according to EN 45545-2, however, this feature is not used.

The test set-up is commonly referred to as the "cone calorimeter" (Figure 11.8).

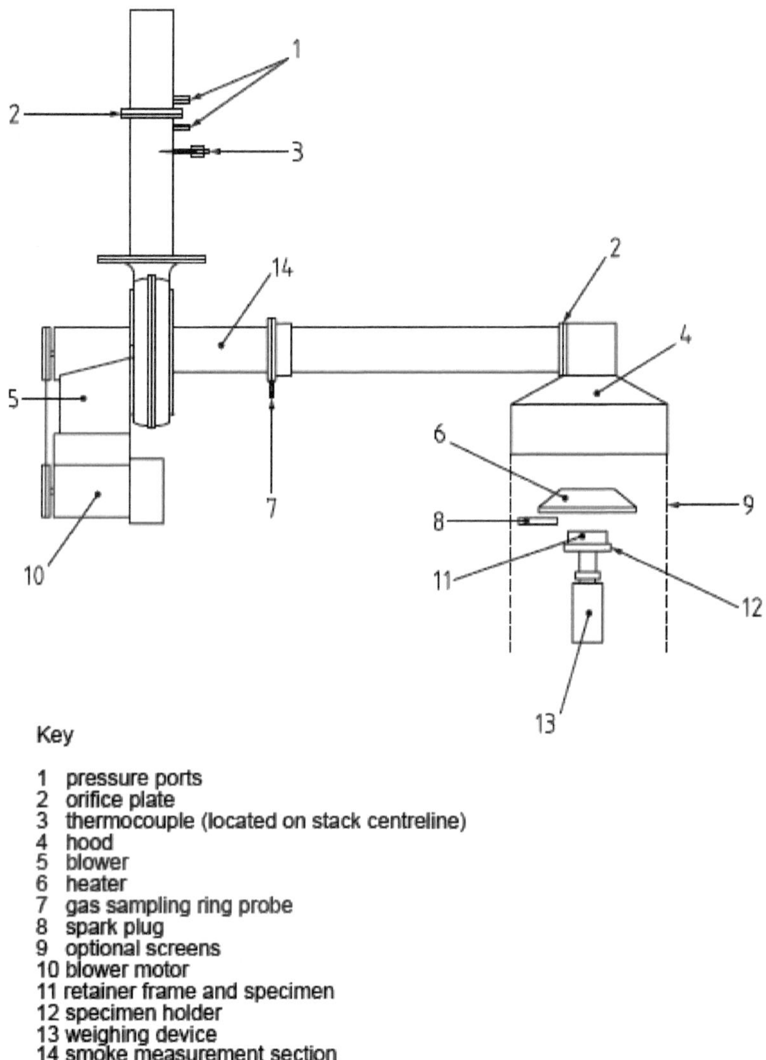

Figure 11.8 Cone calorimeter apparatus

Test samples measure 100 × 100 mm, with a maximum thickness of 50 mm. They are wrapped in a single layer of aluminum foil and placed in the specimen holder. The irradiance level from the cone-shaped radiant electrical heater is either 25 or 50 kW/m^2, depending on the specific requirements in EN 45545-2.

The heat release rate is determined by measurement of the oxygen consumption in the combustion product stream. From this test, the Average Rate of Heat Emission *(ARHE)* is calculated over time, and from the *ARHE* curve the Maximum Average Rate of Heat Emission *(MARHE)* is determined. The obtained value for *MARHE* is used in EN 45545-2 to classify materials according to the respective **"R"** set of requirements. In these cases, a *maximum* value for *MARHE* is listed.

Smoke Density

ISO 5659-2 [7] specifies a method for measuring smoke production from the exposed surface of test specimens of materials and/or products. It is applicable to specimens that have an essentially flat surface and do not exceed 25 mm in thickness. The specimens are placed in a horizontal orientation and subjected to specified levels of thermal irradiance in a closed cabinet with or without the application of a pilot flame. Figure 11.9 shows a typical example of the apparatus.

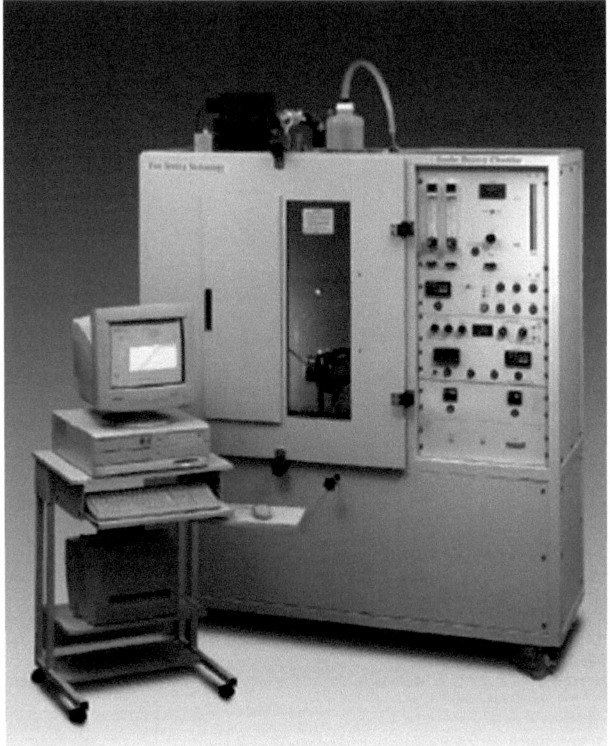

Figure 11.9 Apparatus according to ISO 5659-2

Since smoke production from a material varies according to the irradiance level to which the specimen is exposed, in EN 45545-2 two irradiance levels are listed: 25 kW/m^2 and 50 kW/m^2. The higher irradiance level is typically used for larger surfaces, whereas the lower irradiance level is used for smaller parts.

Test samples have dimensions of 75 × 75 mm, with a maximum thickness of 25 mm. The test specimen is wrapped in aluminum foil. If the thickness of the wrapped specimen is less than 25 mm, then it is backed with a low-density (nominal 65 kg/m^3) refractory fiber blanket. Else, it is tested without a refractory fiber blanket.

The wrapped sample is placed in a holder, which in turn is placed in a horizontal position in the smoke cabinet. The actual test is performed either with a heat flux of 50 kW/m^2 *without* a pilot flame, or with a heat flux of 25 kW/m^2 *with* a pilot flame.

The distance between the sample surface and the cone heater is 25 mm by default. For materials that intumesce more than 10 mm, the distance between the cone heater and the specimen is increased to 50 mm, and the pilot burner is positioned 15 mm below the cone heater bottom edge.

Smoke obscuration is measured by a light beam applied vertically in the chamber. From the light transmission at a given time during the test the specific optical density is calculated:

$$D_s = 132 \log_{10}\left(\frac{100}{T}\right) \quad (11.1)$$

where:
D_s is the specific optical density [-]
132 is a factor specific for the smoke box according to ISO 5659-2
T is the light transmission [%].

From the graph of D_s versus time (t), the following parameters can be derived:

- $D_s(4)$ is the specific optical density after 4 min test time (dimensionless)
- $D_s(\max)$ is the maximum value of the specific optical density reached during the test (dimensionless)
- VOF_4 is the cumulative value of specific optical densities in the first 4 min of the test (in min). VOF_4 is the area under the $D_s(t)$ versus time curve during the test period $t = 0$ to $t = 4$, using a trapezoidal area assumption and a finite element (t) of 1 min:

$$VOF_4 = D_s(1) + D_s(2) + D_s(3) + \frac{1}{2} D_s(4) \quad (11.2)$$

where:
VOF_4 is the cumulative value of specific optical densities [min]
$D_s(i)$ is the specific optical density at $t = i$ [-].

Toxicity

As mentioned before, the toxicity of the combustion gases can be determined in several ways. For general products the test method described in EN 45545-2 Annex C as Method 1 is used. For non-listed products the French test methods NF X 70-100-1 [9] and NF X 70-100-2 [10] are used, described in Annex C as Method 2.

The test apparatus and conditions for Method 1 are described in ISO 5659-2 [7]. During the test, samples of the smoke are taken after 4 and 8 min. These smoke samples are analyzed using Fourier-transform infrared spectroscopy (FTIR). The concentration of eight specific components is determined and compared to reference values, based on the so-called IDLH values (Immediately Dangerous to Life and Health) of the U.S. NIOSH (National Institute for Occupational Safety and Health); see Table 11.14.

Table 11.14 Reference Values of Eight Combustion Components

Combustion component		Reference value [mg/m³]
Carbon dioxide	CO_2	72 000
Carbon monoxide	CO	1380
Hydrogen fluoride	HF	25
Hydrogen chloride	HCl	75
Hydrogen bromide	HBr	99
Hydrogen cyanide	HCN	55
Nitrogen oxide [a]	NO_x	38
Sulfur dioxide	SO_2	262

[a] Includes NO and NO_2.

The reference values of these eight combustion gas components are used to calculate the Conventional Index of Toxicity (CIT_G):

$$CIT_G = 0.0805 \sum \frac{c_i}{C_i} \tag{11.3}$$

where:
CIT_G is the Conventional Index of Toxicity (for general products) [-]
c_i is the concentration of gas component i [mg/m³]
C_i is the reference concentration of gas component i [mg/m³]
0.0805 is a scaling factor specifically calculated for the smoke box according to ISO 5659-2.

The larger value of CIT_G at 4 and 8 min is used for classification according to EN 45545-2.

For non-listed products, the toxicity of the combustion effluents is determined using test methods NF X 70-100-1 (procedure for generation and analysis of fire effluents) [9] and NF X 70-100-2 (test apparatus) [10]. Here too, the same reference values of the eight critical combustion effluents are applied as listed in Table 11.14.

The Conventional Index of Toxicity for non-listed products (CIT_{NLP}) is calculated as follows:

$$CIT_{NLP} = 1 \frac{\text{g}}{\text{m}^3} \sum \frac{Y_i}{C_i} \tag{11.4}$$

where:
CIT_{NLP} is the Conventional Index of Toxicity (for non-listed products) [-]
Y_i is the yield of gas component i in the NF X70-100-1 tube furnace [mg/g]
C_i is the reference concentration of gas component i [mg/m³].

11.2.2.2 Future Developments

Although the EN 45545 series is relatively new, CEN/TC 256-WG01 is already working on first modifications and revisions. As mentioned before, even in 2015 some small modifications to parts 2 and 5 were incorporated into the standards, and published as Amendments.

More recently, work has been undertaken to remove the normative Annexes A ("Standard vandalism test for seat coverings") and B ("Fire test method for seating") from EN 45545-2, as well as the normative Annex C ("Testing methods for determination of toxic gases from railway products"). The aim is to "transform" the EN 45545-2 standard into a true "requirements" standard, without any test method descriptions.

This work has resulted in the publication in 2018 of two new EN standards, namely EN 16989, *Railway applications - Fire protection on railway vehicles - Fire behaviour test for a complete seat* [12], and EN 17084, *Railway applications - Fire protection on railway vehicles - Toxicity test of materials and components* [13].

The "cleaned up" version of EN 45545-2, without the Annexes A–C, has been published in 2020 [14]. In this version reference is made to both new standards (EN 16989 and EN 17084).

In the years to come, more significant updates will be made to all parts of the EN 45545 series, in which some of the as-yet unresolved technical issues will be addressed. This process will most probably take another several years.

References for Section 11.2.2

[1] EN 45545 series, *Railway applications - Fire protection on railway vehicles*
- Part 1: General
- Part 2: Requirements for fire behaviour of materials and components
- Part 3: Fire resistance requirements for fire barriers
- Part 4: Fire safety requirements for railway rolling stock design
- Part 5: Fire safety requirements for electrical equipment including that of trolley buses, track guided buses and magnetic levitation vehicles
- Part 6: Fire control and management systems
- Part 7: Fire safety requirements for flammable liquid and flammable gas installations.

[2] Ebenau, A., Rail Vehicles (Section 11.2), in *Plastics Flammability Handbook* (3rd edition), Troitzsch, J. (ed.), Carl Hanser Verlag, Munich, 2004.

[3] Commission Regulation (EU) No. 1302/2014 of 18 November 2014 concerning a technical specification for interoperability relating to the "rolling stock - locomotives and passenger rolling stock" subsystem of the rail system in the European Union ("TSI-LOC&PAS").

[4] Commission Implementing Regulation (EU) 2019/776 of 16 May 2019 amending Commission Regulations (EU) No. 321/2013, (EU) No. 1299/2014, (EU) No 1301/2014, (EU) No 1302/2014, (EU) No. 1303/2014 and (EU) 2016/919 and Commission Implementing Decision 2011/665/EU as regards the alignment with Directive (EU) 2016/797 of the European Parliament and of the Council and the implementation of specific objectives set out in Commission Delegated Decision (EU) 2017/1474.

[5] ISO 5658-2:2006+A1:2011, *Reaction to fire tests - Spread of flame - Part 2: Lateral spread on building and transport products in vertical configuration.*

[6] ISO 5660-1:2015+A1:2019, *Reaction-to-fire tests - Heat release, smoke production and mass loss rate - Part 1: Heat release rate (cone calorimeter method) and smoke production rate (dynamic measurement).*

[7] ISO 5659-2:2017, *Plastics - Smoke generation - Part 2: Determination of optical density by a single-chamber test.*

[8] ISO 4589-2:2017, *Plastics - Determination of burning behaviour by oxygen index - Ambient-temperature test.*

[9] NF X 70-100-1:2006, *Essais de comportement au feu - Analyse des effluents gazeux - Partie 1: Méthodes d'analyses des gaz provenant de la dégradation thermique.*

[10] NF X 70-100-2:2006, *Essais de comportement au feu - Analyse des effluents gazeux - Partie 2: Méthode de dégradation thermique au four tubulaire.*

[11] IMO Resolution A.653 (16) adopted on 19 October 1989, *Recommendation on improved fire test procedures for surface flammability of bulkhead, ceiling and deck finish materials.*

[12] EN 16989:2018, *Railway applications - Fire protection on railway vehicles - Fire behaviour test for a complete seat.*

[13] EN 17084:2018, *Railway applications - Fire protection on railway vehicles - Toxicity test of materials and components.*

[14] EN 45545-2:2020, *Railway applications - Fire protection on railway vehicles - Part 2: Requirements for fire behavior of materials and components.*

11.2.3 United States of America and Canada

Marc Janssens

11.2.3.1 United States of America

11.2.3.1.1 Federal Regulations

The Federal Railroad Administration (FRA) in the U. S. Department of Transportation (DOT) is responsible for the fire safety regulations for passenger trains. The fire performance requirements for materials, products, and assemblies that are used in the construction of passenger railcars are described in 49 CFR Part 238 [1]. The jurisdiction of the FRA is limited to railroads that are connected to the national network or provide services between locations in different states.

Materials used in constructing a new passenger car or a cab of a locomotive and materials introduced as part of any kind of rebuild, refurbishment, or overhaul of an existing car or cab shall meet the test performance criteria for flammability and smoke emission characteristics as specified in Appendix B to 49 CFR Part 238. In addition, the railroad shall complete a fire safety analysis for new passenger equipment being procured. The purpose of the analysis is to ensure that fire safety considerations and features in the design of the equipment reduce the risk of personal injury caused by fire to an acceptable level in its operating environment. A method for conducting such an analysis has been developed by the American Public Transportation Association (APTA) [2].

The flammability, smoke emission, and fire resistance performance requirements for various types of materials and structural components are given in Table 11.15. The test methods and performance criteria for acceptance are given in columns three and four, respectively. Note that the table refers to specific editions of the standards, e.g., the requirements are based on the 1998 edition of ASTM D3675, the 2001 edition of ASTM E662, etc. These standards may or may not have undergone substantive changes. ASTM E648 and ASTM E119 are briefly discussed in Sections 10.2.1.5 and 10.2.1.8, respectively. The remaining standard test methods are summarized in Section 11.2.3.1.2.

Testing of a seat assembly according to ASTM E1537-99 or a mattress assembly according to ASTM E1590-01 using the pass/fail criteria of CAL TB 133 or CAL TB 129, respectively, are permitted in lieu of testing of seat or mattress components as prescribed in Table 11.15. However, an analysis must be performed to evaluate the risk of vandalism, puncture, cutting, or other acts which may expose the individual components of the assembly to an ignition source. ASTM E1537, CAL TB133, ASTM E1590, and CAL TB 129 are discussed in Section 13.3.

If the surface area of any individual part is less than 100 cm^2 in the end-use configuration, materials used to fabricate such a part may be tested in accordance with ASTM E1354-99 as an alternative to ASTM E162-98 and ASTM E662-01. Testing shall be at 50 kW/m^2 applied heat flux with a retainer frame, and the following performance criteria shall be met: a three-minute average heat release rate less than or equal to 100 kW/m^2, and an average specific extinction area over the same three minutes less than or equal to 500 m^2/kg. ASTM E1354 is described in Section 10.2.1.5.

The ASTM E119 fire resistance period required shall be consistent with the safe evacuation of a full load of passengers from the vehicle under worst-case conditions but shall not be less than 15 minutes. However, a railroad is not required to use the ASTM E119 test method.

Table 11.15 49 CFR, Part 238: Test Procedures and Performance Criteria for the Flammability and Smoke Emission Characteristics of Materials Used in Passenger Cars and Locomotive Cabs

Category	Function of material	Test method	Performance criteria
Cushions Mattresses	All	ASTM D3675-98	$I_s \leq 25$
		ASTM E662-01	$D_s(1.5) \leq 100$
			$D_s(4.0) \leq 175$
Fabrics	Seat upholstery, mattress ticking and covers, curtains, draperies, wall coverings, and window shades	14 CFR 25	Flame time ≤ 10 s
		Appendix F, Part I (vertical test)	Burn length ≤ 155 mm
		ASTM E662-01	$D_s(4.0) \leq 200$
Other Vehicle Components	Seat and mattress frames, wall and ceiling panels, seat and toilet shrouds, tray and other tables, partitions, shelves, opaque windscreens, end caps, roof housings, and component boxes and covers	ASTM E162-98	$I_s \leq 35$
		ASTM E662-01	$D_s(1.5) \leq 100$
			$D_s(4.0) \leq 200$
	Flexible cellular foams used in armrests and seat padding	ASTM D3675-98	$I_s \leq 25$
		ASTM E662-01	$D_s(1.5) \leq 100$
			$D_s(4.0) \leq 175$
	Thermal and acoustic insulation	ASTM E162-98	$I_s \leq 25$
		ASTM E662-01	$D_s(4.0) \leq 100$
	HVAC ducting	ASTM E162-98	$I_s \leq 35$
		ASTM E662-01	$D_s(4.0) \leq 100$
	Floor covering	ASTM E648-00	CRF ≥ 5 kW/m^2
		ASTM E662-01	$D_s(1.5) \leq 100$
			$D_s(4.0) \leq 200$
	Light diffusers, windows and transparent plastic windscreens	ASTM E162-98	$I_s \leq 100$
		ASTM E662-01	$D_s(1.5) \leq 100$
			$D_s(4.0) \leq 200$
Elastomers	Window gaskets, door nosing, inter-car diaphragms, roof mats, and seat springs	ASTM C1166-00	Average flame propagation ≤ 102 mm
		ASTM E662-01	$D_s(1.5) \leq 100$
			$D_s(4.0) \leq 200$
Structural Components	Flooring, other	ASTM E119-00a	Pass

Local Requirements

Rapid transit and light rail vehicles and railcars operated by railroads that are not under the jurisdiction of the FRA have to meet local requirements that are usually based on "NFPA 130: Fixed Guideway Transit Systems" [3]. NFPA 130 includes a table that is very similar to Table 11.15 but has flammability and smoke emission requirements for electrical wires and cables that are not in 49 CFR Part 238.

11.2.3.1.2 Summary of Standard Tests Referred to in 49 CFR Part 238 and NFPA 130

ASTM C1166: Standard Test Method for Flame Propagation of Dense and Cellular Elastomeric Gaskets and Accessories

The test method described in ASTM C1166 measures the flame propagation characteristics of gaskets and accessories. The test specifications are summarized in Table 11.16.

Table 11.16 ASTM C1166 Flame Propagation of Elastomeric Gaskets and Accessories

Specimens	Six specimens, 460 × 25 × 13 mm
Specimen position	Vertical
Ignition source	Bunsen burner with 38 mm high blue flame applied for 15 min (dense materials) or 5 min (cellular materials)
Test duration	Time until the flame ceases to propagate
Conclusions	Average flame propagation based on unburnt lengths from six repeat tests

A 460 × 25 × 13 mm specimen is mounted in the vertical position within the test chamber. It is exposed at the bottom end to a 38 mm blue flame of a Bunsen burner supplied with natural gas for 15 min when evaluating a dense material, or 5 min when a specimen of a cellular material is tested. The velocity of the air flow through the test chamber shall not exceed 18.3 m/min. The test is terminated when the flame ceases to propagate. The flame propagation is equal to the initial specimen length minus the residual unburnt length at the end of the test. The reported flame propagation is the average of the values obtained in six tests.

ASTM E162: Standard Test Method for Surface Flammability of Materials Using a Radiant Heat Energy Source

The test method described in ASTM E162 evaluates the surface flammability characteristics of the test material. The apparatus uses the same radiant heat panel as specified in ASTM E648 (see Section 10.2.1.5). However, the panel is mounted vertically and directed at the specimen, which is inclined at 30° to the panel as shown in Figure 11.10. A pilot flame is located at the top edge of the specimen. The result is an index, I_s, which is calculated from the flame spread rate down the specimen and the temperature rise in the stack. The test specifications are summarized in Table 11.17.

Table 11.17 ASTM E162 Surface Flammability Test Specifications

Specimens	Four specimens, approximately 152 × 457 mm × thickness of use up to 25 mm
Specimen position	Inclined at a 30° angle from vertical
Ignition source	Vertical premixed gas-fired radiant heater, panel surface temperature 805 °C, 305 × 457 mm, top edge at 120 mm from the specimen, 51–75 mm acetylene pilot flame at specimen top edge
Test duration	The test is complete when the stack thermocouples have reached a maximum
Conclusions	The final result is a radiant panel index (I_s), which is equal to the product of a factor derived from the flame spread rate down the specimen (F_s) and a factor derived from the temperature rise in the stack (Q)

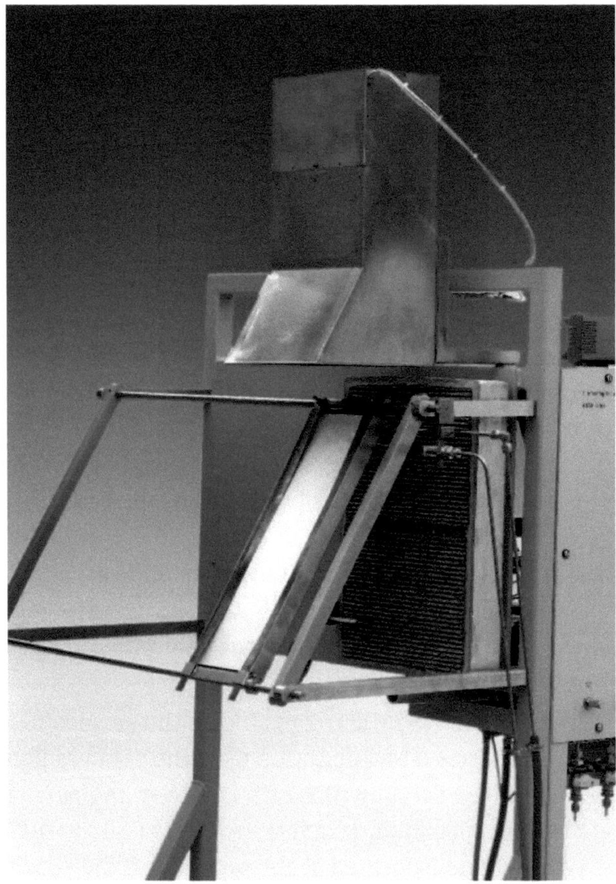

Figure 11.10 ASTM E162 surface flammability test apparatus

ASTM D3675 *(Surface flammability of flexible cellular materials using a radiant heat energy source)* was developed as a modification of ASTM E162 intended specifically for fire testing of flexible cellular materials. For example, the pilot burner in ASTM D3675 is different from the pilot burner in ASTM E162, and ASTM D3675 requires that running or dripping of flaming material be reported.

ASTM E662: Standard Test Method for Specific Optical Density of Smoke Generated by Solid Materials

The NBS smoke chamber is the most commonly used small-scale test apparatus for measuring the optical density of smoke. The apparatus and test procedure are described in ASTM E662. The method was developed in the 1960s at the National Bureau of Standards (currently the National Institute of Standards and Technology). The test specifications are summarized in Table 11.18.

The test apparatus consists of a 0.914 × 0.610 × 0.914 m enclosure (see Figure 11.11a). A radiant heater with a diameter of 76 mm is used to provide the heat flux to the specimen surface (see Figure 11.11b). Smoke density is measured based on the attenuation of a light beam by the smoke accumulating in the closed chamber. A white light source is located at the bottom of the enclosure, and a photomultiplier tube is mounted at the top. The specimen measures 76 × 76 mm and is oriented vertically. Specific optical density (D_s), as defined in ASTM E662, is equal to the optical density measured over unit path length within a chamber of unit volume, produced from a specimen of unit surface area, that is irradiated by a heat flux of 25 kW/m^2 for a specified time.

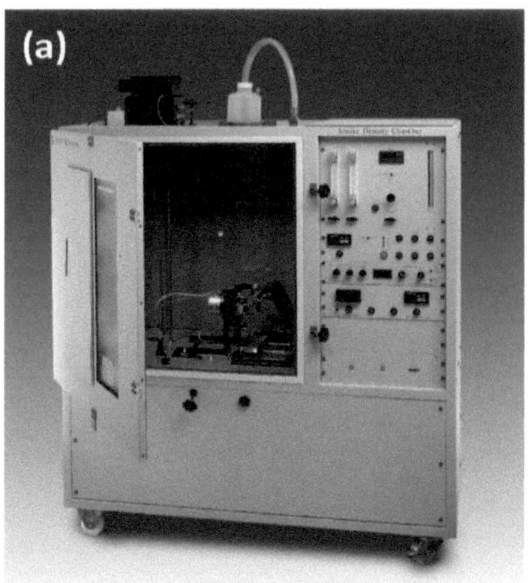

Figure 11.11 (a) ASTM E662 test chamber; (b) ASTM E662 radiant heater

Table 11.18 ASTM E662 Smoke Specific Optical Density Test Specifications

Specimens	Three or six specimens for each mode, 76 × 76 mm × thickness of use up to 25 mm
Specimen position	Vertical
Ignition source	Radiant heater producing an irradiance of 25 kW/m^2, pilot flame burner at specimen bottom edge
Test duration	20 min or 3 min after D_s reaches maximum
Conclusions	The final results are the specific optical density (D_s) as a function of time, and the maximum value of the optical density (D_m)

The ASTM E662 maximum test limits for smoke emission (specific optical density) shall be measured in either the flaming or non-flaming mode, utilizing the mode which generates the most smoke. A 72 W six-tube premixed propane pilot burner is used in the flaming mode. The burner flames impinge on the lower half of the specimen surface. The burner is removed to conduct tests in the non-flaming mode. Three replicate tests are typically conducted in each mode. An additional set of three specimens are tested if there is a high variation in the results from the first set.

A test is terminated 3 min after the light transmittance value reaches a maximum value or 20 min after the start of the test, whichever occurs first. D_s is reported as a function of time. In addition, the maximum value of the specific smoke density, D_m, is reported separately after correction for soot deposits on the windows of the optical system during the test. The correction is equal to the apparent optical density that is recorded after the test is terminated and the smoke has been cleared from the chamber.

The test method described in ASTM E662 appears to have some significant limitations, although it is hard to find objective evidence for this. To make the user aware of these limitations, the ASTM committee that developed the standard added the following caveat to the scope section of the document: "This test method is intended for use in research and development and not as a basis for ratings for regulatory purposes". In spite of this caveat, the method is referenced in several fire safety regulations and specifications for materials used in passenger aircraft, Navy ships and passenger rail transportation vehicles. The acceptance criteria in the FRA regulations for the latter, which refer to the 2001 edition of ASTM E662, are based on limits for D_s at 1½ and 4 min (see Table 11.15).

14 CFR 25 Appendix F, Part I (Vertical Test)

Seat upholstery fabrics, mattress ticking and covers, curtains, draperies, wall coverings, and window shades must be tested according to the vertical method described in 14 CFR 25 Appendix F, Part I. The test specifications are summarized in Table 11.19.

Table 11.19 14 CFR 25 Appendix F, Part I, Vertical Test Specifications

Specimens	Three specimens, minimum 50 mm wide × 300 mm long × thickness of use
Specimen position	Vertical
Ignition source	38 mm tall 843 °C Bunsen or Tirrill burner flame
Test duration	The ignition source exposure time is 60 s or 12 s depending on the type of material; the test is terminated when all flaming ceases
Conclusions	The final results are the flame time and burn length

Specimens are mounted in a metal frame so that the two long edges and the upper edge are held securely during the test. The exposed area of the specimen must be at least 50 mm wide and 300 mm long. A minimum of three specimens must be tested and results averaged. For fabrics, the direction of weave corresponding to the most critical flammability conditions must be parallel to the longest dimension.

Each specimen must be supported vertically. The specimen must be exposed to a Bunsen or Tirrill burner with a nominal 19 mm I. D. tube adjusted to give a flame of 38 mm in height. The minimum flame temperature measured by a calibrated thermocouple in the center of the flame must be 843 °C. The lower edge of the specimen must be 19 mm above the top edge of the burner. The flame must be applied to the center line of the lower edge of the specimen.

For wall coverings, the flame must be applied for 60 seconds and then removed. For seat upholstery fabrics, mattress ticking and covers, curtains, draperies, and window shades, the flame must be applied for 12 seconds and then removed. Flame time and burn length are recorded, the latter to the nearest 2.5 mm (tenth of an inch).

11.2.3.1.3 ASTM Fire Hazard Assessment Guide

In 2000, ASTM Committee E05 on Fire Standards published the first version of ASTM E2061 *(Fire Hazard Assessment of Rail Transportation Vehicles)*. The current version at the time of writing is dated 2020. The standard provides guidance to fire safety engineers who wish to develop a fire hazard assessment of a rail transportation vehicle. This guide is not in itself a fire hazard assessment, and does not provide acceptance criteria. Consequently, it cannot serve as the basis for a regulation. However, a fire hazard assessment developed in accordance with ASTM E2061 can be used in support of a request for a variance from a prescriptive regulation. The latter involves the specification of individual fire-test-response requirements for each material, component, or product deemed important to maintain satisfactory levels of fire safety. Table 11.15 and the equivalent table in NFPA 130 are examples of such prescriptive requirements. In other words, a fire hazard assessment developed following the guidance in ASTM E2061 can be used to demonstrate that the level of safety of a modified or new design of a rail transportation vehicle that does not meet all the requirements of a prescriptive regulation is equal to or greater than the implicit level of safety provided by the regulation.

11.2.3.2 Canada

In 1977, the Canadian federal government created VIA Rail as a Crown corporation with the exclusive mission to organize and provide all intercity passenger train services in the country. Canada is also home to the Rocky Mountaineer, the world's largest privately owned tourist train company, which has train routes through Canada's Rocky Mountains between the provinces of Alberta and British Columbia and into the U.S. Thirteen commuter rail lines operate in Canada, serving the Vancouver, Toronto, Ottawa, and Montreal metropolitan areas, among others.

The regulatory framework for railway safety in Canada encompasses the federal and provincial legislation, regulations, rules, and standards that provide the structure in which railway companies can operate safely. Federal rail regulations in Canada are set by Transport Canada and the Canadian Transportation Agency. Some 34 Canadian railways have interprovincial or Canada-U.S. operations and are therefore regulated by federal law. These include the passenger rail company VIA Rail, and more than 30 short-line companies. Another 62 railways operate entirely within a single province and are, therefore, regulated by provincial governments. However, most provinces have incorporated (by reference into their own legislation) some or all of the provisions of the regulations, ensuring that the same rules apply to provincial railways. Several local railway companies, mainly commuter railways, are not regulated under the corresponding provincial railway safety legislation.

Transport Canada Railway Passenger Inspection & Safety Rules [4] specify that new equipment ordered after April 1, 2001 shall be designed and constructed in accordance with the latest revision in effect at the time of order of the "American Public Transportation Association (APTA) Manual of Standards and Recommended Practices for Passenger Rail Equipment", or an equivalent standard. Consequently, although the requirements are not explicitly stated in the Transport Canada rules and regulations, materials used in the construction of passenger railcars and locomotive cabs generally comply with Appendix B of 49 CFR, Part 238. NFPA 130 is also used in Canada, primarily for light rail systems.

11.2.3.3 Future Trends

In 2018 Lattimer and McKinnon published a comprehensive literature review in which they assessed the state of knowledge of the effects of railcar interior finish materials on fire growth and fully developed fires in passenger railcars [5]. The review included a survey of US and European small-scale test standards and corresponding material fire performance requirements currently used to regulate the flammability of interior finish materials in railcars. The authors noted that the basis for these requirements is largely lacking. They recommended that testing and modeling be conducted to verify that passing railcar materials indeed will not contribute to causing rapid fire growth and that the geometric arrangement of fuels inside a railcar be considered when assessing the appropriateness of the acceptance criteria. Finally, they also noted that international railcar standards have adopted smoke toxicity requirements and suggested that including such requirements in US regulations should be considered.

References for Section 11.2.3

[1] *https://www.govinfo.gov/app/collection/cfr/2017/title49/subtitleB/chapterII/part238.*

[2] "Recommended Practice for Fire Safety Analysis of Existing Passenger Rail Equipment", APTA PR-PS-RP-005-00, The American Public Transportation Association, New York, NY, 2004.

[3] *https://www.nfpa.org/Codes-and-Standards/All-Codes-and-Standards/Free-access.*

[4] *https://www.tc.gc.ca/eng/railsafety/rules-tco26-356.htm.*

[5] Lattimer B. Y., McKinnon, M. A review of fire growth and fully developed fires in railcars. *Fire and Materials,* Vol. 42, pp. 603–619, 2018. *https://doi.org/10.1002/fam.2514*

11.2.4 People's Republic of China

Shuai-Xia Tan

11.2.4.1 Statutory Regulations

The technical legal provisions for rail vehicles in China are issued by the ministry of railway of People's Republic of China (PRC), the ministry of housing and urban-rural development of PRC, and the national standards of PRC.

In practical applications, the fire-preventive performance requirements and the application fields of locomotive vehicles are specified in the relevant standards and regulations, such as:

- GB 6771-2000 – Regulations relating to fire-preventive and firefighting measures for electric locomotives
- TB/T 2640-1995 – Structural design of fire protection for railway passenger cars
- CJ/T 416-2012 – Fire protection requirements for urban rail transit vehicles
- TB/T 3138-2018 – Technical specification of flame retardant materials for railway locomotives and vehicles
- TB/T 3237-2010 – Flame retardant technical specification of decorating materials for multiple unit trains

The specified scope and tested parameters of fire protection are listed in Table 11.20 [1].

Table 11.20 Fire Protection Regulations for Rail Vehicles in China Fire Protection Requirements, Classification, and Tests

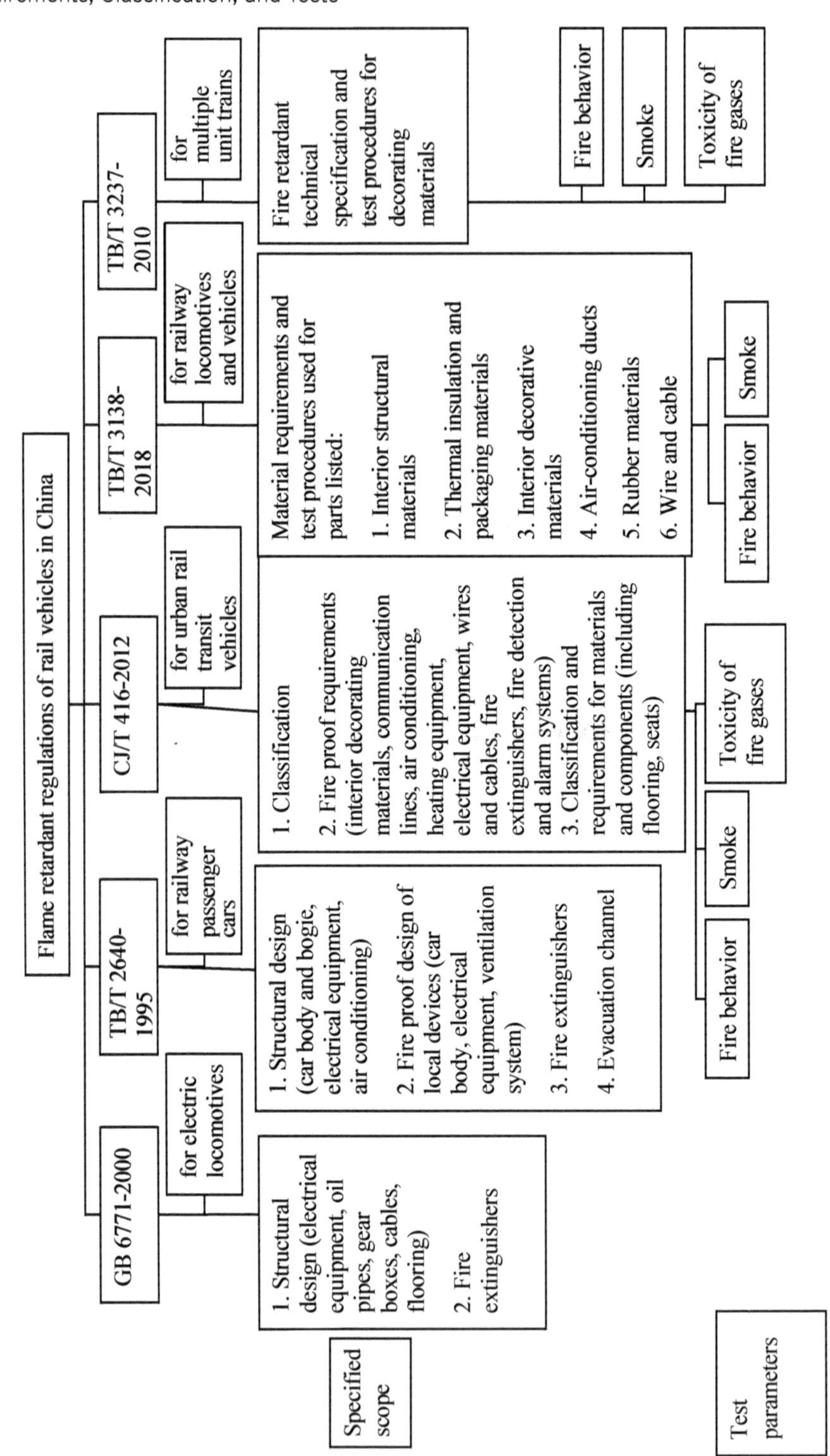

11.2.4.2 Fire Protection Requirements, Classification, and Tests

11.2.4.2.1 GB 6771-2000 – Regulations Relating to Fire-Preventive and Firefighting Measures for Electric Locomotives

GB 6771 is the only national standard of fire protection for locomotives, which is equivalent to UIC 617-1 OR:1979. The standard specifies the regulations related to fire-preventive structural design and firefighting measures for electric locomotives and is applicable to the electric tractive stock, electric multiple units, and new design and manufacture of locomotives in international transport.

The main differences between GB 6771-2000 and UIC 617-1 OR:1979 are listed in Table 11.21.

Table 11.21 Main Differences between GB 6771-2000 and UIC 617-1 OR:1979

	GB 6771-2000	UIC 617-1 OR:1979
Compulsory execution for new manufacture of electric locomotives	Since execution data	Since 1979-1-1
Compulsory execution of firefighting measures for existing electric locomotives	Cancelled, not necessary	Since 1975-1-1
Connection of cable terminals	Screw terminal connection or welded connection	Crimp connection or screw terminal connection

11.2.4.2.2 TB/T 2640-1995 – Rolling Stock. Fire Protection and Firefighting. Design Arrangements

TB/T 2640-1995 mainly refers to French NF F16-103-1988, and is applicable to the passenger-carrying rail vehicles of new design and manufacture. The standard specifies the constructive provisions of the fire-protective structural design, local devices, fire extinguishers, and the evacuation of passengers.

- The fire-protective structural design firstly requires that all parts of structure and decorative non-metallic materials must comply with the specifications of TB/T 2560 (refers to NF F16-101). Moreover, the requirements of structural design for crates and bogies, electrical equipment, and air conditioners (heating) are given individually.
- The local devices for fire safety specify the design requirements of crates, electrical equipment, and ventilation systems individually.
- The fire-extinguishing devices for fire protection specify the instructions, location, and quantities of fire extinguishers.
- The fire protection measures for the evacuation of passengers prescribe that slide doors must be used for entrance and end doors, swing doors for partition doors, and doors with locks must be equipped with tempered glazing.

11.2.4.2.3 CJ/T 416-2012 – Fire Protection Requirements, Classification, and Tests for Urban Rail Transit Vehicles

CJ/T 416-2012 mainly refers to German DIN 5510, and specifies the hazard levels of fire protection, fire protection of structural design, fire performance classes and requirements for material and products, test methods, and so on. The standard is applicable to subway vehicles, light rail vehicles, monorail vehicles, trams, magnetic levitation vehicles, automated guideway transit systems, and urban fast track vehicles.

The hazard levels are based on the operation environment, fire characteristics, and fire hazard as shown in Table 11.22.

Table 11.22 Hazard Levels of Railway Vehicles Related to the Operation Environment

No.	Category	Operation environment
1	1	Ground sections and/or elevated lines
2	2	Underground sections

The structural design and arrangement of fire protection specifies the design requirements for nine categories: general fire protection regulations, interior decorating materials, communication lines, air conditioning, heating equipment, electrical equipment, cables and wires, fire extinguishers, fire detection and alarm systems.

The fire performance requirements for components and materials are specified for three categories, and the test methods are shown in Table 11.23 [2].

1. Structural parts and materials include six classes (car body, interior decorative materials, internal facilities, doors and windows, pipes and hoses, cable ducts, and wire tubes) and 65 kinds of products. Classifications and tests include flammability, smoke generation, smoke toxicity, and dripping. Flammability, smoke generation, and dripping classes must comply with the respective requirements for the fire hazard levels. Smoke toxicity must meet FED ≤ 1.
2. Electrical components and materials include three classes (electrical equipment in enclosures, wires and cables, and other electrical equipment). The enclosures for the protection of electrical equipment must comply with the integrity requirements of GB/T 9978.1 (based on ISO 834-1). Wires and cables must comply with the flame retardant requirements of being low smoke and halogen-free of TB/T 1484. Other electrical equipment in specific installation positions must meet the flammability Class V-1/V-0 to GB/T 5169.16-2008 (identical to IEC 60695-11-10:2003), or have an oxygen index > 28/30 to GB/T 2406.2-2009 (identical to ISO 4589-2:1996).
3. Other parts.
   - The flammability of other parts which can be tested according to DIN 54837 should meet the S2 level.

- The flammability of other parts which cannot be tested according to DIN 54837 due to the small exposed surface and which are in contact with passengers should meet the S1 level.
- The flammability of other parts which cannot be tested according to DIN 54837 and which are not in contact with passengers does not have to be tested.

Table 11.23 Items Classification and Test Methods for Fire Performance in CJ/T 416-2012

Items classification	Test methods/Standards
Flammability class for materials	DIN 54837-2007; GB/T 8626-2007 (identical to ISO 11925-2:2002)
Flammability class for floorings	GB/T 11785-2005 (identical to ISO 9239-1:2002)
Smoke generation class	DIN 54837-2007
Drop forming class	DIN 54837-2007
Smoke toxicity	Annex D of CJ/T 416-2012
Oxygen index	GB/T 2406.2-2009 (identical to ISO 4589-2:1996)

A fire test method for complete passenger seating is also regulated in CJ/T 416-2012. Seats are made from a large number of non-metallic materials and often used in the car body so that it is necessary to test their flammability. CJ/T 416-2012 and EN 45545-2:2013 both provide the test methods and technical requirements for seats, respectively. The comparisons are shown in Table 11.24.

Table 11.24 Comparison of Fire Test Method for Seating in CJ/T 416-2012 and EN 45545-2:2013

Clause	CJ/T 416-2012 Annex B, C	EN 45545-2:2013 Annexes A, B
Preparation of test specimen	The base of the seat is cut off. The distance of 100 mm shall be cut with a blade along the diagonals and the cut depth is about 25 mm. Specifies the cutting area and depth with a blade, it is more random, and has a greater impact on the combustion.	The back and base of the seat shall be cut off. Distances > 50 mm shall be cut along the diagonals beginning at 50 mm from the corners. Specifies the cutting direction and position; the cutting method is more conducive to ensure the consistency of the operation and combustion test result.
Number of tests	3 tests for uncut seats and 3 tests for cut seats	3 tests for cut seats
Ignition source	Paper cushion	Square burner and position as defined in EN 45545-2 of annex B

Table 11.24 Comparison of Fire Test Method for Seating in CJ/T 416-2012 and EN 45545-2:2013 *(continued)*

Clause	CJ/T 416-2012 Annex B, C	EN 45545-2:2013 Annexes A, B
Measured parameters	Temperature, volume flow rate of exhaust system V(t), smoke production rate SPR, smoke density/ transmittance, toxicity index of FED (fractional effective dose)	Temperature, pressure difference between the two chambers of the bi-directional probe (Δp), mass flow rate of propane gas, volume flow of exhaust, smoke density and transmittance, O_2/CO_2 volume fraction in the exhaust flow, total heat release, HRR/MAHRE versus time
Requirements	TSP (total smoke production) ≤ 60 m^3, FED ≤ 1, burning behavior (flame height ≤ 100 cm, flame time from the test beginning ≤ 15 min; flame should not reach the side edges of seats, etc.)	MARHE ≤ 75 kW (HL1), 50 kW (HL2), 20 kW (HL3); HRR peak ≤ 350 kW
Comment	Heat release is not considered	Toxicity is not considered

11.2.4.2.4 TB/T 3138-2018 – Technical Specification of Flame Retardant Materials for Railways: Locomotives and Vehicles

TB/T 3138-2018 mainly refers to UIC 564-2:1991 and NF F 16-101:1988, and is applicable for flame retardant materials for railway locomotives and vehicles with a maximum operating speed of < 200 km/h.

TB/T 3138-2006 specifies the test methods and requirements of flame retardant materials for six categories:

1. Interior structural materials (for interior horizontal downward-facing surfaces, vertical surfaces, horizontal upward-facing surfaces, interior skeleton)
2. Thermal insulation and packaging materials (inorganic materials, polymer materials, packaging materials)
3. Interior decorative materials (leather coverings, foam materials, curtains)
4. Air-conditioning ducts (main materials for air-conditioning ducts)
5. General rubber materials
6. Wire and cable

Most materials must be tested for oxygen index (OI), smoke density, and 45° burning behavior (except wires and cables complying with TB/T 1484).

1. The technical requirements for smoke density are identical for all listed materials. The smoke density at 4 min (Ds_4) (with heat flux 25 kW/m^2 and pilot flame) is ≤ 200.

2. The oxygen index (OI) and flammability class requirements for different materials are listed in Table 11.25.

Table 11.25 Oxygen Index and Flammability Class Requirements for Different Materials

Categories	Location/Specific use	Oxygen index requirement (%)	Flammability class (45° burning behavior)
Interior structural materials	Interior horizontal downward-facing surfaces	≥ 32	Difficult-flammability
	Interior vertical surfaces	≥ 30	Difficult-flammability
	Interior horizontal upward-facing surfaces	≥ 28	Difficult-flammability
	Interior skeleton	≥ 30	Difficult-flammability
Thermal insulation and packaging materials	Inorganic insulation materials	-	Non-combustibility
	Polymer insulation materials	≥ 26	Difficult-flammability
	Packaging materials	≥ 28	Difficult-flammability
Interior decorative materials	Leather coverings	≥ 30	Difficult-flammability
	Foam materials	≥ 26	Difficult-flammability
	Curtains	≥ 30	Difficult-flammability
Main material of air-conditioning ducts		≥ 50	Non-combustibility
General rubber materials		≥ 29	Difficult-flammability

3. The technical requirements for 45° burning behavior are virtually identical for most materials, and must meet the difficult-flammability class as classified in Table 11.26. With the exception of inorganic materials, thermal insulation and air-conditioning duct materials must meet the non-combustibility class. All the other materials (like interior structural materials, polymer insulation materials, packaging materials, interior decorative materials, general rubber materials, as listed in Table 11.25) must meet the difficult-flammability class. The fire-resistant classes of the 45° burning behavior test are summarized in Table 11.26 and the test arrangement is shown in Figure 11.12 of Section 11.2.5.2.

Technical requirements of 45° burning behavior test (similar to UIC 564-2 Appendix 4):

1. Specimen dimensions: 257 mm × 182 mm (prepare two longitudinal and two lateral for anisotropic materials)
2. Specimen held at an angle of 45° to the horizontal
3. Distance from the axis of symmetry of the crucible between the inside bottom surface and the underside of the specimen 25.4 mm
4. Test environment conditions: temperature 15 to 30 °C, relative humidity (RH): 60 to 75%

Table 11.26 Fire Resistant Classes of 45° Burning Behavior Test

Fire-resistant class	Alcohol burning				After alcohol burned out			
	Ignition	Flame	Smoke level	Fire	After flame	After glow	Carbonization	Deformation
Non-combustibility	No	No	Minimal				Discoloration length < 100 mm	Deformation length < 100 mm
Extremely low flammability	No	No	Low				Does not reach the specimen top edge	Deformation length < 150 mm
	Yes	Yes	Low	Weak	No	No	< 30 mm	
Low flammability	Yes	Yes	Normal	Does not reach specimen top edge	No	No	Reaches specimen top edge	Reaches edge/ local perforation

11.2.4.2.5 TB/T 3237-2010 – Flame Retardant Technical Specification for Decorating Materials for Multiple Unit Trains

TB/T 3237-2010 is applicable to decorative flame retardant materials for multiple unit trains with a maximum operating speed ≥ 200 km/h.

TB/T 3237-2010 specifies the test methods and requirements of flame retardant materials for 14 categories of products at different locations:

1) Ceiling paneling including its decorative and sealing/connecting materials
2) Side panels and wall panels including their decorative and sealing/connecting materials
3) Composite materials of doors
4) Curtains and shade cloth
5) Lamp covers
6) Seats and mattresses
7) Floor composites and their connecting materials
8) Non-metallic luggage rack materials
9) Washroom panels
10) Decorative and sealing materials
11) Noise reduction materials with sealed and anti-corrosion effect
12) Cold-resistant materials
13) Air-conditioning ducts and connected sealing materials
14) Others

All 14 product categories must be tested for oxygen index, fire resistance classes, smoke density, and gas toxicity. While the technical requirements for smoke density and gas toxicity are identical for all 14 product categories, the oxygen index and fire resistance requirements are different. The smoke density at 1.5 min ($Ds_{1.5}$) must not exceed 100, and at 4 min (Ds_4), not exceed 200 [3]. The limit gas concentrations are listed in Table 11.27.

Smoke generation, sampling, and analytical methods for gas components of TB/T 3237-2010 refer to EN 45545-2, DIN 5510-2, NF F16-101, BS 6853, and SMP 800-C; the comparisons are given in Table 11.28 [4]. The technical requirements for each gas component in the standards mentioned above are summarized in Table 11.28 as well [5]. The weighted summation of toxic fumes described in TB/T 3237-2010 is not required; however, each single gas component has to meet the concentration limit values indicated.

Table 11.27 Comparison of Reference Gas Components Concentrations

Gas	Standards					
	TB/T3237 [mg/m³] (ppm)	EN 45545-2 [mg/m³]	BS 6853 [mg/m³] (ppm)	DIN 5510 CJ/T 416-2012 [mg/m³]	NF F16-101 [mg/m³]	SMP 800-C (ppm)
CO	< 4000 (3500)	1380	1400 (1200)	1380	1750	3500
CO_2	< 90 000 (50 000)	72 000	73 000 (40 000)	72 000	90 000	90 000
NO_X	< 190 (100)	38	56 (50)	38	-	100
SO_2	< 260 (100)	262	38 (20)	262	260	100
HCl	< 150 (100)	75	76 (50)	75	150	500
HF	< 82 (100)	25	25 (30)	25	17	100
HBr	< 330 (100)	99	101 (30)	99	170	100
HCN	< 110 (100)	55	270 (100)	55	55	100
Weighted summation index	-	CIT	R	FED	ITC	as in NFPA 130 or CFR 238.103

Table 11.28 Smoke Generation, Sampling, and Analytical Methods of Gas Components

Test items	Standards					
	TB/T 3237	EN 45545-2 Annex C (method 1) DIN 5510-2 Annex D.2	NF F16-101 EN 45545-2 Annex C (method 2) BS 6853 Annex B.1	DIN 5510-2 Annex D CJ/T 416-2012 Annexes D, E	BS 6853 Annex B.2/ EN 2826	SMP 800-C
Smoke generation	According to GB/T 8323.2, flaming mode	According to ISO 5659-2 25 kW/m^2, flaming mode (for both) 50 kW/m^2, non-flaming mode (only for EN 45545-2)	Furnace temperature 600 °C	According to ISO 5659-2, 25 kW/m^2, flaming mode	According to ISO 5659-2, 25 kW/m^2, flaming mode	According to NFPA 258/ ASTM E662-01, both flaming and non-flaming mode
Sample	(75 mm × 75 mm) thickness ≤ 25 mm	(75 mm × 75 mm) thickness ≤ 25 mm	(1000 ± 0.050) g	(75 mm × 75 mm) thickness ≤ 25 mm	(75 mm × 75 mm) thickness ≤ 25 mm	(75 mm × 75 mm) thickness ≤ 25 mm
Smoke sampling	Gas collection from 1.5 min to 4 min after test start; gas flow rate ≤ 2 L/min	Gas collection from 225 s to 255 s and 465 s to 495 s to determine the gas concentration at 4 and 8 min after test start	Gas collection from start until the end of the 20 min test; gas flow rate ≤ 2 L/min	Gas collection from 3 min 30 s to 4 min 30 s and from 7 min 30 s to 8 min 30 s to determine the gas concentration at 4 and 8 min after test start	Gas collection starts at the previously determined time to reach 85% of the peak smoke emission	Gas collection from start at 4 min mark until the end of the 20 min test

Test items	Standards					
CO_2/CO	Infrared spectro-metry	FTIR	Non-disper-sive infrared	Non-disper-sive infrared Colorimetric tube	Colorimetric tubes	Follow industry practices such as absorptive or equivalent methods. The gas detection tube method is not acceptable.
NO_X (Includes NO_2 and NO quoted as NO_2)	Spectro-photometry	FTIR	Chemilumi-nescence	Chemilumi-nescence	Colorimetric tubes	
HF	Potentio-metric titration, wet analysis	FTIR	Spectro-photometry Ionometry with specific electrodes	Spectro-photometry Colorimetric tubes	Potentio-metric titration, wet analysis	
HCl/HBr	Potentio-metric titration, wet analysis	FTIR	Titrimetry with Ag electrode Ion liquid chromato-graphy	High-perfor-mance ion chromato-graphy (HPIC)	Colorimetric tubes Potentio-metric titration, wet analysis	
SO_2	Spectro-photometry	FTIR	Ion liquid chromato-graphy	High-perfor-mance ion chromatog-raphy (HPIC) Colorimetric tubes	Colorimetric tubes	
HCN	Colorimetric tubes	FTIR	Spectro-photometry Ion liquid chromato-graphy	Colorimetric tubes	Colorimetric tubes	

Since TB/T 3237-2010 and TB/T 3138-2018 are both technical specifications of flame retardant materials, the test items and methods are compared in Table 11.29.

Table 11.29 Comparison of Test Items and Test Methods in TB/T 3237-2010 and TB/T 3138-2018

Test items	Test methods in TB/T 3237-2010	Test methods in TB/T 3138-2018
Oxygen index	GB/T 2406.2-2009 (identical to ISO 4589-2:1996) for plastics GB/T 8924-2005 for fiber-reinforced plastics GB/T 5454-1997 (not identical to ISO 4589:1984) for textiles	GB/T 10707 for rubber GB/T 2406-1993 (refers to ISO 4589:1984) for plastics GB/T 8924-2005 for fiber-reinforced plastics GB/T 5454-1997 (not identical to ISO 4589:1984) for textiles

Table 11.29 Comparison of Test Items and Test Methods in TB/T 3237-2010 and TB/T 3138-2018 *(continued)*

Test items	Test methods in TB/T 3237-2010	Test methods in TB/T 3138-2018
Smoke density	Smoke density at 1.5 min and 4 min (4 conditions, heat flux 25 kW/m²/50 kW/m² in the presence or absence of a pilot flame) to GB/T 8323.2-2008 (identical to ISO 5659-2)	Smoke density at 4 min (1 condition, heat flux 25 kW/m² in the presence of a pilot flame) according to GB/T 8323.2 (identical to ISO 5659-2)
Fire resistance classes	Fire resistance classes according to UIC 564-2-1991	45° angle burning behavior test according to Annex A
Gas toxicity	Clause 4.4 of TB/T 3237-2010	–
Comments	The differences of fire performance of different materials and parts on vehicles are not fully considered, and the test method of gas toxicity is not very clear.	The test of gas toxicity is not required.

11.2.4.3 Future Developments

High-speed rail vehicles with excellent performance develop rapidly in China, and the standardization administration pays close attention to the development trends of ISO, EN, DIN, NF, and IEC in the field of railways. So, the development of national standards in China for fire protection for railways in the future will be based on Chinese actual features and focus on the following aspects:

1. Develop the operation and design categories of rail vehicles based on rail vehicle types, infrastructure characteristics, and escape routes

2. Formulate a series of systematic fire standards for different vehicle types, and specify the fire protection design requirements for fireproof separation, vehicle design, electrical equipment, etc.

3. Furthermore, regulate the test methods and requirements for different components and materials, and solve the current problems, such as inconsistent test parameters, references, confusion between the standards, and other issues.

References for Section 11.2.4

[1] Xie, Wei; Lei, Yu-Hao; Ning, Jian-Guo. The comparison analysis of fire precaution standards for rail vehicles in China and Abroad, *Railway quality control,* 43 (7), 7–12, 2015.

[2] Wu, Hui-Yong; Wu Shao-Li. The comparison analysis and suggestions of fire precaution standards for materials of rail vehicles, *Railway quality control,* 42 (10), 34–37, 2014.

[3] Jiang, Cui-Xiang. Analysis of fire prevention standard for rail vehicles, *Journal of Shijiazhuang Institute of railway technology,* 12 (2), 37–41, 2013.

[4] Yu, Quan-Lei. Comparison and analysis of fire protection standards for rail vehicles in China and abroad, *Rolling stock,* 55 (1), 24–29, 2017.

[5] Tan, Shuai-Xia; Yu, Xiao-Hui; Wang Jin; Yang Na; Dong, Hong-Hai. Comparison of measurement methods for combustion of materials for rail vehicles, *Rolling stock,* 57 (1), 18–22, 2019.

11.2.5 Japan

Koichi Wada

11.2.5.1 Statutory Regulations

The first technical standard for fire safety for rolling stock was enacted as "Countermeasures against rolling stock fires for electric train" in May 1969 by Ministerial Ordinance No. 81 of the Ministry of Transport (now MILT). In this standard, the test method shown in Figure 11.12 was adopted, and since then has been continuously adapted. In 2001, various technical standards for individual classes of rolling stock were unified as "Technical Regulatory Standards on Japanese Railways" (Ministerial Ordinance No. 151 of December 25, 2001) by the Ministry of Land, Information, Transport and Tourism (MLIT), which had the responsibility for the fire safety regulations for railway rolling stock in Japan.

As a result of the serious subway fire incident at Daegu, South Korea, in February 2003, this "Technical Regulatory Standard" was amended in December 2004 to tighten fire safety regulations for passenger compartments, especially for ceilings, by Ministerial Ordinance No. 124 of MLIT [1].

In this amended standard, Article 83 covers countermeasures against rolling stock fires, as listed below:

1. Electric wires and cables used in rolling stock shall be capable of preventing fire caused by electric malfunction, overheating of equipment, and other reasons.
2. Devices that are likely to generate arcs or heat shall be provided with pertinent protective measures.
3. The body of passenger cars shall be built with the appropriate structure and materials so that they can prevent potential fires from starting and spreading.
4. Locomotives (excluding steam locomotives), passenger cars, and freight cars with crew cabins shall be provided with fire-extinguishing devices to contain fires at an early stage.

The countermeasures against rolling stock fires for the body of passenger cars shall satisfy the specifications shown in Table 11.30 [1].

Table 11.30 Countermeasures against Rolling Stock Fire for Passenger Cars [1]

Part		General passenger cars	Subway etc. (Shinkansen passenger cars excluding superconducting magnetic levitation railways)	Special railways [j]						
					S	G	T	C	N	Su
Roof	Roof [a]	Metal, or equal to or better than the non-combustibility of metal [b]		Noncombustible	Yes	Same as S			Same as S	Same as S
				Metal, or equal to or better than the non-combustibility of metal			Yes			
	Roof top surface	Shall be covered with a flame-retardant electrical insulating material (limited to passenger trains that travel on sections of track with electrified overhead line) (except for extremely high-voltage contact lines)				Yes	Yes			
	Equipment and hardware mounted on the roof	The mounted part shall be insulated from the car body or shall be covered with a flame-retardant insulating material (limited to passenger trains that travel on sections of track with overhead contact lines) (except extremely high-voltage contact lines)				Yes	Yes			
External sheet-ing	End section	Flame-retardant Shall use noncombustible material for the surface paint [f]	Noncombustible Shall use noncombustible material for the surface paint [f]		Yes	Same as S			Same as S	Same as S
	Other sections	Shall be noncombustible, or surface shall be covered with a non-combustible material [c] Shall use noncombustible material for the surface paint [f] Noncombustible	Noncombustible Shall use noncombustible material for the surface paint [f]		Yes	Same as S			Same as S	Same as S

Part		General passenger cars	Subway etc. (Shinkansen passenger cars excluding superconducting magnetic levitation railways)	Special railways [j]						
					S	G	T	C	N	Su
Passen-ger com-part-ment	Ceiling	Shall be noncombustible, or surface shall be covered with a noncombustible material [c] Shall use noncombustible material for the surface paint [f]	Noncombustible Shall have resistance to radiant heat, to melting and dripping. [e] Shall use noncombustible material for the surface paint [f]	Noncombustible Shall use noncombustible material for the surface paint [f]	Same as G	Yes [d]			Same as G	Same as G
	Inside panel	Shall be noncombustible or surface shall be covered with noncombustible material [c] Shall use noncombustible material for the surface paint [f]	Noncombustible Shall use noncombustible material for the surface paint [f]		Yes	Same as S			Same as S	Same as S
Heat insulation and noise insulation			Noncombustible		Yes	Same as S			Same as S	Same as S
Floor	Floor	Structure where there is little risk of smoke and fire flowing			Yes	Same as S			Same as S	Same as S
	Floor covering	Flame-retardant			Yes	Same as S			Same as S	Same as S
	Stuffing under floor covering [g]		Extremely flame-retardant		Yes	Same as S			Same as S	Same as S
	Floor panel		Metal floor sheeting, or equal to or better than the non-combustibility of metal [b]		Yes	Same as S			Same as S	Same as S
	Under-floor surface [h]	Shall be noncombustible, or surface covered with metal	Shall be noncombustible, or surface covered with metal, and shall use noncombustible material for the surface paint [f]	Surface paint [f] shall be noncombustible	Yes	Same as S			Same as S	Same as S

Table 11.30 Countermeasures against Rolling Stock Fire for Passenger Cars [1] *(continued)*

Part		General passenger cars	Subway etc. (Shinkansen passenger cars excluding superconducting magnetic levitation railways)	Special railways [j]					
				S	G	T	C	N	Su
Underfloor equipment box [i]			Noncombustible However, flame-retardant when the need for insulation is unavoidable	Yes	Same as S			Same as S	Same as S
Seat	Fabric	Flame-retardant			Same as C		Y	Same as C	Same as C
	Stuffing		Flame-retardant	Yes	Same as S			Same as S	Same as S
	When electric heater is provided under the seat		Provide noncombustible heat-resistant plate between heating element and seat	Yes	Same as S			Same as S	Same as S
Window shade		Flame-retardant		Yes	Yes	Yes	Yes	Yes	Yes
Gangway bellows			Flame-retardant	Yes	Yes			Same as S	Same as S

[a] "Roof" means the section above the rain trough/gutter of upper car body structure. If the rain trough/gutter is located within one-third of the maximum car body width, as measured from the car body centerline, "roof" shall mean the section up to one-third of the maximum car body width, respectively. However, when part of the roof is integral to the end-section external sheeting, the said section shall be the "end section" of the "external sheeting" shown in Table 11.30.

[b] The expression "equal to or better than noncombustible" for "roof" and "floor" means that the performance is equal to or better than that of the metal used on the existing roof and floor sheeting, which is different from the stipulation for non-combustibility in the combustibility standards for materials used in railway rolling stock.

[c] "Surface shall be covered with a noncombustible material" includes combustible materials combined with noncombustible materials such as metal.

[d] Among rail-guided types of railways, the "ceiling" on rolling stock used for railways running in underground sections and in long tunnels shall be in accordance with stipulations for passenger cars for subways.

[e] Materials that "shall have resistance to burning due to radiant heat and to melting and dripping" shall include main facilities for air conditioning outlets, etc., installed in upper section of passenger compartments, in addition to ceiling materials, excluding small items that will not affect the spreading of the fire. "Resistance to melting and dripping" means that the surface of the material shall maintain its smoothness after ethanol has been burned as described in the "Test Method I for Non-Metallic Materials for Use on Railways".

[f] "Surface paint" means the outermost coat of paint when there are multiple coats.

[g] "Stuffing under floor covering" means the stuffing or filling for floors with keystone construction. The hardboard, water-resistant veneer, etc., sandwiched between metals (or between metal and flooring material), are included in the stipulations for stuffing.

[h] When a metal sheet is attached under the underfloor surface so that hot air from equipment installed underfloor will not affect the underfloor surface, the said metal sheet shall be deemed the "underfloor surface".

[i] The covers for relays, etc., are not included in "underfloor equipment box".

[j] The abbreviations for the types of special railways are: S: Suspended railways, straddled type monorail; G: Rail-guided types of railways; T: Trackless electric vehicles; Cable railways; N: Normal conducting magnetic levitation railways; Su: Superconducting magnetic levitation railways.

11.2.5.2 Classification and Testing of Materials and Components

The stipulated classification as “noncombustible”, “extremely flame-retardant”, and “flame-retardant” used in Table 11.30 is used according to the standards based on the following “Test Method I for Non-Metallic Materials for Use on Railways” in “Technical Regulatory Standards on Japanese Railways”. The classification of combustibility is shown in Table 11.31.

Table 11.31 Classification of Combustibility for Materials for Use on Railways [1]

Classification	During burning of ethanol		After burning of ethanol					
	Ignition	Flame	Smoke	Flame condition	Residual flame	Residual ash	Carbonization [a]	Deformation [a]
Noncombustible	None	None	Negligible	-	-	-	100 mm or less discoloration	100 mm or less surface deformation
Extremely flame-retardant	None	None	Little	-	-	-	Does not reach top edge of test specimen	150 mm or less deformation
	Yes	Yes	Little	Weak	None	None	30 mm or less	-
Flame-retardant	Yes	Yes	-	Flame does not exceed top edge of test specimen	None	None	Reaches top edge of test specimen	Deformation that reaches the edge, localized penetrating holes

[a] The extent of carbonization and deformation is considered as follows: (1) If the specimen burns abnormally, it shall be downgraded to a classification one rank lower; (2) Determination shall be done according to Test Method I.

Test Method I for Non-Metallic Materials for Use on Railways

Test piece
182 mm x 257 mm
Alcohol container
Container Position: the bottom surface center of the test specimen 25.4 mm from container bottom surface
Container holder (cork or other material with low heat conductivity)
45°

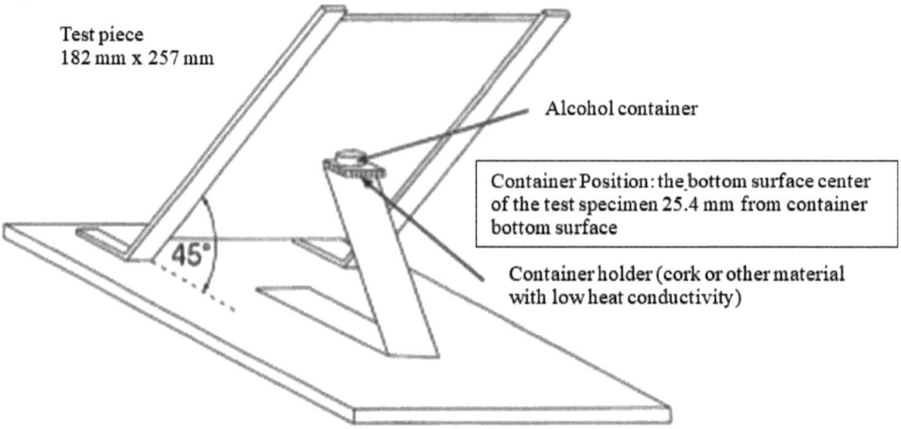

Figure 11.12 Testing apparatus for Method I [1]

Table 11.32 Method I: Test Specifications

Specimen	182 mm × 257 mm, thickness intended for use or more
Specimen position	45° against horizontal bottom plate
Specimen preparation	If test specimen has moisture absorption characteristics, it shall be left in a ventilated room for 5 days, avoiding direct sunshine
Test conditions	15–30 °C, 60–70% relative humidity, no air flow
Ignition source	0.5 mL ethanol with 100% purity in a steel container (17.5 mm diameter, 7.1 mm high, 0.8 mm thick) Bottom of the center of the fuel container placed on a wood platform positioned 25.4 mm (1 inch) perpendicularly below the center of the bottom plane of the test specimen
Test duration	Ethanol ignited and allowed to burn until fully consumed
Conclusions	Ignition, flame, smoke development, and flaming of the test specimen shall be observed during burning of ethanol. Residual flame, residual ash, carbonization, and deformation of the test specimen shall be investigated after burning of ethanol

In Table 11.30, for the ceiling of subway and Shinkansen passenger cars, “resistance to radiant heat” is required, which shall satisfy the specification described in Table 11.33 according to the standard based on the Test Method II for Non-Metallic Materials for Use on Railways.

Test Method II for Non-Metallic Materials for Use on Railways

Test Method II for Non-Metallic Materials for Use on Railways follows ISO 5660-1:2002 (Cone calorimeter method). Test apparatus and specifications are described in Section 9.2.

Table 11.33 Test Method II: Specifications for Non-Metallic Materials for Use on Railways [1]

Total heat release (THR) [MJ/m²]	Ignition time [s]	Maximum Heat Release Rate (MHR) [kW/m²]
< 8	-	< 300
> 8 and < 30	> 60	

Railway Sleepers

In Japan, glass-fiber-reinforced polyurethane rigid foam has been widely adopted as railway sleepers since mid-1980s based on a long-term examination by the Railway Technical Research Institute [4]. JIS E1203 is responsible for the technical requirements of synthetic sleepers in Japan [5].

In this standard, Article 5.12 requires "non-combustibility" for synthetic sleepers, and Article 10.12 covers the test method for fire resistance of synthetic railway sleepers, described in JIS K6911 Article 5.24 Method A [6] as below. Testing apparatus and specifications are shown in Figure 11.13 and Table 11.34.

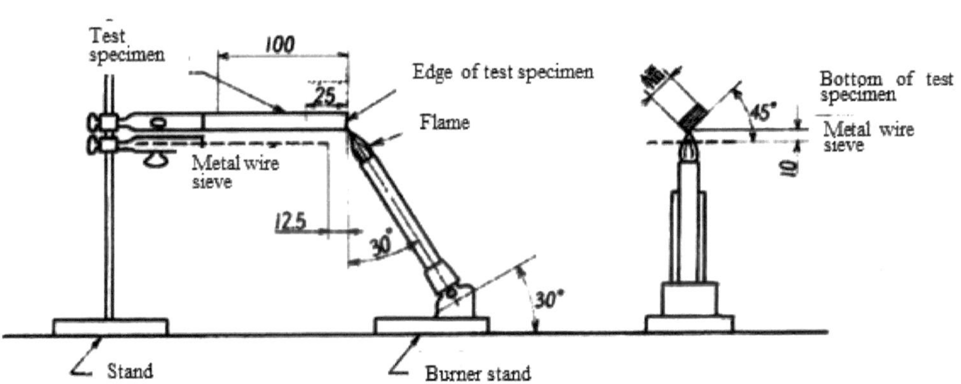

Figure 11.13 Testing apparatus for Method A. All dimensions in mm

Table 11.34 Method A: Test Specifications

Specimen	12.7 mm × 12.7 mm × 12.7 mm, with mark line 25 mm and 100 mm from the edge
Specimen position	Horizontal
Specimen holder	Horizontal in length direction, 45° to horizontal in width direction, height adjusted to contact flame top with bottom edge of test specimen Metal wire grid to JIS Z 8801 (ISO 3310-1), nominal mesh width 0.85 mm, dimension 100 × 100 mm, placed 10 mm below bottom of test specimen
Test conditions	Room temperature, no air flow

Table 11.34 Method A: Test Specifications *(continued)*

Ignition source	Bunsen burner fed with city gas or LPG, nozzle 8.5–11.5 mm, blue flame 25 mm high, inclined 30° to vertical
Test duration	After burner ignition, blue flame adjusted to 25 mm height. Flame top applied to bottom edge of specimen for 30 s. Burner removed after 30 s (distance > 450 mm), time until remaining flame on specimen extinguishes measured
Conclusions	Burning time is time in seconds until the remaining flame extinguishes. If flaming of specimen > 180 s, it shall be extinguished and classified as "Flammable". After the flame extinguishes, the length of the burnt part of the specimen is measured on the bottom side of the specimen

The specification for classification to JIS K6911 Article 5.24 Method A6 is summarized in Table 11.35.

Table 11.35 Specification for Classification to JIS K6911 Article 5.24 Method A[6]

Classification	Flaming after burner removal	Length of specimen burnt part
Noncombustible	Extinguished < 180 s	< 25 mm
Self-extinguishing	Extinguished < 180 s	> 25 mm and < 100 mm
Flammable	> 180 s	–

11.2.5.3 Official Approval

Japan Railway Rolling Stock & Machinery Association (JRMA) is the only certification body to cover fire safety regulation for railway rolling stock in Japan [2]. The procedures to obtain the certificate, details, and Q&A about test methods are covered in the guideline book published by JRMA [3].

11.2.5.4 Future Developments

In Japan, the super magnetic levitation train (Super Maglev) to be operated by Japan Railway Tokai between Tokyo and Nagoya is under construction. Fire safety requirements already apply for Super Maglevs and are part of the "Technical Regulatory Standards on Japanese Railways" of 2004 [1] referred to in Table 11.30 as "Su".

For the future development of fire safety standards for railways in Japan, JRMA is working on the following projects (status mid-2020):

- Development of a new test method for roof tops, where the gas analysis of roof top insulation material according to EN 17084 (ISO 5659-2 + FTIR) is compared to the Railway Technical Research Institute method (Cone calorimeter + FTIR). The study is still ongoing.

- Quantification of smoke and fire performance in a combustion chamber and gas analysis comparison between EN and Japanese standards. It is proposed to set up a new EN standard-based gas analysis method. The report has not been published yet.
- Promotion of a "survey on low-smoke seating materials that meet European fire protection standards". The Tokyo Metro project started in 2017 to collect information on materials that reduce the smoke and flammability of seating materials in order to conduct a survey on improvement measures. In 2019, the survey results were summarized in a report not available to the public.

References for Section 11.2.5

[1] "Technical Regulatory Standards on Japanese Railways", (Ministerial Ordinance No. 124 of 27th December, 2004) Railway Bureau, Ministry of Land, Information, Transport and Tourism (MLIT).

[2] Home page of JRMA: *http://www.rma.or.jp/exam/index.html.*

[3] "The guideline of fire test of materials for use on railways", Japan Railway Rolling Stock & Machinary Association (JRMA), March 2016 (in Japanese).

[4] "Eslon Neolamber FFU, Synthtic Railway Sleeper", Sekisui Chemical, December 2018 (in Japanese).

[5] JIS E1203:2007, Synthetic sleepers - Made from fiber reinforced foamed urethane (in Japanese).

[6] JIS K6911:1995, Testing method for thermosetting plastics (in Japanese).

11.2.6 Republic of Korea

Jungmin Choi

11.2.6.1 Statutory Regulations

The Rail Vehicle Technical Standards (KRTS-VE-Part 1-2016(R1)) [1] (hereinafter "the technical standards") stipulate technical standards regarding the type approval for rolling stock in accordance with Article 26 Clause 3, the approval for manufacturers of rolling stock in accordance with Article 26-3 Clause 2, the completion inspection of rolling stock in accordance with Article 26-6, and the post-management of type approval for rolling stock in accordance with Articles 31 and 32 of the Railroad Safety Law (hereinafter, " the law") [2].

"Type approval inspection" refers to design suitability inspection, conformity inspection, and formal inspection of rolling stock, conducted for verifying whether the rail vehicle design is appropriate for the technical standards in accordance with Article 48 Clause 1 of the Enforcement Regulations of Railway Safety Act (hereinafter, "the regulations"). Fire safety tests for interior materials and other materials used in rail vehicles are part of this type approval inspection.

11.2.6.2 Fire Hazard of Rail Vehicles

Fire safety standards for railroads in Korea have been modified technically based on the concept of the European standard prEN 45545 to be more appropriate for the environment of Korea, including a classification into four hazard levels. For example, a distance limit of 20 km or less has been added for operation category C, in order to better apply the rules for tunnels and elevated roads in Korea.

More specifically, the fire hazard of high-speed rail vehicles, regular rail vehicles, and urban rail vehicles is classified into four levels according to their operation categories and design types, as shown in Tables 11.36 to 11.38.

Table 11.36 Classification of Operation Category of Passenger Cars

Operation category	Evacuation of passenger	Characteristics of operation range
A	Immediate evacuation	Underground section, tunnel, or elevated section is 10% or less, and passenger cars are operated through sections where the maximum length of tunnel or elevated road is 1 km or less
B	Short evacuation time	The distance to the next safety area[a] is 5 km or less, or passenger cars can reach the next safety area in less than 4 minutes when travelled at a regular speed
C	Long evacuation time	The distance to the next safety area[b] is 20 km or less, or passenger cars can reach the next safety area in less than 15 min when travelled at a regular speed
D	Difficult or impossible to evacuate	Regions not classified above or passenger cars operate in areas in which evacuation is not possible[c]

[a] The next safety area is an open area or a ground-level station where passengers are completely unaffected by the fire hazard.

[b] Areas in which evacuation is not possible refer to underground sections, tunnels, or elevated sections of 200 m or longer where no evacuation route is available.

[c] If a rail vehicle belongs to two or more operation categories, the category for which evacuation is more difficult applies.

Table 11.37 Classification of Design Type

Design type	Application
A	Unmanned passenger car
B	Passenger car installed with beds, or two-story passenger car
C	High-speed rail vehicle, diesel vehicle, electric vehicle, or regular passenger vehicle
D	Passenger cars not belonging to type A, B, or C

Table 11.38 Classification of Hazard Levels

Operation Category	Design type hazard level (HL)			
	D	C	B	A
[A]	HL 1	HL 1	HL 2	HL 2
[B]	HL 2	HL 2	HL 3	HL 4
[C]	HL 2	HL 3	HL 4	HL 4
[D]	HL 3	HL 4	HL 4	HL 4

[a] Operation categories and design type symbols must be specified in the manufacturing specifications of passenger cars.

[b] Motor cars with a self load of 10 tons or over and special vehicles (other than small couchette trollies for workers) can be excluded from the fire resistance test.

11.2.6.3 Fire Performance Criteria

11.2.6.3.1 Test Methods

Details of the following test methods and specifications are provided in Section 11.2.2.1. Other terms used in the definitions below are defined in ISO 13943, ISO 3261, and ISO 8421-1.

(1) Heat Release Rate: The amount of heat energy released from the combustion of the specimen per unit area and per unit time, according to ISO 5660-1 or KS F ISO 5660-1.

(2) Limiting Oxygen Index: The minimum concentration of oxygen needed for the sustained combustion of a material at room temperature when tested to ISO 4589-2 or KS M ISO 4589-2.

(3) Density of Smoke (D_s): The amount of smoke generated by the combustion of the specimen measured by the change in light transmittance when tested to ASTM E 662. For the smoke density test, the requirements in both flaming mode and non-flaming mode of the ASTM E 662 test must be satisfied.

(4) Index of Toxicity (R): The comparison between the concentration of major toxic gases (CO_2, CO, HF, HCl, HBr, HCN, NO_2, SO_2) generated during the combustion of the specimen and the reference values. British standard BS 6853 Annex B.2 applies.

(5) Heat for Sustained Burning (Q_{sb}): The average heat energy per unit area needed for the combustion of the specimen to continue. This is tested to ISO 5658-2 or KS F 2844.

(6) Critical Flux at Extinguishment (CFE): The heat energy per hour that the unit area receives at the point where flame spread ends. This is tested to ISO 5658-2 or KS F 2844.

Performance criteria for the body structure and exterior materials besides the interior facilities of passenger car body require items (7) and (8) in addition to items (1)–(6).

(7) Vertical flame spread distance of cable: The extent of damage caused by the combustion of the specimen in the vertical burning test for bunched wires or cables, performed to IEC 60332-3-24.

(8) Fire Resistance Time: The time span in which all requirements for loadbearing capacity, insulation, and integrity of the specimen are satisfied. The requirements of KS F 2257-1 or the general requirements of ISO 834-1 shall be met.

11.2.6.3.2 Items Subject to Testing

Items subject to fire testing are shown in Table 11.39.

Table 11.39 Items Subject to Fire Tests

Number	Item	Detail items[a]	Applicable target	
1	Interior panel	Interior wall, ceiling panel, interior duct, door, window frame, and flame-retardant panels in gangway connections	Interior facilities of passenger cars	Body structure and exterior materials
2	Seat	Covers, cushions, body		
3	Gangway connection	(Type A) When there is no flame-retardant panel at the ends of a cabin or a gangway connection		
		(Type B) When there is a flame-retardant panel at the ends of a cabin or a gangway connection		
4	Flooring	Surface material of flooring		
5	Insulation	Wall, ceiling, interior of cabin duct, interior of exterior air-conditioning unit, bottom of the flooring		
6	Wires (cable)[b]	Interior and exterior wires of the rail vehicle	-	
7	Body Structure[c]	Roof, ends, and floor structure of the body		
8	Body trim	Exterior wall of the body and front mask		

[a] For flooring, paint, adhesive, and coating agents, an aluminum sheet of 3 mm or less can be used as the back board for all tests except for the LOI.

[b] Special wires (power collector high-tension wires, data bus cables) are excluded from the above fire tests.

[c] The entire structure from the base panel of the external flooring to internal floor covering is tested for the fire performance of the floor structure; additional loads are applied for the full capacity of passengers.

The items listed in Table 11.39 are excluded from testing in any of the following circumstances:

- The item is used in the interior of a passenger car, and is manufactured from a material meeting one of the following non-combustibility criteria below:
  1. Materials manufactured from metal, glass, or minerals (such as steel, copper, aluminum, and stainless steel)
  2. Materials having a temperature increase of 50 °C or less in the ISO 1182 or KS F ISO 1182 tests
  3. Materials having an LOI of 70 or more in the ISO 4589-2 or KS M ISO 4589-2 tests.
- The core of the item is completely covered with 1 mm or more of a non-combustible material.
- The item is manufactured from one of the non-combustible materials listed in Table 11.40.

Table 11.40 List of Non-combustible Materials

No.	Material	No.	Material
1	Clay	10	Cement
2	Expanded perlite	11	Limestone
3	Vermiculite	12	Blast furnace slag and pulverized fly ash
4	Mineral wool	13	Mineral
5	Glass fiber	14	Steel, metal, and aluminum metal
6	Concrete (ready-mix type and precast or steel-wire-reinforced cement)	15	Copper and copper alloy
7	Aggregate concrete	16	Zinc and zinc alloy
8	Autoclaved aerated concrete	17	Aluminum and aluminum alloy
9	Fiber cement	18	Lead

The performance criteria are summarized in Table 11.41.

Table 11.41 Performance Criteria

Requirement	Test method	Tested parameter	Criteria for Hazard Levels (HL)			
			HL 1	HL 2	HL 3	HL 4
Interior panel	ISO 4589-2	LOI	≥ 28	≥ 32	≥ 35	≥ 40
	ISO 5658-2	Q_{sb} (MJ/m²)	≥ 1.2	≥ 1.2	≥ 1.5	≥ 1.5
	ISO 5658-2	CFE (kW/m²)	≥ 15	≥ 18	≥ 20	≥ 20
	ASTM E 662	D_s (1.5 min)	≤ 150	≤ 100	≤ 75	≤ 50
	ASTM E 662	D_s (4.0 min)	≤ 300	≤ 200	≤ 150	≤ 100
	ASTM E 662	D_s (10 min)	-	-	≤ 300	≤ 200
	BS 6853 Annex B.2	Index of toxicity (*R*)	≤ 3.6	≤ 2.7	≤ 1.6	≤ 1.6

Table 11.41 Performance Criteria *(continued)*

Requirement		Test method	Tested parameter	Criteria for Hazard Levels (HL)			
				HL 1	HL 2	HL 3	HL 4
Chair	Cover	ISO 4589-2	LOI	≥ 24	≥ 26	≥ 28	≥ 28
		ASTM E 662	D_s (1.5 min)	≤ 150	≤ 125	≤ 100	≤ 100
		ASTM E 662	D_s (4.0 min)	≤ 300	≤ 250	≤ 200	≤ 200
		BS 6853 Annex B.2	Index of toxicity (*R*)	≤ 3.2	≤ 2.7	≤ 2.3	≤ 2.0
	Cushion[b]	ISO 4589-2	LOI	–	≥ 22	≥ 24	≥ 28
		ASTM E 662	D_s (1.5 min)	≤ 175	≤ 175	≤ 125	≤ 100
		ASTM E 662	D_s (4.0 min)	≤ 300	≤ 300	≤ 200	≤ 175
		BS 6853 Annex B.2	Index of toxicity (*R*)	≤ 3.6	≤ 3.6	≤ 3.2	≤ 3.2
	Body	ISO 4589-2	LOI	≥ 28	≥ 28	≥ 32	≥ 32
		ASTM E 662	D_s (1.5 min)	–	–	–	≤ 100
		ASTM E 662	D_s (4.0 min)	≤ 300	≤ 300	≤ 200	≤ 200
		BS 6853 Annex B.2	Index of toxicity (*R*)	≤ 3.2	≤ 2.7	≤ 2.3	≤ 2.0
Gang-way connec-tion	Type A	ISO 4589-2	LOI	≥ 28	≥ 32	≥ 35	≥ 40
		ISO 5658-2	Q_{sb} (MJ/m^2)	–	≥ 1.2	≥ 1.5	≥ 1.5
		ISO 5658-2	CFE (kW/m^2)	≥ 15	≥ 18	≥ 20	≥ 20
		ASTM E 662	D_s (1.5 min)	≤ 150	≤ 100	≤ 75	≤ 50
		ASTM E 662	D_s (4.0 min)	≤ 300	≤ 200	≤ 150	≤ 100
		BS 6853 Annex B.2	Index of toxicity (*R*)	≤ 3.6	≤ 2.7	≤ 1.6	≤ 1.6
	Type B	ISO 4589-2	LOI	≥ 26	≥ 26	≥ 28	≥ 28
		ISO 5658-2	Q_{sb} (MJ/m^2)	–	–	≥ 1.0	≥ 1.0
		ISO 5658-2	CFE (kW/m^2)	≥ 7	≥ 7	≥ 10	≥ 10
		ASTM E 662	D_s (1.5 min)	≤ 200	≤ 150	≤ 100	≤ 100
		ASTM E 662	D_s (4.0 min)	≤ 400	≤ 300	≤ 200	≤ 200
		BS 6853 Annex B.2	Index of toxicity (*R*)	≤ 3.6	≤ 3.6	≤ 2.7	≤ 2.7
Flooring[a]		ISO 4589-2	LOI	≥ 24	≥ 26	≥ 28	≥ 28
		ISO 5658-2	CFE (kW/m^2)	≥ 4.5	≥ 4.5	≥ 7	≥ 7
		ASTM E 662	D_s (1.5 min)	–	–	–	≤ 100
		ASTM E 662	D_s (4.0 min)	≤ 400	≤ 300	≤ 250	≤ 200
		BS 6853 Annex B.2	Index of toxicity (*R*)	≤ 5.0	≤ 4.0	≤ 3.0	≤ 3.0

Requirement		Test method	Tested parameter	Criteria for Hazard Levels (HL)			
				HL 1	HL 2	HL 3	HL 4
Insulation	Wall, ceiling, cabin duct interior	ISO 4589-2	LOI	≥ 24	≥ 28	≥ 32	≥ 35
		ISO 5658-2	CFE (kW/m²)	-	≥ 7	≥ 10	≥ 20
		ASTM E 662	D_s (4.0 min)	≤ 350	≤ 300	≤ 250	≤ 100
		BS 6853 Annex B.2	Index of toxicity (*R*)	≤ 3.0	≤ 2.7	≤ 2.0	≤ 1.6
	Interior of exterior air-conditioning unit	ISO 4589-2	LOI	≥ 24	≥ 24	≥ 28	≥ 32
		ISO 5658-2	CFE (kW/m²)	-	-	≥ 7	≥ 10
		ASTM E 662	D_s (4.0 min)	≤ 400	≤ 350	≤ 300	≤ 250
		BS 6853 Annex B.2	Index of toxicity (*R*)	≤ 4.0	≤ 3.0	≤ 2.7	≤ 2.0
	Flooring	ISO 4589-2	LOI	≥ 24	≥ 24	≥ 28	≥ 28
		ISO 5658-2	CFE (kW/m²)	-	-	≥ 7	≥ 7
		ASTM E 662	D_s (4.0 min)	≤ 400	≤ 350	≤ 300	≤ 250
		BS 6853 Annex B.2	Index of toxicity (*R*)	≤ 4.0	≤ 3.0	≤ 2.7	≤ 2.7
Wires[c]	Internal	IEC 60332-3-24	Vertical flame spread (distance; m)	≤ 2.5	≤ 2.5	≤ 2.5	≤ 2.5
		IEC 61034	Density of smoke (Transmittance: %)	≥ 25	≥ 50	≥ 50	≥ 70
		BS 6853 Annex B.1	Index of toxicity (*R*)	≤ 3.6	≤ 2.7	≤ 1.6	≤ 1.6
	External	IEC 60332-3-24	Vertical flame spread (distance: m)	≤ 2.5	≤ 2.5	≤ 2.5	≤ 2.5
		IEC 61034	Density of smoke (Transmittance: %)	-	≥ 25	≥ 50	≥ 50
		BS 6853 Annex B.1	Index of toxicity (*R*)	-	≤ 3.6	≤ 2.7	≤ 2.7
Body Structure[d]	Flooring	KS F 2257-5 or ISO 834-5	Fire resistance duration (min)	≥ 15	≥ 15	≥ 20	≥ 20
	Ends, Roof[e]	KS F 2257-1 or ISO 834-1	Flame resistance performance duration (min)	≥ 15	≥ 15	≥ 20	≥ 20

Table 11.41 Performance Criteria *(continued)*

Requirement		Test method	Tested parameter	Criteria for Hazard Levels (HL)			
				HL 1	HL 2	HL 3	HL 4
Body Trim[d]	Body, external wall, front mask	ISO 4589-2	LOI	≥ 24	≥ 28	≥ 28	≥ 28
		ISO 5658-2	CFE (kW/m^2)	≥ 10	≥ 10	≥ 15	≥ 15
		ASTM E 662	D_s (1.5 min)	≤ 200	≤ 100	≤ 100	≤ 100
		ASTM E 662	D_s (4.0 min)	≤ 400	≤ 200	≤ 200	≤ 200
		BS 6853 Annex B.2	Index of toxicity (R)	≤ 3.2	≤ 3.2	≤ 3.0	≤ 3.0

[a] Carpet flooring can be used if it meets the French NF F 16 101 standards. In this case, hazard levels 3 and 4 in Table 11.38 are considered to be in category A1, hazard level 2 in A2, and hazard level 1 in B.

[b] Cushions covered with a flame-retardant layer having an LOI of 32 or higher can be exempted from the LOI test.

[c] Wires can be used if they meet the BS 6853 Category Ia or NF F 16 101 Category A1 standards, while wires that penetrate both interior and exterior must satisfy the standards for internal wires.

[d] When the body is made of non-combustible materials, the fire performance test for body structure and body trim specified in this table can be omitted. When the body is not made of non-combustible materials, fire resistance tests for body trim must be performed using the specimen manufactured with the same paint, film, or coating used for actual cars.

[e] If the power supply voltage is 600 V or less, or the feed structure does not penetrate the roof, then the integrity test for the roof is omitted. The integrity test for the ends and roof must be performed for the outermost structure of the body; if the thickness of the structure varies, the test is performed at the thinnest part with a total area of 5 m^2 or more.

11.2.6.4 Future Developments

Objects subject to reaction-to-fire testing have not been subdivided to such a degree as specified in EN 45545. The national standards of Korea follow the Rules of Safety Standards for Urban Railway, which were announced by the Ministry of Land, Infrastructure, and Transportation and enacted as an urgent countermeasure after the Daegu subway arson fire in 2003, but abolished on March 19, 2014 and incorporated into the Urban Rail Vehicle Technical Standards. These designate five major items (interior panel, insulation, flooring, chair, gangway connection) subject to fire testing. Other items must have an oxygen index of at least 24 and are continually being added to the list. Electrical wires, car body, and car structures have been added subsequently, and more items are being added every year based upon discussions with railroad manufacturers.

References for Section 11.2.6

[1] Rail Vehicle Technical Standards, 2016.

[2] Railroad Safety Law, 2016, *https://elaw.klri.re.kr/kor_service/lawView.do?hseq=43441&lang=ENG.*

11.2.7 Australia

Alex Webb

11.2.7.1 Statutory Regulations

The Council of Australian Governments (COAG) drove the establishment of a single national rail safety regulator, the Office of the National Rail Safety Regulator (ONRSR), to administer a nationally consistent rail safety law, the *Rail Safety National Law* [1]. This represented a major change in rail safety regulation, which has historically been delivered by State and Territory regulators. ONRSR is an *independent body corporate* first established under the Rail Safety National Law (South Australia) Act 2012. Each state and territory has passed a law explaining that the *Rail Safety National Law* is the rail safety law in that state or territory or replicates that law. The law establishes the ONRSR as the body responsible for rail safety regulation in that state or territory.

The Commonwealth of Australia consists of six States and two Territories. Each of these joined with ONRSR at dates between 2013 and 2017.

11.2.7.2 Passenger Rolling Stock

Technical specification documents are bespoke for each passenger train project. Historically, they were based on a mix of Australian Standards: state-specific methods such as the FE103 [2] set of tests, and parts of the foreign standards BS 6853 [3] and/or NFPA 130 [4].

In 2014, AS 7529 Australian Railway Rolling Stock - Fire Safety (parts 1 to 4) was published by the Rail Industry Safety and Standards Board (RISSB), a not-for-profit organization funded by its members. This standard now provides a nationally consistent approach and is often the basis of the passenger rolling stock specifications. AS 7529 series is performance-based. It contains a Purpose, Performance Requirements and Deemed to Satisfy (DtS) provisions.

There are two types of controls contained within the Standard:

1. Mandatory (MAN) requirements provided as performance-based requirements
2. Recommended (REC) requirements provided as DtS provisions.

Both of these types of control address hazards that are deemed to require controls on the basis of existing Australian and international rolling stock practices and standards.

Determination of compliance "not through the DtS provisions" is mandated to be a fire engineering assessment of the design carried out in accordance with the methodologies set out in the International Fire Engineering Guidelines (IFEG) [5]. The assessment must also include fire hazard and risk analysis in accordance with AS/NZS ISO 31000 [6].

DtS provisions are non-mandatory, but if met, will result in compliance with the standard. Compliance with the performance requirements is mandatory.

The most relevant standard to the plastics industry is AS 7529.3 Australian Railway Rolling Stock - Fire Safety - Passenger [7], which covers passenger rolling stock.

11.2.7.3 Materials Fire Performance

Materials selected for use must comply with the material fire performance requirements stated in AS 7529.3 Section 5.

In general, this requires testing of each material to the fire performance tests specified in EN 45545-2 [8] hazard level 2 with the following exceptions:

- Hazard level 3 is nominated for ceiling linings and wall panels.
- BS 6853:1999 [3] category 1b is an option for testing and selecting cables.
- Seating has two options:
  - Requirement Set 1 is based on BS 6853 test and requirements.
  - Requirement Set 2 is based on EN 45545-2 with an additional full-scale test, CSIRO's 900 g timber crib ignition source [9]. The crib is similar in construction to a large BS 5852 crib but with dimensions of 300 × 400 × 70 mm and an ignition source with a peak of 250 kW. The crib is placed on or under a set of the seats. The acceptance criterion is flame spread limited to less than one seat width from the seat ignited.
- The performance of fire-rated elements requires testing in accordance with AS 1530.4 [10].
- Many of the technical specifications for passenger trains also require a Duggan's method [11] calculation to determine "fire power" or peak-summed heat release rate. This method is based on summing the product of "exposed surface areas of each combustible material internal to the train cars" and "heat release rate per unit area" (HRRPUA) data from AS 3837 [12] (ISO 5660 [13]).

11.2.7.4 Future Developments

No future developments have been identified.

References for Section 11.2.7

[1] *Rail Safety National Law (South Australia) Act 2012*, Government of South Australia, 2018.

[2] *Flammability requirements for materials used in the construction and fitting out of passenger rolling stock* (Report No.: Specification FE103-99), Maintenance SRPF, 1999.

[3] *BS 6853:1999: Code of practice for fire precautions in the design and construction of passenger carrying trains*, British Standards Institute, 2002.

[4] *NFPA 130: Standard for fixed guideway transit and passenger rail systems*, National Fire Protection Association, USA, 2010.

[5] *International Fire Engineering Guidelines*, Australian Building Codes Board, 2005.

[6] *AS/NZS ISO 31000:2009 Risk Management - Principles and guidelines*, Standards Australia (Superseded), 2009.

[7] *AS 7529.3 Australian Railway Rolling Stock - Fire Safety - Passenger*, Rail Industry Safety Standards Board, 2014.

[8] *BS EN 45545-2:2013: Railway applications. Fire protection on railway vehicles. Part 2: Requirements for fire behaviour of materials and components* British Standards Institute, 2015.

[9] *Selection of seating for land based public transport on the basis of fire performance testing* (EP182585), CSIRO, 2018.

[10] *AS 1530.4-2014: Methods for fire tests on building materials, components and structures - Part 4: Fire Resistance tests for elements of construction*, Standards Australia, 2014.

[11] Duggan, G.J. (editor), *Usage of ISO 5660 data in UK railway standards and fire safety cases. Fire hazards, testing, materials and products*, Rapra Technology Limited, 1997.

[12] *AS/NZS 3837-1998 (R2016) Amdt 1, Method of test for smoke and heat release rates for materials and products using an oxygen consumption calorimeter*, Standards Australia, 1998.

[13] *ISO 5660-1: Reaction-to-fire tests - Heat release, smoke production and mass loss rate - Part 1: Heat release rate (cone calorimeter method) and smoke production rate (dynamic measurement)*, International Organization for Standardization, 2015.

11.3 Ships

Edith Antonatus

International safety standards for merchant ships became a world-wide concern after the tragedy of the passenger ship "Titanic". Consequently, the first version of the International Convention for the Safety of Life at Sea (SOLAS) [1] was adopted in 1914 by an inter-governmental conference, and was revised in 1929 and 1948.

In 1948, an international conference convened by the United Nations established the Inter-Governmental Maritime Consultative Organization (IMCO). The name of the organization was changed to International Maritime Organization (IMO) in 1982. The function of IMO has been, since its establishment as IMCO, to develop international regulations and standards that result in safer ships and cleaner oceans. One of the major works of IMO has been the development and adoption of revisions and amendments to SOLAS, and total revisions of SOLAS were adopted in 1960 and 1974. Since 1974 to the present date, SOLAS has been amended many times to enhance the safety of merchant ships (see *http://www.imo.org*), and several consolidated versions have been published in that time.

In 2013, IMO restructured its sub-committees. The SDC Sub-Committee (Ship Design and Construction) is now responsible for structural fire protection including the IMO FTP Code (Fire Test Procedures Code), and the SSE Sub-Committee (Ship

System and Equipment) defines requirements for fire protection systems (system for fire detection, firefighting, and their fire safety related systems and equipment).

Fire safety of ships engaged in international trade is regulated by Chapter II-2 of SOLAS. The chapter includes detailed fire safety provisions for all ships and specific measures for passenger ships, cargo ships, and tankers. It was published in 1974 and revised several times after major fire disasters on-board ships, and when additional fire safety measures became necessary to avoid such disasters. This accumulation of amendments made the chapter complex, and gave rise to demand for a more user-friendly version. Consequently, the Fire Protection Sub-Committee (FP) of IMO has worked on a comprehensive review of SOLAS Chapter II. The resulting new draft Chapter II-2 was approved by the IMO Maritime Safety Committee (MSC) in May 2000. A further revision was approved in 2010 and came into force in July 2012. In June 2017 new "interim guidelines for use of fiber reinforced plastic (FRP) elements within ship structures: fire safety issues" were issued, in order to also allow the use of combustible pipes on ships, by applying an alternative to the prescriptive approach taken in the past.

11.3.1 Statistics Regarding Fires on Board Passenger Ships

The data published by the Japan Transport Safety Board in Table 11.42 [2] and the overview from the European Maritime Safety Agency [3] show that after flooding, foundering and loss of control fires and/or explosions are the cause of a significant number of casualties. Therefore, fire safety on ships is still a major issue for safety of life at sea.

Table 11.42 Statistics of Marine Accidents [2]

Year	Collision	Contact	Grounding	Sinking	Flooding	Capsizing	Fire	Explosion	Vessel missing	Facility damage	Fatality/ Injury	Others	Total
2019	197	84	174	11	22	57	29	1	0	18	125	0	718
2018	243	87	171	22	26	51	24	2	0	24	179	0	828
2017	200	96	181	14	22	55	27	3	0	23	143	0	764
2016	217	94	163	5	19	46	26	3	0	21	144	0	738
2015	244	102	202	5	12	56	38	3	0	20	122	1	805
2014	265	116	213	7	11	61	35	1	0	37	150	3	899
2013	264	145	210	10	25	49	33	2	0	38	163	2	941
2012	246	132	264	5	21	55	44	2	0	34	155	0	958
2011	282	145	264	12	18	56	32	1	0	23	142	1	977
2010	356	180	369	15	18	50	35	2	0	26	146	0	1197
2009	325	174	431	16	19	58	42	3	0	38	217	2	1325
2008	181	101	255	12	4	28	15	3	0	30	61	0	690
2007	0	1	2	0	0	0	0	0	0	0	0	0	3

11.3.2 SOLAS Chapter II-2: International Regulations for Fire Safety of Ships

11.3.2.1 Development of SOLAS Fire Safety Regulations and Revision Activities

Safety of international trading ships is regulated by the International Convention for the Safety of Life at Sea (SOLAS), adopted in 1974. SOLAS includes a set of fire safety regulations in Chapter II-2. Table 11.43 shows the history of amendments to SOLAS Chapter II-2.

Table 11.43 SOLAS Chapter II-2, History of Amendments

Year of adoption	Amendments	Date of enforcement
1978	Enhancing tanker safety	1 May 1981
1981	Enhancing fire protection of construction	1 September 1984
1983	Enhancing means of escape and tanker safety	1 July 1986
1989	Introducing new fire safety measures	1 February 1992
1992	Enhancing passenger ship safety (after a fire on the passenger ferry "Scandinavian Star")	1 October 1994
1994	Enhancing protection of fuel lines	1 July 1998
1995	Enhancing passenger ship safety (after foundering of the passenger ferry "Estonia")	1 July 1997
1996	Introducing Fire Test Procedures Code (FTP Code) and providing interpretation of vague expressions	1 July 1998
2010	Introducing revised FTP Code 2010 [4]	1 July 2012
2017	Interim guidelines for fiber-reinforced plastic (FRP) elements within ship structures: fire safety issues	9 June 2017

SOLAS Chapter II-2 specifies the fundamental fire safety requirements in a particular regulation and the functional requirements in every regulation, specifying fire safety measures precisely. Precise requirements and technical standards for each type of fire protection system and equipment, such as sprinkler systems and CO_2 fire extinguishing systems, are now included in the Fire Safety Systems Code, which was introduced in 2014 [4].

Fundamental Principles for Fire Protection on Board of Ships

The fundamental objectives for fire safety for ships are specified in SOLAS Chapter II-2 as:

1. to prevent the occurrence of fire and explosion
2. to reduce the risk to life caused by fire

3. to reduce the risk of damage caused by fire to the ship, its cargo, and the environment
4. to contain, control, and suppress fire and explosion in their compartment of origin
5. to provide adequate and readily accessible means of escape for passengers and crew.

In order to achieve these fire safety objectives, the following functional requirements have been derived and are embodied in the regulations of SOLAS Chapter II-2:

1. division of the ship into main vertical or horizontal zones by thermal and structural boundaries
2. separation of accommodation spaces from the remainder of the ship by thermal and structural boundaries
3. restricted use of combustible materials
4. detection of any fire in the zone of origin
5. containment and extinction of any fire in the space of origin
6. protection of means of escape or access for fire fighting
7. ready availability of fire-extinguishing appliances
8. minimization of possibilities of ignition of flammable cargo vapors.

11.3.2.2 Structure of Chapter II-2 of SOLAS

Chapter II-2 of SOLAS is structured based on the following principles:

1. Level of fire safety of the new Chapter II-2 shall be the same as that of the existing Chapter II-2
2. Each regulation shall clearly state functional requirements
3. Each regulation shall have a clear relationship to regulation 2 "Fire safety objectives and functional requirements"
4. Amendments already agreed shall be incorporated, but new ideas for amendments not yet agreed shall not be incorporated
5. Vague expressions in existing regulations shall be clarified
6. Precise requirements and technical standards for each type of fire protection system and equipment shall be moved from SOLAS to the Fire Safety Systems Code
7. A new regulation for evaluation of alternative fire safety design, arrangements and solutions shall be developed.

Table 11.44 shows the structure of Chapter II-2. Each regulation specifies its functional requirement at the beginning of the regulation.

Table 11.45 shows examples of such functional requirements.

Table 11.46 shows the contents of the newly developed Fire Safety Systems Code (FSS Code).

Table 11.44 Structure of Chapter II-2

Part	Regulations
Part A: General	Regulation 1: Application
	Regulation 2: Fire safety objectives and general requirements
	Regulation 3: Definitions
Part B: Prevention of fire and explosion	Regulation 4: Probability of ignition
	Regulation 5: Fire growth potential
	Regulation 6: Smoke generation potential and toxicity
Part C: Suppression of fire	Regulation 7: Detection and alarm
	Regulation 8: Control of smoke spread
	Regulation 9: Containment of fire
	Regulation 10: Fire fighting
	Regulation 11: Structural integrity
Part D: Escape	Regulation 12: Notification of crew and passengers
	Regulation 13: Means of escape
Part E: Operational requirements	Regulation 14: Operational readiness and maintenance
	Regulation 15: Instruction, onboard training and drills
	Regulation 16: Operations
Part F: Alternative design and arrangements	Regulation 17: Alternative design and arrangements
Part G: Special requirements	Regulation 18: Helicopter facilities
	Regulation 19: Carriage of dangerous goods
	Regulation 20: Protection of vehicle, special category and ro-ro spaces

Table 11.45 Examples of Functional Requirements

Regulation 4 – Probability of ignition
The purpose of this regulation is to prevent the ignition of combustible materials or flammable liquids. For this purpose, the following functional requirements shall be met: 1. means shall be provided to control leakage of flammable liquids 2. means shall be provided to limit the accumulation of flammable vapors 3. the ignitability of combustible materials shall be restricted 4. ignition sources shall be restricted 5. ignition sources shall be separated from combustible materials and flammable liquids 6. the atmosphere in cargo tanks shall be maintained out of the explosive range.
Regulation 5 – Fire growth potential
The purpose of this regulation is to limit the fire growth potential in every space of the ship. For this purpose, the following functional requirements shall be met: 1. means of control for the air supply to the space shall be provided 2. means of control for flammable liquids in the space shall be provided 3. the use of combustible materials shall be restricted.

Table 11.46 Contents of the Fire Safety Systems Code (FSS Code)

Chapter	Contents
1	General, definitions, etc.
2	International shore connections
3	Personnel protection
4	Fire extinguishers
5	Fixed gas fire-extinguishing systems
6	Fixed foam-extinguishing systems
7	Fixed pressure water-spray and water-mist fire-extinguishing systems
8	Automatic sprinkler systems
9	Fixed fire detection and fire alarm systems
10	Sample extraction smoke detection systems
11	Low location lighting for escape routes
12	Fixed emergency fire pumps
13	Arrangement of means of escape on passenger ships
14	Fixed deck foam systems
15	Inert gas systems
16	Fixed hydrocarbon gas detection systems

11.3.2.3 Alternative Design and Arrangements for Fire Safety

Each regulation in Parts B, C, D, E, and G of the drafted new Chapter II-2 contains prescriptive requirements. However, designers and/or owners of ships may wish to use new fire safety design and/or arrangements which are not explained in the regulations but may have a fire safety level equal to or better to that specified by the prescriptive requirements. Therefore, FP has developed a regulation (new Chapter II-2 Part F Regulation 17) for evaluation of the level of fire safety provided by alternative design and arrangements [5], based on standards [6] and the SFPE Engineering Guide to Performance-Based Fire Protection [7]. The principles are specified in Regulation 17 as follows:

1. The designer may deviate from the prescriptive requirements in the regulations and design the fire safety measures differently to meet the fire safety objectives defined in Regulation 2 or other regulations in Part B, C, D, E, and G of Chapter II-2.
2. The designers, when deviating from the prescriptive requirements, shall provide proof that the alternative design or arrangements have a level of safety equivalent to that achieved by a design in accordance with the prescriptive requirements. For proving the safety level, an engineering analysis shall be carried out.

The fundamentals of such an engineering analysis are also specified in Regulation 17. It is required that:

1. The proposal states which prescriptive requirements are not used
2. The proposal specifies fire safety performance and its criteria addressed by the prescriptive requirements
3. Details of alternative design and arrangements, and the assumptions used in the design, are described, which may lead to operational restrictions of the ship (such as limiting operation under certain conditions)
4. Proof is given that the alternative design and arrangements meet the required performance and criteria derived from point 2 above.

In order to specify details of the method of proof, "Guidelines for the analysis of alternative design and arrangement for fire safety" have been developed.

Following this approach, in 2017 "interim guidelines for use of fiber-reinforced plastic (FRP) elements within ship structures: fire safety issues" [8] were published.

11.3.3 Fire Test Procedures Code (FTP code)

In 1996, IMO FP developed the Fire Test Procedures Code (FTP Code) which contains fire test procedures for fire-safe constructions and materials used on board ships. This Code has become a mandatory instrument by regulations in SOLAS Chapter II-2 (as revised in 1996 and entered into force on July 1st, 1998). This test system and related regulations are moved into the new Chapter II-2 without any change. In 2010 a new version was published, and adopted in 2012 [9]. The 2010

FTP Code provides the international requirements for laboratory testing, type-approval and fire test procedures for products referenced under SOLAS Chapter II-2 (which includes regulations on fire protection, fire detection and fire extinction). Table 11.47 shows the contents of the FTP Code 2010.

Table 11.47 Contents of the IMO FTP Code 2010 (Valid From July 2012)

- Scope
- Application
- Definitions
- Testing
- Approval
- Products which may be installed without testing and/or approval
- Use of equivalents and modern technology
- Period of grace for type approvals issued in accordance with the previous FTP Code
- List of reference
- Annex 1: Fire test procedures
  - Part 1 - Non-combustibility test
  - Part 2 - Smoke and toxicity test
  - Part 3 - Test for “A”, “B” and “F” class divisions
  - Part 4 - Test for fire door control systems
  - Part 5 - Test for surface flammability
  - Part 7 - Test for vertically supported textiles and films
  - Part 8 - Test for upholstered furniture
  - Part 9 - Test for bedding components
  - Part 10 - Test for fire-restricting materials for high-speed craft
  - Part 11 - Test for fire-resisting divisions of high-speed craft
- Annex 2: Products which may be installed without testing and/or approval
- Annex 3: Fire protection materials and required approval test methods
- Annex 4: Interpretation of SOLAS Chapter II-2, regulations 5.3 and 6.2 (MSC/Circ.1120)

Some of the parts describe the test procedures to be used. However, many of the parts refer to ISO standards or IMO test standards that have been already established.

Non-Combustibility Test (FTP Code Part 1)

Materials used for fire-resistant divisions of ships must be non-combustible. According to Annex 2 of the FTP Code, products made only of glass, concrete, ceramic products, natural stone, masonry units, common metals, and metal alloys are considered non-combustible and may be installed in ships without testing and approval. Other materials that need to be classified as non-combustible shall be tested according to Part 1 of FTP Code, which refers to ISO 1182/2010 as the test procedure. The pass/fail criteria for marine use are different from those specified in ISO 1182. The criteria are:

1. the average temperature rise of the furnace thermocouples shall not exceed 30 K
2. the average temperature rise of the surface thermocouples shall not exceed 30 K
3. the mean duration of sustained flaming shall not exceed 10 s
4. the average mass loss shall not exceed 50%.

Further details of the test apparatus and test specifications are available under Section 9.2 ISO.

Figure 11.14 shows the equipment for the ISO 1182 non-combustibility test.

Figure 11.14 Equipment for the ISO 1182 non-combustibility test (FTP code part 1)

Smoke and Toxicity Test (FTP Code Part 2)

According to the requirements of SOLAS Chapter II-2, interior surface (deck, wall and ceiling) materials shall not be capable of producing excessive quantities of smoke and toxic effluents. For this purpose, products to be used in interior of ships shall be tested and approved according to the smoke and toxicity test of Part 2 of the FTP Code. Specimens shall be tested according to ISO 5659-2/2016, and the specific optical density shall be measured under three test conditions:

1. Irradiance of 25 kW/m^2 in the presence of a pilot flame
2. Irradiance of 25 kW/m^2 in the absence of a pilot flame
3. Irradiance of 50 kW/m^2 in the absence of a pilot flame

When the optical smoke density reaches the maximum value, gases in the smoke chamber shall be sampled and the concentration of defined gas components measured with FTIR (Fourier Transform Infrared Spectrometry). Table 11.48 shows the requirements.

Table 11.48 Requirements Regarding Smoke and Toxic Gas Generation

Product application	Smoke generation	Toxicity	
	Average of max. optical density not to exceed D_s	Gas component	Max. acceptable concentration
Surface of wall and ceiling	200	CO	1450 ppm
Primary deck covering (the first layer applied to the steel deck plate)	400	HBr	600 ppm
		HCl	600 ppm
		HCN	140 ppm
Surface of floor	500	HF	600 ppm
		SO_2	120/200[a] ppm
		NO_x	350 ppm

[a] For floor coverings

Recognizing that the evaluation of toxicity using this test is not fully based on fire safety engineering and that such techniques are under development within ISO/TC92, the test method and the criteria have been adopted as an interim solution for SOLAS. In future, the test method and the criteria will be reviewed. Figure 11.15 shows a test apparatus according to ISO 5659-2, which is referred to in Part 2 of the FTP Code. Further details are given in this book in Section 9.2.

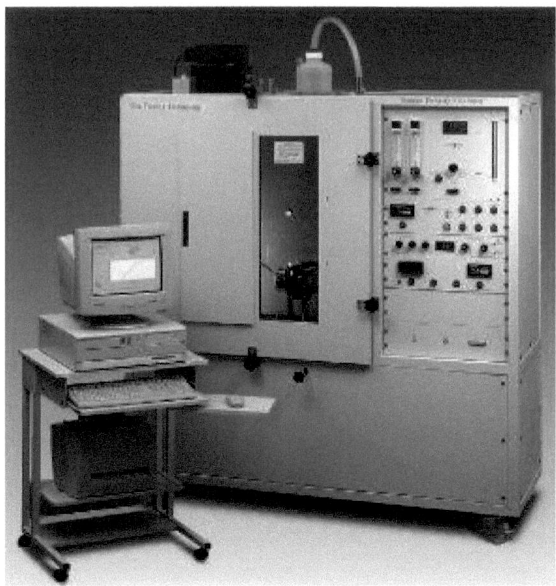

Figure 11.15
ISO 5659-2 smoke chamber

Test for Surface Flammability (FTP Code Part 5)

Fire safety regulations in SOLAS require interior surface materials to show "low flame spread". This has to be verified in accordance with Part 5 of the FTP Code.

The test method is based on ISO 5658-2. Further details are given in this book in Section 9.2. In addition, Part 5 of the FTP code requires measurement of heat release using thermocouples. A rectangular box type stack is fitted above the specimen position, and combustion fumes flow into the stack. Five thermocouples fitted at the outlet of the stack measure the temperature of the exhaust gases, and their signals are converted into a heat release rate.

The specimen (155 mm × 800 mm) is supported vertically with the longer edges in horizontal orientation, and is exposed to a heat flux generated by a gas-fueled radiant panel heater. The maximum heat flux imposed to one end of the specimen is about 50 kW/m^2. The heat flux decreases along the specimen.

Flame spread time along the specimen is measured in 50 mm intervals. Heat for sustained burning (kJ/m^2) is the average of the flame spread time (s) multiplied by the incident heat flux (kW/m^2) at the respective position on the specimen.

The maximum flame spread distance is used to determine the Critical Heat Flux at Extinguishment (CFE), which is the irradiance at the point of the specimen where the flame front ceased its progress.

The principle of the apparatus is shown in Figure 11.16. Table 11.49 shows the criteria for interior surface materials.

FTP Code Part 5 also specifies that the test method in ISO 1716 is recommended for determining the gross calorific value, if there is a requirement regarding the maximum gross calorific value for surface materials (see Section 9.2).

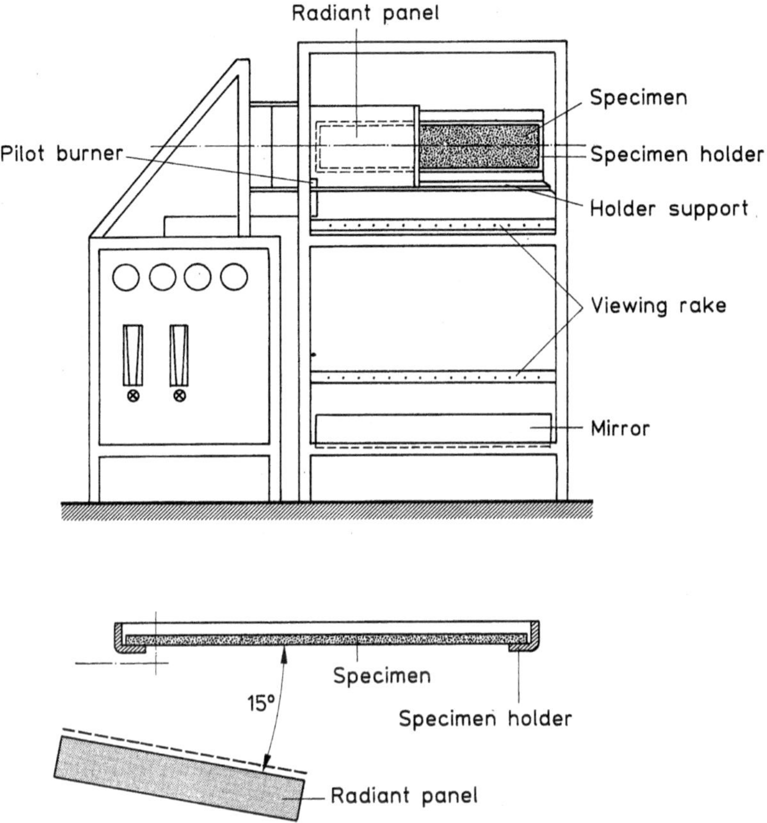

Figure 11.16 Surface flammability test to FTP code part 5

Table 11.49 Criteria for Surface Flammability

Parameter	Ceiling and wall surface	Deck surface
Critical Heat Flux at Extinguishment (kW/m^2)	≥ 20.0	≥ 7.0
Heat for Sustained Burning (MJ/m^2)	≥ 1.5	≥ 0.25
Maximum Heat release rate (kW)	≤ 4.0	≤ 10.0
Total Heat Release (MJ)	≤ 0.7	≤ 2.0

Test for Primary Deck Coverings (FTP Code Part 6)

Steel deck plates usually have many welding beads and the surface is uneven. "Primary deck covering" is a system of coating which is applied directly on the steel deck plate to make the surface flat enough to install floor coverings such as floor tiles, flooring coatings, and carpets.

A primary deck covering is required by the fire safety regulations of SOLAS, to meet the criteria for floor surfaces of Part 5 of the FTP Code. Specimens shall be made in a way that the primary deck covering is applied to a steel plate with a thickness of 3.0 ± 0.3 mm. In addition, the contribution to smoke development and toxicity have to be tested according to FTP code part 2, if the total heat release (Q_t) is > 0.2 MJ and the maximum heat release (Q_p) is > 1.0 kW.

Test for Vertically Supported Textiles and Films (FTP Code Part 7)

Draperies, curtains and other suspended textile materials in rooms of restricted fire risk shall be tested according to Part 7 of the FTP Code.

The FTP Code specifies a flammability test method for vertically supported textiles and films. A burner according to DIN 50051, type KBN, is used. The size of the test specimens is 220 mm × 170 mm.

The surface of one specimen is exposed at a central point for 5 s to the flame. If no ignition occurs the next specimen is exposed for 15 s. If again no sustained ignition occurs, the lower edge of the specimen is exposed to the flame in the same way.

The ignition condition to be used for testing the specimens shall be that at which sustained ignition is first achieved when the order of tests listed above is followed. In the absence of sustained ignition, the specimens shall be tested under conditions showing the greatest char length. The method of flame application for warp and weft specimens shall be determined using the ignition sequence given above. After determining the appropriate burner position and flame application time, further five specimens cut in both warp and weft directions shall be tested.

The classification is based upon the following criteria:

- After-flame time ≤ 5 s for any of the 10 or more specimens tested with a surface application of the pilot flame
- No burn-through to any edge of any of the 10 or more specimens tested with a surface application of the pilot flame
- No ignition of the cotton wool below the specimen for any of the specimens tested
- No average char length ≥ 150 mm in any of the batches of five specimens tested by either surface or edge ignition
- No surface flash propagating more than 100 mm from the point of ignition, with or without charring of the base fabric.

Test for Upholstered Furniture (FTP Code Part 8)

Upholstered furniture and seating in rooms of restricted fire risk and in high speed craft have to meet the requirements of Part 8 of the FTP Code regarding resistance to the ignition and propagation of flames and smoldering. Part 8 of the FTP Code defines the test method.

Surface fabric and filler are attached to a test rig 450 mm wide, 150 mm deep and 300 mm high. An ignited cigarette and a small flame of a propane gas burner are applied to the junction of seating part and back part of the test set-up. No smoldering or flaming shall propagate beyond the vicinity of the applied ignition source. Details of the test rig and specifications are shown in Section 13.2.1.1.

Type approval for "upholstered furniture" might be applied by the combination of the covering and filling. But if both materials (cover and filling material) pass the criteria of this standard and have sufficient test reports for each individual material as evidence of the independent test, an additional test for the actual combination is not required.

Test for Bedding Components (FTP Code Part 9)

Bedding components, such as mattresses, pillows, and blankets, in rooms of restricted fire risk shall meet the requirements of Part 9 of the FTP Code regarding resistance to the ignition and propagation of flames.

The specimen size is 450 mm × 350 mm × thickness as used. A mock-up of the mattress shall have the actual structure except reduced length and width. An actual pillow can be tested. Blankets should be cut to the size of 450 mm × 350 mm. Test rig and detailed specifications are described in Section 13.2.1.1.

An ignited cigarette and a small flame from a propane gas burner are applied to the top surface. The cigarette shall be covered by a piece of cotton wool. No smoldering or flaming shall propagate beyond the vicinity of the applied ignition source, and weight loss by burning has to be limited.

Test for "A", "B", and "F" Class Divisions (FTP Code Part 3)

One of the fundamental requirements regarding fire safety of ships in SOLAS is "containment and extinction of any fire in the space of origin". In order to meet this requirement, the interior of a ship shall be divided into "Main Vertical Zones" every 40 m or less by steel walls (bulkheads), and most of the decks shall be constructed of steel. These decks and bulkheads shall have a fire resistance of 60 minutes.

Materials or systems other than steel may be used for bulkheads and decks if the material or system is made of non-combustible materials and has the same degree of fire resistance as a steel construction. Other boundaries such as walls and ceilings of cabins, rooms and corridors shall have, in general, a resistance to fire of 30 minutes.

The grade of fire resistance of these bulkheads, decks, ceilings and walls (including windows) shall be verified by a standard fire resistance test specified in FTP Code Part 3. Details for these tests and the related requirements are not relevant for plastics, and therefore not explained in detail in this book.

Tests for Fire-Restricting Materials for High-Speed Craft (FTP Code Part 10)

Appendix 1 specifies the full-scale room test for surface materials on bulkheads, walls and ceiling linings, including their supporting structure, of high-speed craft.

The ISO 9705 room corner test shall be used as test method (for details, see Section 9.2). The ignition source shall have 100 kW heat output for 10 min and thereafter 300 kW heat output for another 10 min. The total testing time shall be 20 min.

The classification criteria are defined as follows:

- The average heat release rate (HRR) (excluding the HRR from the ignition source) does not exceed 100 kW
- the maximum HRR averaged over any 30 s period of time during the test (excluding the HRR from the ignition source) does not exceed 500 kW
- The average smoke production rate does not exceed 1.4 m^2/s
- the maximum value of the smoke production rate averaged over any 60 s period of time during the test does not exceed 8.3 m^2/s
- flame spread shall not reach any further down the walls of the test room than 0.5 m from the floor excluding the area, which is within 1.2 m from the corner where the ignition source is located
- no flaming drops or debris of the test sample may reach the floor of the test room outside an area which is within 1.2 m from the corner where the ignition source is located

Appendix 2 defines the fire test procedures for heat release, smoke production and mass loss rate for materials used for furniture and other components of high-speed craft.

The tests are based on:

- ISO 5660, Reaction-to-fire tests - Heat release, smoke production and mass loss rate
  - Part 1: Heat release rate (cone calorimeter method); and
  - Part 2: Smoke production rate (dynamic measurement). (see Section 9.2).

The irradiance level shall be set at 50 kW/m^2.

Materials used for furniture and other components are qualified as "fire-restricting material" if the following four criteria are fulfilled:

- the time to ignition (TIG) is greater than 20 s
- the maximum 30 s sliding average heat release rate ($HRR_{30,max}$) does not exceed 60 kW/m^2
- the total heat release (THR) does not exceed 20 MJ/m^2
- the time average smoke production rate (SPR_{avg}) does not exceed 0.005 m^2/s.

Appendix 3 defines fire test procedures for fire-resisting divisions of high-speed craft.

Details for these tests and the related requirements are not relevant for plastics, and therefore not explained in detail in this book.

11.3.4 Provisions for High-Speed Craft

High-speed passenger craft have been operated in commercial service routes since the 1980s. Hovercraft and hydrofoils were the first generation of high-speed passenger craft. In the early 1990s, there was a demand to develop larger high-speed passenger craft that could carry also cars (high-speed passenger ferry). The service speed exceeded 40 knots, and the capacity exceeded 500 passengers. The safety of such large high-speed passenger ferries became a matter of public concern. Consequently, IMO developed an international code for safety of high-speed craft (HSC Code) and introduced it as mandatory instrument alternative to SOLAS in 1996. It was amended several times, published as new version in 2000 and a further consolidated version [10] was introduced in 2008.

The current HSC Code applies to high-speed craft engaged on international voyages, including passenger craft that do not proceed for more than four hours at operational speed from a place of refuge when fully laden, and cargo craft of ≥ 500 tonnes gross weight that do not go more than eight hours from a port of refuge. It also includes air-cushion boats.

The code is based on the following principles regarding fire safety:

- maintenance of the main functions and safety systems of the craft, including propulsion and control, fire-detection, alarms and extinguishing capability of unaffected spaces, after fire in any one compartment on board
- division of the public spaces for category B craft, in such a way that the occupants of any compartment can escape to an alternative safe area or compartment in case of fire
- subdivision of the craft by fire-resisting boundaries
- restricted use of combustible materials, and materials generating smoke and toxic gases in a fire
- detection, containment and extinction of any fire in the space of origin
- protection of means of escape and access for fire fighting
- immediate availability of fire-extinguishing appliances.

Light-weight construction is the most important design goal for high-speed craft. Fire safe constructions are also required to be light-weight. Therefore, the HSC Code allows using fire-resistant boundaries, which restrict spread of fire by two ways:

- The construction has to meet the requirements of FTP Code Part 3
- The materials used do not create a flashover in case of fire, if tested according to FTP Code Part 10 Annex 1.

Additional requirements are defined referring to the test and classification methods in the FTP Code, for example:

- case furniture (FTP Code, Annex 1, Parts 1 and 10)
- frames of all other furniture (FTP Code, Annex 1, Parts 1 and 10)
- draperies, textiles and other suspended textile materials (FTP Code, Annex 1, Part 7)
- upholstered furniture, e.g., passenger seating (FTP Code, Annex 1, Part 8)
- bedding components (FTP Code, Annex 1, Part 9)
- deck finish materials (FTP Code, Annex 1, Parts 2 and 6)
- Further requirements based on the FTP code - for example, for defined surfaces, insulation, pipes, etc.

11.3.5 Future Developments

IMO is trying to use ISO standards as far as possible. In 2019, representatives of EU member state governments submitted a new work item proposal for a review of the IMO Fire Test Procedures Code (FTP Code). The intention is the introduction of the new ISO 1182 [11] and ISO/TS 19021 [12]. New evaluation criteria for toxicity of fire effluents might be considered. In addition, it is planned to further develop in cooperation with ISO more detailed specifications regarding fire testing of plastic pipes and fiber-reinforced plastic structural parts on ships.

References for Section 11.3

[1] 1974 International Convention for the Safety of Life at Sea (74 SOLAS).

[2] Statistics of Marine Accident, Japanese Transport Safety Board, 2019, *http://www.mlit.go.jp/jtsb/statistics_mar.html#p0.*

[3] Annual overview of marine casualties and incidents 2017, European Maritime Safety Agency, *https://www.isesassociation.com/wp-content/uploads/2017/11/Annual-overview-of-marine-casualties-and-incidents-2017_final.pdf.*

[4] FSS Code, International Code for Fire Safety Systems, IMO Publishing, 2015.

[5] Guidance notes on alternative design and arrangements for fire safety, January 2004, updated 2010, *https://ww2.eagle.org/content/dam/eagle/rules-and-guides/current/design_and_analysis/122_altdesignandarrangforfiresafety/fire_safety_guidance_e-july10.pdf.*

[6] ISO TR 13387-1 through 13387-8 "Fire safety engineering", ISO/TC92/SC4, 1999.

[7] SFPE Engineering Guide to Performance-Based Fire Protection. Society of Fire Protection Engineers and National Fire Protection Association, 2nd Edition, 2007.

[8] Guidelines for use of fiber reinforced plastic (FRP) elements within ship structures: Fire safety issues (MSC.1/Circ.15749), June 2017; and Corrigendum, March 2018 (MSC.1/Circ.1574/Corr.1).

[9] FTP Code 2010, International Code for Application of Fire Test Procedures, IMO Publication, 2012.

[10] HSC Code, International Code of Safety for High-Speed Craft, 2008 Edition.

[11] ISO 1182:2020-06, Reaction to fire tests for products – Non-combustibility test.

[12] ISO/TS 19021:2018, Test method for determination of gas concentrations in ISO 5659-2 using Fourier transform infrared spectroscopy.

11.4 Aircraft[1]

Torben Kempers

11.4.1 Introduction

Modern wide-bodied jets contain a large amount of plastics in structural applications, but also in compartments occupied by passengers and crew. As an example, the Boeing 787 Dreamliner is claimed to be 50% composite by weight; a majority of the primary structure is made of composite materials, most notably the fuselage. In addition, large quantities of combustible materials including maintenance materials, provisions and hand baggage are stored in the cabins. Finally, the freight (in metal containers) and the airplane's fuel in wing and fuselage tanks all add up to a significant amount of combustible material.

Surveys of transport aircraft accidents indicated that approximately 75% of accidents occur on the runway or within a radius of 3,000 m of the airport. By far the largest number of fatalities occurs because of crash landings in this area, i.e., when planes land short of or overshoot the runway on take-off.

Large quantities of fuel can be released during crashes, and are ignited by short circuits or heat. Such fires can develop fully within 1 to 3 min, and because of the low fire resistance of the aircraft fuselage, occupants have only a few minutes in which to escape.

In-flight fires occur mainly in the galleys as a result of electrical faults or overheating. In the past, smokers' materials were a frequent cause of cabin fires. Other commonly cited causes include faults in the electrical and oxygen supply systems.

Fires started by small ignition sources in the cockpit or cabin can generally be successfully tackled with on-board equipment. Exceptions include fires, which develop unnoticed in inaccessible compartments or toilets.

[1] **Note:** Portions of the text have been taken from – Plastics Flammability Handbook, 3rd Edition – Federal Aviation Regulation, Part 25 – Aircraft Materials Fire Test Handbook

Fires also occur in aircraft while on the ground, mainly during maintenance work on the electrical and oxygen supply systems. Such incidents usually do not result in casualties and can be brought under control relatively easily.

In 1980, the authorities started to improve regulations regarding the fire behavior of materials and parts used in aircraft, and achieved a significant contribution to risk reduction. In addition, most airlines no longer allow smoking on board their aircrafts.

In fires, passengers should be protected as long as necessary from flames, heat, smoke, and toxic gases by suitable measures such as division into fire compartments, and the use of fire-resistant partitions in front of concealed fuselage areas and spaces where there is a high fire risk. These measures should be supplemented by fire alarm systems and effective automatic extinguishing equipment.

The only effective way of reducing fire risk, particularly in air transport, appears to be the introduction of safety systems that take adequate account of all aspects of fire protection. Large-scale fire tests under operating conditions are a prerequisite for more effective fire regulations in aviation. They must take into account the types of accident that pose a particular hazard for occupants. Most tests involve small ignition sources simulating the start of a fire during flight. As a rule, they do not enable any predictions to be made on the behavior of structures exposed to large ignition sources.

11.4.2 Statutory Regulations

Although airplanes are normally built based on specifications agreed between manufacturers and purchasers (typically the operators), they always need to comply with national and international regulations. The national air transport authority of the country in which the aircraft is registered monitors compliance. However, because of multinational manufacturers, cross-border aircraft leasing, and the liberalization of air transport economic regulations, wide international co-operation of safety requirements is necessary.

Most countries have adopted the Federal Aviation Regulations (FAR) [1] prepared by the U.S. Federal Aviation Administration (FAA), as the basis for ensuring airworthiness. These FARs are part of Title 14, "Aeronautics and Space" of the Code of Federal Regulations (CFR) [2], and are often also referred to as "14 CFR Part xx".

In Europe, the European Aviation Safety Agency (EASA), formerly known as the Joint Aviation Authorities (JAA), have harmonized their Joint Aviation Requirements (JAR) to be in line with FAA's FARs. EASA and the FAA are working closely together to harmonize all relevant regulations.

From time to time, the Regulations are amended to reflect the latest technical developments. Experts from various industry associations and regulators advise the FAA in this process.

The following FAR regulations are most common, and apply to most countries worldwide:

- FAR Part 23 - Airworthiness Standards: Normal category airplanes
- FAR Part 25 - Airworthiness Standards: Transport category airplanes
- FAR Part 27 - Airworthiness Standards: Normal category rotorcraft
- FAR Part 29 - Airworthiness Standards: Transport category rotorcraft
- FAR Part 33 - Airworthiness Standards: Aircraft engines.

Each Part is divided into subparts, which are composed of several sections. The most relevant sections of Part 25 related to plastics flammability are the following:

- Section 853 ("FAR 25.853") - Compartment interiors
- Section 855 ("FAR 25.855") - Cargo or baggage compartments
- Section 856 ("FAR 25.856") - Thermal/Acoustic insulation materials
- Section 857 ("FAR 25.857") - Cargo compartment classification.

11.4.3 Fire Behavior Testing

In FAR Part 1 (Definitions and Abbreviations), the terminology of technical fire protection is given. However, many terms employed such as "fireproof", "fire-resistant", and "flash-resistant" can often have different meanings in other standards and regulations.

Requirements for materials and parts used in crew and passenger compartments in transport aircraft are defined in FAR Part 25, Section 853 ("FAR 25.853") [3]. They also apply to other aircraft where appropriate. See Table 11.50 and Table 11.51.

Table 11.50 FAR 25.853 - Test Requirements for All Aircraft - Compartments Occupied by Crew and Passengers

Application	Test requirements
Interior ceiling panels, interior wall panels, partitions, galley structure, large cabinet walls, structural flooring, and materials used in the construction of stowage compartments (other than underseat stowage compartments and compartments for stowing small items such as magazines and maps)	▪ 60 s Vertical Bunsen Burner test
Seat cushions, except those on flight crewmember seats	▪ 60 s Vertical Bunsen Burner test ▪ Oil Burner test

Table 11.50 FAR 25.853 - Test Requirements for all Aircraft - Compartments Occupied by Crew and Passengers *(continued)*

Application	Test requirements
Floor covering, textiles (including draperies and upholstery), seat cushions, padding, decorative and non-decorative coated fabrics, leather, trays and galley furnishings, electrical conduit, air ducting, joint and edge covering, liners of Class B and E cargo or baggage compartments, floor panels of Class B, C, E, or F cargo or baggage compartments, cargo covers and transparencies, molded and thermoformed parts, air ducting joints, and trim strips (decorative and chafing)	▪ 12 s Vertical Bunsen Burner test
Clear plastic windows and signs, parts constructed in whole or in part of elastomeric materials, edge-lighted instrument assemblies consisting of two or more instruments in a common housing, seat belts, shoulder harnesses, and cargo and baggage tiedown equipment, including containers, bins, pallets, etc., used in passenger or crew compartments	▪ Horizontal Bunsen Burner test (max. 2.5 in/min burn rate)
Materials in items not specified above - *except* for small parts (such as knobs, handles, rollers, fasteners, clips, grommets, rub strips, pulleys, and small electrical parts) that would not contribute significantly to the propagation of a fire, and for electrical wire & cable insulation	▪ Horizontal Bunsen Burner test (max. 4.0 in/min burn rate)

Table 11.51 FAR 25.853 - Test Requirements for Aircraft with Passenger Capacities of 20 or More

Application	Test requirements	Notes
1. Interior ceiling and wall panels, other than lighting lenses and windows 2. Partitions, other than transparent panels needed to enhance cabin safety 3. Galley structure, including exposed surfaces of stowed carts and standard containers, and the cavity walls that are exposed when a full complement of such carts or containers is not carried 4. Large cabinets and cabin stowage compartments, other than underseat stowage compartments for stowing small items such as magazines and maps	▪ 60 s Vertical Bunsen burner test ▪ OSU 65/65 ▪ Smoke density test	The interiors of compartments, such as pilot compartments, galleys, lavatories, crew rest quarters, cabinets and stowage compartments, need not meet the OSU and smoke density test requirements, provided the interiors of such compartments are isolated from the main passenger cabin by doors or equivalent means that would normally be closed during an emergency landing condition

In FAR Part 25, Section 855 [4], the test requirements for each cargo or baggage compartment are given, whereas in FAR Part 25, Section 857 [5] a cargo compartment classification is defined.

For thermal/acoustic insulation materials, FAR 25 Part 856 [6] prescribes the following test requirements:

- Thermal/acoustic insulation material installed in the fuselage must meet the flame propagation test requirements; this requirement does not apply to "small parts"
- For airplanes with passenger capacities of 20 or greater, thermal/acoustic insulation materials (including the means of fastening the materials to the fuselage) installed in the lower half of the airplane fuselage must meet the flame penetration resistance test requirements; this requirement does not apply to thermal/acoustic insulation installations that the FAA finds would not contribute to fire penetration resistance.

Often additional requirements are listed in airline operators' and/or manufacturers' specifications, on top of the basic FAR requirements. Examples are additional requirements related to smoke emission and toxicity of combustion products. As an example, Boeing and Airbus have specifications on smoke toxicity.

In Appendix F to FAR Part 25 [7], the methods used to test the materials and components in the cabins and holds of transport aircraft are described in some detail. These tests are typically carried out on test specimens taken from the final product, or simulating the components used. Composites must be tested as such.

Detailed test descriptions and procedures can be found in the Aircraft Materials Fire Test Handbook [8], published by the FAA and publicly available through the National Technical Information Service (NTIS), or via the website of the FAA *(https://www.fire.tc.faa.gov/Handbook).*

11.4.3.1 Bunsen Burner Flammability Testing

In Chapters 1–4 of the Aircraft Materials Fire Test Handbook, the four different options of the Bunsen Burner test are described: vertical, horizontal, 45° and 60° testing. These test methods apply to most parts used in the interior. Depending on the type and area of application, additional tests are required. Windows, small parts, and electrical components are tested differently.

Vertical Bunsen Burner Test

This test is used to demonstrate that materials are "self-extinguishing", and is schematically shown in Figure 11.17.

The 60-second Vertical Bunsen Burner test is described in FAR Part 25, Appendix F, Part 1(a)(1)(i), and the 12-second test in Part 1(a)(1)(ii).

Test requirements and specifications are summarized in Table 11.52.

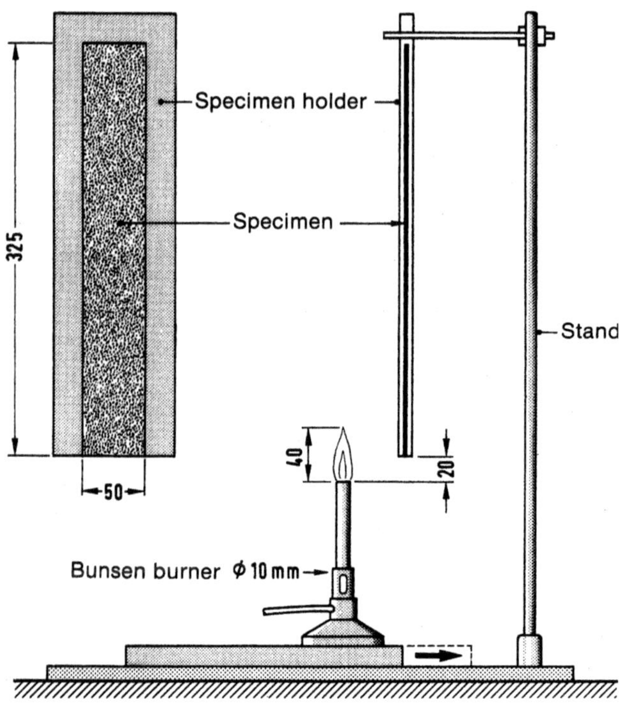

Figure 11.17 Vertical Bunsen burner test according to FAR Part 25, Appendix F. Dimensions in mm

Table 11.52 Requirements for Vertical Bunsen Burner Test

Specimens	▪ Three specimens, minimum 76 mm (3 in) × 305 mm (12 in) ▪ Thickness: minimum thickness in actual use; foam parts cut to 13 mm		
Position	Vertical		
Ignition source	Bunsen or Tirrill-type burner, with 10 mm inside diameter barrel		
Flame	38 mm (1.5 in) high flame; top of burner 19 mm (¾ in) below lower edge of specimen		
Requirements	FAR Part 25, App. F, Part 1	(a)(1)(i)	(a)(1)(ii)
	Flame application	60 s	12 s
	Burn length	≤ 152 mm (6 in)	≤ 203 mm (8 in)
	Flame time	≤ 15 s	≤ 15 s
	Flame time drippings	≤ 3 s	≤ 5 s

Horizontal Bunsen Burner Test

Windows, signs, parts made from elastomeric materials, small parts and a number of other parts, which are not installed permanently in the aircraft, are tested in the horizontal position; see Figure 11.18 for a schematic.

Proof must be obtained that a maximum permitted rate of flame spread (burn rate) is not exceeded; see Table 11.53 for exact requirements.

This test corresponds to ISO 3795 [9] and FMVSS 302 [10].

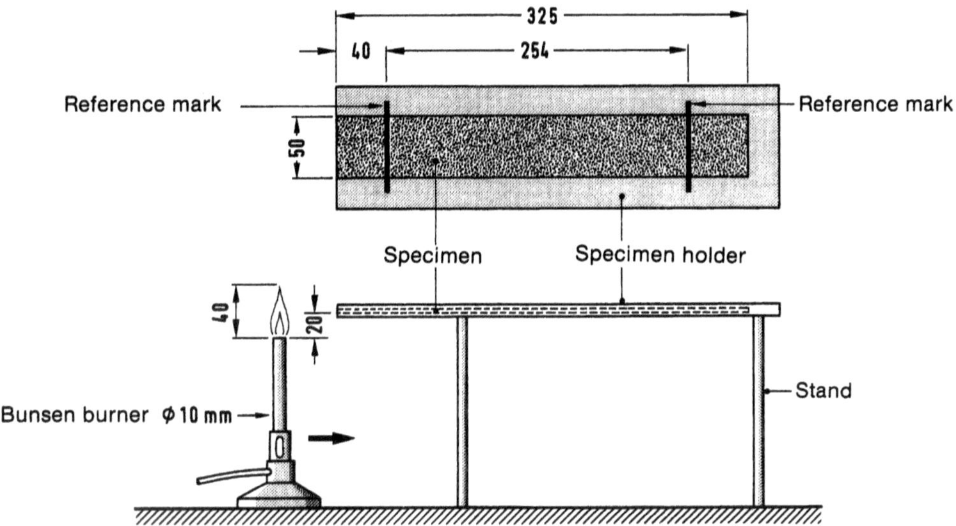

Figure 11.18 Horizontal Bunsen burner test according to FAR Part 25, Appendix F. Dimensions in mm

Table 11.53 Requirements for the Horizontal Bunsen Burner Test

Specimens	▪ Three specimens, minimum 76 mm (3 in) × 305 mm (12 in) ▪ Thickness: minimum thickness in actual use, max. 3 mm ▪ Reference marks at 38 mm (1.5 in) and 292 mm (11.5 in)		
Position	Horizontal		
Ignition source	Bunsen or Tirrill-type burner, with 10 mm inside diameter barrel		
Flame	38 mm (1.5 in) high flame; top of burner 19 mm (¾ in) below edge of specimen		
Flame application	15 s		
Requirements	FAR Part 25, App. F, Part 1	(a)(1)(iv)	(a)(1)(v)
	Burn rate	≤ 64 mm/min (2.5 in/min)	≤ 102 mm/min (4.0 in/min)

Note: A minimum of 254 mm (10 in) of specimen must be used for timing purposes; approximately 38 mm (1.5 in) must burn before the burning front reaches the timing zone.

45° Bunsen Burner Test

Cargo compartment liners and waste stowage compartment materials (see FAR 25.855) must be subjected to the 45° Bunsen Burner test, in addition to the Vertical test. See Figure 11.19 for a schematic of the test, and Table 11.54 for the exact requirements.

This test will prove that a fire starting in the cargo compartment will not easily pass through the walls of the compartment.

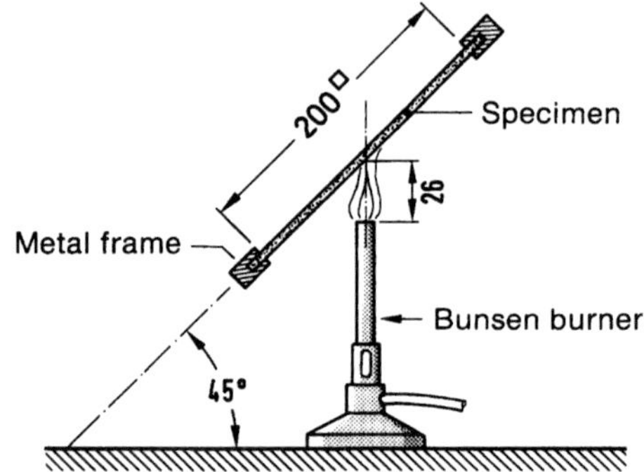

Figure 11.19 45° Bunsen burner test according to FAR Part 25, Appendix F. Dimensions in mm

Table 11.54 Requirements for the 45° Bunsen Burner Test

Specimens	▪ Three specimens, 254 mm (10 in) × 254 mm (10 in); exposed area 203 mm (8 in) × 203 mm (8 in) ▪ Thickness: minimum thickness in actual use	
Position	45°	
Ignition source	Bunsen or Tirrill-type burner, with 10 mm inside diameter barrel	
Flame	38 mm (1.5 in) high flame; top of burner 25 mm (1 in) below the center of the bottom surface of the specimen	
Flame application	30 s	
Requirements	FAR Part 25, App. F, Part 1	(a)(2)(ii), (a)(2)(iii)
	Flame time	≤ 15 s
	Glow time	≤ 10 s
	Flame penetration	Not allowed [a]

[a] The flame may not penetrate (pass through) the material during application of the flame or subsequent to its removal.

60° Bunsen Burner Test

The 60° Bunsen Burner test is used to determine the resistance of electric wires and cables to flame. See Figure 11.20 for a schematic, and Table 11.55 for exact requirements.

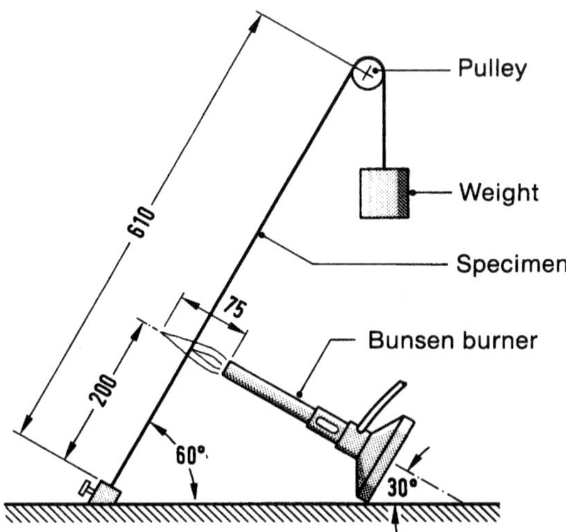

Figure 11.20 60° Bunsen burner test according to FAR Part 25, Appendix F. Dimensions in mm

Table 11.55 Requirements for the 60° Bunsen Burner Test

Specimens	▪ 3 specimens, 762 mm (30 in) in length; exposed length 610 mm (24 in) ▪ A weight is attached to keep the specimen taut	
Position	60°	
Ignition source	Bunsen or Tirrill type burner, with 10 mm inside diameter barrel	
Flame	76 mm (3 in) high flame; top of burner 25 mm (1 in) from the specimen; burner perpendicular to the specimen	
Flame application	30 s	
Requirements	FAR Part 25, App. F, Part 1	(a)(3)
	Flame time	≤ 30 s
	Drip flame time	≤ 3 s
	Burn length	≤ 76 mm (3 in)

Note: It will not be considered a failure if the wire breaks during the test.

11.4.3.2 Heat Release Testing

For large parts inside the passenger cabin of aircraft, a heat release test was developed at Ohio State University (OSU). It was introduced around 1990 [11] and resulted in major developments of new materials, because most of the materials tested did not pass.

The intent of the test is to model a post-crash, fuel-fed fire. The purpose is to extend the time to cabin flashover, so that passengers inside the cabin will have enough time to exit the aircraft. See Figure 11.21 for a schematic of the test apparatus.

The heat release rate is measured for the duration of the test from the moment the specimen is injected into the controlled exposure chamber, and encompasses the period of ignition and progressive flame involvement of the surface. At the end of the test, both the maximum heat release rate (in kW/m^2) during the test, and the total heat released (in $kW \cdot min/m^2$) during the first 2 minutes of the test, are recorded. See Table 11.56 for exact requirements.

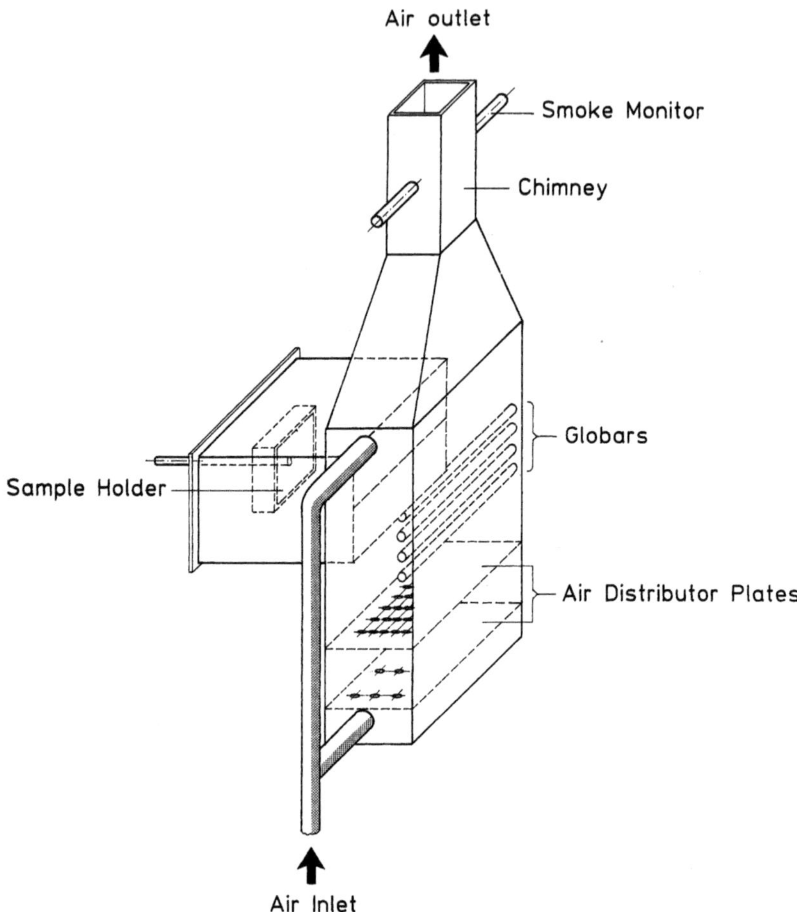

Figure 11.21 Ohio State University (OSU) calorimeter

Table 11.56 Requirements for Heat Release Testing with OSU Calorimeter

Specimens	▪ Three specimens, 150 × 150 mm (5.94 × 5.94 in) ▪ Thickness as in actual use
Position	Vertical
Ignition source	35 kW/m² radiant heat flux, plus pilot flames located near the bottom and the top of the specimen
Test duration	5 min
Requirements	FAR Part 25, App. F, Part IV, (g)
	▪ Maximum heat release rate (5 min) ≤ 65 kW/m² ▪ Total heat released (first 2 min) ≤ 65 kW·min/m²

11.4.3.3 Smoke Emission and Toxicity Testing

Many years ago, the major aircraft manufacturers such as Boeing and Airbus introduced additional requirements regarding the optical density of the smoke emitted ("smoke density"), and the toxicity of the fire combustion products. Since 1990, the smoke density test method has been included in FAR Part 25, Appendix F [12].

Currently discussions are ongoing as to whether the smoke density requirements need to be kept in the Regulation, or removed. Since 2014, Airbus has introduced two options for purchasers in their Airbus Company Directive ABD0031, Issue G [13]:

1. Set A (standard): Smoke and toxicity requirements beyond applicable certification requirements for parts located in areas non-accessible during taxi, take-off, flight, and landing.
2. Set B (optional): Smoke and toxicity requirements beyond applicable certification requirements for all parts in the pressurized section of the fuselage.

Smoke Density

The test is performed according ASTM F814-83 [14]. The instrument and test method described in this standard are similar to ASTM E662 [15], but differ in procedures and apparatus, as specifically required for aerospace applications. Three test specimens are measured in flaming mode with pilot burners on, over the course of 4 minutes. This time is regarded as sufficient for the evacuation of passengers in case of a fire.

See Figure 11.22 for a schematic view of the NBS smoke chamber, and Figure 11.23 for the principle of the NBS smoke chamber. In Table 11.57, the exact requirements for smoke density are listed.

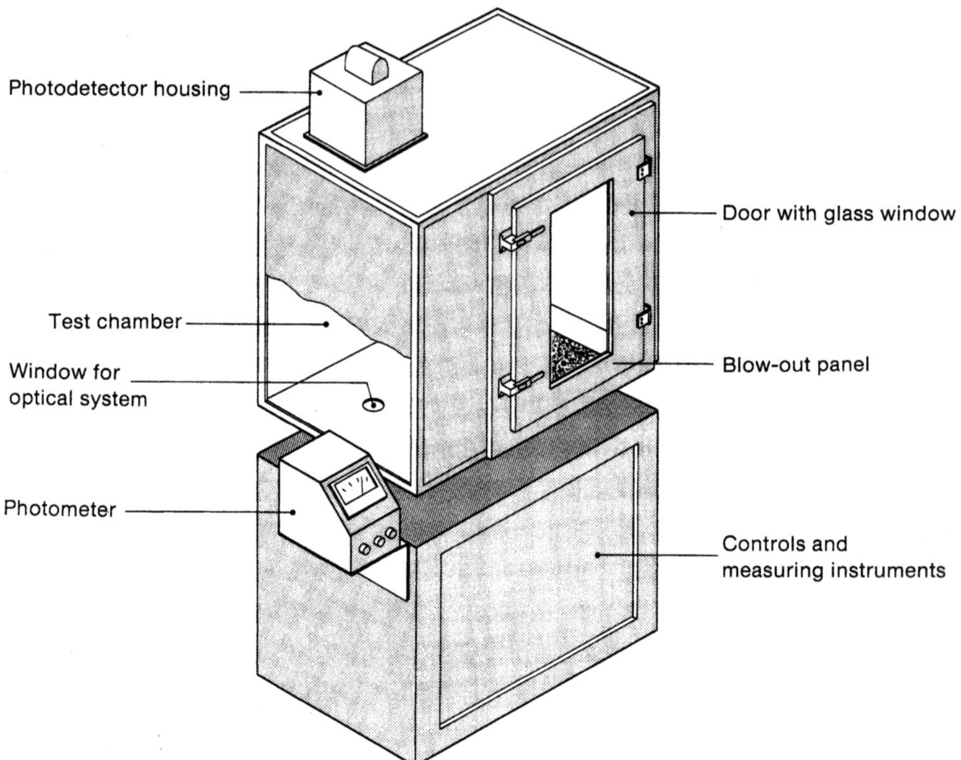

Figure 11.22 Schematic view of the NBS smoke chamber to ASTM E662

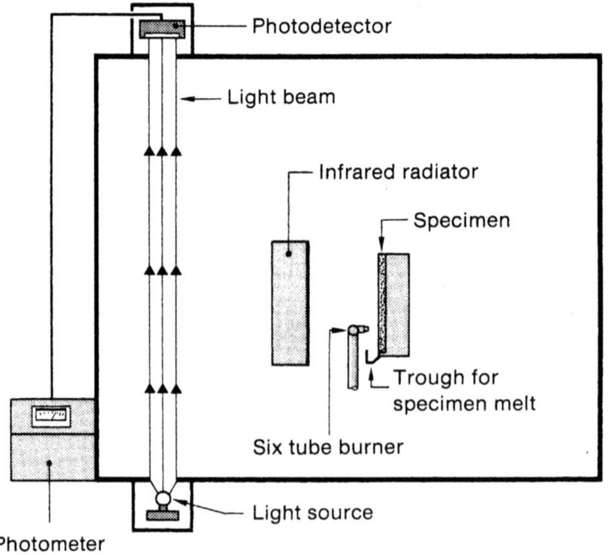

Figure 11.23 Principle of the NBS smoke chamber to ASTM E662

Table 11.57 Requirements for Smoke Density Testing with NBS Smoke Chamber

Specimens	▪ Three specimens, 73 × 73 mm (2.9 × 2.9 in) ▪ Thickness as in actual use
Position	Vertical
Ignition source	25 kW/m^2 radiant heat flux, with pilot flames located near the bottom of the specimen
Test duration	Minimum 4 min
Requirement	FAR Part 25, App. F, Part V, (b)
	▪ Maximum specific optical density ≤200 (during 4 min test)

Toxicity

The test method to determine the concentration of toxic components in the combustion gases of a burning material is described in Airbus standard AITM 3-0005 [16] and Boeing standard BSS 7239 [17]. The test is run in combination (but not simultaneously) with the test for smoke density as described above. The gas sampling procedure starts immediately after the 4-minute smoke density test.

In Table 11.58, the maximum allowed concentrations of specific toxic gas components are listed. These maximum concentrations are listed in Airbus Company Directive ABD0031 [13], and Boeing Document No. D6-51377 [18].

Table 11.58 Maximum Allowed Concentrations of Toxic Gas Components

Gas component	Concentration limits [ppm]	
	ABD0031	D6-51377
Hydrogen fluoride (HF)	100	200
Hydrogen chloride (HCl)	150	500
Hydrogen cyanide (HCN)	150	150
Sulfur dioxide (SO_2)	100	100
Nitrous gases (NO/NO_2)	100	100
Carbon monoxide (CO)	–	3500

11.4.3.4 Additional Testing

For many years, the Bunsen Burner type tests were the only measures taken by the authorities to ensure reasonable fire behavior for interior materials used in aircraft. Some major accidents initiated research programs and subsequent developments, and resulted in amendments of the Federal Aviation Regulations. Apart from the OSU Heat Release test [11] and the Smoke Emission test [12], the Seat Cushion Flammability test [19] was also introduced, as well as the Flame Penetration Resistance test for cargo compartment linings [20].

Seat Cushion Flammability Test

This test simulates a post-crash fire burning through the fuselage. The intent is to limit the contribution of aircraft seats to fire spread and smoke emission. By limiting the allowed burn length on the seat, fire spread is limited. Furthermore, the contribution of the seat to heat development and smoke emission is also restricted, as only 10% of the mass of the seat is allowed to be burnt. Test results can be positively influenced by using a fire-blocking layer, which protects the seat foam from the flame attack.

In Figure 11.24, a schematic of the Seat Cushion Flammability test is given, and the requirements are listed in Table 11.59.

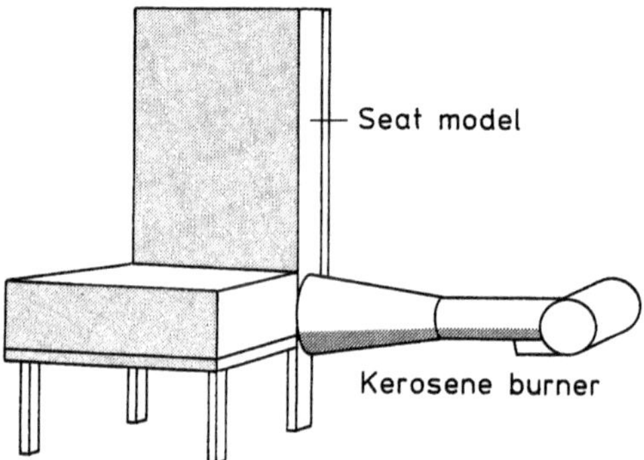

Figure 11.24 Seat cushion flammability test

Table 11.59 Requirements for Seat Cushion Flammability Testing

Specimens	▪ Three sets of seat bottom and seat back cushion samples ▪ Bottom: 457 mm (18 in) wide × 508 mm (20 in) deep × 102 mm (4 in) thick ▪ Back: 457 mm (18 in) wide × 635 mm (25 in) high × 51 mm (2 in) thick
Position	Model seat mounted on a steel stand
Ignition source	Kerosene burner, horizontally mounted, delivering nominal 2 gallons/hour kerosene
Flame application	2 min, to the side of the bottom seat cushion
Requirements	**FAR Part 25, App. F, Part II, (a)(5)**
	▪ Average weight loss ≤ 10% ▪ Burn length ≤ 432 mm (17 in) on at least two-thirds of the total number of samples tested

Flame Penetration Resistance Test

This test has been developed in order to prevent a fire originating in a cargo compartment, breaking through the linings and spreading further through the aircraft. Integrity of the linings is also important, because a fire in a cargo compartment can only be successfully extinguished when the concentration of the extinguishing agent can be maintained for a certain time, and no additional oxygen can come into the area of the origin of the fire.

In Figure 11.25, a schematic is given, and in Table 11.60 the requirements are listed.

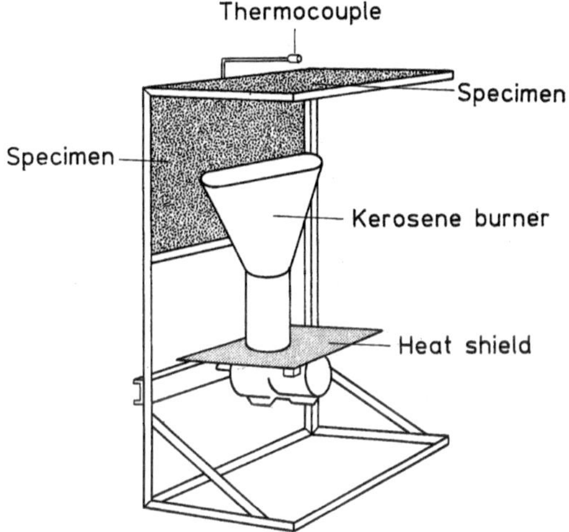

Figure 11.25 Flame penetration resistance test for cargo compartment linings

Table 11.60 Requirements for Flame Penetration Resistance Testing

Specimens	■ Three sets of ceiling and sidewall cargo liner panels ■ Ceiling panel: 406 mm (16 in) × 610 mm (24 in) ■ Sidewall panel: 406 mm (16 in) × 610 mm (24 in)
Position	Vertically mounted (sidewall) and/or horizontally mounted (ceiling) on a steel frame
Ignition source	Kerosene burner, vertically mounted, delivering nominal 2 gallons/hour kerosene
Flame application	5 min or until flame penetration occurs
Requirements	FAR Part 25, App. F, Part III, (a)(3)
	■ No flame penetration of any specimen ■ Sample backside temperature during flame exposure ≤ 204 °C (400 °F) – measured 102 mm (4 in) above the ceiling panel

Note: Specimens that pass in the ceiling orientation may be used as a sidewall panel without further testing.

11.4.3.5 Thermal/Acoustic Insulation Testing

The entire pressurized section of the aircraft is completely lined with thermal/acoustical insulation, which is by far the largest volume of non-metallic material in an aircraft. In FAR 25.856 [6] the requirements for thermal/acoustic insulation materials are described:

a) Thermal/acoustic insulation material installed in the fuselage must meet the flame propagation test requirements of part VI of Appendix F to this part.

b) For airplanes with a passenger capacity of 20 or greater, thermal/acoustic insulation materials (including the means of fastening the materials to the fuselage) installed in the lower half of the airplane fuselage must meet the flame penetration resistance test requirements of part VII of Appendix F to this part.

Flame Propagation Test

This test method evaluates the flammability and flame propagation characteristics of thermal/acoustic insulation when exposed to both a radiant heat source and a flame. In Figure 11.26, a schematic of the radiant panel test chamber is given, and in Table 11.61 the requirements are listed.

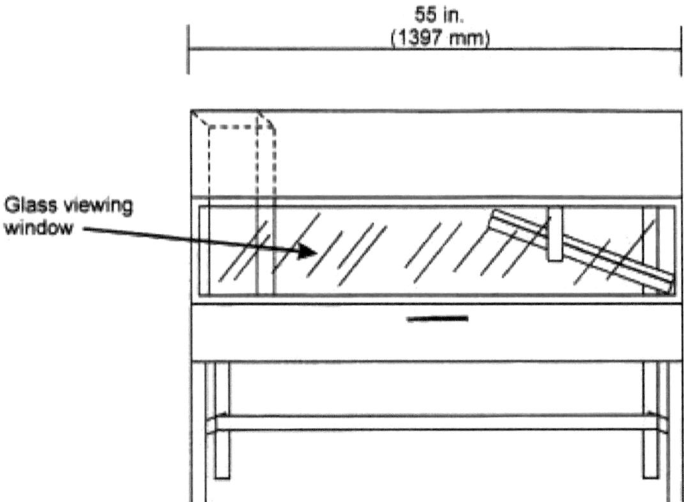

Figure 11.26 Radiant panel test chamber

Table 11.61 Requirements for Flame Propagation Testing

Specimens	▪ Three test specimens ▪ Non-rigid core materials: 318 mm (12½ in) × 584 mm (23 in) ▪ Rigid materials: 292 mm (11½ in) × 584 mm (23 in)
Position	Horizontal
Ignition source	▪ Electric radiant panel at 704 °C (1300 °F) ▪ Propane pilot burner with 127 mm (5 in) flame
Flame application	15 s
Requirements	**FAR Part 25, App. F, Part VI, (h)(1) & (h)(2)**
	▪ No flame propagation beyond 51 mm (2 in) ▪ Flame time after removal of the pilot burner ≤ 3 s

Flame Penetration Resistance Test for Thermal/Acoustic Insulation Materials

This test method evaluates the burn-through resistance characteristics of aircraft thermal/acoustic insulation materials when exposed to a high-intensity open flame. See Figure 11.27 for a schematic, and Table 11.62 for the requirements.

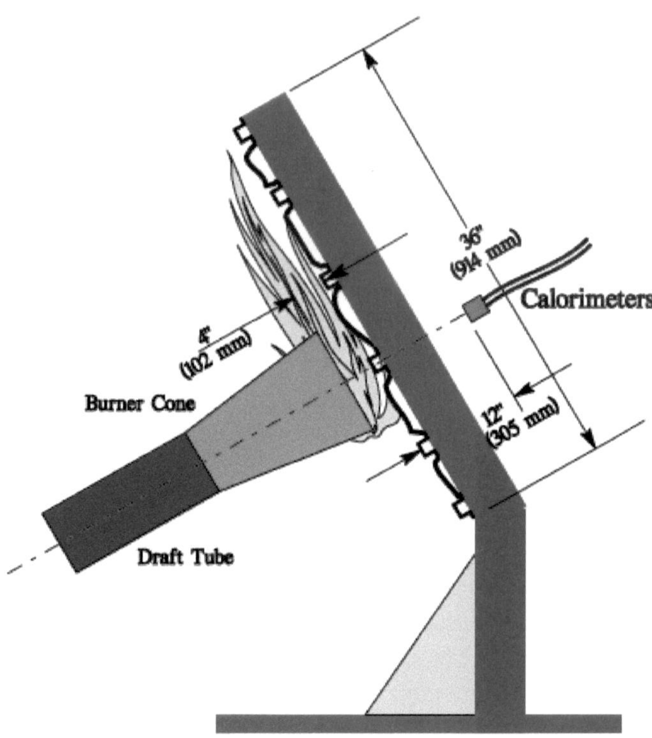

Figure 11.27 Flame penetration resistance test for thermal/acoustic insulation materials

Table 11.62 Requirements for Flame Penetration Resistance Testing

Specimens	▪ Three sets of insulation blanket test specimens, with dimensions 813 mm (32 in) × 914 mm (36 in)
Position	30° angle from vertical
Ignition source	Kerosene burner, delivering nominal 6 gallons/hour kerosene
Flame application	4 min
Requirements	FAR Part 25, App. F, Part VII, (h)(1) & (h)(2)
	▪ No fire or flame penetration in less than 4 min ▪ ≤ 2.27 W/cm^2 on the cold side of the insulation specimens, measured at 305 mm (12 in) from the face of the test rig

11.4.4 Future Developments

The FAA and EASA, together with the experts from industry and regulators, are continuously investigating ways to improve the fire testing capabilities, either by optimizing existing test methods, or by developing completely new ones.

This work is typically done in FAA's William J. Hughes Technical Center, and in the International Aircraft Materials Fire Test Forum (IAMFTF) – formerly known as the International Aircraft Materials Fire Test Working Group (IAMFTWG). The FAA facilitates this Forum, with strong participation from EASA.

Within the Forum, several Task Groups are being formed around specific topics, to develop new test methods, test similarities, help in writing advisory materials, etc. Examples of such Task Groups are the ones working on the new test methods Vertical Flame Propagation (VFP) and Heat Release Rate 2 (HR2).

Vertical Flame Propagation Test

New designs of commercial transport airplanes more frequently include primary and secondary structures constructed from carbon fiber composites, since these provide increased strength, lower density, and better corrosion resistance. Current regulations do not require flammability testing for fuselage skins or structural components, as traditional designs are inherently non-flammable.

The FAA has therefore started the development of a flammability test for fuselage skins and structural components, as well as for other materials extensively used in inaccessible areas. These could include wiring, sleeving, and ducting material. This test method under development is currently referred to as the "Vertical Flame Propagation" (VFP) test.

Specific requirements for testing and specification limits have not yet been defined.

Heat Release Rate 2 Test

The goal of the development of the new Heat Release Rate 2 test (HR2) is to define a robust test method to determine peak and total heat release in a way that results in better repeatability and reproducibility when compared to the current OSU Heat Release test.

Many variables are being investigated, such as the airflow split, thermopile optimization, heat flux, inlet airflow, inlet temperature, etc. The project is currently in the "reproducibility" phase, where the robustness of the equipment and test procedure is validated; initial results look promising. As part of the process to determine the reproducibility performance, extensive round-robin testing will be performed.

Major changes to the reaction-to-fire regulations of materials used in airplanes are not expected in the near future. The work of the last 20+ years has resulted in considerable reduction in the ignitability, heat release, and flame spread of the aircraft materials, as well as smoke emission and toxicity, and consequently a satisfactory level of safety has been reached.

References for Section 11.4

[1] Federal Aviation Regulations, Airworthiness Standards - Department of Transportation - Federal Aviation Administration.

[2] Code of Federal Regulations *(https://www.ecfr.gov)*.

[3] FAR Part 25, *Airworthiness standards: Transport category airplanes*, Subpart D, *Design and Construction*, § 25.853, *Compartment interiors*.

[4] FAR Part 25, *Airworthiness standards: Transport category airplanes*, Subpart D, *Design and Construction*, § 25.855, *Cargo or baggage compartments*.

[5] FAR Part 25, *Airworthiness standards: Transport category airplanes*, Subpart D, *Design and Construction*, § 25.857, *Cargo compartment classification*.

[6] FAR Part 25, *Airworthiness standards: Transport category airplanes*, Subpart D, *Design and construction*, § 25.856, *Thermal/Acoustic insulation materials*.

[7] FAR Part 25, *Airworthiness standards: Transport category airplanes*, Appendix F, Part I, *Test criteria and procedures for showing compliance with § 25.853 or § 25.855*.

[8] Aircraft Materials Fire Test Handbook, U.S. Department of Transportation, Federal Aviation Administration *(https://www.fire.tc.faa.gov/Handbook)*.

[9] ISO 3795, *Road vehicles, and tractors and machinery for agriculture and forestry - Determination of burning behaviour of interior materials*.

[10] Federal Motor Vehicle Safety Standard No. 302 ("FMVSS 302"), *Flammability of interior materials - Passenger cars, multipurpose passenger vehicles, trucks and buses*.

[11] FAR Part 25, *Airworthiness standards: Transport category airplanes*, Appendix F, Part IV, *Test method to determine the heat release rate from cabin materials exposed to radiant heat*.

[12] FAR Part 25, *Airworthiness standards: Transport category airplanes*, Appendix F, Part V, *Test method to determine the smoke emission characteristics of cabin materials*

[13] Airbus Company Directive ABD0031, Issue G, *Fireworthiness requirements for the pressurized section of fuselage*.

[14] ASTM F814-83, *Standard test method for specific optical density of smoke generated by solid materials for aerospace applications* (withdrawn).

[15] ASTM E662, *Standard test method for specific optical density of smoke generated by solid materials.*

[16] AITM 3-0005, *Determination of specific gas components of smoke generated by component parts or sub-assemblies of aircraft interior.*

[17] Boeing Specification Support Standard BSS 7239, *Test method for toxic gas generation by materials on combustion.*

[18] Boeing Document No. D6-51377, *Airplane fireworthiness design criteria – Pressurized compartments.*

[19] FAR Part 25, *Airworthiness standards: Transport category airplanes,* Appendix F, Part II, *Flammability of seat cushions.*

[20] FAR Part 25, *Airworthiness standards*: *Transport category airplanes,* Appendix F, Part III, *Test method to determine flame penetration resistance of cargo compartment liners.*

12 Electrical Engineering and Cables

12.1 Electrical Engineering

Jun Haruhara, Mattia Ferraris, and Jürgen Troitzsch

Significant technological advances in electrical engineering would not have been possible without the contribution of plastics. Plastics are well known for their durability, corrosion resistance, strength, and outstanding insulation properties. They are therefore the material of choice for many applications in electrical and electronic engineering, such as appliances, consumer and office equipment, electrical parts, and electrical cables. The diversity of polymers available as well as their aesthetic design qualities gives the development engineer, designer, and manufacturer of electrotechnical products tailor-made solutions to meet a variety of needs, including function and aesthetics, safety, and technical and environmental performance.

The production of plastics in 2017 amounted to 348 million tons worldwide, and to 64.4 million tons in Europe. In Europe, the electrical and electronic (E&E) sector accounted for 6.2% (3.17 million tons) of the total plastics converter demands in 2017 [1].

It cannot be denied that the use of plastics in electrotechnical products demands special fire precautions, because a fire hazard is inherent in any energized circuit. The primary objective of circuit, component, and equipment design and the choice of materials are to minimize the risk of electrically induced ignition, even in the event of foreseeable abnormal use, malfunction, or failure. If, despite precautions, ignition and fire do occur, a fire should be controlled within the area of the equipment's enclosure.

Potential ignition sources within electrical equipment result primarily from overheated current-carrying parts, electrically overloaded parts or components, overheated terminals due to loosening and faulty installation of connections (bad connections), sparks, and arcs. Like all other organic substances, plastics are combustible. Therefore, plastics are a potential fire hazard in electrical equipment. Flames may develop and impinge on adjacent combustible materials, if, for example, a fault current flows over a tracking path onto the surface of the insulation or a bad connection causes insulation to ignite. Special fire tests described below have been developed to simulate as closely as possible real ignition risks.

In cases where surfaces of E&E products like large enclosures, cables, and cable management systems are exposed to an external fire, they must not contribute more to fire growth than other combustible materials, building components, or building structures in the immediately surrounding areas.

Electrical equipment can only be operated if it complies with regulatory requirements. The same is true for plastics used in electrical insulation or equipment housings. In many applications, the polymers used and/or the final component made from them have to meet one or more flammability standards. Historically, the development of standards for fire testing of electrical equipment was the responsibility of national organizations or regional standardization commissions. Numerous fire tests methods have been developed, which often differ from country to country. Depending on the test method, the relevant importance of the fire parameters being assessed is different. In standardization, be it on national or international level, there are two philosophies of testing.

Side by side with hazard-oriented *end-product testing*, which is based on the view of fire hazard technologists that only the as-operated assembly in the as-installed state can provide a valid hazard assessment, *preselection testing of standard specimens* is requested. These tests are often small (bench-scale) tests of materials, which examine the reaction-to-fire of standardized specimens under defined conditions. Mostly, they are used to provide data on combustion characteristics, like flammability, ignitability, and flame spread rate. The results of preselection tests on materials are generally quoted as numerical values or as classifications in flammability classes. Flammability classes based on the results of preselection tests on materials are often cited in material specifications. Data from preselection tests alone are insufficient to predict the performance of the plastic material in the end-product, unless the small-scale test has been validated by a fire hazard test reflecting the real fire scenario. The interaction of all related material and design factors can only be assessed by fire simulation tests on assemblies, electrotechnical end-products, or parts taken out of them.

However, combustion characteristics tests can be quite useful in product development and quality control. The data from these tests can help the plastics industry to compare the performance of different materials, particularly in research and material development, or in assisting the development engineer to select materials during the design of an electrotechnical product. In quality control, material tests serve to evaluate the uniformity of a particular property of a product, in this case flammability.

Much work has been done in recent decades by international standardization bodies, such as the International Electrotechnical Commission (IEC) and the European Electrotechnical Standardization Commission (CENELEC), to develop small-scale and large-scale tests for electrotechnical products and cables that more accurately model real fire conditions. In these tests, the real conditions of use - including the surroundings of an electrotechnical product - are simulated as closely as possible, and the design of the test procedure is related to the real risk. Special test ignition sources have been developed to simulate the conditions of a fault, improper use, or exposure to an external fire. This concept of fire hazard testing is well accepted by international and national standardization committees.

Due to the increased need for worldwide free movement of electrotechnical equipment, national standards and classification systems for fire hazard assessment in the E&E sector are being increasingly replaced by harmonized systems. Advances in the understanding and use of fire safety engineering will have a positive impact on the standardization process and will help to integrate the two philosophies of preselection testing of materials and hazard-oriented end-product testing into an overall systematic approach to fire hazard assessment.

12.1.1 International Fire Safety Requirements and Standards (IEC)

12.1.1.1 Objectives and Organization

Harmonization reduces trade barriers, promotes safety, allows inter-operability of products, systems, and services, and promotes common technical understanding. The development of international standards is the task of organizations such as ISO and IEC. ISO (the International Organization for Standardization) is responsible for all technical fields with the exception of electrical technology [2]. International electrical and electronic standards are established by the IEC (International Electrotechnical Commission) [3], which was founded in 1906.

The object of the IEC is to promote international cooperation on all questions of standardization and related matters in the fields of electrical and electronic engineering. The IEC is composed of 86 national committees (members and associate members) representing all the industrial countries in the world. The preparation of standards, technical specifications, and technical reports is entrusted to Technical Committees (TCs).

In order to promote international understanding, IEC National Committees undertake to apply IEC International Standards as often as possible instead of national standards. IEC cooperates with numerous other international organizations, particularly with ISO, and in Europe with CENELEC (European Committee for Electrotechnical Standardization) [4]. The supreme authority of the IEC is the Council. The Council delegates the management of technical work to the Committee of Action. ACOS (the Advisory Committee on Safety) supervises the coordination of IEC safety standards and their coherence. It assigns to certain TCs a "horizontal safety function" (preparation of basic safety publications applicable to many electrotechnical products) or a "group safety function" (preparation covering all safety aspects of a specific group of products within the scope of two or more product TCs) according to the principles laid down in ISO/IEC Guide 51 [5] and IEC Guide 104 [6]. More than half of the 3500 IEC publications are safety-related. The ACEA (Advisory Committee on Environmental Aspects) considers all aspects of the protection of the natural environment against detrimental impacts from a product, group of products, or a system using electrical technology.

IEC publications are known as International Standards (IS), Technical Specifications (TS), Technical Reports (TR), Publicly Available Specifications (PAS), or IEC Guides. Full information on published standards is available on the IEC website [7].

12.1.1.2 Fire Hazard Testing, IEC TC 89

IEC Technical Committee (TC) No. 89, Fire Hazard Testing, was formed in 1988 by a decision of the IEC Council to change subcommittee 50D into a full technical committee. TC 89 is a horizontal committee with a basic safety function within the IEC to give guidance and develop standards related to fire hazards for use by other IEC Product Committees responsible for finished components or electrotechnical products [8].

The scope of TC 89 is to prepare international standards, technical specifications, and technical reports in the areas of:

1. Fire hazard assessment, fire safety engineering, and terminology as related to electrotechnical products
2. Measurement of fire effluents (e.g., smoke, corrosivity, toxic gases, and abnormal heat), and reviews of the state of the art of current test methods as related to electrotechnical products
3. Widely applicable small-scale test methods for use in product standards and by manufacturers and regulators
4. Horizontal safety function: Guidance and test methods for assessing fire hazards of electrotechnical equipment, their parts (including components), and electrical insulating materials.

At present, the work of TC 89 is executed through two working groups (WG) consisting of experts from 15 participating countries (P-members), two project teams (PT) and one advisory group consisting of the P-members and 18 product or horizontal TCs and their subcommittees (SCs) liaised to TC 89:

- WG 11 – Fire effluent (smoke, heat, corrosivity, and toxicity), fire hazard assessment, fire safety engineering, flame spread, general guidance, and terminology
- WG 12 – Test flames and resistance to heat. Small-scale heat and flame test methods
- PT 60695-2-15 – Fire containment test on finished units
- PT 60695-2-20 – Hot-wire ignition test
- AG 13 – Chairman's Advisory Group.

Standardization is expensive and time-consuming. Wherever possible, IEC works with other standardization organizations. In order to support and advise IEC TC 89, various active liaisons to other IEC and ISO Technical Committees exist. These liaisons are listed in Table 12.1.

Table 12.1 IEC TC 89 Liaisons

IEC	ISO	Title
TC 10	–	Fluids for electrotechnical applications
TC 14	–	Power transformers
TC 20	–	Electric cables
SC 22F	–	Power electronics for electrical transmission and distribution systems
TC 23	–	Electrical accessories
SC 23 A	–	Cable management systems
TC 46	–	Cables, wires, waveguides, RF connectors, RF and microwave passive components and accessories
TC 61/MT4	–	Safety of household and similar electrical appliances
SC 86A	–	Fibers and cables
TC 99	–	System engineering and erection of electric power installations in systems with normal voltages above 1 kV_{ac} and 1.5 kV_{dc}, particularly considering safety aspects
TC 104	–	Environmental conditions, classification and methods of test
TC 108	–	Safety of electronic equipment within the field of audio/video, information technology, and communication technology
TC 112	–	Evaluation and qualification of electrical insulating materials and systems
SC 121A	–	Low-voltage switchgear and controlgear
–	TC 61/SC 4	Plastics – Burning behaviour
–	TC 92	Fire safety
–	TC 92/SC 1	Fire initiation and growth
–	TC 92/SC 3	Threats to people and environment

The fire hazard of batteries is not currently in the scope of IEC TC 89. Considering the big differences between batteries and related applications, the IEC product committees are specifically drafting requirements for their products (e.g., IEC 60335-1 Annex B, where IEC TC 61 in Table 12.1 is responsible for [34]).

12.1.1.2.1 Fire Hazard Assessment

As any electrical circuit has a fire risk, the objectives of component circuit, equipment design, and the choice of materials are to reduce the likelihood of a fire, even in the case of foreseeable abnormal use, malfunction, or failure.

In its basic guidance documents IEC 60695 Parts 1-10, 1-11, and 1-12, IEC TC 89 provides an overall framework for assessing the hazard that an electrotechnical product poses to life, property, and the environment in a fire situation [9]. These documents are intended to provide guidance to IEC product committees when assessing the fire hazard of their products. In the context of the standard, electrotechnical products are materials, components, and sub-assemblies as well as the complete, fully-assembled end-use products.

The basic process in IEC 60695 Parts 1-10, 1-11, and 1-12 is to identify the class or product range to which the assessment applies, the kind of fire scenarios associated with the product, and to define the fire safety objectives and acceptable safety levels. The standard includes guidance on how to incorporate preselection tests on materials and end-products, service conditions, fire scenarios, and environmental aspects into the overall hazard assessment.

In IEC 60695-1-11 Annex B, an example of hazard assessment on a hypothetical installation of a rigid plastic conduit is demonstrated to illustrate the hazard assessment process and provide guidance on how to use results of materials tests and fire growth models for assessing whether products are likely to provide any significant contribution to the overall fire hazard.

Electricity is a potential fire hazard. Fire statistics demonstrate that electrical equipment is one of the primary causes of fire. In particular, the benefits of using plastics in electrical and electronic appliances go hand-in-hand with the need to manage the risk of fire. Fires can result from faulty product design, but are often due to misuse or abuse of the appliance. Educating the consumer, for instance, to remove dust from the appliance backplate, not to place candles or smoking materials on the appliance, or to avoid do-it-yourself repairs, is a long-term process. The increased use of plastics in electrical and electronic appliances over the years has only been possible by the parallel development and application of flame retardants. As a result, plastics containing flame retardants or inherently flame-retardant polymers can meet fire safety requirements for combustible materials.

One basic safety objective is to influence the course of a fire, which always follows the same pattern of three stages – initiation, growth to full development, and decay.

The main "reaction-to-fire" parameters and side effects of a fire that determine the course of a fire and (first of all) the initiating fire, are:

- Reaction-to-fire parameters:
  - Combustibility
  - Ignitability
  - Spread of flame
  - Heat release
- Side effects of a fire:
  - Smoke development
  - Toxic potential of fire gases
  - Corrosive potential of fire gases.

Fire protection measures for combustible materials and electrotechnical products can only apply to the early stage of a fire (i.e. the initiation stage), by preventing ignition as a first fire safety objective, or if ignition occurs, by preventing or reducing fire propagation and minimizing heat release. Common causes of ignition encountered in electrotechnical products can be found in Table 1 of IEC 60695-1-10.

According to UL 746 C edition 7, Section 3.34 [72], a fire risk exists at any two points in a circuit where:

a) The open-circuit voltage is > 42.4 V peak and the energy available in the circuit under any conditions on load, including short-circuit, results in a current of 8 A or more after 1 min of operation, or

b) A power of more than 15 W can be delivered into an external resistor connected between the two points.

Certain electrotechnical products such as large enclosures, insulated cables, and conduits, may cover large areas in building construction or may penetrate fire-resisting walls. Here, electrotechnical products must be exposed to an external fire and their fire hazard contribution compared to materials or structures without them. Thus, the secondary fire safety objective is to make sure that electrotechnical products do not contribute to a greater fire growth than building products and structures in their immediate vicinity.

The fire hazard of an electrotechnical product depends on its characteristic service conditions and the environment in which it is used. Therefore, a fire hazard assessment procedure for a particular product shall describe the product, its conditions of operation, and its environment. The contribution of a material, product, or system to an assumed fire scenario is assessed as follows:

- Is the product likely to be the source of ignition?
- Is the product likely to be the secondary ignited item?

- Is the product a potentially significant fuel source, even if it is not the first item ignited?
- What is the potential product's contribution to hazard and risk?
- How close are occupants and/or critical equipment to the origin of the fire?

Testing and calculations are used to determine whether the fire safety objectives will be met. For the fire hazard assessment procedure to be valid, it is necessary that the characteristic fire test responses used produce valid estimates of success or failure in achieving the fire safety objectives given in the specified fire scenario (see Table 12.2). A schematic representation of the principles of a fire hazard assessment procedure can be found in Figures 1–5 of IEC 60695-1-11. In fire safety engineering guidelines such as IEC 60695-1-12, the flowcharts may be followed, but in national and international fire test standards for electrotechnical products this strategy is at the current time only partly taken into account.

Table 12.2 Fire Parameters Related to Classification and Testing

Stage	Parameter		
Developing fire	Reaction to fire	Ignitability Flame spread Heat release	Smoke Toxicity Corrosivity
Fully developed fire	Resistance to fire	Load-bearing insulation and integrity capacity, e. g., fire resistance of protective systems for cables; circuit integrity of cables	Smoke Toxicity Corrosivity

12.1.1.2.2 Fire Hazard Assessment Tests for Electrotechnical Products

The characteristic fire test responses for electrotechnical products or their components, or the materials that may be required for such an assessment, are ignition and flame spread, currently investigated in a set of small-scale fire tests. Realistic ignition sources are used to ignite a specimen, and are applicable to a variety of potential fire scenarios.

The IEC TC 89 guidance publications and their equivalent EN standards are summarized in Table 12.3.

Table 12.3 IEC TC 89: Guidance Publications

Reference IEC 60695-	Reference EN 60695-	Description	Subject	End-product test	Preselection test for materials
1-10 [9]	1-10	General guidelines	Fire hazard	Yes	Yes
1-11 [9]	1-11	Fire hazard assessment		Yes	Yes
1-12 [9]	-	Fire safety engineering		Yes	Yes
1-14 [10]	-	Guidance on the different levels of power and energy related to the probability of ignition and fire in low-voltage electrotechnical products		Yes	Yes
1-40 [13]	1-40	Guidance for insulating liquids		-	-
1-20 [11]	1-20	General guidance	Ignitability	Yes	Yes
1-21 [12]	1-21	Summary and relevant test methods		Yes	Yes
1-30 [21]	1-30	General guidelines	Preselection testing process	-	Yes
4 [14]	4	Terminology	Terminology	Yes	Yes
5-1 [59]	5-1	General guidance	Corrosion damage	Yes	Yes
5-2 [60]	-	Summary and relevance of test methods		Yes	Yes
6-1 [53]	6-1	General guidance	Smoke obscuration	Yes	Yes
6-2 [54]	6-2	Summary and relevance of test methods		Yes	Yes
7-1 [56]	7-1	General guidance	Toxicity of fire effluent	Yes	Yes
7-2 [57]	7-2	Summary and relevance of test methods		-	-
7-3 [58]	7-3	Use and interpretation of results		Yes	Yes
8-1 [49]	8-1	General guidance	Heat release	-	-
8-2 [50]	8-2	Summary and relevance of test methods		Yes	Yes
9-1 [43]	9-1	General guidance	Surface spread of flame	Yes	Yes
9-2 [44]	-	Summary and relevance of test methods		Yes	Yes

Table 12.4 contains the relevant standards for the fire hazard tests for end-products and the preselection tests for materials.

Table 12.4 IEC TC 89: Fire Hazard Assessment Tests for End-Products and Preselection Tests for Materials

Reference IEC 60695-	Description	Subject	End-product test	Preselection test for materials
2-10 [17]	Glow-wire apparatus and common test procedure	Ignitability and flammability	Yes	Yes
2-11 [18]	Glow-wire flammability test on end-products	Flammability	Yes	-
2-12 [22]	Glow-wire flammability test on materials		-	-
2-13 [23]	Glow-wire ignitability test on materials	Ignitability	-	Yes
11-5 [19]	Needle-flame test	Flammability	Yes	-
11-2 [20]	1 kW nominal premixed test flame and guidance		Yes	-
10-2 [15]	Method for testing products made from non-metallic materials for resistance to heat using the ball pressure test	Abnormal heat	Yes	Yes
10-3 [16]	Mold stress relief distortion test		Yes	-
11-10 [39]	50 W horizontal and vertical flame test methods	Flammability	-	Yes
11-20 [40]	500 W flame test methods		-	Yes

Ball pressure test, IEC 60695-10-2

The properties of insulating materials in electrotechnical appliances or parts may change when exposed to heat from internal or external sources. As a consequence, the insulating material's physical properties may be changed, impairing the mechanical or electrical suitability of the material. The IEC 60695-10-2 Ball pressure test is a method for testing parts of non-metallic materials for resistance to heat [15]. A steel ball of 5 mm diameter is pressed at (20 ± 0.2) N for 1 hour against the surface of the test specimen at (80 ± 3) °C or (125 ± 5) °C or at a temperature (40 ± 2) °C in excess of the temperature rise of the relevant part. The test specimen has to be taken from the electrotechnical equipment, its sub-assemblies, and components. If this is not feasible, a panel of an identical material (3.0 ± 0.5) mm thick, either square with > 10 mm sides or circular with > 10 mm diameter, may be used. After the ball is removed, the test specimen is cooled down for 10 s by immersion in water at room temperature. The diameter of the indentation is measured. The test is passed if the diameter caused by the ball does not exceed 2.0 mm.

Glow-wire test for end-products, IEC 60695-2-11 (GWEPT)

IEC 60695-2-10 (EN 60695-2-10) specifies the glow-wire test apparatus and the test procedure to assess the fire hazard of electrotechnical materials and end-products in simulating the short-term effect of thermal stress, which may be produced by heat sources such as glowing wires or overloaded resistors [17]. The glow-wire test for end-products is prescribed in IEC 60695-2-11 [18]. The test objective is to ensure that the glow-wire does not cause parts to ignite, or if so, that their burning time is limited with no flame spread or falling burning/glowing particles. The glow-wire coil consists of a nickel/chromium (> 77% Ni/(20 ± 1)% Cr) wire (4 ± 0.07) mm in diameter. A sheathed fine-wire thermocouple (e.g., chromel-alumel) of 1.0 mm diameter is used for measuring the temperature of the glow-wire. The test apparatus positions the electrically heated glow-wire in a horizontal plane while applying a force of (0.95 ± 0.10) N to the test specimen. The force is maintained when the glow-wire is moved horizontally towards the test specimen. The depth of penetration of the glow-wire is limited to (7 ± 0.5) mm. The glow-wire is heated by a simple electric circuit and, once the temperature is stable, no further adjustment may be made to the current. The glow-wire is applied to the test specimen for a period of (30 ± 1) s. To evaluate the possibility of flame spread by burning or glowing particles, a tissue-covered wooden board is placed underneath the test specimen. The test specifications are summarized in Table 12.5 and the test apparatus is shown in Figure 12.1.

Table 12.5 IEC 60695-2-11: Glow-Wire Flammability Test Method for End-Products (GWEPT)

Test specimen	Complete electrotechnical equipment, its sub-assemblies and components, solid electrical insulating materials, and other combustible materials. Only one test is carried out. If ambiguous data is obtained, two further specimens must be tested.				
Test layout	Test specimen position should correspond to worst case in use, glow-wire coil horizontal, penetration depth limited to 7 mm, pressing force (0.95 ± 0.1) N.				
Ignition source	Glow-wire coil made of chrome-nickel alloy with built-in miniature thermocouple				
Suggested test severities	Preferred test temperatures: (550 ± 10) °C	(650 ± 10) °C	(750 ± 10) °C	(850 ± 15) °C	(960 ± 15) °C
Ignition source application time	(30 ± 1) s				
Conclusions	Test passed if there is no flaming or glowing, or: a) if flames or glowing of the test specimen extinguish < 30 s after removal of the glow-wire, and b) no ignition of the wrapping tissue under the specimen occurs.				

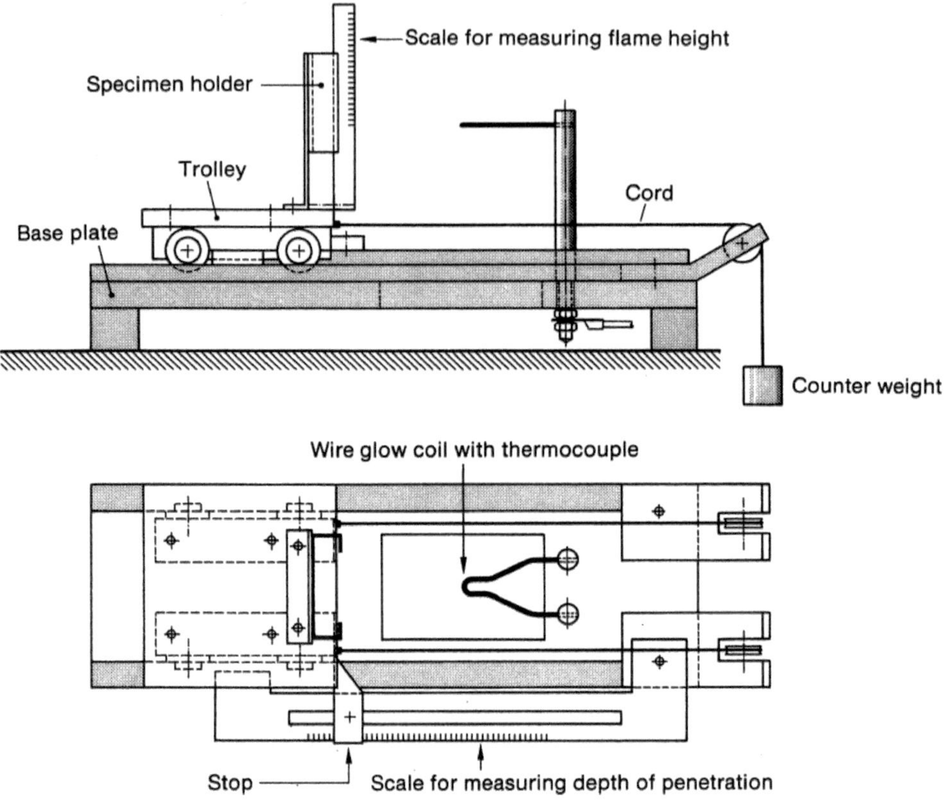

Figure 12.1 IEC 60695-2-10 glow-wire test apparatus

Examples for glow-wire tests specified in standards for electrotechnical products are shown in Table 12.6.

Table 12.6 Use of Glow-Wire Flammability Test Methods in Standards for Electrotechnical Products

Reference		Description	Test object	Test temperature [°C]
IEC 60998-1 [24]	DIN/EN 60998-1, VDE 0613-1 [25]	Connecting devices for low-voltage circuits for household and similar purposes; General requirements	Parts of insulating material in contact with current-carrying parts and for enclosures retaining in position only earthing clamping units	650 [a]
			Parts of insulation material retaining current-carrying parts and parts of the earthing circuit in position	850 [a]

Reference		Description	Test object	Test temperature [°C]
IEC 60400 [26]	DIN/EN 60400, VDE 0616-3 [27]	Lamp-holders for tabular fluorescent lamps and starterholders	External parts of non-metallic material	650
IEC 60730-1 [28]	DIN/EN 60730-1, VDE 0631-1 [29]	Automatic electrical controls for household and similar use; General requirements	External parts of non-metallic material; Parts of insulating material supporting connections in position and parts in contact with them; Enclosures retaining in position only earthing clamping units	550 Category A: 550 Category B: 550 Category C: 750 Category D: 850
IEC 60669-1 [30]	DIN/EN 60669-1, VDE 0632-1 [31]	Switches for household and similar fixed electrical installations; General requirements	Parts of insulating material retaining current-carrying parts and parts of the earthing circuit in position	850
IEC 61439-3 [32]	DIN/EN 61439-3, VDE 0660-600-3 [33]	Switchgear and control-gear; Particular requirement for low-voltage switchgear; Distribution boards	Parts of insulating materials retaining current-carrying parts	960
IEC 60335-1 [34]	DIN/EN 60335-1 [35]	Safety of household and similar electrical appliances; General requirements	External parts of non-metallic material; Parts of insulating materials supporting connections in position; $I \geq 0.2$ A. Operating while unattended (1); operating while attended (2)	550 IEC: (1) 850; (2) 750 EN: (1) 850; (2) 750

[a] Application of the glow-wire: 5 s.

The glow-wire test is also used to test the flammability (IEC 60695-2-12) or ignitability (IEC 60695-2-13) of electrical insulation or other materials (see Section 12.1.1.2.3).

Flame Tests – General

In general, flames are not the primary cause of fires in electrotechnical equipment, but may occur due to malfunction within the equipment, i.e. a leakage current flowing over a tracking path. Such small flames should not lead to a safety hazard; they are simulated by the needle-flame test to IEC 60695-11-5 [19]. Larger flames may arise from other ignited items in the vicinity of electrotechnical products and act on combustible parts, e.g., cables or enclosures. They are simulated by the 1 kW premixed test flame to IEC 60695-11-2 [20].

Needle-flame test, IEC 60695-11-5

IEC 60695-11-5 is applicable to complete electrotechnical equipment, its sub-assemblies and components, and to solid electrical insulating materials or other combustible materials [19]. The test objective is to ensure that the test flame does not cause parts to ignite, or if so, that they burn to a limited extent with no flame spread or falling burning/glowing particles. The test specifications are summarized in Table 12.7 and the test arrangement is shown in Figure 12.2. Examples for the use of the needle-flame test in standards for electrotechnical products are given in Table 12.8.

Table 12.7 IEC 60695-11-5: Needle-Flame Test Specifications

Test specimen	Complete electrotechnical equipment, its sub-assemblies, and components. Three tests are carried out.	
Test specimen position	Should correspond to worst case in use. Flame application to specimen part, where flames caused by malfunction may arise.	
Ignition source	Needle-flame burner:	
	Tube length	≥ 35 mm
	Inner diameter	(0.5 ± 1) mm
	Outer diameter	≤ 0.9 mm
	Flame length	(12 ± 1) mm
	Fuel gas	Butane with a purity of at least 95%
Preferred flame application time	5, 10, 20, 30, 60, and 120 s	
Conclusions	Test passed if: a) No flame, no glowing of test specimen. No ignition of wrapping tissue or wooden board scorching b) Flames or glowing of the test specimen, and the layer below extinguish < 30 s after removal of needle-flame. No ignition of wrapping tissue or wooden board scorching.	

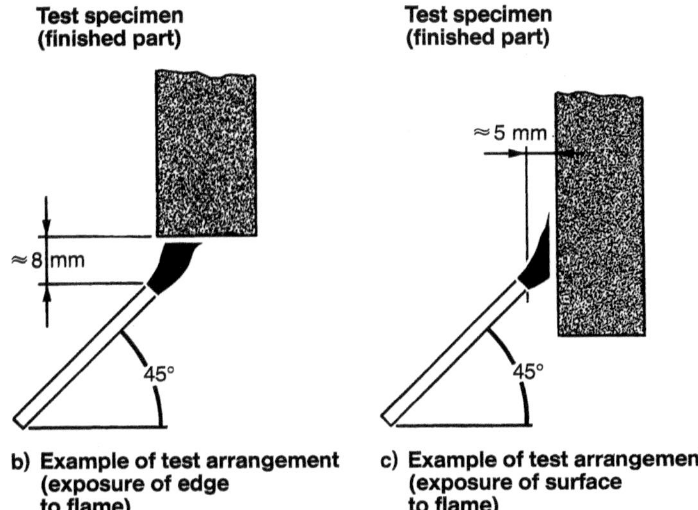

Figure 12.2 Needle-flame test to IEC 60695-11-5: Examples of test arrangement

Table 12.8 Use of Needle-Flame Test in Standards for Electrotechnical Products

Reference		Description	Test object	Requirements
IEC 60598-1 [36]	DIN/EN 60598-1, VDE 0711-1 [37]	Luminaries Part 1: General requirements and tests	Parts of insulating material supporting electrical connections in position.	Needle-flame test [a]. Application time: 10 s. Afterburning time: ≤ 30 s (no burning droplets).
IEC 60730-1 [28]	DIN/EN 60730-1, VDE 0631-1 [29]	Automatic electrical controls for household and similar use; Part 1: General requirements	Parts in the vicinity of 50 mm from current-carrying parts.	Needle-flame test [a]. Application time: 30 s. Afterburning time: ≤ 30 s (no burning droplets).
IEC 60400 [26]	DIN/EN 60400, VDE 0616-3 [27]	Lamp-holders for tabular fluorescent lamps and starter-holders	Parts of insulating material supporting live parts, other than connections, in position.	Needle-flame test [a]. Application time: 10 s. Afterburning time: ≤ 30 s (no burning droplets).

Table 12.8 Use of Needle-Flame Test in Standards for Electrotechnical Products *(continued)*

Reference		Description	Test object	Requirements
IEC 60335-1 [34]	EN 60335-1, [35]	Safety of household and similar electrical appliances; General requirements	External parts of non-metallic material; Parts of insulating material supporting connections in position; I ≥ 0.2 A. Operating while unattended (1); operating while attended (2); printed circuit boards.	Needle-flame test [a]. Application time: 30 s. Afterburning time: ≤ 30 s; printed circuit boards ≤ 15 s.

[a] Needle-flame test according to IEC 60695-11-5.

1 kW nominal premixed-flame test, IEC 60695-11-2

The detailed requirements for the production of the 1 kW nominal, propane-based premixed-type test flame with an approximate overall flame height of 175 mm and guidance for the test arrangements, are given in IEC 60695-11-2 [20].

The standard specifies the hardware for the 1 kW propane burner and the related confirmatory test. The test specimen should be a complete electrotechnical product. Care should be taken to ensure that the test conditions are not significantly different from those occurring in normal use with regard to shape, ventilation, effect of possible flames occurring as well as burning or glowing particles falling in the vicinity of the test specimen. The recommended distance from the top of the burner tube to the test specimen is 100 mm. Examples for test arrangements are given in IEC 60695-11-2.

12.1.1.2.3 Combustion Characteristics Tests for Preselection of Materials, IEC 60695-1-30

The selection of the right material is driven by the appearance, functionality, and serviceability of an electrotechnical product. Production and overall costs are mainly influenced by the choice of the material, its design, and processing. The techniques outlined in IEC 60695-1-30 [21] help the producers to use material tests for better predicting the fire performance of the finished products, and thus to save costs.

General guidance for the IEC product committees is given to assess the significance, relevance, and limitations of IEC and ISO preselection fire tests compared to fire hazard assessment tests. From an economic point of view, combustion characteristics tests are mainly small in scale and low-cost, so that they can be advantageously used for routine laboratory testing. In addition, they can serve as valuable tools for quality testing, production control, and for screening processes in research and development.

The next section contains detailed information about a selection of IEC tests listed in IEC 60695-1-30 that are commonly used today to compare the various combustion characteristics of materials. They are only valid as performance measurements under specific test conditions and cannot reflect real fire conditions. As a single material test may lead to an incorrect fire hazard assessment, the aim should be to simulate the end-use conditions of an electrotechnical product as closely as possible.

Glow-wire flammability test method for materials, IEC 60695-2-12 (GWFI)

The IEC 60695-2-12 glow-wire flammability test is applied for solid electrotechnical insulating materials or other solid combustible materials [22]. The test results allow comparison of the ability of various materials to extinguish after removal of the heated glow-wire and to prevent flaming or glowing particles from igniting a tissue-covered wooden board.

Table 12.9 IEC 60695-2-12: Glow-Wire Flammability Test Method for Materials (GWFI)

<table>
<tr><td rowspan="4">Test specimen</td><td colspan="4">Test specimens having a sufficiently large plane section with fixed dimensions:</td></tr>
<tr><td colspan="2">Length</td><td colspan="2">≥ 60 mm</td></tr>
<tr><td colspan="2">Width (inside clamps)</td><td colspan="2">≥ 60 mm</td></tr>
<tr><td colspan="2">Thickness</td><td colspan="2">Preferred values 0.1, 0.2, 0.4, 0.75, 1.5, or 3.0 mm</td></tr>
<tr><td>Test specimen position</td><td colspan="4">Typically, a set of 15 test specimens is used in vertical orientation.</td></tr>
<tr><td>Test layout</td><td colspan="4">Test specimens may be any shape, provided that the diameter of the specimen is a minimum of 60 mm. The test apparatus is described in IEC 60695-2-10.</td></tr>
<tr><td>Ignition source</td><td colspan="4">Glow-wire coil, made of chrome-nickel alloy with built-in thermocouple.</td></tr>
<tr><td rowspan="3">Test temperatures</td><td>(550 ± 10) °C</td><td>(600 ± 10) °C</td><td>(650 ± 10) °C</td><td>(700 ± 10) °C</td></tr>
<tr><td>(750 ± 10) °C</td><td>(800 ± 15) °C</td><td>(850 ± 15) °C</td><td>(900 ± 15) °C</td></tr>
<tr><td>(960 ± 15) °C</td><td></td><td></td><td></td></tr>
<tr><td>Ignition source application time</td><td colspan="4">30 s</td></tr>
<tr><td>Conclusions</td><td colspan="4">Test passed if there is no ignition of the test specimen or if the following applies:
a) No flames or glowing of the test specimen < 30 s after glow-wire removal
b) No ignition of the wrapping tissue.</td></tr>
</table>

The test apparatus is identical to that described in IEC 60695-2-10 (Figure 12.1). The test specimen is fixed on the trolley in a vertical position. The tip of the glow-wire is brought into contact with the test specimen for (30 ± 1) s. Nine test temperatures ranging from 550 °C to 960 °C can be used. Duration of burning of the test specimen and ignition of the wrapping tissue are noted. The test specifications are summarized in Table 12.9.

The GWFI (glow-wire flammability index) is the highest temperature of the glow-wire at which, during three subsequent tests, flames or glowing of the test specimen extinguish within 30 s after removal of the glow-wire. Ignition of the tissue-covered wooden board by burning droplets or particles is not allowed. The GWFI will usually vary depending on the thickness of the material tested.

Glow-wire ignitability test method for materials, IEC 60695-2-13 (GWIT)

The IEC 60695-2-13 glow-wire ignitability test is applied to solid electrotechnical insulating materials or other solid combustible materials [23]. The test results allow comparison of the temperatures at which the various materials ignite during the glow-wire application. The test apparatus is identical to that described in IEC 60695-2-10 (Figure 12.1) and the test specifications are summarized in Table 12.10. The ignitability will usually vary depending on the thickness of the material tested. The tip of the glow-wire is brought into contact with the surface of the test specimen for 30 s. Repeated tests at different glow-wire temperatures determine the GWIT (glow-wire ignition temperature) of the material, defined as the temperature 25 °C higher than the maximum test temperature not causing ignition.

Table 12.10 IEC 60695-2-13: Glow-Wire Ignitability Test method for Materials (GWIT)

<table>
<tr><td rowspan="4">Test specimen</td><td colspan="4">Test specimens having a sufficiently large plane section with fixed dimensions:</td></tr>
<tr><td colspan="2">Length</td><td colspan="2">≥ 60 mm</td></tr>
<tr><td colspan="2">Width (inside clamps)</td><td colspan="2">≥ 60 mm</td></tr>
<tr><td colspan="2">Thickness</td><td colspan="2">Preferred values 0.1, 0.2, 0.4, 0.75, 1.5, or 3.0 mm</td></tr>
<tr><td>Test specimen position</td><td colspan="4">Typically, a set of 15 test specimens is used in a vertical orientation.</td></tr>
<tr><td>Test layout</td><td colspan="4">Specimens may be any shape provided that the diameter of the specimen is a minimum of 60 mm. The description of the test apparatus is given in IEC 60695-2-10.</td></tr>
<tr><td>Ignition source</td><td colspan="4">Glow-wire coil, made of chrome-nickel alloy with built-in thermocouple.</td></tr>
<tr><td rowspan="3">Test temperatures</td><td>(500 ± 10) °C</td><td>(550 ± 10) °C</td><td>(600 ± 10) °C</td><td>(650 ± 10) °C</td></tr>
<tr><td>(700 ± 10) °C</td><td>(750 ± 15) °C</td><td>(800 ± 15) °C</td><td>(850 ± 15) °C</td></tr>
<tr><td>(900 ± 15) °C</td><td>(960 ± 15) °C</td><td></td><td></td></tr>
</table>

Ignition source application time	(30 ± 1) s
Evaluation of test results	Ignition defined as visible flame > 5 s during glow-wire application. Temperature 25 °C higher than the maximum test temperature not causing ignition during three subsequent tests, reported as GWIT.

Flammability of Solid Non-Metallic Materials

One aim of the TC 89 work is to make available standardized test flames and test methods described in IEC 60695-11-10 and IEC 60695-11-20 [39, 40]. These harmonized standards include the UL 94 HB, V-2, V-1, V-0, 5VA, and 5VB classification criteria [38], and some specifications for considering colors, densities, melt flow, and impact strength of test specimens. They are considered essential by the fire testing community because they satisfy the needs of material suppliers, end-product designers, and safety regulators.

50 W horizontal flame test method, IEC 60695-11-10

IEC 60695-11-10 [39] describes the application of a 20 ± 1 mm test flame for 30 s to a horizontally mounted test specimen. Successful completion of this test leads to the classifications in classes HB 75 or HB 40. The classification depends on the burning rate of the test specimen. The test layout is shown in Figure 12.3, and the test specifications are summarized in Table 12.11. They are similar to UL 94 with the exception of the classification for the HB test.

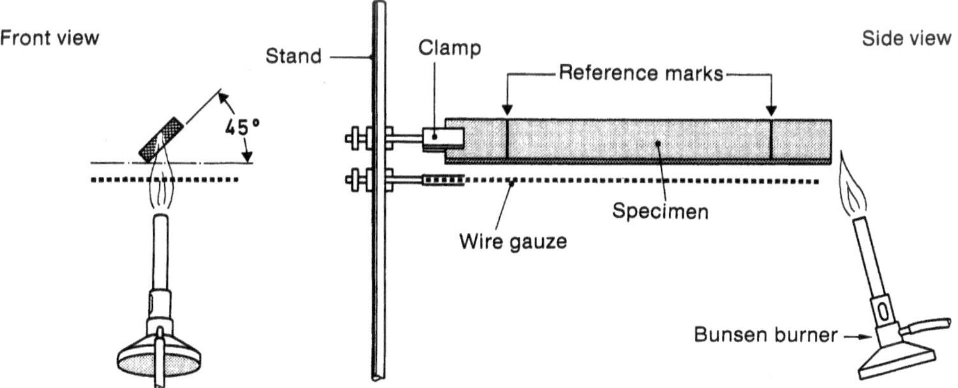

Figure 12.3 Test layout for the classes HB 40 and HB 75 to IEC 60695-11-10

Table 12.11 IEC 60695-11-10: 50 W Horizontal Flame Test Method

Test specimens	Three specimens 125 mm × 13 mm with two reference marks 25 mm and 100 mm from the end to be ignited. Maximum thickness 13 mm. Depending on test results, a further set of three samples may be tested.
Test specimen conditioning	48 h, 23 °C, and 50% relative humidity.
Test specimen position	Longitudinal axis horizontal, traverse axis inclined 45° to horizontal.
Ignition source	Laboratory burner according to IEC 60695-11-4. The longitudinal axis of the burner is tilted at 45° towards the horizontal during flame application.
Flame application	30 s. If the flame front reaches the 25 mm mark on the test specimen before 30 s, the burner is removed.
Conclusions	Classification in class HB 40 if: a) No visible flame after burner removal b) Specimen extinguishes before the 100 mm mark is reached c) Burning rate between reference marks < 40 mm/min. Classification in class HB 75 if burning rate between reference marks < 75 mm/min.

50 W vertical flame test method, IEC 60695-11-10

The vertical flame test method to IEC 60695-11-10 is well-known as UL 94, and has been used for more than 60 years in North America. The test classifies plastic test specimens as V-0, V-1, and V-2. It is a small-scale laboratory screening procedure for comparing the burning behavior of vertically oriented test specimens exposed to a small flame, which is applied twice to the lower end of the test specimen for 10 s. Burning, glowing, dripping, and whether the surgical cotton beneath is ignited, are recorded. The test specifications are summarized in Table 12.12 and the test layout is shown in Figure 12.4.

Table 12.12 IEC 60695-11-10: 50 W Vertical Flame Test Method

Test specimens	Two sets of five specimens 125 mm × 13 mm × minimum thickness normally supplied. The thickness shall not exceed 13 mm. Depending on test results, additional sets may be tested.
Test specimen conditioning	a) Storage of one set of specimens for at least 48 h at 23 °C and 50% relative humidity. b) Storage of a second set of test specimens for 168 h in an air-circulating oven at 70 °C followed by a 4 h cooling to room temperature in a desiccator.
Test specimen position	Vertical. Surgical cotton placed under specimen.
Ignition source	Laboratory burner according to IEC 60695-11-4.

Flame application	Two applications 10 s for each specimen. Second application as soon as specimen extinguishes.
Conclusions	a) Classification in class V-0 if: Single specimen afterflame time ≤ 10 s, total set afterflame time ≤ 50 s Single specimen afterflame + afterglow time ≤ 30 s after burner removal No ignition of surgical cotton Specimens do not burn up completely. b) Classification in class V-1 if: Single specimen afterflame time ≤ 30 s, total set afterflame time ≤ 250 s Single specimen afterflame + afterglow time ≤ 60 s after burner removal Other criteria as for a). c) Classification in class V-2 if: Ignition of surgical cotton Other criteria as for b).

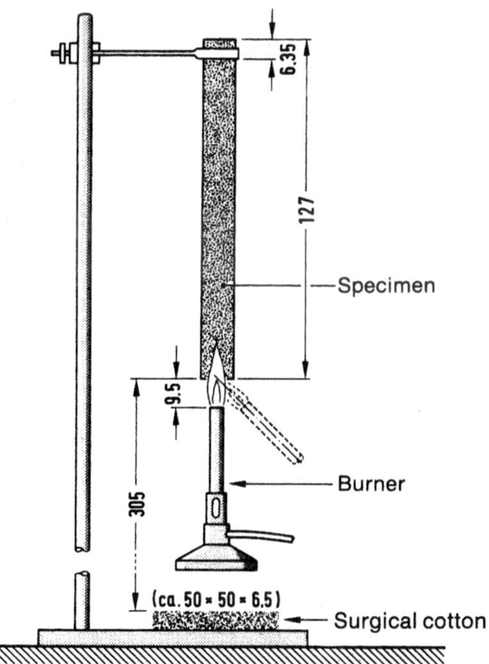

Figure 12.4 Test layout for the classes V-0, V-1, and V-2 to IEC 60695-11-10. Dimensions in mm

500 W flame test method, IEC 60695-11-20

IEC 60695-11-20 specifies a small-scale laboratory screening procedure for comparing the relative burning behavior of test specimens made from plastics as well as their resistance to burn-through when exposed to a flame ignition source of 500 W nominal power [40]. This method requires the use of two test specimen configurations to characterize material performance. Rectangular bar-shaped test specimens 125 mm long × 13 mm wide in the minimum supplied thickness are used to assess ignitability and burning time (bar test). Square plate test specimens 150 mm × 150 mm in the minimum supplied thickness are used to assess the resistance of the material to burn-through (plate test). The test specifications are summarized in Table 12.13 and the test layout for the bar test is shown in Figure 12.5.

Table 12.13 IEC 60695-11-20: 500 W Flame Test Methods

Test specimens	Bar test specimens: Two sets of five bar-shaped test specimens 125 mm × 13 mm × minimum thickness normally supplied. The thickness shall not exceed 13 mm. Plate test specimens: Two sets of three plates 150 mm × 150 mm × minimum thickness normally supplied. Depending on test results, additional sets may be tested.
Test specimen conditioning	a) Storage of one set of test specimen for at least 48 h at 23 °C and 50% relative humidity. b) Storage of a second set of test specimens for 168 h in an air-circulating oven at 70 °C. 4 h cooling to room temperature in a desiccator.
Test specimen position	Bar test specimens: Vertical. Plate test specimens: Horizontal. Surgical cotton placed under specimen.
Ignition source	Laboratory burner to IEC 60695-11-3 with axis inclined at 20° to vertical; flame length 125 mm with 40 mm blue inner cone.
Flame application	5 × 5 s with 5 s intervals.
Conclusions	a) Burning category 5 VA: Individual afterflame time ≤ 60 s after fifth flame application on vertical specimen No ignition of surgical cotton No hole developed in plate specimens. b) Burning category 5 VB: Individual afterflame time ≤ 60 s after fifth flame application on vertical specimen No ignition of surgical cotton Hole developed in plate specimens.

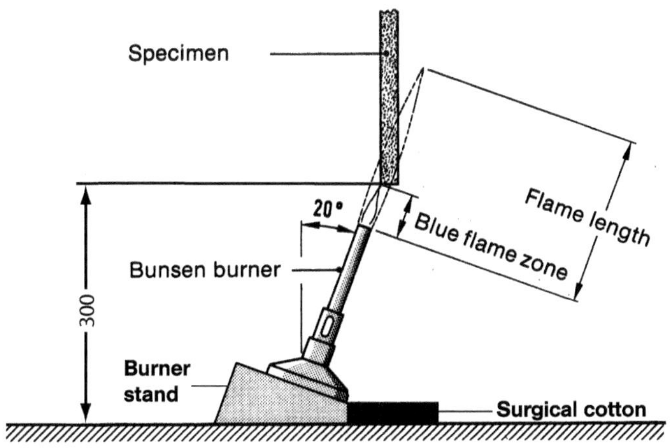

Figure 12.5 Test layout for the bar test to IEC 60695-11-20. Dimension in mm

Confirmational Test Methods

IEC 60695-11-3 [41] gives the detailed requirements for the production of a nominal 500 W, premixed-type test flame, with an overall approximate height of 125 mm. These test flames are used (for example) for the IEC 60695-11-20 small-scale laboratory screening procedure or the UL 94 5V tests. IEC 60695-11-4 [42] provides detailed requirements for the production of a 50 W nominal, premixed-type test flame, with an overall height of approximately 20 mm.

Both IEC standards prescribe confirmatory tests based on copper block calorimetry. These confirmatory tests require a thermocouple to be placed in a copper block under which the flame is located at a specified distance.

The flame is considered to be confirmed if the time taken for the temperature of the copper block to increase from 100 °C to 700 °C is within the time required.

Surface spread of flame, IEC 60695-9-1

IEC 60695-9-1 provides guidance on the assessment of surface spread of flame for electrotechnical products and the materials from which they are formed [43].

IEC TS 60695-9-2 presents a summary of test methods that are used to determine the surface spread of flame of electrotechnical products and the materials from which they are formed [44]. It reflects the current state of the art and includes special observations on their relevance and use where available.

12.1.1.2.4 Side Effects of a Fire

Electrotechnical products, when involved in a fire, contribute to fire hazard due to the release of heat and smoke. Smoke may cause loss of vision and disorientation, and impede escape and firefighting. The production of toxic effluents may be a significant factor contributing to the overall fire hazard. Fire effluents from burning plastics may be to some degree corrosive, and thus damage electrotechnical equipment and building structures.

Heat release, smoke density and toxicity, as well as corrosivity, require fire safety engineering methods for the hazard assessment concerning rise of temperature, visibility, and the production of toxic, and/or corrosive effluents. The standardized test method (the fire model) chosen for decomposing materials or products must allow comprehensive simulation of the many different fire situations.

ISO has published a general classification of fire stages in ISO 19706 [45]. Conditions for use in the fire model can be derived from this table in order to correspond, as far as possible, to real-scale fires. Important factors affecting heat release rate and smoke production are oxygen concentration and irradiance/temperature.

Heat release

Fire calorimetry is the technology used to measure heat release rate from a fire of a product or a whole system. Oxygen consumption calorimetry is the usual method used to determine the heat release rate, described, for example, in ISO 5660 Parts 1, 3, and 4 [46–48]. Testing ranges from bench-scale calorimetry to intermediate-scale, full-scale room, and open calorimeter tests. This technology finds use in a quickly growing field. The measurements provide basic fire characteristics of materials and products that help the fire scientist to better understand the performance of a material or electrotechnical equipment in real fire situations. Heat release characteristics are used as an important factor to determine fire hazard and also used in fire safety engineering calculations.

IEC 60695-8-1 provides guidance in the assessment of heat release from electrotechnical products and materials [49]. IEC 60695-8-2 provides a summary of relevant test methods that are used to determine heat release. It represents the current state of the art of the test methods and includes special observations on their relevance and use [50].

One further example for fire calorimetry in an open-product calorimeter test is the UL 2043 test method for determining the fire performance response of discrete products and their accessories in air-handling spaces [51]. The purpose of this test is to determine the rate of heat and smoke release of burning product samples to the requirements of the U.S. National Electrical Code NEC (NFPA 70) [99]. The test rig and the test specifications are shown in Table 12.14 and Figure 12.6. Further details of fire calorimetry and an introduction to this field are given in [52].

Table 12.14 UL 2043 Test Specifications for Determining Heat and Visible Smoke Release

Test specimen	Discrete products (electrical equipment) installed in air-handling spaces. Test specimens must be representative of the construction components and design. Examples: Loudspeakers, loudspeaker assemblies, ceiling-mounted products suspended in air-handling space products.

Test specimen conditioning	At least 24 h in standard atmosphere (23 °C and 50% relative humidity).
Test layout	Products mounted on test frame enclosure, subjected to a gas flame, and evaluated using a product calorimeter. All combustion products collected in the hood (face dimensions at least 2.4 m × 2.4 m). Nominal exhaust flow rate: 0.71 m^3/s.
Ignition source	Propane gas diffusion burner 0.318 m wide with square surface 0.203 m above bottom of test frame enclosure.
Flame application	10 min
Conclusions	Passed if: ▪ Peak heat release rate (HRR) < 100 kW ▪ Peak normalized optical density < 0.5 ▪ Average optical density < 0.15.

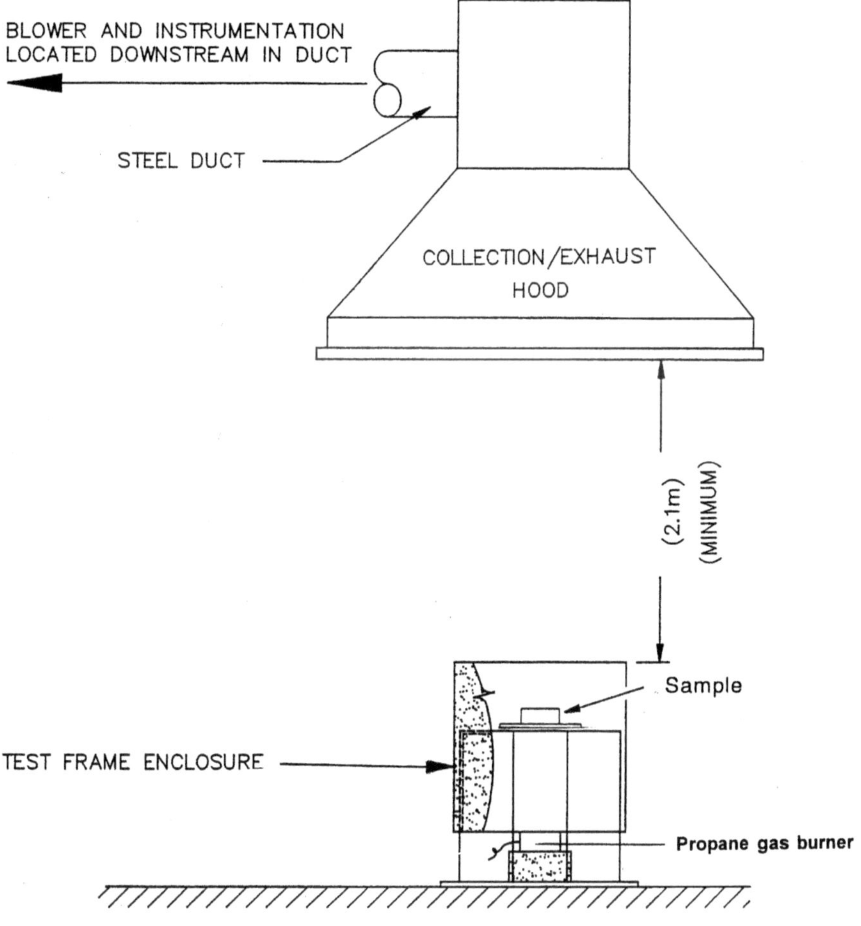

Figure 12.6 Product calorimeter to UL 2043

Smoke obscuration to IEC 60695-6-1 and opacity to IEC 60695-6-2

IEC 60695-6-1 [53] gives guidance on the optical measurement of smoke obscuration, the general aspects of optical smoke test methods, and the relevance of optical smoke data to hazard assessment. IEC 60695-6-2 gives a summary of the test methods that are used in the assessment of smoke opacity [54] as shown in Table 12.15. It presents a brief summary of static and dynamic test methods in common use. For electrotechnical products and materials it includes special observations on their relevance for real fire scenarios and gives guidance on their use. It is intended to support product committees wishing to incorporate test methods for smoke obscuration in product standards.

Toxic hazard assessment

Fire does not discriminate between electrotechnical and non-electrotechnical products. IEC TC 89 works closely with the fire experts in ISO/TC 92 [55] on a basis for publishing guidance for toxic hazards in fires involving electrotechnical products. The basic philosophy developed by ISO/TC 92 is that toxic hazard cannot be taken in isolation, but forms part of an assessment of the total fire hazard. Other hazards can include smoke, heat, and structural endurance. The total combination of all hazards leading to lethality or incapacitation must be considered as the fire grows and spreads. One task of IEC TC 89 was to translate this guidance into terms that would be useful for IEC product groups.

Toxicity is a function of the exposure dose and the toxic potency of fire effluents. The main questions concerning the reduction in toxic hazards from fires in which electrotechnical products have been involved are:

a) Potency: How toxic is the fire effluent?

b) Dose: How much of the fire effluent is inhaled?

c) Escape: How is escape impeded? Does the time available to escape exceed the time needed to escape?

Table 12.15 IEC 60695-6-2: Smoke Opacity – Summary and Relevance of Test Methods[a]

Type of test method	Test method reference		Limitations on test specimen	Relevance to stage of fire[b]						Limitations on use for regulatory purposes	Suitability of data format for input to fire safety engineering
	Section			1(a)	1(b)	1(c)	2	3(a)	3(b)		
Static	6.1	Determination of opacity of smoke in an NBS chamber	Essentially flat only, and less than 75 mm square Not suitable for liquids or some thermoplastics	No	Yes	No	Yes	No	No	Not recommended	No
	6.2	Determination of opacity of smoke by a single-chamber test	Essentially flat only, and less than 75 mm square Possibly suitable for products					Yes		Only for products with suitable geometry, and for appropriate fire stage	Not as currently reported
	6.3	Determination of opacity of smoke in the "three meter cube" smoke chamber	Self-supporting products only, approx. 1 m long		No			No		Only for self-supporting products, and for appropriate fire stage	
Dynamic	7.1	Determination of smoke opacity generated by electric cables mounted on a horizontal ladder	Essentially flat building products or cables				No		Yes	Only for products with suitable geometry, and for appropriate fire stage	Yes
	7.2	Determination of smoke opacity from electrical cables mounted on a vertical ladder	Self-supporting products only				Yes			Only for self-supporting products, and for appropriate fire stage	
	7.3	Determination of smoke opacity using the cone calorimeter	Intended for essentially flat materials Possibly suitable for products							Only for products with suitable geometry, and for appropriate fire stage	

[a] The suitability of data for fire safety engineering purposes is limited to applications where the test method is relevant to the appropriate stage of fire.

[b] to ISO 19706 [45]: 1(a): Self-sustained smoldering, 1(b): Oxidative radiation, 1(c): Anaerobic external radiation; 2: Well-ventilated flaming; 3(a): Low ventilation room fire, 3(b): Post flashover

IEC TC 89 addresses all these questions, which have resulted in the list of publications summarized in Table 12.16.

Table 12.16 IEC List of Publications Concerning Toxicity of Fire Effluent

Publications	Title
IEC 60695-7-1 [56]	Toxicity of fire effluent - General guidance
IEC 60695-7-2 [57]	Toxicity of fire effluent - Summary and relevance of test methods
IEC 60695-7-3 [58]	Toxicity of fire effluent - Use and interpretation of test results

IEC 60695-7-1 [56] gives general guidance for IEC product committees.

Key messages to IEC product committees incorporating requirements for the assessment of toxic hazard in product standards are:

- Toxic potency data should not be used directly in product specifications to give individual pass/fail criteria, and only be used in an integrated system analysis, within a total fire hazard assessment
- Toxic potency should not be confused with toxic hazard
- Toxic hazards from fires can best be minimized by the control of ignition, rates of flame spread, fire growth, and smoke development
- Carbon monoxide is by far the most significant agent contributing to toxic hazard. Other agents of major significance are hydrogen cyanide and carbon dioxide, as well as irritants, heat, and hypoxia caused by oxygen depletion.

Prediction of toxic potency is based on mathematical models taking into account analytical results of fire effluents for acute toxicity by inhalation in animal tests. This requires the generation of thermal decomposition products and analytical determination of the concentration of those components of effluents responsible for toxicity.

IEC 60695-7-2 provides a summary of test methods which are in common use [57]. For electrotechnical products and materials it includes special observations on their relevance for real fire scenarios and gives guidance on their use. It is intended to support product committees wishing to incorporate test methods for toxicity in product standards.

IEC 60695-7-3 provides guidance on the use and interpretation of results from laboratory tests for the determination of toxic potency in assessing the toxic hazard as a part of the total fire hazard [58]. It discusses currently available approaches to toxic hazard assessment consistent with the approach of ISO/TC 92 SC 3, as described in ISO 19706 [45]. The detailed methodology is directly applicable to data produced by tests measuring the lethal effects of fire effluents.

Corrosion damage: IEC 60695-5 series

About fifty years ago, the corrosive behavior of smoke began to be investigated systematically, and the work concentrated mainly on determining the effects of chlorine. Fires in power stations, telecommunication facilities, ships, and underground railways, etc. have shown that exposure of sensitive electronic or mechanical equipment to fire effluents has the potential to cause extensive damage. In combination with cleaning, equipment replacement, and temporary operating failure, great financial losses may occur. Work in ISO/TC 61 and IEC TC 89 led to the development of universally valid testing and evaluation criteria and the preparation of guidance documents.

IEC 60695-5-1 [59] gives guidance on the assessment of corrosion damage to electrotechnical equipment, systems, and building structures from fire effluent emitted from electrotechnical equipment and systems.

IEC TS 60695-5-2 [60] provides a summary of test methods, and advices which are suitable for use in fire hazard assessment and fire safety engineering. For electrotechnical products and materials it includes special observations on their relevance for real fire scenarios and gives guidance on their use. It is intended to support product committees wishing to incorporate test methods for corrosion damage in product standards.

Prescribed are three approaches to testing differentiated by the type of corrosion target:

- Product testing: The corrosion target shall be a manufactured product, e.g., printed wiring board, switchboards
- Simulated product testing: The corrosion target shall be a reference material simulating a product. Effects can be assessed, e.g., by mass loss, physical characteristics, or electrical characteristics
- Indirect assessment: No corrosion target is used, but characteristics of the gases and vapors evolved are measured, e.g., the pH value and the conductivity of a solution in which the effluents have been dissolved.

12.1.1.3 International Certification

The CB Scheme

Companies wishing to enter the global market have to overcome certain hurdles. In the past, manufacturers had to obtain national safety approvals for every country they sold their products in. To overcome these problems, the Certified Body (CB) Scheme has been established by IEC. The system is based on the principle of mutual recognition by means of certification through the use of internationally accepted standards. By virtually eliminating duplicate testing, the CB Scheme facilitates the international exchange and acceptance of product safety test results.

The International Electrotechnical Committee for Conformity Testing to Standards for Electrical Equipment (IECEE) is a global network of National Certification Bodies (NCBs) that has agreed to mutual acceptance of CB test certificates and reports. The Scheme is based on the use of international (IEC) Standards. If some national standards of a member country adhering to the CB Scheme are not yet completely harmonized to IEC standards, national differences are permitted, if clearly declared to all other members. Currently, there are 54 member countries in the IECEE. Further details about the 77 participating NCBs, the over 500 CB Testing Laboratories, and an introduction to the CB Scheme are available [63].

The CB Scheme applies to many specific product categories established by the IECEE, for example IEC 60065: Safety requirements for audio, video, and similar electronic apparatus [102], or IEC 60335: Safety of household and similar electrical appliances [34].

If a product is already UL-Listed, Classified, or Recognized (Section 12.1.2.1) or has a test mark, e.g., a German GS- or VDE-Mark (Section 12.1.3.3), the required testing for the CB Scheme is mostly already completed and accepted. In some cases, additional testing is necessary to comply with national deviations required in countries to which the products are to be marketed.

12.1.2 Fire Safety Regulations in North America (UL, CSA)

Electrotechnical products are tested worldwide according to established safety standards. While in many countries the testing regulations and performance requirements are based on standards developed by the IEC, in the USA and Canada, national standards that sometimes deviate from the IEC standards are commonly used and recognized. In the USA, safety testing of electrotechnical products is often based on Underwriters' Laboratories Inc. (UL) [61]. In Canada, the standards issued by the Canadian Standards Association (CSA) are applied [64].

Following the harmonization of UL and CSA testing standards and the corresponding alignment of existing standards and regulations, the two organizations entered into a Memorandum of Understanding (MoU), granting each other acceptance of test data, with verification, for the purpose of listing or recognition.

12.1.2.1 Test and Approval Procedures of the Underwriters' Laboratories

Underwriters' Laboratories Inc. was founded in 1894 with the aim of carrying out technical safety tests for the protection of people and goods. As a "not-for-profit" organization without capital stock, UL must use the profits to establish, maintain, and operate laboratories for examination and testing of devices, systems, and materials to determine their relation to hazards to life and property, and to develop standards, classifications, and specifications for (for example) devices, products, and equipment.

UL started their work by carrying out technical fire protection tests. Later, their field of activity expanded, and today it includes product safety testing and certification in the field of (for example) fire protection, electrical engineering, protection from burglary, heating and air conditioning, chemical safety, as well as marine and automotive.

UL has fire testing laboratories, numerous international, affiliate, and representative offices, as well as field representatives located throughout the world. The UL website [61] gives detailed information about (for example) global resources, UL news, testing and certification services, and information about UL-certified products.

UL Standards for Safety are developed with the participation and comments from manufacturers, consumers or consumer-oriented organizations, governmental officials, inspection authorities, and insurers. They provide input to UL before UL draws up a final version of the UL Standard taking into account the comments made. The UL Standards are often taken over by other bodies and standards organizations. The UL Standards Department has developed a separate website that contains information on UL Standards - for example, the Catalog of Standards and the scopes of the standards [62].

Three types of UL certifications exist: Listing, Classification, and Recognition. Successful completion of a Listing Certification allows a finished product to carry the UL Listing Mark, which is similar to the German VDE Test Mark. In contrast, Classification covers specific characteristics of the finished product designated by the manufacturer. It results in a UL Mark incorporating wording for a specific hazard, or indicating that it was evaluated to a non-UL-standard (ASTM, OSHA, etc.). Recognition is a preselection certification for materials and components, and can expedite the issuing of a Listing Mark to the finished product.

The UL categories of materials are very important in connection with the use of plastics in electrotechnical products, because specific ratings are often a pre-requirement of the Listing of a product by UL. By itself, neither a part nor a material can gain a UL Mark, but the testing of a certified product is simplified if materials that are already UL-Recognized are used.

On behalf of UL, field representatives, and authorized representatives periodically visit (unannounced) manufacturers' facilities, in order to inspect and monitor production, and to verify that the product still complies with the UL-Report requirements. The field representative randomly selects production samples for Follow-up testing at UL.

The UL product directories help to find products that meet UL requirements. They contain the names of companies that manufacture products, components, devices, materials, or systems in accordance with UL safety requirements. These are products eligible to carry UL's Listing Mark, Recognized Component Mark or Classification Marking. The Plastics Recognized Component Directory contains the category of polymeric materials components QMFZ2. Recognition cards, also known as "Yellow Cards", are an excerpt of the Recognized Component Directory. They give the manufacturers the documentary evidence that their product meets the relevant UL requirements.

12.1.2.1.1 UL 94 Flammability Tests

UL 94 "Test for flammability of plastic materials for parts in devices and appliances" is one of the most important UL standards relating to fire safety test methods and requirements, and contains several fire tests [38].

The various flammability tests to UL 94, which have been adopted by national or international standardization bodies like IEC, are exclusively material tests. They involve standard-size specimens and are intended to serve as a preliminary indication of their acceptability with respect to their flammability in a particular application. These methods are not intended to provide correlation with performance under real end-use conditions. The final acceptance of the material is dependent upon its use in complete equipment that conforms with the standards applicable to such equipment.

The requirements of UL 94 together with a few other tests described in UL 746 A, B, and C form the basis for the "Recognition" of plastics as summarized in the "Recognized component directory" [68, 71, 72]. UL 94 applies not only to electrical parts, appliances, consumer, and office equipment, but also to all areas of application except the use of plastics in buildings. UL 94 is particularly significant for the use of plastics in electrotechnical products, since a UL Listing of the product frequently requires a specific flammability classification of the materials used.

Depending on the flammability requirements of the UL standards for the end-product, materials have to meet the horizontal burning test (class HB) or the more stringent vertical burning tests (class V-0, V-1, or V-2). These test methods and their criteria are identical to IEC 60695-11-10 (see Section 12.1.1.2.3).

Testing for classification in class HB

The test method describes the application of a 20 mm test flame for 30 s to a horizontally mounted test specimen. Classification in class HB requires that the burning rate of the test specimen should not exceed a maximum value dependent on its thickness or that the specimen extinguishes after removal of the flame. Successful completion of this test leads to the classifications in classes HB 75 or HB 40. The test specifications and the test layout have already been shown in Table 12.11 and Figure 12.3 of Section 12.1.1.2.3.

Testing for classification in classes V-0, V-1, and V-2

This well-known test for classifying solid plastic specimens in classes V-0, V-1, and V-2 is a small-scale laboratory screening procedure for comparing the relative burning behavior of vertically oriented specimens, exposed to a small-flame ignition source. For the vertical test the same specimens are used as for the HB-test. A flame is applied twice to the lower end of the test specimen for 10 s. Burning and glowing times, dripping, and whether or not the surgical cotton beneath is ignited, are noted. The best rating, V-0, is achieved if the mean afterflame time of five samples after 10 flame applications does not exceed 5 s. The material is classified in V-1 if the mean afterflame time is less than 25 s. If flaming particles ignite the surgical cotton, the material is classified in V-2. The test specifications and the test layout have already been shown in Table 12.12 and Figure 12.4 of Section 12.1.1.2.3.

Testing for classification in classes VTM-0, VTM-1, and VTM-2

UL 94 contains a method for testing materials that (due to their thinness) distort, shrink, or are consumed up to the holding clamp when tested using the UL 94 Vertical Burning Test. The test method and criteria are identical to ISO 9773 [65].

Instead of solid test specimens, cylindrically wound film rolls are used. Oriented specimens are tested vertically. The film roll is produced by winding a sample strip around a metal rod. A piece of adhesive is used to prevent unrolling. After the rod is withdrawn and the film roll clamped vertically, a burner flame is applied twice to the lower end of the film roll. The test specifications are summarized in Table 12.17. The main assessment criteria are the same as in the vertical test for solid specimens, i. e. afterflame time and igniting of the surgical cotton.

Table 12.17 UL 94 Test Specifications for Classification in Classes 94 VTM-0, 94 VTM-1, and 94 VTM-2

Specimens	Two sets of five film rolls 200 mm long, 12.7 mm internal diameter. Marking 125 mm from one end. Film rolls prepared by wrapping a specimen 200 mm long by 50 mm around the axis of a 12.7 mm diameter mandrel. Depending on test performance, further sets of samples may have to be tested.
Specimen conditioning	a) Storage of first set of samples for at least 48 h in a standard atmosphere (23 °C and 50% relative humidity). b) Storage of second set of samples for 168 h in air-circulating oven at 70 °C. 4 h cooling to room temperature in desiccator.
Specimen position	Film roll clamped vertically with reference mark 125 mm above sample bottom. Film roll closed by spring clamp at top and open at bottom. Surgical cotton placed under specimen.
Ignition source	Bunsen or Tirrill burner with 20 mm high non-luminous flame.
Flame application	Twice, for a duration of 3 s each time, for each sample. Second application as soon as the sample extinguishes.

Table 12.17 UL 94 Test Specifications for Classification in Classes 94 VTM-0, 94 VTM-1, and 94 VTM-2 *(continued)*

Conclusions	a) Classification in class 94 VTM-0 if: Afterflame time ≤ 10 s Sum of afterflame times for 10 flame applications ≤ 50 s Single specimen afterflame + afterglow time ≤ 30 s after second flame application No surgical cotton ignition No material consumed beyond 125 mm mark. b) Classification in class 94 VTM-1 if: Afterflame time ≤ 30 s Sum of afterflame times for 10 flame applications ≤ 250 s Single specimen afterflame + afterglow time ≤ 60 s after second flame application Other criteria as for a). c) Classification in class 94 VTM-2 if: Ignition of surgical cotton Other criteria as for b).

Testing for classification in classes 5 V

Testing for classification in classes 5 V is the most stringent of all UL small-scale burning tests. This method requires the use of two test specimen configurations to characterize material performance. Rectangular bar-shaped test specimens 125 mm long × 13 mm wide in the minimum supplied thickness are used to assess ignitability and burning time (bar test). Square plate test specimens 150 mm × 150 mm in the minimum supplied thickness are used to assess the resistance of the material to burn through (plate test). The test specifications are summarized in Table 12.13 and the test layout for the bar test is shown in Figure 12.5 of Section 12.1.1.2.3. The 5 VA classification is specified for fire enclosures in office machines (see UL Standard 1950 in Section 12.1.2.1.2).

Testing for classification in classes HBF, HF-1, or HF-2

UL 94 contains a method for testing foam materials in a horizontal position. The test method and criteria are identical to ISO 9772 [66]. A flame from a special burner is applied to one side of the foam sample lying horizontally on a wire grid. Class HBF is achieved if the burning rate does not exceed 40 mm/min. If the specimen extinguishes within 2 s after removal of the ignition source, it is classified in classes HF-2 or HF-1, depending on whether burning drips are produced. The test specifications are summarized in Table 12.18 and the test layout is shown in Figure 12.7.

Table 12.18 UL 94 Test Specifications for Classification in Classes 94 HBF, 94 HF-1, and 94 HF-2

Specimens	Two sets of five specimens 150 mm × 50 mm in minimum and maximum thickness of intended application with three reference marks 25 mm, 60 mm, and 125 mm from the end to be ignited. Maximum thickness 13 mm. Depending on performance, additional sets may be tested.
Specimen conditioning	Storage of first set of samples at least 48 h in standard atmosphere (23 °C and 50% relative humidity). Storage of second set of samples for 168 h in air-circulating oven at 70 °C. 4 h cooling to room temperature in desiccator.
Specimen position	Horizontal on wire grid of specified mesh. Surgical cotton placed under specimen.
Ignition source	Burner with special wing tip to give 38 mm non-luminous flame.
Flame application	60 s
Conclusions	a) Classification in class HBF if: Burning rate between 25 mm and 125 mm marks ≤ 40 mm/min or Specimen extinguishes before 125 mm mark Classes HF-1 and HF-2 requirements not fulfilled. b) Classification in class HF-1 if: Afterflame time ≤ 2 s for at least four of the five specimens Afterflame time ≤ 10 s for any specimen No material consumed beyond 60 mm mark No incandescence > 30 s for each specimen after burner removal No ignition of surgical cotton. c) Classification in class HF-2 if: Ignition of surgical cotton Other criteria as for b).

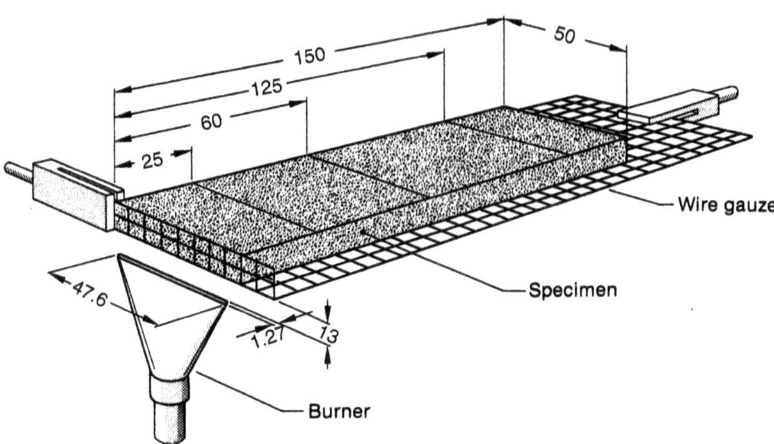

Figure 12.7 Layout for testing foam materials to UL 94. All dimensions in mm

Radiant panel flame spread test to ASTM E162

UL 94 requires evaluation of the potential contribution of large sections of material used as electrical housing to flame spread with the radiant panel test to ASTM E162 [67]. The surface flammability characteristics of the test material are evaluated. The end-product criteria establish the level of flame spread accepted. The test specimen is exposed to a radiant panel and ignited at its upper hot end, and the extent of combustion towards the bottom of the specimen is evaluated. The test is described in Section 11.2.3, with the test specifications summarized in Table 11.17 and the test layout shown in Figure 11.10.

12.1.2.1.2 Fire Safety Requirements in Other UL Standards

UL 746A

The safety of electrical equipment depends on the right selection of materials, design, and processing, as well as on the assembly, mounting, and relative positions of parts. The properties of a part are defined by its function. UL 746A contains short-term test procedures to be used for the evaluation of materials used for parts intended for specific applications in electrical end-products [68].

The UL 746A evaluations provide data with respect to the physical, electrical, flammability, thermal, and other properties of materials, and provide guidance for the material manufacturer and other interested parties (Table 12.19).

Programs for the evaluation of material modifications, such as plating of plastics or the use of flame-retardant paints, are contained in the standard UL 746C [72]. As thickness and orientation of the material may influence its burning behavior, these parameters have to be considered when testing specimens representative of the finished article.

Table 12.19 UL 746A Polymeric Materials Short-Term Property Evaluations

High-voltage, low-current, dry arc resistance	Comparative tracking level (CTI)	High-voltage arc-tracking rate (HVTR)	Hot-wire ignition (HWI)	High-current arc ignition (HAI)	High voltage arc resistance to ignition (HVAR)	Assigned performance level category (PLC)
ASTM D495 range-mean time of arc resistance [s]	Range-tracking index [V]	Range-tracking rate [mm/min]	Range-mean ignition time [s]	Range-mean number of arcs to cause ignition	Range-mean time to ignition [s]	
420 ≤ TAR	600 ≤ TI	0 < TR ≤ 10	120 ≤ IT	120 ≤ NA	300 ≤ TI	0
360 ≤ TAR < 420	400 ≤ TI < 600	10 < TR ≤ 25.4	60 ≤ IT < 120	60 ≤ NA < 120	120 ≤ TI < 300	1
300 ≤ TAR < 360	250 ≤ TI ≤ 400	25.4 < TR ≤ 80	30 ≤ IT < 60	30 ≤ NA < 60	30 ≤ TI < 120	2
240 ≤ TAR < 300	175 ≤ TI ≤ 250	80 < TR ≤ 150	15 ≤ IT < 30	15 ≤ NA < 30	0 ≤ TI < 30	3
180 ≤ TAR < 240	100 ≤ TI ≤ 175	150 < TR	7 ≤ IT < 15	0 ≤ NA < 15		4
120 ≤ TAR < 180	0 ≤ TI ≤ 100		0 ≤ IT < 7			5
60 ≤ TAR < 120						6
0 ≤ TAR < 60						7

Comparative tracking index (CTI)

A fire risk may develop within electrical equipment as a result of electrical tracking of insulating material that is exposed to various contaminating environments and surface conditions. The comparative tracking index (CTI) provides a comparison of the performance of insulating materials under wet and contaminated conditions. The CTI is the numerical value of the highest voltage (max. 600 V) at which an electrical insulation material will withstand 50 drops of electrolytic test solution A (398 Ω cm) being dripped onto it without permanent electrically conductive tracking. The test is carried out under the conditions specified in ASTM D3638-12 [69] or IEC 60112 [70]. Based on the comparative tracking index, the comparative tracking performance level category (PLC) is assigned (Table 12.19).

High-voltage tracking resistance (HVTR)

The high-voltage tracking resistance (HVTR) test method is used to assess the susceptibility to tracking of insulating materials that are exposed to high voltages outdoors. The test determines the tracking resistance that defines the dielectric strength of the insulating material surface and the maximum allowable leakage current (tracking). High-voltage arcs are ignited on the surface of a specimen for 2 min. A note is made if ignition on the material takes place. Thereafter, the length of the tracking path is measured and divided by two. Table 12.19 indicates the tracking path generated in mm/min and the corresponding TR values and PLCs.

Hot-wire ignition test (HWI)

Under certain operational conditions or in the event of electrical equipment malfunction, wires, conductors, resistors, or other parts will become abnormally hot. When these overheated parts are in intimate contact with insulating materials, the latter may ignite. The hot-wire ignition test to UL 746A [68], Section 30, determines the relative resistance of insulation materials to ignition. For a given material, the hot-wire performance level is to be assigned based on the determined mean time for ignition (Table 12.19). Hot-wire ignition tests are carried out on five bar-shaped specimens 125 mm long, 130 mm wide, and 3 mm thick.

High-current arc ignition (HAI)

Under certain normal or abnormal operational conditions, insulating material used in electrical equipment might be in proximity to arcing. If the intensity and duration of the arcing are severe, the insulating material may ignite. UL 746A, Section 31, simulates such arcing. The specimen is located in the direct vicinity of a high-current arc. The HAI rating identifies the mean number of arcs required to cause ignition, and this is used to assign a PLC as shown in Table 12.19.

High-voltage, low current, dry arc resistance

The ASTM D495 high-voltage, low-current, dry arc resistance test is intended to simulate service conditions as exist in alternating current circuits operating at high voltage, but at currents limited to units and tens of milliamperes. The specimen is exposed to a high-voltage arc for a maximum of 7 min, while the load is increased. The time (in s) is measured until the arc extinguishes. The results are used to assign a PLC as shown in Table 12.19. The results obtained with 3 mm thick material are considered representative of the material's performance in any thickness.

High-voltage arc resistance to ignition

A high-voltage arc is generated on the surface of the specimen for a maximum of 5 min. The time (in s) is measured until a hole forms or the specimen ignites. The results are used to assign a PLC as shown in Table 12.19.

Glow-wire ignitability test

The test method for the determination of ignitability of an insulating material from an electrically heated wire is described in the glow-wire ignitability test method on materials, IEC 60695-2-13 (Section 12.1.1.2.3). The glow-wire ignition temperature (GWIT) is defined as 25 °C higher than the temperature of the tip of the glow-wire not causing ignition during three consecutive tests.

UL 746B

UL 746B contains long-term test procedures to be used for the evaluation of materials used for parts intended for specific applications in end-products [71]. The relative thermal index (RTI) is an indication of a material's ability to retain a particular property when exposed to elevated temperatures for an extended period of time, and is based on an evaluation of long-term thermal-aging data. The tests include mechanical, electrical, and flammability (vertical burning UL 94) requirements.

UL 746C

UL 746C contains requirements for mechanical, electrical, thermal, and flammability properties of plastics, depending on their application in electrical equipment, devices, and appliances – for example, as materials for polymeric enclosures or internal barriers, or as carriers for live parts [72]. The service conditions of the electrotechnical products, e.g., portable or fixed equipment, and the interaction of the material's properties, are considered. Individual requirements are selected in such a way that in combination they ensure an adequate safety level for the polymeric part in the finished electrotechnical product. The test procedures refer to the data obtained from UL 94 flammability tests, UL 746A standard property tests, as well as from other evaluation means. The standard also contains programs for the examination of material part modifications, such as plating of plastics or the use of flame-retardant paints.

While the methods described in UL 94 measure and describe the flammability properties of materials in response to heat and flame under controlled laboratory conditions, UL 746C includes tests for finished components, to assess the influence of the structural make-up of the finished part on the actual response to heat and flame of the electrotechnical product.

For example, enclosures for electrotechnical appliances, as a finished product, must pass a fire performance test (with a flame ignition source) if the material does not fulfill the relevant UL 94 requirements 5 VA, 5 VB, V-0, V-1, or V-2 for testing flammability of plastic materials. The test conditions depend on the use of the appliance. Portable appliances that are usually attended are tested with a 12 mm needle-flame or a 20 mm flame applied twice for 30 s with a break of 1 min. The results are acceptable if all flames have ceased within 1 min after the ignition source is removed. Electrical equipment that is fixed or stationary and not easily carried or conveyed is tested with a 127 mm flame applied five times for 5 s with intervals of 5 s. After the fifth flame application, the after-burning time must not exceed 1 min. In addition, in the area of the test flame, the housing must not be damaged to such an extent that its fire containment integrity is affected. Flaming drops, and flaming or glowing particles, that ignite the surgical cotton below the test specimen are not allowed.

UL 746C also requires flame-retardant coatings to pass the enclosure flammability tests with the flaming ignition sources mentioned above if the coated test specimens do not fulfill the requirements of the relevant UL 94 material specifications. In Section 50, UL 746C contains the enclosure flammability 746-5VS-test for coated materials. Test plaques of dimensions 150 mm × 150 mm are tested in their minimum and maximum thicknesses with a 125 mm flame with the inner blue cone applied for 60 s.

UL uses several test methods to evaluate the burning behavior of external plastic materials of appliances and devices, depending on their type and size, the likelihood of ignition of the material due to its proximity to an internal or external ignition source, and the amount (surface area) of plastic involved.

For intermediate-size and larger-size products, the material's ignition probability by internal or external sources should be considered. UL primarily uses two larger laboratory-scale fire tests to assess the burning behavior of such materials.

Generally, intermediate-size products with surface plastic are exposed to radiant heat and flaming pilot ignition of the ASTM E162 Radiant Panel Test [67]. The test specimen is exposed to a radiant panel and ignited at its upper hot end, and the extent of combustion towards the bottom of the specimen is evaluated. The rate of flame spread and the temperature rise in the exhaust stack of the apparatus are used for calculating the Flame Spread Index. A maximum value of 200 is allowed (see also Section 12.1.2.1.1).

The surface plastic materials used in larger-size products (surface > 0.93 m^2 or single dimension > 1.83 m) are exposed to the large flame of the UL 723 (ASTM

E84) 25-foot tunnel test [73]. Details of the test set up and specifications are described in Section 10.2.1. Larger-size products shall have a maximum flame spread index of 200.

UL 746C also requires polymeric materials used in electrical equipment to pass two tests with electrical ignition sources, the hot-wire ignition test and the high-current arc ignition test. The requirements in both tests depend upon the UL 94 flammability classification of the polymeric material. According to UL 746C, polymeric materials used in enclosures for portable appliances should not ignite before 7 s when tested by the hot-wire ignition test. Materials that do not comply with the minimum hot-wire ignition levels may be evaluated by an abnormal overload test or the glow-wire end-product test.

The abnormal overload test involves passing abnormally high currents through current-carrying parts of the equipment or representative sections of it. Overcurrent values and times depend on the overcurrent protective device rating. The glow-wire temperature requirements are based on the functional end-use application of the product. As an example, fixed or stationary equipment is tested at 750 °C.

In the high-current arc test described in UL 746A, Section 31, the number of arc ignitions required to set the specimen on fire is determined. According to UL 746C, polymeric materials in contact with uninsulated live parts < 0.8 mm thick should survive at least 15 cycles without igniting.

UL 746D

UL 746D Polymeric Materials - Fabricated Parts, contains requirements for traceability and performance of parts molded and fabricated from polymeric materials, as well as requirements for materials that have been modified to match the requirements of a specific application, including the use of recycled and "regrind" materials, the use of additives (e. g., flame retardants and colorants), and the blending of two or more materials [74].

UL 1694

Small components, i. e. encapsulated integrated circuits, which contain materials that cannot be fabricated into standardized specimens in minimum-use thickness and subjected to preselection tests such as UL 94, are tested with a needle-flame to simulate the effect of the small flames that may result from equipment malfunction. The burner and the confirmatory test procedure are in accordance with IEC 60695-11-5, referenced in Section 12.1.1.2.2.

The 12 mm needle-flame is applied to that part of the surface of the specimen likely to be affected by flames resulting from malfunction or accidentally exposed to any ignition source. Whether a small component ignites depends to a large extent on its mass or volume, and on flame exposure time. Depending on component size, four different exposure times (5 to 30 s) are scheduled in UL 1694 [75]. The assessment criteria are the same as in the UL 94 vertical test, i. e. afterflame time and burning drops. There are additional material classifications for small components that (after ignition by the test flame) are totally consumed, but have a limited burning time. The test specifications are summarized in Table 12.20. Small molded components from materials classified as 5 VA, 5 VB, V-0, V-1, or V-2 in minimum-use thickness are exempted from the needle-flame test.

Table 12.20 UL 1694: Flammability Test Specifications for Small Components

Specimens	Five samples of the small component.	
Specimen conditioning	Storage 24 h in air-circulating oven. 2 h cooling to room temperature in a desiccator.	
Specimen position	Sample arranged in its most unfavorable position of normal use. Surgical cotton placed under specimen.	
Ignition source	12 mm needle-flame to IEC 60695-11-5.	
Flame application	Two flame applications for each sample. Second application as soon as specimen extinguishes.	
	Volumetric range [mm^3]	**Flame application time [s]**
	V ≤ 250	5
	250 < V ≤ 500	10
	500 < V ≤ 1750	20
	1750 < V ≤ 2500	33
Conclusions	a) Classification in class SC-0 if: Afterflame time ≤ 10 s Total afterflame time for 10 flame applications ≤ 50 s, no burning drops Single sample afterflame + afterglow time ≤ 30 s after burner removal No ignition of surgical cotton Samples do not burn up completely. b) Classification in class SC-1 if: Afterflame time ≤ 10 s Total afterflame time for 10 flame applications ≤ 250 s, no burning drops Single sample afterflame + afterglow time ≤ 60 s after burner removal Other criteria as a). c) Classification in class SC-2 if: Ignition of surgical cotton Other criteria as b). d) Classification in SC-TC 0, SC-TC 1, and SC-TC 2 if: Small component is totally consumed Other criteria as a), b) and c).	

UL 60950-1

The common UL and CSA standards for Safety of Information Technology Equipment, including Electrical Business Equipment UL 60950-1 and CSA 60950-1 [76], have been harmonized to IEC 60950-1 [77]. The standard contains basic requirements for products covered by UL under its follow-up service. It aims to prevent injury or damage due to electric shock, mechanical and heat hazards, and fire. Temperatures causing a fire risk in information technology equipment, i.e. visual display units and data processing equipment, may result from overloads, component failure, insulation breakdown, high resistance or loose connections. Fires within the equipment must not spread beyond the fire source or damage the whole piece of equipment.

To meet these targets, the following design objectives should be achieved:

- Taking all reasonable steps to avoid high temperatures that might cause ignition
- Controlling the position of combustible materials in relation to possible ignition sources
- Limiting the quantity of combustible material used
- Ensuring that, if combustible materials are used, they have the lowest flammability practicable
- Using enclosures or barriers, if necessary, to limit the spread of fire within the equipment
- Using suitable materials for the outer enclosures of the equipment.

Materials and components inside a "fire enclosure" and air-filter assemblies should meet the UL 94 flammability classes V-2 or HF-2 (or better) [38], or pass the flammability test for fire enclosures of movable equipment ≤18 kg described below.

The standard contains many exemptions, for which these requirements do not apply. Examples are materials and components that either meet the flammability requirements of the relevant IEC component standard or which are encapsulated within a metal enclosure < 0.06 m^3 in volume without any ventilation openings, or within a sealed unit filled with an inert gas.

Materials used for equipment and decorative enclosures should minimize the risks of ignition and flame spread. For movable equipment > 18 kg in weight, fire enclosures are exempted from a finished product flammability test if, in the smallest thickness used, the material fulfills class V-1 or better. Alternatively, the complete enclosure, a part of it, or a test plaque representing the thinnest thickness, is tested with a 20 mm Bunsen burner flame, or a 12 mm needle-flame to IEC 60695-11-5 (Section 12.1.1.2.2) applied to the inner sample surface likely to ignite because of its proximity to potential ignition sources. The flame is applied twice for 30 s with a break of 1 min after the second flame application; the afterburning time must not exceed 1 min. Flaming drops or glowing particles must not ignite the surgical cotton below the sample.

Materials fulfilling class 5 V are also exempted. Alternatively, three samples, each consisting each either of a complete fire enclosure or a section representing the lowest wall thickness including any ventilation openings, are tested with a 130 mm flame. The flame is applied five times for 5 s with intervals of 5 s to the inner sample surface considered to be most critical because of its proximity to potential internal ignition sources. After the fifth flame application, the afterburning time must not exceed 1 min. Here too, flaming drops or glowing particles must not ignite the surgical cotton below the sample.

UL 60950-1 and CSA 60950-1 will be withdrawn on 20 December 2020, and replaced by UL 62368-1 and CSA 62368-1, also harmonized in Europe as EN 62368-1 [100].

12.1.2.2 CSA Test and Approval Procedures

The Canadian Standards Association (CSA) describes the fire performance of products similar to the UL Yellow Card, and rates polymeric materials or finished parts of electrical equipment based on a number of standardized tests. Flammability standards are summarized in CSA C22.2 No. 0.17: Evaluation of properties of polymeric materials [78]. This standard consists of ten test procedures (A–J). Most of them are identical to the various UL 94, IEC 60695-11 series, or UL 746A test methods.

The CSA C22.2 test standard contains two flame tests for the housings of live parts. In the first one, a 19 mm Bunsen burner flame is applied twice for 30 s with an interval of 1 min to a test component or a corresponding thick sheet of insulating material. Depending on the type and use of the electrical equipment, the test specimen is positioned horizontally or vertically. As a rule, the test is passed if after both flame applications the afterflame time does not exceed 1 min, and the test specimen does not show major damage. The second one, CSA C22.2 Test A, is similar to UL 94 5V, but more rigorous: each application of the 127 mm Bunsen burner flame to the finished housing or to a part of insulating material sheet of the same thickness is 15 s instead of 5 s. In addition, during the first four flame applications the afterflame time must be < 30 s, and < 60 s after the fifth application.

12.1.3 Fire Regulations and Standards in Europe

12.1.3.1 European Committee for Electrotechnical Standardization (CENELEC)

CENELEC is the European Committee for Electrotechnical Standardization [4]. It was set up in 1973 as a non-profit-making organization under Belgian Law, and has been officially recognized as the only European Standards Organization in its field by the European Commission. CENELEC supports the policies of the European Union and EFTA for free trade, worker and consumer safety, and public procurement. CENELEC membership encompasses 34 countries. In addition to the members, there are 13 other "National Electrotechnical Committees" that participate in the work of CENELEC as "Affiliates" and/or "Companion Standardization Bodies" (CSBs). All interested parties are consulted during the drafting of CENELEC standards, through involvement in technical meetings at national and European level, and through enquiries conducted by the members. The Technical Board (BT) coordinates the work of the technical bodies, e.g., Technical Committees (TC) or Sub-Committees (SC). The TCs are responsible for the preparation of the European Standards (EN). TCs take into account any ISO/IEC work coming within their scope, together with such data as may be supplied by members and by other relevant international organizations, and work on related subjects in any other Technical Committee. CEN, CENELEC, and ETSI, the three European Standards Organizations, have jointly developed a common website [79] for the European Committee for Electrotechnical Standardization.

Reporting Secretariats (SR) exist to provide information to the Technical Board on any ISO or IEC work which could be of concern to CENELEC. Once CENELEC has started work by selecting an international standard or any other document to develop into a European Standard, all national work on the same subject is stopped. This suspension of national work is called "Standstill". New standardization initiatives originating in Europe are offered to IEC with the request that they will be undertaken at international level.

Only if IEC does not want to undertake the work, or if it cannot meet CENELEC target dates, does the work continue at European level. Afterwards, the result is offered to IEC again. All drafts of an IEC standard are the object of a parallel voting procedure in CENELEC and IEC. Globalization of the markets affects the volume of CENELEC work due to the progressively shifting emphasis towards international standardization in IEC. Today, in more than 80% of cases, the initial document comes from the International Electrotechnical Commission (IEC). European Standards are established as a general rule because it is important that members' national standards become identical wherever possible. Members are obliged to implement European Standards by giving them the status of a national standard.

12.1.3.2 Electrotechnical Products and EU Directives

The primary purpose of the several Directives and Regulations laid down by the European Commission was to ensure the free movement of goods [80]. To support this objective, each Directive established "essential requirements" (ER) and required these requirements to be incorporated into European harmonized standards.

In the frame of the future European Technical Approval, every product standard shall have, as far as possible, a harmonized indication related to the level of performance or classes under the scope of the individual Directives. Where necessary, existing standards may be amended for fully meeting the ER of a European Directive. The following EU Regulation and Directives are important for electrotechnical products:

- Construction Products Regulation (CPR) [82]
- Low Voltage Directive (LVD) [81]
- Machinery Directive (MD) [83]
- Directive on the interoperability of the rail system within the Community [85].

Construction Products Regulation and Low Voltage Directive

At the moment, the Member States of the EU are in the process of changing over to a harmonized reaction-to-fire classification system based on the Construction Products Regulation (CPR) [82], relating to construction products. In general, products permanently installed in buildings and constructions are governed by the CPR. Section 10.3.1 shows the Euroclasses of the uniform classification system together with the corresponding test methods and classification criteria for construction products.

In general, low-voltage cables, optical fibers, and optical cables that are permanently installed in buildings and constructions are governed by the Low Voltage Directive (LVD) [81], which essentially ensures the electrical safety of products. It applies to products with 50–1000 V_{ac} or 75–1500 V_{dc} input. Products may include computers, information technology equipment, and household products. This directive is only a requirement for the EEA member nations, and not required for products sold outside this community.

The nature and use of power, control, and communication cables are sufficiently different from other construction products to warrant separate treatment. This consists of a separate classification for such products, based on test methods specifically adapted to take account of these differences. Details are found in Section 10.3.1.

Machinery Directive

The Council Directive on the approximation of the laws of the EU member states relating to machinery (2006/42/EC) demands in its Annex I, Section 1.5.6, that machinery shall be designed and constructed to avoid any risk of fire [83]. CEN and CENELEC are producing a set of standards to assist designers, manufacturers, and other interested bodies to interpret this essential safety requirement in order to achieve conformity with European legislation.

The European Standard EN ISO 19353 describes methods of identification of the fire hazard from machinery and the performance of a corresponding risk assessment [84]. It describes the basic concepts and methodology of technical measures for fire prevention and protection during design and construction of machinery. Risk reduction is primarily achieved by design/engineering measures that eliminate or minimize the fire hazard, e. g., use of flame-retardant materials in the construction of the machine, minimization of the risk of overheating, or the limitation of fire effects by using proper shields or enclosures. The safety of machinery against fire involves fire prevention, protection, and fighting. In general, these include technical, structural, organizational, workplace, and public firefighting measures. Effective fire safety of machinery may require the implementation of a single measure or a combination of measures.

Fire hazard assessment of electrotechnical equipment of railway vehicles

The railway industry is generally concerned with the movement of people as well as goods. It is therefore essential that safety of rolling stock is achieved when failures, including fires, occur. A mandate (M024) was given to CEN by the Commission of the European Communities and the European Free Trade Association to prepare a European standard in order to support essential requirements of the Directive on the interoperability of the European high-speed train network [85].

The European Standard EN 45545-1 “Fire protection on railway vehicles: General” has been prepared by CEN/TC 256 “Railway applications” in cooperation with CENELEC/TC 9X “Electrical and electronic applications for railways”. It is a part of a series of seven interlinked standards [86] relating to “Railway applications - Fire protection on railway vehicles”. The standard specifies the measures on railway vehicles for fire protection, fire safety objectives and general requirements for fire protection measures. It is a general document which defines operation categories, design categories, and vehicle classifications to satisfy the fire protection measures. It takes into account potential hazards and operation categories in an effort to ensure fire protection, limit fire spread, and limit the effects of fire and smoke (thermal, toxic fumes, obscuration of escape routes). Part 2 specifies the fire behavior requirements for selecting materials and products for railway vehicles. Part 5 prescribes the fire safety requirements for electrical equipment. Details are found in Sections 11.2.2 and 12.2.5.

12.1.3.3 Approval Procedures in Europe

CE Marking

CE marking indicates that the manufacturer of a specific product claims, under its sole responsibility, compliance with all applicable EU directives and regulations.

In order to use CE marking, the manufacturer must carry out a conformity assessment of the product in relation to the safety principles laid down in each of the applicable Directives or Regulations, draw up a Technical File and issue under its name the Declaration of Conformity.

Although CE marking is applied under the sole responsibility of the manufacturers, certain EU directives and regulations do not allow manufacturers to carry out self-certification of the product but require that the conformity assessment is performed by a notified body. In these specific cases the affixed CE marking is also followed by the registration code of the notified body.

Since fulfilling all the requirements to apply the CE marking can be quite complex, even if not strictly required by the applicable EU directives or regulations, the involvement of a notified body during all the necessary steps can be helpful, and ensures that all assessments and verifications are properly performed.

Part of the process to apply CE marking involves testing of the goods to demonstrate compliance with the safety principles of the EU directives or regulations. Since directives and regulations do not provide specific testing specifications, manufacturers can refer to the list of harmonized Standards for each specific directive or regulation.

Harmonized Standards are those EN standards developed by CEN, CENELEC, or ETSI technical committees whose application is considered adequate to demonstrate compliance with the EU directives and regulations. The list of harmonized standards for most of the EU directives and regulations is available on the European Commission website [105].

For proof of the overall safety of a product by virtue of compliance with the relevant harmonized standards, manufacturers often apply for testing and certification by independent institutions such as the VDE Institute. Only marks for safety and quality attestations such as the VDE Mark License Certificate or the GS-Mark reassure (a) the manufacturer that the product complies with legal requirements of product safety and environmental issues, and (b) the end user that the product complies with the acknowledged rules of technology.

VDE Regulations and Approval Procedures

Electrical safety requirements in Germany are laid down by the VDE (Association for Electrical, Electronic & Information Technologies) [87], a nonprofit association for electrical science and technology, founded in 1893. One of VDE's main tasks is the promotion of technical progress and the application of electrical engineering/

electronics, information technology and associated technologies. VDE is one of the largest technical and scientific associations in Europe.

The German Electrotechnical Commission (DKE) [88] is responsible for the development of standards and safety requirements in the field of electrical engineering. DKE represents Germany in the International and European Standardization Organizations IEC, CENELEC, and ETSI (European Telecommunications Standards Institute). The results of the DKE standardization work are harmonized as European and international standards, and published as DIN EN standards, VDE regulations, and VDE guidelines. Since the VDE regulations are based on civil law, they do not have the status of legally binding provisions such as state laws or the safety regulations of the underwriters for statutory accident insurance. However, they are accepted by presumption of law as standard technical practice, and are thus often referred to in legal provisions and safety regulations. In the event of a failure or accident involving electrical equipment that does not comply to the VDE regulations, it has to be shown that the design is at least equal to accepted standard technical practice.

Within DKE, Committee K 133 is responsible for the development of fire safety standards. It is a horizontal committee with a pilot safety function within the DKE to give guidance on the fire hazard assessment and testing to other DKE Committees responsible for finished components and electrotechnical products. DKE/K 133 is the German mirror committee to IEC Technical Committee (TC) 89 Fire Hazard Testing.

As a neutral and independent institution, the VDE Testing and Certification Institute provides testing of electrotechnical products, components, and systems, and awards the VDE test mark recognized worldwide [89]. The VDE Institute is accredited on a national and international level. The VDE Mark indicates conformity with VDE standards and European or internationally harmonized standards, and confirms compliance with protective requirements of the applicable EC Directive. The VDE Mark is a symbol for testing based on the assessment of electrical, mechanical, thermal, toxic, fire, and other hazards.

For cables, insulated cords, installation conduits, and ducts, a special VDE Cable Mark is applicable. In addition, there are many other VDE Marks such as the VDE Component Mark for electronic components. The VDE Institute also certifies test results obtained from other certified bodies, e.g., CB Test Certificates, CCA Notifications of Test Results, or EC Type-Examination Certificates, and is a Notified Body for the Low-Voltage Directive and the Medical Devices Directive. Through partner organizations worldwide and the presence of VDE Authorized Office and VDE experts, the VDE Institute is an international partner for safety and quality.

The GS-Mark

The GS-Mark, the German safety approval mark, shows conformity with the German Equipment Safety Law. It is a recognized mark for electrotechnical products, such as office equipment, household appliances, and industrial equipment, and is widely accepted throughout Europe. The GS-Mark informs the buyer, customer, and consumer that product and production process have been checked by an authorized institution such as TÜV Rheinland [90] and that regular surveillance audits are made.

12.1.4 Fire Regulations and Standards in Asia

Most of the consumer goods to be sold in Asian markets (Japan, Philippines, China, Malaysia, Singapore, Indonesia, Thailand, Taiwan, and Korea) need an approval. Because many Asian states are in the process of implementing new testing and classification systems, obtaining appropriate safety compliance licenses and approvals can be complex and confusing. This is particularly true for safety and security regulations for E&E products. Although these regulations are to some extent in line with IEC standards, the written rules and current practices are often at odds and require some interpretation. In many cases, the safety tests have to be carried out by "Notified Bodies" of the corresponding Asian state, such as national or semi-national institutes. Some Asian countries (e.g., Japan, Korea, and Singapore) have joined the CB scheme, which simplifies the testing and certification process. Special companies, e.g., FEMAC [91], offer their assistance to companies trying to enter Asian markets. In the following, some flame-retardant regulations and test standards in Japan and China are exemplified.

The Electrical Appliance and Material Safety Act in Japan

The purpose of the Electrical Appliance and Material Safety Act is to prevent hazards and disturbances resulting from electrical appliances. To accomplish this purpose, the Act regulates the manufacture, sale, etc. of electrical appliances and materials.

All persons engaged in manufacturing electrical appliances in Japan, or importing them into the country (referred to as "Notifying Suppliers") must notify the Ministry of Economy, Trade and Industry of their name, etc. Therefore, importers based in Japan must make such a notification when starting their business.

For electrical appliances, it is the importers or the manufacturers in Japan who have the obligation to secure the conformity of electrical appliances to the mandatory technical requirements, and which have the right to attach the PSE marks including the name of the importer, once their conformity to the mandatory technical requirements is confirmed [92].

Particularly for "specified electrical appliances", the importers or the manufacturers in Japan shall either undergo third-party conformity assessment for the product they import (obtaining and keeping the corresponding certificate) or obtain an official copy of "the equivalent of a certificate" through manufacturers of the specified electrical appliances concerned and keep the official copy of the equivalent of a certificate.

The Japanese industrial standards (JIS) that are relevant for the characterization of the flammability and fire safety assessment of plastics, electrotechnical parts, and equipment are based on the relevant tests developed by IEC TC 89 and relevant IEC technical committees responsible for individual types of electronic and electrical products. The fire hazard test methods of materials were set up on the basis of the corresponding IEC standards shown in Table 12.21.

Table 12.21 Fire Hazard Test Methods for Electric and Electronic Products in Japan (Examples) [93]

Test method		Equivalent IEC standard
Glow-wire flammability index (GWFI)	JIS C 60695-2-12 [94]	IEC 60695-2-12
Glow-wire ignition temperature (GWIT)	JIS C 60695-2-13 [94]	IEC 60695-2-13
Flammability of the 50 W test flame	JIS C 60695-11-10 [94]	IEC 60695-11-10
Comparative tracking index (CTI)	JIS C2134 [95]	IEC 60112

Certification According to the Fire Safety Regulations and Test Standards in China

Since China participates in the CB scheme, the evaluation process is straightforward; testing is based on Chinese national standards (GB) or the technical specifications (GB/T), which are translated from IEC standards. There are currently two certification schemes for the Chinese market, the CCC mark (China Compulsory Certification) and CQC mark (China Quality Certification). The CCC certification is compulsory for a wide range of product categories, including data processing or household products. CQC is a voluntary certification. For some products (e.g., refrigerators and televisions), both CCC and CQC certifications are required.

Most of the fire safety regulations in China have been issued by the relevant ministries of the central government, such as the Ministry of Public Security. At the current time, there are few mandatory regulations for electric and electronic products. The fire hazard test methods of materials were set up on the basis of the corresponding IEC standards shown in Table 12.22.

Table 12.22 Examples of Fire Hazard Test Methods for Electric and Electronic Products in China [96]

Test method		Equivalent IEC standard
Glow-wire flammability Index (GWFI)	GB/T 5169.12-2013 [97]	IEC 60695-2-12
Glow-wire ignition temperature (GWIT)	GB/T 5169.13-2013 [97]	IEC 60695-2-13
Flammability of the 50 W test flame	GB/T 5169.16-2017 [97]	IEC 60695-11-10
Comparative tracking index (CTI)	GB/T 4207-2012 [98]	IEC 60112

12.1.5 Other Fire Safety Test Methods

Office Equipment

The explosion in global communication during the 1980s and 1990s has been made possible through the use of plastics. Plastics used in office equipment generally have to meet fire performance requirements. These are part of standards like IEC 60950 [77], EN 62368-1 [100], UL 60950, and CSA 60950 [76]. The UL and CSA standards are identical and described in Section 12.1.2.1.2.

The IEC, EN, and UL standards essentially have the same flammability requirements, with minor differences. The standards based on IEC 60950 are used worldwide, whereas UL/CSA 60950 is more commonly used in the USA, and the CSA standard is used in Canada.

IEC 62368-1 [100], which has been developed using hazard-based principles, is expected to replace IEC 60065 [102] and IEC 60950 series over the next few years. IEC 62368-1 includes in its scope those products included in the scopes of the IEC 60065 and IEC 60950 series, and will therefore cover a wide variety of products in the audio/video, information, and communication technology market. Adoption of IEC 62368-1 is already progressing and it is expected to be used extensively, with an overall expected acceptance equal to or greater than the IEC 60065 and IEC 60950 series.

Consumer Equipment

The fire performance requirements for plastic material used in the manufacture of consumer products (e.g., TV-sets) are different in US, IEC, and European standards. In the USA, plastics used for enclosures are virtually always required to pass one of the vertical UL 94 flammability tests. The requirements for TV sets in the USA are laid down in UL 1492 [101]. IEC 60065 [102] and EN 62368-1 [100] allow major plastic parts in TV-set backplates and housings to be made from materials that fulfill the requirements of the UL 94 horizontal test. No flammability requirements are necessary for plastic materials, which are well protected from internal ignition sources using fire enclosures or specified distances from potential ignition sources. Several studies were carried out to assess the role of flame retardants in plastics for preventing or delaying fires. In studies focusing on TV-sets, it was found that flame-retarded plastics guarantee high resistance to external ignition sources [103].

Household Appliances

An elegant appearance and high performance are essential for household appliances and commodities. Safety requirements, including those of fire safety, are set out in standards, recommendations and guidelines.

IEC TC 61 aims to produce and maintain international standards relating to the safety of household and similar electrical appliances in a manner that is timely and efficient, and which keeps pace with modern technology [104]. MT 4 of TC 61 deals with the revision of IEC 60335 series Clause 30, which contains requirements for resistance to heat and fire. The basic discussion in TC 61 and in other committees like TC 74 is whether new approaches in testing or other means of fire protection can assist in minimizing the fire hazard presented by end-products. Special discussion points in IEC TC 61 and TC 89 are how to use statistically relevant ignition sources such as the bad-connection test, the glow-wire test, and the needle-flame test to simulate the flames emitted by the test specimen on surrounding parts, or to use other fire-protective measures.

12.1.6 Future Developments

IEC TC 89 has started a new challenge to develop fire containment tests on finished units. Reactive analysis of failed finished units from the market often showed evidence of the ignition of internal flammable parts, caused by overheating of high-resistance electrical connections and consequent evolution of the fire, culminating in flames bursting out of the finished unit enclosure, with the attendant risk to users. This demonstrated a gap in current fire hazard testing and evaluation methodologies, which mainly focus on the fire performance of single materials.

Assessment of the evolution of fires in complex environments has become of crucial importance. Examples are finished units and the evaluation of their fire containment ability, where interactions between different flammable materials enclosed in a limited space can easily lead to different results from those expected on the basis of the characteristics of single materials. Current work done in the Italian mirror committee to IEC TC 89, on forced ignition by bad connections inside electrical appliances, shows the need for a new part in the standard series of end-product tests. A project team responsible for this was established in 2016 under IEC TC 89/WG12, and currently more than 40 experts from 12 countries have met for discussions, initiated two round-robin tests, and developed normative and informative texts. The new proposal has been circulated as IEC TS 60695-2-21, and has been under discussion since 2019.

A hot-coil ignition test for materials in IEC TC 89, equivalent to the hot-wire ignition test (HWI) of IEC 60695-2-20, was first published in 1995. It was later revised in view of improving the repeatability and reproducibility, and published as IEC 60695-2-20 Edition 2 in 2004. However, the anticipated improvement did not materialize, and this publication was withdrawn in 2007. In 2008, in response to requests from various IEC Product Committees, a TC 89 Project Team (PT 60695-2-20) was formed to analyze the fundamental physics of the test method and try again to improve its repeatability and reproducibility. Recent work of this project team, including several round-robin tests and drafting of an Edition 3, indicates that these desired improvements have been achieved. Publication as a technical specification is expected early 2021.

In addition, IEC TC 89 has developed IEC TS 60695-11-11, a test method to determine the characteristic heat flux for ignition (CHFI) from a non-contacting flame source for materials used in electrical and electronic products and sub-assemblies. It provides a relationship between ignition time and incident heat flux. This different approach uses direct and quantitative exposure data to evaluate material or product ignition, in contrast to the existing approach, which applies test flames or heat sources of nominal values to test specimens in order to simulate fire hazards by testing. This technical specification has been upgraded to an international standard published in 2020.

References for Section 12.1

[1] *Plastics - the Facts 2018:* Plastics Europe *http://www.plasticseurope.org/.*

[2] International Organization for Standardization (ISO) *http://www.iso.org/home.html.*

[3] International Electrotechnical Commission (IEC) *http://www.iec.ch/index.htm.*

[4] European Committee for Electrotechnical Standardization (CENELEC) *http://www.cenelec.eu.*

[5] ISO Guide 51, Ed.3 Apr. 2014: Safety aspects - Guidelines for their inclusion in standards.

[6] IEC Guide 104, Ed.5.0 Jul. 2019: The preparation of safety publications and the use of basic safety publications and group safety publications.

[7] IEC Publications *http://www.iec.ch/standardsdev/publications/?ref=menu.*

[8] IEC TC 89, Scope *http://www.iec.ch/dyn/www/f?p=103:7:0::::FSP_ORG_ID:1283.*

[9] IEC 60695-1-10, Ed.2.0, Nov. 2016: Fire hazard testing - Part 1-10: Guidance for assessing the fire hazard of electrotechnical products - General guidelines; IEC 60695-1-11, Ed.2.0, Oct. 2014: Fire hazard testing - Part 1-11: Guidance for assessing the fire hazard of electrotechnical products - Fire hazard assessment; IEC 60695-1-12, Ed.1.0, Jan. 2015: Fire hazard testing - Part 1-12: Guidance for assessing the fire hazard of electrotechnical products - Fire safety engineering.

[10] IEC 60695-1-14, Ed.1.0, Dec. 2017: Fire hazard testing - Part 1-14: Guidance on the different levels of power and energy and energy related to the probability of ignition and fire in low voltage electrotechnical products.

[11] IEC 60695-1-20, Ed.1.0, Jan. 2016: Fire hazard testing - Part 1-20: Guidance for assessing the fire hazard of electrotechnical products - Ignitability - General guidance.

[12] IEC 60695-1-21, Ed.1.0, Sep. 2016: Fire hazard testing - Part 1-21: Guidance for assessing the fire hazard of electrotechnical products - Ignitability - Summary and relevance of test method.

[13] IEC 60695-1-40, Ed.1.0, Nov. 2013: Fire hazard testing - Part 1-40: Guidance for assessing the fire hazard of electrotechnical products - Insulating liquids.

[14] IEC 60695-4, Ed.4.0, May 2012: Fire hazard testing - Part 4: Terminology concerning fire tests for electrotechnical products.

[15] IEC 60695-10-2, Ed.3.0, Feb. 2014: Fire hazard testing - Part 10-2: Abnormal heat - Ball pressure test method.

[16] IEC 60695-10-3, Ed.2.0, Sep. 2016: Fire hazard testing - Part 10-3: Abnormal heat - Mould stress relief distortion test.

[17] IEC 60695-2-10, Ed.2.0, Apr. 2013: Fire hazard testing - Part 2-10: Glowing hot/wire based methods - Glow-wire apparatus and common test procedure.

[18] IEC 60695-2-11, Ed.2.0, Feb. 2014: Fire hazard testing - Part 2-11: Glowing/hot-wire based test methods - Glow-wire flammability test method for end-products (GWEPT).

[19] IEC 60695-11-5, Ed.2.0, Dec. 2016: Fire hazard testing - Part 11-5: Test flames - Needle-flame test method - Apparatus, confirmatory test arrangement and guidance.

[20] IEC 60695-11-2, Ed.3.0, Jun. 2017: Fire hazard testing - Part 11-2: Test flames - 1 kW pre-mixed flame - Apparatus, confirmatory test arrangement and guidance.

[21] IEC 60695-1-30, Ed.3.0, Feb. 2017: Fire hazard testing - Part 1-30: Guidance for assessing the fire hazard of electrotechnical products - Preselection testing process - General guidelines.

[22] IEC 60695-2-12, Ed.2.1, Feb. 2014 (Ed.2.0:2012+AMD1:2014): Fire hazard testing - Part 2-12: Glowing/hot-wire based test methods - Glow-wire flammability index (GWFI) test method for materials.

[23] IEC 60695-2-13, Ed.2.1, Feb. 2014 (Ed.2.0:2010+AMD1:2014): Fire hazard testing - Part 2-13: Glowing/hot-wire based test methods- Glow-wire ignition temperature (GWIT) test method for materials.

[24] IEC 60998-1, Ed.2.0, Dec. 2002: Connecting devices for low-voltage circuits for household and similar purposes - Part 1: General requirements.

[25] DIN EN 60998-1, VDE 0613-1, Mar. 2005 (IEC 60998-1:2002 modified): Connecting devices for low-voltage circuits for household purposes.

[26] IEC 60400, Ed.8.0, Jun. 2017: Lampholders for tubular fluorescent lamps and starterholders.

[27] DIN EN 60400, VDE 0616-3, Jun. 2018: Lampholders for tubular fluorescent lamps and starterholders.

[28] IEC 60730-1, Ed.5.1, Dec. 2015 (Ed.5.0:2015+AMD1:2015): Automatic electrical controls - Part 1: General requirements.

[29] DIN EN 60730-1, VDE 0631-1, May 2017 (IEC 60730-1:2013 modified + Corrigendum 1:2014): Automatic electrical controls - Part 1: General requirements.

[30] IEC 60669-1, Ed.4.0, Dec. 2017: Switches for household and similar fixed-electrical installations - Part 1: General requirements.

[31] DIN EN 60669-1, VDE 0632-1, Mar. 2019 (IEC 60669-1:2017 modified): Switches for household and similar fixed-electrical installations - Part 1: General requirements.

[32] IEC 61439-3, Ed.1.0., Mar. 2019 (Ed.1.0:2002+COR2:2019): Low-voltage switchgear and controlgear assembled - Part 3: Distribution boards intended to be operated by ordinary persons (DBO).

[33] DIN EN 61439-3, VDE 0660-600-3, Feb. 2013 (IEC 61439-3:2012): Low-voltage switchgear and controlgear assemblies - Part 3: Distribution boards intended to be operated by ordinary persons (DBO).

[34] IEC 60335-1, Ed.5.2, May. 2016 (Ed.5.0:2010+AMD1:2013+AMD2:2016): Household and similar electrical appliances - Safety - Part 1: General requirements.

[35] DIN EN 60335-1, 2012 (Ed.2012 + AMD13:2017): Household and similar electrical appliances - Safety - Part 1: General requirements.

[36] IEC 60598-1, Ed.8.1, Sep. 2017 (Ed.8.0:2014+AMD1:2017): Luminaries - Part 1: General requirements and tests.

[37] DIN EN 60598-1 VDE 0711-1, Sep. 2018 (IEC 60598-1:2014 modified + AMD1:2017): Luminaries - Part 1: General requirements.

[38] UL 94, 6th Ed., 2013: Test for Flammability of Plastic Materials for Parts in Devices and Appliances, with supplements up to May 2018.

[39] IEC 60695-11-10, Ed.2.0, Apr. 2013: Fire hazard testing Part 11-10: Test flames - 50 W horizontal and vertical flame test methods.

[40] IEC 60695-11-20, Ed.2.0, Apr. 2015: Fire hazard testing - Part 11-20: Test flames - 500 W flame test methods.

[41] IEC 60695-11-3, Ed.1.0, Aug. 2012: Fire hazard testing – Part 11-3: Test flames – 500 W flames – Apparatus and confirmational test methods.

[42] IEC 60695-11-4, Ed.1.0, Sep. 2011: Fire hazard testing – Part 11-4: Test flames – 50 W flame – Apparatus and confirmational test method.

[43] IEC 60695-9-1, Ed.3.0, Apr. 2013: Fire hazard testing – Part 9-1: Surface spread of flame – General guidance.

[44] IEC 60695-9-2, Ed.1.0, Mar. 2014: Fire hazard testing – Part 9-2: Surface spread of flame – Summary and relevance of test methods.

[45] ISO 19706, Ed.2.0, Sep. 2011: Guidelines for assessing the fire threat to people.

[46] ISO 5660-1, Ed.3.1, Aug. 2019 (Ed.3:2015+A1:2019): Reaction-to-fire tests – Heat release, smoke production and mass loss rate – Part 1: Heat release rate (Cone calorimeter method) and smoke production rate (Dynamic measurement).

[47] ISO TS 5660-3, Ed.1.0, Dec. 2012: Reaction-to-fire tests – Heat release, smoke production and mass loss rate – Part 3: Guidance on measurement.

[48] ISO TS 5660-4, Ed.1.0, Dec. 2016: Reaction-to-fire test – Heat release, smoke production and mass loss rate – Part 4: Measurement of low levels of heat release.

[49] IEC 60695-8-1, Ed.3.0, Nov. 2016: Fire hazard testing – Part 8-1: Heat release – General guidance.

[50] IEC 60695-8-2, Ed.1.0, Nov. 2016: Fire hazard testing – Part 8-2: Heat release – Summary and relevance of test methods.

[51] UL 2043, Ed.4, Oct. 2013 + Supplements up to Jul. 2018: Fire test for heat and visible smoke release for discrete products and their accessories installed in air-handling spaces.

[52] Babrauskas, V., Grayson, S. J., eds., 1992: *Heat Release in Fires.* Elsevier Applied Science Publishers, London.

[53] IEC 60695-6-1, Ed.2.1, Sep. 2010 (Ed.2.0:2005+AMD1:2010): Fire hazard testing – Part 6-1: Smoke obscuration – General guidance.

[54] IEC 60695-6-2, Ed.2.0, Feb. 2018: Fire hazard testing – Part 6-2: Smoke obscuration – Summary and relevance of test methods.

[55] ISO/TC 92 Fire safety, *http://www.iso.org/committee/50492.html*

[56] IEC 60695-7-1, Ed.3.0, Jun. 2010: Fire hazard testing – Part 7-1: Toxicity of fire effluent – General guidance.

[57] IEC 60695-7-2, Ed.1.0, Aug. 2011: Fire hazard testing – Part 7-2: Toxicity of fire effluent – Summary and relevance of test methods.

[58] IEC 60695-7-3, Ed.1.0, Aug. 2011: Fire hazard testing – Part 7-3: Toxicity of fire effluent – Use and interpretation of test results.

[59] IEC 60695-5-1, Ed.2.0, Nov. 2002: Fire hazard testing – Part 5-1: Corrosion damage effects of fire effluent – General guidance.

[60] IEC TS 60695-5-2, Ed.2.0, Sep. 2002: Fire hazard testing – Part 5-2: Corrosion damage effects of fire effluent – Summary and relevance of test methods.

[61] Underwriters Laboratories (UL), *https://www.ul.com/aboutul/.*

[62] UL Standards, *https://ulstandards.ul.com/access-standards/.*

[63] International Electrotechnical Committee for Conformity Testing to Standards for Electrical Equipment (IECEE), Geneva, Switzerland, *https://www.iecee.org/about/cb-scheme/.*

[64] Canadian Standards Association (CS), *http://www.csagroup.org.*

[65] ISO 9773, Ed.1.1, Sep. 2003 (Ed.1.0:1998+AMD1:2003): Plastics – Determination of burning behaviour of flexible vertical specimens in contact with a small-flame ignition source.

[66] ISO 9772, Ed.3.0, Sep. 2012: Cellular plastics – Determination of horizontal burning characteristics of small specimens subjected to a small flame.

[67] ASTM E162-16: Standard test method for surface flammability of materials using a radiant heat energy source.

[68] UL 746A, Ed.6, Sep. 2012 + Supplements up to Jul. 2019: Polymeric materials – Short term property evaluations.

[69] ASTM D3638-12: Test method for comparative tracking index of electrical insulating materials.

[70] IEC 60112, Ed.4.1, Oct. 2011 (Ed.4.0:2003+AMD1:2009): Method for the determination of the proof and the comparative tracking indices of solid insulating materials.

[71] UL 746B, Ed.5, Aug. 2018 + Supplements up to Aug. 2019: Polymeric materials – long term property evaluations.

[72] UL 746C, Ed.7, Feb. 2018: Polymeric materials – use in electrical equipment evaluations.

[73] UL 723/ASTM E84-19b: Standard test method for surface burning characteristics of building materials.

[74] UL 746D, Ed.8, Jan. 2018: Polymeric materials. Fabricated parts; with supplements up to Jan. 2018.

[75] UL 1694, Ed.3, Feb. 2002 + Supplements up to Jul. 2015: Tests for flammability of small polymeric component materials, supplements up to Feb. 2002.

[76] UL 60950-1; CSA 60950-1.

[77] IEC 60950-1, Ed.2.2, May 2013 (Ed.2.0:2005+AMD1:2009+AMD2:2013): Information technology equipment – Safety – Part 1: General requirements.

[78] CSA C22.2 No. 0.17: Evaluation of properties of polymeric materials.

[79] New Approach standardization in the European internal market, *http://www.NewApproach.org.*

[80] The New Approach Directives, *http://www.NewApproach.org/directiveList.asp.*

[81] Low Voltage Directive (LVD) 2014/35/EU.

[82] Commission Delegated Regulation (EU) 2016/364 of 1 July 2016 on the classification of the reaction to fire performance of construction products pursuant to Regulation (EU) No. 305/2011 of the European Parliament and of the Council (text with EEA relevance) C/2015/4394, Ref. OJ L 65, 15 Mar. 216, pp. 4–11, *http://eur-lex.europa.eu/legal-content/en/TXT/?uri=CELEX:32016R0364.*

[83] Safety of Machinery Directive 2006/42/EC.

[84] EN ISO 19353, Mar. 2019: Safety of machinery – Fire prevention and protection.

[85] Directive 2008/57/EC of the European Parliament and the Council of 17 June 2008 on the interoperability of the rail system within the Community (recast) (text with EEA relevance). Ref. OJ L 191, 18 Jul. 2008, pp. 1–45.

[86] EN 45545-1, Jul. 2013 + Cor.:2015: Railway applications – Fire protection on railway vehicles: General

EN 45545-2, 2020: Railway applications – Fire protection on railway vehicles: Requirements for fire behaviour of materials and components

EN 45545-3, Jul. 2013 + Cor.:2015: Railway applications – Fire protection on railway vehicles: Fire resistance requirements for fire barriers

EN 45545-4, Jul. 2013 + Cor.:2013: Railway applications – Fire protection on railway vehicles: Fire safety requirements for railway rolling stock design

EN 45545-5, Jul. 2013 + AMD1:2015: Railway applications – Fire protection on railway vehicles: Fire safety requirements for electrical equipment including that of trolley buses, track guided buses and magnetic levitation vehicles

EN 45545-6, Jul. 2013 + Cor.;2015: Railway applications – Fire protection on railway vehicles: Fire control and management systems

EN 45545-7, Jul. 2013 + Cor.:2015: Railway applications – Fire protection on railway vehicles: Fire safety requirements for flammable liquid and flammable gas installations.

[87] Association for Electrical, Electronic & Information Technologies (VDE), *http://www.vde.com/vde/html/e/online/online.htm.*

[88] German Electrical Commission (DKE), *http://www.dke.de.*

[89] VDE Testing and Certification Institute, Merianstrasse 28, D-63069 Offenbach (Main), Germany.

[90] TÜV Rheinland, Am Grauen Stein, D-51101 Köln, Germany, *http://www.us.tuv.com.*

[91] FEMAC (Far East Market Access Services) GmbH, Friedrichstrasse 10, D-70174 Stuttgart, Germany, *http://www.femac.com.*

[92] Japan Electrical Safety & Environment Technology Laboratories, Support for compliance with Japanese legal regulations, *https://www.jet.or.jp/en/import/index.html.*

[93] Japan Electrical Safety & Environment Technology Laboratories, Certification of Components and Materials/CMJ Registration, *https://www.jet.or.jp/en/products/parts/index.html.*

[94] JIS C 60695-2-12: Glow-wire flammability index (GWFI)

JIS C 60695-2-13: Glow-wire ignition temperature (GWIT)

JIS C 60695-11-10: Flammability of 50W test flame.

[95] JIS C 2134: Comparative tracking index (CTI).

[96] Ou, Y., *Flame Retardant Regulations and Test Standards in China,* National Laboratory of Flame-retarded Materials, Institute of Technology, Beijing 100081, China.

[97] GB/T 5169.12-2013: Glow-wire flammability index (GWFI)

GB/T 5169.13-2013: Glow-wire ignition temperature (GWIT)

GB/T 5169.16-2017 Flammability of 50W test flame.

[98] GB/T 4207-2012 Comparative tracking index (CTI).

[99] NFPA 70, 2017: National Electrical Code, NEC, National Fire Protection Association, Batterymarch Park, Quincy, MA, USA.

[100] EN 62368-1, Aug. 2018 (Ed. 2014+AMD:2018): Audio/video, information and communication technology equipment – Safety requirements.

[101] UL 1492, Apr. 2002: Safety audio-video products and accessories.

[102] IEC 60065, Ed.8.0, Dec. 2018 (Ed.8.0:2014+COR3:2018): Audio, video and similar electronic apparatus – Safety requirements.

[103] Blais, M., Carpenter, K., Combustion characteristics of flat panel televisions with and without fire retardants in the casing, *Fire Technology*, Volume 51, Issue 1, pp. 19–40, 2015.

[104] IEC TC 61: Strategic Business Plan, *https://www.iec.ch/dyn/www/f?p=103:7:0::::FSP_ORG_ID,FSP_LANG_ID:1236,25.*

[105] European Commission. Single market and standards. List of Harmonized Standards, *https://ec.europa.eu/growth/single-market/european-standards/harmonised-standards_en.*

12.2 Fire Hazard Assessment of Cables

Esther Hild

A variety of standards for the fire hazard testing of cables and wires are in use around the world. In the past, most product specifications quoted national standards for fire testing. For example, in the USA, the National Fire Protection Agency (NFPA) regulates electrical cables through NFPA 70 [1], which is the National Electrical Code (NEC).

In view of the globalization of the markets, the standardization bodies increasingly refer to ISO/IEC work. Particularly in Europe, a comprehensive conversion from international to European Standards (EN) took place in the field of fire hazard testing of cables and wires.

IEC TC 20 "Electric cables" and specifically WG18 "Burning characteristics of electric cables" is responsible for the international standardization of fire hazard testing and test methods. In fire testing of cables and wires, a section of the finished end-product is exposed to a small flame under defined conditions. The degree to which it burns during and after exposure, and the rate of combustion, are assessed. In most cases, testing is restricted to small-scale laboratory tests.

Increased attention has been devoted to real fires, where the fire behavior of cables and wires may be greatly influenced by the installation conditions and preheating caused by the combustion of long vertical runs of bundles of cables on cable trays. As a consequence, the classification of single cables or wires is not generally transferable to bundled cables. For this reason, many national and international standards for intermediate or large-scale tests have been developed, especially dealing with the fire behavior of cable bundles.

Note: Although IEC has decided that the decimal sign to be used shall be a comma[1] instead of a point, in this book the use of the point is maintained to avoid confusion.

12.2.1 Small-Scale Testing

Vertical Burning Test on a Single Insulated Wire or Cable

A non-exhaustive list of the most important small-scale test methods for wires and cables, including the test specifications and classifications, is shown in Table 12.23. The principle of these tests is demonstrated by the flame propagation tests

[1] *https://www.iec.ch/standardsdev/resources/draftingpublications/directives/principles/numbers_quantities.htm*

according to IEC 60332-1 and IEC 60332-2, or their technically equivalent European standards, the EN 60332-1 series and the EN 60332-2 series.

In the flame propagation test according to IEC 60332-1 [2], a vertically orientated piece of finished cable or wire 600 ± 25 mm long, is exposed to a Bunsen burner flame with a height of 175 mm and a 40 mm inner blue cone. The test layout is illustrated in Figure 12.8. The flame is continuously applied for a period of time that is related to the overall diameter of the test specimen (Table 12.23). The damaged length is a measure of its resistance to flame propagation. After all burning has ceased, the charred or affected portion must not have reached a point within 50 mm of the lower edge of the top clamp.

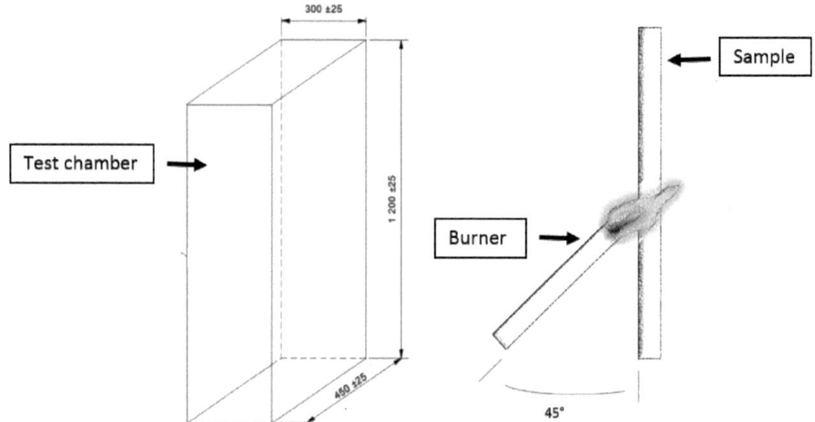

Figure 12.8 Test layout for the flame propagation test according to IEC 60332-1; dimensions in mm

Table 12.23 Summary of Test Conditions for Cable and Wire Test Methods at Laboratory Scale

Test method	Ignition source				Test enclosure specified	No. of test runs[a]	Specimen			Performance requirements / Maximum allowable		
	Burner	Power [kW]	Flame length [mm]	Flame application time t [s]			Conditioning	Length [mm]	Position	Char length [mm]	Burning time [s]	Burning time of drips [s]
EN 60332-1-1 (IEC 60332-1-1)	Describes the test apparatus				Metallic screen 1200 ± 25 mm high, 300 ± 25 mm wide, 450 ± 25 mm deep	–	–			–		
EN 60332-1-2 (IEC 60332-1-2)	According to IEC 60695-2-4/1	1	–	Depending on outer diameter D of the sample: $D \leq 25$ mm: $t = 60$ s; $25 < D \leq 50$: $t = 120$ s; $50 < D \leq 75$: $t = 240$ s; $D > 75$: $t = 480$ s		1	50 ± 20% 23 ± 5 °C [b] Min. 16 h	600±25	Vertical	Distance between the lower edge of the top support and the onset of charring is > 50 mm; burning does not extend downwards to a point > 540 mm from the lower edge of the top support	–	–

Table 12.23 Summary of Test Conditions for Cable and Wire Test Methods at Laboratory Scale *(continued)*

Test method	Ignition source				Test enclosure specified	No. of test runs[a]	Specimen			Performance requirements / Maximum allowable		
	Burner	Power [kW]	Flame length [mm]	Flame application time t [s]			Condi-tioning	Length [mm]	Posi-tion	Char length [mm]	Burning time [s]	Burning time of drips [s]
EN 60332-2-1	Describes the test apparatus				Metallic screen 1200 ± 25 mm high, 300 ± 25 mm wide, 450 ± 25 mm deep	–	–			–	–	
EN 60332-2-2 (IEC 60332-2-2)	Pro-pane burner	–	125 ± 25	20 ± 1 (should the conductor prematurely melt at < 20 ± 1 s, the test shall be repeated on sample no. 2, to a duration of (t – 2) s	–	2	[b]	600 ± 25	Vertical	Distance between the lower edge of the top clamp and the charred portion is > 50 mm	–	–
FAR Part 25 Appendix F Part 1	Bunsen or Tirill burner	–	76 mm	30	Draft free cabi-net with suffi-cient oxygen supply	3	–	500±25	–	76	30	3

[a] If failure is recorded, two more tests must be carried out.

[b] For a single core with paint or lacquer coating, the sample is stored 4 h at 60±2 °C before conditioning.

Vertical Burning Test on a Single Small Copper Wire and Cable

IEC 60332-2 [3], which has been prepared by IEC TC 20 and IEC TC 46, specifies a fire test method on small insulated copper cables or wires when the method specified in IEC 60332-1 is not applicable because the specimen melts during the application of the flame. The test specimen consists of a piece of finished copper cable or wire 600 ± 25 mm long with a range of application of 0.4–0.8 mm for solid copper conductors and 0.1–0.5 mm^2 cross-section for stranded conductors. The test specimen is held in a vertical position and a load of 5 N/mm^2 of conductor cross-sectional area is attached to the lower part of the specimen. A calibrated propane burner is used to ignite the specimen. The 125 ± 25 mm flame is continuously applied for a maximum period of 20 s. After all burning has ceased, the charred or affected portion must not have reached the zone within 50 mm of the lower edge of the top clamp (Figure 12.9).

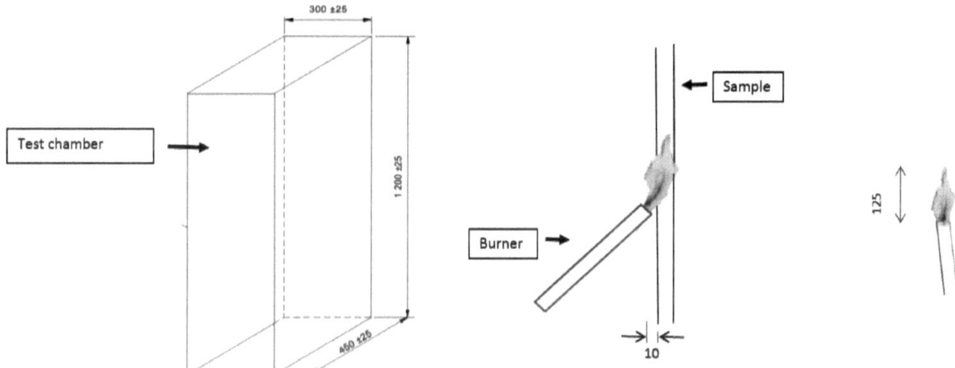

Figure 12.9 Vertical burning test on a single copper wire and cable according to IEC 60332-2. Dimensions in mm

12.2.2 Large-Scale Testing

Small-scale fire tests carried out according to IEC 60332-1 or 60332-2 are not appropriate for scenarios with cables or wires of high or concentrated fire load and with ignition sources of very high energy. To meet the requirements of such scenarios, several national and international standards with more stringent large-scale flame spread tests that better reflect end-use conditions have been developed. They can be divided into vertical and horizontal ladder test methods.

Vertical Ladder Test Methods

A non-exhaustive list of the most popular vertical ladder test methods, including a comparison of the test specifications and classifications, is shown in Table 12.24. The principle of these large-scale tests is demonstrated by the bundled wires or cables tested to IEC 60332-3, EN 50399, and UL 1666.

Table 12.24 Summary of Test Conditions of Vertical Ladder Cable Test Methods

Test method	Heat source			Application time of heat source [min]	Test enclosure		No. of test runs	Specimen				Conclusion		
	Burner	Power [kW]	Alternate source		Specified	Air flow rate [m^3/s]		Length [m]	Width of test sample [m]	Mounting techniques	Thin size cables to be bundled	Maximum allowable char length from bottom [m]	Peak smoke release rate [m^2/s]	Total smoke released [m^2]
IEC 60332-3	Yes [a]	21	No	Category A + B: 40 min Category C + D: 20 min	Yes	0.083	1	3.5	0.3/0.6	Front (A to D) or front and back (AF/R)	Mounted flush with no spaces	2.5 m above the bottom edge of the burner	No requirement	
EN 50399	Yes	20.5/30	No	20	Yes	0.7–1.2 m^3/s	1	3.5	0.5	Front only	If *D* ≤ 5 mm	1.75 (B1ca) 1.5 (B2ca) 2.0 (Cca) > 2.0 (Dca)	≤ 0.25 (s1) ≤ 1.5 (s2)	≤ 50 (s1) ≤ 400 (s2)
UL 1581-1160	Yes	21	No	20	No	Not applicable	3	2.4	0.15	Front only	No	2.4	No requirement	

Test method	Heat source			Application time of heat source [min]	Test enclosure		No. of test runs	Specimen				Conclusion		
	Burner	Power [kW]	Alternate source		Specified	Air flow rate [m^3/s]		Length [m]	Width of test sample [m]	Mounting techniques	Thin size cables to be bundled	Maximum allowable char length from bottom [m]	Peak smoke release rate [m^2/s]	Total smoke released [m^2]
UL 1581-1164 CSA FT 4/ IEEE 1202	Yes [b]	20	No	20	Yes	≥ 0.17	2	2.3	0.25	Front only	If D < 13 mm	1.5[d]	No requirement	
UL 1685	Yes [a]	20.6	No	20	Yes	0.6–0.7	2 [e]	2.44	0.15	Front only	No	2.44	0.25	95
CSA FT 4/ IEEE 1202	Yes [b]												0.40	150
UL 1666	Yes [c]	154.5	No	30	Yes	–	2 [f]	5.33	0.31	Single layer	No	3.66	No requirement	
NF C 32-070 Class C 1 UIC 895 VE	No	–	Electrical ring furnace [g,h]	–	–	0.5	2	1.6	–	–	Yes	0.25 [f]	No requirement	

[a] Angle of the burner parallel to the horizontal.
[b] Angle of the burner 20° up.
[c] Diffusion burner plate at the bottom of the cable tray.
[d] Char length measured from horizontal line of burner.
[e] Second set depending on the variance of the test results.
[f] Second and third set depending on the variance of the test results.
[g] Two 20 mm long tangential pilot flames are located above the top of the furnace (upper position).
[h] Temperature of the furnace after stabilization 780 °C to 880 °C.

Cable Test According to IEC 60332-3

IEC 60332-3 consists of a series of international standards published in six separate parts [4]. These standards are technically equivalent to the European standards EN 60332-3-10 to EN 60332-3-25. IEC 60332-3-10 contains details of the apparatus, its arrangement and calibration.

The test chamber is 1 m wide, 2 m deep, and 4 m high. The test rack is placed vertically in the chamber, with the cables and wires pointing to the burner. The distance of the test rack to the rear chamber wall is 150 mm.

The ribbon gas burner (one or two as requested in the relevant part), specified as the ignition source, is placed horizontally in front of the test rack with its front edge at 75 ± 5 mm from the specimen's surface and at least 500 ± 5 mm from the lower end of the testing rack. A scheme of the test bench with an installed cable ladder and ribbon burner is shown in Figure 12.10. The energy of the ribbon gas burner, which is supplied with a mixture of propane/air according to IEC 60332-3, is 73.7 MJ/h. Combustion air is admitted at the base of the test chamber at a rate of 5 ± 0.5 m^3/min at 20 ± 10 °C. The test sample consists of a number of cable or wire test pieces with the same length, each having a minimum length of 3.5 m. The number of 3.5 m test pieces depends on the volume per meter of non-metallic material of one test piece.

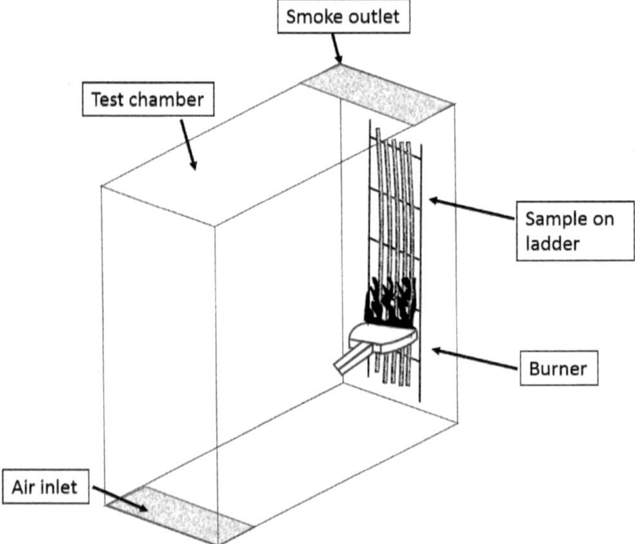

Figure 12.10 Test layout for the cable test according to IEC 60332-3

The categories A, B, C, and D correspond to a nominal volume of non-metallic material of 7, 3.5, 1.5, and 0.5 L/m. Two methods of mounting are applicable to category A. A summary of the test conditions is given in Table 12.25. Depending on the category, the flame application time is 20 min or 40 min. After burning has ceased, the maximum charred length of the test sample should not exceed 2.5 m above the bottom edge of the burner, neither at the front nor the rear of the ladder. If burning has not ceased 1 h after completion of flame application, the flames are extinguished and the extent of the damage is measured.

Table 12.25 Summary of IEC 60332 Test Conditions [a]

Category	A F/R (IEC 60332-3-21)	A (IEC 60332-3-22)		B (IEC 60332-3-23)		C (IEC 60332-3-24)		D (IEC 60332-3-25)
Range of conductor cross-sections [mm^2]	> 35 [b]	≤ 35 [c]	> 35 [b]	≤ 35 [c]	> 35 [b]	≤ 35 [c]	> 35 [b]	-
Non-metallic volume per meter of test samples [L]	7	7		3.5		1.5		0.5
Number of layers [d]	2 (standard ladder) ≥ 1 (wide ladder)	1		≥ 1	1	≥ 1	1	≥ 1
Position of test pieces	Spaced	Touching	Spaced	Touching	Spaced	Touching	Spaced	Touching
Flame application time [min]	40	40		40		20		20
Number of burners	1	1		2	3	3		3

[a] From IEC 60332 (2nd edition).
[b] At least one conductor > 35 mm^2.
[c] No conductor cross-section > 35 mm^2.
[d] For the standard ladder: Max. width of test sample 300 mm. For the wide ladder: Max. width of test sample 600 mm.

Cable Test According to EN 50399

The test EN 50399 [5] is used to classify the reaction to fire of cables used as construction products in the European Union (see Section 10.3.1.2.3). The test rig consists of a test chamber which is based upon that of EN 60332-3-10 but with additional instrumentation to measure heat release and smoke production during the test. The cables or bundles are mounted on a standard ladder of 500 ± 5 mm width, which is vertically mounted on the back panel of the chamber. The ignition source is a ribbon-type propane gas burner. A scheme of the test bench with an installed cable ladder, ribbon burner, and the hood above the test chamber is shown in Figure 12.11. The cables are ignited by a 20.5 kW burner for the classification of classes $B2_{ca}$, C_{ca}, and D_{ca}, and with a 30 kW burner for class $B1_{ca}$. The ignition duration is 1200 s. During the test there is an air supply by a system that blows air into the chamber at a rate of 8000 ± 400 L/min.

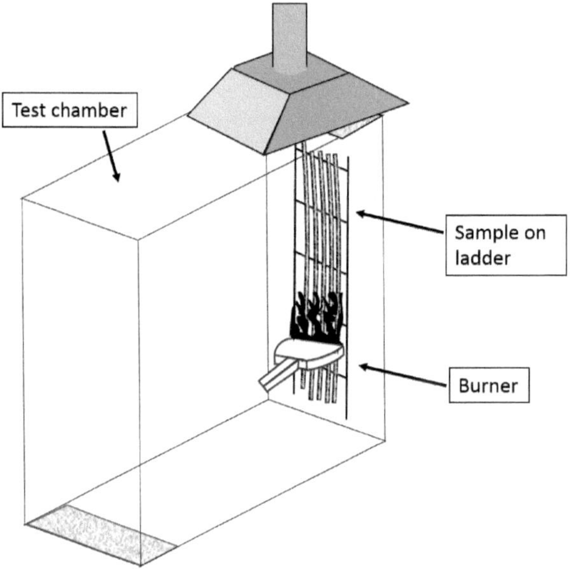

Figure 12.11 Test layout for the cable test according to EN 50399

For the classification of $B1_{ca}$ an additional board made of calcium silicate is mounted on the back panel of the ladder.

The following parameters are determined during the test:

- Flame spread
- Heat release rate
- Total heat release
- Smoke production rate
- Total smoke production
- Fire growth rate index
- Occurrence of flaming droplets/particles.

All data are measured every 3 s, and these measurements are averaged every 30 s for the heat release parameters, and every 60 s for the smoke production.

The test conditions for EN 50399 are shown in Table 12.26.

Table 12.26 Test Specifications for EN 50399

Specimen	Cables or bundles of specimens of electrical or optical-fiber cables. The number of test specimens depends on the diameter of the cable.
Specimen conditioning	Conditioning for at least 16 h at a temperature of 20 ± 10 °C.
Test layout	Fire test chamber 4 m high and 1 m wide. One air inlet entering from the rear of the chamber, parallel to the floor, and along the burner center line. A hood with a minimum length of 1.50 m and a minimum width of 1.00 m is centered above the outlet of the test chamber. Connected to the hood is the exhaust duct. At the end of the exhaust duct, an extraction ventilator is installed. The number (N) of the cables is calculated as follows: ▪ $N = (300 + 20)/(D + 20)$ for cables with $D \geq 20$ mm ▪ $N = (300 + D)/(2D)$ for cables with 20 mm > D > 5 mm ▪ $N = 15$ for cables with $D \leq 5$ mm, with number of cables in each bundle $n = 100/D^2$. D = Outside diameter of the round cable in mm. For non-circular cables, D is replaced by the smallest major axis dimension of the cable
Ignition source	Ribbon-type propane gas burner in accordance with EN 60332-3-10: ▪ 20.5 kW for classes $B2_{ca}$, C_{ca}, and D_{ca} ▪ 30 kW for class $B1_{ca}$. The test flame is applied for 1200 s, after which it is extinguished.
Test duration	20 min
Conclusion	Depending on classification (see Section 10.3.1.2.3)

Cable Tests According to UL 1666 (Riser Test)

The test UL 1666 [6] is mainly used in North America. The test rig consists of two chambers over two floors, connected by a slot. A simplified scheme for the flame test is shown in Figure 12.12. The total height is 5.7 m and the inner space 2.44 m × 1.22 m. The cables are mounted on a metal frame running vertically through the two floors and a burner diffusion plate on the floor ignites the cables for 30 min. The test conditions are shown in Table 12.27.

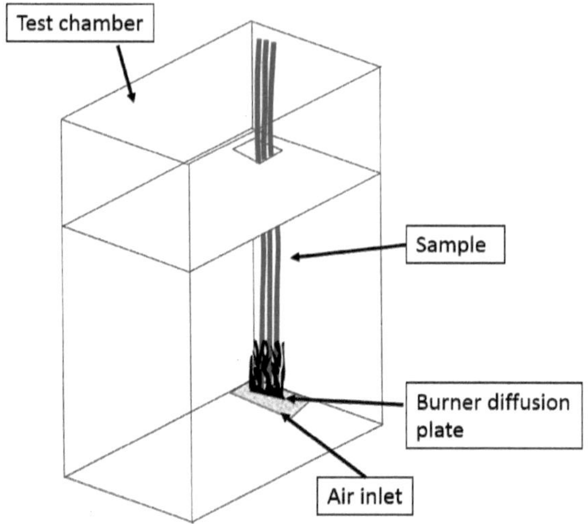

Figure 12.12 Test layout for the cable test according to UL 1666

Table 12.27 Test Specifications for Determining the Flame Propagation Height of Cables in Shafts According to UL 1666

Specimen	Two sets of specimens of electrical or fiber-optic cables. Each set consists of multiple specimens of cable, 5.33 m in length.
Specimen conditioning	Examples: Storage 48 h in a standard atmosphere (23 °C, 50% relative humidity).
Test layout	Fire test chamber 5.79 m high and 1.22 m wide. One slot 0.305 m × 0.61 m in the first floor. An identical slot directly above in the first floor. The cable lengths are to be installed through both slots. The number (N) of cables is computed by: $N = (12 \times 25.4)/D$. D = Outside diameter of the round cable in mm. For non-round cables, D is the smallest dimension of the cable diameter.
Ignition source	Burner apparatus consisting of a piping and a steel diffuser plate, 76 mm above the bottom of the first test chamber, propane gas flame, power 154.5 kW.
Test duration	Until non-metallic cable parts are completely consumed for the full length, or for 30 min, whichever is soonest.
Conclusion	Passed if: ▪ Flame propagation height (combustible material having been softened, partially or completely consumed) of each set has not equaled or exceeded 3.66 m ▪ The difference between the propagation heights of cable specimens (two sets) does not exceed 15% (if yes, a third set has to be tested).

Cable Test According to NF C 32-070

The fire test methods of the French standard NF C 32-070 [7] differ in some cases considerably from those in other countries or those prescribed by IEC TC 20.

The standard contains test methods for the reaction to fire of the cable and its fire resistance. The cables are divided into three classes of combustibility (C1, C2, and C3) and two classes of fire resistance (CR1 and CR2). In order to differentiate between classes C2 and C3, a test is carried out with one Bunsen burner (or two in the case of thicker cables), according to IEC 60695-2-4/1. If the test is passed, the cable is classified as C2, otherwise it is classified in the least favorable class C3. Cables used in applications involving stringent fire safety requirements must be of class C1, and must pass a special test developed in France using the fire test rig illustrated in Figure 12.13.

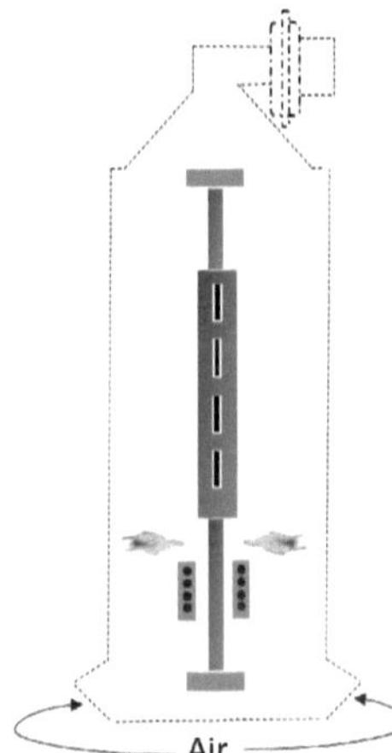

Figure 12.13
C1 test according to NF C 32-070

A 160 cm long sample of the finished cable (or bundle of cables in the case of thin cables) is suspended vertically from a frame in a test chamber with internal dimensions of 0.8 m × 0.7 m × 2.0 m. An annular furnace that can be slid up and down is fixed to the lower part of the frame. In its top position, it surrounds the lower part of the cable, and in its lowest position it surrounds a probe for checking its heat output (the secondary calibration device). Two 20 mm long, tangential pilot flames are located above the top furnace position, 10 to 15 mm from the surface of the cable. Above the pilot flames, the cable is enclosed concentrically by an 80 mm long metallic chimney, which serves to conduct the flames upwards from the cable below. An air stream flowing upwards at approximately 2 m/s is generated with a fan in the flue of the chamber. Before the start of the test, the radiation output of the furnace is first adjusted at its upper position by placing a measuring probe instead of the cable. This is the primary calibration device. Once the output of the furnace as determined by the temperature rise of this measuring probe has reached the required value, the furnace is slid to the lower position. The cable is then suspended into the test chamber. The actual test starts when the furnace is brought to its upper position again, and lasts 30 min, during which ventilation is interrupted for 1 min after 10 min. The cable achieves class C1 if in two tests the section of cable above the chimney remains undamaged.

Horizontal Ladder Test Methods

In the USA, horizontal plenums, which include the area above a suspended ceiling and under a raised floor, are very popular to place electrical and optical-fiber communication cables. The rapid growth of computers on local area networks (LAN) has resulted in an increased mass of communication cables in such plenum areas. To address the fire hazard from the accumulation of combustibles in return-air plenums, cables laid there must be listed as having passed the flame and smoke requirements of NFPA 262 [8]. The test is performed in the Steiner Tunnel, in which cable bundles are placed on a horizontal tray and exposed for 20 min to two burners with 87.9 kW output. Any resulting fire in the cable must not spread more than 162 cm, with strict limitations on the amount of smoke permitted. Figure 12.14 shows the test layout for cables in the Steiner Tunnel (see also Section 10.2.1). The test specifications are summarized in Table 12.28.

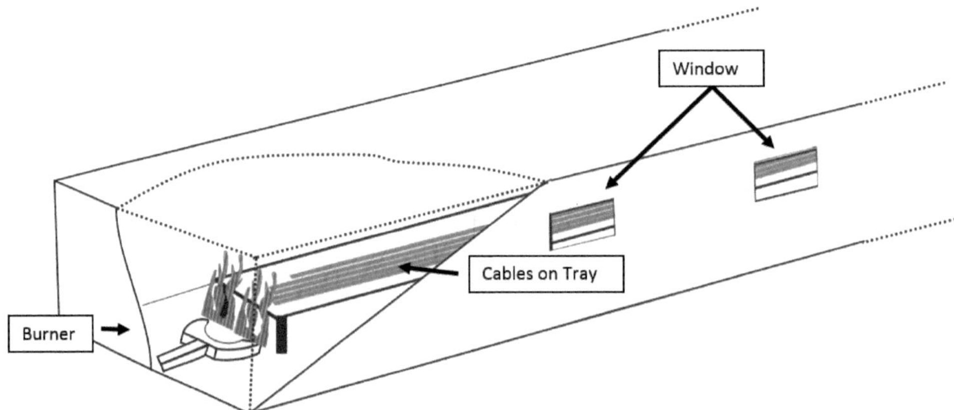

Figure 12.14 Test layout for the Steiner Tunnel test for cables according to NFPA 262

Table 12.28 Test Specifications for Determining Flame Propagation and Smoke Density of Cables to NFPA 262

Specimen	Specimen length 7.32 m, installed without any space between adjacent specimens in a single layer across the bottom of a cable tray
Specimen position	Horizontal
Ignition source	Two gas burners (methane gas) with 87.9 kW output located 47.6 mm below the cable tray

Test duration	20 min
Conclusion	Passed if: ▪ Maximum flame-propagation distance does not exceed 162 cm beyond the initial (137 cm) test flame ▪ Peak optical smoke density ≤ 0.5 (32% light transmission) and average optical density ≤ 0.15

12.2.3 Side Effects of Cable Fires

12.2.3.1 Smoke

Smoke from burning cables is an identified risk in the early stages of a fire, and is addressed by IEC 61034 or EN 61034 [9]. The standard describes the smoke density measurement of optical cables burned under defined conditions in a static smoke test, and consists of two parts. Part 1 gives details of the test apparatus, and includes details of a test cube of 27 m^3 volume, a photometric system for light measurement, a qualification procedure (which also defines the standard fire source), and a smoke mixing method. The details of the test rig are shown in Figure 12.15. The test method was first developed in the UK by scientists on behalf of London Underground Ltd.

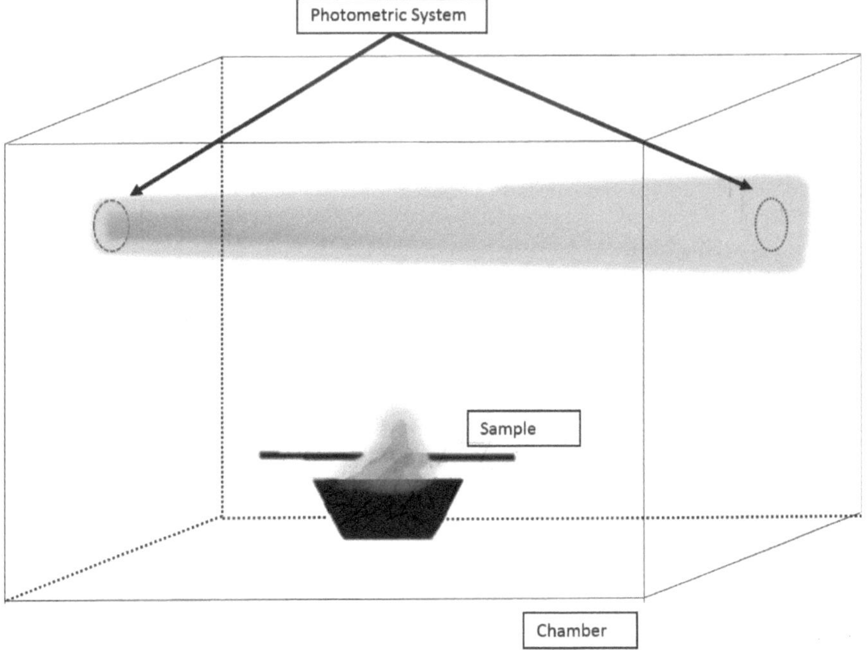

Figure 12.15 Test layout for smoke density of cables to IEC 61034/EN 61034

IEC or EN 61034 Part 2 describes the test procedure and recommended requirements for use, when no specific requirements are given in the particular cable standard or specification. The details for the preparation and assembly of the cables for the test are shown in Table 12.29. The standard fire source consists of 1 L alcohol placed below a 1 m long, horizontally oriented test piece consisting of cables or cable bundles. Observation of light absorbance within the cubic room is made by measuring the smoke emission over time.

Table 12.29 IEC or EN 61034 Part 1 and Part 2 Cable Smoke Density Test Specifications

Specimen	Electric or optical cables: Cable test pieces shall consist of one or more samples of cable 1.00 ± 0.05 m long		
Number of test pieces	Overall diameter of the cable (D):	Number of cables tested	Number of bundles tested [a]
	$D > 40$	1	–
	$20 < D \leq 40$	2	–
	$10 < D \leq 20$	3	–
	$5 < D \leq 10$	$N_1 = 45/D$ cables [b]	–
	$2 \leq D \leq 5$	–	$N_2 = 45/3D$ [b]
Test piece position	Cables and bundles in horizontal position above the tray		
Conditioning	At least 16 h at 25 ± 5 °C		
Assembly of test pieces	Cables or bundles shall be bound together at the ends and at 0.3 m from each end		
Test layout	Cubic enclosure; inside dimensions 3000 ± 30 mm; photometric system		
Ignition source	Standard fire source 100 ± 0.01 L alcohol with the composition: 90 ± 1% ethanol, 4 ± 1% methanol, 6 ± 1% water		
Conclusion	Recommended minimum value of the light transmittance: 60%		

[a] Each bundle shall consist of seven cables twisted together.
[b] Values of N_1 and N_2 to be rounded downwards to the nearest integer.

12.2.3.2 Flaming Droplets

The term “flaming droplets” refers to material separating from the specimen during the test and continuing to burn for a certain period. According to EN 13501-6 [11], classification d0 applies when during the test there are no flaming droplets and classification d1 applies when the droplets continue to flame for 10 s or less after reaching the floor of the test chamber.

12.2.3.3 Acidity

The acidity is measured according to the standard EN 60754-2 [10]. This standard describes the method for the determination of potentially corrosive gases evolving from burning materials of electrical and optical cables. The corrosivity is measured by the pH value, and by the conductivity of an aqueous solution absorbing the gases developed during the burning process.

The measurement procedure below applies to single cable materials. The effective length of the heating zone of the furnace must be 480–620 mm, and the internal diameter must be 38–62 mm. The furnace must be equipped with an electrically controlled heating system. The cable material, lying on a combustion vial made of porcelain, quartz glass, or soapstone, has to be inserted into a quartz glass tube, which runs through the oven.

The inner diameter of the quartz glass tube must be 30–46 mm.

For the classification to EN 13501-6 [11], the following values apply:

- a1: Conductivity: < 2.5 μS/mm; pH value: > 4.3
- a2: Conductivity: < 10 μS/mm; pH value: > 4.3

12.2.4 Fire Resistance Characteristics of Cables

In the event of a fire, it may be vital for the safety of the building occupants that certain electrical systems remain functioning until the rescue has been completed. Examples are:

- Fire detectors and alarm systems
- Emergency escape lighting
- Smoke extraction systems
- Power supplies for extinguishing systems.

IEC 60331-1, -2, -3

The standard IEC 331-1970 has seen two major steps in its development. In 1999, IEC 60331 was extended and a range of devices introduced to allow tests to be carried out on large and small power cables, control cables, data cables, and optical-fiber cables. Part 11 described the test apparatus and a flame temperature of at least 750 °C, while Parts 21, 23, and 25 addressed the procedures and requirements for cables of rated voltage up to and including 0.6/1 kV, including the requirements for electric data cables and for optical-fiber cables.

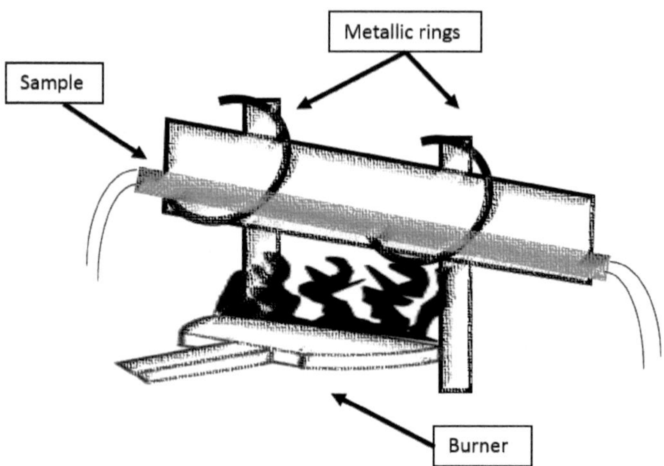

Figure 12.16 Test layout for the circuit integrity test for cables in accordance with IEC 60331-11, -21, and -23

The next revision in 2009 (and its 2nd edition in 2018) contained a further step in the development of the circuit integrity test for electric cables. The test temperature was increased to at least 830 °C and a shock treatment during flame application was introduced for cables with an overall diameter exceeding 20 mm (60331-1), cables with an overall diameter not exceeding 20 mm (60331-2), and cables tested in a metal enclosure (60331-3). This series of standards covers the recommended test procedures, and Parts 11, 21, 23, and 25 are no longer used.

The new series of standards is applicable to cables of rated voltage not exceeding 0.6/1 kV, including those of rated voltage below 80 V, metallic data and telecom cables, and optical fiber cables. The parts of IEC 60331 [12] specify the test method for cables that are required to maintain circuit integrity when subject to fire and mechanical shock under specified conditions. They include details for the specific point of failure, continuity checking arrangements, test samples, test procedures and test reports relevant to electric power and control cables with rated voltage up to and including 0.6/1 kV. The procedure may be used, with the agreement of the manufacturer and the purchaser, for cables with rated voltage up to and including 1.8/3 (3.3) kV, provided that suitable fuses are used. Details for the specific point of failure, continuity checking arrangements, test samples, test procedures, and test reports relevant to metallic data and telecom cables and optical-fiber cables are not provided, and subject to further development of the standard. The test is essentially the same as the test described in EN 50200 [13].

The test layout is shown in Figure 12.17. The apparatus is used for testing cables required to maintain circuit integrity when subjected to a flame with a controlled heat output corresponding to a temperature of at least 830 °C, and to mechanical shock under specified conditions. It shall be placed in a draught-free environment within a suitable chamber with a minimum volume of 20 m^3. For Part 1 and Part 2, the 1500 mm long cable sample shall be bent to form an approximately circular arc with the minimum bending radius of cables in normal use, and held by supports. In Part 3, the metal enclosure shall be fixed in the middle of the test setup. The heat source, a ribbon-type propane burner with a nominal burner face length of 500 mm with a Venturi mixer, is aligned with the test sample. Annex A of the standard contains a verification procedure to check the burner and control system, and the determination of the exact burner location during the cable testing. During the test, a current for continuity checking is applied to all cable conductors.

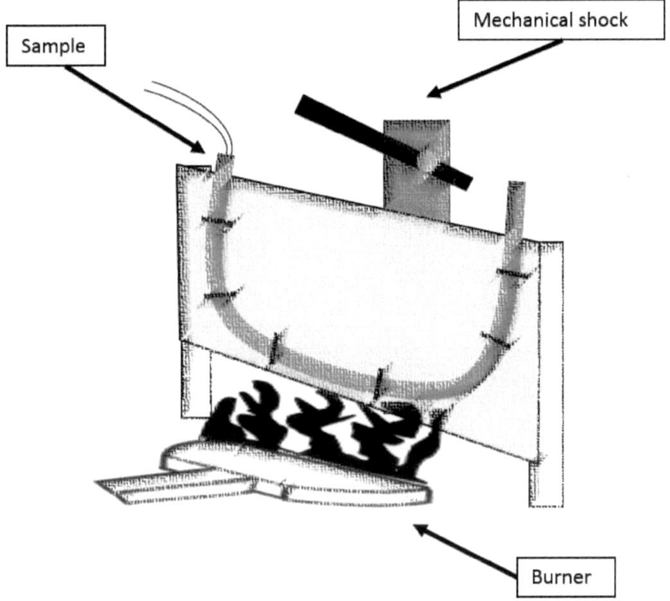

Figure 12.17 Test layout for the circuit integrity test for cables in accordance with IEC 60331-2/EN 50200

The connection of the conductors to the transformer and the test voltage is prescribed in the standards. The shock-producing device must impact the ladder after 5 min ± 10 s from activation, and subsequently at 5 min ± 10 s intervals. If not specified in the relevant cable standard, a 30 min, 60 min, 90 min, or 120 min flame application may be chosen. A cable possesses the characteristics for providing circuit integrity as long as the voltage is maintained during the course of the test, no fuse fails, and no lamp is extinguished. The points of failure for data cables are given in EN 50289-4-16, that for optical-fiber cables in EN 50582.

EN 50577

The standard EN 50577 [14] is used to test low voltage power cables and control cables with a rated voltage up to and including 0.6/1.0 kV, electric data cables and optical-fiber cables. The fire resistance of the cable is tested in a furnace to the EN 1363-1 standard time/temperature curve (Figure 12.18) until its point of failure.

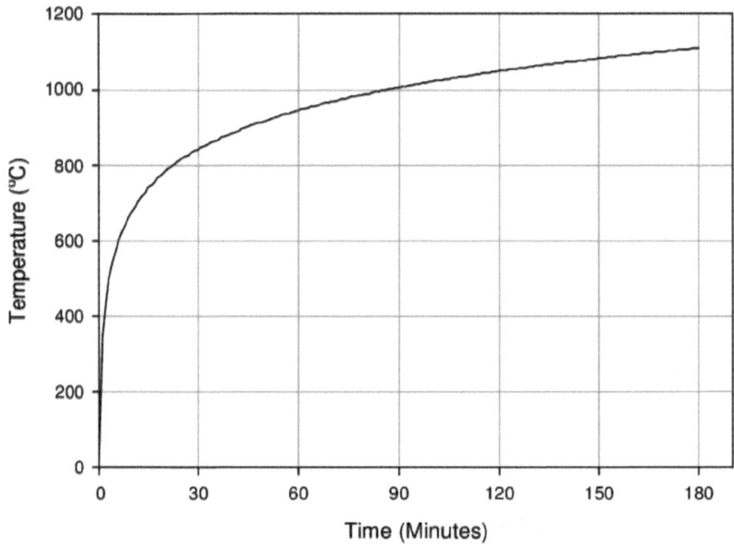

Figure 12.18 EN 1363-1 standard time/temperature curve

The dimensions of the furnace are 3 m in length, 2 m in depth, and 2.5 m in height. The test layout consists of:

- A cable tray made of galvanized steel with a width of 400 ± 20 mm and a maximum working load of 200 N/m
  - Suspensions and horizontal supports (U-shaped and perforated) made of galvanized steel with a cross-section of 220 ± 10 mm² and a minimum working load of 10 kN
- 90° fittings made of galvanized steel from the same system as the cable tray
- Steel chains (used as loading) made of uncoated mild steel with a weight of approximately 3.5–4.0 kg/m.

The test specimens have to be conditioned for at least 16 h at a temperature of 20 ± 10 °C. The point of failure for the cable occurs when the test voltage is interrupted, indicated either by a failing fuse or an extinguished lamp.

The points of failure for data cables are given in EN 50289-4-16; that for optical-fiber cables, in EN 50582.

The test result is determined by the duration of the current flow up to 120 min.

DIN 4102-12

The standard DIN 4102-12 [15] is a German standard to test low-voltage power cables and control cables with a rated voltage up to and including 0.6/1.0 kV. In contrast to the insulation and integrity tests of a single cable according to IEC 60331-series or the related harmonized EN standards, the test to DIN 4102-12 is designed to establish whether a complete electrical cable system is able to maintain circuit integrity over a defined period of time in the heating regime in the test chamber under conditions of the EN 1363-1 standard time/temperature curve (Figure 12.18).

The dimensions of the furnace are 3 m in length, 2 m in depth, and 2.5 m in height. The cable samples are mounted to the walls and the ceiling inside the fire test chamber in an end-use application, using the recommended fixing devices, cable trays, ladders, etc. The test can also be run on a standardized installation, but if the installation is of a "practical" design that deviates from the standardized one, the installation as it will be used in practice has to be assessed. For the measurement of circuit integrity, the cables are protruded through the outer walls of the fire test chamber and connected to the relevant test voltage (110 V for telecommunication cables; 400 V for power cables). The point of failure for the cable occurs when the test voltage is interrupted, indicated either by a failing fuse or an extinguished lamp.

The test result is determined by the duration of the current flow up to 90 min. DIN 4102-12 classifies the cable system into the following categories depending on the duration of the current flow: E30 (≥ 30 min); E60 (≥ 60 min); E90 (≥ 90 min).

12.2.5 Fire Hazard Assessment on Railway Rolling Stock Cables

Railway rolling stock cables are intended for transmission and distribution of electricity in monitoring, control, and power circuits. The European standard EN 50355 [16] gives guidance in the safe use of rolling stock cables specified in the EN 50306 [17], EN 50264 [18], and EN 50382 [19] series. It was prepared for Technical Committee CENELEC TC 20 (Electric cables), specifically by Working Group 12 (Railway cables), as part of the overall program of work in CENELEC TC 9X (Electrical and electronic applications for railways).

EN 50264 covers a range of cables, with standard wall thickness of insulation and sheath rated at 3.6/6 kV with conductor sizes 1.0 mm^2 up to 400 mm^2. EN 50306 covers cables with thin wall insulation, restricted to a rating of 300 V to earth and a maximum conductor size of 2.5 mm^2. Both standards require the use of cables based on halogen-free materials. EN 50355 gives recommendations on how to select, locate, and install cables so that they do not present a fire hazard to adjacent materials.

Based on the protective goal of the standard, insulating and sheathing materials referred to in EN 50264 and EN 50306 are specifically selected and tested to make spread of flame, emission of smoke and toxicity consistent with the required performance of the cable to suit the operation categories and hazard levels given in EN 45545-1 [20].

These standards include specifications according to reaction-to-fire, flame propagation, smoke emission, emission of corrosive and acid gases, toxic potency of the combustion gases, and resistance to fire. EN 50305 [21] gives particular test methods applicable to the cables covered by EN 50264 and EN 50306. The standard describes detailed reaction-to-fire tests for determining flame propagation according to EN 60332-2-1 for single cables, and to the relevant part of EN 60332-3 depending on the cable diameter and the nominal total volume of non-metallic material.

12.2.6 Future Developments

In recent years, numerous developments have taken place regarding fire test methods for cables. Among other things, this is due to the introduction of the Construction Products Regulation for cables, which has led to the development of a new test standard, EN 50399. EN 50399 includes many methods from other test standards, and was developed by recourse to empirical values. This process of change is ongoing, and the more experience is gained with the testing methods, the better they will be. It is expected that standards will continue to be updated regularly so that the highest level is maintained.

In addition, due to the European regulations, a unification of the test methods for power cables and communication cables, including fiber-optic cables, has taken place. It is expected that this will lead to some new developments regarding the use of tests for very thin cables, and fiber-optic cables that so far have only applied to power cables.

References for Section 12.2

[1] NFPA 70:2020. National Electrical Code (NEC).

[2] IEC 60332-1:2004+A1:2015. Tests on electric and optical fibre cables under fire conditions. Part 1: Test for vertical flame propagation for a single insulated wire or cable – Apparatus. Part 2: Test for vertical flame propagation for a single insulated wire or cable – Procedure for 1 kW pre-mixed flame. Part 3: Test for vertical flame propagation for a single insulated wire or cable – Procedure for determination of flaming droplets/particles.

[3] IEC 60332-2:2004. Tests on electric and optical fibre cables under fire conditions. Part 1: Test for vertical flame propagation for a single small insulated wire or cable – Apparatus. Part 2: Test for vertical flame propagation for a single small insulated wire or cable – Procedure for diffusion flame.

[4] IEC 60332-3:2018. Tests on electric and optical fibre cables under fire conditions. Part 10: Test for vertical flame spread of vertically-mounted bunched wires or cables – Apparatus. Part 21: Test for vertical flame spread of vertically-mounted bunched wires or cables – Category A F/R. Part 22: Test for vertical flame spread of vertically-mounted bunched wires or cables – Category A. Part 23: Test for vertical flame spread of vertically-mounted bunched wires or cables – Category B. Part 24: Test for vertical flame spread of vertically-mounted bunched wires or cables – Category C. Part 25: Test for vertical flame spread of vertically-mounted bunched wires or cables – Category D.

[5] EN 50399:2011 + A1:2016. Common test methods for cables under fire conditions – Heat release and smoke production measurement on cables during flame spread test – Test apparatus, procedures, results.

[6] UL 1666: Standard for test for flame propagation height of electrical and optical-fiber cables installed vertically in shafts. Edition 5, 16 February 2007.

[7] NF C32-070/A1 November 2005, *https://www.boutique.afnor.org/norme/nf-c32-070-a1/conducteurs-et-cables-isoles-pour-installations-essais-de-classification-des-conducteurs-et-cables-du-point-de-vue-de-leur-compo/article/728160/fa141993* (+ F1, F2 November 2007, F3 April 2009, F4 July 2010, F5 April 2011).

[8] NFPA 262:2019. Standard method of test for flame travel and smoke of wires and cables for use in air-handling spaces.

[9] EN 61034-1:2005 + A1:2014. Measurement of smoke density of cables burning under defined conditions. Part 1: Test apparatus (IEC 61034-1:2005 + A1:2013). Part 2: Test procedure and requirements (IEC 61034-2:2005 + A1:2013).

[10] EN 60754-2:2015. Test on gases evolved during combustion of materials from cables. Part 2: Determination of acidity (by pH measurement) and conductivity.

[11] EN 13501-6:2019. Fire classification of construction products and building elements. Part 6: Classification using data from reaction to fire tests on power, control and communication cables.

[12] IEC 60331:2018. Tests for electric cables under fire conditions. Part 1: Test method for fire with shock at a temperature of at least 830 °C for cables of rated voltage up to and including 0,6/1,0 kV and with an overall diameter exceeding 20 mm. Part 2: Test method for fire with shock at a temperature of at least 830 °C for cables of rated voltage up to and including 0,6/1,0 kV and with an overall diameter not exceeding 20 mm. Part 3: Test method for fire with shock at a temperature of at least 830 °C for cables of rated voltage up to and including 0,6/1,0 kV tested in a metal enclosure.

[13] EN 50200:2006. Method of test for resistance to fire of unprotected small cables for use in emergency circuits.

[14] EN 50577:2015. Fire resistance test for unprotected electric cables (P classification).

[15] DIN 4102-12:1998. Brandverhalten von Baustoffen und Bauteilen. Teil 12: Funktionserhalt von elektrischen Kabelanlagen – Anforderungen und Prüfungen [Fire behavior of building materials and building components. Part 12: Circuit integrity maintenance of electric cable systems – Requirements and testing].

[16] EN 50355:2013. Railway applications – Railway rolling stock cables having special fire performance – Guide to use.

[17] EN 50306. Railway applications – Railway rolling stock cables having special fire performance – Thin wall. Part 1: General requirements (EN 50306-1:2002). Part 2: Single core cables (prEN 50306-2:2018). Part 3: Single core and multicore cables screened and thin wall sheathed (prEN 50306-3:2018). Part 4: Multicore and multipair sheathed cables (prEN 50306-4:2018).

[18] EN 50264. Railway applications - Railway rolling stock power and control cables having special fire performance. Part 1: General requirements (EN 50264-1:2008). Part 2-1: Cables with crosslinked elastomeric insulation - Single core cables (EN 50264-2-1:2008). Part 2-2: Cables with crosslinked elastomeric insulation - Multicore cables (EN 50264-2-2:2008). Part 3-1: Cables with crosslinked elastomeric insulation with reduced dimensions - Single core cables (EN 50264-3-1:2008). Part 3-2: Cables with crosslinked elastomeric insulation with reduced dimensions - Multicore cables (EN 50264-3-2:2008).

[19] EN 50382. Railway applications - Railway rolling stock high temperature power cables having special fire performance. Part 1: General requirements (EN 50382-1:2008/A1:2013). Part 2: Single core silicone rubber insulated cables for 120 °C or 150 °C (EN 50382-1:2008/A1:2013).

[20] EN 45545-1:2013/A1:2015. Railway applications - Fire protection on railway vehicles - General.

[21] EN 50305:2021. Railway applications - Railway rolling stock cables having special fire performance - Test methods.

13 Furniture and Furnishings

13.1 Introduction

Edith Antonatus

Fires often start due to ignition of building contents. The amount of combustible materials used in buildings has greatly increased over the past decades. Building contents include all moveable items such as cupboards, wardrobes, chests, tables, desks, seating, and bedding, as well as household textiles (decorations, curtains) or floor coverings (the use of which is controlled by the building regulations in many countries). Building contents such as furniture and furnishings may constitute a major fire hazard because, even if building regulations are strict in most countries, the technical fire safety requirements for the building contents may be low. As a result, the fire hazard from building contents may be much higher than that from building materials.

This chapter only deals with upholstered furniture and mattresses in building. Upholstered furniture used in transportation (mainly as seating) is dealt with in Chapter 11. Special regulations for textiles are covered in Chapter 4.

Upholstered furniture and mattresses have been identified as the most hazardous building content in terms of ease of ignition, fire load, and rapid fire growth. Most research and development into test methods has therefore focused on these products. In buildings, the main ignition sources for furniture and mattresses are lighters, cigarettes, candles, and electrical blankets. A significant hazard in bedding and furniture fires in private dwellings arises from people being trapped by a rapidly developing fire, with the consequential smoke spread which gives less time to escape.

In Europe, furniture and mattresses are regulated by the European General Product Safety Directive. In the early 1990s, a research program was initiated by the Commission of the European Communities Directorates-General XII to investigate the combustion behavior of upholstered furniture (CBUF) and to develop methods for measuring the burning behavior of upholstered furniture needed for the implementation of possible EU or national legislation, or for European standardization [1]. A draft directive concerning the fire behavior of upholstered furniture was prepared [2], but has never been finalized. The European General Product Safety directive [3] only contains general statements regarding safety, but it does not give specific instructions on how to control fire properties of upholstered furniture and mattresses. Nevertheless, European standards for testing the ignition resistance of furniture (EN 1021-1 and EN 1021-2) [4] and mattresses (EN 597-1 and EN 597-2) [5] by cigarettes and small flames have been developed.

Some countries have introduced regulations for the fire safety of furniture and mattresses. The USA has developed a broad set of requirements for upholstered furniture and mattresses, which are explained in Section 13.3. In Europe, the UK and Ireland are the only countries with mandatory fire safety requirements for all furniture, including private use (see Section 13.2). They were introduced in 1989 (UK) and 1995 (Ireland). In addition, a number of European countries have introduced limited requirements for furniture and mattresses used in public spaces or other critical areas (for example prisons, hospitals etc.). Examples are Belgium, France, Italy, and Spain, where requirements have been set for furniture in public spaces (see the respective sections under 10.3.2). Other countries (for example Sweden and Finland) have introduced voluntary recommendations or guidelines.

In the Asia/Pacific region, fire safety requirements for furniture in public spaces apply in China (see Section 10.4.1). In Australia, there are no specific flammability requirements for furniture, but the presence of furniture has to be considered when fire risks in a building are assessed. In 2019, a new Product Safety Policy Statement for foam-filled furniture [6] was introduced in New Zealand, which aims to reduce the fire-related risk of foam-filled furniture and refers to Australian/New Zealand and UK standards.

There is an ongoing discussion whether fire safety requirements for furniture are effective for the improvement of safety. The UK Department of Trade and Industry (DTI) published a detailed assessment of the effectiveness of the Furniture and Furnishings (Fire) (Safety) Regulations 1988 [7]. These data had been corrected to allow for the effect of smoke alarms, which in the UK has been relatively small, with 1–2% of fires being detected by alarms in 1995 and 10–12% in 1997. The DTI report shows that there were 247 fire deaths caused by upholstered furniture in 1988, and 95 in 1998. In 2009, a second report was published by the Department for Business, Innovation and Skills (BIS) in the UK [8], stating that, in the period 2002–07, furniture fires accounted for:

- 54 fewer deaths per year;
- 780 fewer non-fatal casualties per year;
- 1,065 fewer fires each year.

This reduction was valued at about £140m per year, of which:

- 50% was attributed to the cigarette test;
- 10% to the match test;
- 40% to the Crib 5 test.

Newer statistics from the UK show that, in 2018/19, textiles, upholstery, and furnishings were mainly responsible for fire development (24% of all dwelling fires), and that they were the item first ignited in 26% of cases. They also caused 58% of all fire-related fatalities in dwellings [9, 10]. So, even with strict regulations in place, upholstered furniture and mattresses still make a significant contribution to fire fatalities in dwellings. But the absolute numbers of fatalities caused by furniture and beddings have become quite low in the UK. The number of the relevant fatalities in the last years is shown in Figure 13.1.

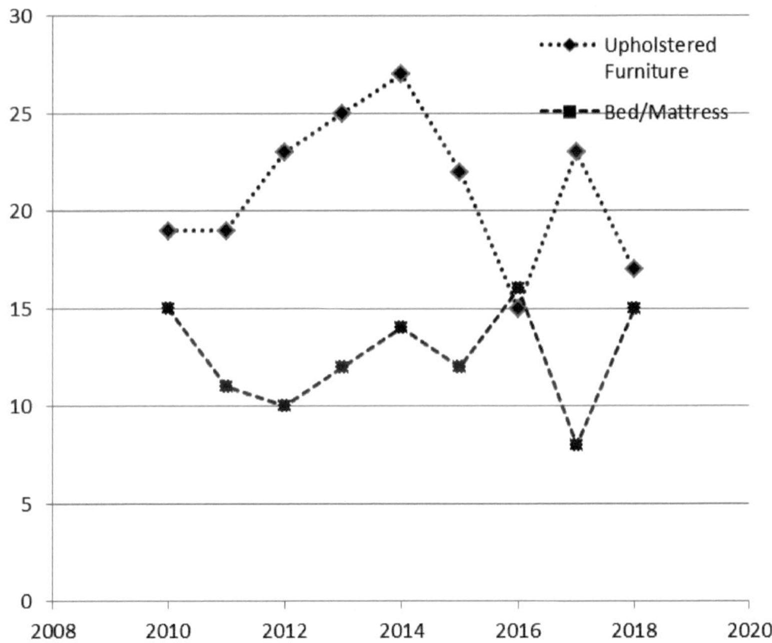

Figure 13.1 Fire-related fatalities in dwellings in the UK in relation to the first-ignited item (data from [9])

In the USA as well, statistical studies show that the introduction of fire safety requirements for furniture has contributed to improved fire safety. A NFPA (National Fire Protection Association) report published in 2017 on the subject of "home fires that began with upholstered furniture" [11] states that: "Since 1980, home upholstered furniture fires have fallen. Home structure fires beginning with upholstered furniture fell 85% from a high of 36,900 in 1980 to a low of 5,400 in 2014, the lowest point since 1980. This is a much larger decrease than the 50% drop seen for total home structure fires over the same period. From 2013 to 2014, upholstered furniture fires fell 2% while total home fires fell 1%."

13.2 UK and Ireland

Edith Antonatus

13.2.1 Regulations and Tests for Furniture Suitable for Domestic Use

The Upholstered Furniture (Safety) Regulations introduced in 1980 in Great Britain [12] required that upholstered seating should pass the BS 5852 Part 1 cigarette test. Furniture that did not also pass the BS 5852 part 1 flame test was labeled accordingly. More stringent requirements for furniture suitable for domestic use were introduced in 1988 and amended in 1993 and 2010, with the Furniture and Furnishings (Fire) (Safety) Regulations defined in the Statutory Instruments SI 1324 (1988), SI 2358 (1989), SI 207 (1993), and SI 2205 (2010) [13]. These Statutory Instruments are acts of parliament, which define the application, enforcement and test procedures, and requirements for upholstered furniture suitable for domestic purposes within the UK. The tests used are essentially BS 5852 Part 1 [14], BS 5852 Part 2 [15], and BS 6807 [16]. The Regulations do not apply to sleeping bags, bedclothes (including duvets), loose covers for mattresses (i.e. mattress protectors), pillowcases, curtains, and carpets. However, it is important to note that these products are covered under the General Product Safety Regulations 2005 (GPSR).

Ireland has introduced almost identical regulations, with slight changes regarding the ignition time for the match test [17].

13.2.1.1 Test Methods Applied in the UK Regulations for Upholstered Furniture and Corresponding European Test Methods

The European Standards Organization CEN has published four ignition tests for upholstered furniture. EN 1021-1 and EN 1021-2 [4] are cigarette and small-flame tests for upholstered seating, and EN 597-1 and EN 597-2 [5] are cigarette and small-flame ignition tests for mattresses. The seating tests for the ignitability of upholstered seating are based on BS 5852: Part 1 (1979), and use a glowing cigarette and a small flame intended to simulate a match.

The test apparatus is shown in Figure 13.2, and consists of two steel frames hinged together, capable of being locked at right-angles to each other. The frames are equipped with expanded metal mesh. The horizontal frame measures 450 mm × 150 mm, while the vertical frame is 450 mm × 300 mm. The test apparatus is opened out so that both parts are horizontal for inserting the covering material, which is stretched over the entire frame and behind the retaining hinge rod. The covering material measures 800 mm × 650 mm. The upholstery fillings are laid under the cover and measure 450 mm × 150 mm × 75 mm, and 450 mm × 300 mm × 75 mm for the seat and back, respectively. The cover material is held on the steel frames with clips. Finally, the frames are returned to their original perpendicular positions. An important aspect of the test rig is that it enables fabric tension and filling compression to be simulated. It also simulates the crevice, which occurs between the seat and back or seat and arm of actual furniture, which is considered to be a critical feature for the ignition of seating. The test specimen comprises the outer cover, the interliner (if used), and the soft upholstery fillings. It does not include trims such as piped or ruched edges, or features such as deep buttoning.

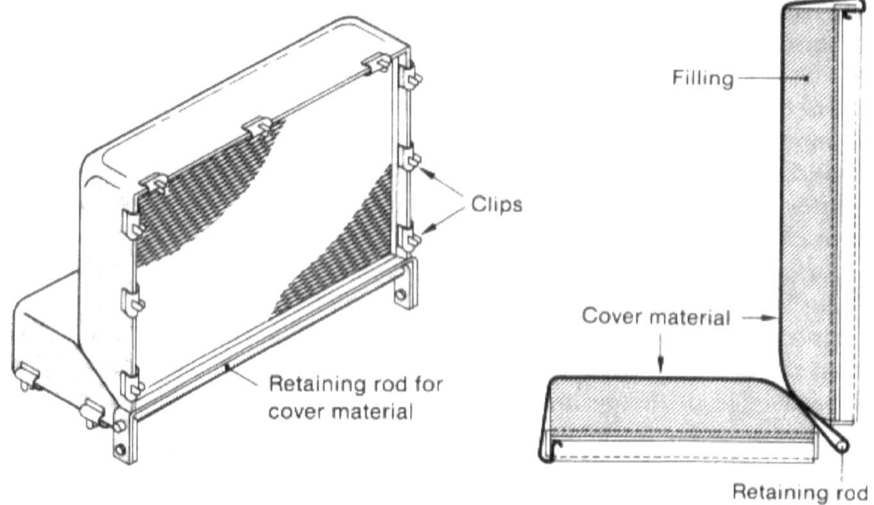

Figure 13.2 Apparatus for testing the ignitability of upholstered seating according to BS 5852: Part 1, ISO 8191-1, ISO 8191-2, EN 1021-1, EN 1021-2, and IMO resolution A.652

The glowing cigarette (68 mm long, 8 mm diameter, mass 1 g) is placed at the junction between the vertical and horizontal components at least 50 mm from the nearest side edge. If progressive smoldering takes place for more than 1 hour, or if flaming occurs, the test is failed. If no such phenomena occur, the test is repeated with a new cigarette placed at least 50 mm away from the position of the first cigarette. If no progressive smoldering or flaming is observed in both tests, a pass result is recorded.

The small-flame test uses a 65 mm internal diameter tube, 200 mm long, connected to flexible tubing as a burner, which produces a butane flame approximately 35 mm high. The burner tube is laid along the joint between the vertical and horizontal components of the test rig and at least 50 mm from the nearest edge (and the damage from any previous test). It is applied for 15 s (the BS 5852 Part 1 test uses 20 s). If flaming or glowing continues for more than 120 s after removing the ignition source or if progressive smoldering occurs, the test is failed. If this does not occur with two tests, a pass result is recorded. A water soak is applied to covers to prevent the use of easily removed flame retardant treatments.

The procedure described above is essentially the same as for ISO 8191-1 and ISO 8191-2, and IMO resolution A.652 for international shipping. The ignition flame time varies between tests (15 s or 20 s), which may give different results in some cases.

EN 597-1 and EN 597-2 are cigarette and small-flame ignition tests for mattresses. A test specimen of minimum size 350 mm × 450 mm × nominal thickness is used, which reproduces the features of the full-size mattress. In the tests, the ignition sources are applied to the mattress surface, to tufts, along quilt lines and also to tape edges if present. The test rig is shown in Figure 13.3.

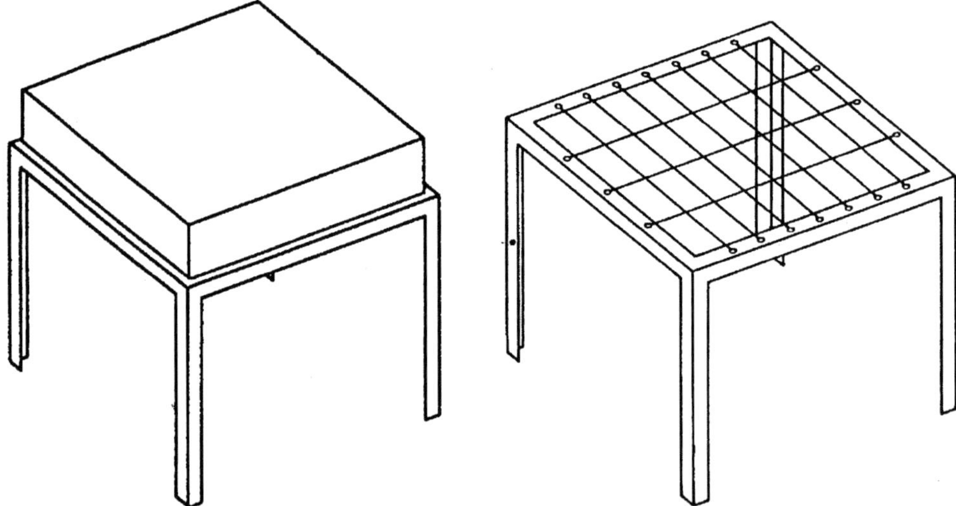

Figure 13.3 Test rig and specimen used for EN 597-1 and EN 597-2 mattress tests

13.2.1.2 Classification of Soft Upholstery in the UK

A very important aspect of the UK Regulations is that the tests are used to classify materials used for soft upholstery. Upholstery components that pass the relevant test can be interchanged and used with other components that pass their relevant test. This applies to fabrics, interliners, and to all fillings. The importance of this is that manufacturers can classify their materials without excessive testing. The exception is the cigarette test, which is applied to the actual upholstery composite. In practice, the cigarette test may be carried out with the fabric over the standard PU foam used for the flame test, since this represents a worst-case PU foam filling when compared to the specified combustion-modified PU foams. This enables a further reduction in testing. The various test schedules are summarized in Table 13.1, and are discussed in the next section.

Table 13.1 Component Tests within SI 1324 (1988), SI 2358 (1989), SI 207 (1993), and SI 2205 (2010)

Test schedule	Test component
SI 1324 (1988)	
Schedule 1 Part I	PU slab or cushion foam
Schedule 1 Part II	PU crumb foam
Schedule 1 Part III	Latex rubber foam
Schedule 2 Part I	Single non-foam fillings
Schedule 2 Part II	Composite fillings other than mattresses
Schedule 2 Part III	Composite test for ignitability of pillows and cushions with primary covers
Schedule 2 Part IV	Composite test for bed bases, cushions, pillows
Schedule 3	Interliners, ignition resistance
Schedule 4	The cigarette test
Schedule 5 Part I	The match test for covers
Schedule 5 Part I	The match test for stretch covers
SI 2358 (1989)	
Schedule 4 Part II	Cigarette test for invisible parts (no water soak)
Schedule 5 Part III	Match test for invisible parts of covers
SI 207 (1993)	Extends the scope, does not alter the tests
SI 2205 (2010)	Fabrics and interliners in Schedules 1, 2, and 3 are defined in more detail

13.2.1.2.1 Fillings

Essentially all fillings are tested using a specified flame-retarded polyester cover using the test method of BS 5852 Part 2 with ignition source 2. This is applied to the fillings listed with the exception of PU block foams, which is tested to crib 5 with modified pass/fail criteria of a weight loss of less than 60 g without the flame spread and penetration criteria. Composite fillings of mattresses and bed bases are tested to the following:

- Schedule 2 Part IV uses ignition source 2 with the BS 6807 mattress-type test specimen;

- Schedule 1 Part II PU crumb foam also requires that the crumb is made from foam, which complies with Schedule 1 Part I.

13.2.1.2.2 Covers and Interliners

Schedule 3 tests interliners for use with designated types of non-flame-retarded and natural fabrics over a standard (non-flame-retarded) PU foam.

Schedule 5 Part I tests coverings and Part II stretch covers with a standard PU foam filling using the small flame (20 s duration) of BS 5852 Part 1 (1979). This essentially means that satisfactory fabrics are ignition-resistant and protective, i. e. non-melting. Tests for covers and interliners require that these are first subjected to a water soak procedure to ensure that easily soluble treatments that could be removed by simple cleaning or spillage are not used in the manufacture of actual furniture.

SI 2358 (1989) requires that "invisible" fabric (i. e. fabric used beneath cushions, on seat platforms, beneath chairs, etc.) is tested to Schedule 4 Part II (cigarette) and Schedule 5 Part III (match flame), which are the equivalent of SI 1324 (1988) Schedule 4 and Schedule 5 respectively. Neither test requires the water soak for the covering, while the small-flame test is carried out using combustion-modified PU foam complying with Schedule 1 Part I.

13.2.1.2.3 Composites

Schedule 4 is unique in that it tests the actual furniture composite with the cigarette of BS 5852 Part 1 (1979). Tests are frequently carried out using a standard PU filling to represent a worst-case PU foam to reduce the testing load. This often gives a more reasonable test than using PU foam complying with Schedule 1 Part I, but is not used as an alternative to more easily ignitable fillings, e. g., cotton wadding.

13.2.2 UK Regulations and Tests for Furniture Used in Public Areas

The fire safety of persons in public areas in the UK is essentially controlled by a number of measures directly relating to the assessment of the threat to life from fire, which is assessed by local authorities or fire prevention officers. Legislation on this was first published in 1971. It was replaced by the regulatory reform (fire safety) order 2005 [18], which is now the main legislation governing fire safety precautions in non-domestic premises in England and Wales. To support the Order, the Ministry of Housing, Communities and Local Government (MHCLG) has published a number of guidance documents. These guides apply to places of work, certain multiple occupancies, hotels and boarding houses, places of entertainment and similar premises, etc., which are also published on the above-mentioned website.

The required fire performance for the medium and high hazard levels is defined in BS 7176 [19] and BS 7177 [20]. In addition, it is usual to require that the fillings of upholstered furniture for public area use should comply with the requirements of SI 1324.

Here, larger ignition sources are applied. The test rig dimensions are 450 mm × 450 mm for the vertical and 450 mm × 350 mm for the horizontal component. The ignition sources comprise two gas flames and four wood cribs. The number 5 crib, together with the cigarette and the small flame of EN 1021-1 and EN 1021-2, respectively, are used for upholstered furniture for use in public buildings. The number 7 wood crib (126 g) is used with the cigarette and small-flame tests to define upholstery for high-risk areas. Section 5 of this test is applied to actual upholstered furniture and the sources are applied to the upper surface at the seat/back junction and below or against the furniture depending on the distance between the base of the furniture and the floor. The number 5 wood crib (17.5 g) is used to define block PU fillings, while the number 2 gas flame is used to define all other fillings within the UK Furniture and Furnishings (Fire) (Safety) Regulations (see previous section).

BS 6807 is similar to the EN 597 standards, but uses the two large gas flames and the four wood cribs described above for BS 5852. These sources are used to test mattresses for similar end-use applications as those described for seating tests. The BS 6807 test method has a number of sections, and may be used to test mattresses on their own and also made-up beds with different combinations of bedding and mattress. Tests may be carried out on the test mattress test specimen of EN 597 or on actual mattresses.

BS 7175 [21] is similar to BS 6807 [16], but tests bedding combinations over a non-combustible fiber mat, which is used to simulate a mattress. This test is used with the cigarette and small flame, and also with the two large gas flames and the four wood cribs of BS 5852 (1990). Textile working groups have published cigarette and small-flame ignition resistance tests for bedding [22, 23].

BS 7176 and BS 7177 list ignition test performance requirements for upholstered seating and mattresses, respectively. Furniture is specified according to low, medium, high, and very high hazard end-use applications. The standards also contain examples of the hazard applications, for example:

- Low-hazards applications are typically domestic, office, some hotels, and require cigarette and small-flame resistance to EN 1021-1 and EN 1021-2, respectively (note that the regulations of SI 1324 and amendments also apply to furniture suitable for domestic use);
- Medium-hazard applications apply to public areas, and typically require a cigarette test and a No. 5 crib;

- High-hazard applications require the cigarette test and a No. 7 crib;
- Very high hazard applications do not have an ignition source specified in the British Standards, but a test for mattresses for use in locked accommodation uses four No. 7 cribs to test a fully vandalized mattress, and is published by the Home Office Prison Service.

13.3 United States of America

Marc Janssens

13.3.1 Statutory Regulations

A distinction has to be made between regulations that apply to upholstered furniture and mattresses at the point of sale, and those that apply at the point of use. The former apply to all mattresses or upholstered furniture that are sold in the geographical area where the regulation is in effect. The latter only apply to mattresses or upholstered furniture when they are used for the purpose specified in the regulation.

At the national level, flammability of upholstered furniture and mattresses is regulated by the U.S. Consumer Product Safety Commission (CPSC). The federal flammability standards for mattresses described in 16 CFR Part 1632 and 16 CFR Part 1633 are examples of regulations that apply at the point of sale. In other words, every mattress that is sold in the U.S. has to comply with the requirements of these standards. Part 1632 describes a full-scale test procedure to determine the smoldering ignition resistance of a mattress or mattress pad prototype when exposed to a lighted cigarette. Part 1633 describes another full-scale test that is used to measure the heat release rate of a mattress ignited with the flames of a pair of T-shaped gas burners designed to represent burning bedclothes. The test procedures and minimum performance requirements for both parts are discussed in subsequent sections. The text for the two standards can be downloaded from the e-CFR website [24].

In 2008, CPSC issued a notice of proposed rulemaking based on a standard for residential upholstered furniture [25]. The proposed standard consisted of a small-scale smoldering cigarette test for fabrics and a small-scale open-flame test to evaluate barriers, which are used to prevent smoldering (and ultimately flaming) combustion of the filler material. However, CPSC subsequently determined that performance in the proposed small-scale tests is not sufficiently representative of fire behavior of actual furniture, and the standard was never formally issued.

Consequently, there are no federal point-of-sale flammability regulations for upholstered furniture at this time. A detailed discussion of CPSC's historical and current rulemaking activity and supporting research pertaining to furniture flammability is available [26].

The Home Furnishings and Thermal Insulation Act issued by the Bureau of Household Goods and Services (BHGS, formerly Bureau of Electronic and Appliance Repair, Home Furnishings, and Thermal Insulation, or BEARHFTI), regulates the flammability of upholstered furniture in the state of California. These regulations include specific testing requirements which are specified in California Technical Bulletin (CA TB) 117-2013 [27], which describes the requirements, test procedure, and test apparatus to determine the smolder resistance of components of furniture upholstery.

It is worth noting that older editions of CA TB 117 included an open-flame test, which was removed in the 2013 version due to concerns related to the use of flame retardants. In the summer of 2014, the NFPA Fire Test Committee started development of an open-flame test for upholstered furniture, NFPA 277. However, nearly four years later the NFPA Standards Council concluded that "there is a fundamental lack of consensus on how to test and evaluate residential upholstered furniture flammability exposed to a flaming ignition source" and decided to cease the activity.

In the absence of national flammability requirements for upholstered seating, the voluntary Upholstered Furniture Action Council (UFAC) set up by furniture manufacturers developed a series of small-scale tests for determining the resistance of materials used in upholstered furniture to smoldering initiated by a lighted cigarette [28]. The UFAC cigarette smoldering standard for upholstered seating composites was introduced in the 1970s, and revised in 1990.

According to Table 10.5 in Section 10.2.1.3, the International Fire Code (IFC) and NFPA 101: Life Safety Code® (LSC) refer to the following standard test methods for evaluating upholstered furniture and mattress flammability:

- ASTM E1537: *Standard test method for fire testing of upholstered furniture*
- ASTM E1590: *Standard test method for fire testing of mattresses*
- NFPA 260: *Standard methods of tests and classification system for cigarette ignition resistance of components of upholstered furniture*
- NFPA 261: *Standard method of test for determining resistance of mock-up upholstered furniture material assemblies to ignition by smoldering cigarettes.*

ASTM E1537 is functionally identical to CA TB 133 [29], which, prior to its repeal on January 22, 2019, was used in the state of California to regulate the flammability of upholstered furniture for use in public occupancies. ASTM E1590 is similar to (and now superseded by) the test procedure in 16 CFR 1633 but uses a single

T-shaped 18 kW gas burner that is applied at the crevice between the bottom of the mattress and top of the foundation or frame. NFPA 260 is similar to TB 117-2013, except for some differences in the procedure and specifications for the standard component materials. NFPA 261 is identical to NFPA 260, except that a complete mock-up comprising all components of the furniture upholstery is tested. To determine where the IFC requires these tests, the reader can view the online version of the code for free [30]. The LSC and NFPA 260 and 261 can also be viewed online for free [31]. ASTM standards, however, have to be purchased from ASTM International [32].

By August 26, 2011, all 50 U.S. states and the District of Columbia had passed legislation mandating the exclusive sale of cigarettes with reduced ignition propensity. These cigarettes exhibit a greater likelihood of self-extinguishing when tested according to ASTM E2187: *Standard test method for measuring the ignition strength of cigarettes* and may not be suitable for use in fire testing. The aforementioned smoldering ignition tests are therefore conducted with a special cigarette (Standard Reference Material® or SRM 1196), which can be obtained from the National Institute of Standards and Technology (NIST) in Gaithersburg, MD.

13.3.2 U.S. Smoldering Tests for Upholstered Furniture and Mattresses

16 CFR 1632: Standard for the Flammability of Mattresses and Mattress Pads (FF 4-72, Amended)

The standard was first introduced in 1972 and is now enforced by CPSC. It was implemented because of the number of fire fatalities caused by mattress fires, which were largely started by smoking materials. The test method for determining the smoldering ignition resistance of a mattress to a lighted cigarette is used for products that are in development and not yet on the market, as well as for quality control of commercially available mattresses. Test specifications are summarized in Table 13.2. Detailed guidance for preparing specimens, conducting tests, and reporting the results can be found in the Laboratory Test Manual [33].

Table 13.2 16 CFR 1632 Mattress Smoldering Test Specifications

Specimens	One mattress; testing is first performed on one surface of the bare mattress and subsequently on the opposite surface partly covered with two laundered cotton bedsheets
Specimen position	Horizontal
Ignition source	Lighted cigarette (NIST SRM 1196)

Test duration	Test duration varies and depends on the time it takes for all cigarettes to burn their full length: ▪ For the first part of the test, lighted cigarettes are placed at a minimum of nine different locations on the surface of the bare mattress ▪ For the second part of the test, the mattress is turned over and lighted cigarettes are placed at a minimum of nine different locations in between two laundered sheets covering half of the surface of the mattress
Conclusions	The mattress fails the test if char length on the surface or char depth into the mattress exceeds 51 mm (2 in.) at any location where a lighted cigarette was placed

The mattress meets specified performance requirements if lighted cigarettes, placed in certain spots on the horizontally positioned mattress illustrated in Figure 13.4, do not cause charring of its surface or into the mattress extending more than 51 mm (2 in.) in any direction from the nearest point of the cigarette. A lighted cigarette is first laid on a smooth surface of the bare mattress, then at the edge of the mattress (where in the absence of a depression, it is held in place by three pins). Finally, various spots such as quilted surfaces are tested. The mattress is then turned over and the tests are repeated using two sheets covering half of the mattress. The lighted cigarette is laid on a stretched sheet tucked under the mattress, and a loose sheet laid immediately over it. At least 18 cigarettes are burned on each mattress, nine on the bare mattress and nine under the cotton sheeting material. The test is also used to determine the smoldering ignition resistance of mattress pads, which generally need to be laundered (ten cycles) or dry-cleaned prior to testing.

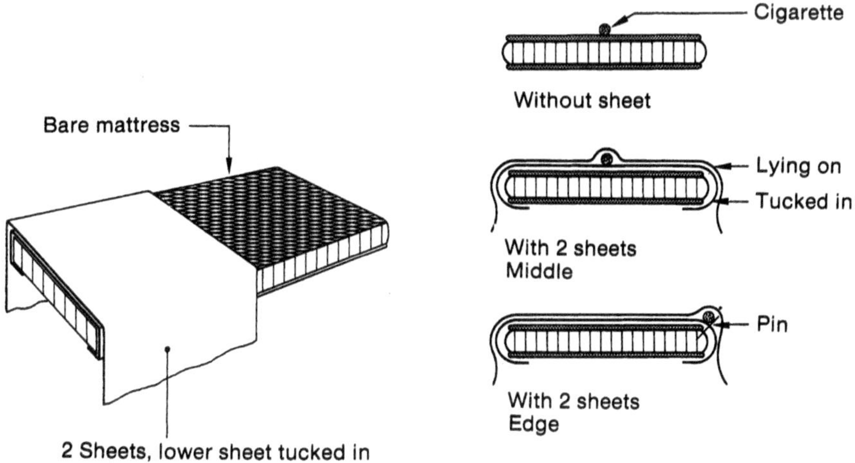

Figure 13.4 Various arrangements for the 16 CFR Part 1632 flammability test for mattresses

Flammability Tests to Meet UFAC Construction Criteria

Furniture manufacturers that wish to participate in UFAC's voluntary action program are required to produce furniture in accordance with UFAC construction criteria [28]. To meet these criteria, UFAC has developed voluntary test methods for six components of upholstered furniture:

- Fabric Classification Test Method (1990)
- Barrier Test Method (1990)
- Filling/Padding Component Test Method, Part A or B (1990)
- Welt Cord Test Method (1990)
- Decking Materials Test Method (1990)
- Standard Test Methods for Decorative Trims, Edging, and Brush Fringes (1993).

The tests on fabric covers, barrier materials, filling/padding components, and welts are carried out with the apparatus illustrated in Figure 13.5. The test rig in this figure is made of plywood. The test specimen consists of a miniature seat assembly comprising the component to be tested, in combination with the remaining components made of specified materials. Three tests are carried out with a lighted cigarette, 85 mm long, weighing 1.1 g, laid in the gap or on the welt between the back panel and seat cushion so that it is in contact with both. A piece of cotton sheeting, 125 × 125 mm, is laid over the cigarette.

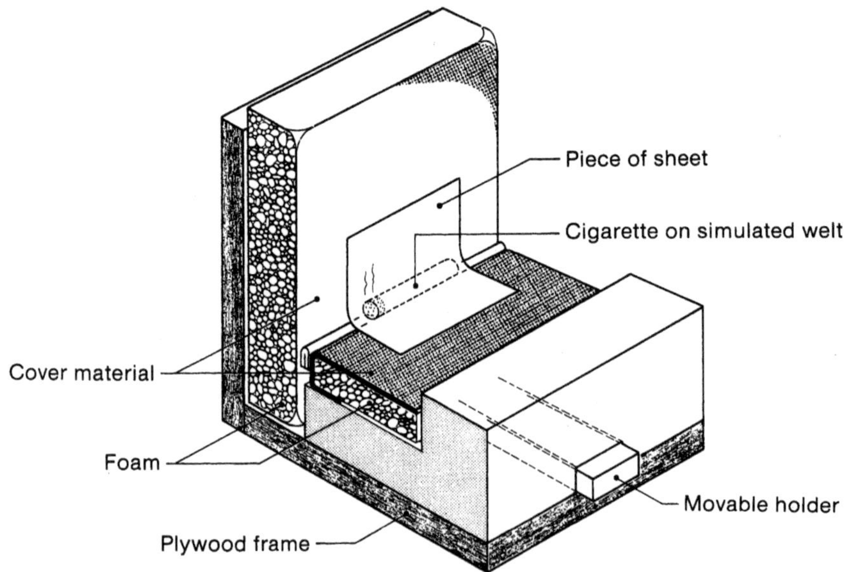

Figure 13.5 UFAC test rig for fabrics, barriers, filling/padding components, and welt cords

For tests on fabrics and barriers, a specified PUR foam, 203 × 203 × 51 mm, is used as cushioning for the miniature back panel. For the welt cord tests, a cotton filling is used. The miniature seat cushion consists of the same filling/padding material as the back panel and measures 203 × 127 × 51 mm (for the welt cord test 203 × 114 mm). The covering fabric used in the tests is 203 × 381 mm for the back panel, except with the barrier test and filling/padding component test where it is 305 × 305 mm and 203 × 305 mm, respectively. The miniature seat cushion cover is 203 × 203 mm.

The barrier and filling/padding component tests are successful if the charred length above the cigarette does not exceed 51 mm. Depending on the test performance, covering materials are classified as Class I or II. The former is achieved if the charred length above the cigarette is less than 38 mm. If this length is exceeded, Class II applies. In the welt cord test, the cotton filling acting as a backrest must not be charred for more than 38 mm above the cigarette.

Decking materials (upholstered support materials under the seat cushion in a loose seat construction) are tested using the apparatus illustrated in Figure 13.6. The decking is laid on a plywood base and the cover fabric laid over it. The dimensions of all three components are 533 × 343 mm. A plywood frame with the same external measurements and internal dimensions of 406 × 216 mm is laid over this assembly. Three glowing cigarettes are placed on the remaining exposed surface of the cover fabric at equal intervals (see Figure 13.6) and covered with a piece of sheeting. The decking material passes the test if the charred length on the base is less than 38 mm measured from each cigarette.

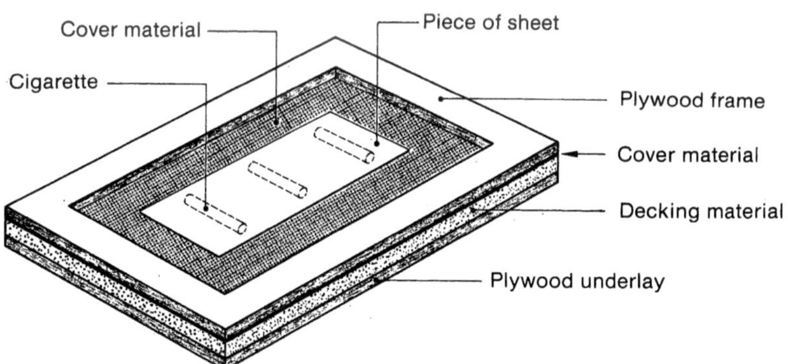

Figure 13.6 UFAC test rig for decking materials

CA TB 117-2013: Requirements, Test Procedure, and Apparatus for Testing the Smolder Resistance of Materials Used in Upholstered Furniture

CA TB 117-2013 [27] consists of several tests used to evaluate the cigarette ignition resistance of upholstery cover fabrics, barrier (interliner) materials, resilient filling materials, and decking materials used in the manufacture of upholstered furniture. CA TB 117-2013 makes frequent reference to the sections on specimen preparation and test procedures in ASTM E1353: *Standard test methods for cigarette ignition resistance of components of upholstered furniture.* In addition, the test rigs are identical to those developed by UFAC and described in ASTM E1353. However, there are some differences between the three methods, e.g., the pass/fail criteria in CA TB 117-2013 are different from the UFAC construction criteria.

Each test on a fabric, barrier, or filler material involves a miniature seat assembly consisting of the component to be tested in combination with the remaining components made of specified materials, mounted on a plywood mockup that resembles a small chair seat and back. The assembly is exposed to a lighted cigarette as an ignition source, which is placed in the crevice and against the back panel. A piece of sheeting material is used to cover the cigarette. A schematic of the assembly is shown in Figure 13.5 (except for the welt cord, which is not used in CA TB 117-2013). CA TB 117-2013 test specifications, including pass/fail criteria for fabrics, barriers, and filling materials, are summarized in Table 13.3.

Table 13.3 CA TB 117-2013 Test Specifications for Fabrics, Barriers and Filling Materials

Specimen	Miniature seat assembly consisting of the component to be tested, in combination with the remaining components made of specified materials
Specimen position	The specimen consists of a horizontal seat panel, approximately 203 × 127 × 51 mm in size, and a vertical back panel, approximately 203 × 127 × 51 mm in size
Ignition source	A lighted cigarette (NIST SRM 1196) placed on top of the fabric in the crevice between the seat and back panels and covered with a piece of cotton sheeting
Test duration	Test duration varies and depends on the time it takes for the cigarettes to burn their full length, tests are conducted in triplicate
Conclusions	A fabric specimen fails the test if any of the following occurs: ▪ The specimen continues to smolder for more than 45 min ▪ Char length on the fabric exceeds 45 mm ▪ The specimen transitions to open flaming The criteria are the same for barriers and filling materials except that the maximum char length is 38 and 51 mm, respectively A specimen passes if passing results are obtained for the three initial tests If one and only one of the three initial specimens fails, a second set of three tests is conducted, and the specimen passes if passing results are obtained for each of the three additional tests

The setup for testing decking materials is shown in Figure 13.6. In this case, a specimen of the decking material to be tested is placed on a plywood base and covered by a specified fabric. Lighted cigarettes are placed on top of the fabric at three locations and are covered with a piece of sheeting material. CA TB 117-2013 test specifications, including pass/fail criteria for decking materials, are summarized in Table 13.4.

Table 13.4 CA TB 117-2013 Test Specifications for Decking Materials

Specimen	533 × 343 mm piece of decking material placed on a plywood base and covered with standard fabric
Specimen position	Horizontal
Ignition source	A lighted cigarette (NIST SRM 1196) covered by a piece of sheeting placed on top of the fabric Tests can be combined for efficiency by placing cigarettes at up to three locations, provided they are equally spaced and at least 152 mm from each other and from the short edges of retaining ring that is placed on top of the fabric (see Figure 13.6)
Test duration	Test duration varies and depends on the time it takes for the cigarette(s) to burn their full length
Conclusions	A decking material fails the test if any of the following occurs: ▪ The specimen continues to smolder for more than 45 min ▪ Char length on the fabric exceeds 38 mm ▪ The specimen transitions to open flaming A specimen passes if passing results are obtained for the three tests initially conducted. If one and only one of the three initial tests results in a failure, a second set of three tests is conducted, and the specimen passes if passing results are obtained for each of the three additional tests

The 2018 IFC requires that the upholstery of new furniture used in unsprinklered spaces of selected public occupancies (ambulatory care facilities, correctional and detention facilities, college and university dormitories, etc.) meet the requirement for a Class I when its components are tested according to NFPA 260, or have a char length of 38 mm or less when the assembly is tested according to NFPA 261. The 2018 LSC has the same requirements for the same types of public occupancies.

13.3.3 U.S. Open-Flame Tests for Upholstered Furniture and Mattresses

This section describes the two open-flame tests for upholstered furniture and mattresses that are used in the U.S. for regulatory purposes. Both are full-scale tests. The first test is ASTM E1537 [34], which is specified in the IFC and LSC to control the flammability of upholstered furniture used in specific occupancies. The second test is described in 16 CFR 1633 [35], which every mattress sold in the U.S. has to comply with.

ASTM E1537: Standard Test Method for Fire Testing of Upholstered Furniture

ASTM E1537 describes a fire test procedure for measuring the heat release rate of seating furniture for use in public occupancies. The test can be run either on a finished piece of upholstered furniture or a smaller-scale mock-up. The test specimen is placed in a rear corner or against the back wall of a room (dimensions: 3.05 × 3.66 × 2.44 m high) with an open door (dimensions: 0.97 × 2.06 m) in the front wall. Alternatively, the specimen can be tested in the standard NFPA 286 room (see Section 10.2.1.6) or placed directly under the hood of a furniture calorimeter (see Section 13.4.1.2). If a room is used, the hood of the oxygen consumption calorimeter is located outside the room above the open door. The specimen is placed on a weighing platform to continuously measure the mass loss during the test.

ASTM E1537 specifies a severe ignition source for upholstered furniture, which is identical to the burner specified in the 1991 version of CA TB 133. CA TB 133 was originally developed and issued in 1984, and specified an ignition source of crumpled newspaper, and requirements relating to temperature level increase, smoke opacity, CO concentration in the fire effluents, and weight loss. The 1991 version of CA TB 133 prescribes a square propane ignition burner and oxygen consumption calorimetry test criteria. The burner, illustrated in Figure 13.7, generates a flame with a heat release rate of 19 kW and is applied to the top surface of a seat cushion for a period of 80 s.

Figure 13.8 illustrates the application of the burner in a test on a mock-up conducted in general accordance with ASTM E1537. The mockup is placed on a weighing platform, directly under the hood of a furniture calorimeter.

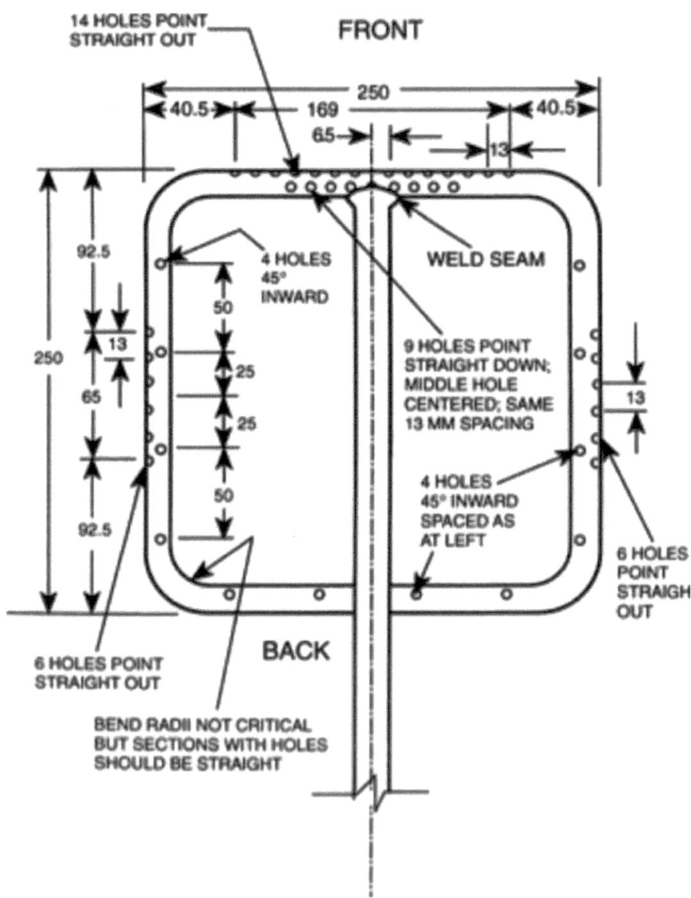

Figure 13.7 Top view of ASTM E1537 gas burner

Figure 13.8 Furniture mock-up test in general accordance with ASTM E1537

The 2018 IFC requires that new furniture used in unsprinklered spaces of selected public occupancies be tested according to ASTM E1537 and meet the following requirements:

- Peak heat release rate of 80 kW or less.
- Total heat release in the first 10 min of the test of 25 MJ or less.

The 2018 LSC has the same requirement for new furniture used in the same types of facilities.

16 CFR 1633: Standard for the Flammability (Open-Flame) of Mattress Sets

CPSC published a draft of 16 CFR Part 1633 as part of a notice of proposed rulemaking in the Federal Register of January 13, 2005. The fire test method for mattresses and mattress/box spring sets described in this document was developed at NIST in a research program sponsored by the Sleep Products Safety Council (SPCS), an affiliate of the International Sleep Product Association (ISPA) [36–38]. As of July 1, 2007 all mattresses for sale in the U. S. have had to comply with this regulation.

A detailed discussion of the test procedure can be found in the Laboratory Test Manual [39]. The test specimen is placed under the hood of an oxygen consumption calorimeter. At the start of the test, the mattress is exposed to the flames of two T-shaped gas burners that simulate the thermal exposure from burning bedclothes. A 19 kW flame from a horizontal line burner is applied for 70 s to the top surface of the mattress and a 10 kW flame from a vertical line burner is applied for 50 s to the side of the mattress. To pass the test, the following criteria must be met:

- Peak heat release rate during the 30 min test period does not exceed 200 kW.
- Total heat release during the first 10 min does not exceed 15 MJ.

Figure 13.9 and Figure 13.10 show pictures of a failing mattress at the start of the test and 2 min into the test, respectively. Figure 13.11 and Figure 13.12 show pictures of a passing mattress at the start of the test and 10 min into the test, respectively.

Figure 13.9
Failing mattress at start of test

Figure 13.10
Failing mattress at 2 min

Figure 13.11
Passing mattress at start of test

Figure 13.12
Passing mattress at 10 min

ASTM E1590 (and CAL TB 129, which is functionally identical) is similar to (and now superseded by) the test procedure in 16 CFR 1633 but uses a single T-shaped 18 kW gas burner that is applied at the crevice between the bottom of the mattress and top of the foundation or frame.

The 2018 IFC requires that new mattresses used in unsprinklered spaces of selected public occupancies be tested according to ASTM E1590 and meet the following requirements:

- Peak heat release rate of 100 kW or less.
- Total heat release in the first 10 min of the test of 25 MJ or less.

The 2018 LSC has the same requirement for new mattresses used in the same types of facilities.

13.4 Measurements and Predictions of the Burning Rate of Furniture

Marc Janssens

13.4.1 Measurement Methods

13.4.1.1 Cone Calorimeter

The cone calorimeter is the most commonly used bench-scale fire test apparatus to obtain heat release rate and related flammability data on planar materials. The apparatus and test procedure are standardized internationally as ISO 5660-1, and in the U.S. as ASTM E1354. A schematic of the apparatus is shown in Section 9.2.3.4, Figure 9.7. ASTM E1474, *Standard test method for determining the heat release rate of upholstered furniture and mattress components or composites using a bench-scale oxygen consumption calorimeter* specifies the same apparatus as ASTM E1354, but provides detailed instructions for specimen preparation and testing of furniture upholstery and mattress components or composites. ASTM E1474 test specifications are summarized in Table 13.5.

Consistent specimen preparation of furniture upholstery and mattress composites is important for achieving reproducible and representative test results. The procedure for preparing specimens in ASTM E1474 is based on that developed in the CBUF program [40]. Specimens of furniture upholstery and mattresses consist of the cover fabric, intermediate layers (if present), and the filling material(s). Specimen size is 100 × 100 × 50 mm. Fabric and intermediate layers cover the top surface and the sides of the specimen. The specimen is wrapped in aluminum foil, except for the exposed upper surface. ASTM E1474 also provides a less tedious specimen preparation procedure that can be used for screening tests.

Table 13.5 ASTM E1474 Test Specifications

Specimens	A minimum of three specimens, 100 × 100 × thickness of use (≤ 50 mm)
Specimen position	Horizontal
Ignition source	Truncated conical electrical heater producing an initial pre-set heat flux of 35 kW/m^2, electric spark ignition pilot
Test duration	20 min or until flame-out, whichever occurs first
Conclusions	This test method provides for measurements of the time to sustained flaming (s), heat release rate curve (kW/m^2), average heat release rate over the first 180 s following ignition (kW/m^2), peak heat release rate (kW/m^2), total heat released by the specimen (MJ/m^2) and average effective heat of combustion over the entire test (MJ/kg). There are no pass/fail criteria and the data are used for fire safety engineering applications.

13.4.1.2 Furniture Calorimeter

Furniture calorimeters were initially developed in the 1980s in several laboratories to measure the burning rate of chairs, sofas, and mattresses [41, 42]. They have since been used to test a wide variety of combustibles such as electrical cabinets and cables in trays, motor vehicles, commodities stacked on pallets, etc.

A furniture calorimeter consists of a weighing platform that is located on the floor of the laboratory beneath the standard hood (see Figure 13.13). The object is placed on the platform and ignited with the specified ignition source. The products of combustion are collected in the hood and extracted through the exhaust duct. Measurements of oxygen and carbon oxides concentrations, flow rate, and light transmission in the exhaust duct are used to determine the heat release rate and smoke production rate from the object as a function of time. Yields of toxic gases can be calculated based on the mass loss rate of the specimen and the flow rate and toxic gas concentrations measured in the exhaust duct (typically using Fourier Transform Infrared Spectroscopy).

The first furniture calorimeter test standard was published in 1987 in the Nordic countries as NT Fire 032. More recently, furniture calorimeter test standards have been developed in ASTM for single chairs and stacked chairs. The corresponding designations are ASTM E1537 and ASTM E1822, respectively. ASTM E1537 is functionally identical to CA TB 133 [29], which, prior to its repeal on January 22, 2019, was used in the state of California to regulate the flammability of upholstered furniture for use in public occupancies.

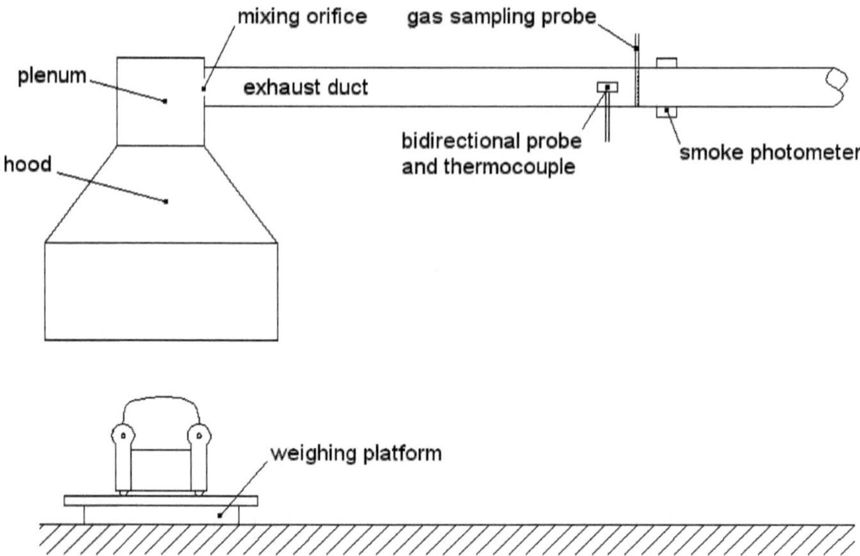

Figure 13.13 Furniture calorimeter

13.4.1.3 Room Calorimeter

The furniture calorimeter is very well ventilated, and tests have also been carried out in which furniture is burned within the ISO 9705 room (see Section 9.2) or an enclosure of comparable size, where the ventilation is restricted by the doorway. A room calorimeter is essentially a small room with an open door that is positioned beneath a hood of the type used for the furniture calorimeter. An example of a room calorimeter is the test arrangement shown in Figure 13.14.

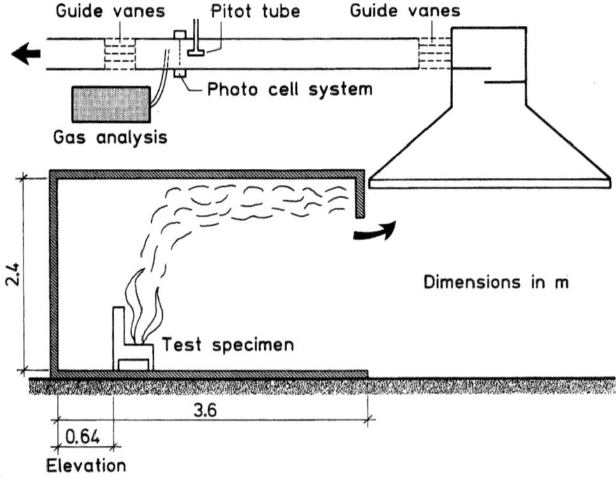

Figure 13.14 ISO 9705 test room used as a room calorimeter to test furniture

The original version of CA TB 133 specified that the test specimen be located in a corner against the back wall of a 3.7 × 3.0 × 2.4 m room with a 1.0 × 2.1 m open doorway in the front wall. The ignition source consisted of five double sheets of loosely wadded newspaper inside a chicken-wire cage placed on the back of the seat. Acceptance was based primarily on a maximum temperature rise of 111 °C just below the ceiling above the specimen. Subsequent research at NIST resulted in the development of the presently-used gas burner as an alternative ignition source [43]. The gas burner flame generates an equivalent thermal insult but is much more repeatable and reproducible than the original ignition source. The research project at NIST also resulted in an equivalent heat release rate criterion and demonstrated that the room effects are negligible for heat release rates below 600 kW [44]. Later studies found significant effects at lower heat release rates and proposed a threshold of 460 kW [45]. Because this is still much higher than peak heat release rate limits specified in regulations, ASTM E1537 permits the use of an open furniture calorimeter configuration as well as two room configurations: the original CA TB 133 room and the smaller 3.6 × 2.4 × 2.4 m room specified in the ISO 9705 and NFPA 265 room/corner tests.

A more recent literature review of data from furniture fire tests in small rooms concluded that enclosure effects are negligible below 200 kW, and that, above 400 kW, the peak heat release rate is increased by approximately 32% over the furniture calorimeter value, as shown in Figure 13.15 [46]. A Danish/Canadian study found no significant enclosure effects for a slab of flexible polyurethane foam in a small room lined with cement board, but measured a dramatic increase in the peak heat release rate (from approximately 500 kW to more than 900 kW) when the room was lined with mineral wool insulation [47].

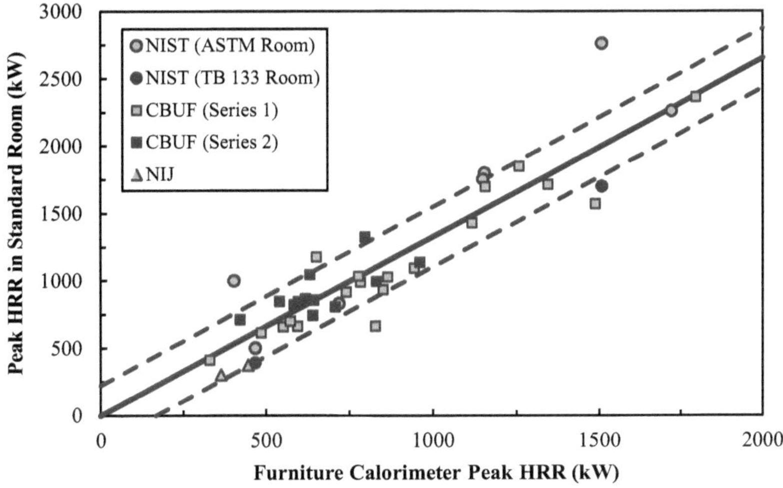

Figure 13.15 Enclosure effects based on heat release rates of upholstered furniture tested in a furniture calorimeter and a small room calorimeter
NIJ = National Institute of Justice funded project done at Southwest Research Institute

13.4.2 Model Predictions

Since the mid-1980s, several empirical correlations and models have been developed to estimate the heat release rate of upholstered furniture on the basis of small-scale fire test data of the furniture components and generic characteristics of the item. Perhaps the most comprehensive research program on flammability of upholstered furniture was conducted in Europe in the early to mid-1990s. The program was referred to as the CBUF program [1]. The CBUF program developed three models to predict the heat release rate versus time curve of upholstered furniture. The models were validated with an extensive database of furniture calorimeter measurements, and provide reasonably accurate predictions of the heat release rate of the furniture items that were tested, as shown in Figure 13.16.

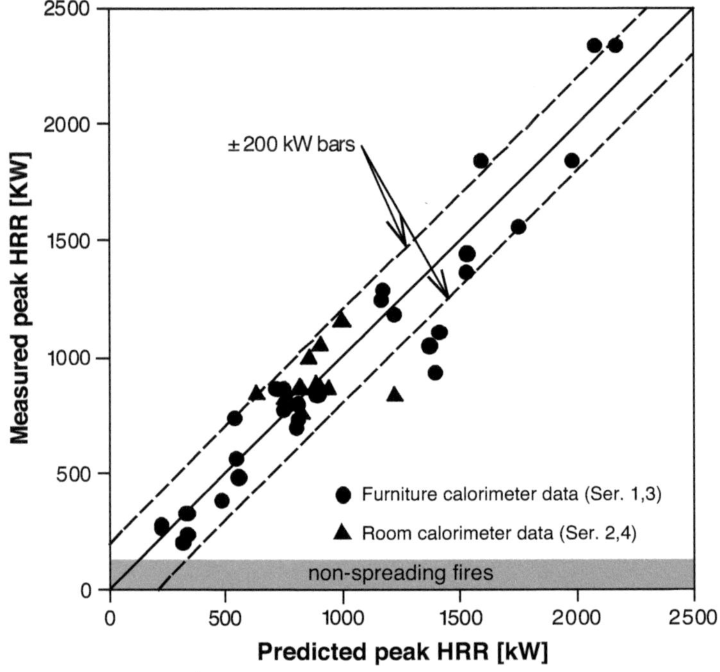

Figure 13.16 ISO 9705 test room used as a room calorimeter to test furniture

However, neither the CBUF models nor any of the other correlations and models of similar nature that have been developed account for the effect of the severity (intensity and duration) of the ignition source and the location where it is applied. For example, Figure 13.17 compares the heat release rate curve measured in the test shown in Figure 13.8 of Section 13.3.3 to the heat release rate measured in a furniture calorimeter test of the same furniture mockup with the ignition burner applied to a side cushion instead of the center cushion. As shown in Figure 13.17, the peak heat release rate for the three-seat sofa mockup ignited with the burner in the center is approximately 2.5 times the peak observed in the test with ignition of one of the side cushions (approximately 1000 kW vs. 400 kW). The CBUF Model I agrees well with the center cushion ignition test.

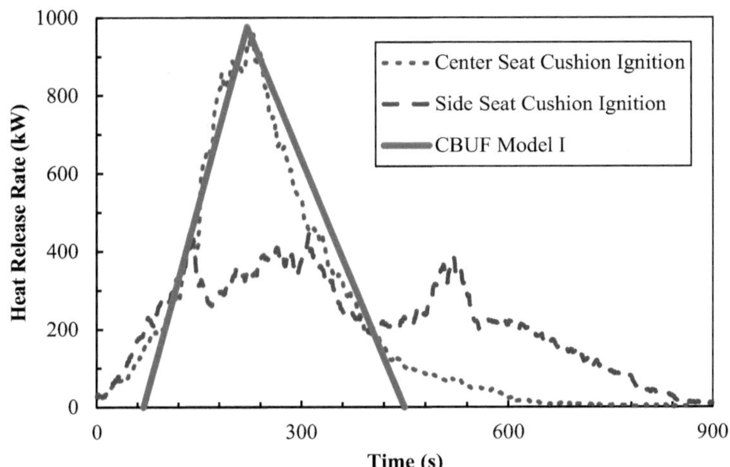

Figure 13.17 Effect of burner location on the heat release rate of a three-seat sofa mockup

A much more detailed approach to model furniture fires involves the use of the Fire Dynamics Simulator (FDS), a Computational Fluid Dynamics (CFD) code developed at NIST in Gaithersburg, MD, specifically to model fires. The FDS user's guide [48] describes an example to illustrate how the heat release rate of an upholstered chair or sofa can be calculated. FDS can account for the intensity, duration, and location of the ignition source, but underestimates the flame spread rate over a horizontal surface and therefore underestimates the heat release rate. If this problem could be solved, FDS would be the most versatile tool to predict the heat release rate of furniture fires.

13.5 Conclusions and Future Developments

For some time now, the widespread use of flame retardants to meet fire safety requirements in furniture has increasingly been questioned. Starting in the U.S., a worldwide debate about risks related to flame retardants in furniture developed. It includes the potential release of toxic products from flame-retarded furniture into dust and air in buildings. Another concern is that the use of flame retardants may lead to increasingly toxic combustion gases and post-fire contamination. This discussion has already led to some changes in Californian regulations, in which open-flame tests were eliminated (CA TB 117 in 2013, and CA TB 133 in 2019) because flame retardants had to be used to meet them (see Section 13.3).

In the UK and Ireland, similar activities to revise furniture regulations have taken place. In 2014 and 2016, the UK government published public consultations regarding a revision of the fire safety regulations for furniture. The aim of these consultations was to "further enhance safety provisions, enable newer and innovative materials to come onto the market and reduce the use of hazardous flame-retardant chemicals as a means of making furniture fire resistant - ensuring the UK continues to have the strongest product safety standards" [49].

The outcome of the latest 2019 results of the 2016 consultation was (as for the one before) the decision not to change the current legislation, but it was announced that the Office for Product Safety and Standards will work with stakeholders and businesses to share information about technological developments and alternative approaches to fire safety [50]. In Ireland, a similar public consultation has been launched and was closed in April 2020.

These consultations are part of the current discussion in the UK and Ireland about the need to reduce the use of toxic chemicals. This has to be seen in the context of an inquiry launched by the UK Environmental Audit Committee to assess the impact of toxic chemicals in everyday life on human health and the environment. The inquiry focused on how toxic chemicals are used in everyday products, such as furniture, food, and toys, on current government regulation of these substances, and on the environmental and human health problems associated with them.

This discussion is going on world-wide and there are different solutions being discussed. While some countries are hesitating to maintain existing or to introduce new flammability requirements for furniture, there are also other solutions to maintain fire safety without accepting the above-mentioned negative effects.

In some regions, strict regulations apply, defining whether flame retardants are considered as safe in use and considering emissions during fires. For Europe, the REACH approach already gives a basic framework for this approach. This leads to the application of flame retardants that do not raise concern for the health of building occupants and firefighters.

Different concepts have also been developed. Instead of flame-retardant foams, protective layers (interliners) are used for upholstery and mattresses, protecting the upholstery foam from ignition. This concept has been used for many years in cases where the impact of severe ignition sources is expected (for example in aircraft); it is now increasingly applied in other areas. In this case the concept is not to test the single furniture components (as required for example in the UK), but instead to test the fire performance of the complete piece of furniture or mattress, as the mattress tests in the UK and the USA do.

In parallel to the flame retardants discussion, research on furniture flammability is still an important topic in the USA. CPSC has been involved in research and rulemaking pertaining to furniture flammability since 1994. In 2019, Lock and Miller presented a historical overview of CPSC work in this area at the Interflam conference [26]. In this paper, which does not necessarily represent the Commission's opinion, CPSC staff made several important observations. From research conducted by CPSC and other organizations they concluded that small-scale testing conducted according to methods such as CAL TB 117-2013 and the smoldering/open-flame test proposed in 2008 [51] does not accurately predict full-scale fire behavior. Furthermore, identifying standard materials that lead to repeatable and reproducible results from small-scale component testing was found to be a major challenge. Finally, full-scale fire testing conducted by CPSC and other organizations determined that the effect of fire barriers on heat release rate was unclear and that, in most cases, barriers actually increased the likelihood of a fire developing from a smoldering ignition source. These issues will be important topics for future research.

Upholstered furniture is typically ignited by smoking materials. Initially the fire involves smoldering, which eventually transitions to flaming combustion, resulting in rapid growth in intensity due to the contribution of the padding material(s). Stoliarov et al. developed a bench-scale apparatus and test procedure to assess the propensity of a cushioning/fabric composite to transition from smoldering to flaming when exposed to a small ignition source that simulates a burning cigarette [52]. The validity of the method still needs to be established based on experiments with actual furniture items. A standard that describes the apparatus and test procedure is under development in ASTM Committee E05.

References for Chapter 13

[1] CBUF Fire safety of upholstered furniture - The final report of the CBUF research programme. Edited by B. Sundström. European Commission, Measurements and Testing, Report EUR 16477 EN, 1995.

[2] Gravigny, L., "Draft European directive on the fire behaviour of upholstered furniture", *Flame Retardants '92* [Conference], pp. 238–242. Elsevier Science Publishers Ltd, ISBN 1-85166-758-x.

[3] Directive 2001/95/EC of The European Parliament and of The Council of 3 December 2001 on general product safety.

[4] EN 1021-1:2014, Furniture. Assessment of the ignitability of upholstered furniture. Ignition source smouldering cigarette; EN 1021-2:2014, Furniture. Assessment of the ignitability of upholstered furniture. Ignition source match flame equivalent.

[5] EN 597-1:2016, Furniture - Assessment of the ignitability of mattresses and upholstered bed bases. Ignition source smouldering cigarette; EN 597-2:2015 Furniture - assessment of the ignitability of mattresses and upholstered bed bases. Ignition source match flame equivalent.

[6] Ministry of Business, Innovation & Employment (MBIE), "Product safety policy statement foam-filled furniture", 2019.

[7] Effectiveness of the Furniture and Furnishings (Fire) (Safety) Regulations 1988, Consumer Affairs Directorate, Department of Trade and Industry, London, June 2000.

[8] A statistical report to investigate the effectiveness of the Furniture and Furnishings (Fire) (Safety) Regulations 1988, Research commissioned by Consumer and Competition Policy Directorate, BIS, 2009.

[9] *https://assets.publishing.service.gov.uk/government/uploads/system/uploads/attachment_data/file/831136/detailed-analysis-fires-attended-fire-rescue-england-1819-hosb1919.pdf*

[10] Detailed analysis of fires attended by fire and rescue services, England, April 2018 to March 2019, Home Office, UK.

[11] Ahrens, M., Home fires that began with upholstered furniture, NFPA Research, 2017.

[12] The Upholstered Furniture (Safety) Regulations, 1980, and (Amendment) Regulations 1983. Department of Trade and Industry, London, UK.

[13] Furniture and Furnishings (Fire Safety) Regulations 1988/1989, 1993 and 2010 and The Statutory Instrument 1988 No. 1324, the Statutory Instrument 1989 No. 2358, the Statutory Instrument 1993 No. 207 and Statutory Instrument 2010 No. 2205.

[14] BS 5852-1:1979, Fire tests for furniture. Methods of test for the ignitability by smokers' materials of upholstered composites for seating.

[15] BS 5852:Part 2:1982, Fire tests for furniture. Methods of test for the ignitability of upholstered composites for seating by flaming sources.

[16] BS 6807:2006, Assessment of the ignitability of mattresses, divans and bed bases with primary and secondary ignition sources.

[17] S.I. No. 316/1995 – Industrial Research and Standards (Fire Safety) (Domestic Furniture) Order, 1995 and Irish Standard IS 419:2011.

[18] *https://www.firesafe.org.uk/regulatory-reform-fire-safety-order-2005/.*

[19] BS 7176:2007+A1:2011, British Standard Specification for resistance to ignition of upholstered furniture for non-domestic seating by testing composites.

[20] BS 7177+ A1:2011, British Standard Specification for resistance to ignition of mattresses, divans and bed bases.

[21] BS 7175:1989, Method of test for the ignitability of bedcovers and pillows by smouldering and flaming ignition sources.

[22] BS EN ISO 12952: Parts 1 and 2, 2010, Parts 3 and 4: 1999. Textiles – Burning behaviour of bedding items.

[23] BS 5852:2006, Methods of test for assessment of the ignitability of upholstered seating by smouldering and flaming ignition sources.

[24] *https://www.govinfo.gov/app/collection/cfr.*

[25] *https://www.cpsc.gov/s3fs-public/pdfs/blk_pdf_furnflamm.pdf.*

[26] Lock, A. and Miller, D., "Progress and status of the U.S. consumer product safety commission upholstered furniture flammability project", *Interflam 2019* (15th International Conference and Exhibition on Fire Science and Engineering), Royal Holloway College, London, UK, July 1–3, 2019, pp. 1223–1232.

[27] *https://bhgs.dca.ca.gov/about_us/tb117_2013.pdf.*

[28] *https://ufac.org/technical-specifications/.*

[29] *https://bhgs.dca.ca.gov/industry/tb133.pdf.*

[30] *https://codes.iccsafe.org/.*

[31] *https://www.nfpa.org/Codes-and-Standards/All-Codes-and-Standards/Free-access.*

[32] *https://www.astm.org/.*

[33] *https://www.cpsc.gov/s3fs-public/pdfs/blk_media_testmatt.pdf.*

[34] *https://www.astm.org/Standards/E1537.htm.*

[35] *https://www.govinfo.gov/content/pkg/CFR-2012-title16-vol2/pdf/CFR-2012-title16-vol2-part1633.pdf.*

[36] Ohlemiller, T., Shields, T., McLane, A., and Gann, R., "Flammability assessment methodology for mattresses", NIST, Gaithersburg, MD, NISTIR 6497, 2000.

[37] Ohlemiller, T. and Gann, R., "Effect of bed clothes modifications on fire performance of bed assemblies", NIST, Gaithersburg, MD, NIST Technical Note 1449, 2003.

[38] Ohlemiller, T., "Flammability tests of full-scale mattresses: Gas burners versus burning bedclothes", NIST, Gaithersburg, MD, NISTIR 7006, 2003.

[39] *https://www.cpsc.gov/s3fs-public/pdfs/blk_media_labmanual.pdf.*

[40] Babrauskas, V. and Wetterlund, I., "Fire testing of furniture in the cone calorimeter–The CBUF test protocol", SP Report 1994:32, Research Institutes of Sweden (RISE, formerly SP Technical Research Institute of Sweden), Borås, Sweden, 1994.

[41] Babrauskas, V., Lawson, J., Walton, W., and Twilley, W., "Upholstered furniture heat release rates measured with a furniture calorimeter", National Bureau of Standards, Gaithersburg, MD, NBSIR 82-2604, 1982.

[42] Ames, S. and Rogers, S., "Large and small scale fire calorimetry assessment of upholstered furniture", *5th Interflam Conference,* Canterbury, UK, 1990, pp. 221–232.

[43] Ohlemiller, T. and Villa, K., "An investigation of the California Technical Bulletin 133 Test, Part II: Characteristics of the ignition source and a comparable gas burner", National Bureau of Standards, Gaithersburg, MD, NBSIR 90-4348, 1990.

[44] Parker, W., Tu, K., Nurbakhsh, S., and Damant, G., "An investigation of the California Technical Bulletin 133 Test, Part III: Full scale chair burns", National Bureau of Standards, Gaithersburg, MD, NBSIR 90-4375, 1990.

[45] Krasny, J. and Parker, W., "Impact of the room enclosure on the peak heat release rates of upholstered furniture", *4th Fire and Materials Conference,* Crystal City, VA, 1995, pp. 181–190.

[46] Hirschler, M. and Janssens, M., "Effect of enclosure on heat release rate of furniture fires", *Fire and Materials '13* (13th International Conference), San Francisco, CA, 2013, pp. 343–354.

[47] Poulsen, A., Bwalya, A., and Jomaas, G., "Experimental study on the influence of thermal feedback on the burning behavior of flexible polyurethane", *13th Interflam Conference,* London, UK, 2013, pp. 669–674.

[48] McGrattan, K., Hostikka, S., McDermott, R., and Vanella, M., "Fire dynamics simulator user's guide", NIST, Gaithersburg, MD, NIST Special Publication 1019, Sixth Edition, 2019.

[49] *https://www.gov.uk/government/consultations/furniture-and-furnishing-fire-safety-regulations-proposed-changes-2016.*

[50] *https://assets.publishing.service.gov.uk/government/uploads/system/uploads/attachment_data/file/822072/furniture-fire-regulations-2016-consultation-government-response-july-2019.pdf.*

[51] *https://www.cpsc.gov/s3fs-public/pdfs/blk_pdf_furnflamm.pdf.*

[52] Stoliarov, S., Zeller, O., Morgan, A., and Levchik, S., "An experimental setup for observation of smoldering-to-flaming transition on flexible foam/fabric assemblies", *Fire and Materials,* Vol. 42, pp. 128–133, 2018, *https://doi.org/10.1002/fam.2464.*

Part III
Appendices

Appendix 1 – Suppliers of Flame Retardants and Smoke Suppressants

Manufacturers and suppliers	Products
Adeka Corporation 7-2-35 Higashiogu, Arakawaku 116-8554 Tokyo, Japan *https://www.adeka.co.jp/en/chemical*	Phosphorus and P/N compounds Intumescent systems Brominated compounds
and	
Adeka Polymer Additives Europe SAS 13 rue du 17 Novembre 68100 Mulhouse, France *www.adeka-pa.eu*	Phosphorus and P/N compounds Intumescent systems Brominated compounds
Albemarle Corporation 451 Florida Blvd. Baton Rouge, LA 70801, USA *https://www.albemarle.com/businesses/bromine-specialties/fire-safety-solutions*	Brominated compounds
BASF SE 67056 Ludwigshafen, Germany *https://plastics-rubber.basf.com/global/en/plastic_additives/special_effect/flameretardancy.html*	Melamine cyanurate and derivatives N-alkoxy hindered amines (NOR HALS)
Campine n. v. Nijverheidsstraat 2 B-2340 Beerse, Belgium *https://www.campine.com/en/specialty-chemicals/flame-retardant-solutions/*	Antimony trioxide, masterbatches
Chemische Fabrik Budenheim Rheinstrasse 27 D-55257 Budenheim, Germany *https://www.budenheim.com/en/solutions/flame-protection/*	Phosphorus and nitrogen compounds Ammonium polyphosphate Melamine cyanurate

Manufacturers and suppliers	Products
Clariant Plastics & Coatings (Deutschland) GmbH, Germany *www.clariant.com/en/Business-Units/Additives/Flame-Retardants*	Organic phosphorus compounds, phosphinates Red phosphorus, ammonium polyphosphate
Daihachi Chemical Industry Co. Ltd. Hiranomachi Yachiyo Bldg. 8-13 Hiranomachi 1-chome, Chuo-ku Osaka 541-0046, Japan *https://www.daihachi-chem.co.jp/eng/technology/flame_retardant.html*	Halogenated and halogen-free phosphate esters Phosphorus compounds
FRX Polymers, Inc. 200 Turnpike Road Chelmsford, MA 01824, USA *https://www.frxpolymers.com*	Polyphosphonates
Grafitbergbau Kaisersberg GmbH Bergmannstraße 39 A-8713 St. Stefan ob Leoben, Austria *https://www.grafit.at/en/homepage/*	Expandable graphite
ICL-IP Maklef Building, Kreutzer Street 1 Beer Sheva 84101, Israel *http://www.icl-ip.com/flame-retardants/*	Brominated compounds Phosphorus compounds Magnesium hydroxide
Italmatch Chemicals S.p.A. Via Magazzini del Cotone, 17, Modulo 4 Genova (GE) 16128, Italy *https://www.italmatch.com/home/product-lines/flame-retardants-and-plastic-additives/*	Phosphorus compounds Red phosphorus
J.M. Huber Engineered Materials 4401 Northside Parkway, Suite 600 Atlanta, GA 30327, USA *www.hubermaterials.com*	Aluminum trihydrate Magnesium hydroxide Molybdate smoke suppressants
Kisuma Chemicals BV Billitonweg 7 9641 KZ Veendam, The Netherlands *https://www.kisuma.com*	Magnesium hydroxide
and	
Kyowa Chemical Ind. Co., Ltd. 4035, Hayashida-machi Sakaide-shi, Kagawa 762, Japan *https://kyowa-chem.jp/en/*	Magnesium hydroxide

Manufacturers and suppliers	Products
Lanxess Deutschland GmbH Kennedyplatz 1 50569 Köln, Germany *https://flameretardants.lanxess.com*	Phosphorus compounds Brominated compounds
Martin Marietta Magnesia Specialties, LLC 8140 Corporate Drive, Suite 220 Baltimore, MD 21236, USA *https://magnesiaspecialties.com*	Magnesium hydroxide
Martinswerk GmbH, a subsidiary of J. M. Huber Engineered Materials Kölnerstrasse 110 D-50127 Bergheim, Germany *https://www.martinswerk.de*	Aluminum trihydrate Magnesium hydroxide
Metadynea Hafenstraße 77 3500 Krems, Austria *www.metadynea.com*	Phosphorus compounds, DOPO
Nabaltec GmbH Alustrasse 50-52 D-92421 Schwandorf, Germany *https://nabaltec.de/en*	Aluminum trihydrate Aluminum monohydrate (Boehmit) Magnesium hydroxide
Nordmann, Rassmann GmbH Kajen 2 D-20459 Hamburg, Germany *https://nordmann.global/en*	Phosphorus compounds, zinc borate Nitrogen compounds, expandable graphite
Oceanchem Group Limited 9th Floor, Future Plaza No.88 Fenghuang Street, Fangzi District Weifang, Shandong 261200, China *http://www.oceanchem-group.com*	Brominated compounds Phosphorus compounds, DOPO Zinc borate Zinc stannate, zinc hydroxy stannate Smoke suppressants
Otsuka Chemical Co., Ltd. 3-2-27, Ote-Dori, Chuo-Ku Osaka 540-0021, Japan *https://www.otsukac.co.jp/en/*	Phosphazenes
PCC Rokita, PCC Group 56-120 Brzeg Dolny Sienkiewicza 4, Poland *https://en.pcc.rokita.pl*	Phosphorus compounds Phosphonates

Manufacturers and suppliers	Products
Rinkagaku Kogyo Co. Ltd. 34 Shinbori, Imizu-shi, Toyama 934-8534, Japan *http://www.rinka.co.jp/english*	Red phosphorus
Schill + Seilacher GmbH Schoenaicher Str. 205 D-71032 Boeblingen, Germany *https://www.schillseilacher.de/en/*	Phosphorus compounds Phosphonates Phosphorus and nitrogen compounds
Shandong Moris Technology Co., Ltd. *http://www.techmoris.com/en*	Brominated compounds Phosphorus compounds
Shouguang Weidong Chemical Co.,Ltd *http://www.wdchem.com/en/*	Brominated compounds Phosphorus compounds Nitrogen compounds
Thor GmbH PO Box 19 09 67329 Speyer, Germany *www.thor.com*	Phosphorus and nitrogen compounds
Tolsa Group 2da Planta Edificio 4 P. E. Las Mercedes, Calle Campezo 1 28022 Madrid, Spain *http://adins.tolsa.com* *https://www.tolsa.com/en*	Clays as synergists
Tor Minerals International 722 Burleson Street Corpus Christi, Texas 78402, USA *https://torminerals.com/products-services/flame-retardants*	Aluminum trihydrate Aluminum monohydrate (Boehmite)
Tosoh Corporation 3-8-2, Shiba, Minato-ku Tokyo 105-8623, Japan *www.tosoh.com*	Brominated compounds
Trovotech GmbH Chemiepark Bitterfeld-Wolfen, Areal A Edisonstrasse 3 06766 Bitterfeld-Wolfen, Germany *https://www.trovotech.de/index.php/en/*	Inorganic Si-based synergists

Manufacturers and suppliers	Products
U.S. Borax Inc. 14486 Borax Road Boron, CA 93516, USA *https://www.borax.com/products/firebrake*	Zinc borate
William Blythe Ltd Church, Accrington Lancashire, BB5 4PD, UK *www.williamblythe.com*	Zinc stannate, zinc hydroxy stannate Smoke suppressants
Zhejiang Wansheng Co., Ltd. No. 8 Rd. Jujing Liangshui Development Zone Linhai, Zhejiang, 317000, China *http://www.ws-chem.com/en/index.aspx*	Phosphorus compounds

Appendix 2 – Abbreviations for Plastics and Flame Retardants

Abbreviations for Plastics

Thermoplastics

ABS	acrylonitrile-butadiene-styrene terpolymer
ASA	acrylonitrile-styrene-acrylate
COC	polycycloolefin
EMA	ethylene-methylacrylate copolymer
EPS	expanded polystyrene
ETFE	ethylene-tetrafluoroethylene copolymer
FEP	perfluoroethylene-propylene
HIPS	high-impact polystyrene
LCP	liquid crystal polymer
PA	polyamide
PAA	poly(acrylic acid)
PA-6	polyamide-6
PAI	polyamideimide
PAN	polyacrylonitrile
PBD	polybutadiene
PBI	polybenzimidazole
PBO	poly(*p*-phenylene-2,6-benzobisoxazole)
PBS	polybutylene succinate
PBT	polybutylene terephthalate
PBZT	polybenzothiazole
PC	polycarbonate
PC/ABS	polycarbonate/ABS blend
PE	polyethylene

PEEK	poly(ether ether ketone)
PE-HD (HDPE)	high density polyethylene
PEK	poly(ether ketone)
PE-LD (LDPE)	low-density polyethylene
PE-LLD (LLDPE)	linear low-density polyethylene
PES	polyether sulfone
PET	polyethylene terephthalate
PI	polyimide
PIP	polyisoprene
PIPD	poly(2,6-diimidazol[4,5-b: 4',5'-e]pyridinylene-1,4(2,5-dihydroxy)phenylene)
PLA	polylactide
PMAA	poly(methacrylic acid)
PMA	polymethacrylonitrile
PMMA	poly(methyl methacrylate)
POM	polyoxymethylene
PP	polypropylene
PPE	polyphenylene ether
PPE/HIPS	polyphenylene ether/high-impact polystyrene blend
PPO	polyphenylene oxide
PPS	polyphenylene sulfide
PPTA	*p*-phenylene terephthalamide (*p*-aramid)
PS	polystyrene
PSU	polysulfone
PTFE	polytetrafluoroethylene
PVC	polyvinyl chloride
PVDC	polyvinylidene chloride
PVDF	polyvinylidene fluoride
SAN	styrene-acrylonitrile copolymer
XPS	extruded polystyrene

Thermosets

DGEBA	diglycidyl ether of bisphenol A
EP	epoxy resin
MF	melamine formaldehyde resin
PF	phenol formaldehyde resin
PIR	polyisocyanurate
PUR	polyurethane
TGDDM	tetraglycidyl diaminodiphenylmethane
UF	urea formaldehyde resin
UP	unsaturated polyester resin

Inorganic Resins

PCS	polycarbosilane copolymer
PDMS	polydimethylsiloxane
POSS	polyhedral oligomeric silsesquioxane
PSS	polysilsesquioxane
VTS	vinyltriethoxysilane

Elastomers/Thermosets

ACM	acrylic rubber
AFMU	carboxy-fluoro-nitroso rubber
AU/EU	polyurethane
BIIR	brominated isobutylene isoprene rubber (bromobutyl rubber)
BR	1,4-polybutadiene rubber
CIIR	chlorinated isobutylene isoprene rubber (chlorobutyl rubber)
CM	chlorinated polyethylene
CO (ECO)	epichlorohydrin rubber
CR	neoprene rubber (polychloroprene)
CSM	chlorosulfonated polyethylene
EAM	ethylene-acrylate rubber
EPM (EPDM)	ethylene-propylene rubber
EVM (EVA)	ethylene-vinyl acetate rubber
FPM	fluorocarbon rubber
FFKM	perfluoro rubber

FMQ (FVMQ)	fluoromethyl(vinyl)silicone rubber
FZ	polyfluoroalkoxyphosphazene rubber
GPO	propyleneoxide rubber
HNBR	hydrogenated acrylonitrile-butadiene rubber
IIR	isobutylene isoprene rubber (butyl rubber)
IR	1,4-polyisoprene rubber
MQ (VMQ)	methyl(vinyl)silicone rubber
NBR	acrylonitrile-butadiene rubber
NR	natural rubber
PNR	polynorbornene rubber
PZ	polyaryloxyphosphazene rubber
SBR	styrene-butadiene rubber
TM(T)	a polysulfide rubber
XNBR	carboxylated acrylonitrile-butadiene rubber

Elastomers/Thermoplastics

CPE	polyether-ester copolymer
EVA/VC	EVA/chlorinated polyolefin blend
MPR	melt-processible rubber
NBR/PVC	NBR/PVC blend
PEBA	polyether-amide copolymer
PP/NR-VD	polypropylene/cross-linked NR blend
PP/EPDM-VD	polypropylene/cross-linked EPDM blend
PP/IIR-VD	polypropylene/cross-linked IIR blend
PP/NBR-VD	polypropylene/cross-linked NBR blend
PUR	polyurethane
SBS	styrene-butadiene-styrene block copolymer
SEBS	styrene-ethylene-butylene-styrene block copolymer
SIS	styrene-isoprene-styrene block copolymer
TPO	thermoplastic polyolefin elastomer
TPU	thermoplastic polyurethane elastomer
TPV	thermoplastic vulcanizate

Abbreviations for Flame Retardants

AlO(OH)	boehmite
AlPi	aluminum diethylphosphinate
APH	aluminum hypophosphite
APP	ammonium polyphosphate
ATH	aluminum trihydrate (aluminum hydroxide)
ATO	antimony trioxide
BDP	bisphenol A bis(diphenyl phosphate)
BEO	brominated epoxy resins
BrPBPS	polystyrene–brominated polybutadiene–polystyrene block copolymer
BrPol	brominated polyols
BrPS	brominated polystyrene
CNT	carbon nanotube
CP	chloroparaffin
DBDPO	decabromodiphenyl oxide
Deca	decabromodiphenyl ether
DEPAL	diethyl aluminum phosphinate
Dicumyl	2,3-dimethyl-2,3-diphenylbutane
DICUP	dicumyl peroxide
DMMP	dimethyl methylphosphonate
DMPP	dimethyl propylphosphonate
DOPO	9,10-dihydro-9-oxa-10-phosphaphenanthrene-10-oxide
DOPO-HQ	2-(6-oxido-6*H*-dibenz[c,e][1,2]oxaphophorin-6-yl)-1,4-benzenediol
EBTPI	ethylene bis(tetrabromophthalimide)
EG	expanded graphite
HBCD (HBCDD)	hexabromocyclododecane
HET	hexachloro-*endo*-methylenetetrahydrophthalic acid (chlorendic acid)
LDH	layered double hydroxide
MC	melamine cyanurate
MDH	magnesium dihydroxide (magnesium hydroxide)
Mel	melamine

MPP	melamine polyphosphate
MWNT	multi-walled nanotube
NOR	*N*-alkoxy-hindered amine
P red	red phosphorus
PBB-PA	poly(pentabromobenzyl acrylate)
PER	pentaerythritol
RDP	resorcinol bis(diphenyl phosphate)
RXP	biphosphate resorcinol bis(dixylenyl phosphate)
SWNT	single-walled nanotube
TBBA (TBBPA)	tetrabromobisphenol A
TBNPP	tris(bromoneopentyl) phosphate
TCEP	tris(chloroethyl) phosphate
TCPP	tris(chloropropyl) phosphate
TDCP	tris(dichloropropyl) phosphate
THP	tetrakis(hydroxymethyl)phosphonium salts
TPP	triphenyl phosphate
ZB	zinc borate
ZHS	zinc hydroxystannate
ZS	zinc stannate

Appendix 3 – Flame Retardants: Chemical Names and CAS Numbers

Rudolf Pfaendner

Chemical Name	CAS Number
Alkoxyamines, i. e., 1,3,5-Triazine-2,4,6-triamine, *N*2-[3-[[4,6-bis[butyl[1-(cyclohexyloxy)-2,2,6,6-tetramethyl-4-piperidinyl]amino]-1,3,5-triazin-2-yl]amino]propyl]-*N*2-[2-[[3-[[4,6-bis[butyl[1-(-cyclohexyloxy)-2,2,6,6-tetramethyl-4-piperidinyl]amino]-1,3,5-triazin-2-yl]amino]propyl]amino]ethyl]-*N*4,*N*6-dibutyl-*N*4,*N*6-bis[1-(cyclohexyloxy)-2,2,6,6-tetramethyl-4-piperidinyl]-	[122587-07-9]
1,3,5-Triazine-2,4,6-triamine, *N*2,*N*2'-1,2-ethanediylbis[*N*2-[3-[[4,6-bis[butyl[1-(cyclohexyloxy)-2,2,6,6-tetramethyl-4-piperidinyl]amino]-1,3,5-triazin-2-yl]amino]propyl]-*N*4,*N*6-dibutyl-*N*4,*N*6-bis[1-(cyclohexyloxy)-2,2,6,6-tetramethyl-4-piperidinyl]]-	[217084-73-6]
Aluminum diethyl phosphinate	[225789-38-8]
Aluminum hypophosphite	[7784-22-7]
Aluminum methylmethyl phosphonate	[35851-62-8]
Aluminum polymelaminephosphate	[1111732-49-0]
Ammonium polyphosphate, APP	[68333-79-9]
Ammonium sulfamate	[7773-06-0]
Antimony oxide, ATO	[1309-64-4]
Bis(2,3-dibromopropylether) of tetrabromobisphenol-A	[21850-44-2]
4,4'-Bishydroxydeoxybenzoin-polyphosphonate	[911127-04-3]
Bisphenol-A-bis(dicresyl phosphate)	[93981-32-9]
Bisphenol-A-bis(diphenyl phosphate)	[5945-33-5]
Boehmite, AlO(OH)	[1318-23-6]
Brominated carbonate oligomers	[94334-64-2, 71342-77-3]
Brominated epoxy polymers	[68928-70-1]
Brominated epoxy oligomers, end-capped	[135229-48-0]
Brominated polyphenylene ether	[1367795-87-6]
Brominated polystyrene	[88497-56-7]
Calcium borate	[13840-55-6, 12007-56-6]

Chemical Name	CAS Number
Calcium carbonate	[471-34-1]
2-Carboxyethyl methyl phosphinic acid	[15090-23-0]
2-Carboxyethyl phenyl phosphinic acid	[14657-64-8]
Chitosan	[9012-76-4]
Condensate of *N,N'*-bis(2-aminoethyl)phenyl phosphorodiamidate with a piperidine-substituted triazine	[1643591-76-7]
Condensate from tris(2-hydroxyethyl)isocyanurate with condensate from tris(2-hydroxyethyl)isocyanurate and spirocyclic pentaerythritoldiphosphoryl dichloride	[714-87-4]
Decabromodiphenyl ethane	[84852-53-9]
Decabromodiphenyl ether	[1163-19-5]
Dibromoneopentyl glycol	[3296-90-0]
Dicumyl(2,3-dimethyl-2,3-diphenyl butane)	[1889-67-4]
Dicumylperoxide	[80-43-3]
Diethyl-*N,N'*bis(2-hydroxyethyl) aminomethylphosphonate	[2781-11-5]
Diglycidylether of tetrabromo bisphenol A	[3072-84-2]
2,3-Dimethyl-2,3-diphenylbutane	[1889-67-4]
Dimethyl propyl phosphonate	[18755-43-6]
Dimethylspirophosphonate	[3001-98-7]
Dipentaerythritol	[126-58-9]
Diphenylcresyl phosphate	[26444-49-5]
DOPO reaction product with itaconic acid, i.e., butanedioic acid, 2-[(6-oxido-6*H*-dibenz[c,e][1,2]oxaphosphorin-6-yl) methyl]-	[63562-33-4]
DOPO zinc derivative, i.e., 6*H*-dibenz[c,e][1,2]oxaphosphorin, 6-hydroxy-, 6-oxide, zinc salt (2:1)	[69151-14-0]
Ethylene bis(tetrabromophthalimide)	[32588-76-4]
Ethylene diamine phosphate salt	[52657-34-8]
Expandable or expanded graphite	[7782-42-5]
Glycerol tri(DOPO-acrylate)	[1947313-19-0]
Graphene	[1034343-98-0]
Halloysite	[12298-43-0, 12068-18-0]
Hexabromocyclododecane, HBCD	[3194-55-6]
Huntite, $Mg_3Ca(CO_3)_4$	[19569-21-2]
Hydromagnesite, $Mg_5(CO_3)_4(OH)_2 \cdot 4H_2O$	[12275-04-6, 12072-90-1]
Hydroquinone bis(diphenyl phosphate)	[51732-57-1]
Lignin	[9005-53-2]
Magnesium dihydroxide, MDH	[1309-42-8]
Magnesium poly melamine phosphate	[1621136-11-5]

Chemical Name	CAS Number
Melamine	[108-78-1]
Melamine cyanurate	[37640-57-6]
Melamine hypophosphite	[1609368-04-8]
Melamine phosphate	[20208-95-1]
Melamine phosphite	[120505-16-0]
Melamine polyphosphate	[41583-09-9]
Mica	[12001-26-2]
Molybdic oxide	[7782-91-4]
Montmorillonite	[1318-93-0]
2-(6-Oxido-6*H*-dibenz[c,e][1,2]oxaphophorin-6-yl)-1,4-benzenediol (DOPO-HQ)	[99208-50-1]
1-Oxo-2,6,7-trioxa-1-phosphabicyclo-[2.2.2]-octane-methyl diallyl phosphate	[1453115-53-1]
Pentaerythritol, PER	[115-77-5]
Phosphaphenanthrene oxide, i.e., 6*H*-dibenz[c,e][1,2] oxaphosphorin, 6-oxide, DOPO	[35948-25-5]
Phosphonium sulfonates	[106503-53-1]
Phytic acid	[83-86-3]
Piperazine pyrophosphate	[52492-62-3]
Polycondensation products with ethylene glycol and phosphoric acid ester	[184538-58-7]
Poly(pentabromobenzyl) acrylate	[59447-57-3]
Poly(1,3-phenylene methylphosphonate)	[850072-32-1]
Polyphosphonates	[68664-06-2]
Polystyrene-brominated butadiene-polystyrene block copolymer	[694491-73-1DP], [1195978-93-8]
Polytetrafluorethylene, PTFE	[9002-84-0]
Poly(triazinyl, morpholonyl)triazine	[1078142-02-5]
Poly(1,3,5-triazine-2-aminoethanol) diethylenetriamine	[1352945-31-3]
Potassium perfluorobutane sulfonate	[29420-49-3]
Potassium-*p*-toluene sulfonate	[16106-44-8]
Potassium diphenylsulfone sulfonate	[63316-43-8]
Red phosphorus	[7723-14-0]
Resorcinol bis(di-2,6-xylylphosphate)	[139189-30-3]
Sepiolite	[63800-37-3]
Silica	[7631-86-9]
Tetrabromobisphenol A	[CA 79-94-7]
Tetrabromobisphenol-A-bis(2,3-dibromopropylether)	[21850-44-2]
Tetrabromophthalic acid derivatives	[77098-07-8]

Chemical Name	CAS Number
1,3,5,7-Tetrakis(phenyl-4-sodium sulfonate) adamantane	[1346770-66-8]
Triazine derivative with morpholine groups, i. e., 1,3,5-triazine, 2,4,6-trichloro-, polymer with piperazine, reaction products with morpholine	[1078142-02-5]
Triphenyl phosphate	[101-02-0]
Tris(1-chloro-2-propyl) phosphate	[13674-84-5]
Tris(2-chloroethyl) phosphate	[115-96-8]
1,3,5-Tris(2,3-dibromopropyl) isocyanurate	[52434-90-9]
Tris(hydroxyethyl)isocyanurate	[839-90-7]
Tris(tribromoneopentyl) alcohol	[36483-57-5]
Tris(tribromoneopentyl) phosphate	[19186-97-1]
4,6-Tris(2,4,6-tribromophenoxy)-1,3,5-triazine	[25713-60-4]
Vermiculite	[1318-00-9]
Wollastonite	[13983-17-0]
Zinc borate	[1332-07-6]
Zinc diethyl phosphinate	[284685-45-6]
Zinc hydroxystannate	[21172-59-8]

Appendix 4 – Journals and Books

The following list contains only those journals and monographs devoted exclusively to fire protection. No claim is made regarding completeness, and sources concerned with fire fighting and the fire services have been disregarded. For relevant papers in plastics journals and pertinent sections in monographs on plastics, the reader should consult the references at the end of each section in this book.

Journals

Combustion and Flame

https://www.journals.elsevier.com/combustion-and-flame

Fire and Materials

https://onlinelibrary.wiley.com/journal/10991018

Fire Safety Journal

https://www.journals.elsevier.com/fire-safety-journal

Fire Technology

https://www.springer.com/journal/10694

Journal of Fire Sciences

https://journals.sagepub.com/home/jfs

Books

Babrauskas, V., *Ignition Handbook.* (2003) Fire Science Publishers, Issaquah, WA.

Babrauskas, V.; Gann, R.; Grayson, S. (Eds.), *Hazards of Combustion Products.* (2008) Interscience Communications Ltd., London, UK.

Cullis, C. F.; Hirschler, M. M., *The Combustion of Organic Polymers.* (1981) Clarendon Press, Oxford, UK.

Grand, A. F.; Wilkie, C. A. (Eds.), *Fire Retardancy of Polymeric Materials.* (2000) Marcel Dekker, Inc., New York, NY.

Horrocks, A. R.; Price, D. (Eds.), *Fire Retardant Materials.* (2001) CRC Press, Boca Raton, FL.

Horrocks, A. R.; Price, D. (Eds.), *Advances in Fire Retardant Materials.* (2008) Woodhead Publishing Limited, Cambridge, UK.

Kuryla, W. C.; Papa, A. J. (Eds.), *Flame Retardancy of Polymeric Materials,* five volumes (1973–1979). Marcel Dekker, Inc., New York, NY.

Le Bras, M.; Camino, G.; Bourbigot, S.; Delobel, R., *Fire Retardancy of Polymers. The Use of Intumescence.* (1998) The Royal Society of Chemistry, Cambridge, UK.

Lewin, M.; Atlas, S. M.; Pearce, E. M. (Eds.), *Flame Retardant Polymeric Materials,* three volumes (1975–1982). Plenum Press, New York, NY.

Morgan, A. B.; Wilkie, C. A., *Flame Retardant Polymer Nanocomposites.* (2007) John Wiley & Sons, Inc.

Nelson, G.; Wilkie, C. A. (Eds.), *Fire and Polymers.* (2001) Oxford University Press, UK.

Quintiere, J. G., *Fundamentals of Fire Phenomena.* (2006) Wiley, Chichester, UK.

Weil, E. D.; Levchik, S. V., *Flame Retardants for Plastics and Textiles,* Second Edition. (2016) Hanser, Munich.

Wilkie, C. A.; Morgan, A. B., *Fire Retardancy of Polymeric Materials,* Second Edition. (2009) CRC Press, Boca Raton, FL.

Authors Index

Only the authors mentioned in the text are listed here. Most of them can be found in the respective references of the individual chapters and sections.

Babrauskas 170, 273
Backer 133
Benisek 145
Blockley and Speitel 227
Bourbigot 81
Butcher and Parnell 169
Camino 60, 125 f., 128
Crane 237
Diamantes 294
Döring 75
Ellison 75
Fenimore 64
Garvey 133
Hall 8
Hartzell 237
Hendrix 133
Horrocks 145
Iji 80
Jin 169, 170
Jin and Yamada 170
Kaplan 237
Kaprinidis 73
Kashiwagi 61
Lattimer 578
Lock 757
Malhotra 169
McKinnon 578
Miller 133, 757
Nachtigall 57
Nakhmanovich 75
Ning 79
Pauluhn 225, 227, 236, 237
Purser 237, 240
Rasbash 163, 169
Reyniers 75
Schartel 76
Speitel 237
Stoliarov 757
Sweeney 237, 243
Troitzsch 8
van Krevelen 32, 148
Wilén 73 f.
Wilkie 60

Standards Index

14 CFR 25 Appendix F, Part I **577**
16 CFR 1500.3 **308**
16 CFR 1500.44 **300, 308**
16 CFR 1630 **300, 307**
16 CFR 1632 **302, 740**
16 CFR 1633 **740, 746, 748, 750**
16 CFR Part 1632 **738**
16 CFR Part 1633 **738**
49 CFR Part 238 **555, 570, 572, 573**
95/28/EC **542**
ABD0031 **639**
Airbus Company Directive ABD0031 **637**
Airbus standard AITM 3-0005 **639**
AS 1366 **533**
AS 1530.1 **519, 520, 528, 532**
AS 1530.2 **519, 521, 528, 529, 532, 534**
AS 1530.4 **519, 520, 528, 529, 608**
AS 2122.1 **528**
AS 3837 **525, 608**
AS 3959 **519, 526, 529**
AS 4072.1 **520, 528, 529**
AS 5113 **526**
AS 5637.1 **519, 525, 529**
AS 7529 **555, 607–609**
AS ISO 9239.1 **519**
AS/NZS 1530.3 **519, 523, 528, 529, 532**
AS/NZS 3013 **520, 528–530**
AS/NZS ISO 31000 **607**
ASTM 2387 **329**
ASTM C1166 **572, 573**
ASTM D495 **684**
ASTM D635 **290, 300, 303, 304**
ASTM D1929 **300, 304, 305, 307, 323**
ASTM D2843 **301, 309, 310, 456, 495**
ASTM D2859 **300, 306**
ASTM D3014 **496, 528**
ASTM D3638-12 **684**
ASTM D3675 **571, 572, 575**
ASTM D5424 **163, 172**
ASTM D6413 **136**
ASTM E84 **165, 291, 301, 315, 317, 318, 337, 340, 341**
ASTM E108 **303, 329–331, 333, 337**
ASTM E119 **302–303, 325–329, 333, 519, 571, 572**
ASTM E136 **289, 300, 307, 308, 337**
ASTM E162 **571–575, 682, 686, 703**
ASTM E648 **301, 310–312, 528, 571–573**
ASTM E662 **165, 571, 572, 575, 576, 588, 601, 603–606, 637, 638, 645**
ASTM E814 **302, 303, 327, 328**
ASTM E970 **301, 312**
ASTM E1353 **744**
ASTM E1354 **201, 301, 312, 313, 338, 571, 750**
ASTM E1474 **750, 751**
ASTM E1529 **302, 327, 328**
ASTM E1537 **302, 571, 739, 746–748, 751, 753**
ASTM E1590 **302, 571, 739, 750**
ASTM E1678 **200**
ASTM E1966 **302, 303, 327, 328**
ASTM E2058 **201**
ASTM E2187 **740**
ASTM E2307 **302**

ASTM E2387 **302, 329**
ASTM E2404 **317**
ASTM E2573 **317**
ASTM E2579 **317**
ASTM E2599 **317**
ASTM E2652 **300, 308**
ASTM E2768 **317**
ASTM E2965 **301, 313**
ASTM E9060 **137**
ASTM F814-83 **637, 645**
ASTM F1930 **141**
BS 476 **467**
BS 476-3 **467**
BS 476-6 **467, 468**
BS 476-7 **429, 467, 468, 470**
BS 476-11 **466–468, 472**
BS 476-20 **519**
BS 2782 **473, 478**
BS 5438:1989 **136**
BS 5651:1989 **138**
BS 5722 **139**
BS 5722:1991 **136**
BS 5852 **137, 732–737, 758**
BS 5852-1 **137**
BS 5852-2 **137**
BS 5867-2:1980 **136**
BS 6249-1:1982 **136**
BS 6341 **131**
BS 6561 **140**
BS 6807 **137, 732, 735, 737, 758**
BS 6853 **555, 587, 588, 601, 603–609**
BS 7175 **137, 737, 758**
BS 7176 **737, 758**
BS 7177 **737, 758**
BS 7990 **202**
BS 8414 **412, 475, 478, 497, 517, 526**
BS AU 169 **542**
BS EN 71-2:2011+A1:2014 **135**
BS EN 367:1992 **138**
BS EN 407:2004 **138**
BS EN 469:2014 **138, 140**
BS EN 597-1 & -2:2015 **137**
BS EN 702:1995 **138**
BS EN 1021 **131**
BS EN 1021-1:2014 **137**
BS EN 1021-2:2014 **137**
BS EN 45545-2:2013+A1:2015 **135**
BS EN ISO 11611:2015 **138**
BS EN ISO 469:2014 Annex E **141**
BS EN ISO 6330:2012 **138**
BS EN ISO 6940:2004 **136**
BS EN ISO 6941:2003 **136**
BS EN ISO 6942 **138, 141**
BS EN ISO 9151:2016 **141**
BS EN ISO 9185 **131, 138**
BS EN ISO 9239-1 **131, 137**
BS EN ISO 10528:1995 **138**
BS EN ISO 11612:2015 **138, 140**
BS EN ISO 12138:1997 **138**
BS EN ISO 12952 **131**
BS EN ISO 12952-1 & -2:2010 **137**
BS EN ISO 13934-1:2013 **141**
BS EN ISO 14116:2015 **138**
BS EN ISO 15025:2016 **136, 138**
BS ISO 13506 **138, 140**
BSS 7239 **639**
CA TB 129 **571**
CA TB 133 **571**
CAN/ULC-S101 **336, 346**
CAN/ULC-S102 **337, 340, 341, 345–347**
CAN/ULC S 102.2 **317**
CAN/ULC-S104 **336, 337, 346, 347**
CAN/ULC-S105 **336, 346**
CAN/ULC-S106 **336, 346**
CAN/ULC-S107 **337, 347**
CAN/ULC-S110 **336, 347**
CAN/ULC-S111 **336, 347**
CAN/ULC-S112 **337, 347**
CAN/ULC-S113 **337, 347**
CAN/ULC-S114 **337, 339, 347**
CAN/ULC-S115 **337, 347**
CAN/ULC-S124 **337, 347**
CAN/ULC-S126 **337, 347**
CAN/ULC-S134 **338, 347**
CAN/ULC-S135 **338, 347**
CAN/ULC-S138 **338, 347**
CAN/ULC-S139 **338, 347**
CAN/ULC-S143 **338, 347**
CAN/ULC-S144 **337, 347**

CA TB 117 739, 743–745, 756
CA TB 133 739, 746, 751, 753, 756
CEN/TR 16988:2016 285
CEN/TS 1187 285, 291, 366–368, 397, 403, 405, 409, 429, 432, 433, 467
CEN/TS 15117:2005 285
CEN/TS 15447:2006 285
CEN/TS 15912 510, 512
CEN/TS 16459:2013 285
CFR-2012-16-2 136
CJ/T 416 582–584, 587, 588
CSA 60950 698
CSA 60950-1 689
CSA 62368-1 690
CSA C22.2 690
CSA FT 4/IEEE 1202 711
D6-51377 639
D45 1333 542
DASMA 107 301, 320
DBL 5307 542
DIN 4102 384, 405–407, 409, 412, 429, 455–457
DIN 4102-1 74
DIN 4102-12 725
DIN 4102-15 165
DIN 4102-20 409, 410
DIN 5510 555, 582, 587, 588
DIN 5510-2 165
DIN 50051 622
DIN 53436 192, 218, 227, 496, 497
DIN 54837 582, 583
DIN 75200 542
DIN/EN 60335-1 659
DIN/EN 60400, VDE 0616-3 659, 661
DIN/EN 60598-1, VDE 0711-1 661
DIN/EN 60669-1, VDE 0632-1 659
DIN/EN 60730-1, VDE 0631-1 659, 661
DIN/EN 60998-1, VDE 0613-1 658
DIN/EN 61439-3, VDE 0660-600-3 659
DIN SPEC 4102-23 409
EN 469:2011 141
EN 597-1 730, 732, 734, 757
EN 597-2 730, 732, 734, 757
EN 1021 392, 445
EN 1021-1 730, 732, 733, 737, 757
EN 1021-2 730, 732, 733, 737, 757
EN 1363-1 724
EN 13238 285, 358
EN 13501-1 285, 350, 356, 364, 380, 381, 383, 385, 391, 407, 415, 417, 429, 430, 432, 438, 441, 444, 446, 452, 458, 466, 467, 487
EN 13501-5 285, 365, 392, 397, 403, 409, 452, 460, 467
EN 13501-6 369, 392, 432, 452, 461, 720
EN 13772 381, 385
EN 13773 381, 382, 385, 445
EN 13823 165, 285, 291, 351, 352, 354, 355, 359–363, 377, 415, 417, 493
EN 14390 285, 358–361
EN 14878 135
EN 15619 444
EN 16733 364
EN 16989 569, 570
EN 17084 569, 570, 598
EN 45545 555, 556
EN 45545-1 558, 559, 693
EN 45545-2 271, 543, 556–561, 563–569, 583, 584, 587, 588, 608, 609
EN 50200 722, 723, 727
EN 50264 725
EN 50289-4-16 723
EN 50306 725
EN 50355 725
EN 50382 725
EN 50399 369–373, 376, 709, 710, 713–715, 726, 727
EN 50577 724, 727
EN 50582 723
EN 60332-1 369, 370, 375
EN 60332-3-10 713
EN 60335-1 662, 701
EN 60695-2-10 657
EN 60754-2 721, 727
EN 61034 719, 720, 727
EN 61034-1 164, 165, 172
EN 61034-2 370, 375

EN 62368-1 **698**
EN ISO 1421:1998 **141**
EN ISO 5659-2 **165**
EN ISO 6941 **542**
EN ISO 6942:2002 **141**
EN ISO 9239-1 **165**
EN ISO 13934-1 **141**
EN ISO 14116:2015 **141**
EN ISO 19353 **693**
FAR 25.853 **629, 630**
FAR 25.853(b) **136**
FAR 25.853 Part 4, App F **137**
FAR/JAR/CS 25.853 **543**
FAR/JAR/CS 25.855 **543**
FAR Part 23 **629**
FAR Part 25 **629**
FAR Part 25, App. F **631–635, 637, 639–641, 643, 644**
FAR Part 29 **629**
FAR Part 33 **629**
FLTM-BN 24-2 **542**
FM 4450 **301, 320, 323**
FM 4880 **301, 321–323, 329**
FM 4996 **301, 323**
FM 6921 **302**
FMVSS 217 **541**
FMVSS 302 **136, 260, 542, 633**
FMVSS 304 **541**
FTP Code **610, 612, 617–621, 626**
FTP Code Part 1 **617**
FTP Code Part 2 **618**
FTP Code Part 3 **623**
FTP Code Part 5 **620**
FTP Code Part 6 **621**
FTP Code Part 7 **622**
FTP Code Part 8 **622**
FTP Code Part 9 **623**
FTP Code Part 10 **624**
GB 6771 **581**
GB 8624 **487, 492, 497, 499**
GB 17927.1 **492, 497**
GB 50016 **479, 499**
GB 50222 **480, 482–484, 486, 497, 499**
GB/T 2406.2 **491, 494, 582, 583**
GB/T 2408 **491, 494**
GB/T 4207-2012 **697**
GB/T 5169.12-2013 **697**
GB/T 5169.13-2013 **697**
GB/T 5169.16 **491, 495, 582, 697**
GB/T 5454 **491, 495**
GB/T 5455 **491, 495**
GB/T 5464 **488–490, 492**
GB/T 8332 **496**
GB/T 8333 **491, 496**
GB/T 8625 **496**
GB/T 8626 **488–490, 494**
GB/T 8627 **491, 495**
GB/T 11785 **489, 493, 583**
GB/T 14402 **488–490, 493**
GB/T 14403-2014 **496**
GB/T 16172 **491, 496**
GB/T 20284 **488, 490, 493**
GB/T 20285 **496**
GB/T 27904 **492, 497**
GB/T 29416 **497, 499**
GM 6090 M **542**
GS 97038 **542**
IEC 60065 **676**
IEC 60112 **684, 697, 703**
IEC 60331-2 **723**
IEC 60331-11, -21, and -23 **722**
IEC 60332-1 **706, 707, 709, 726**
IEC 60332-2 **706, 708, 709, 727**
IEC 60332-3 **709, 710, 712, 727**
IEC 60332-3-21 **713**
IEC 60332-3-22 **713**
IEC 60332-3-23 **713**
IEC 60332-3-24 **602, 605, 713**
IEC 60332-3-25 **713**
IEC 60335 **676, 699**
IEC 60335-1 **652, 659, 662, 701**
IEC 60400 **659, 661, 701**
IEC 60598-1 **661, 701**
IEC 60669-1 **659, 701**
IEC 60695-1-11 **652, 654, 700**
IEC 60695-1-12 **654, 700**
IEC 60695-1-30 **662, 663, 701**
IEC 60695-2-4/1 **717**
IEC 60695-2-10 **657, 658, 663, 664, 701**
IEC 60695-2-11 **657, 701**

IEC 60695-2-12 659, 663, 664, 697, 701
IEC 60695-2-13 659, 664, 685, 697, 701
IEC 60695-2-20 699
IEC 60695-5-1 675, 702
IEC 60695-6-1 672, 702
IEC 60695-6-2 672
IEC 60695-7-1 674, 702
IEC 60695-7-2 674, 702
IEC 60695-7-3 674, 702
IEC 60695-8-1 670, 702
IEC 60695-8-2 670, 702
IEC 60695-9-1 669, 702
IEC 60695-10-2 656, 700
IEC 60695-11-2 660, 662, 701
IEC 60695-11-4 666, 669, 702
IEC 60695-11-5 660–662, 687–689, 701
IEC 60695-11-10 494, 495, 582, 665–667, 678, 697, 701
IEC 60695-11-20 665, 668, 669, 701
IEC 60695 Parts 1-10, 1-11, and 1-12 652
IEC 60730-1 659, 661, 701
IEC 60950 698
IEC 60950-1 689, 703
IEC 60998-1 658, 701
IEC 61034 719, 727
IEC 61439-3 659, 701
IEC TS 60695-2-21 699
IEC TS 60695-5-2 675
IEC TS 60695-9-2 669
IEC TS 60695-11-11 700
IMO FTP code *See* FTP code
IMO Resolution 653 543
IMO resolution A.652 733, 734
ISO 834-1 582, 602, 606
ISO 1182 47, 265, 268, 269, 285, 289, 308, 337, 351–354, 357, 358, 420, 424, 429, 438, 492, 501, 511, 513, 520, 530, 532, 603, 617, 618, 626, 627
ISO 1210 494, 495
ISO 1716 265, 268, 285, 289, 351–354, 357, 369, 493, 620
ISO 3261 601
ISO 3795 542, 543, 552, 633, 645
ISO 4589 100, 290, 494, 495
ISO 4589-2 562, 570, 582, 583, 589, 601, 603–606
ISO 5657 265
ISO 5658-2 265, 271, 422, 543, 547, 551, 561–564, 570, 601, 603–606, 620
ISO 5658-4 265, 272
ISO 5659-2 265, 267, 275, 276, 510, 543, 561, 562, 566–568, 570, 588, 590, 598, 618, 619, 627
ISO 5660 670
ISO 5660-1 163, 165, 172, 201, 265, 273, 274, 291, 496, 500, 501, 504, 507, 510, 513, 517, 528, 543, 561, 562, 564, 570, 596, 601, 609, 624
ISO 5725 217, 218
ISO 6722 548
ISO 6940/41 543
ISO 6941 543
ISO 8191 543
ISO 8191-1 137, 497, 733, 734
ISO 8191-2 137, 733, 734
ISO 8421-1 601
ISO 9239-1 266, 272, 273, 285, 353, 365, 526, 528, 532, 543, 583
ISO 9239-2 266, 272, 273
ISO 9705 165, 205, 207, 218, 291, 323, 358, 360, 374, 377, 502, 504, 507, 511, 525, 528, 532, 534, 752, 754
ISO 9705-1 266, 276–278
ISO 9705-2 277, 543
ISO 9772 680, 702
ISO 9773 679, 702
ISO 11925-2 266, 269, 270, 285, 290, 351–355, 363, 420, 421, 494, 543, 583
ISO 11925-3 266, 269
ISO 12136 201, 266, 275
ISO 12828 215–218
ISO 12863 266, 270
ISO 12949 266, 280
ISO 13344 187, 217
ISO 13571 153, 154, 171, 172, 187, 193, 217, 221, 225, 227, 232, 234, 240–243
ISO 13784-1 165, 266, 277, 278
ISO 13784-2 266, 278, 279, 526
ISO 13785 505, 511

ISO 13785-1 **267, 279**
ISO 13785-2 **267, 279**
ISO 13943 **47, 601**
ISO 14696 **267, 275**
ISO 14697 **267, 280**
ISO 14934-1 **267, 280**
ISO 14934-2 **267, 280**
ISO 14934-3 **267, 280**
ISO 14934-4 **267, 280**
ISO 16312-1 **193, 218**
ISO 16405 **267**
ISO 17493 **141**
ISO 17554 **267**
ISO 19701 **193, 204, 211, 215, 218**
ISO 19702 **193, 211, 218**
ISO 19706 **186, 195, 198, 199, 217, 670, 674, 702**
ISO 20632 **267, 277, 278**
ISO 21367 **165**
ISO 24473 **267, 280**
ISO 29473 **267, 268**
ISO 29904 **155, 162, 166–168, 172**
ISO/IEC 17020 **297, 335**
ISO/IEC 17025 **292, 296, 335, 410**
ISO/IEC 17065 **292, 410**
ISO/TR 9122 **496, 497**
ISO/TR 9705-2 **266, 276**
ISO/TR 11696-1 **266, 268**
ISO/TR 11696-2 **266, 268**
ISO/TR 11925-1 **266, 269**
ISO/TR 13571-2 **186**
ISO/TR 16312-2 **194, 501, 502, 511, 513**
ISO/TR 17252 **267**
ISO/TR 17755-1 **10**
ISO/TS 3814 **265, 268**
ISO/TS 5658-1 **265, 270**
ISO/TS 5660-3 **265, 273, 274**
ISO/TS 5660-4 **265, 273, 274**
ISO/TS 5660-5 **266, 273, 274**
ISO/TS 12828 **217**
ISO/TS 17341 **502**
ISO/TS 17431 **267, 501, 503, 504, 507, 508, 511**
ISO/TS 17755-2 **11, 21**
ISO/TS 19021 **200, 267, 626, 627**
ISO/TS 19700 **202, 510**
ISO/TS 22269 **267**
JIS 1091 **495, 496**
JIS A 1310 **504–507, 510, 511**
JIS A 1320 **504, 507, 508, 511**
JIS A 1321 **496, 497, 500, 510**
JIS A 1326 **510, 512**
JIS C 2134 **697**
JIS C 60695-2-12 **697**
JIS C 60695-2-13 **697**
JIS C 60695-11-10 **697**
JIS D 1201 **542**
JIS E 1203 **597, 599**
JIS K 6911 **598**
JSTM J 7001 **510, 512**
KRTS-VE-Part 1-2016(R1) **599**
KS F 2257-1 **602**
KS F 2271 **513**
KS F 2844 **601**
KS F ISO 1182 **603**
KS F ISO 5660-1 **601**
KS M ISO 4589-2 **601**
LEPIR2 **397**
MES DF 050D **542**
method 1231 of the Japanese Ministry of Construction **200**
MSZ 14800 **413**
MSZ 14800-6 **414**
MSZ 14890 **413, 415–417**
NBN EN 597-1 **392**
NBN EN 597-2 **392**
NBN EN 1101 **393**
NBN EN 1102 **393**
NBN S21-203 **391**
NBN S21-204 **390, 391**
NBN S21-208-1 **391**
NBN S21-208-2 **391**
NEN 1775 **429, 430, 432**
NEN 6064 **429, 430, 432**
NEN 6065 **429, 430, 432**
NEN 6069 **429, 432**
NES 713 **200**
NF C 32-070 **711, 717**
NF D 60-013 **397, 398, 403**
NF F 16-101 **555, 581, 584, 606**

NF F 16-103-1988 **581**
NF P 92501-7 **137**
NF P 92505 **542**
NFPA 70 **295, 670, 704, 705, 726**
NFPA 101 **294, 739**
NFPA 130 **555, 573, 587, 607, 609**
NFPA 252 **302, 303, 328**
NFPA 253 **301, 312**
NFPA 257 **302, 303, 328**
NFPA 259 **300, 308**
NFPA 260 **302, 739, 740, 745**
NFPA 261 **302, 739, 740, 745**
NFPA 262 **718, 727**
NFPA 265 **301, 319, 320**
NFPA 266 **497, 498**
NFPA 268 **303, 331**
NFPA 269 **200, 295**
NFPA 270 **200**
NFPA 271 **201**
NFPA 272 **201**
NFPA 275 **302, 328**
NFPA 276 **301, 320, 321**
NFPA 277 **739**
NFPA 285 **303, 329, 332, 333, 338, 533**
NFPA 286 **301, 318–320, 329**
NFPA 286, UL 1715 **301**
NFPA 287 **201**
NFPA 288 **303, 328**
NFPA 289 **302**
NFPA 415 **303, 328**
NFPA 701 **301, 313–315**
NFPA 2112 **141**
NFPA 5000 **294, 295, 308**
NFPA (National Fire Protection Association) **294**
NF X 10-702 **165**
NF X 70-100 **202**
NF X 70-100-1 **562, 567, 568, 570**
NF X 70-100-2 **562, 567, 568, 570**
NT FIRE 032 **751**
NT FIRE 053 **510, 512**
ÖNORM A 3800-1 **382, 385, 387**
ÖNORM B 3800-1 **382**
ÖNORM B 3800-5 **383–385, 388**
ÖNORM B 3822 **385**
ÖNORM B 3825 **385**
PTL 8501 **542**
SMP 800-C **555, 587, 588**
STD 5031.1 **542**
TB/T 1484 **582, 584**
TB/T 2560 **581**
TB/T 2640 **581**
TB/T 3138 **584, 589, 590**
TB/T 3237 **586–590**
TL 1010 **542**
TSI-LOC&PAS **557, 558, 569**
UIC 564 **555, 584, 585, 590**
UIC 564-2 **584, 585, 590**
UIC 617-1 **581**
UIC 895 VE **711**
UL 9 **302, 303, 328**
UL 10 **302**
UL 94 **38, 41, 685**
UL 94-5V **669**
UL 263 **302, 303, 325**
UL 723 **301, 315, 323, 686**
UL 746 **678**
UL 746A **682, 684**
UL 746B **685**
UL 746C **682, 685**
UL 746D **687**
UL 790 **303, 329, 333, 334**
UL 1040 **301, 323, 329**
UL 1072 **101**
UL 1256 **301, 323, 324**
UL 1277 **101**
UL 1315 **302**
UL 1479 **302, 303, 327**
UL 1492 **698**
UL 1581-1160 **710**
UL 1581-1164 **711**
UL 1666 **101, 709, 711, 715, 716, 727**
UL 1685 **711**
UL 1694 **687, 688, 703**
UL 1703 **303, 333**
UL 1715 **301, 320, 329**
UL 1975 **302**
UL 2043 **670, 671, 702**
UL 2079 **302, 303, 327**

UL 2196 **303, 328, 338**
UL 60950 **698**
UL 60950-1 **689**
UL 62368-1 **690**
ULC-S102.2 **301**
ULC-S109 **337, 343, 347**
ULC-S127 **337, 340, 341, 343, 347**
UNE 23102 **447**
UNE 23721 **447–449**
UNE 23723 **447, 448**
UNE 23724 **447–449**
UNE 23725 **447–449**
UNE 23726 **447, 449**
UNE 23735 **447**
UN-ECE-R 34 **548**
UN-ECE-R 36 **540**
UN-ECE-R 67 **540, 550**
UN-ECE-R 107 **540, 550, 551**
UN-ECE-R 110 **540, 550**
UN-ECE-R 118 **540–548, 550, 551**
UNI 8456 **420, 421, 424, 426, 428**
UNI 8457 **421, 422, 424, 426, 428**
UNI 9174 **422–424, 426–428**
UNI 9175 **425–428**
US Federal Standard 191, Method 5136(3) **136**
U.T.A.C. 18-502/1 **542**
U.T.A.C. 18-502/2 **542**

Keywords Index

1,2-bis(tetrabromophthalimide)ethane 63
1,3,5-tris(2,3-dibromopropyl) isocyanurate 101
1 kW nominal premixed-flame test, IEC 60695-11-2 662
3 m cube test 375
4,6-tris(2,4,6-tribromophenoxy)-1,3,5-triazine 104
9,10-dihydro-9-oxa-10-phosphaphenanthrene-10-oxide 62, 65, 106, 116
50 W horizontal flame test method, IEC 60695-11-10 665
50 W vertical flame test method, IEC 60695-11-10 666
500 W flame test method, IEC 60695-11-20 668
ABS 57, 63, 68, 79, 81, 97, 104, 105, 114, 118
absorption 27, 45
acrylic 132, 142, 147
acute effects of fire effluents 186
acute toxicity 204, 674
AEGL 229, 238
aerosols 153, 154, 166, 171, 174, 220
AFNOR (Association Française de Normalisation) 261
afterglow 139, 143
afterglow suppressant 79
agglomeration 154
air-conditioning ducts 584
aircraft 627
alkoxyamines 100
aluminum diethyl phosphinate 76, 108, 110, 113, 116
aluminum hypophosphite 77, 99, 105, 110
aluminum trihydroxide (*See* also ATH) 53, 59
Americas 293
ammonium borate 176
ammonium octamolybdate 175
ammonium phosphate 143, 149
ammonium polyphosphate (APP) 66, 72, 77, 99, 113, 117, 143, 178
analytical methods 211, 215
analytical methods for gas components 588
analyzer 204, 208
animal bioassays 223
animal exposure 193
ANSI (American National Standards Institute) 261
anti-dripping agent 61, 106, 110–112, 114
antimony trioxide (*See* also ATO) 62, 80, 102, 115, 118, 143, 147
appliances 686
Approved Document B, UK 466
APTA (American Public Transportation Association) 571, 578
ARHE (Average Rate of Heat Emission) 565
aromatic nuclei 155
Asia/Pacific 479
asphyxiants 187, 222, 226, 229, 234, 235, 240

ASTM (American Society for Testing and Materials) **298**
ATH **53, 55, 58, 59, 77, 100, 101, 113, 117–119**
ATO **62, 99, 102, 105–107, 109, 113, 115, 118, 119**
Australia **518, 607**
Austria **378**
automobiles **538**
average rate of heat emission *See* ARHE
avis de chantier **400**
avis technique **399**
azoalkanes **72, 73**
AZONOR **73**
back-coatings **143, 144, 146, 147**
ball pressure test, IEC 60695-10-2 **656**
barrier effect **56**
barrier materials **742**
BDP **67**
bedding **270**
bedding components **623**
Belgium **388**
binder, polymeric **48**
bio-based chemicals **81**
biphosphate resorcinol bis(dixylenyl phosphate) **67**
bis(pentabromophenyl) ethane **63, 144**
bisphenol A bis(diphenyl phosphate) **41, 67**
bis(tribromophenoxy)ethane **63**
blowing agent **48, 56, 72**
boehmite **58, 99, 100, 108, 110, 116**
bomb calorimeter **33, 47**
boron-based flame retardants **78**
Bouwbesluit **428**
brominated butadiene-styrene block copolymers **57**
brominated polystyrene **64, 107, 109, 112**
bromine-containing flame retardants **61, 62**
brucite **98**
BSI (British Standards Institution) **261**
Building Code of Australia **518**
Bunsen burner test, 45° **634**
Bunsen burner test, 60° **634**
buses **537–539**
bushfire **519, 526**
cable materials **101**
cables **369**
- acidity **721**
- flaming droplets **720**
- large-cale testing **709**
- small-scale testing **705**
- smoke **719**
calcium borate **100, 176**
calorimeter
- micro combustion **47**
Canada **335**
carbonaceous layer **56**
carbon dioxide *See* CO_2
carbonized acrylic fibers **148**
carbonized acrylics **143**
carbon monoxide (*See* also CO) **187, 198, 220, 674**
carbon nanotubes **59, 61, 101, 110, 114, 119**
carboxyhemoglobin **187, 224**
catastrophic fires **7**
CB scheme **675, 697**
CBUF program **754**
CCC mark **697**
cellulosics **143**
CE marking **694**
CENELEC (European Committee for Electrotechnical Standardization) **259, 375, 650, 691**
CENELEC TC 20 **725**
CEN (European Committee for Standardization) **259, 282, 349, 556**
CEN/TC 127 (fire safety in buildings) **284**
CEN/TC 256 **557, 569, 693**
CFD (Computational Fluid Dynamics) **755**
CFE (critical heat flux at extinguishment) **547, 561, 562, 564, 601, 603–606, 620**
CFR (Code of Federal Regulations, USA) **259, 299, 628**
chain reaction **43**
chain scission **35**

chain stripping 36
chain transfer 34
chair, railways Korea 604
char 31
- char layer 32
- char yield 31
char-forming 148
characteristic heat flux for ignition *See* CHFI
charring 27, 34, 37
chemical analysis of fire effluents 204
CHFI (characteristic heat flux for ignition) 700
China 479
chlorinated cycloaliphatics 62
chlorinated hydrocarbons 62
chlorinated paraffin 62, 102, 119, 177
chlorine-containing flame retardants 61, 62
chronic effects of fire effluents 188
CIB (International Council for Research and Innovation in Building and Construction) 263
cigarette test 732, 734, 735, 744
CIT (Conventional Index of Toxicity) 568
CO (*See* also carbon monoxide) 187, 220, 221, 224, 227, 228, 235, 240
CO_2 220, 221, 226, 240
coagulation and agglomeration 159
coating, intumescent 48
coatings 149
code de la construction et de l'habitation 394
code of design on building fire protection and prevention, China 479
COHb 224, 225, 227, 228
combustibility 255, 519, 520, 595
combustion
- efficiency 23, 33, 44
- flaming 24
- heat 27, 32, 47, 131
combustion-modified foam 736
combustion process 54
combustion products 153
combustion toxicology 219
comparative tracking index *See* CTI
condensed phase 41, 55, 56, 58, 60, 64, 65, 67, 76, 79, 103, 112
conduction 24, 27
- conductive heat transfer 28
cone calorimeter 38, 47, 132, 163, 176, 273, 291, 312, 496, 500, 501, 517, 596, 750
- time-to-ignition TTI 132
confirmational test methods 669
consumer equipment 698
convection 27
conventional index of toxicity *See* CIT
corrosion damage 675
corrosivity 110
cotton 132, 142
coupling agents 57, 100
CPR (Construction Products Regulation) 259, 284, 287, 348, 692
CPSC (Consumer Product Safety Commission, USA) 738
CQC mark 697
critical heat flux 271, 620
critical heat flux at extinguishment *See* CFE
critical radiant flux 310, 493
cross-linking 37
CSA (Canadian Standards Association) 690
CTI (comparative tracking index) 684
curtains 381
curtains and drapes 393, 445, 491
cushions mattresses 572
decabromodiphenyl ethane 99, 101, 105, 107, 109, 116, 118, 119
decabromodiphenyl oxide 144
decking materials 742, 745
decomposition
- PET and PA 6 37
- polymers 34
- polyolefins 35
- thermal 30
- thermo-oxidative 29, 40
decorating materials 586
Denmark 432

DEPAL 68, 76
depolymerization 34
design category 559
design type 600
diethylphosphinic acid salts 68
DIN (Deutsches Institut für Normung) 261
discotheque fires 6
disulfides 74
divisions, ships 623
DKE (German Electrotechnical Commission) 695
dodecachloropentacyclooctadeca-7,15-diene 62
DOPO 62, 65, 69, 106, 108, 110, 113–118, 120
DOPO-HQ 69
dose-response/effect model 219
DOT (U.S. Department of Transportation) 570
dwelling fire 130
dwellings 731
early fire hazard test 523
EASA (European Aviation Safety Agency) 628
EIFS (exterior insulation finish systems) 332
elastomers 97, 119
electrical appliance and material safety act Japan 696
electrical cables 548
electrical engineering 647
electrical equipment 648
electric locomotives 581
electric wires and cables 634
electrotechnical equipment 562
endothermic decomposition 58
end-product testing 648
EN (European standard) 283
environmental effects of fire effluents 189
environmental pressure 147
EPDM rubber 103
epoxy resin 63, 68, 97, 115, 116
EPS 105
equivalence ratio 220
ERPG 229
établissements recevant du public 394
ETICS (External Thermal Insulation Composite System) 385, 387, 409, 414
ethylene bis(tetrabromophthalimide) 101, 112
ethylenediamine 99
euroclasses 350
Europe 348
European General Product Safety Directive 729
European Union 348
EVA (ethylene-vinyl acetate) 58, 59, 66, 96, 100, 101, 119
EXAP (extended application) 349
expandable graphite 77, 105, 118
exposure 187, 221, 238
exposure concentration 231
exterior cladding 533
exterior fire exposure tests 329
exterior wall assemblies 331, 332
exterior walls 438
external fire exposure 408
external fire spread 445
external sheeting 592
external wall covering systems 463
extinction coefficient 164, 169, 170
FAA (Federal Aviation Administration) 259, 628
fabric 133, 572, 742
- cotton 133
fabrics and films 343, 344, 521
façades 383, 396, 409, 413, 445, 463, 474, 497, 504, 510
- intermediate scale test 279
- large-scale test 279
FAR (Federal Aviation Regulations) 628
FDS (Fire Dynamics Simulator) 755
FEC (fractional effective concentration) 235
FED (fractional effective dose) 234, 240, 584
FIGRA (fire growth rate) 133, 360, 362

filling/padding component 742
fillings 735
filtering system 206
finishing materials 513
Finland 432
fire
- fully developed 47
- well-ventilated 47
fire behavior 27
fire cause 6
fire classification 350
fire effluents 153, 171, 185, 215, 219, 650
fire enclosure 689, 690
fire fatalities 11, 14, 16, 17
firefighters 10, 11
firefighters' clothing 141
fire growth rate *See* FIGRA
fire hazard 23
fire hazard assessment 652
fire hazard assessment of cables 705
fire hazard testing 705
fire hazard testing, IEC TC 89 650
fire hazard tests for end-products 655
fire injuries 17
fire load 47
fire losses 19
fire modeling 154, 281
fire performance 25
fire prevention 3
fire propagation apparatus 275
fire propagation test 468
fire protection 3, 4, 9
fire regulations and standards in Asia 696
fire resistance 47
fire resistance tests 325
fire-resistive construction 500
fire retardant materials 500, 514
fire risk 653
fire risk assessment 222
fire safety engineering 198, 203, 268
fire safety regulations and test standards in China 697
fire scenario 25, 44, 221, 653
fire spread 45, 255
FIRESTARR 557
fire statistics 7, 8, 10–14, 16, 17, 21, 731
- fatalities 129
- ships 610
- UK 129
- vehicles 538
fire threat 171
fire ventilation 220
first item ignited 6
flame
- diffusion 42
- premixed 42
flame inhibition 44
flame inhibitor 41
flame penetration test, aircraft 641, 643
flame propagation test, aircraft 642
flame retardant 65, 118, 143, 147
flame-retardant cellulosic blends 144
flame retardant classes 57
flame-retardant cottons 143
flame retardant efficiency 54
flame retardant materials 514
flame retardant reactive 67, 115, 117
flame retardants 53, 173, 652
flame retardants, chemical action 55
flame retardants in furniture 755
flame retardants, mode of action 54
flame retardants, physical action 55
flame-retardant synthetic fibers 146
flame-retardant textiles 142
flame-retardant viscose 144
flame-retarded plastics 53, 95
flame-retarded wool and blends 145
flame spread 42, 269
- flow 46
- velocity 46
flame-spread rating 345
flame zone 43, 55, 61
flammability 26
flammability class 585
flammability test 140
flammability testing, aircraft 631
flammable solids 308
floor covering 306, 310, 312, 340, 347, 365, 405, 532

flooring radiant panel test **574**
floorings **272, 352, 489, 493, 519, 526, 604**
fluoropolymers **80**
FMVSS (Federal Motor Vehicle Safety Standards) **541**
fogging **62**
Fourier transform infrared spectroscopy *See* FTIR
fractional effective concentration *See* FEC
fractional effective dose *See* FED
FRA (Federal Railroad Administration) **570**
France **394**
FSS Code (Fire Safety Systems Code **614**
FTIR (Fourier Transform Infrared Spectroscopy) **567, 618**
functional requirements, ships **615**
furnishings **325, 729**
furniture **497, 729**
- burning rate **750**
furniture and bedding **392**
furniture calorimeter **746, 751**
gas analysis **193**
gas bags **211**
gas burners **254**
gas concentration FTIR **275**
gas phase **43, 54–56, 61, 64, 65, 76**
gas-solution absorbers **208, 209**
General Product Safety Directive **135**
generation of fire effluents **193**
Germany **403**
glowing and smoldering combustion **255**
glowing cigarette **733**
glow wire flammability index *See* GWFI
glow-wire flammability test **658, 663**
glow-wire ignitability test **664, 685**
glow wire ignition temperature *See* GWIT
glow wire test **38, 657**
glow wire test for end products *See* GWEPT
GPSR (General Product Safety Regulations 2005) **732**
graphene **61, 101**
GS-mark **694–696**
GWEPT (glow wire test for end products) **657, 701**
GWFI (glow wire flammability index) **663, 664, 697, 701, 704**
GWIT (glow wire ignition temperature) **664, 665, 685, 697, 701, 704**
HAI (high-current arc ignition) **684**
halogenated dibenzodioxins and furans **189**
halogenated flame retardants **53, 55, 57, 61, 174, 177**
halogen-free flame retardants **178**
HALS **73**
hazard level, rail vehicles **559, 601**
HBCD (hexabromocyclododecane) **57, 105, 144**
HBr *See* hydrogen bromide
HCl *See* hydrogen chloride
HCN *See* hydrogen cyanide
heat capacity **39, 58**
heat evolved, total **47**
heat flux **28, 55, 98**
- effective heat flux **31**
- external **40, 44**
heat flux meters **280**
heat release rate *See* HRR
heat release parameter **44**
heat-resistant fibers **148**
heat shielding effect **45**
heat transfer **27**
hemoglobin **187, 224**
HET acid **62**
hexachloro-endo-methylenetetra-hydrophthalic acid **62**
high-current arc ignition test **687**
high-performance polymers **114**
high-rise building **4, 11**
high-speed craft **624, 625**
high-voltage arc resistance to ignition **685**
high-voltage, low current, dry arc resistance **684**
high-voltage tracking resistance *See* HVTR

HIPS (high-impact polystyrene) 63, 67, 104, 105, 115
Home Furnishings and Thermal Insulation Act 739
horizontal Bunsen burner test 633
horizontal ladder test methods 718
hot-wire ignition test 687
household appliances 698
HR (heat release) 273, 312, 505, 561, 564, 597, 635, 670, 751
HRR (heat release rate) 32, 38, 44, 45, 59–61, 131, 273, 277, 584, 751
Hungary 413
huntite 100
HVTR (high-voltage tracking resistance) 684
HWI (hot-wire ignition) 684
hydrogen bromide 61, 64, 188
hydrogen chloride 61, 113, 188, 238
hydrogen cyanide 187, 220, 221, 224, 235, 240
hydrogen fluoride 188
hydrogen halides 56, 61, 63
hydromagnesite 100
hypoxia 188, 224
IAMFTF (International Aircraft Materials Fire Test Forum) 644
IC_{50} 219
ICAO (International Civil Aviation Organization) 135, 260
ICC (International Code Council) 294
Iceland 432
IEC 258, 649
IEC (International Electrotechnical Commission) 257, 649
IEC TC 20 705, 709, 717
IEC TC 46 709
IEC TC 61 699
IEC TC 89 699
IEC TC 89 guidance 654
IEC TC 89 Liaisons 651
IFC (International Fire Code) 739
ignitability 255
ignition 38, 269, 312
- time to ignition 28, 39
ignition source 38, 254
ignition temperature 304
immeubles de grande hauteur 394
IMO (International Maritime Organization) 135, 260, 263, 268, 271, 273, 275, 277, 609
IMO MSC (IMO Maritime Safety Committee) 610
IMO Resolution A.653 (16) 563
incapacitation 190, 219, 222, 232, 237, 238, 240
industrial facilities 443, 446
inhalation exposure 219
inhalation toxicity 224, 226
inhalation toxicology 234
inherently flame retardant 146
inherently flame-retardant synthetic fibers 147
inherently flame retarded 23
inorganic fillers 57, 60
inorganic flame retardants 57, 58, 108, 113
insulation 519, 593, 605
- aircraft 642
interior ceiling and wall panels
- aircraft 630
interior decorative materials 584
interior finish materials 321, 345
interior panel 603
interior products 350
interior structural materials 584
interior surfaces 561
interliners 736
intermediate-scale calorimeter (ICAL) 275
international certification 675
interoperability 557
intoxication 187
intumescence 56, 64, 71, 77
intumescent 99, 102, 119
intumescent systems 148
intumescent textiles 149
Ireland 732
irradiation 39
irritants 188, 220, 222, 235, 238

ISO (International Organization for Standardization) **10, 257, 261, 649**
ISO/TC 61 **675**
ISO/TC 92 **10, 262, 619, 672**
ISO/TC 92/SC1 **264**
ISO/TC 92/SC3 **674**
ISPA (International Sleep Product Association) **748**
Italy **417**
Japan **499, 591**
JAR (Joint Aviation Requirements) **259**
JISC (Japanese Industrial Standards Committee) **261**
JRMA (Japan Railway Rolling Stock & Machinery Association) **598**
Korea **512, 599**
laboratory-scale tests **254**
Landesbauordnung **404**
large-scale tests **256, 276, 291**
lateral flame spread **271**
layered double hydroxides **60, 99, 119, 179**
LC_{01} **220, 222, 224, 229, 235, 238, 240**
LC_{50} **192, 198, 219, 230, 231, 235**
lethality **190, 191, 237**
lignin **81**
limiting oxygen index *See* LOI
limit of detection **216**
linear pipe thermal insulation products **351**
lining materials **532**
liquid smoke aerosols **160**
listed products **560**
LOI (limiting oxygen index) **38, 80, 99, 103, 115, 116, 118, 131, 133, 136, 139, 145, 148, 176, 290, 494, 495**
LSC (Life Safety Code) **739**
LVD (Low Voltage Directive) **692**
magnesium dihydroxide *See* MDH
magnesium hydroxide (*See* also MDH) **59, 79**
manikin **141**
MARHE (maximum average rate of heat emission) **565, 584**
mattresses **732, 734, 740, 746, 748**
mattress smoldering test **740**
maximum average rate of heat emission *See* MARHE
MDH **58, 77, 98, 100, 103, 105, 108, 113, 117, 119**
MD (Machinery Directive) **692**
Meeker burner (also Meker burner) **314**
Meker burner (also Meeker burner) **515, 516**
melamine **72, 77, 78, 117**
melamine condensates **78**
melamine cyanurate **78, 108, 110, 113, 117**
melamine phosphate **149**
melamine poly(metal phosphates) **116**
melamine polyphosphate **66, 68, 77, 103, 108, 110, 116, 117, 178**
melting behavior **545**
metal hydroxides **57**
methenamine pill test **306**
MHCLG (Ministry of Housing, Communities and Local Government, China) **736**
mineral fillers **57**
MLIT (Ministry of Land, Information, Transport and Tourism, China) **591**
modacrylics **147**
model box test **502, 503**
model predictions **754**
MOLIT (Ministry of Land, Infrastructure and Transport, China) **512**
molybdenum compounds **79**
molybdenum oxide **177**
molybdenum trioxide **175**
montmorillonite **103, 119**
montmorillonite clay **60**
motor vehicles **537**
MPP *See* melamine polyphosphate
Musterbauordnung, Germany **404**
MVV TB (Muster-Verwaltungsvorschrift Technische Baubestimmungen), Germany **412**
N-alkoxy hindered amines **72**
nanoclays **100**

nanocomposites 45, 59, 99, 104, 106, 110, 121, 128
NBCC (National Building Code of Canada) 335
NCC (National Construction Code, Australia) 518
needle-flame test 660, 686
Netherlands 428
NFPA (National Fire Protection Association) 6, 299
NHTSA (National Highway Traffic Safety Administration) 541
nightwear 135
NIST (National Institute of Standards and Technology), USA 8, 9, 740, 748
nitrogen-containing flame retardants 77
nitrogen oxides 188
non-combustibility 269, 339, 357, 617
non-combustible 289, 513
non-combustible materials 501
non-harmonized construction products 382
NOR 72, 73
- NOR 116 142
Nordic countries 432
NORDTEST 434
Norway 432
novoloid 143
NT FIRE 434
NTIS (National Technical Information Service) 631
nucleation 157
number of fires 12
nylon fabrics 146
office equipment 698
Ohio State University calorimeter 132, 636
OIB (Österreichisches Institut für Bautechnik, Austrian Institute of Construction Engineering) 378
OIB (Österreichisches Institut für Bautechnik, Austrian Institute of Construction Engineering)-Guidelines 379
opacimeters 163
open corner test 321
open flame tests 746
operation category 558, 600
optical density 164, 575
organoclays 59
OSNR (Office of the National Rail Safety Regulator), Australia 607
OTSZ (Hungarian Fire Safety Code) 413
oxygen consumption calorimeter 312
oxygen index (*See* also LOI, limiting oxygen index) 562, 585
oxyhemoglobin 187
oxyimides 72, 100
pallets 323
parallel panel test 323
particle oxidation 160
particle size 59
partitions, aircraft 630
partition walls 561
passenger compartment 593
PC/ABS 41, 112
pentabromobenzyl acrylate 144
pentaerythritol 71, 149
perfluorinated sulfonate salts 114
perimeter fire barriers 329
peroxide compounds 73
phenolic formaldehyde resins 118
phenoxyphosphazenes 68
phosphate esters 67
phosphorus-containing flame retardants 57, 62, 64, 112, 113
photovoltaic 333
PHRR (peak heat release rate) 60, 75, 81, 99, 101, 132, 748
physical fire models 193, 195
plasticizing effect 56, 68, 112
plenum areas 718
Poland 435
polyacrylates 106
polyacrylonitrile 105
polyamide 62, 97, 107
- polyamide 6 63, 79, 107, 108
- polyamide 6.6 107
polyaramide 143, 148
poly(aramid-arimid) 143, 148

polybenzimidazole **143, 148**
polybrominated diphenyl ethers **63**
polybutylene terephthalate **62, 63, 109**
polycarbonate **63, 64, 67, 68, 79, 80, 97, 111, 115**
polycarbonate blends **111**
polychloroprene **119**
polycyclic aromatic hydrocarbons (PAHs) **155, 189**
polydibromophenylene oxide **64**
polyester **142, 147**
polyester/cotton **132, 144**
polyesters **97, 109**
polyethersulfones **115**
polyethylene **96, 100**
polyimides **115**
polylactic acid **109, 111**
polymeric brominated polybutadiene-polystyrene (BrPBPS) **63**
polymeric flame retardants **53, 56, 68, 70**
polymeric metal polymelaminephosphate **108**
polymeric methylphosphonates **68**
polymethacrylate **97**
poly(methyl methacrylate) **106, 155**
polyoxymethylene **97, 106, 155**
poly(pentabromobenzyl acrylate) **64, 101**
polyphthalamides **108**
polypropylene **96, 98, 142, 155**
polypropylene fabrics **146**
polystyrene **155**
polystyrene-brominated butadiene-polystyrene block copolymers **101**
polystyrene-brominated polybutadiene-polystyrene block copolymer **105**
polystyrene foam **41**
polystyrene **96, 104**
polysulfone **115**
polytetrafluoroethylene (*See* also PTFE) **41, 61, 114**
polyurethane **62, 97, 117**
polyurethane foams **117**
polyvinylchloride (*See* also PVC) **97, 113, 142**
polyvinylidene chloride **114, 142**
polyvinylidene fluoride **114**
POSS **80**
potassium perfluorobutane sulfonate **111**
potential heat **308**
PRC (People's Republic of China) **579**
preselection of materials **662**
preselection testing **648**
preselection tests for materials **655**
primary deck coverings **621**
printed circuit boards **63, 68, 115**
probes **205**
property insurers **251**
protective clothing **140**
protective layer **55, 56, 61, 77**
PTFE **97, 104, 105, 109–112, 114**
public areas **736**
PVC **113, 174**
pyrolysis **24, 27, 44, 130**
- anaerobic pyrolysis **29**
- polymer **24, 38, 43**
- zone **25**
pyrolysis products **58**
pyrotechnics **6**
quantitative risk assessment of fire effluents **190**
quasi-non-combustible materials **500, 502**
radiant ignition sources **255**
radiant panel **272**
radiant panel flame spread test to ASTM E162 **682**
radiation **24**
radical **34, 56, 61, 65, 66, 72**
- forming agents **57, 72, 74**
- free **43**
- generator **35, 41, 100**
- hydroxyl **43**
- mechanism **55**
rail vehicles **135, 554, 693**
railway rolling stock cables **725**
railway sleepers **597**
RD (Royal Decree, Belgium) **388**

REACH (Registration, Evaluation, Authorisation and Restriction of Chemicals) **756**
reaction to fire **289, 357, 361, 377, 391, 405, 408, 500**
reactive flame retardants **53**
recycling **54, 80**
red phosphorus **66, 101, 105, 113, 117**
reference gas components concentrations **587**
reference scenario for cables **371**
Regulation (EU) No. 305/2011 **293**
relative thermal index *See* RTI
repeatability **203, 217**
reproducibility **203, 217**
resorcinol bis(diphenyl phosphate) **67, 112, 115**
respiratory tract irritants **229**
respiratory ventilation **228**
retreat effect **56, 75**
Riser test UL 1666 **715**
roof coverings **329, 330, 366, 433, 460, 467**
roof deck construction **323**
roofing assemblies **320**
roofs **291, 365, 366, 397, 408, 429**
room calorimeter **752**
room corner test **273, 276, 318, 360, 502**
RTI (relative thermal index) **685**
rubber materials **584**
SAC (Standardization Administration of the People's Republic of China) **261**
sampling conditions **208**
sampling line **205, 207**
sandwich panels **488, 507**
- large room test **278**
- small room test **278**
SBI (single burning item) **291, 360, 361, 493**
Schlyter test **382**
seats **560, 583, 584, 594**
seat shells **562**
selectivity **215**
semi-non-combustible **514**
sensitivity **215**
Setchkin furnace **304**
ships **609**
side effects of a fire **669**
side effects of cable fires **719**
silicon-containing compounds **79**
silicone rubbers **120**
silk **132**
silsesquioxanes **80**
SI (Statutory Instrument, UK) **732**
single burning item *See* SBI
small flame test **269, 732**
small room test **266**
- pipe insulation **278**
small scale tests **193**
SMOGRA (smoke growth rate) **362**
smoke **256**
smoke and toxicity **618**
smoke and toxicity from fire effluents **185**
smoke box **375**
smoke density **309, 495, 561, 566**
smoke developed classification **345**
smoke development and measurement **153**
smoke emission
- aircraft **637**
smoke formation **179**
smoke generation **154, 173, 275, 588**
smoke inhalation **185**
smoke measurement **165**
smoke obscuration **234, 274, 564, 672**
smoke opacity **162, 672**
smoke production **161, 273**
smoke production rate *See* SPR
smoke suppressants **78, 79, 114, 173**
smoke suppressants for PVC **174**
smoke suppression **61, 79, 147**
smoke toxicity **185, 193, 497**
smoldering **38, 40, 364, 740**
smolder resistance **739**
soft upholstery **734**
SOLAS (International Convention for the Safety of Life at Sea) **609, 612**
solid sorption tubes **210**

soot 171
- particles 155, 156, 160
- production rate 170
- yield 154
Spain 442
SPCS (Sleep Products Safety Council) 748
SPR (smoke production rate) 277, 312
spread of flame 271, 561, 563
Steiner tunnel 315, 317, 718
stowage compartments
- aircraft 630
styrene butadiene rubber 119
styrene copolymers 105
substantial component, building 356
sulfenamides 72, 100
sulfur-based compounds 72
sulfur dioxide 188
surface burning characteristics 315, 340
surface flammability 573, 574, 620
surface spread of flame, IEC 60695-9-1 669
surface spread of flame test BS 476 part 7 470
Sweden 432
Switzerland 451
synergist 62, 72, 77, 79, 108, 109, 116, 120, 147
synergistic 101
synergistic effect 56, 58, 75, 79
TBBA (tetrabromobisphenol A) 63, 64, 116
technical building code Spain 442
test mark 251
testo unico di prevenzione incendi, Italy 418
tetrabromobisphenol-A-bis(2,3-dibromo-propylether) 63, 101
tetrabromophthalate diol 63
tetrabromophthalate ester 64
tetrabromophthalic anhydrides 63
tetrakis(hydroxymethyl)phosphonium (THP) salt 143
textile fibers and fabrics
- burning behavior 130
textiles 129
- fire regulations 134
- flammability testing 134
- test categorization 136
textiles and films 313, 622
The Building Regulations UK 465
thermal conductivity 55, 58
thermal decomposition 36
thermal degradation 196
thermal inertia 28
thermal insulation 47, 584
thermally thin 25, 29
thermal stability 36
thermal transitions fibers 131
thermofixation 146
thermogravimetry 30
thermoplastic polyurethanes 112
thermoplastics 98
thermosets 115
THR (total heat release) 44, 132, 748
timber 334, 346, 533
time-dose-response 235
total heat release *See* THR
toxicants 199, 220, 228
toxic combustion gases 755
toxic effect 191, 193
toxic endpoints 229
toxic gases 751
toxic hazard assessment 672
toxic hazards 221
toxicity 256, 502, 510, 514, 561, 567
- aircraft 637
- fire effluent 674
- testing 199, 200, 202
toxic load model 236
toxicological hazard assessment 222
toxicological risk assessment 234
toxic potency 194, 198, 214, 674
TRANSFEU 557
translational toxicology 223
transportation 259
transportation fires 7
trihydroxymethylphosphine oxide 70
triphenyl phosphate 67, 118
tris(1,3-dichloro-2-propyl)phosphate 62

tris(1-chloro-2-propyl)phosphate 117
tris(2-chloroethyl)phosphate 62, 117
tris(chloroisopropyl)phosphate 62
tris(tribromoneopentyl)phosphate 99, 101, 142
tris(tribromophenyl)cyanurate 105
tube furnace 398
UFAC construction criteria 742
UFAC test rig 742
UFAC (Upholstered Furniture Action Council, USA) 739
UIC (Union Internationale des Chemins de Fer) 260
UL 94 flammability tests 678
- 5 V 680
- HB 666, 678
- HB 40, HB 75 665, 678
- HBF, HF-1, or HF-2 680
- V-0 59, 67, 68, 73, 99, 100, 102, 114, 118
- V-0, V-1, and V-2 679
- V-2 59
- VTM-0, VTM-1, and VTM-2 679
UL 746 C 653
UL listing 678
UL listing mark 677
UL product directories 677
UL recognized component directory 678
UL standards for safety 677
UL (Underwriters' Laboratories) 297, 676
under-ventilated fire 171
UNECE (United Nations Economic Commission for Europe) 260, 539
United Kingdom 465, 732
United States of America 293, 738
unsaturated polyesters 97, 118
upholstered furniture 129, 270, 302, 385, 397, 425, 426, 445, 491, 492, 497, 622, 732
Upholstered Furniture (Safety) Regulations 732
VDE mark 694, 695
VDE regulations and approval procedures 694
vehicle fires 537
ventilated fire 171
ventilation-controlled fires 220
vertical burning test single insulated wire or cable 705, 709
vertical flame propagation test 644
vertical ladder test methods 709
vertical tube furnace 307
visas de façades 401
viscose 133, 142
viscosity 25, 41
visibility 153, 162, 169, 170
VKF Fire Protection Register 464
VKF Guidance 452
VKF (Vereinigung Kantonaler Feuerversicherungen) 451
volatile fuel production 27
volatile hydrocarbons 189
wall and ceiling linings 519, 525
well-ventilated fires 199, 220
welt cord 742
wire and cable 491, 584
wires 605
wood crib 255, 737
wool 132, 133, 142
XPS 105
yarn 133
yellow card 678, 690
zeolites 72
zinc borate 78, 99, 119, 176, 179
zinc hydroxystannate 79, 99, 147, 175, 177, 179
zinc stannate 79, 108, 147, 175, 177, 178